T0295232

Introduction to Energy Systems

Introduction to Energy Systems

Ibrahim Dincer and Dogan Erdemir
Ontario Tech. University, ON, Canada

Library of Congress Cataloging-in-Publication Data
Names: Dincer, İbrahim, 1964- author. | Erdemir, Dogan, author.
Title: Introduction to energy systems / Ibrahim Dincer, University of Ontario ON, CA, Dogan Erdemir, University of Ontario Institute of Technology, ON, CA.
Description: Hoboken, New Jersey : John Wiley & Sons Inc., [2023] | Includes bibliographical references and index. | Summary: "Energy is critical commodity for every society, and nothing is technically possible without energy. Energy studies are carried out by every discipline, and many academic programs cover energy related courses. During the past decade it has been clear that developing an introductory level undergraduate textbook is badly needed. That has been a kind of motivation to prepare a book about conventional and innovative energy systems and applications and their linkages to the environment and sustainable development with numerous analysis methods, including energy and exergy approaches, illustrative examples and cases studies to provide a solid knowledge source for undergraduate students in engineering programs, potentially ranging from mechanical to chemical and electrical to industrial engineering programs"-- Provided by publisher.
Identifiers: LCCN 2023009320 (print) | LCCN 2023009321 (ebook) | ISBN 9781119825760 (hardback) | ISBN 9781119825777 (epdf) | ISBN 9781119825784 (epub)
Subjects: LCSH: Power resources.
Classification: LCC TJ163.2 .D556 2023 (print) | LCC TJ163.2 (ebook) | DDC 621.042--dc23/eng/20230323
LC record available at https://lccn.loc.gov/2023009320
LC ebook record available at https://lccn.loc.gov/2023009321

Cover Images: © kamisoka/Getty Images; Yaorusheng/Getty Images; zorazhuang/Getty Images; Sylvain Sonnet/Getty Images; Ron and Patty Thomas/Getty Images; RelaxFoto.de/Getty Images; Red ivory/Shutterstock
Cover Design: Wiley

Set in 9.5/12.5pt STIXTwoText by Integra Software Services Pvt. Ltd., Pondicherry, India

SKY10053743_082223

Contents

Preface

Energy is considered a critical commodity for every society, and we know that nothing is technically possible without energy. Energy studies are carried out by every discipline, and many academic programs cover energy-related courses. During the past decade it has been clear that developing an introductory level undergraduate textbook is badly needed, which has been a kind of motivation to prepare such a book about conventional and innovative energy systems and applications and their linkages to the environment and sustainable development with numerous analysis methods, including energy and exergy approaches, illustrative examples, and cases studies to provide a solid knowledge source for undergraduate students in engineering programs, potentially ranging from mechanical to chemical and electrical to industrial engineering programs. Furthermore, theory and concepts along with analysis and assessment are emphasized throughout this comprehensive book, reflecting new techniques, models and applications, together with complementary materials and recent information.

This book starts with some introductory perspectives on energy and the environment. Importance of energy, energy-related issues and environmental problems are discussed with some important solutions. The role of engineering in energy and environment-related problems is also provided. The concepts of life cycle assessment and industrial ecology are presented with illustrative examples. Finally, the need for energy labeling is discussed. Second chapter deals with energy sources and sustainability. First, the key instruments, drivers and indicators are introduced. Then, the historical perspectives of the energy consumption and generation are presented with developments and landmark type achievements. Next, regression analysis and future projection techniques for energy systems are introduced. Finally, dimension of sustainability and sustainable development indexes are discussed with illustrative examples. In Chapter 3, the thermodynamic analyses of energy systems are introduced in detail with illustrative examples. Thermodynamic balance equations, which are mass, energy, entropy, and exergy balance equations, are written for common components that are commonly used in energy systems. Some illustrative examples are in this regard presented. Chapter 4 provides the background for combustion processes. Fossil fuels and their impacts are discussed. Then, combustion of fuels is introduced. Thermodynamic analysis for combustion in a closed system and an open system are analyzed with illustrative examples. As a following topic, nuclear energy systems are provided in Chapter 5. First, historical perspectives and types of nuclear energy are introduced. Nuclear fission and nuclear fusion processes are presented. Then, nuclear fuel production techniques are introduced, and types of nuclear reactors are discussed. Importance of small modular reactors and their potential contribution in electricity generation are discussed. This is followed by nuclear cogeneration and hydrogen production processes. Finally, integrated nuclear energy systems for communities are provided over some case studies. Chapter 6 presents the solar energy systems. First, atmospheric and direct solar radiations are described with the solar radiation measurement methods. Then, solar radiation distribution in the world is presented. Next, solar energy applications are discussed with illustrative examples. Solar thermal and PV applications are introduced with their thermodynamic analysis techniques. Finally, photovoltaic thermal hybrid solar panels and their applications are given with the practical examples. Wind energy systems are introduced in Chapter 7. This chapter starts with historical development of the wind energy systems. This is followed by wind effect and global wind patterns which help create the wind power. Then, the types of wind turbines are introduced. Next, thermodynamic analyses of wind energy systems are provided with illustrative examples. Finally, some case studies and exergy maps for wind energy systems are shared with the readers. Chapter 8 deals with geothermal energy technologies. In this chapter, after introductory information about geothermal energy systems, their advantages and disadvantages are discussed with the readers. Then, power generation methods from geothermal energy are presented with illustrative examples. Finally, other applications of geothermal energy, such as heat pump, district heating, absorption cooling and hydrogen production are provided with practical examples. Biofuels and biomass energy systems are discussed in Chapter 9. This chapter starts with CO_2 balance which makes biofuel and biomass are renewable sources. Details about combustion, gasification and pyrolysis are provided with illustrative examples. Waste-to-energy concept is introduced with a case study. Finally, biodigestion and micro-gas turbine applications are presented. Chapter 10 covers the water-related energy systems which are essentially hydro and ocean (marine) energies. First, hydro power generation systems are presented with illustrative examples. Types of hydro turbines are introduced. Then, ocean energy systems are discussed, which cover energy production with tides, waves, currents, ocean thermal energy, and salinity gradients. Here, energy sources and production techniques are presented in detail with practical example. In Chapter 11, energy storage methods are presented with the need of energy storage systems and their importance. Mechanical, thermal, chemical, electrochemical, magnetic and electromagnetic and biological energy storage techniques are discussed with their working principles. Chapter 12 presents the hydrogen energy system which are hydrogen production, storage and utilization. First, hydrogen production systems are introduced briefly, and then electrolysis process is discussed in detail with illustrative examples. Next, hydrogen storage techniques are presented. Finally, utilization of hydrogen is provided based on fuel cell applications. Chapter 13 introduces the integrated energy concept. System integration and analysis techniques are presented with illustrative examples. Multigeneration concept is also discussed with case studies. Chapter 14 aims to provide background for life cycle assessment of the energy systems. First, goals and scope of life cycle assessment are defined. Then, life cycle inventory analysis for energy systems is presented with illustrative examples. Finally, impact assessment and improvement analysis are given with practical examples.

Incorporated throughout are many analysis methodologies, illustrative examples, and case studies, which provide the students with a substantial learning experience, especially in areas of practical applications of energy systems. There are appendices included where unit conversion factors and tables and charts of thermophysical properties of various materials and substances in the International System of units (SI) are covered.

In closing, the assistance provided by of two doctoral students, Mert Temiz and Ali Karaca, of Prof. Ibrahim Dincer are highly acknowledged.

Ibrahim Dincer
Dogan Erdemir

Nomenclature

Symbols

A   Area (m^2)

c   Specific heat (kJ/kg K)

D   Diameter (m, mm)

e   Specific energy (kJ/kg)

E   Energy (kJ)

$\dot{E}$ Energy transfer rate (kW)

ex   Specific exergy (kJ/kg)

Ex   Exergy (kJ)

$\dot{Ex}$ Exergy rate (kW)

F   Shape factor

g   Gravity (m/s^2)

h   Specific enthalpy (kJ/kg)

$\dot{I}$ Solar radiation (W/m^2)

m   Mass (kg, ton)

$\dot{V}$ Mass flow rate (kg/s, kg/h)

$\dot{n}$ Molar flow rate (mol/s)

N   Number of moles

V   Velocity (m/s)

$\dot{V}$ Volumetric flow rate (m^3/s, L/min)

P   Pressure (kPa, bar), Power (kW)

Q   Heat transfer (kJ)

$\dot{Q}$ Heat transfer rate (kW)

R   Universal gas constant or solar radiation (kW/m^2) or radius (m)

t   Time (s, min, h)

T   Temperature (°C, K)

s   Specific entropy (kJ/kg K)

S   Entropy (kJ/K)

u   Internal energy (kJ/kg)

v   Specific volume (m^3/kg)

V   Volume (m^3, L)

$\dot{W}$ Work rate (W, kW)

x   Vapor quality

z   Elevation (m)

Greek Letters

Δ	Difference
ε	Emitted radiation
η	Efficiency
ρ	Density (kg/m^3)
λ	Equivalence ratio
Ψ	Exergy index, exergy coefficient
τ	Transmitted radiation

Subscripts

0	Reference
abs	Absolute or adsorbed
ave	Average
b	Boundary
boil	Boiler
c	Compressor or compression or combustion
cc	Combustion chamber
ch	Charging period
com	Combustion
comp	Compressor
con	Condenser
d	Destruction
dest	Destruction
disch	Discharging period
down	Downward flow
dry	Dryer
elect	Electric
en	Energy
eva	Evaporator
ex	Exergy
g	Generator
gen	Generation
f	Final or fluid
fg	Fluid/gas mixture
H	High temperature
Hx	Heat exchanger
HRSG	Heat recovery steam generator
i	Inlet or initial or any stream
is	Isentropic
loss	Loss
in	Inlet
L	Low temperature
max	Maximum
min	Minimum
net	Net

o	Outlet
o	Outlet
p	Constant pressure or pump or piping
prod	Products
react	Reactant
ref	Reference
s	Source
sf	Solid/fluid mixture
sh	Shaft
t	Turbine, total
tur	Turbine
up	Upward flow
v	Constant volume

Abbreviations

ADP	Abiotic depletion potential
AC	Air conditioner
AP	Acidification potential
AFR	Air fuel ratio
BiPV	Bifacial photovoltaic
COP	Coefficient of performance
COVID	Corona Virus Disease
CSP	Concentrated solar power
EBE	Energy balance equation
EnBE	Entropy balance equation
EP	Eutrophication potential
ExBE	Exergy balance equation
ES	Energy storage
GC	Gas consumption
CC	Coal consumption
CNG	Compressed natural gas
COMM	Commercializability
DMFC	Direct methanol fuel cells
OC	Oil consumption
EMC	Electromagnetic Suspension System
EPS	Electromagnetic Pulse System
FC	Fuel cell
FLT	First law of thermodynamics
GDP	Gross domestic product
GFR	Gas-cooled fast reactor
GHG	Greenhouse emissions
GWP	Global warming potential
HE	Heat engine
HHV	High heating value
HP	Heat pump

HEX	Heat exchanger
HRSG	Heat recovery steam generator
HVAC	Heating, ventilation, and air conditioning
JC	Job creation
KE	Kinetic energy
LHV	Lower heating value
LNG	Liquid natural gas
LCA	Life cycle assessment
LCC	Life cycle costing
LED	Light-emitting diode
LFR	Lead-cooled fast reactor
LNG	Liquid natural gas
MBE	Mass balance equation
MED	Multi-effect desalination
MSR	Molten salt reactor
NG	Natural gas
NGC	Natural gas consumption
ODP	Ozone depletion potential
ORC	Organic Rankine cycle
PE	Potential energy
PEC	Photoelectrochemical
PEM	Polymer Electrolyte Membrane
PV	Photovoltaic
PVT	Photovoltaic and thermal
SB	Solid body
SCWR	Supercritical water-cooled reactor
SFR	Sodium-cooled fast reactor
SLT	Second law of thermodynamics
SMES	Superconducting Magnetic Energy Storage
SMR	Small modular reactor
SNG	Synthetic natural gas
SSSF	Steady state, steady flow
TC	Total consumption
TES	Thermal energy storage
TLT	Third law of thermodynamics
TR	Technology readiness
TSR	Tip-speed ratio
USUF	Uniform-state, uniform flow
VHTR	Very high temperature reactor
WAM	Weighted arithmetic mean
WGM	Weighted geometric mean
ZLT	Zeroth law of thermodynamics

About the Companion Website

This book is accompanied by companion website:

www.wiley.com\go\dincer\energysystems

This website includes

- Appendices.

1

Energy and Environment Perspectives

1.1 Introduction

Humanity has been struggling with many global issues such as energy, water, food, education, pandemic and diseases, terrorism and wars, population increase, immigration and refugees, poverty, and environment as illustrated in Figure 1.1 that makes the world cry. This is primarily because of the anthropogenic activities where fossil fuels in global energy portfolio have played a critical, but damaging role. There have been various researchers and institutions making various lists of problems or ranking the global challenges, and they have more or less ended up with the same or similar items, ranging from energy to water and from food to environment. Almost all countries have been affected by one or more, or all of these items as presented in Figure 1.1. Among these, energy is of course listed one of the top-ranked challenges as it is essential to drive the sectors, economies, and hence societies for their activities. During the past a few years, it has been crystal clear to all of us that the COVID-19 pandemic has changed many tangible and intangible things in our daily life, personal relations, institutional arrangements, country affairs, energy matters, business, trade, politics, economy, education, social life, etc. Although each item has a unique impact on people's lives, some of them, for example, energy, are affecting the rest of the items significantly. Energy is recognized as a significant necessity which is considered responsible for many issues in particular related to ecosystem, air, water, and food. For instance, traditional energy systems which use fossil fuels cause air pollution and water pollution issues. The emissions, in particular greenhouse gas (GHG) emissions, cause major environmental challenges, such as global warming (or greenhouse effect) and stratospheric ozone depletion. For example, due to the global warming we face water scarcity in many parts of the world causing crises in many sectors, including agricultural sector.

Furthermore, each item given in Figure 1.1 is, no matter, of great importance. However, one may extract four out of them, namely clean energy, clean air, clean water, and clean food, that affected the past, are affecting the present, and will probably affect the future of humanity. Clean energy plays a key role among these four key humanity-needs as the nature and cleanliness of energy may badly influence the other three, namely air, water, and food. Here, clean energy primarily refers to renewable energy where we use renewable energy sources, such as solar, wind, geothermal, hydro, marine and biomass to generate clean outputs (for example, electricity, heat, and cooling) for daily sectoral applications. Energy is essentially needed for almost every operation or application or process in our daily life. Therefore, clean energy has a crucial role in providing clean air, clean food and clean water which will result in a more sustainable community.

Since the food chain from harvesting to the shelf is so diverse and there are many processes involved which are energy intensive and fuel consuming, they require huge amounts of energy and fuels,

Introduction to Energy Systems, First Edition. Ibrahim Dincer and Dogan Erdemir.
© 2023 John Wiley & Sons, Inc. Published 2023 by John Wiley & Sons, Inc.
Companion Website: www.wiley.com/go/Dincer/Introduction_to_Energy_Systems

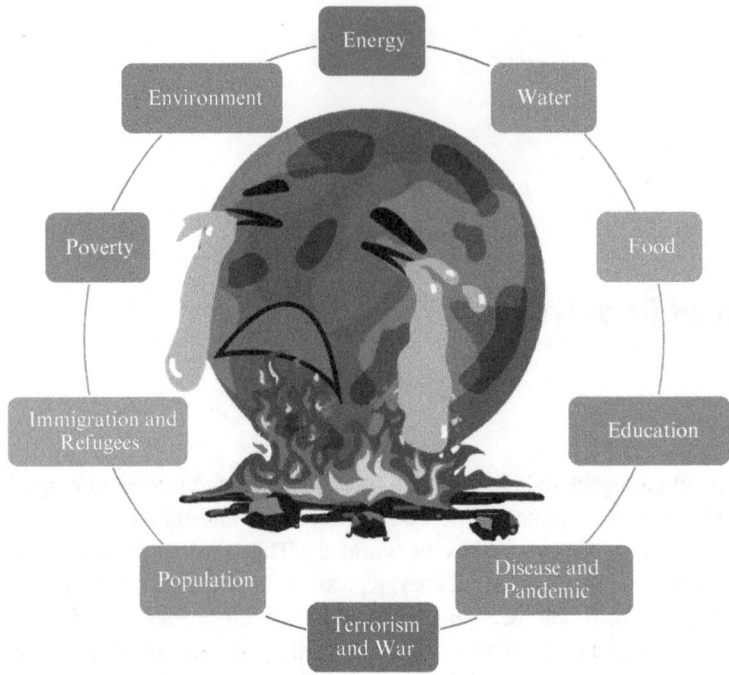

Figure 1.1 Global challenges affecting the people's past, present and future.

accounting for over 30% of the world's total energy consumption. Of course, having more and more processed foods coming to our tables, due to the advanced preservation and processing technologies as well as rising population, will even increase this much more. For example, ammonia is one of the most essential inputs for agriculture since it is used as fertilizer, or agricultural machinery consumes a voluminous fuel for tillage. Also, water is the most significant input for agriculture and aquaculture. Pumps used in watering or oxygen generators used in aquaculture consume a considerably high amount of energy. On the other hand, the transportation of the agricultural products is an indispensable part of the food-supply chain. Clean transportation options will help achieve clean food and clean air targets. Moreover, electricity is consumed to preserve the food in the refrigerator. Consequently, energy, food, environment and water are really connected very much to each other. So, we need to find newly developed and innovative options to operate all these steps in a sustainable manner.

Although the Earth is called as the water planet, almost 99% of the water resources on the Earth are not usable by people. It is widely known that about 1% of the water sources is fresh water in a drinkable form which is largely found in lakes, rivers, and underground sources. In order to obtain fresh water from the remaining 99% of the unusable forms of water, water should be cleaned and hence desalinated accordingly, which again requires a significant amount of energy for operation. So, clean energy will really be the key to clean water, too. Finally, this shows us that it may not take too much time to be faced with the water crisis.

Energy itself is, no doubt, an essential need for people, and the amount of energy consumption has been ever-increasing in the world due to the increasing population, the rising living standards, and the comfort level requirements in almost all sectors. Today, most of the energy demands are met by fossil fuel-based systems, and the consumption of fossil fuels is also increasing day by day. Fossil fuels can be assumed to be one of the major contributors to environmental problems. The increasing energy consumption and environmental problems have compelled people to use energy more efficiently, especially in existing energy systems and to benefit from clean energy sources (renewables and nuclear).

We begin this chapter with the importance of energy and how to solve energy issues effectively. Next, the smart solutions to solve these issues are discussed. The role of engineering is essentially introduced and discussed for solving energy-based problems. Then, the introductory information on environmental issues, industrial ecology, and life-cycle assessment is presented. Finally, the importance of energy labeling is discussed along with some examples.

1.2 Importance of Energy

It is a common fact that life is impossible without energy which is an essential driver for anything and everything. Energy is vital for our daily life and sectoral activities. Some examples may be introduced as follows:

- In our home for lighting, appliances, televisions, computers, air conditioners, etc.
- In factories to power the manufacturing processes; and
- In transportation for cars, trucks, ships, airplanes to transport people and goods.

The energy demand in the world has been ever-increasing due to the growing population and enhanced living standards. Figure 1.2 demonstrates the global changes in the world energy consumption (in Figure 1.2a), world gross domestic product (Figure 1.2b), and world population with respect to years (Figure 1.2c). The world's population is also projected to be around 10 billion by the year 2050, as seen in Figure 1.2a. Most of the population will eventually require food, energy, shelter, water, etc. more than what people have today. The increase in the population definitely results in a substantial amount energy consumption. As seen from Figure 1.2a, by 2050, it is

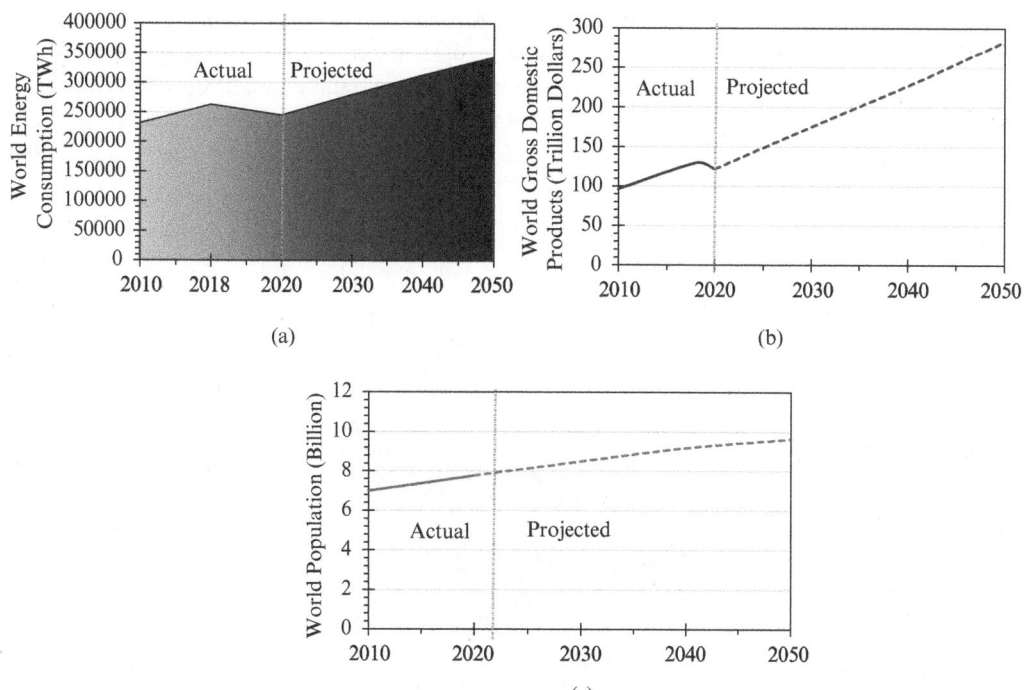

Figure 1.2 The projections of (a) the world energy consumption, (b) gross domestic product, and (c) population (data from [1]).

expected to drastically increase in the global energy use approximately by 50%. The biggest contribution in this increase is done by non-OECD economic growth and population. Also, the increasing population will require more domestic products. Figure 1.2b shows the historical and projected data for world gross domestic product changes. It is expected to increase more than double. These numbers clearly show that urgent action is really required for a sustainable future, especially in energy production and consumption.

1.3 Energy Issues

All sectors have relied on energy as a constant source of power. Over time, energy has become more important, especially after the industrial revolution. Many economic domains and areas have been governed by it, such as relationships, business deals, wars, terrorist activities, etc. At present, it dominates more than ever and will certainly dominate far more in the future. The present civilization has a fundamental responsibility to deal fairly and diligently with this issue.

The increasing world population and the rising living and working standards in all sectors have been the reasons behind the worldwide ever-rising energy consumption. Figure 1.3 illustrates the total amount of energy consumption around the world by source from 1965 to 2020. As mentioned previously, the global energy consumption has increased significantly every year. Figure 1.3 also highlights that fossil fuels cover nearly 65% of the global energy demand. Additionally, while the decrease in energy consumption in 1972–1975, 1979–1983, and 2008 have occurred due to economic recession, it has occurred due to the COVID-19 pandemic occurred in 2020. From 1965 to 2020, energy consumption increased approximately four times.

Fossil-based fuels essentially meet almost two-third of the global energy demand. They also have some risks due to their limited and nonhomogeneous reserves, environmental impacts, and energy security concerns. These risk factors have forced people toward alternative energy sources and systems such as renewable energy sources. Traditional energy systems dramatically pollute the environment, so, today, we are talking about the net zero emissions target, clean energy technologies, etc. Also, with the oil crisis that emerged in October 1973, the interest in alternative energy sources has

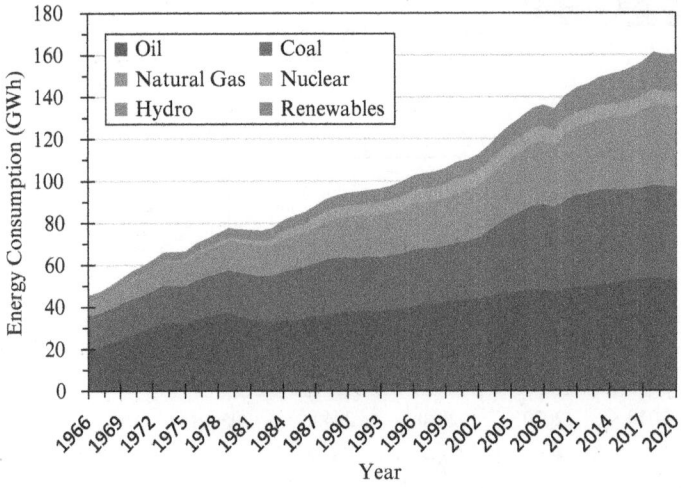

Figure 1.3 Total energy consumption of the world by sources (data from [2]).

increased significantly. Especially, solar domestic hot water systems and other solar space heating systems have become prevalent after the oil crisis. The use of renewable energy sources has increased significantly over the past decade. Despite the magnificent advantages of renewables in terms of energy security and environmental impact, the biggest disadvantage in their use is that they are mostly non-continuous energy sources, such as solar and wind, due to their fluctuating nature [3].

Another issue with energy is the difference between the energy supply and demand profiles. The energy demand profiles vary with respect to the sectors (residential, industrial, commercial, etc.) and time frame (daily, weekly, seasonal, yearly etc.). For instance, when the weekly energy demand profiles of the industrial and commercial sectors are considered, their energy demands tend to decrease on the weekends due to lowering activities in these sectors. In the residential buildings, against to the commercial and industrial sectors, the energy demands are higher on the weekends due to increasing time spent in residential units and indoor activities. When the yearly energy demand profile is taken into consideration for all sectors, it is quite higher during the summer due to the intensive air conditioner (AC) usage. That is why the highest electricity consumption rates are observed in the summer months.

In addition to these two issues, the peak loads are another issue faced in the energy field. Fluctuating trend in energy demand means peak energy loads, which requires a substantially higher amount energy according the off-peak and average loads. Peak loads are also seen in the short time depending on the time frame considered. For example, the highest cooling loads which are almost 50% higher than the average load, are seen 2–4 hours in a day. On a yearly basis, the highest electricity consumption is seen in the few hottest weeks in summer in many countries. In order meet the peak load, high-capacity devices should be included in the buildings for micro scale (buildings, factories, etc.). For macro scale such as communities, regions, countries, extra power plants or energy imports are required to meet the peak energy demands. Extra power plants and energy imports bring with it a high cost for energy supply. The different tariff structures, called triple tariff or multi tariff, are used to cover a part of this high-cost energy. Also, the time-of-use tariffs aim to promote to reduce energy consumption for savings during peak periods and to shift the peak demand from peak hours to off-peak hours. In order to avert those issues in the energy use and reduce the environmental impact due to energy consumption, we need to find smart solutions which will be presented forthcoming sections.

1.4 Environmental Issues

Energy supply and demand are related not only to problems such as global warming but also to such environmental concerns as air pollution, ozone depletion, forest destruction, and emissions of radioactive substances. All of them are called environmental impact categories. All environmental impact categories must be taken into consideration if human society is to develop in the future while maintaining a healthy and clean environment. Much evidence suggests that the future will be negatively impacted if people and societies continue to degrade the environment. Figure 1.4 shows the energy related CO_2 emissions in the world. CO_2 emissions have exponentially increased, and it is therefore expected that it will continue increasing with the increasing energy consumption. If current policy and technology trends continue, global energy consumption and energy-related carbon dioxide emissions will increase through 2050 as a result of population and economic growth. Renewables will be the primary sources for new electricity generation, but natural gas, coal, and increasingly batteries will be used to help meet load and support grid reliability. Oil and natural gas production are still expected to continue in a growing manner, mainly to support increasing energy consumption in developing countries, including Asian economies.

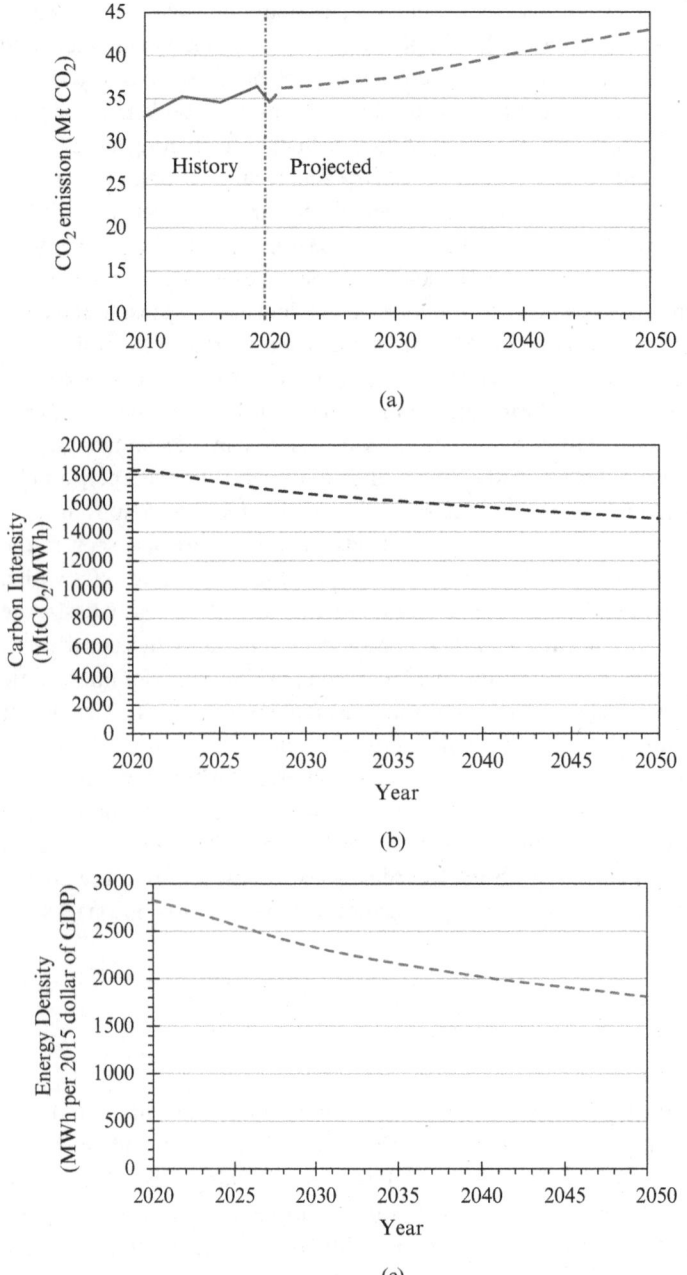

Figure 1.4 Energy related (a) carbon dioxide (CO_2) emissions, (b) carbon intensity and (c) energy intensity (data from [1]).

There is an intimate connection between energy, the environment, and sustainable development. A society seeking sustainable development ideally must utilize only energy resources that cause no environmental impact (e.g., which release no emissions or only harmless emissions to the environment). However, since all energy resources lead to some environmental impact, it is reasonable to suggest that some (but not all) of the concerns regarding the limitations imposed on sustainable development by environmental emissions and their negative impacts can be overcome through increased energy efficiency. A strong correlation clearly exists between energy efficiency and environmental impact since, for the same services or products, less resource utilization and hence pollution are normally associated with higher efficiency processes.

Achieving solutions to the environmental problems that we face today requires long-term planning and actions, particularly if we are to approach sustainable development. In this regard, renewable energy resources appear to represent one of the most advantageous solutions. Hence, a strong connection is often reported between renewable energy and sustainable development. Environmental considerations have been given increasing attention in recent decades by energy industries and the public. The concept that consumers share responsibility for pollution and its impact and cost has been increasingly accepted. In some jurisdictions, the prices of many energy resources have increased over the last 20 years, in part to account for environmental remediation costs. It is really necessary to look at some major environmental issues which are listed below:

- Acid rain: It results when sulfur dioxide and nitrogen oxides are emitted into the atmosphere and transported by wind and air currents. The combustion of the fossil fuels is the main contributor of acid rain.
- Stratospheric ozone depletion: it is used for defining the thickness of the ozone later in the stratosphere.
- Global warming and climate change: Global warming and climate change occur due to greenhouse emissions created by human activities.
- Hazardous air pollutants: include carbon monoxide, lead, nitrogen oxide, ozone, sulfur dioxide etc., which are mainly formed by burning fossil fuels.
- Poor ambient air quality: Air quality is defined over oxygen, carbon monoxide, carbon dioxide, and nitrogen oxide and hazardous air pollutants in the air. Combustion of fossil fuels and husbandry results in a poor ambient air quality.
- Water and maritime pollution: occur when the mixture of chemical and trash are disposed to the lakes, rivers, sea, and oceans. This pollution causes damages in the environment and living organisms in the water. It breaks down the balance of the world.
- Land use and siting impact: it is related to the economic and cultural activities which happen at a location. It can cause impervious surfaces, point source discharges, and development patterns. It affects quality of water and watersheds, loss of native habitats and spread of invasive species.
- Radiation and radioactivity: Radiation is the particles or energy released during radioactive decay. Radioactivity is the natural process by which some atoms spontaneously disintegrate, emitting both particles and energy as they transform into different, more stable atoms. Everyone is exposed to radiation on a daily basis, primarily from naturally occurring cosmic rays, radioactive elements in the soil, and radioactive elements are incorporated in the body. Man-made sources of radiation, such as medical X-rays or fallout from historical nuclear weapons testing also contribute, but to a lesser extent. About 80% of background radiation originates from naturally occurring sources, with the remaining 20% resulting from man-made sources.
- Solid wastes: Solid wastes are produced with almost every activity done by people. Their collection for disposal requires energy. Diverting waste by recycling and composting can help reduce the impact of solid waste on the environment.
- Environmental accidents: Anthropogenic environmental disasters such as nuclear accidents, oil leakages, chemical waste dumps, forest fires, and many more can be included in major environmental accidents. Better engineering, engineering materials, and people's education can help to reduce the number of such accidents.

- Thermal pollution: It is the degradation of water quality by any process that changes ambient water temperature. It occurs when cooling of the thermal and nuclear power plants are done by water sources such as lake, sea, and river. Thermal pollution, unlike chemical pollution, results in a change in the physical properties of water.

Various potential solutions to the current environmental problems associated with harmful pollutant emissions have recently evolved, including:

- use of renewable energy technologies,
- use of advanced nuclear energy technologies,
- integration of energy storage systems into existing systems for better management practices,
- efficient energy use and energy conservation,
- development of district energy systems, including heating and cooling,
- use of clean energy sources and fuels for transportation vehicles,
- energy-source switching from fossil fuels to more environmentally benign-energy forms,
- use of clean coal technologies,
- optimum monitoring and evaluation of energy indicators,
- policy and strategy development and deployment,
- recycling and waste management practices,
- development of waste to energy (i.e., power) options,
- process improvement and sectoral rehabilitation studies,
- acceleration of forestation,
- implementation of carbon and/or fuel taxes,
- greener material substitutions,
- promoting public transportation,
- changing lifestyles,
- increasing public awareness of energy-related and environmental problems,
- increased ethical responsibilities related to resources and their utilization,
- environmentally-focused education and training.

1.5 Smart Solutions

As mentioned previously, energy is recognized as an essential part of our life for everything, but causes the serious problems for environment, economy, development and sustainability if it is not clean enough. Here, the key question will be "How to meet future energy needs in a sustainable manner?" In order to meet the future's energy demand in a sustainable way, we really need the smart and reasonable solution covering clean energy sources, energy storage technologies, multi-objective optimizations, and real-time controlling. In order to reach smart solutions in the energy field, first, we need to identify our targets. Figure 1.5 demonstrates the key targets for the smart solutions to reach a sustainable future.

It is widely experienced that things are getting smarter and better day by day. That's why we have smart materials, smart devices, smart technologies, smart grid, etc. Also, we really need smart energy solutions. The smart energy portfolio is apparently expected to provide nine solution areas as illustrated in Figure 1.6. In order to reach the smart solutions, the following technologies, methods and strategies should be included into the systems, processes and services accordingly:

- Renewable energy technologies: Renewables are one of the significant solutions to generate energy in a clean way for local and global activities. Therefore, they should be the first option in order to generate power in any location where we have any renewable source available.

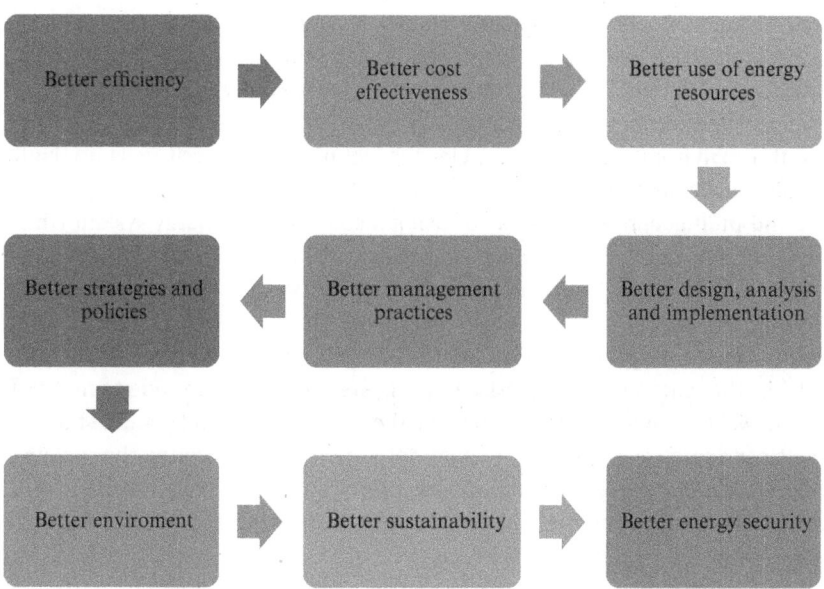

Figure 1.5 The targets to reach the smart solutions.

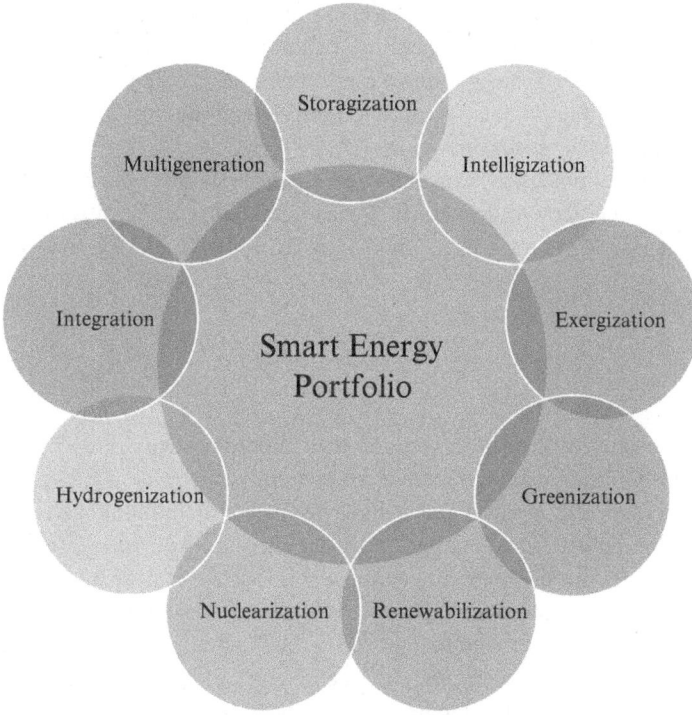

Figure 1.6 Nine branches of smart energy portfolio (modified from [4]).

- Clean fuel technologies (hydrogen, ammonia, etc.): Fuels are critical especially for the mobility. The systems should be developed, where we use carbon free fuels, such hydrogen and ammonia, instead of fossil fuels.
- Efficient energy use: Energy efficiency is called as the one best "energy source" to reduce the energy consumption and emissions since many systems are still using inefficient energy

conversion systems. Using energetically efficient systems, devices, processes and services will help to reduce energy consumption and emission depending on them.

- Cleaner technologies for fossil fuels: Fossil fuels are still expected to continue covering the main part of our daily life activities for some years, which may range from one decade to two decades. In order to reduce their environmental impacts, cleaner techniques for fossil fuels are badly needed to reduce their carbon footprints in the near future.
- System integration and multigeneration: Instead of using a conventional energy system which consists of a source, a system, and service, we need to obtain more useful outputs from a source in order to increase the efficiency. Therefore, multigeneration systems play a key role for sustainability.
- Energy storage technologies: Energy storage systems are critically needed and recognized as one of the essential parts of the renewable-energy based multigenerational systems due to intermittent availability of renewables. They also help to manage the energy supply and demand profiles.
- Nuclear energy: Nuclear energy is expected to play an important role in meeting the societies' energy need in a clean way. In this regard, small-modular nuclear systems are expected to meet the electricity demand of the communities along with various other types of useful commodities, including heat, cooling, hydrogen, clean fuels, etc.
- Waste to energy technologies: Wastes should first be recycled as much as possible at their sources and converted in useful forms for further utilization. If they are not recycled accordingly, waste to energy technologies are necessary to convert such wastes into useful forms of energy and fuels by deploying various methods ranging from gasification to pyrolysis.
- Multi-objective optimization studies: Building multigenerational and renewable energy-based systems requires high costs. Also, in order to design the systems, their operation periods, conditions should be adjusted well. For all of them, it is required to perform a multi objective optimization to better design and build the systems for operation.
- Real-time automation and control technologies: In order to operate the systems in the optimum desirable conditions, there is a strong need to develop real-time automation and control technologies to implement accordingly and manage the demands and supplies in an effective and efficient manner.

1.6 3S Concept

Energy has been, is, and will be a significant need in people's lives. Energy consumption in the world is ever-increasing significantly. Today, energy demands are generally met by fossil-fuel-powered systems. Since fossil-based fuels bring some risks, have limited and nonhomogeneous reserves, and results in environmental impacts and energy security concerns, people have oriented toward alternative energy sources. In the deployment of renewables as alternative and clean energy sources, despite their major advantages in terms of energy security and environmental friendliness as well as sustainability, there is still a challenge, due to the fluctuating nature of some renewables, namely solar and wind. In order to overcome such a challenge, there is strong need to develop right energy storage (ES) solutions for implementation. ES techniques primarily have a great potential to solve the discontinuity of renewables. The second biggest issue in the use of energy is an unbalanced energy demand profile and hence fluctuating nature of it. In this regard, ES methods have the potential to balance energy supply and demand profile periods.

ES has many methods and different applications. A conventional energy conversion system consists of a source, a system, and a service. Over time, each of the above elements has improved significantly with the developing technology. In order to reduce carbon emission and environmental impact, renewable energy sources have been integrated the energy conversion systems. Also, multigeneration systems have been started to use in the system for increasing the useful outputs of the system.

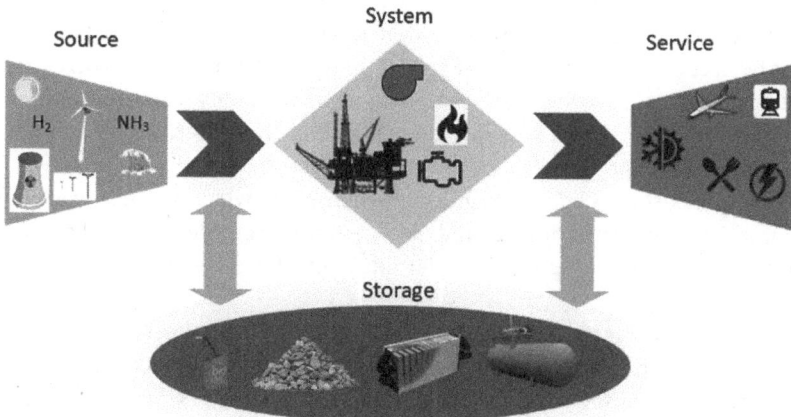

Figure 1.7 The rule of 5S (5S = 3S + 2S) in the respect of energy domain.

These changes have played a very critical role in better sustainability. However, energy production and demand profiles do not always match. The energy demand profile shows fluctuating nature over a time frame. As expressed before, to meet peak loads require additional power plants or the energy imports, and so it requires a high cost. Also, the ever-increasing use of wind turbines and their increasing contribution to nighttime electricity generation will bring about a rising nighttime surplus of electricity generation. This causes waste energy, which is produced by wind turbines. Nuclear power plants also have a similar situation. ES has a high impact on recovering waste energy. Figure 1.7 demonstrates a novel aspect of energy conversion systems. Integrating ES to the conventional energy conversion system, between the source and system, and the system and service reduce the waste energy and solves the mismatch between the energy production and demand periods. This rule is called as 3S + 2S = 5S. The rule of 3S + 2S presents a higher efficiency for energy systems and a better sustainability for communities.

1.7 Role of Engineering

Although there are many definitions of engineering, we define engineering as "creating the optimum solutions for physical problems and challenges with new ideas or innovative techniques." As highlighted earlier, the biggest problems of today are energy and environmental related as both are affecting the other parameters such as economy, transportation, etc. as mentioned in Section 1.1. Although engineers are responsible for solving the problems in all sectors, they are firstly responsible for finding the optimum and cost effective solutions to energy and environmental related problems, considering each step of the rule of 5S. They need to find a secure, reasonable, efficient, and reliable way to benefit from any clean energy source. Then, they develop systems considering life cycle, economy, and efficiency. Finally, they will meet the demands considering energy recovery and demand side management tools. It is really needed to perform those in a smart way as demonstrated in Figure 1.6.

Note that everything engineers do as practicing technical staff can have an impact on the environment. Engineers are further responsible for, or are involved in technology development and deployment as they do designs, analyses, developments, buildings, and implementations. For an engineering system, the following are three key activities of work.

Material selection: One may ask the following questions for material selection: Can we use environmentally friendly alternative materials? Can we use less amount of materials without compromising the functionality or reliability? Can we use recyclable materials? Can we obtain the raw materials in a green way? For example, aluminum-based materials are used in many sectors from automotive to construction, energy to electronics. Regrettably, most of them are not adequately recycled after they complete their lifetime or use period. More importantly, one may need to note that aluminum may not be needed for every targeted application or process or system. Therefore, correct material selection should be performed by considering its recyclability, cost, advantages, and disadvantages. The key to doing this is knowing the features of the materials. As another example, nanoparticles are used in many applications to enhance system performance. However, nanoparticle production requires some chemicals and high energy. So, when a nanoparticle is used in any application, its environmental impact and cost should be considered. The benefits expected from nanoparticles should be higher than the ones consumed during nanoparticle production.

Process selection: Here, process selection is specifically targeted for production and manufacturing stages where processing the raw materials to transform them into a final product is a key objective. In most cases, every step releases waste materials to the environment as air pollutants and solid wastes. All sectoral processes and activities require energy and material input(s) and produce some outputs with wastes. Therefore, an optimum process should be designed and implemented to reduce the energy and material inputs, maximize the useful outputs, and minimize the wastes.

Energy selection: All raw-material obtaining and production stages require energy along with all production activities. Therefore, it plays the most crucial role in environmental impact. The quantities and the types of energy we use directly affect environmental quality. Energy input should be selected depending on the type of demand. For example, when hot water is needed, boiling water by burning natural gas will not be the logical technique in an environmental manner. Or, when you have exhaust gasses with a temperature of 300 °C, exhausting it without recovering heat will not be the effective solution. Consequently, both needs and energy input(s) should be considered carefully by taking into consideration energy recovery options.

For example, let's investigate ammonia production, which is one of the most produced chemicals all around the world. Hydrogen and nitrogen should be provided as the raw materials, with a couple of chemicals used as catalysts for the ammonia synthesizer. Today, nearly all hydrogen for ammonia production is provided by a fossil-based source, mainly natural gas. So, to be engineers, we need to include green hydrogen in our ammonia production facility. Nitrogen is generally obtained via air separation techniques which require high energy input. So, in order to reduce the environmental impact of air separation, we need to use renewables and/or develop a method for air separation that requires less energy input. All ammonia production facilities use a traditional technique called the Haber-Bosch, which has low efficiency. So, again, as to be engineers, we are responsible for developing the cycles and chemical reaction chains for more efficient ammonia production. Civil engineering can be part of this facility with sustainable materials. Software engineering can work on better management and autonomous control coding. Consequently, there are many opportunities in all engineering fields for more sustainable ammonia production.

As today's systems, machines, and devices require energy input and management, many programs specialize in energy-related studies in higher education, from civil to electronic, environmental to planning, and mechanical to software engineering programs. Today, energy and the environment should be part of all programs as our globe requires urgent acting.

1.8   Life Cycle Assessment

Impact of the energy systems on the environment is today clear for everyone. However, not only energy systems cause the environmental problems. Almost all activities in all sectors have impact on the environmental impact categories. In order to assess systems in terms of their environmental impacts, life cycle assessment (LCA) is an essential tool. LCA provides a picture of how engineering decisions in any particular field for building and producing goods to serve society affect the environment. All stages of a product's life cycle must be considered in finding ways to reduce environmental impacts by generating clearer and more efficient manufacturing operations (with less energy and materials inputs) and recovering energy and materials during the waste management. LCA is a significant tool in implementing the concepts of green design, green power use, and waste minimization. Therefore, it can be used in any sectors and any applications.

Figure 1.8 illustrates a process layout of the life cycle assessment. In LCA analysis, the system boundaries should be defined carefully. In other words, it must be determined which parameters, processes, inputs and outputs will be included in the analysis. Normally, it is expected that to include everything for a system/process/manufacturing in the LCA, from raw-material obtaining to waste disposal, the lifetime of the product is considered. Thus, we can see the full picture of the system in terms of environmental impacts. Then, we can choose what can be implemented to the system in order to enhance their environmental impacts.

Figure 1.9 demonstrates the conventional (linear) economy and circular economy for the vehicles. In the linear production and consumption economy (Figure 1.9a), raw materials are used for production. Then, products are reached to user with a distribution network. Users use the vehicles until the vehicles complete their lifespan. The vehicles go to junkyard as the waste. In circular economy (Figure 1.9b), every item in production and consumption periods have a great impact on the LCA of the vehicle. A circular economy aims less waste output and emissions releases. Each step should be carefully studied in terms of LCA.

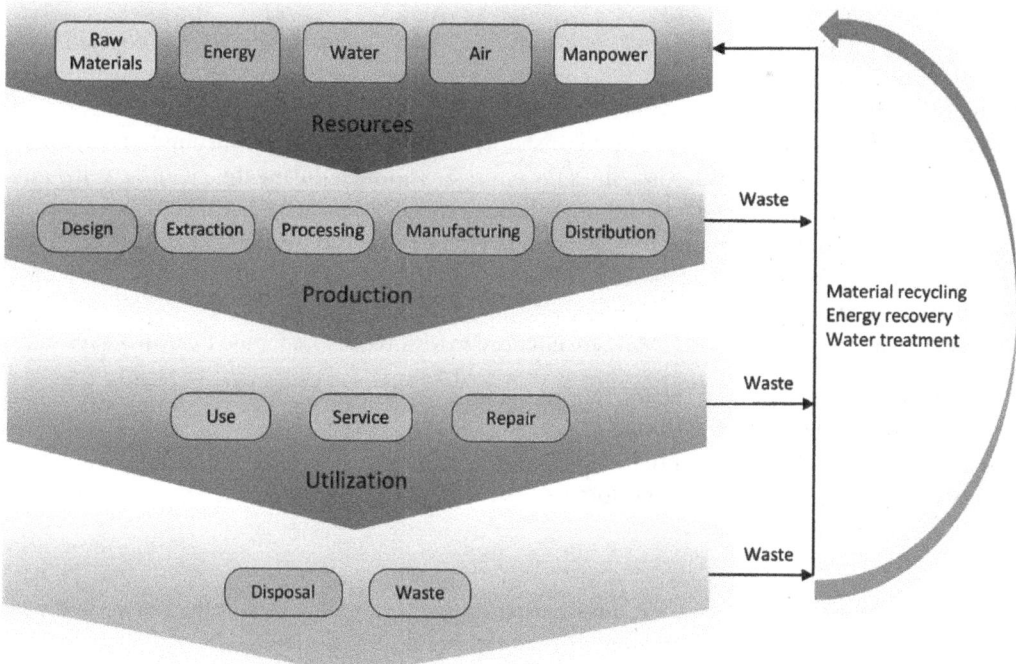

Figure 1.8 A system layout for the life cycle assessment.

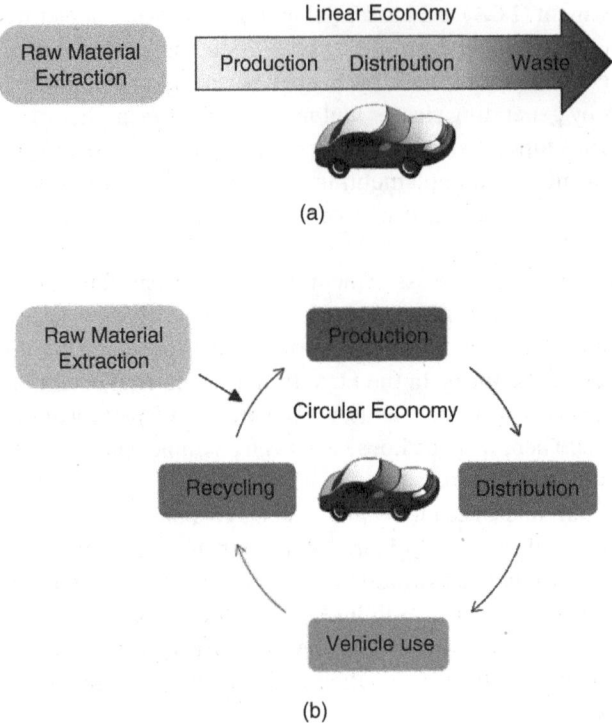

(a)

(b)

Figure 1.9 Two specific views of the life cycle of the vehicles: (a) linear economy and (b) circular economy.

As another interesting example, the heat transfer enhancement techniques are considered a common way to improve the thermo-fluid performance of heat exchangers and burners. In order to enhance the heat transfer, many methods have been introduced and implemented in practical applications. It is clear that enhanced heat transfer will help reduce energy consumption and emissions depending on it. However, sometimes the emissions emitted during the manufacturing and operating the technique developed to improve the heat transfer may be higher than the cases without enhancement or with another option. So, it is really needed to perform a complete analysis considering all performance criteria such as thermodynamics, heat transfer, fluid flow, economy, environmental impacts, etc.

In order to perform a complete LCA, there is a need to seek responses for the following questions:

- What raw materials to use?
- Where to order these materials from?
- What types of energy to use in production?
- How to transport the products to customers?
- How to recycle the materials?
- How to dispose of the wastes?

While seeking the responses for these questions, the key criteria must be minimizing the cost and environmental impacts and maximizing the product quality and efficiency.

The following key steps, shown in Figure 1.10, should be followed in the LCA analyses and studies:

- Goal and scope definition
- System boundary identification
- Inventory analysis

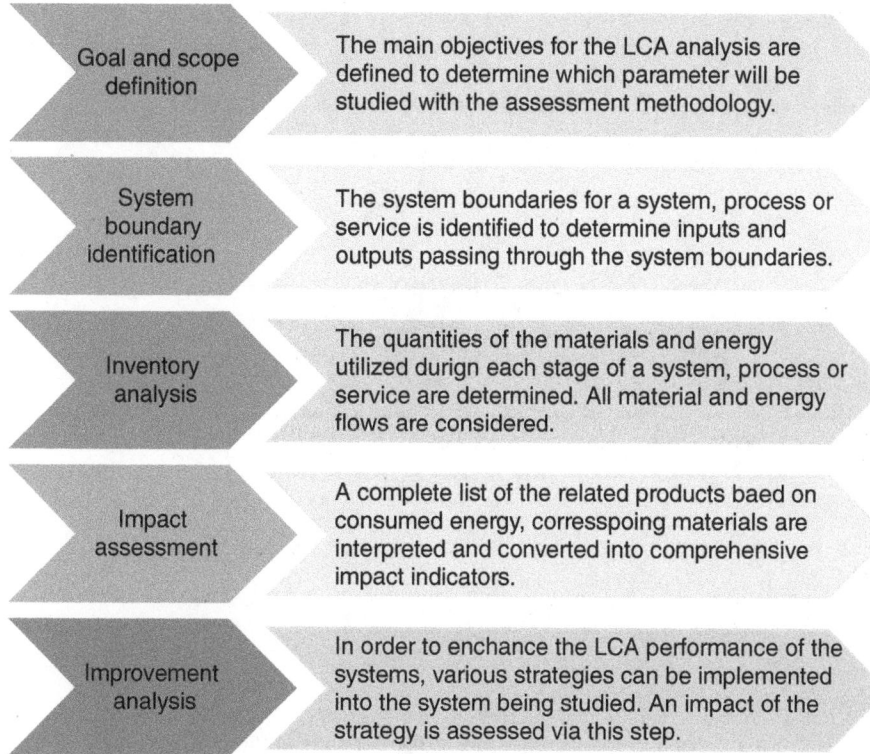

Figure 1.10 The steps of a complete life cycle assessment.

- Impact assessment
- Improvement analysis

The first two steps are more introductory, but considered important ones for achieving a consistent LCA analysis as they define the objectives of the analysis and the boundary of the system. The boundary of the system can be mainly defined as cradle-to-grave and cradle-to-gate. Some conventional LCA studies may include these two first steps combined. In either way, they are important to handle properly depending on the system chosen. The next one is inventory analysis which covers the identification of materials, systems, equipment, devices, etc., used in the system. For this step, some libraries and databases, which are readily available in the literature and the LCA software, are used. This is then followed by the impact assessment which is performed based on the inventory analysis by considering the objectives. Finally, the improvement strategies are implemented to enhance the system's performance and reduce the environmental impact categories.

Illustrative Example 1.1 Let consider the life cycle costing (LCC) analysis and comparative assessment of the lighting, incandescent, compact fluorescent and LED (light-emitting diode) bulbs. Note that incandescent bulbs are less expensive (but converting 95% to heat and 5% to usable light). Compact fluorescent bulbs are obviously much more energy efficient. In addition, we have LED bulbs which are even more efficient than compact fluorescent bulbs. So, let's look at which type of lighting comes out top in the LCC analysis.

Results and Discussion
Table 1.1 demonstrates the benchmark for both lighting bulbs. It is clear from Table 1.1 that only unit price of the bulbs is considered, incandescent bulbs are three times cheaper than compact fluorescent ones and six times cheaper than LED ones. However, when we consider the lifespan of the bulbs, for the same lifespan period which is 50,000 hours, while fifty incandescent bulbs

Table 1.1 The benchmarking for incandescent bulbs and compact fluorescent bulbs. *Source:* Feit Electric.

	Incandescent	Compact Fluorescent	LED
Cost of Buying Bulbs			
Lifetime of one bulb (hours)	1,000	10,000	50,000
Bulb price ($)	2	6	12
Number of bulbs required for 50,000 operating hours	50	5	1
Cost of bulbs ($)	$50 \times \$2 = \$100$	$5 \times \$6 = \$30$	$1 \times \$12 = \$12$
Energy cost			
Equivalent wattage (W)	75	20	12
Watt-hour (Wh) needed for lighting for 10,000 hours	$75 \times 50,000 = 3,750,000$Wh $=3,750$ kWh	$20 \times 50,000 = 1,000,000$Wh $=1,000$ kWh	$12 \times 50,000 = 600,000$Wh $=600$ kWh
Cost at $0.05 per kWh	3,750 kWh $\times \$0.120 = \$450$	1,000 kWh $\times \$0.120 = \$120$	600 kWh $\times \$0.120 = \$72$
Total Costs ($)	$\$100 + \$450 = \$550$	$\$30 + \$120 = \$150$	$\$12 + \$72 = \$84$
Environmental impact	3,750 kWh $\times$ 1 kgCO$_2$/kWh = 3,750 kg CO_2	1,000 kWh $\times$ 1 kgCO$_2$/kWh = 1,000 kg CO_2	600 kWh $\times$ 1 kgCO$_2$/kWh = 600 kg CO_2

and ten compact fluorescent bulbs are required, one LED bulb is needed. Also, when their energy consumption is taken into consideration for the operating hours of 50,000 hours, they are $100, $30 and $12, respectively. Namely, while the compact fluorescent bulb reduces the lighting costs by 3.33 times, LED bulb reduces it 8.33 times. Lastly, thanks to the decreasing electricity consumption, the CO_2 emissions therefore reduce significantly. Consequently, despite the fact that the unit price of the LED bulbs are higher than the compact fluorescent and incandescent bulbs to build, the operating costs and environmental impacts of LED bulbs are drastically lower than compact fluorescent and incandescent bulbs.

Illustrative Example 1.2   Compare the amount of CO_2 emission for transportation to the Ontario Tech University campus with different types of vehicles:

Let us consider a student who daily comes to the Ontario Tech University campus from the residence/home which is about 20 km away from the campus. Calculate the amounts of CO_2 emissions of the transportation per day for the following cases: vehicle I: gasoline, vehicle II: hybrid (gasoline and electric), vehicle III: plug-in hybrid, and vehicle IV: electric (either battery or fuel cell (hydrogen) based).

Results and Discussion

As we are consider the Ontario Tech University campus as final destination, the actual operational emission values for these four common types of passenger cars used in the Ontario, Canada are shown in Figure 1.11. As expected, gasoline-fueled car emits the highest CO_2 with the value of 158.2

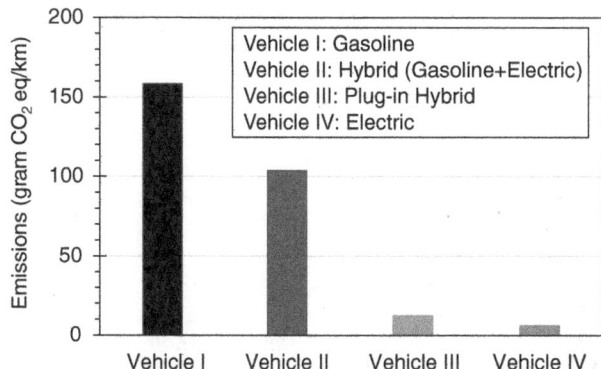

Figure 1.11 The emission values of the various vehicles using in the Ontario, Canada (data from [5]).

grams CO_2e/km. The hybrid vehicle (gasoline and electric) (HEV) emits 103.6 grams CO_2e per km. Plug-in hybrid vehicle emits 12.4 grams CO_2e per km. Finally, full electric powered one emits 6.2 grams CO_2e per km. Note that these values are for 45% highway and 55% city driving values.

For the student considered in this problem, total daily commuting emissions will be 6336, 4144, 496, and 248 grams CO_2e/km for each vehicle, respectively. Consequently, upgrading the vehicle from gasoline to hybrid vehicle will reduce the emissions 34%. It will be 92% for plug-in hybrid vehicle and 96% for electric vehicle. Here, it should be noted that the electricity production emission will be critical for plug-in and full electric vehicles. In this problem, Ontario's electric emission values is taken into consideration. For the hydrogen fueled vehicle, the main concern will be the method of the hydrogen production and energy source. When hydrogen is produced via electrolysis by using renewable energy sources, as hydrogen fueled vehicle doesn't emit any CO_2 emission, the emission for the student's car will be zero.

1.9 Industrial Ecology

The field of industrial ecology is now recognized as an emerging and challenging discipline for scientists, engineers, and policy-makers. Often termed the "science of sustainability." The multi-disciplinary nature of industrial ecology makes it difficult to provide a consistent and universally accepted definition. Figure 1.12 depicts an illustration of the industrial ecology. It is promoted as an approach to close industrial production loops and reduce waste, thereby making better use of resources and preventing the overuse of raw materials. Industrial ecology aims at transforming industries to resemble natural ecosystems where any available source of material or energy is consumed by some organism. The managerial approach of industrial ecology essentially involves analyzing the interaction between industry and the environment, through the use of tools such as life cycle analysis (LCA).

Industrial ecology provides a comprehensive view of design for the environment. It is very popular for engineering design since it combines it with environment principles. It aims to minimize and eliminate the overall environmental consequences of engineering design decisions. Industrial ecology includes:

- Circulating and reusing the materials flows within the system with a minimum cost and effort effectively and efficiently,
- Reducing the amount materials used in products to achieve a particular function,
- Protecting living organisms by minimizing or eliminating the flow of harmful substances, and
- Minimizing the energy consumption and the waste heat to the surroundings which will also help reduce thermal pollution.

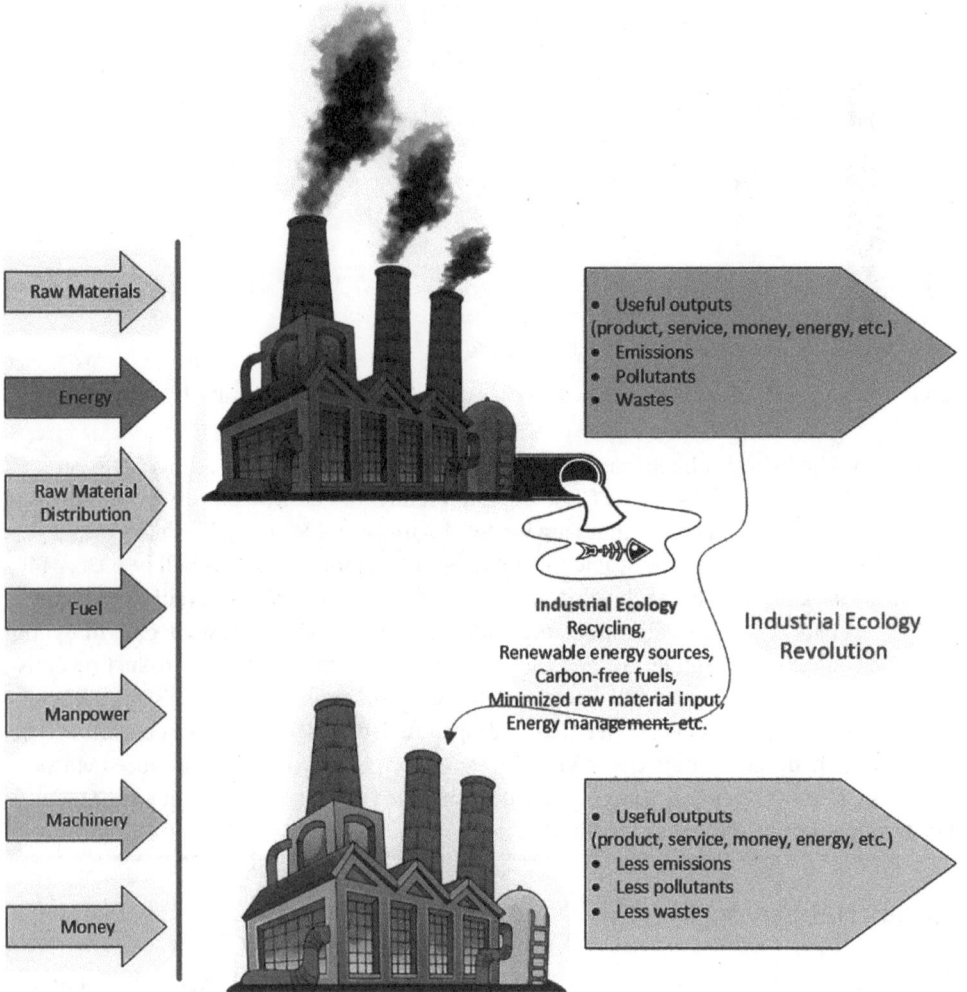

Figure 1.12 Illustration of industrial ecology.

In order to analyze any industrial ecology, performing environmental analyses based on LCA is not enough; energetic, exergetic, and economic analyses should be performed with it. Thus, it is possible to see an entire picture of the sustainability for any system, device, production process, service, etc.

1.10 Energy Labeling

In order to increase the use of energy-efficient devices and guide the customer in the market, many applications have been carried out such as energy efficiency labels. In Figure 1.13, the energy labels used in refrigerators, washing machines, and vehicles are shown. Such practices are both informative and encouraging to people. Since devices used in daily life such as TV, computer, refrigerator, air conditioner, washing machine, heater, etc. can be easily selected according to their energy-efficient situation, they have wide usage. Thus, it is possible to reduce energy consumption in residential buildings. Extending these labeling and grading applications to devices and applications which are consumed energy will play an important role in reducing energy consumption.

In the USA, the Energy Star symbol, which is the government-backed symbol for energy efficiency, is used for providing simple, credible, and unbiased information that consumers and businesses rely on to make well-informed decisions. Energy Star, and its partners, which are thousands

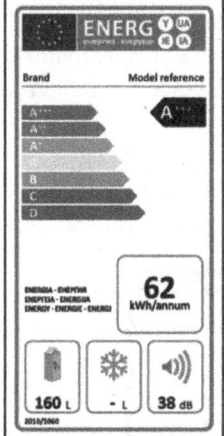

Current energy labels used in European Union countries.

These labels use in Fridges and freezers, dishwashers, washing machines, washer-dryers, electronic displays including televisions and lighting.

Source: Ref. [6]

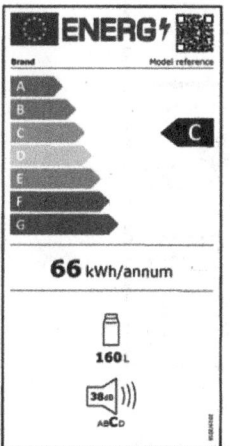

New energy labels will be used in European Union countries.

The label given above has been updated by European Commission as a result of the development of more and more energy efficient products, and because the difference between A++ and A+++ is less obvious to the consumer, the EU energy labels categories will be gradually adjusted to reintroduce the simpler A to G scale.

For example, a product showing an A+++ energy efficiency class could become a class B or lower after rescaling without any change in its energy consumption. The class A will initially be empty to leave room for more energy efficient models to be developed.

Source: Ref. [6]

A label for a gasoline-fueled vehicle, which indicated CO_2 emission, the different consumption if fuel (urban, extra urban, mixed)

Source: [7]

Energy Star symbol used in the USA

Source: [8]

Figure 1.13 Some common energy labels practically used in various countries around the world.

of industrial, commercial, utility, state, and local organizations, have provided saving in residential and commercial sectors more than 4 trillion kilowatt-hours of electricity and achieve over 3.5 billion metric tons of greenhouse gas reductions, equivalent to the annual emissions of more than 750 million cars. In only 2018, the Energy Stars and its members have helped to save $35 billion in energy costs in the USA [8].

In order to reduce emissions in the transportation sector in Canada, they have started to use energy labels on vehicles. Figure 1.14 shows the energy labels for gasoline- and hydrogen-powered vehicles. The label indicates the fuel consumption rate, yearly fuel cost and emission values. Thus,

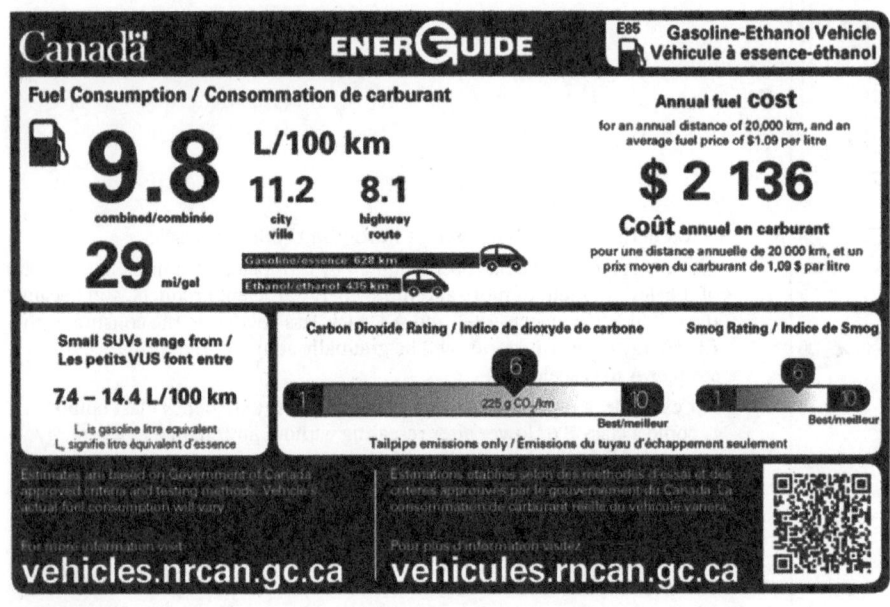

(a)

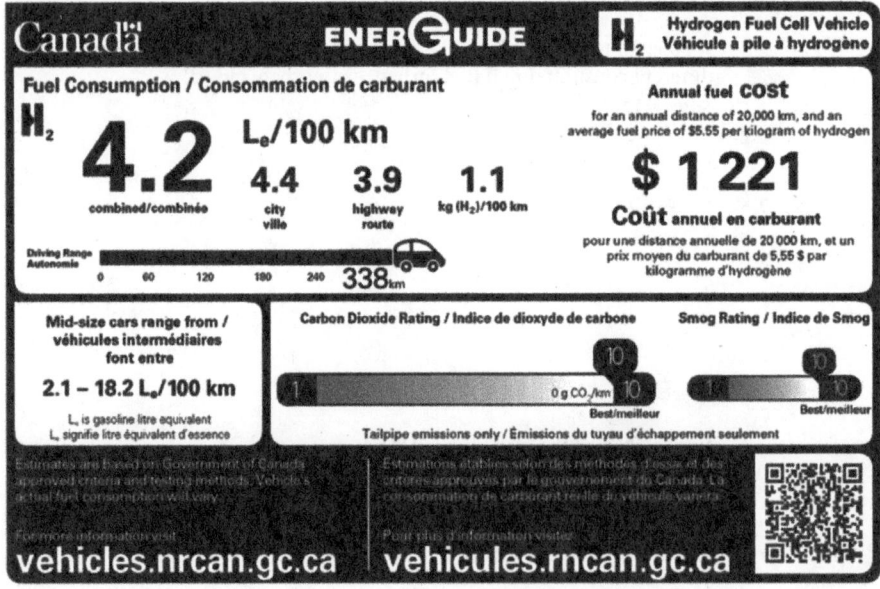

(b)

Figure 1.14 EnerGuide labels for (a) gasoline-fueled and (b) hydrogen-fueled vehicles in Canada (adapted from [9]).

they have created an awareness of the environmental effects caused by vehicles, by including encouraging impacts on fuel consumption and cost. Here, the potential of the hydrogen-fueled vehicles to reduce the emissions and cost is clearly seen in Figure 1.14.

The labeling is technically critical for every sector and provides a significant recognition and a meaningful information for selection of the right products as needed. Such a labeling is also very common in the food sector. Figure 1.15 shows the energy and contents label for a frozen lasagna for four people. It is clear for everyone that this kind of labeling provides the important information for the people to understand the contents and possible implications. These labels increase awareness to choose the right product for consumption. The bottom line here is that labelling has to be a prime responsibility for every product, system, service or application. For a sustainable future, the use of energy labels should be expanded for energy systems, buildings, products, services, etc. With the carbon taxes, the use of energy labels will probably be further increased. Figure 1.16 demonstrates a potential energy label for a thermal power plant. A label for a power plant summarizes all critical parameters for the entire system such as fuel type, fuel consumption rate, waste, energy recovery option, etc. Critical environmental impact categories should be included in the label. Thus, it is possible to understand how power generation systems work effectively. An energy label may consist of the following for per W, kW, or MW production, transportation, conversion, consumption, etc., but not limited to:

- Fuel type
- Fuel consumption rate
- Total energy input
- Total energy output
- Cooling/heating requirements
- Energy or exergy efficiency (or performance coefficient)
- Emissions (or environmental impact categories)
- Green category (i.e., out of 10)

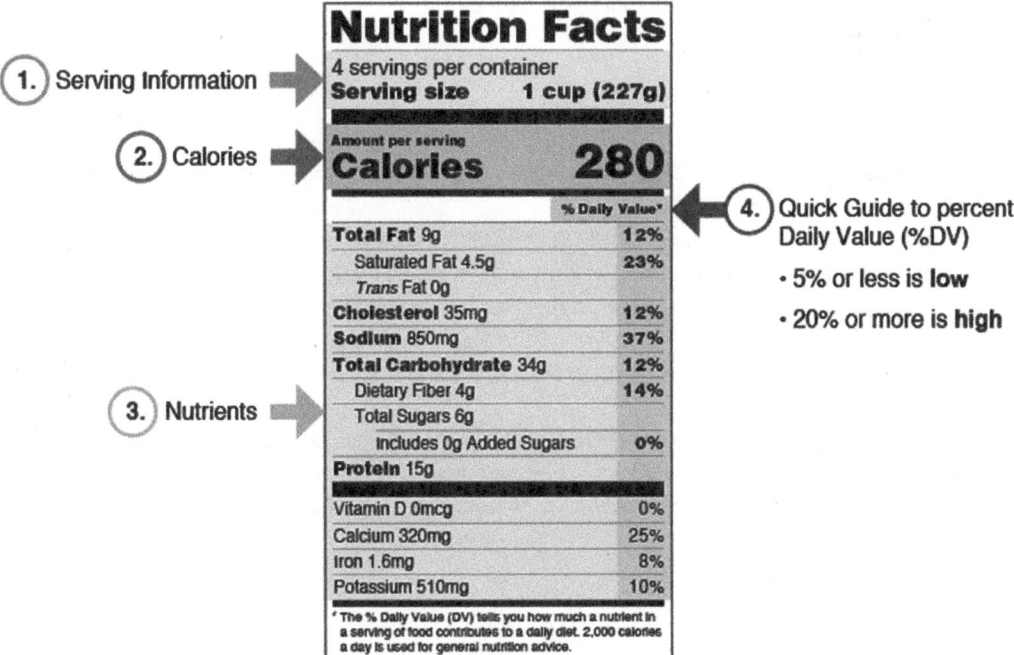

Figure 1.15 Nutrition facts label (adapted from [10]).

Thermal Power Plant Facts	
Fuel type:	......
Fuel consumption rate (kg/s):	......
Total energy input (MW):	......
Power output (MW):	......
Cooling requirement (MW):	......
Energy efficiency (%):	......
Heat recovery option (Y/N):	......
Heat rejection rate (MW):	......
CO_2 emissions (kg/MWh):	......
SO_2 emissions (kg/MWh):	......
NO_x emissions (kg/MWh):	......
VOC emissions (kg/MWh):	......
PM emissions (kg/MWh):	......
Ash (kg/MWh):	......
Ozone depletion potential (ODP):	......
Global warming potential (GWP):	......
Acidification potential (AP):	......

Figure 1.16 A potential energy label for a power plant.

1.11 Closing Remarks

This chapter deals with the energy and environmental perspectives. First, the importance of energy is presented with illustrative examples. Next, energy issues and environmental impacts are introduced and discussed from various perspectives. Then, some smart solutions to overcome energy and environment related challenges are introduced and discussed. The role of engineers in providing solutions is discussed. Finally, energy labels are introduced potentially for energy sector same as what is done in other sectors, such as food sector.

References

1 International Energy Outlook 2021 (2022). https://www.eia.gov/outlooks/ieo (accessed 28 July 2022).
2 BP. Statistical review of world energy. (2020). https://www.bp.com/en/global/corporate/energy-economics/statistical-review-of-world-energy.html (accessed 26 April 2020).
3 Dincer, I. and Erdemir, D. (2021). *Heat Storage Systems for Buildings*, 1e. Amsterdam: Elsevier. doi: 10.1016/C2019-0-05405-2.
4 Dincer, I. (2016). Smart energy solution. *International Journal of Energy Research* 40: 1741–1742. doi: 10.1002/er.3621.
5 Market Snapshot: how much CO2 do electric vehicles, hybrids and gasoline vehicles emit? (2018). https://www.cer-rec.gc.ca/en/data-analysis/energy-markets/market-snapshots/2018/market-snapshot-how-much-co2-do-electric-vehicles-hybrids-gasoline-vehicles-emit.html (accessed 22 September 2022).

6 About the energy label and ecodesign. (2022). European Commission. https://ec.europa.eu/info/energy-climate-change-environment/standards-tools-and-labels/products-labelling-rules-and-requirements/energy-label-and-ecodesign/about_en (accessed 22 September 2022).

7 European Union Energy Label (n.d.). https://en.wikipedia.org/wiki/European_Union_energy_label#:~:%20text=The%20energy%20efficiency%20of%20the,%20they%20choose%20between%20various%20models (accessed 2 August 2020).

8 Energy Star (n.d.). https://www.energystar.gov (accessed 5 August 2020).

9 Natural Resources Canada. Comprehensive energy use database. (n.d.). https://oee.nrcan.gc.ca/corporate/statistics/neud/dpa/menus/trends/comprehensive_tables/list.cfm (accessed 9 August 2020).

10 How to understand and use the nutrition facts label. (2022). U.S. Food & Drug Administration. https://www.fda.gov/food/new-nutrition-facts-label/how-understand-and-use-nutrition-facts-label (accessed 29 July 2022).

Questions/Problems

Questions

1 Discuss the importance of energy with some examples.
2 Explain how global issues affect people's life?
3 Discuss the major environmental issues.
4 Discuss the projected CO_2 emissions from 2020 to 2050.
5 What is energy intensity? Explain with an example.
6 What are the key requirements in obtaining clean air, clean water and clean food?
7 Define each item in the concept of 3S+2S=5S.
8 Discuss the importance of the storage techniques in 5S rule with some practical examples.
9 What are the nine specific targets to achieve the smart solutions?
10 List the nine branches specific smart energy portfolio branches and discuss each in brief.
11 Discuss the role of engineering and its consequences.
12 What is life cycle assessment? and explain how to complete its key steps.
13 What are the six questions to perform a complete LCA analysis?
14 What is life cycle costing? and why is it required?
15 What is industrial ecology? Discuss its role and implications.
16 What are the main differences between linear and circular economy?
17 Why is energy labeling important? Discuss it with examples coming from our daily life.

Problems

1 There are various light bulb types where they have different energy consumptions and costs. Perform a comparative life cycle cost analysis and assessment for spotlight bulbs, halogen lamps, led strip lights. Use the information provided in the table below and fill the missing sections to make a life cycle cost analysis.

	Spotlight	Halogen	LED strip
Lifetime of one bulb (hours)	5,000	7,500	15,000
Bulb price ($)	5	4	12
Number of bulbs required for 30,000 hours	3	2	1
Cost of bulbs ($)	...	...	...
Energy cost			
Equivalent wattage (W)	20	60	20
Watt-hour (Wh) needed for 30,000 hours	...	...	...
Cost at $0.05 per kWh	...	...	...
Total Costs ($)	...	...	...
Environmental impact (1 kgCO$_2$/kWh)	...	...	...

Sources for table images: OSRAM GmbH/Koninklijke Philips N.V./BAZZ SMART HOME.COM

2 Make a comparative emissions analysis for the four modes of daily transportation from home to school, such as diesel fueled vehicle, pneumatic vehicle, electric vehicle and hydrogen vehicle (similar to the example given illustrative example 2). Discuss how more carbon dioxide emissions reduction will be possible if there are public gasoline fueled and electric buses used for about 50 students instead of using individual cars.

3 Consider that the diesel-fueled bus consumes 20 liters of diesel per 100 km and emits 400 grams CO_2eq/km. The hydrogen based electric bus consumes 2 kWh/km where hydrogen is produced by renewable energy sources in a clean manner. Compare the calculated results with the results presented in illustrative Example 1.2.

4 Make an example about industrial ecology based on the criteria presented for a product selected and discuss pros and cons.

5 Identify energy labels available for at least five different commodities (including household appliances) and see what criteria each label considers. Discuss them comparatively.

2

Energy Sources and Sustainability

2.1 Introduction

Sources are critical for everything that humanity needs to cover the key necessities of life for survival. For example, sources, namely food, water, and energy are badly needed. Having clean sources are essential to better health and better welfare for humanity. Since the prime subject is energy, we need to look at the energy sources and availabilities. There are mainly two key points in energy topic, which are supply and demand. Figure 2.1 demonstrates how humanity uses energy sources, such as fossil fuels, nuclear, and renewables, to cover their demands which are badly needed for the activities of people and economic sectors. Making demand side management requires careful planning on the sources side to ensure that the useful outputs needed by people, sectors, and communities as illustrated and listed in Figure 2.1. It is also a well-known fact that getting clean outputs as demanded requires clean sources and their management cleanly.

As mentioned in Chapter 1, energy consumption is primarily considered responsible for many global issues, particularly air pollution, global warming, water pollution, stratospheric ozone depletion, etc. In order to reduce those impacts, the recipe is clear for the solution. Because we will not reduce energy demands as they increase with exponentially with population and life standards, it is required to reduce the use of fossil fuels and utilize clean sources. Even though this sounds easy to implement, they have various complications. Therefore, the sustainability of the energy sources is the main critical topic for the future of the energy systems. Since energy is required for anything everything, it has great impact on the world's future as well.

Figure 2.1 depicts the main energy sources: fossil fuels, nuclear energy, and renewables. The use of energy resources has started since the beginning of humanity. People first started to use fossil fuels, especially wood for heating and cooling purposes. Then, they started to use coal. Coal took significant attention after the industrial revolution with the invention of steam engines due to their relatively higher heating values than biomass-based sources and the discovery of large reserves. Coal was the main fuel for the steam engines and primarily used in the transportation and industrial activities. Also, it is still the main fuel source of thermal power plants. It was widely used for meeting the heating demands in buildings until a few decades. Now, in many countries, it is prohibited to use coal for building heating purposes due to air pollution in the communities. It is expected that the use of coal in the energy sector will reduce due to its increasing environmental impacts such as CO_2 emissions, acid rain, and air pollution.

With the electrification in the world, nuclear power plants have been a part of our life since December 1951. Today, approximately 10% of the gross electricity generation is actualized by nuclear sources. In some countries, the share of nuclear power plants in electricity generation

Introduction to Energy Systems, First Edition. Ibrahim Dincer and Dogan Erdemir.
© 2023 John Wiley & Sons, Inc. Published 2023 by John Wiley & Sons, Inc.
Companion Website: www.wiley.com/go/Dincer/Introduction_to_Energy_Systems

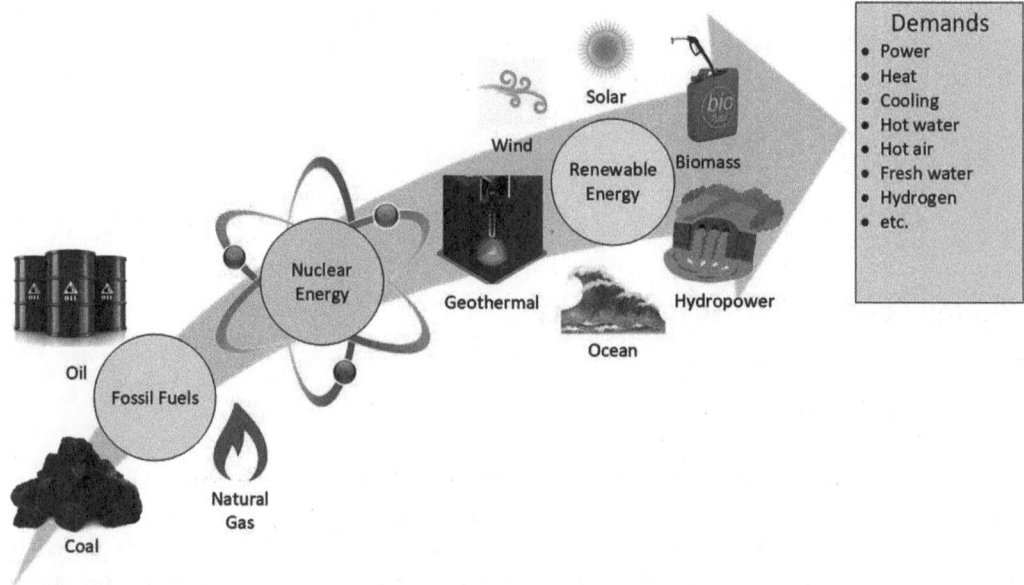

Figure 2.1 An illustration of common energy sources and ultimate demands.

reaches up to 70%. Since the dependence on electricity will increase, safe, reliable, and especially small modular nuclear technologies are really needed for our future. Additionally, despite the practical applications of the fusion reactors have not been appeared yet, there is a significant expectation on this technology for clean energy production. It is expected that a big milestone may going to be achieved, in order to make fusion reactors possible in upcoming a few years by researchers.

Although people were utilizing renewables via passive solar thermal systems, wind and water mills for many years, their use in engineering and technological manner has started with solar water heaters and space heating systems since the 1970s after the petrol crisis. Due to the challenges in the supply and distribution of fossil fuels, the countries that do not have fossil fuel sources started to find alternative solutions. With the trend of renewabilization, solar PV electricity, solar thermal power plant and wind turbines have taken part in energy sector. Many governments have supported and encouraged the investment of these energy power plants. Today, we are practically utilizing renewables via many techniques. They are one of the key sources of sustainable future.

Selection of the energy resources are one of the critical points for the design the energy systems. Energy resource should be selected according to the economic conditions, environmental impacts, energy contents, efficiency, reliability, safety, their availability, their transportability, etc. Energy resources have a great impact on the sustainability of the energy systems while meeting the needs. Main issue regarding energy resources is that the distribution of the energy resources is not uniform all around the world, especially for the fossil fuels. It is also similar with the renewables. While solar energy is available in the certain location in the world due to geometrical condition of the globe, geothermal is available in some locations. Wind energy depends on the air streams in the world. Therefore, it is required to perform a comprehensive analysis to choose energy source for any energy systems.

In this chapter, three significant points, five key dimensions, and historical perspectives of energy use are presented. Then, exponential growth dynamic and its importance in energy sector is discussed. Next, dimension of the sustainability is expressed. Finally, the sustainability assessment indicator is presented with an illustrative example.

2.2 Three Key Points

Increasing energy demand and hence related fossil fuel consumption have brought the environmental issues into the picture. Although there are various environmental impact categories, CO_2 emissions is the critical one among them, especially for the fossil fuels. The minimizing the CO_2 emissions is often associated with clean energy systems.

The CO_2 emissions in the world is ever-increasing with the increasing population. Figure 2.2 shows the CO_2 emissions per capita for the world and some selected countries. The CO_2 emission per capita is recorded to be 3.88 tCO_2/capita in 1990 and 4.39 tCO_2/capita in 2019. When the data of 2010, 2015, and 2019 is observed to be 4.42, 4.41, and 4.39 tCO_2/capita, respectively. When the data for 2019 is considered, the Saudi Arabia, Canada, and United States has the higher emissions per capita. Energy consumption is mainly responsible for the emissions.

As energy demands depend on the sectoral activities, it is not possible to change it significantly. Therefore, managing energy sources is the only way to reduce the emissions and other issues depend on energy consumptions. Here, as energy sources are used for meeting the energy demands, we need to consider the demand profiles and their changes. Three key points are critical for energy supplies and demands. Figure 2.3 depicts these key points accordingly as summarized below:

1) As shown in Figure 2.3a, at any given growth rate of the population, the total energy consumption will grow at a greater rate. There is obviously a strong correlation between them.
2) Life is mainly consisting of peoples demands, parallel to the population, the demands, items, and their rates will increase, as indicated in Figure 2.3b. Fundamental human goals include both the desire for abundant energy on demand and a clean and safe environment.
3) All activities of the people are irreversible which cause disorders and chaos. It is impossible to turn back to the past using any reversible way. Therefore, irreversibilities increase with the increase population along with people's needs, as shown in Figure 2.3c. The future of humanity will continue to follow a one-way and irreversible path.

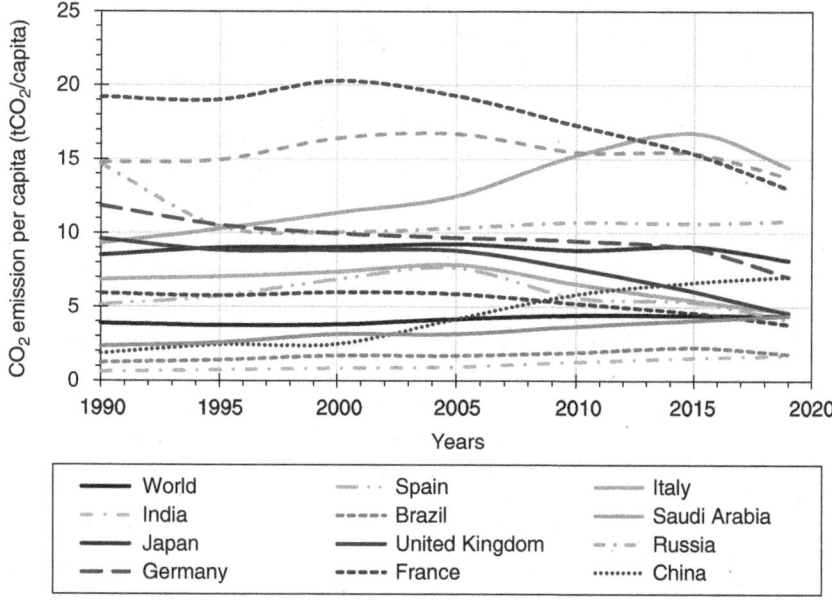

Figure 2.2 Carbon emission per capita for the world and selected countries (data from [1]).

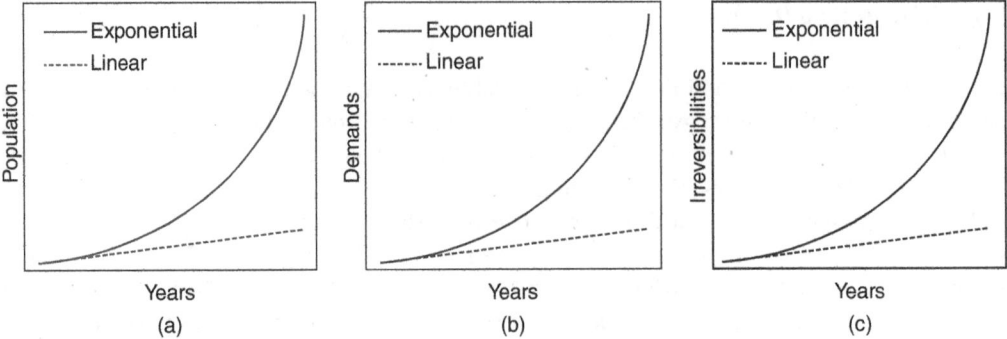

Figure 2.3 Demonstration of the three key points (a) population trend, (b) demand trend and (c) irreversibility trends.

2.3 Five Major Economic Drivers

Energy, environment, and sustainability; those three are very critical for our future. It is really required to identify the key drivers, key criteria, and key elements significantly affecting those. Kruger [2] has presented five major economic drivers associated with P as population, production, power, pollution, and partnership. However, current developments in the world have changed that picture. We are defining the five major drivers over population, production, power, pollution, and partnership, as illustrated in a logical order in Figure 2.4 which are summarized as follows:

- **Population:**

The first driver is population and people make the population. It is the first item because everything we do or aim is for people. Everything is coming down to people. So, increasing population means more demand in all sectors and services. In order to sustain life with increasing demands, we need the development and technical tools to provide a continuous food supply in the form of agriculture and livestock. These are going to be very critical for the population. Today, the main driver for the future projections for all sectors is the population. Engineers and technology developers design their systems according to the populations. Here, it should be noted that the population density may be critical in terms of communities. Communities can be defined as subdivisions such as neighborhood, region, city, province, country, etc. according to the subject dealt with. When we consider population and individuals, it is easy to determine the needs such as food supply, energy, etc. When we raise the level to community stage, houses, buildings, infrastructures, roads, bridges, anything else, you may bring it down to traffic lights. Those are things what will make the community. Population density affects the size, cost, capacity of each item. In order to reach a sustainable future, we need to adjust uniform population density to manage the demand and supply profiles.

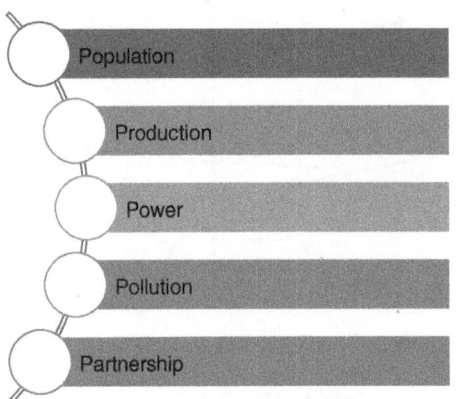

Figure 2.4 Five major economic drivers playing role in energy, environment and sustainability.

- **Production:**

The next driver will be the production. Production is going to come down to various groups, such as food, goods, energy, fuel, feedstock, etc. Production depends on the population and population density as the demands increase with the increasing population. Also, it is very critical how production processes are done as they affect the inputs, outputs, and wastes in the production. For better sustainability, the inputs such as raw materials, energy, water, etc. should be minimized, while keeping the output the same or better. Also, wastes should be reduced, and recycling should be integrated into production activities.

- **Power:**

As one of the major demands for people, communities, and sectors is energy, and power is needed for almost every device, plant, community, sector, etc. which makes the next economic driver power. In order to meet the demands for a community accordingly, the power demand of the population and production activities should be met by sustainable power generation systems. In order to achieve this, integration of renewable energy systems, green nuclear, hydrogen and energy storage is the only option, as mentioned in Chapter 1.

- **Pollution:**

The next P brings pollution into the picture. Almost all human activities cause irreversibilities through pollution and hence environmental impacts. Since one of the significant needs for population is clean air, we start with the air pollution. Today, HVAC systems, conventional energy generation systems, transportation and production process cause emissions which cause air pollution. In order to continue all our activities in a sustainable manner, we really need a clean and safe environment. In order to reach this, clean energy systems will be the most significant driver, as it affects the rest of it directly. For examples, the electric and hydrogen powered vehicles are the promising option for reduce the environmental impact of the transportation section. However, it should be highlighted that it is only possible when electricity and hydrogen is produced in green techniques.

- **Partnership:**

The last P is the partnership. Today's challenges are complicated and depend heavily on many factors. While fighting with challenges, partnership between organizations, governments, countries, and many more must be built between them to manage the processes. The partnership is also required in the professional disciplines to bring all parties together for a common goal to achieve clean environment and sustainable development. In order to build sustainable energy systems, mechanical, energy, software, electrical, and electronical engineering disciplines should work together coordinately with finance and environmental engineering disciplines.

As a summary for 5Ps, everything comes down to people to meet their needs, provide clean and healthy environment, achieve sustainable development. Recognizing the critical role of these 5s in a society becomes a necessity.

2.4 Historical Perspectives

Energy has always been a major demand item for the humanity since ancient times. In the ancient times, energy demand items are only limited with the human activities. Because all activities had been done by manpower. With the invention of the fire, wood was used as the fuel for heating

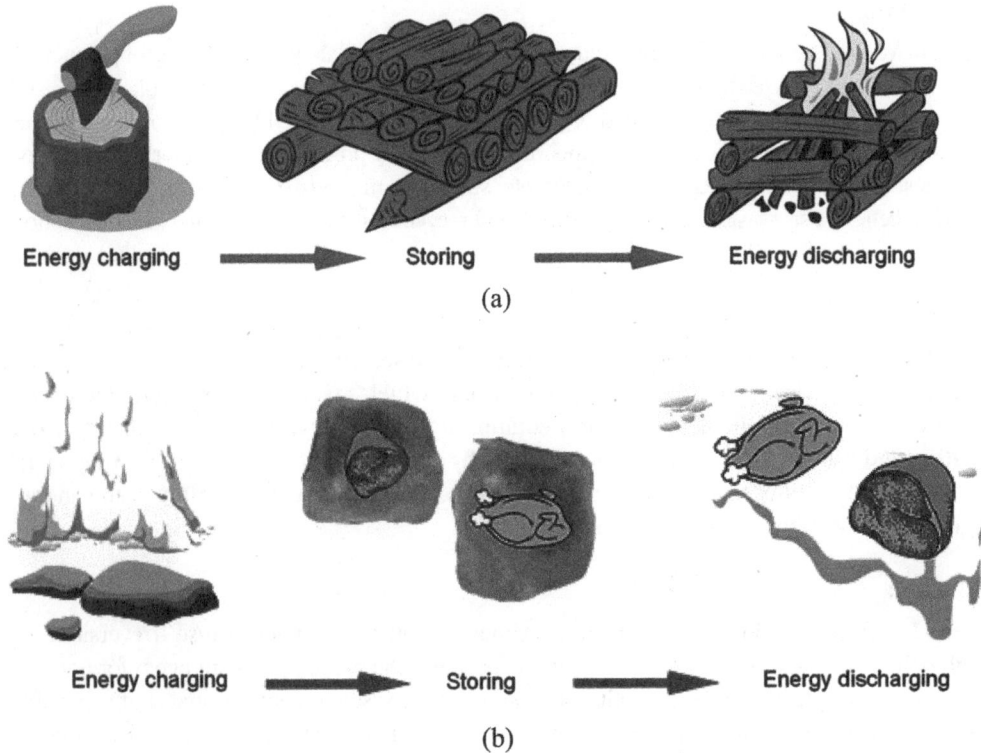

Figure 2.5 Two types of illustrations of ancient way of energy charging to discharging (a) a wood example and (b) a snow example (Reproduced with permission from [3].

purposes. For cooling, snow was used for cooling purposes. For later cooling requirements, the snow and ice were stored in wells and caves. These systems are the first energy storage systems.

Energy storage techniques are, by the way, not new, but had been used by humanity for various purposes since ancient times. Probably, energy storage is as old as civilization itself. From past to present, energy has been an important need for people in any form of it such as thermal, mechanic, electricity, etc. Since recorded times, people have harvested ice and stored it for later use for cooling purposes and food preservation. They have collected wood for the later heating purposes. An illustration of the ancient energy storage methods is given in Figure 2.5. Parallel to developing technology, energy needs have changed and increased considerably. Today, energy is an essential need for people's life and takes part in everything from mobile devices to vehicles, from thermal comfort devices to space technologies. Increasing and changing energy use has forced people to use energy efficiently. Thus, multi-generation systems, which generate more than one useful output, have appeared. In addition, discontinuous energy sources like renewables, and the differences between the energy supply and demand periods forced people to seek alternative methods to manage the energy supply and demand. Herein, energy storage methods play a critical role to solve problems in energy use, to benefit efficiently from the energy source to the recovery of lost energy, and manage the balance betwee energy demand and supply.

2.5 Exponential Growth in Energy Dynamics

To understand the limitations imposed on the quest for abundant energy, it is helpful to understand the arithmetic of exponential growth. Several types of growth are on interest in regard to both personal and energy considerations. Three major ones are linear, polynomial, and

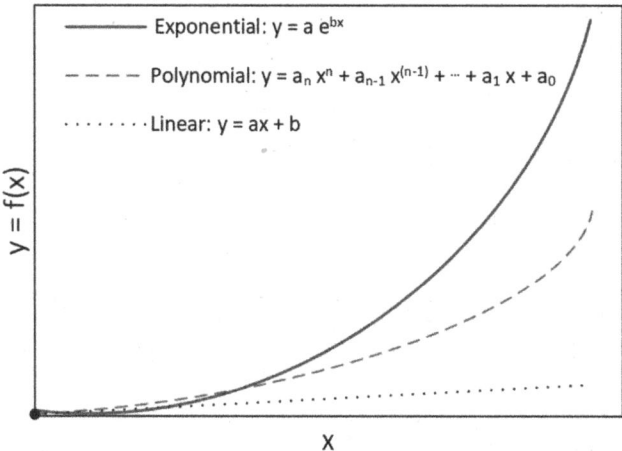

Figure 2.6   The illustrations of various trendlines illustrating the growths.

exponential growths. Also of interest is the form of saturation growth and, from the concept of what goes up must come down, the growth (and decline) rates of extraction of a finite resource. Each of these four types of growth history is reviewed here; we examine resource statistics.

Linear growth is generally used for any dynamics that is increasing or decreasing proportional to any independent variables. Linear growth model is useful for the engineering system such as the relation between gas pedal and acceleration and fuel consumption for a car. There is a linear proportioning between those two parameters.

Although parabolic growth dynamics had advantages to modeling the existing situation since they can almost fit anything and everything, it can cause a substantial mistake in the forecasting exceeding its existing independent value range.

Exponential growth is a process that increases quantity over time. It occurs when the instantaneous rate of change (that is, the derivative) of a quantity with respect to time is proportional to the quantity itself. Figure 2.6 shows the various growth models. It is clear from Figure 2.6 that any item growing exponentially requires attention. Today, parallel to the growing population all demand items growing exponentially (see Figure 2.3). However, here it should be noted that along with the exponentially growing population and demand items, the pollutions and their environmental impacts is increasing exponentially.

When we look at the energy demands over the years plus CO_2 emissions, we can see the exponential growth. The main parameter for this exponential growth is the population. When population increases exponentially, all demand items (energy, product, service, water, etc.) rise exponentially.

2.5.1   Regression Analysis

Forecasting, also called projection of future values of a variable from a recent history, generally is based on extrapolation of the most recent sequence of prior values of the variable that has a well-established historical growth-rate. Forecasting or future projection is widely used for estimating future demands such as energy demand, weather forecasting, emissions, etc. A useful tool for estimating a mean growth rate is the statistical method of regression analysis. This method is especially useful for evaluating the relationship between two variables, for example, chronological behavior of a parameter such as population as a function of time. The data may follow a linear, exponential, logarithmic, or power series relationship. The relationship of interest in forecasting

Table 2.1 General equation for the linear, polynomial, and exponential.

Type	Expression
Linear	$y = ax + b$
Polynomial	$y = a_n x^n + a_{n-1} x^{n-1} + a_{n-2} x^{n-2} + \ldots + a_1 x + a_0$
Exponential	$y = ae^{bx}$ which may be turned into linear form as $\ln(y) = \ln(a) + bx$

future energy demand are linear, polynomial, and exponential relationships. Table 2.1 gives the three types of general expressions for trendlines in the form of linear, polynomial, and exponential relations.

Illustrative Example 2.1 Table 2.2 shows the actual Canadian fossil fuel consumption data which covers the period from 1965 to 2020.

a) Utilize these actual data sets to make projections for short-term till 2030, mid-term till 2050, and long-term till 2075 through curve fitting (trend line analysis) in EXCEL. This will be done for coal, oil, natural gas, and total fossil fuels.
b) Draw the data distribution graphs for all three types of fuels and show regression coefficients and correlations for each of these analyses.

Results and Discussion
These data can be processed in the Microsoft Excel software. The Excel has useful tools for plotting and curve fitting. The first step will be taking these data to the Excel for processing. After that scatter plots can be drawn for each data set. Then, curve fitting can be applied to create correlation for these data. After this step, forecast to forward and backward can be done. In this example, future forecast is performed. Also, future attempts in the fossil fuel consumption is discussed.

Table 2.2 Fossil fuel consumption data for Canada for a period between 1965 and 2020.

Year	Gas Consumption (GC) – TWh	Coal Consumption (CC) – TWh	Oil Consumption (OC) – TWh	Total Canadian (TC) – TWh
1965	216.7	180.0	642.8	1039.5
1966	237.2	176.5	677.3	1090.9
1967	251.9	174.8	723.1	1149.8
1968	280.5	189.9	769.8	1240.2
1969	313.5	183.4	801.5	1298.4
1970	346.3	196.9	854.8	1398.0
1971	369.7	187.1	877.9	1434.6
1972	415.8	176.6	924.3	1516.7
1973	442.2	181.9	1009.6	1633.7
1974	447.0	184.8	1027.0	1658.8
1975	461.6	179.8	1013.2	1654.6
1976	465.7	212.8	1037.9	1716.4
1977	488.4	272.0	1046.0	1806.4

Table 2.2 (Continued)

Year	Gas Consumption (GC) – TWh	Coal Consumption (CC) – TWh	Oil Consumption (OC) – TWh	Total Canadian (TC) – TWh
1978	493.1	223.6	1079.7	1796.4
1979	500.6	211.7	1121.6	1834.0
1980	497.5	257.9	1101.6	1857.0
1981	482.7	263.3	1043.4	1789.4
1982	511.2	278.2	947.3	1736.7
1983	487.5	291.1	890.7	1669.3
1984	541.1	324.3	899.1	1764.5
1985	567.6	311.7	903.0	1782.3
1986	546.3	288.9	923.1	1758.2
1987	552.9	310.5	952.6	1816.0
1988	608.5	333.4	1002.8	1944.7
1989	657.8	332.7	1041.9	2032.4
1990	637.6	315.6	962.8	1916.0
1991	642.0	305.5	912.7	1860.2
1992	680.6	311.2	936.5	1928.3
1993	722.1	276.3	948.6	1947.0
1994	747.5	293.0	968.9	2009.4
1995	772.5	306.2	1016.6	2095.4
1996	785.6	312.5	1041.8	2139.9
1997	801.1	326.6	1073.7	2201.4
1998	787.6	349.2	1094.4	2231.2
1999	841.9	335.7	1115.6	2293.2
2000	892.2	355.2	1109.2	2356.6
2001	840.4	369.1	1140.5	2350.0
2002	874.7	371.3	1183.1	2429.0
2003	897.8	370.4	1227.2	2495.4
2004	878.2	349.9	1280.9	2509.0
2005	875.9	348.3	1245.7	2469.9
2006	870.7	337.4	1256.6	2464.8
2007	936.6	352.8	1290.2	2579.6
2008	925.4	335.2	1245.3	2506.0
2009	897.3	280.1	1214.8	2392.1
2010	915.7	287.0	1269.0	2471.7
2011	1005.5	258.7	1271.2	2535.4
2012	994.2	245.4	1289.1	2528.8
2013	1054.4	239.3	1283.1	2576.8
2014	1098.4	227.6	1281.9	2607.9
2015	1103.3	229.0	1285.0	2617.3

(Continued)

Table 2.2 (Continued)

Year	Gas Consumption (GC) – TWh	Coal Consumption (CC) – TWh	Oil Consumption (OC) – TWh	Total Canadian (TC) – TWh
2016	1049.9	213.1	1281.8	2544.8
2017	1099.0	221.6	1270.7	2591.3
2018	1156.3	181.7	1315.2	2653.3
2019	1172.9	170.2	1304.9	2648.1
2020	1132.6	146.0	1142.5	2421.1

Source: data from Ref [4].

Figure 2.7 shows projections for the gas, coal, oil, and total fossil fuel consumptions in Canada. For the gas consumption (Figure 2.7a), the following exponential curve is plotted.

$$NGC = \left(3 \times 10^{-20}\right) e^{0.0258 \times y}$$

where NGC indicates gas computation and y represents year. R^2 value is calculated to be 0.9534. Gas consumptions are estimated to be 1671 TWh for 2030, 2799 TWh for 2050, and 5334 TWh for 2075. The curve is only considering the relation between year and natural gas consumption. Therefore, it shows an exponential increase. However, many governments are taking action to reduce fossil fuel consumption. In Figure 2.7a, the effect of this decrease is illustrated in orange. The natural gas consumption between 2020 and 2075 is defined as:

$$NGC = -0.3492 \times y^2 + 1420.9 \times y - 10^6$$

Here, R^2 for this equation is found to be 0.9391. In 2075, it is expected that natural gas consumption will be 784 TWh, which is the same level in 1995.

For the coal consumption (Figure 2.7b), there is an increase from 1965 to 2002, after 2002, there is a reduction in the coal consumption. So, coal consumption is divided into two sections. Between 1965 and 2022:

$$CC = 8 \times 10^{-17} e^{0.0215 \times y}$$

Here, CC is coal consumption. R^2 is found to be 0.8384 for this period. After 2022, the coal consumption can be modeled as follows:

$$CC = 4 \times 10^{21} e^{0.021 \times y}$$

This equation is a good fit for the values of 0.9704 of R^2.

Oil consumption (OC) is modeled with linear function with the value of R^2 of 0.753, as shown in Figure 2.7c. The oil consumption as function of year is given below:

$$OC = 9.3302 \times y - 17527$$

It is expected that oil consumption in 2030 will be 1400 TWh. It is found as 1600 TWh for 2050 and 1810 TWh for 2075. Only, yearly change is considered as curve fitted in Figure 2.7c, it will

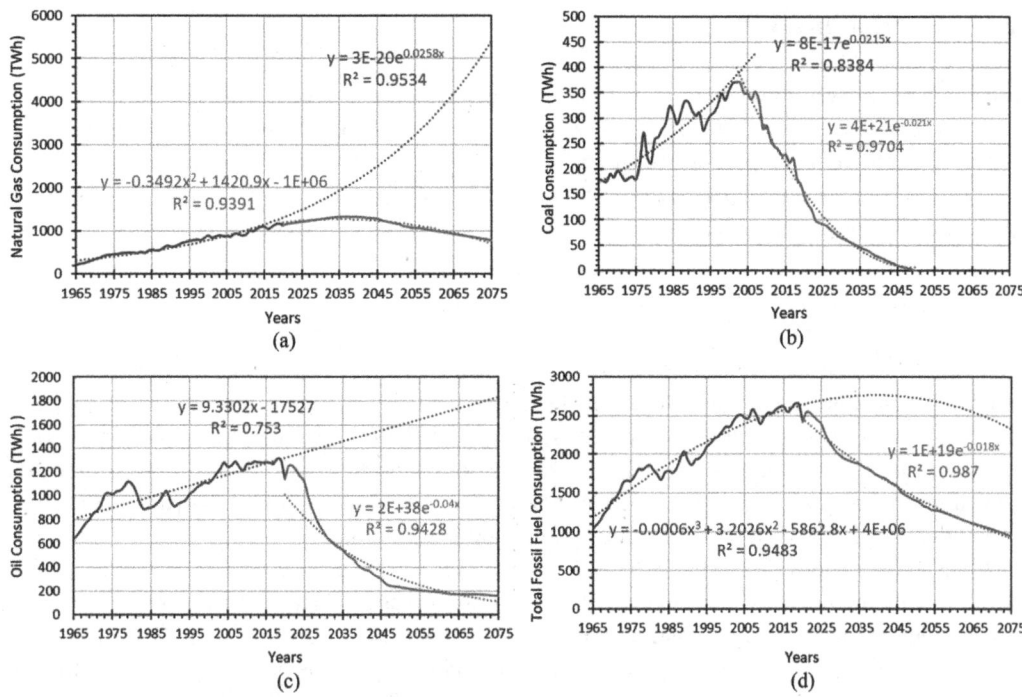

Figure 2.7 The actual data sets in blue and their projections in orange color for (a) natural gas, (b) coal, (c) oil and (d) total fossil fuels consumption in Canada.

increase. However, in the future, it is expected that the OC will decrease with the increasing number of electricity vehicles in the transportation. The oil consumption in Canada can be modeled as:

$$OC = 2 \times 10^{28} e^{0.04 \times y}$$

Here, R^2 value is calculated to be 0.9428. Oil consumption will reduce by 160 TWh in 2075.

Figure 2.7d shows the projection of total fossil fuel consumption (TC) in Canada. Its correlation is the following:

$$TC = -0.0006 \times y^3 + 3.2026 \times y^2 - 5862.8 \times y + 4 \times 10^{-6}$$

Here, R^2 is calculated to be 0.9483. As seen from Figure 2.7d, the increase in the fossil fuel consumption increases until 2046 and it will reduce after this year. In 2075, it is expected that the fossil fuel consumption will reach back its level in 2000. As mentioned previously, the equation only considers the relation between year and the total fossil fuel consumption. The decrease in the coal, oil, and natural gas consumption will reduce the total fossil fuel consumption, too. So, the total fossil fuel consumption between 2020 and 2075 can be modeled exponential as given below:

$$TC = 10^{19} \times e^{-0.018 \times y}$$

where R^2 is found to be 0.987.

2.6 Energy Intensity

Energy intensity was used to be a more important as it was linked to economic development in the part. It is one of the indicators for assessing the development level of the countries. It has similar trend with gross national product. Economic development was requiring a higher energy intensity. And then what they were doing in the past, they were making it. Energy intensity for a country can be defined as energy use per capita.

$$EI = \frac{\text{Total Energy Consumption}}{\text{Population}} (\text{kWh per person})$$

Figure 2.8 demonstrates the energy density for the selected countries. Iceland is by far the largest per capita consumer of electricity in the world due to a low-cost electricity production, a high heating demand and energy-intensive industries along with the low population of 370,000 in 2020. Qatar has high energy density due to heavy cooling demands. Although China is the country that consumes the most energy in the world, its energy density is lower than other counties that consume less energy, due to the China's huge population contribution.

When the developed countries are considered, we can clearly see the relation between economic development and energy consumption. All industrial processes which create an economic worth require energy input. Energy consumptions of the transportation, utility, and commercial sectors are quite high in these countries due to high life quality factors. However, the picture for the energy density will probably get interesting with energy efficient application deployed in these countries. While the rate of increase in the energy consumption is getting lower and lower over the years in the OECD countries, it is increasing in the non-OECD countries. Figure 2.9 shows the

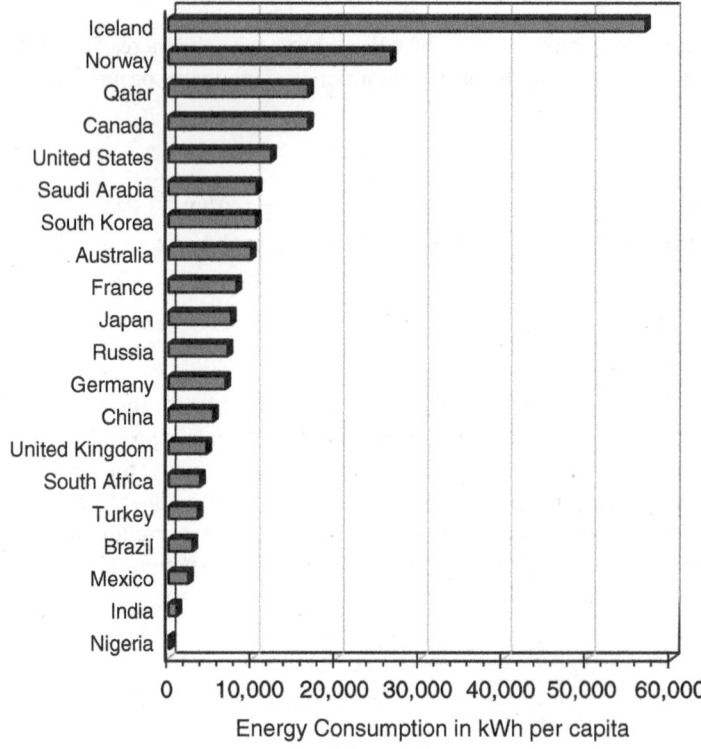

Figure 2.8 Energy intensity for countries in 2021 (data from [5]).

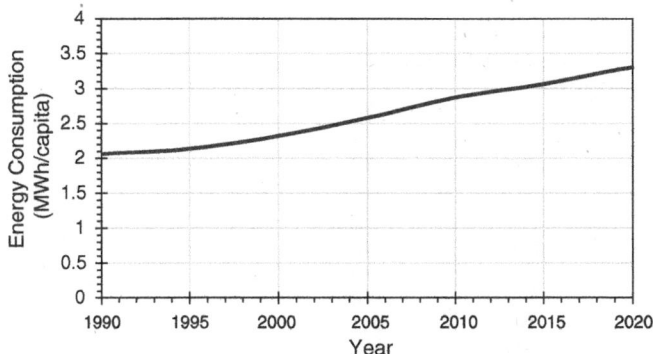

Figure 2.9 Energy consumption per capita for all over the world (data from [1]).

energy consumption per capita for the world. It is clear from the Figure 2.9 that the energy density is ever-increasing due to growing population and rising living conditions.

In addition to the energy consumption per capita, the gross domestic product (GDP) is also used for assessing the level of the development for the countries. Figure 2.10 shows the counties with the highest gross domestic product. It is clear from Figure 2.10 that the United States of America, China, and Japan have the largest value in the world. In 2021, while the sum of USA and China is 40,455 billion US$, the rest of them is 37,370 billion US$. It is clear from Figure 2.10 that USA and China is leading the world in terms of gross domestic production.

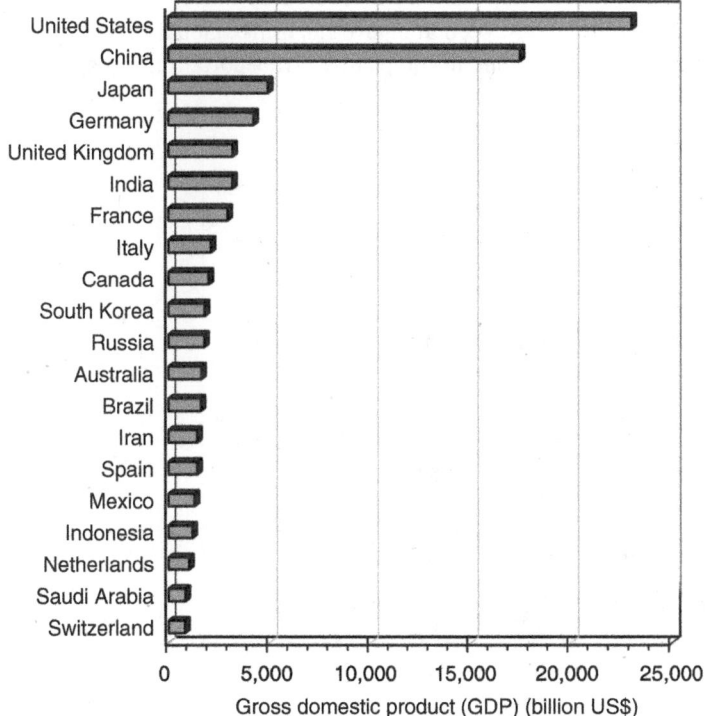

Figure 2.10 Countries which has the highest gross domestic product (GDP) in 2021 (data from [5]).

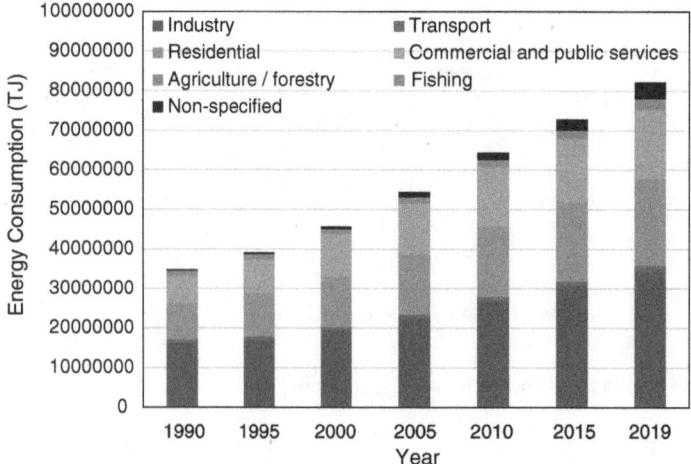

Figure 2.11 Electricity consumption by sector in the world (data from [1]).

Figure 2.11 illustrates the electricity consumption in the world by classifying common sectors. It is clear from Figure 2.11 that industrial, residential, and commercial and public services are mainly responsible for the electricity consumption. Also, as mentioned previously, the energy consumption has grown over the years. At the end of 2019, it is recorded that the electricity consumption of 82,251,569 TJ has occurred globally. Also, in the near future, it is expected that the share of the transportation in the electricity production will increase with the increasing deployment of electricity vehicle. So, the energy consumption trends has changed with the new technologies.

When the total final energy consumption by sectors is taken into consideration, even the increasing trends is the same, the shares are different. As seen in Figure 2.8, unlike global electricity consumption in Figure 2.12, transportation is included in the picture. Transportation is responsible

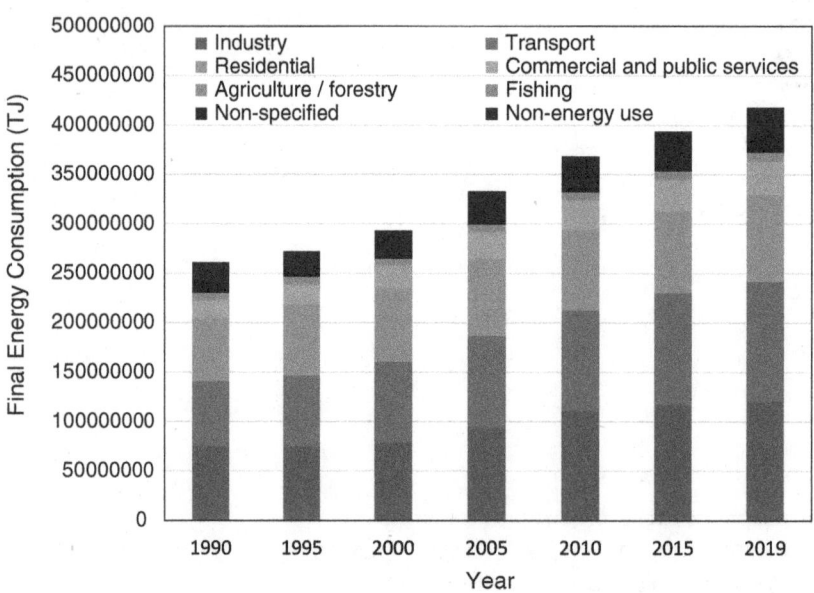

Figure 2.12 Total final energy consumption by sector in the world (data from [1]).

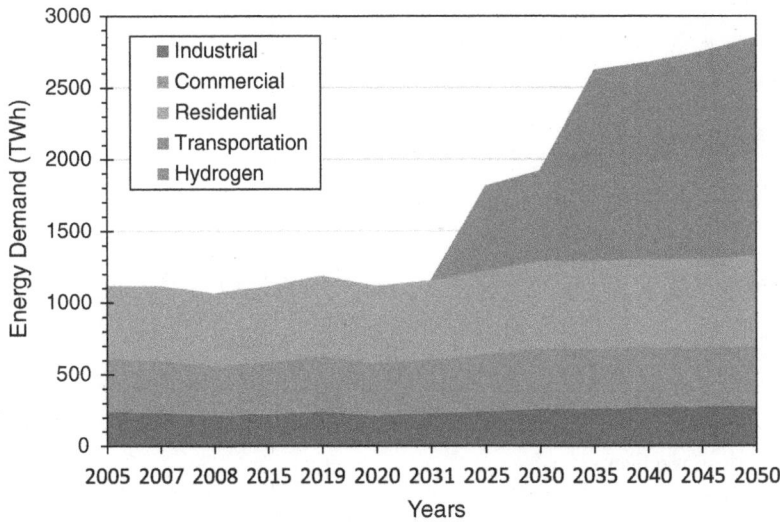

Figure 2.13 Energy demand by sector in Canada (data from [6]).

for one third of the total energy consumption in the world. So, the industrial, transportation, and residential sectors are dominating the global energy consumption. They are covering 80% of the total final energy consumption.

As mentioned previously, while the additional energy demand items take into our lives, many of them exit from our lives. As we mention in Chapter 1, we are in the new era called the hydrogen era. So, it is expected that the energy consumption depending on the hydrogen production will increase with the increase in hydrogen systems to be deployed in the near future. Energy demand of the Canada is seen with future projection in Figure 2.13. In 2050, it is expected that hydrogen production will cover one-tenth of the total energy consumption. Also, the total energy consumption may increase about 30% according to the 2021 data.

2.7 Dimensions of Sustainability

Energy resources needed for societal and sustainable developments require a supply of energy resources that are sustainably available at reasonable cost with no negative social impacts. Energy resources, e.g., fossil fuels, are finite lack of sustainability, while others such as renewable energy sources are sustainable over the relatively long term. Environmental concerns are also a major factor in sustainable development as activities which degrade the environment are not sustainable. The Brundtland Commission defined sustainable development as "development that meets the needs of the present without compromising the ability of future generations to meet their own needs."

In order to maintain sustainability development, we need to achieve the sustainability indexes given in Figure 2.14. Sustainability has seven key dimensions associated with E which are energy, education, economy, environment, ethics, and energy. Energy is of course one of the main drivers in sustainability. Today, everything in people's lives depend on the energy. Almost all devices need an energy input, such as mobile phone, tablet, laptop, vehicle, lights, microwave oven, water heater, etc. In order to manage reaching the sustainability targets, a substantial investment is required; so, economy is also another key dimension for the sustainability. The main objective of sustainability targets is to leave a

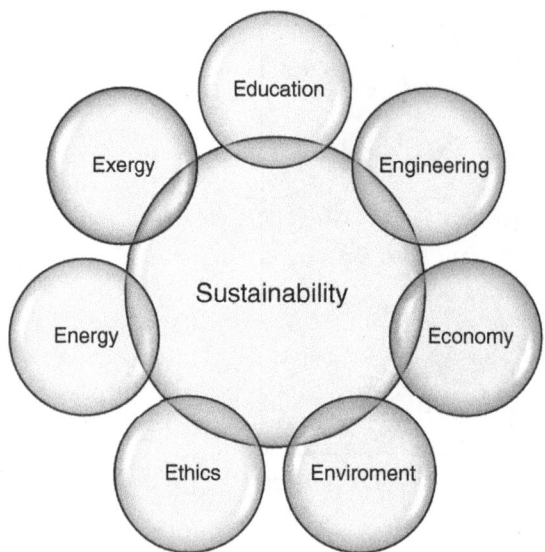

Figure 2.14 Dimensions of sustainability and sustainability as a function of education, engineering, economy, environment, ethics, energy, and exergy (7Es).

clean world to our children and grandchildren through clean air and clean environment. Therefore, the environment is another critical dimension of the sustainability. In order to create and operate energy systems, economies, and environment, people is the key factor. Therefore, well-educated people are really needed to achieve each dimension of the sustainability. Of course, while performing each of these dimensions, ethics plays a critical and ethical principles should be considered to better fulfil responsibilities.

Education at all levels is one of the most critical components in achieving sustainable development. This may even cover information education and training activities to further educate individuals and communities for a better sustainability. Better education will obviously make people more conscious about the environment and hence sustainability. Engineering which is really considered the science of converting the theory to practical solutions for individuals, systems, communities, societies, etc. In this regard, engineering disciplines appear to be very important. The next one is economy which can be classified into numerous sectors, including: food, water, and agricultural, industrial, chemical and petrochemical, commercial, residential, and transportational utility. It is essential to make all these sectors and their subsectors sustainably driven for a better future.

The next one is the environment which is actually the prerequisite for sustainable development. So, if we cannot achieve better environment, we cannot accomplish sustainable development either. We used to think in the past that the nature cleans itself and that we should not worry about it. The end result has been very damaging locally and globally. Although there have been natural cycles taking place over many centuries, resulting in various catastrophic events, the prime damage was originated by a heavy use of hydrocarbon fuels since industrial revolution and obviously caused by human activities locally and globally. We are now in an era where clean environment is more than a need. Trying to implement some partial mitigation solutions locally is good, but quite insufficient.

The next is the component of ethics, which is technically defined as a set of moral principles and values that help govern the behavior and activities of people, which is mostly attributed to individuals or personal acts. However, I want to take it beyond this meaning and make it more societal. Thus, it is more critical to look at societal ethics in harmony with the personal ethics. I also cover social dimensions under this domain so that is critical in achieving sustainable development. The final component is energy, but its magnitude is huge. Clean energy is a key requirement for human wealth and welfare and hence ultimately for sustainable development.

Considering energy evolution historically one needs to note that people started with wood as their primary energy source for their daily activities and continued the use of this energy source through the centuries up until reaching the coal era (particularly with industrial revolution), oil era (primarily after the first world war) and natural gas era (increasingly after 1980s). There is now an ultimate destination in this energy journey towards hydrogen as carbon free fuel, energy carrier, and feedstock.

A large portion of the environmental impact in a society is associated with its utilization of energy resources. Ideally, a society seeking sustainable development utilizes only energy resources that cause no environmental impact (e.g., which release no emissions to the environment). However, since all energy resources lead to some environmental impact, it is reasonable to suggest that some (not all) of the concerns regarding the limitations imposed on sustainable development by environmental emissions and their negative impacts can be in part overcome through increased efficiency. Clearly, a strong relation exists between efficiency and environmental impact since, for the same services or products, less resource utilization and pollution is normally associated with increased efficiency.

In the light of the 6Es, humanity can find optimum to reach the following benefits: better environment, better ecosystem, better efficiency, better economy, and economic development, better synchronization with renewable energy options, better health and healthier societies, and hence better sustainable development.

In order to provide for the needs of humanity in a sustainable way, The Sustainable Development Goals were announced by the UN in 2015; Sustainable Development Goals are shown in Table 2.3.

Table 2.3 UN Sustainable Development Goals for 2015–2030.

Goal Number		Goal Description
1		End poverty in all its forms everywhere
2		End hunger, achieve food security and improved nutrition, and promote sustainable agriculture

(Continued)

Table 2.3 (Continued)

Goal Number		Goal Description
3		Ensure healthy lives and promote well-being for all at all ages
4		Ensure inclusive and equitable quality education and promote lifelong learning opportunities for all
5		Achieve gender equality and empower all women and girls
6		Ensure availability and sustainable management of water and sanitation for all
7		Ensure access to affordable, reliable, sustainable, and modern energy for all

Table 2.3   (Continued)

Goal Number		Goal Description
8		Promote sustained, inclusive, and sustainable economic growth, full and productive employment, and decent work for all
9		Build resilient infrastructure, promote inclusive and sustainable industrialization, and foster innovation
10		Reduce inequality within and among countries
11		Make cities and human settlements inclusive, safe, resilient, and sustainable
12		Ensure sustainable consumption and production patterns

(Continued)

Table 2.3 (Continued)

Goal Number		Goal Description
13	**13** PROTECT THE PLANET	Take urgent action to combat climate change and its impacts
14	**14** LIFE BELOW WATER	Conserve and sustainably use the oceans, seas, and marine resources for sustainable development
15	**15** LIFE ON LAND	Protect, restore, and promote sustainable use of terrestrial ecosystems, sustainably manage forests, combat desertification, and halt and reverse land degradation and halt biodiversity loss
16	**16** PEACE AND JUSTICE	Promote peaceful and inclusive societies for sustainable development, provide access to justice for all, and build effective, accountable, and inclusive institutions at all levels
17	**17** PARTNERSHIPS FOR THE GOALS	Strengthen the means of implementation and revitalize the global partnership for sustainable development

Source: The United Nations' Sustainable Development Goals (adapted from [1]).

The main objective of the sustainable development goals is to achieve a better and more sustainable future the globe. It states the seventeen worldwide issues including climate change, environmental problems, health, poverty, justice, and peace. Sustainable Development Goals invite the countries to start to action reaching the seventeen goals which are presented in Table 2.3. They are important to address the people's needs in both developed and developing countries.

Economy, communication, transportation, population, natural resources, shelter, food supply, energy, etc. those are really critical global issues for counties. There are complex relations between them. However, although each of them is very important, energy is one of the critical one as it affects the rest of them significantly. Therefore, it is a multidimensional problem.

Various parameters are important to achieving sustainable development in a society, some of which are as follows [7]:

- Public awareness. A fundamental step in making a sustainable energy program successful is improving public awareness of its need. This step should be carried out through the media and by public and/or professional organizations.
- Information. Necessary information on energy utilization, environmental impact, renewable energy resources, and so on should be provided to the public through public and government channels.
- Environmental education and training. This activity complements the provision of information. Any approach that does not include as integral education and training is likely to fail, so this activity can be considered as necessary to a sustainable energy program. For this reason, a wide scope of specialized agencies and training facilities should be made available to the public.
- Innovative energy strategies. Such strategies should be included where appropriate in an effective sustainable energy program. In parallel, efficient dissemination of information is required of the new methods through public relations, training, and counseling.
- Renewable energy resources and cleaner technologies. In developing an environmentally benign sustainable energy program, renewable energy sources and cleaner technologies (including heat storage) should be promoted at every stage. Such activities form a basis for short- and long-term policies.
- Financing. Financing is an important tool for achieving the main goal of sustainable energy development in a country and accelerating the implementation of environment friendly energy technologies.
- Monitoring and evaluation tools. In order to assess how successfully a program has been implemented, it is of great importance to monitor each step and evaluate the data and findings obtained. In this regard, appropriate monitoring and evaluation tools should be used.

2.8 Sustainable Development

Sustainability assessment is a complicated procedure which has many dimensions, as shown in Figure 2.14. A sustainability assessment methodology is proposed base on indexes given in Figure 2.15 by Abu-Rayash and Dincer [8]. This model provides an estimation based on energy, exergy and environment along with some economic and social indicators. Although sustainability assessment is crucial, there is no universal standard for sustainability assessment. An accurate, measurable, comparable, and reliably sustainability assessment is required. For energy systems, eight indicators shown in Figure 2.15 can be used for energy systems for sustainability assessment.

Energy is a vital dimension of sustainability assessment as it indicates the performance of the systems considering energy inputs and outputs. It can be defined over energy efficiency. Exergy is

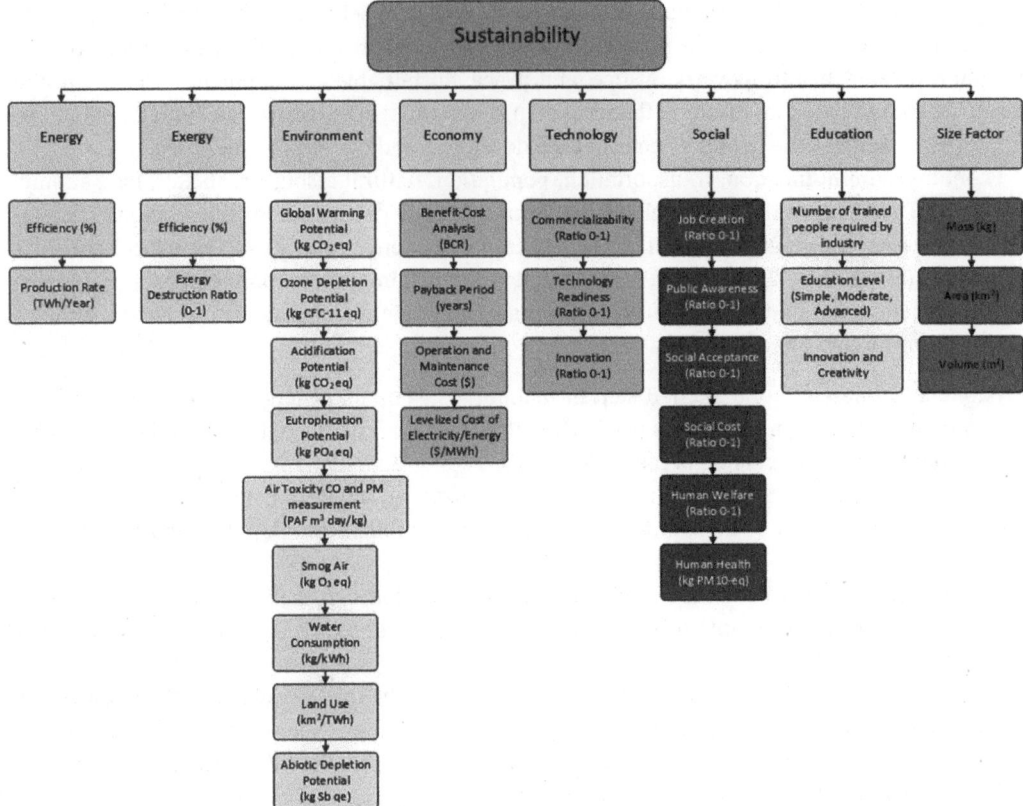

Figure 2.15 Integrated sustainability assessment model for energy systems (modified from [8]).

related to the second law of the thermodynamics. As exergy defines the reversibilities in the system, it can provide more reliable information about the system considering all energy and exergy flows. Exergy efficiency can be used as the exergy indicator for sustainability assessment as it measures the quality of the energy used as inputs and outputs.

The environmental impact categories, which are defined in Chapter 1, can be used for assessing sustainability based on the environmental indicators. The environmental impact categories are, in this regard, listed as:

- Global warming potential
- Ozone depletion potential
- Acidification potential
- Eutrophication potential
- Air toxicity potential
- Water ecotoxicity potential
- Smog
- Water consumption
- Abiotic depletion potential

Economy is a critical indicator for every sector, system, process, and service. So, it is an essential part of the sustainability assessment in the energy sector, too. All attempts require a substantial amount of cost in energy sector for performing better sustainability. Benefit cost ratio, payback period, operation, and maintenance cost can be used for economic indicators in sustainability assessment.

Energy systems can be assumed economically sustainable when the following criteria are achieved [8]:

- The economic benefit of the energy generation outweighs operational, capital, and maintenance cost. Simply, the project is economically viable.
- Energy systems with shorter payback periods are preferred over systems with longer payback periods. This attracts investors.
- Lower levelized cost of energy/electricity. Energy available for everyone at a relatively lower cost.

Many technologies are developing all around the world by researchers and technology developers. Despite many of them presents promising results, the main concern about them their technology level. Technological indicators, including commercializability, technology readiness, and innovation, can be used for assessing the energy systems. For better sustainability, we need commercializable and technologically ready systems, methods, processes, and services.

Social aspects of energy systems are crucial for their sustainability. Social indicators help assess the impacts on the social system, which is composed of the beneficiaries of the energy system, whether directly or indirectly. Job creation, public awareness, social acceptance, social cost, human welfare, and human health are count as the social indicators.

Education within the various stakeholders involved in the construction, operation, and maintenance of the energy system is vital to the sustainability and performance of that system. For example, staff that are more educated reflect more competent and skilled talents, which increases the sustainability score. On the other hand, poorly trained or educated staff could conduct the project in an unsustainable manner. The following can be assumed as education indicators:

- Training
- Educational level
- Innovation and creativity

Size factor is the last sustainability assessment indicator, which covers mass and volume of the system and land use. They are critical since the dimensions of the proposed energy system could be a limiting factor.

The critical step for performing a sustainability assessment is to collect data. After creating dataset, multi-criteria decision analysis is performed for assessment. Non-compensatory aggregation is used in this model to ensure that each dimension is valued accordingly without undermining any important criteria.

The equal weighting is used to aggregate values within one dimension. The following equation shows the weighted arithmetic mean calculation:

$$WAM_{(Y.m)} = \sum_{i=0}^{n} w_i Y_i$$

where w_i is the weight associated with each indicator, Y_i represents the non-dimensional value of the indicator and n is the number of indicators in each dimension. This means that a very high value of an indicator can be compensated by a very low value of another indicator.

The following equation represents the weighted geometric mean calculation as follows:

$$WGM_{(Y.w)} = \prod_{i=0}^{m} Y_i^{w_i}$$

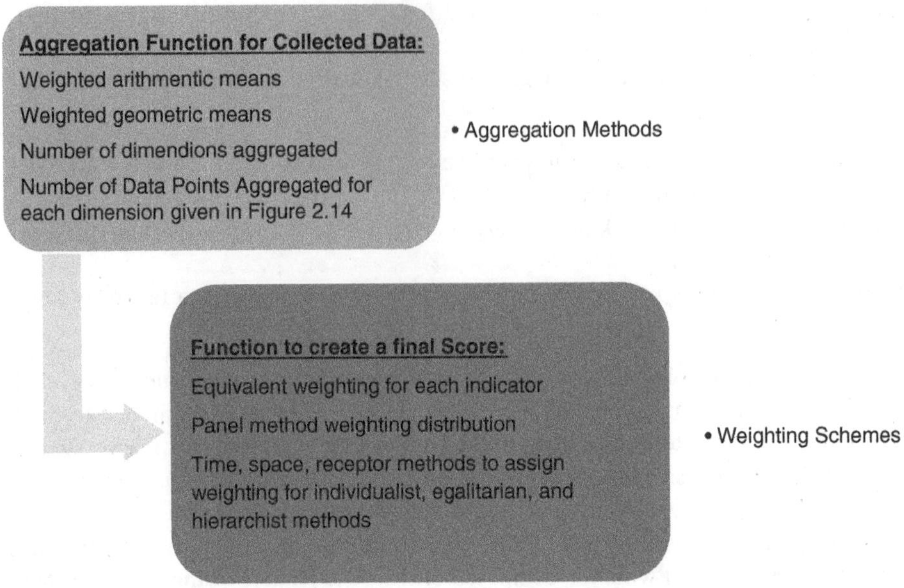

Figure 2.16 The process flow for coming to a final sustainability index (modified from [8]).

Here, w_i is the weight associated with each dimension, Y_i shows the non-dimensional value of the dimension and m is the number of dimensions used in this model.

Weighting is a critical parameter for the sustainability assessment model at stake for biases and inaccuracies. Weighting can be done by various techniques, includes individualist, hierarchist, egalitarian, panel, and equal weighting methods. All those indicators are used for creating the sustainability assessment model panel of panel method, as illustrated in Figure 2.16.

Illustrative Example 2.2 Sustainability assessment of a net zero energy house (Ref: [8])

In this illustrative example, a net zero energy house integrated with solar PV as energy input and a geothermal heat pump is modeled. The systems proposed is evaluated over energetic and exergetic efficiencies and sustainability indicators.

System Description
The view of the proposed system depicts in Figure 2.17. The house is two-storied which has 177 m^2 of area. It is located in Bowmanville, Ontario, Canada. Yearly natural gas and electricity consumptions are 37,771 kWh and 7,331 kWh, respectively. Solar energy is used as the energy source, backed up with battery-based energy storage. The heat pump meets 90% of the heating demand of the building during the winter. The heat pump uses glycol ethylene solution in the vertical ground loop as illustrated in Figure 2.17. The system is designed to meet the demand of a 7.7 kW of load.

In order to calculate the ratio of sustainability, each sustainability index should be calculated. The energy index is calculated by the following formula:

$$Y_{ER} = \left(\eta \times W_\eta\right) + \left(Y_{Pr} \times W_{Pr}\right)$$

Here, Y_{ER} is the total score of energy index. η is the energy efficiency. W_η is the weight of the energy efficiency. Y_{Pr} is the score of the productivity of the energy system and W_{Pr} is its weight.

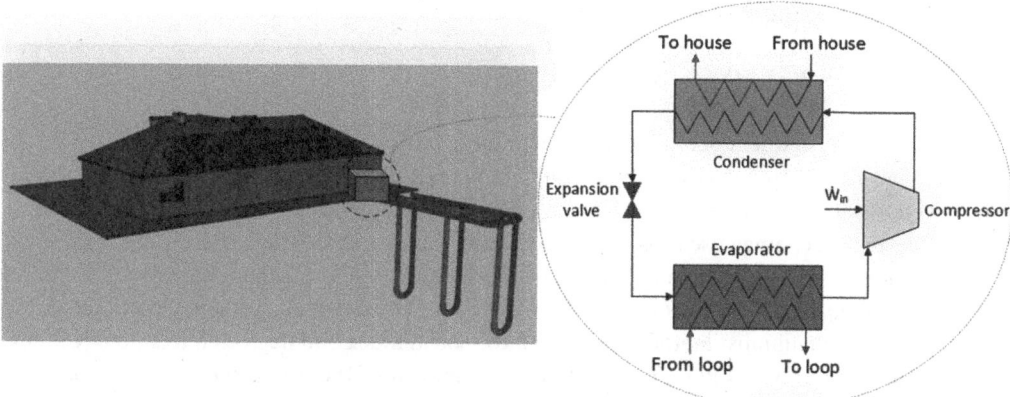

Figure 2.17 Schematic illustration of the proposed system (modified from Ref. [8]).

For the exergy index, the following equation is used:

$$Y_{EX} = (\Psi \times W_{\Psi}) + (Y_{ED} \times W_{ED})$$

where Y_{EX} indicates the total score of the exergy index. Ψ is the exergy efficiency of the system. Y_{ED} is the score of the exergy destruction indicator.

The environmental friendliness index is calculated by the following equation:

$$Y_{ENV} = (Y_{GWP} \times W_{GWP}) + (Y_{ODP} \times W_{ODP}) + (Y_{AP} \times W_{AP}) + (Y_{EP} \times W_{EP}) + (Y_{AT} \times W_{AT})$$
$$+ (Y_{WE} \times W_{WE}) + (Y_{SA} \times W_{SA}) + (Y_{WC} \times W_{WC}) + (Y_{LU} \times W_{LU}) + (Y_{ADP} \times W_{ADP})$$

Each subscription in this equation denotes an environmental impact category.

The score of technology index is calculated as the following:

$$Y_{TECH} = (Y_{COMM} \times W_{COMM}) + (Y_{TR} \times W_{TR}) + (Y_{IN} \times W_{IN})$$

Here, COMM stands for the commercializability, TR is the technology readiness and IN is the innovation.

The score of the social index is found as follows:

$$Y_{SOC} = (Y_{JC} \times W_{JC}) + (Y_{PA} \times W_{PA}) + (Y_{SA} \times W_{SA}) + (Y_{SC} \times W_{SC})$$
$$+ (Y_{HW} \times W_{HW}) + (Y_{HH} \times W_{HH})$$

where JC is the job creation, PA is the public awareness, SA is the social acceptance, CS is the social cost, HW is the human welfare and HH is the health respectively.

The score of the education index can be calculation as the following:

$$Y_{EDU} = (Y_{TRAIN} \times W_{TRAIN}) + (Y_{EL} \times W_{EL}) + (Y_{EI} \times W_{EI})$$

Here, TRAIN denotes the number of trained people required by the industry, EL is the education level and EI represents the educational innovation.

Lastly, the score of the sizing factor is found with the following equation:

$$Y_{MF} = (Y_M \times W_M) + (Y_{LU} \times W_{LU}) + (Y_V \times W_V)$$

Here, M, LU, and V denote the mass, land, and volume utilizations.

In the analysis, the time-space-receptor method is used unappointing suitable values for each index given in Table 2.4. Table 2.4 gives the sustainability assessment indexes and their associated weights as per the schemas used: panel method, individualist, egalitarian, hierarchist, and equal weighting methods. There are slight changes between the different schematic as given in Table 2.4 and shown in Figure 2.18. the panel method is prioritized the exergy index and neglected the sizing index while the individualist method prioritized the social index and neglected technology index. The panel schema provides less priority for education, rest of them indicated a higher priority. When overall system performance is taken into consideration, the sustainability indexes vary between 0.58 and 0.66, respectively.

Table 2.4 Priority factor of the eight main indexes of the sustainability assessment.

Index	Individualist	Egalitarian	Hierarchist	Panel	Equal Weighting
Energy	0.13	0.12	0.13	0.10	0.13
Exergy	0.13	0.12	0.13	0.17	0.13
Environmental friendliness	0.13	0.13	0.17	0.18	0.13
Economic	0.13	0.17	0.13	0.14	0.13
Technology	0.09	0.12	0.12	0.12	0.13
Social	0.17	0.12	0.12	0.17	0.13
Educational	0.13	0.12	0.12	0.09	0.13
Sizing	0.10	0.10	0.12	0,07	0.13

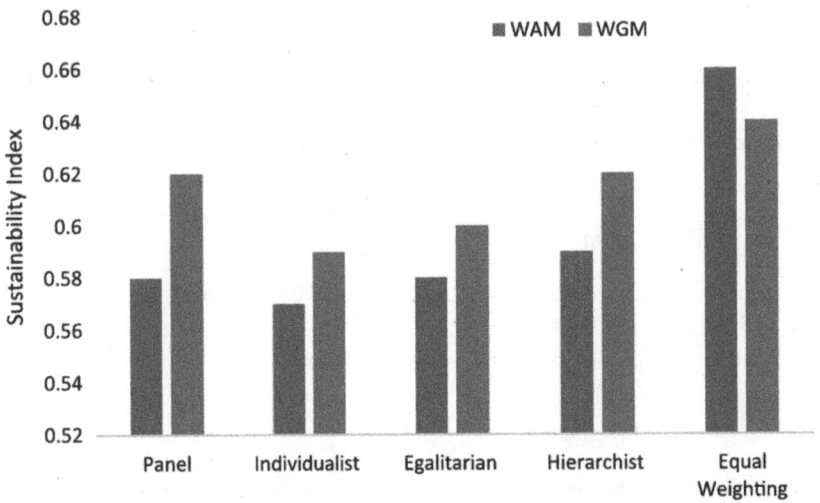

Figure 2.18 Distribution of the sustainability index results based on the various aggregation method and characterization scheme (data from Ref. [8]).

2.9 Closing Remarks

This chapter deals with the sustainability of energy resources. Three key points are critical for energy management systems: increasing energy consumption, need for a clean and safe environment, and irreversible path for humanity. When energy related issues are considered, three key issues are included: environmental problems, sustainability of needed resources, and energy security. The mathematics of exponential growth in energy demands are the critical for understand the growth rates of energy consumption today and into the future. Dimensions of the sustainability is discussed with practical examples. Indicators for sustainability assessment is defined accordingly.

References

1 U.S. Energy Information Administration. International Energy Outlook 2021. https://www.eia.gov/outlooks/ieo (accessed 28 July 2022).
2 Kruger, P. (2006). *Alternative Energy Resources*. John Wiley & Sons.
3 Dincer, I. and Erdemir, D. (2021). *Heat Storage Systems for Buildings*, 1e. Elsevier. doi: 10.1016/C2019-0-05405-2.
4 Our World in Data 2021. https://ourworldindata.org/grapher/fossil-fuel-consumption-by-fuel-type (accessed 10 October 2022).
5 OWID. Electricity consumption per capita worldwide in 2020, by selected country. *Statista*. https://www.statista.com/statistics/383633/worldwide-consumption-of-electricity-by-country (accessed 11 August 2022).
6 Canada Energy Regulator. *Canada's energy future 2021*. https://www.cer-rec.gc.ca/en/data-analysis/canada-energy-future/2021/canada-energy-futures-2021.pdf (accessed 11 August 2022).
7 Dincer, I. and Rosen, M. (2022). *Thermal Energy Storage Systems and Applications*, 3e. John Wiley & Sons.
8 Abu-Rayash, A. and Dincer, I. (2019). *Energy Sustainability*, 1 ed. Elsevier, London.

Questions/Problems

Questions

1 Discuss the three key points considered energy supply and demand.
2 Discuss the five major economic drivers from the environment and sustainability perspectives.
3 Explain how ancient people used to perform energy charging, storage and discharging through examples.
4 Explain what types of trendlines become important in studying the energy consumption growth and highlight the importance of exponential growth with an example.
5 Explain why regression analysis is important and what purposes one can employ it for.
6 Discuss the importance of R^2 in the regression analysis and what to do to improve its value close enough to 1.
7 Discuss the role of energy intensity for countries. And explain how it is related to economic development of any country.
8 Discuss the dimensions of sustainability in 7s and explain each with a practical example.

9 Discuss the importance of United Nation's 17 sustainable development goals with a specific example.

10 List the integrated sustainability model indicators and explain how they are utilized in sustainability assessment.

Problems

1 The following table shows the World's actual fossil fuel consumption data (in TWh).

 a Draw the data distribution graphs for all three types of fuels and show regression coefficients and correlations for each of these analyses.

 b Utilize these actual data sets to make projections for short-term till 2030, mid-term till 2050 and long-term 2075 through curve fitting (trend line analysis) in the Excel for each parameter.

 c Discuss the results of the above requested analysis thoroughly and their possible consequences.

 d Discuss how energy consumption relates to standard of living by comparing energy consumption statistics with either Gross National Product (GNP) or Gross Domestic Product (GDP) for Canada.

Year	Natural Gas Consumption – TWh	Coal Consumption – TWh	Oil Consumption – TWh	Total World – TWh
1965	6303.8	16140.2	17989.6	40433.6
1966	6868.8	16324.1	19341.1	42534.0
1967	7374.0	16060.8	20732.1	44167.0
1968	8044.3	16301.1	22488.3	46833.8
1969	8833.4	16798.7	24353.0	49985.2
1970	9614.8	17058.6	26505.2	53178.6
1971	10293.2	16965.8	27994.8	55253.8
1972	10861.7	17158.6	30112.3	58132.6
1973	11377.8	17667.8	32534.8	61580.5
1974	11659.9	17682.1	32078.2	61420.1
1975	11659.7	18024.5	31715.8	61400.0
1976	12354.1	18688.5	33752.5	64795.1
1977	12759.8	19241.3	34928.1	66929.3
1978	13294.0	19457.9	36509.9	69261.8
1979	14118.0	20363.5	37064.6	71546.1
1980	14237.0	20857.6	35514.4	70609.0
1981	14395.9	21149.7	34251.6	69797.1
1982	14469.7	21385.5	33185.3	69040.5
1983	14703.8	22046.3	32937.6	69687.8
1984	15902.7	23001.1	33680.1	72583.8
1985	16262.2	23987.8	33660.1	73910.2

Year	Natural Gas Consumption – TWh	Coal Consumption – TWh	Oil Consumption – TWh	Total World – TWh
1986	16421.1	24258.0	34700.8	75380.0
1987	17281.9	25212.4	35393.5	77887.9
1988	18088.8	25967.1	36552.6	80608.6
1989	18867.3	26215.7	37156.8	82239.8
1990	19481.1	25904.5	37608.0	82993.6
1991	19972.8	25654.3	37610.6	83237.7
1992	20063.5	25556.6	38151.8	83771.9
1993	20265.5	25685.6	37960.6	83911.7
1994	20389.6	25784.8	38793.5	84967.9
1995	21104.5	25963.7	39442.2	86510.4
1996	22159.3	26582.1	40314.3	89055.7
1997	22030.4	26536.7	41413.2	89980.3
1998	22434.0	26366.4	41644.4	90444.8
1999	23071.9	26479.5	42370.1	91921.5
2000	23994.3	27427.5	42880.9	94302.7
2001	24316.8	27860.2	43255.7	95432.7
2002	25030.1	28955.1	43608.9	97594.1
2003	25729.6	31499.2	44580.6	101809.5
2004	26721.1	33673.0	46270.7	106664.8
2005	27438.0	36169.1	46672.7	110279.8
2006	28161.1	38069.0	47172.6	113402.8
2007	29315.4	40228.4	47758.9	117302.7
2008	30021.3	40773.6	47338.7	118133.6
2009	29401.2	40175.4	46371.6	115948.1
2010	31588.9	41996.4	47895.1	121480.3
2011	32340.2	44017.3	48307.7	124665.1
2012	33194.0	44184.4	49010.7	126389.1
2013	33729.7	44842.6	49520.2	128092.5
2014	33944.3	45161.2	49832.4	128937.9
2015	34768.7	44054.4	50796.3	129619.4
2016	35560.8	43504.2	51760.8	130825.7
2017	36529.2	43751.7	52515.5	132796.4
2018	38356.3	44315.9	53250.6	135922.8
2019	39062.9	43699.9	53368.6	136131.5
2020	38455.7	41964.0	48380.7	128800.4

Source: data from Ref. [4].

2 Find a dataset for any country for natural gas, coal, oil, and total fossil fuel consumption, as in Illustrative Example 2.1. You can use Ref. [4] for finding data. After creating your dataset,

   a Draw the data distribution graphs for all three types of fuels and show regression coefficients and correlations for each of these analyses.

   b Utilize these actual data sets to make projections for short-term till 2030, mid-term till 2050 and long-term 2075 through curve fitting (trend line analysis) in EXCEL for each parameter.

3 Use the data in Figure 2.9 for the energy consumption per capita for all over the world. Utilize this actual data sets to make projections for short-term till 2030, mid-term till 2050 and long-term 2075 through curve fitting (trend line analysis) in EXCEL for each parameter.

4 Create a dataset by using Figure 2.12 for total final energy consumption by sector in the world. Use your dataset to

   a Draw the data distribution graphs for each sector

   b Estimate their yearly values until 2075 by curve fitting (trend line analysis) in EXCEL for each sector.

5 Consider the system given in Illustrative Examples 2.2. Now, use the wind energy via wind turbines as energy input beside of solar energy. Calculate the sustainability indexes in the case of wind energy input. It will be assumed the net power output from wind turbine will be equal to solar power output given value in Illustrative Examples 2.2.

3

System Analysis

3.1 Introduction

In this chapter, system analysis means thermodynamic analyses of energy systems and their components. One may see this as the hearth of the book, and understanding the concept of thermodynamics is therefore of paramount importance. Everything in the universe is one way another connected to thermodynamics either through the phenomena in or changes caused by the effects of external interactions. Humanity did not realize the power of thermodynamics as a scientific tool up until the industrial revolution where there were many engines (particularly steam engines), machines, systems, etc. developed. In this regard, many inventions were achieved only by understanding the phenomena of each process, learning the scientific principles of thermodynamics, and applying them to develop many things (including steam engines) accordingly. Industrial revolution technically opened the door for inventions and discoveries for energy systems and applications where thermodynamic principles were instrumental.

It is really important to understand that the science of thermodynamics is an essential tool for technical things around us, but not limited to. There are many social, economic, and environmental events which are naturally described by the help of thermodynamics. Since the primary focus in this book is about introduction to energy systems and their applications, it is really a necessity to understand the role of thermodynamics correctly as it is needed for system design, analysis, construction, assessment, and performance improvement thoroughly. In this regard, one can define thermodynamics precisely as *the science of energy and exergy* accordingly while remembering the much earlier definitions as *the science of heat* or *the science of work* or *the science of energy* and more recently *the science of energy and entropy*.

Thermodynamics with four laws (the zeroth law, first law, second law, and third law of thermodynamics) is like a human being who has two legs and two hands. Here, two legs refer to the first and second law of thermodynamics while two hands may refer to the zeroth and third law of thermodynamics. Thermodynamics essentially stands on both first and second law of thermodynamics, just like human beings standing on both legs. While the first and second laws of thermodynamics cover methods to assess the systems with a set of rules, the zeroth and third laws of thermodynamics provide the guide to the first and second law. These are illustrated in Figure 3.1 with a soccer player where the ball is like a thermodynamic system governed by the legs which are further referred to the first and second laws of thermodynamics. Further to quote from Dincer [1] about these laws and their roles:

> *The first and second laws of thermodynamics are recognized as governing laws like the constitutional laws for a state or country or institution which are known as the primary rules for regulating the functioning of a state or country or institution. When we look at the*

Introduction to Energy Systems, First Edition. Ibrahim Dincer and Dogan Erdemir.
© 2023 John Wiley & Sons, Inc. Published 2023 by John Wiley & Sons, Inc.
Companion Website: www.wiley.com/go/Dincer/Introduction_to_Energy_Systems

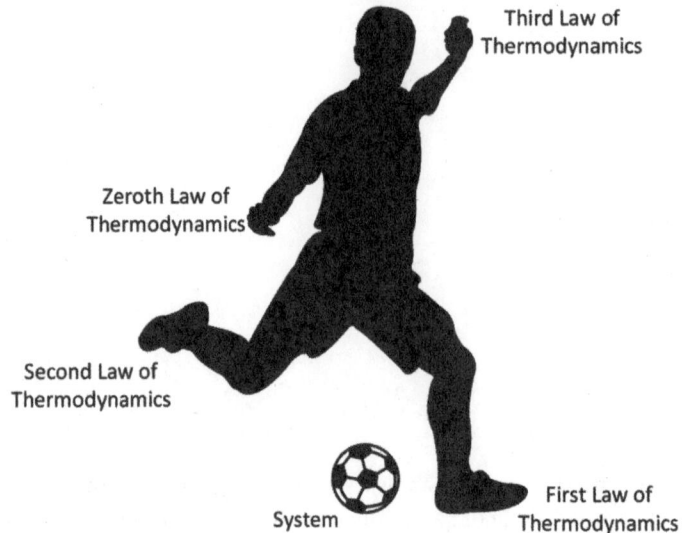

Figure 3.1 An illustration of the laws of the thermodynamics.

zeroth and third laws of thermodynamics, these are seen more as guiding policies for any state or country or institution. After this linkage, one may clearly understand that the first and second laws of thermodynamics are governing laws and the zeroth and third laws of thermodynamics are guiding laws depending on special/specific situations.

One may explain the main objectives of using thermodynamics as a potential tool through the following actions:

- to comprehend the system phenomena and interactions with its surroundings,
- to learn how the systems operate and what processes take place in them,
- to understand thermodynamics laws, concepts and principles, and their roles,
- to design the systems based on sources available and useful outputs needed,
- to identify all inputs and outputs for the systems,
- to analyze the systems based on the balance equations and relevant concepts and principles,
- to address the operating strategies for the systems,
- to evaluate the system performance accordingly based on the thermodynamic performance criteria,
- to make recommendations accordingly.

To further note that in order to design and operate energy systems, all criteria given above should be considered carefully to maximize the useful outputs and minimize the energy or fuel inputs. Today, as thermodynamics part of the any system which has energy flows, thermodynamic approaches are deployed in almost all sectors from energy to environment, from economy to social sciences. The ultimate aim of the thermodynamic analysis is to better understand the energy systems and processes. A better understanding helps to improve the systems further. Thermodynamically improved systems can provide more effective results in many ways like less fuel consumption, power generation, the number of useful outputs, economy, environmental impacts, etc.

Thermodynamics, as mentioned in the beginning, is directly linked to many things, including energy, environment, economy, and hence sustainability as a big picture. Thermodynamics is then located at the intersection of these four key domains, as illustrated in Figure 3.2. Better thermodynamic performance can provide better performance indexes for energy, environment, economy, and sustainability. Therefore, in order to design optimum energy systems, thermodynamic analysis is the first step. It is an essential tool to assess and optimize systems and system components.

In order to better understand thermodynamics, Dincer [1] has introduced a seven-step approach. These steps are important to really comprehend with the spectrum of thermodynamics in a logical manner as illustrated in Figure 3.3. As the clock does not move from 1 to 3 by skipping 2, one should follow all these steps accordingly for true analysis and performance assessment in thermodynamics. The figure shows that the starting point is the property, which is defined as the specific characteristic of a system (for example, pressure, temperature, etc.). The next stop is the state, which is defined as the condition at any system (for example, inlet condition, exit condition, etc.), and it essentially requires minimum two of the properties, so the states depend essentially upon the properties. The third one is the process, which is defined as a change from one state to another (for example, isothermal process, isobaric process, etc.), and it essentially requires minimum two state points to make a process. The fourth one is the cycle, which is defined as the meaningful set of processes to generate something useful before reaching back to the original starting point (for example, steam cycle, air-standard cycle, etc.), and it essentially requires minimum two of the thermodynamic processes to construct it. The next

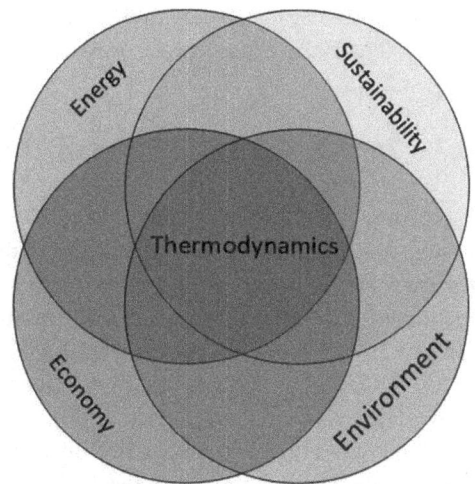

Figure 3.2 Thermodynamics as an intersection of energy, environment, economy, and sustainability.

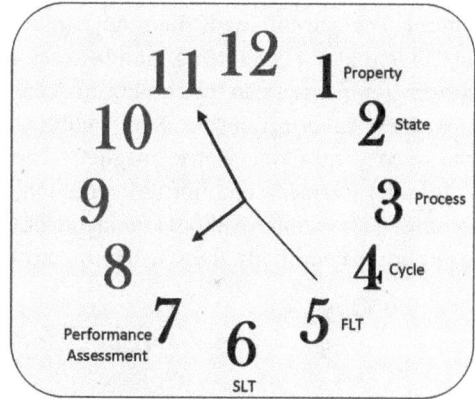

Figure 3.3 Seven steps to follow in thermodynamic system analysis and assessment.

stop is the first law of thermodynamics (FLT), which is defined as the conservation of energy law, referring to the fact that neither is created nor destroyed, but changes form (for example, from heat to work in thermal power plant and from work to heat or cold in a heat pump or refrigeration unit). The sixth stop is the second law of thermodynamics (SLT), which is defined as the non-conservation of exergy law, referring to the fact that there are irreversibilities taking place in any practical system in the form of entropy generations and exergy destructions. It is also an indication of how much deviation takes place from reversible to irreversible behavior. The last stop is the performance assessment where one needs to evaluate the system performance through energy and exergy efficiencies or energetic or exergetic coefficient of performance values or other energy and exergy performance criteria.

In this chapter, first, the zeroth and third laws of thermodynamics are introduced. Then, a smart approach for performing a complete thermodynamic analysis is presented. For this purpose, a seven-step approach is described. Next, the first and second laws of thermodynamics are defined.

This is followed by the presentation of four thermodynamic balance equations for common devices. Finally, illustrative examples are provided to readers.

3.2 Zeroth and Third Laws of Thermodynamics

The zeroth law of the thermodynamics (ZLT) is often defined over an illustrative sketch, as shown in Figure 3.4. The zeroth law of thermodynamics is really about how states reach and stay in the thermal equilibrium (referring to the fact that they have the same temperature) for the objects (or materials). If the object A is in thermal equilibrium with the object B and the object B is in equilibrium with the object C, the object A is also in equilibrium with the object C. Here, the objects A, B, and C indicate that the three thermodynamic systems are in thermal equilibrium. This may look simple, but has played an important role in developing the thermometers for temperature measurements. There are also cases where thermodynamic systems remain in mechanical equilibrium based on the fact that they have reached the same pressure.

Here, the critical question is what the equilibrium is or how the equilibrium conditions are to be defined. In the literature, the zeroth law of thermodynamics is often defined over the thermal equilibrium. Nonetheless, here, we should note that the equilibrium conditions should be defined through all parameters that have the potential to produce power, movement, flow, flux, or useful output. The zeroth law of thermodynamics provides information regarding the potential of the useful output or power generation between two objects/bodies/environments. If two objects/bodies/environments are in the equilibrium conditions, there will be no power generation, movement, flux, flow, or useful output. Some additional equilibrium conditions may be listed as follows: phase, concentration, electric, magnetic, chemical, force, sonic, etc.

In order to ensure equilibrium conditions between the two systems/environments/bodies, the conditions listed above should be identical in each system/environment/body. When they are in equilibrium conditions, there will be no movement, flow, flux between them. In other words, there

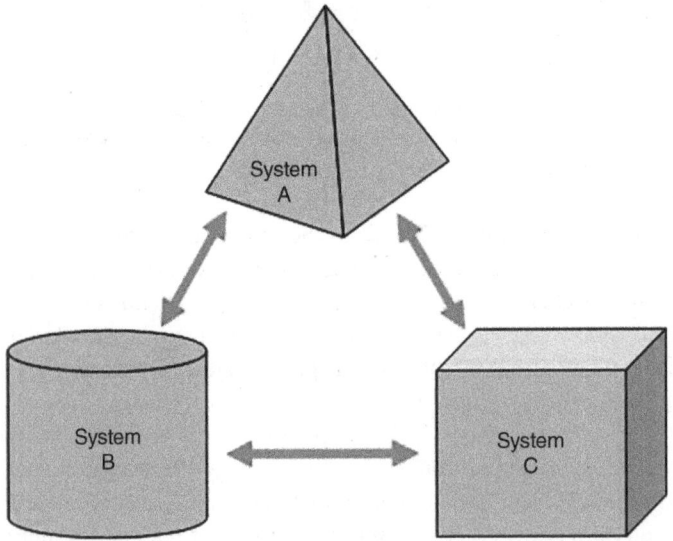

Figure 3.4 A demonstration of the zeroth law of the thermodynamics where thermal equilibrium is achieved.

is no potential to produce work or useful output with a system or process to work between those systems. For instance, batteries are able to produce power due to non-equilibrium electron conditions even they are at the same temperature. It is possible to obtain a cooling effect with a water/ ice mixture despite being at the same temperature.

The third law of thermodynamics (TLT) states that the entropy and entropy generation are zero at the 0 K that called absolute zero. The TLT was developed by Walther Nernst, so it is called Nernst's theorem. In 1923, the TLT was modified by Gilbert N. Lewis and Merle Randall by defining the perfect crystalline substances. They presented that the perfect crystalline substances could occur at the absolute zero. In other words, it explains that there is no way to avoid entropy generation.

3.3 First Law of Thermodynamics

The first law of thermodynamics (FLT), also known as the principle of energy conservation, defines that energy cannot be created nor destroyed, but it can change from one form to another. The general formulation of the first law of thermodynamics is written as follows:

$$\sum_i \dot{Q}_i + \sum_i \dot{W}_i + \sum_i \dot{m}_i \left(h_i + \frac{V_i^2}{2} + gz_i \right) = \sum_o \dot{Q}_o + \sum_o \dot{W}_o + \sum_o \dot{m}_o \left(h_o + \frac{V_o^2}{2} + gz_o \right) \quad (3.1)$$

In Eq. 3.1, $\dot{Q}$ and $\dot{W}$ denote the heat and work transfers, the parameter in the parenthesis is called the total flow energy. h is the enthalpy, $\frac{V^2}{2}$ is the kinetic energy and gz is the potential energy.

For example, imagine an object of mass m falling from above to the bottom, the velocity of the object will increase with the distance toward the ground. So, while the potential energy of the object reduces, its kinetic energy increases. The changes in the kinetic and potential energies are the same. In other words, the potential energy transforms into kinetic energy. Here it should be noted that the air friction (air drag) is neglected. If it is considered, the air friction will turn to heat.

The FLT defines energy as one of the thermodynamics properties and allows analyzing the energy interactions and exchanges in the systems. Thus, it is possible to assess the system by using energy inputs, outputs, and losses across the border of the system. The FLT enables us to understand the systems and interactions between the system and its surroundings or other systems. We can define how much useful output to be obtained from a system or how much fuel/energy input to be needed. The FLT is one of the essential items of thermodynamic analysis. It is also an essential tool for engineers to clearly understand the systems and their interactions with their surrounding environment or other objects. On the other hand, the FLT remains incapable of defining the irreversibilities, losses, inefficiencies, and quality destructions.

3.4 Second Law of Thermodynamics

The second law of the thermodynamics (SLT) defines the irreversibilities, losses, inefficiencies, and quality destructions inside the systems. Therefore, it is often described as a measure of the quality of energy. The SLT is related to both quantity and quality, while the FLT is only related to quantity. Therefore, the SLT gives detailed information regarding the systems. The difference between the FLT and SLT is expressed over the x-ray and tomography images. Although both are critical tools to examine the human body, the x-ray provides limited information according to the tomography. Here, while the FLT is the x-ray, the SLT is the tomography. In order to understand and evaluate the system clearly, the SLT is a powerful tool as it defines the maximum potential useful outputs to be obtained from the systems. The SLT is defined through entropy and exergy approaches.

The SLT can be defined over entropy and exergy. The general entropy and exergy balances are written as follow:

$$\sum_i \dot{m}_i s_i + \sum_{net} \frac{\dot{Q}_{net}}{T_S} + \dot{S}_{gen} = + \sum_o \dot{m}_o s_o \tag{3.2}$$

$$\sum_i \dot{Ex}_{\dot{Q}_i} + \sum_i \dot{W}_i + \sum_i \dot{m}_i ex_i = \sum_o \dot{Ex}_{\dot{Q}_o} + \sum_o \dot{W}_o + \sum_o \dot{m}_o ex_o + \dot{Ex}_d \tag{3.3}$$

where $\dot{S}_{gen}$ is the entropy generation, and $\dot{Ex}_d$ is the exergy destruction. Both represent the magnitudes of irreversibilities inside the system using different approaches.

3.5 Six-Step in System Analysis

In order to deploy a proper thermodynamic analysis, the six-step approach can be followed up, as shown in Figure 3.5. This six-step approach is called "a smart approach to thermodynamic analysis" by Dincer [1]. To begin with a thermodynamic analysis for a system, the system boundaries should be determined for entire system and each component in the system. System and system boundaries should be defined as they are essential part of the thermodynamic aspect. A thermodynamic system is defined as a mass or volume of matter, which is separated from its surroundings with an imaginary or physical border. The system and its border may be fixed in a stationary form or moving in a nonstationary form. Then, all inputs, output and also losses if considered, have to be determined accordingly. Next, the mass balance equation, also called conservation of the mass law, will be written for each component of the system. After that, the energy balance equation will be deployed for the system component considering inputs, outputs, and losses passing through the system boundaries. The entropy balance equation is then written to determine the entropy generation for each component of the system. Finally, the exergy balance equation is written for each component of the system to determine the exergy destructions.

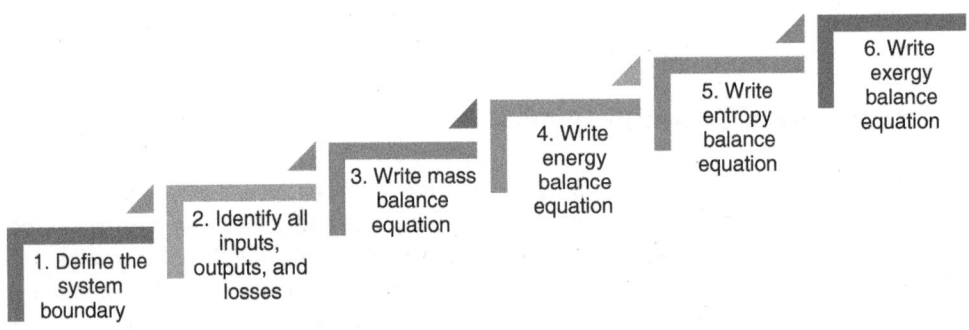

Figure 3.5 Six steps used in system analysis.

In addition to the six-step approach, it should be noted that some assumptions are required in any type of thermodynamic analysis. The following are just some examples about commonly made assumptions in thermodynamic analysis:

- The changes in kinetic and potential energies are negligible.
- The system components operate in steady-state and steady-flow process.
- The system is an adiabatic one.
- The reference conditions are 25°C and 100 kPa.
- The pressure drops within the system are negligible.
- The heat losses from the system are negligible.
- Air is treated as an ideal gas.
- No chemical reactions take place within the system.

Here, it should be note that, some specific conditions or systems can require specific assumptions. Although all assumptions provide easier analysis to find energy and exergy efficiencies, they tend to decrease entropy generations and exergy destructions as they bring the systems closer to ideal. Another common application of thermodynamic analysis is parametric analysis in order to understand how the system works under different operating, design or ambient conditions. Besides, assumptions tend to increase energetic and energetic efficiencies, not their relative values. Therefore, it can help to optimize the system and operating conditions.

3.6 Closed Systems

Thermodynamic systems can generally be classified into two main categories, depending on whether or not mass crosses the boundary of the system. If there is no mass crossing the system boundary, such a system is called a closed system. They are often called as a control mass. Closed systems are in general categorized into two main categories depending on the type of the boundary of the system. These two categories are named as fixed (stationary), as illustrated in Figure 3.6 with (a) hot water tank where the system boundary does not change with time, and moving boundary with (b) through a piston-cylinder mechanism where the system boundary change during the process, as shown in Figure 3.6. Thermodynamic balance equations for common closed systems will then be introduced accordingly for system analysis.

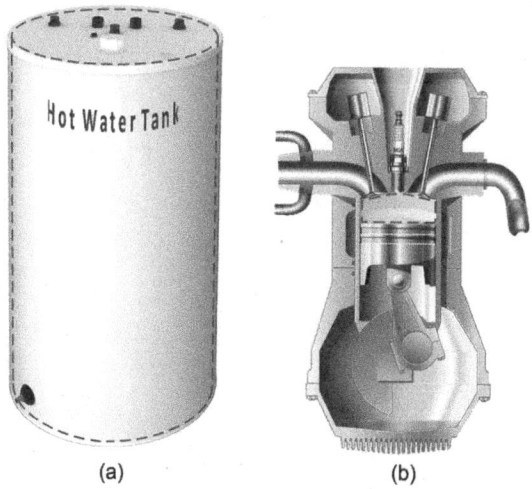

(a)   (b)

Figure 3.6   Types of the closed systems: (a) fixed boundary and (b) moving boundary.

3.6.1 A Rigid Tank Filled with a Fluid

Consider a rigid tank which is filled with a fluid and subject to heat and work transfer, as shown in Figure 3.7. The fluid is mixing with a fan and work input of the fan can be defined over the **shaft work**. There is an electricity resistant to heat the fluid up. The electricity is provided into the system via this resistant, so it is called **electricity work**. Also, heat directly send to control mass as a **heat input**. In this specific case, it is assumed that the temperature of the fluid is higher than the surrounding temperature. So, **heat loss** takes place in the system as it leaves the system.

One may place the system boundary and identify all energy flow crossing the system boundaries, as demonstrated in Figure 3.7. Now, we will write the mass, energy, entropy, and exergy balance equations for the system. The mass balance equation can be written as follows:

$$m_1 = m_2 = m = constant \tag{3.4}$$

where 1 and 2 indicate the initial and final state of the fluid, respectively. As there is no mass input or output. So, they are equal.

The energy balance equation of the system can be written as follows:

$$m_1 u_1 + W_e + Q_{in} + W_{sh} = m_2 u_2 + Q_{loss} \tag{3.5}$$

where u is the internal energy of the fluid. W_e is the electricity input via the resistant. W_{sh} is the shaft work. Q_{in} is the heat input from any source to the system. Q_{loss} is the heat loss from system to environment.

The entropy balance equation can be written as follows:

$$m_1 s_1 + \frac{Q_{in}}{T_s} + S_{gen} = m_2 s_2 + \frac{Q_{loss}}{T_b} \tag{3.6}$$

where s is the specific entropy of the fluid. S_{gen} denotes the entropy generation inside the system boundaries. T_s is the source temperature of the heat storage and T_b is the boundary temperature between the system and ambient. Here, it should be noted that it is hard to determine the boundary temperature. Therefore, surface temperature of the system (T_{sur}), the average temperature for

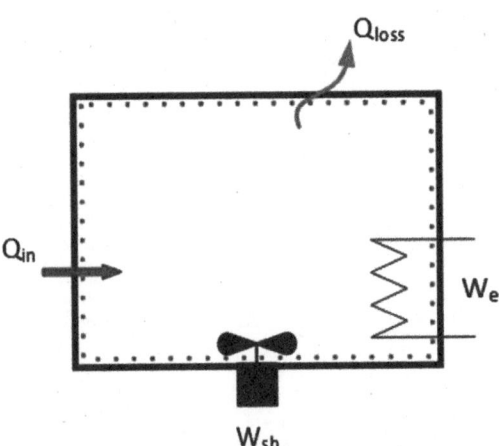

Figure 3.7 A rigid tank filled with a fluid.

the system and ambient (T_{ave}) or directly ambient temperature (T_{amb}) can be used instead of boundary temperature. Also, the entropy balance equation is not associated with work input and outputs.

The exergy balance equation of the system shown in Figure 3.7 can be written as follows:

$$m_1 ex_1 + W_e + Ex^{Q_s} + W_{sh} = m_2 ex_2 + Ex^{Q_{loss}} + Ex_{dest} \tag{3.7}$$

where ex denotes the specific exergy of the fluid. Ex^{Q_s} and $Ex^{Q_{loss}}$ denote the exergy contents of the heat transfers which are heat input and heat lost. Ex^{Q_s} and $Ex^{Q_{loss}}$ are written as:

$$Ex^{Q_s} = \left(1 - \frac{T_0}{T_s}\right) Q_{in} \tag{3.8}$$

$$Ex^{Q_{loss}} = \left(1 - \frac{T_0}{T_b}\right) Q_{loss} \tag{3.9}$$

As the boundary temperature is hard to measure, instead of it, the surface temperature of the system (T_{sur}), the average temperature for the system and ambient (T_{ave}) or directly ambient temperature (T_{amb}) can be used. Exergy content of the any state for the closed system can be calculated to be:

$$ex_i = (u_i - u_0) - T_0 (s_i - s_0) \tag{3.10}$$

where "i" denotes any state or stream in the system. "0" represents the reference conditions.

3.6.2 A Cooling Bath with Two Solid Blocks

Oil bath and other fluid baths are widely used for cooling purposes and heat treatment processes. Now, we will write the thermodynamic balance equations for a cooling bath where two solid blocks are quenched in, as shown in Figure 3.8. Here, the air gap at the top of the rigid tank is neglected. The system boundary only considers the fluid section.

The mass balance equation is then written as follows:

$$m_1 = m_2 = m \quad : constant \ for \ each \ material \tag{3.11}$$

where 1 and 2 indicate the initial and final state of each material. While "sb" denotes the solid block, "f" represents the fluid in the control mass.

The energy balance equation is expressed as follows:

$$m_{f,1} u_{f,1} + m_{sb1,1} u_{sb1,1} + m_{sb2,1} u_{sb2,1} = m_{f,2} u_{f,2} + m_{sb1,2} u_{sb1,2}$$
$$+ m_{sb2,2} u_{sb2,2} + Q_{loss} \tag{3.12}$$

The entropy balance equation can be written as:

$$m_{f,1} s_{f,1} + m_{sb1,1} s_{sb1,1} + m_{sb2,1} s_{sb2,1} + S_{gen} = m_{f,2} s_{f,2}$$
$$+ m_{sb1,2} s_{sb1,2} + m_{sb2,2} s_{sb2,2} + \frac{Q_{loss}}{T_b} \tag{3.13}$$

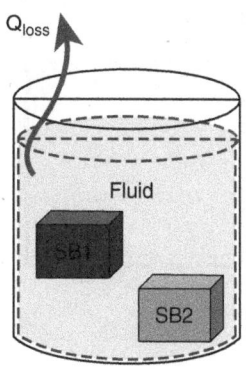

Figure 3.8 A cooling bath with two solid blocks.

Finally, the exergy balance equation for the system is written as follows:

$$m_{f,1}ex_{f,1} + m_{sb1,1}ex_{sb1,1} + m_{sb2,1}ex_{sb2,1} = m_{f,2}ex_{f,2} + m_{sb1,2}ex_{sb1,2} + m_{sb2,2}ex_{sb2,2}$$
$$+ Ex^{Q_{loss}} + Ex_{dest} \qquad (3.14)$$

Here, $Ex^{Q_{loss}}$ is determined as mentioned in Eq. 3.9.

3.6.3 Piston-Cylinder Mechanism (with Moving Boundary)

Piston-cylinder mechanism is common the system used in engineering applications. Figure 3.9 shows a piston-cylinder mechanism subjected to the heat input and lost.

The mass balance equation for the piston-cylinder mechanism can written as follows:

$$m_1 = m_2 = m : constant \qquad (3.15)$$

In the piston-cylinder mechanism, though there is no mass change in the system, the volume of the system changes due to pressure changes inside the system. This volume change produces useful output.

In this regard, the energy balance equation is written as follows:

$$m_1 u_1 + Q_{in} = m_2 u_2 + Q_{loss} + W_b \qquad (3.16)$$

where W_b can be defined as:

$$W_b = P\Delta V \qquad (3.17)$$

and

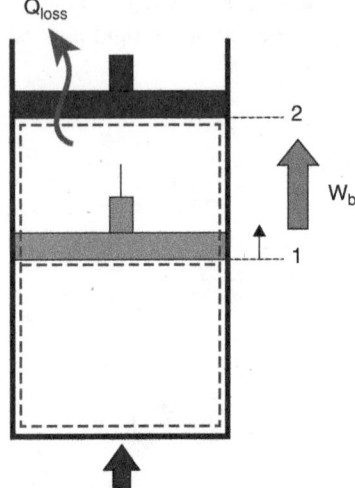

Figure 3.9 Piston cylinder mechanism with the moving boundary.

$$W_b = mP\Delta v \qquad (3.18)$$

Thus, the energy balance equation results in

$$m_1 u_1 + m_1 Pv_1 + Q_{in} = m_2 u_2 + m_2 Pv_2 + Q_{loss} + W_b \qquad (3.19)$$

Here, we should define the enthalpy as

$$h = u + Pv \qquad (3.20)$$

where "u" is the specific internal energy and Pv is called the flow energy. So, enthalpy is the sum of the internal energy and flow energy. Then, energy balance equation is written to be:

$$m_1 h_1 + Q_{in} = m_2 h_2 + Q_{loss} + W_b \qquad (3.21)$$

The entropy balance equation is written as:

$$m_1 s_1 + \frac{Q_{in}}{T_s} + S_{gen} = m_2 s_2 + \frac{Q_{loss}}{T_0} \qquad (3.22)$$

The exergy balance equation is written as:

$$m_1 ex_1 + Ex^{Q_{in}} = m_2 ex_2 + Ex^{Q_{loss}} + W_b + Ex_d \qquad (3.23)$$

where $Ex^{Q_{in}}$, $Ex^{Q_{loss}}$, and W_b are defined as follows:

$$Ex^{Q_{in}} = \left(1 - \frac{T_0}{T_s}\right) Q_{in} \qquad (3.24)$$

$$Ex^{Q_{loss}} = \left(1 - \frac{T_0}{T_b}\right) Q_{loss} \qquad (3.25)$$

$$W_b = m\left(P_1 v_1 - P_2 v_2\right) \qquad (3.26)$$

Also, the exergy content of any steam or state which are subjected to the enthalpy flow:

$$ex_i = \left(u_i - u_0\right) - T_0\left(s_i - s_0\right) \qquad (3.27)$$

The exergy balance equation is also can be written without W_b can be written as:

$$m_1 ex_1 + Ex^{Q_{in}} = m_2 ex_2 + Ex^{Q_{loss}} + Ex_d \qquad (3.28)$$

In this case, the specific exergy is calculated as follows:

$$ex_i = \left(h_i - h_0\right) - T_0\left(s_i - s_0\right) \qquad (3.29)$$

Illustrative Example 3.1 Consider a rigid tank filled with water which has the volume of 150 L, as shown in Figure 3.10. There is a 10 kW electric heater heating the water in the tank. The heater is operated only for 1.5 hours. The following parameters are known:

$T_1 = 5°C$, $\rho = 1000$ kg/m^3, $c_p = 4.18$ kJ/kg°C, $T_s = 400°C$, $T_b = 35°C$, $T_0 = 23°C$, $u_1 = 21.019$ kJ/kg, $u_2 = 230.24$ kJ/kg, $s_1 = 0.0763$ kJ/kg K and $s_2 = 0.768$ kJ/kg K

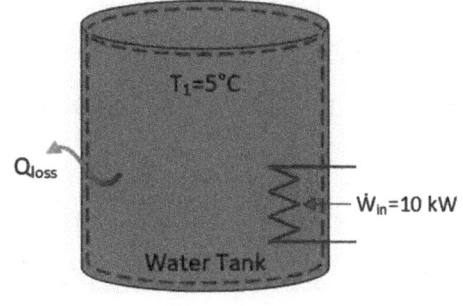

Figure 3.10 Schematic diagram of a rigid tank in illustrative Example 3.1.

- Write the balance equations for mass, energy, entropy, and exergy.
- Find the final temperature of water inside the tank.
- Calculate $\dot{Q}_{loss}$
- Determine the entropy generation (S_{gen}) and exergy destruction (Ex_d) for the system.

Solution:

In order to analyze the system, thermodynamic balance equations should be written as follows:

MBE: $m_1 = m_2 = m = 150$ kg $=$ constant

EBE: $m_1 u_1 + W_{in} = m_2 u_2 + Q_{loss}$

EnBE: $m_1 s_1 + S_{gen} = m_2 s_2 + \dfrac{Q_{loss}}{T_b}$

ExBE: $m_1 ex_1 + W_{in} = m_2 s_2 + \left(1 - \dfrac{T_0}{T_b}\right) Q_{loss} + Ex_d$

where W_{in} is the electric work for heater and can be calculated as follows:

$$W_{in} = \dot{W}_{in} t = 10\,(kW) \times 1.5\,(h) \times 3600\left(\dfrac{s}{h}\right) = 54000\ kJ$$

From EBE:

$$150 \times 21.019 + 54000 = 150 \times 230.24 + Q_{loss}$$

$$Q_{loss} = 22616.85\ kJ = 6.28\ kWh$$

$$\dot{Q}_{loss} = \dfrac{Q_{loss}}{\Delta t} = \dfrac{6.28\,(kWh)}{1.5\,(h)} = 4.15\ kW$$

Also,

$$m(u_1 - u_2) = mc_p(T_2 - T_1)$$

From this equation:

$$(230.24 - 21.019) = 4.18(T_2 - 5)$$

$$T_2 = 55°C$$

From EnBE:

$$m_1 s_1 + S_{gen} = m_2 s_2 + \dfrac{Q_{loss}}{T_b}$$

$$150 \times 0.0763 + S_{gen} = 150 \times 0.768 + \dfrac{22616.85}{35 + 273}$$

$$S_{gen} = 96.94\ kJ\,/\,K$$

The ExBE for the system is written as

$$m_1 ex_1 + W_{in} = m_2 s_2 + \left(1 - \dfrac{T_0}{T_b}\right) Q_{loss} + Ex_d$$

where the specific exergies can be calculated by using the Eq. 3.10.

$$ex_1 = (u_1 - u_0) - T_0(s_1 - s_0) = 11.76\ kJ\,/\,kg$$

$$ex_2 = (u_2 - u_0) - T_0(s_2 - s_0) = 27.12\ kJ\,/\,kg$$

$$m_1 ex_1 + W_{in} = m_2 s_2 + \left(1 - \frac{T_0}{T_b}\right) Q_{loss} + Ex_d$$

$$Ex_d = 28,904 \ kJ$$

In the case of the specific exergies are not given in the problem, exergy destruction cannot be found from energy balance equation. So, exergy destruction can be found from

$$Ex_d = S_{gen} T_0$$

$$Ex_d = 96.94 \times (25 + 273) = 28,904 \ kJ$$

Illustrative Example 3.2 Consider a piston-cylinder mechanism filled with 10 kg of argon gas, as illustrated in Figure 3.11. At the beginning, the temperature and pressure of the argon are 30°C and 200 kPa, respectively. The piston compresses the argon up to 1400 kPa. There is no mass inlet and outlet during the compression process. Take n=1.2, c_v=0.3122 kJ/kg K.

- Write the balance equations for mass, energy, entropy, and exergy.
- Find the boundary work, W_b.
- Calculate the temperature of the argon at the final state (T_2).

Solution:

MBE: $m_1 = m_2$

EBE: $m_1 u_1 + W_b = m_2 u_2 + Q_{loss}$

EnBE: $m_1 u_1 + S_{gen} = m_2 u_2 + \dfrac{Q_{loss}}{T_b}$

ExBe: $m_1 ex_1 + W_b = m_2 ex_2 + Ex^{Q_{loss}} + Ex_d$

The system is a closed system. So, mass balance equation can be written as follow:

$$m_1 = m_2 = 10 \ kg = constant$$

Let's write the balance equation for the system shown in Figure 3.11:

$$m_1 u_1 + W_b = m_2 u_2 + Q_{loss}$$

where W_b can be calculated as:

$$W_b = w_b \times m = \left\{ -\left[\frac{RT}{n-1}\left[\left(\frac{P_2}{P_1}\right)^{\frac{(n-1)}{n}} - 1\right]\right] \right\} \times m$$

$$= \left\{ -\left[\frac{0.2081 \times 303.5}{1.2 - 1}\left[\left(\frac{1400}{200}\right)^{\frac{(1.2-1)}{1.2}} - 1\right]\right] \right\} \times m$$

$$= 436\left(\frac{kJ}{kg}\right) \times 10(kg) = 4360 \ kJ$$

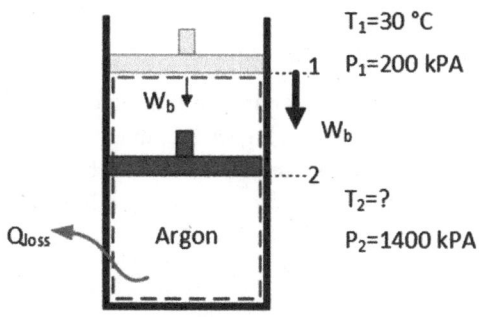

Figure 3.11 Schematic illustration of the piston-cylinder mechanism filled with argon.

T₁=30 °C
1 P₁=200 kPA
W_b
W_b
2
T₂=?
Q_loss   Argon   P₂=1400 kPA

where T_2 is found via the following equation:

$$\frac{T_2}{T_1} = \left(\frac{P_2}{P_1}\right)^{\frac{(n-1)}{n}}$$

$$T_2 = 49°C$$

From the energy balance equation, Q_{loss} is then found as follows:

$$m_1 u_1 + W_b = m_2 u_2 + Q_{loss}$$

$$Q_{loss} = W_b - m_1 (u_2 - u_1)$$

Since $(u_2 - u_1) = c_v \Delta T$

$$Q_{loss} = 4360 - 10 \times 0.3122 \times (49 - 30) = 4300 \; kJ$$

3.7 Open Systems

Many thermodynamic devices are defined as open system as they are subjected to the mass inlet and outlet. Open systems are classified as steady-state, steady flow process (SSSF) and uniform-state, uniform process (USUF). For an open system, the total flow energy is defined as follows:

$$\text{Total flow energy} = h + ke + pe \tag{3.30}$$

$$\text{Total flow energy} = h + \frac{1}{2}V^2 + gz \tag{3.31}$$

Here, it should be noted that exergy contents of the kinetic and potential energies are identical.

$$\text{kinetic exergy} \equiv \text{kinetic energy}$$

$$\text{potential exergy} \equiv \text{potential energy}$$

In this section, thermodynamic balance equations are written for the common systems used in the practical engineering applications.

3.7.1 Nozzle and Diffuser

Nozzle and diffuser are used for adjustment of the flow velocity and pressure by changing cross-section of the flow. Figure 3.12 demonstrates the schematic views of the nozzle and diffuser. They are adiabatic devices, which means there is no heat transfer crossing the boundary.

The mass balance equation for both nozzle and diffuser is written as

$$\dot{m}_1 = \dot{m}_2 = \dot{m} \tag{3.32}$$

where $\dot{m}$ is the flow rate in the system. Since nozzle and diffuser have one inlet and outlet, $\dot{m}_1$ and $\dot{m}_2$ is the equal. The mass balance equation can also be used for calculating the velocities.

$$\dot{m}_1 = \rho \dot{V}_1 A_1 = \rho \dot{V}_2 A_2 = \dot{m}_2 \qquad (3.33)$$

The energy balance equation for both devices can be written as follows:

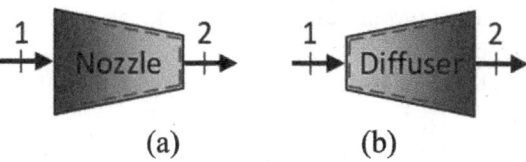

Figure 3.12 Schematic illustrations of (a) nozzle and (b) diffuser.

$$\dot{m}_1 \left(h_1 + \frac{1}{2} V_1^2 + gz_1 \right) = \dot{m}_2 \left(h_2 + \frac{1}{2} V_2^2 + gz_2 \right) \qquad (3.34)$$

Here, the potential energy changes in both nozzle and diffuser, gz_1 and gz_2 can be neglected. So, the final energy balance equation for both:

$$\dot{m}_1 \left(h_1 + \frac{1}{2} V_1^2 \right) = \dot{m}_2 \left(h_2 + \frac{1}{2} V_2^2 \right) \qquad (3.35)$$

The entropy balance equations for both units are written as:

$$\dot{m}_1 s_1 + \dot{S}_{gen} = \dot{m}_2 s_2 \qquad (3.36)$$

The exergy balance equation can be written as:

$$\dot{m}_1 ex_1 = \dot{m}_2 ex_2 + \dot{Ex}_d \qquad (3.37)$$

where ex_1 and ex_2 are the specific flow exergies and can be calculated as:

$$ex_1 = \left(h_1 - h_0 \right) - T_0 \left(s_1 - s_0 \right) + \frac{1}{2} V_1^2 \qquad (3.38)$$

$$ex_2 = \left(h_2 - h_0 \right) - T_0 \left(s_2 - s_0 \right) + \frac{1}{2} V_2^2 \qquad (3.39)$$

3.7.2 Compressor and Turbine

Compressor and turbine are considered two of the main components of the power and refrigeration cycles. Figure 3.13 shows the schematic views for the compressor and turbine. While compressor compresses the gases, turbine expands it for power generation. Therefore, in order to analyze such power generating cycles, writing thermodynamic balance equations for

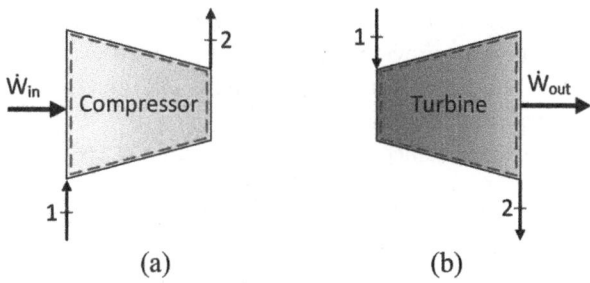

Figure 3.13 Schematic illustrations of (a) a compressor and (b) a turbine.

compressor and turbine and analyzing them carefully are considered very important tasks.

The mass balance equation for both compressor and turbine becomes

$$\dot{m}_1 = \dot{m}_2 = \dot{m} \qquad (3.40)$$

For most turbine and compressor, kinetic and potential energy changes can be neglected as there is no significant changes.

Therefore, the energy balance equation for the compressor is written as:

$$\dot{m}_1 h_1 + \dot{W}_{in} = \dot{m}_2 h_2 + \dot{Q}_{loss} \tag{3.41}$$

The energy balance equation of the turbine becomes:

$$\dot{m}_1 h_1 = \dot{m}_2 h_2 + \dot{W}_{out} + \dot{Q}_{loss} \tag{3.42}$$

The entropy balance equation for compressor and turbine can be written as follows, respectively:

$$\dot{m}_1 s_1 + \dot{S}_{gen} = \dot{m}_2 s_2 + \frac{\dot{Q}_{loss}}{T_b} \tag{3.43}$$

$$\dot{m}_1 s_1 + \dot{S}_{gen} = \dot{m}_2 s_2 + \frac{\dot{Q}_{loss}}{T_b} \tag{3.44}$$

The entropy balance equations are the same for both compressor and turbine, as entropy is not associated with the work.

The exergy balance equation for compressor and turbine can be written as follows:

$$\dot{m}_1 ex_1 + \dot{W}_{in} = \dot{m}_2 ex_2 + \dot{Ex}^{Q_{loss}} + \dot{Ex}_d \tag{3.45}$$

$$\dot{m}_1 ex_1 = \dot{m}_2 ex_2 + \dot{W}_{out} + \dot{Ex}^{Q_{loss}} + \dot{Ex}_d \tag{3.46}$$

where exergy content of the any stream can be determined with the following equation:

$$ex_i = (h_i - h_0) - T_0 (s_i - s_0) \tag{3.47}$$

3.7.3 Pump

Pumps are commonly used devices from pumping the liquids from a lower pressure side to a higher one by providing work input (so-called: pressurizing the liquids). Figure 3.14 demonstrates a practical pump water pump with the inlet and exit sections. The pumping work required comes from the electrical power.

The mass balance equation for the pump is written as follows:

$$\dot{m}_1 = \dot{m}_2 = \dot{m} \tag{3.48}$$

The energy balance equation for the pump is written as:

$$\dot{m}_1 h_1 + \dot{W}_p = \dot{m}_2 h_2 + \dot{Q}_{loss} \tag{3.49}$$

The entropy balance equation is written as the following for the pump.

$$\dot{m}_1 s_1 + \dot{S}_{gen} = \dot{m}_2 s_2 + \frac{\dot{Q}_{loss}}{T_b} \tag{3.50}$$

Figure 3.14 Schematic illustration for the pump.

The exergy balance equation is expressed as follows:

$$\dot{m}_1 ex_1 + \dot{W}_p = \dot{m}_2 ex_2 + \dot{Ex}^{Q_{loss}} + \dot{Ex}_d \qquad (3.51)$$

In order to further calculate the pump power, it is common to use the pressure changes at the inlet and outlet of the pump; the following equation can be used for determining the pumping work depending on the pressure difference which is so called: shaft work:

$$\dot{W}_p = \dot{m} \dot{w}_p \qquad (3.52)$$

$$\dot{w}_p = v_f \left(P_2 - P_1 \right) \qquad (3.53)$$

3.7.4 Heat Exchanger

Heat exchangers are used for heat transfer purposes for either heating or cooling purposes in energy systems. If it is not provided, potential and kinetic energy changes can be neglected. Figure 3.15 demonstrates a heat exchanger.

Since the flows do not mix, the mass balance equation can be written separately for each steam in the heat exchanger.

$$\dot{m}_1 = \dot{m}_2 \ and \ \dot{m}_3 = \dot{m}_4 \qquad (3.54)$$

The energy balance equation is written as follows:

$$\dot{m}_1 h_1 + \dot{m}_3 h_3 = \dot{m}_2 h_2 + \dot{m}_4 h_4 + \dot{Q}_{loss} \qquad (3.55)$$

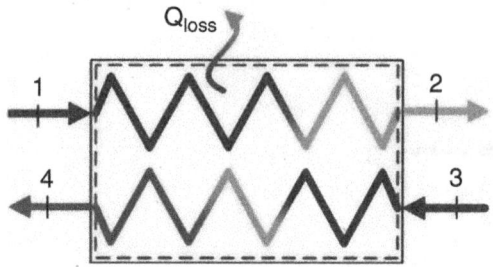

Figure 3.15 Schematic illustration for a heat exchanger.

The entropy balance equation is expressed in the following form:

$$\dot{m}_1 s_1 + \dot{m}_3 s_3 + \dot{S}_{gen} = \dot{m}_2 s_2 + \dot{m}_4 s_4 + \frac{\dot{Q}_{loss}}{T_0} \tag{3.56}$$

The exergy balance equation is written as

$$\dot{m}_1 ex_1 + \dot{m}_3 ex_3 = \dot{m}_2 ex_2 + \dot{m}_4 ex_4 + \dot{E}x^{\dot{Q}_{loss}} + \dot{E}x_d \tag{3.57}$$

It is should be noted that if flow velocities and elevations are given, the total flow energy is to be evaluated.

3.7.5 Open Feedwater Heater (Mixing Chamber)

Consider a mixing chamber as shown in Figure 3.16 where two streams come in to be mixed to obtain a specific stream at state point 3. Open feed water heaters are treated like mixing chambers where there are two or more streams come in to get one desired output. The thermodynamic balance equations for mass (MBE), energy (EBE), entropy (EnBE), and exergy (ExBE) are written as follows:

$$\text{MBE: } \dot{m}_1 + \dot{m}_2 = \dot{m}_3 \tag{3.58}$$

$$\text{EBE: } \dot{m}_1 h_1 + \dot{m}_2 h_2 = \dot{m}_3 h_3 \tag{3.59}$$

$$\text{EnBE: } \dot{m}_1 s_1 + \dot{m}_2 s_2 + \dot{S}_{gen} = \dot{m}_3 s_3 \tag{3.60}$$

$$\text{ExBE: } \dot{m}_1 ex_1 + \dot{m}_2 ex_2 = \dot{m}_3 ex_3 + \dot{E}x_d \tag{3.61}$$

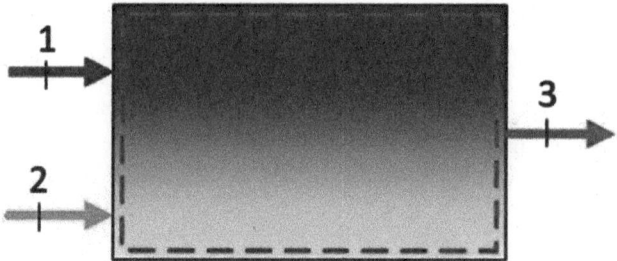

Figure 3.16 Schematic illustration for a heat exchanger.

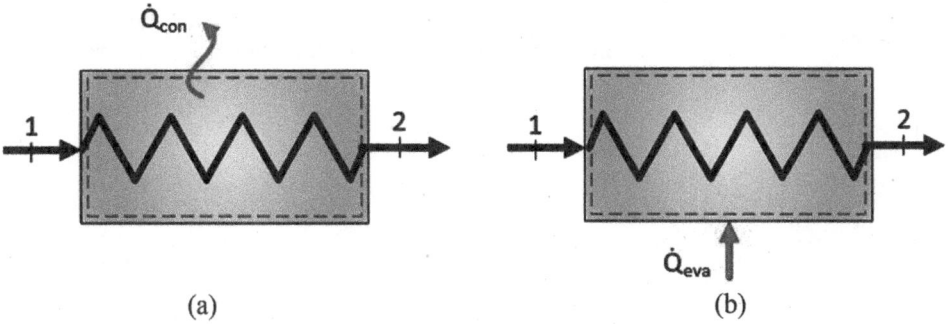

Figure 3.17 Schematic illustration for (a) condenser and (b) evaporator.

3.7.6 Condenser and Evaporator

Condenser and evaporator are the common component of the refrigeration and power cycles. Schematic illustrations of the condenser and evaporator are shown in Figure 3.17. The thermodynamic balance equations for mass (MBE), energy (EBE), entropy (EnBE), and exergy (ExBE) are written as follows:

MBE for both: $\dot{m}_1 + \dot{m}_2 = \dot{m}_3$ (3.62)

EBE for condenser: $\dot{m}_1 h_1 = \dot{m}_2 h_2 + \dot{Q}_{con}$ (3.63)

and

EBE for evaporator: $\dot{m}_1 h_1 + \dot{Q}_{eva} = \dot{m}_2 h_2$ (3.64)

EnBE for condenser: $\dot{m}_1 s_1 + \dot{S}_{gen} = \dot{m}_2 s_2 + \dfrac{\dot{Q}_{con}}{T_b}$ (3.65)

and

EnBE for evaporator: $\dot{m}_1 s_1 + \dot{S}_{gen} + \dfrac{\dot{Q}_{eva}}{T_s} = \dot{m}_2 s_2$ (3.66)

EnBE for condenser: $\dot{m}_1 ex_1 = \dot{m}_2 ex_2 + \dot{Ex}_{Q_{con}} + \dot{Ex}_{dest}$ (3.67)

and

EnBE for evaporator: $\dot{m}_1 ex_1 + \dot{Ex}_{Q_{eva}} = \dot{m}_2 ex_2 + \dot{Ex}_{dest}$ (3.68)

For condenser and evaporator, $\dot{Q}_{con}$ and $\dot{Q}_{eva}$ denote heat lost and heat input.

3.7.7 USUF Process

Filling or emptying a tank, vessel, and chamber can be an example for uniform-state, uniform process (USUF) systems. Figure 3.18 shows a tank which is filled and emptied with a fluid. The rate of fill and empty is different. So, the volume of the fluid will change with respected to the time.

The mass balance equation for the tank shown in Figure 3.18 is written as follows:

$\dot{m}_1 \Delta t + m_i = \dot{m}_2 \Delta t + m_f$ (3.69)

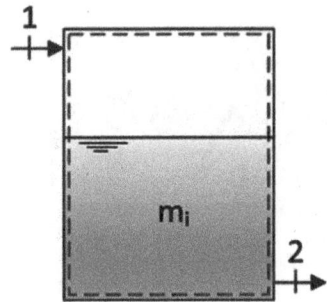

Figure 3.18 Schematic illustration a tank has an input and output.

where Δt is the duration of the process. "i" and "f" denote the initial and final state of the fluid in the tank, respectively.

The energy balance equation is written as:

$$\dot{m}_1 h_1 \Delta t + m_i u_i = \dot{m}_2 h_2 \Delta t + m_f u_f \tag{3.70}$$

The entropy balance equation can be written as:

$$\dot{m}_1 s_1 \Delta t + m_i s_i + \dot{S}_{gen} \Delta t = \dot{m}_2 s_2 \Delta t + m_f s_f \tag{3.71}$$

The exergy balance equation then becomes

$$\dot{m}_1 ex_1 \Delta t + m_i ex_i = \dot{m}_2 ex_2 \Delta t + m_f ex_f + \dot{Ex}_d \Delta t \tag{3.72}$$

Illustrative Example 3.3   Consider a pump as shown in Figure 3.19. This pump is used to increase the pressure of the water from 50 kPa to 800 kPa at the flowrate of 10 kg/s. Take $v_{f@50kPa}$=0.001012 m³/kg, and h_1=340.5 kJ/kg. Calculate the pump work rate and final specific enthalpy.

Solution:

Pump operates as an open system. The MBE for this pump is calculated as:

$\dot{m}_1 = \dot{m}_2 = 10 \ kg/s = constant$

$\dot{m}_1 h_1 + \dot{W}_p = \dot{m}_2 h_2$

$\dot{W}_p = \dot{m} \dot{w}_p$

$\dot{w}_p = v_f (P_2 - P_1) = 0.001012 \times (800 - 50) = 0.759 \ kJ/kg$

$\dot{W}_p = \dot{m} \dot{w}_p = 10 \left(\dfrac{kg}{s}\right) \times 0.759 \left(\dfrac{kJ}{kg}\right) = 7.59 \dfrac{kJ}{s} \ or \ kW$

Figure 3.19 A practical pump.

from MBE:

$$10 \times 340.5 + 7.59 = 10 \times h_2$$

$$h_2 = 341.259 \frac{kJ}{kg}$$

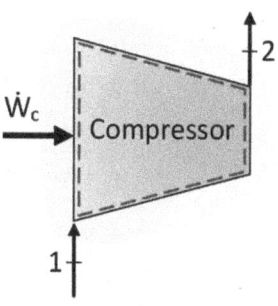

Figure 3.20 Schematic illustration of the hydrogen compressor.

Illustrative Example 3.4 Consider a compressor used for hydrogen compression, as shown in Figure 3.20. It compresses hydrogen from 100 kPa to 1000 kPa. The temperature of hydrogen is 20°C at the compressor inlet. Isentropic efficiency of the compressor is 0.85. Find the compressor power and outlet temperature of hydrogen. Consider the flowrate of the hydrogen as 1 kg/s.

Take h_1=3860 kJ/kg, s_1=53.19 kJ/kg K.

Solution:

MBE for the compressor:

$$\dot{m}_1 = \dot{m}_2 = 1 \; kg/s = constant$$

EBE for the compressor is:

$$\dot{m}_1 h_1 + \dot{W}_c = \dot{m}_2 h_2$$

In order to calculate the compressor power, h_2 should be known. In order to determine the h_2, we need to know two properties of the hydrogen. However, we only have the pressure (P_2=1000 kPa).

We first assume that the compressor works isentropic. That means:

$$s_1 = s_{2s} = 53.10 \frac{kJ}{kg \; K}$$

where s_{2s} denotes the isentropic efficiency at the output of the compressor for hydrogen. Therefore, we know the following information for outlet.

$$s_{2s} = 53.10 \frac{kJ}{kg \; K} \; and \; P_2 = 1000 \; kPa$$

Now, it is possible to determine temperature of the outlet by using thermodynamic tables. So, outlet temperature under the isentropic conditions:

$$T_{2s} = 292.2 \; °C$$

Now, as we know temperature and pressure value at the outlet, we can find the specific enthalpy of hydrogen at the outlet:

$$h_{2s} = 7806 \frac{kJ}{kg}$$

From mass balance equation for isentropic conditions:

$$\dot{m}_1 h_1 + \dot{W}_{p.s} = \dot{m}_2 h_2$$

$$1 \times 3860 + \dot{W}_{p.s} = 1 \times 7806$$

$$\dot{W}_{p.s} = 3946 \ kW$$

As provided in the problem, $\eta_{\text{isentropic}} = 0.85$:

$$\eta_{\text{isentropic}} = \frac{\dot{W}_{p.s}}{\dot{W}_{p.actual}} \rightarrow 0.85 = \frac{3946}{\dot{W}_{p.actual}}$$

$$\dot{W}_{p.actual} = \frac{3946}{0.85} = 4642 \ kW$$

Illustrative Example 3.5 Consider a heat exchanger shown in Figure 3.21. The following information is provided:

T_1=20°C, T_2=250°C, T_3=400°C, T_4=350°C, $\dot{m}_1$ 25 kg/s, h_1=84.01 kJ/kg, h_2=2975 kJ/kg, h_3=3278 kJ/kg

h_4=3175 kJ/kg, s_1=0.2965 kJ/kg K, s_2=8.035 kJ/kg K, s_3=8.357 kJ/kg K, s_4=8.198 kJ/kg K

- Find the flowrate for the second stream.
- Calculate the entropy generation and exergy destruction for the heat exchanger.

Solution:

MBE: $\dot{m}_1 = \dot{m}_2 = 25 \ kg/s/$ and $\dot{m}_3 = \dot{m}_4 = ? \ kg/s$

EBE: $\dot{m}_1 h_1 + \dot{m}_3 h_3 = \dot{m}_2 h_2 + \dot{m}_4 h_4$

$$\dot{m}_1 (h_1 - h_2) = \dot{m}_3 (h_4 - h_3)$$

$$25 \times (84.01 - 2975) = \dot{m}_3 (3175 - 3278)$$

$$\dot{m}_3 = 701 \ kg/s$$

For entropy generation from EnBE:

$$\dot{m}_1 s_1 + \dot{m}_3 s_3 + \dot{S}_{gen} = \dot{m}_2 s_2 + \dot{m}_4 s_4$$

$$\dot{S}_{gen} = 82.01 \ kW/K$$

For exergy destruction from ExBE:

$$\dot{m}_1 ex_1 + \dot{m}_3 ex_3 = \dot{m}_2 ex_2 + \dot{m}_4 ex_4 + \dot{Ex}_d$$

$$\dot{Ex}_d = 24449.3 \ kW$$

Figure 3.21 Schematic illustration of the pump.

3.8 Performance Assessment

The thermodynamic analyses are performed to assess the performance of the energy systems. Before providing the information and equations related to thermodynamic efficiencies, we need to refer the Carnot cycles and Carnot efficiencies. The Carnot cycle is related to the Concept of Ideality for energy systems, which was first exhibited by Sadi Carnot. It aims to define the limit to power generating and refrigeration cycles. In conjunction with this, the Carnot concept is the concept of ideality, which shows the maximum potential performance of thermodynamic cycles can achieve. There are two Carnot cycles which are Carnot heat engine and Carnot refrigerator/heat pump which are reversible devices and run on reversible processes. Both have four identical processes:

- isothermal expansion
- adiabatic expansion
- isothermal compression
- adiabatic compression

Figure 3.22 shows the basic demonstrations of the Carnot principles for heat engine and refrigerator. The Carnot principle draws the theoretical limits for the energy systems, which is subjected to a source and sink which have different temperatures.

The energy balance equation of the Carnot heat engine can be written as follows:

$$Q_H = W - Q_L \quad or \quad W = Q_H - Q_L \tag{3.73}$$

Under the Carnot principle, one may proceed with temperature equivalency ratio and obtain the following ratio:

$$\frac{Q_H}{Q_L} = \frac{T_H}{T_L} \tag{3.74}$$

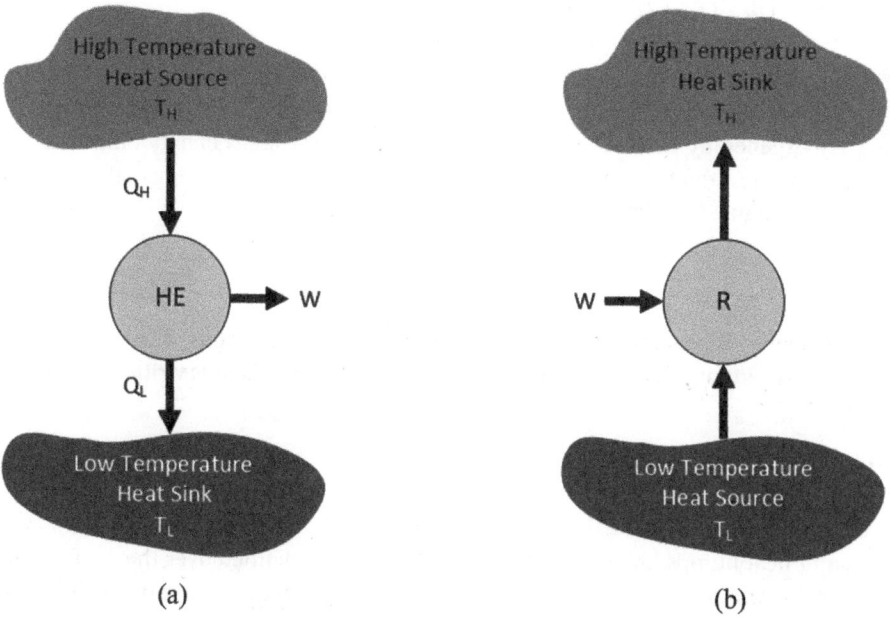

Figure 3.22 Demonstration of the Carnot principle for (a) heat engine and (b) refrigerator.

The efficiency of a Carnot heat engine is written as:

$$\eta_c = \frac{W}{Q_H} = \frac{(Q_H - Q_L)}{Q_H} = \frac{(T_H - T_L)}{T_H}$$

(3.75)

Here, the work output, W, is called the reversible work, W_{rev}.

$$W_{rev} = \left(\frac{T_H - T_L}{T_H}\right) Q_H$$

(3.76)

and then

$$W_{rev} = \left(1 - \frac{T_L}{T_H}\right) Q_H$$

(3.77)

When T_L is the surrounding environment temperature, Eq. 3.77 becomes

$$W_{rev} = \left(1 - \frac{T_0}{T_H}\right) Q_H = Ex^Q$$

(3.78)

which is the exergy content of the heat transfer, which is the theoretical limit of the heat transfer from a source to sink.

For the Carnot refrigerator or heat pump, the energy balance equation is written as follows:

$$Q_L + W = Q_H \quad or \quad W = Q_H - Q_L$$

(3.79)

The performance of the refrigeration cycle is performed over the coefficient of performance (COP). The COP of the Carnot refrigeration and heat pump can be written as:

$$COP_{c,R} = \frac{Q_L}{W} = \frac{Q_L}{Q_H - Q_L} = \frac{T_L}{T_H - T_L}$$

(3.80)

$$COP_{c,HP} = \frac{Q_H}{W} = \frac{Q_H}{Q_H - Q_L} = \frac{T_H}{T_H - T_L}$$

(3.81)

The thermodynamic performance of the system is evaluated through efficiencies.

Furthermore, both energy and exergy efficiencies are used for assessing the systems.

Energy efficiency is generally defined as the ratio of the useful (desired) energy output to be obtained from the system to the total energy input given to the system. The energy efficiency is then written for a heat engine as follows:

$$\eta_{en} = \frac{W_{net}}{Q_{in}}$$

(3.82)

Similarly, exergy efficiency is generally defined as the ratio of the desired exergy output from the system to the total exergy input given to the system. The exergy efficiency is thus written for a heat engine as follows:

$$\eta_{ex} = \frac{W_{net}}{Ex^{Q_{in}}}$$

(3.83)

For refrigerators and heat pumps, the performance of the system is defined over the coefficient of performance (COP). The general definition of the COP for heat pump is the ratio of the heat output in the condenser to work input of the compressor. The COP for heat pump is written as follows:

$$COP_{HP,en} = \frac{Q_{heating}}{W_{in}} \qquad (3.84)$$

The exergetic COP for the heat pump is

$$COP_{HP,ex} = \frac{Ex^{Q_{heating}}}{W_{in}} \qquad (3.85)$$

The COP for the refrigerator is defined as the ratio of the useful cooling capacity in the evaporator to the compressor work input. The energetic and exergetic COPs for the refrigerator are defined as:

$$COP_{R,en} = \frac{Q_{cooling}}{W_{in}} \qquad (3.86)$$

and

$$COP_{R,ex} = \frac{Ex^{Q_{cooling}}}{W_{in}} \qquad (3.87)$$

Illustrative Example 3.6 Someone claims to develop a new power plant operating with an efficiency of 65% at a boiler temperature of 650°C and condenser temperature of 35°C. Is this claim possible?

Solution:
As mentioned earlier, the Carnot efficiency shows the theoretical limit for a heat engine. So, there is no heat engine which has higher efficiency than the one working with Carnot principles. Therefore, we can check the claim by calculating the Carnot efficiency:
From Eq. 3.75:

$$\eta_c = \frac{W}{Q_H} = \frac{(Q_H - Q_L)}{Q_H} = \frac{(T_H - T_L)}{T_H} = \frac{973 - 308}{973} = 0.6834$$

The claim was 65% and the Carnot efficiency is 68.34%. As the efficiency of the claim is lower than the Carnot efficiency, the claim is possible to achieve it.

Illustrative Example 3.7 Engineer discovered a new heat pump and claims that it works by taking outdoor air at –15°C and heating the house at 25°C. The engineer declared the COP of heat pump to be 10. Is this heat pump performance practically possible?

Solution:
We can check its possibly over Carnot efficiency. From Eq. 3.81:

$$COP_{c,HP} = \frac{Q_H}{W} = \frac{Q_H}{Q_H - Q_L} = \frac{T_H}{T_H - T_L} = \frac{298}{298 - 258} = 7.45$$

The engineer mentioned that the COP of the heat pump was 10, despite the Carnot efficiency is 7.45. So, this claim is totally wrong since no heat pump can work with a higher COP than the Carnot heat pump.

Illustrative Example 3.8 Consider an actual Rankine cycle, shown in Figure 3.23a, steam enters the turbine 6 MPa and 450°C. The pressure of the condenser is 100 kPa. The temperature at the condenser outlet 90°C. The mass flow rate in the cycle is 15 kg/s. Consider the isentropic efficiencies of turbine and pump as 90%.

a) Write the thermodynamic balance equations in terms of mass, energy, entropy, and exergy for each component of the system
b) Calculate the heat needed in the boiler
c) Calculate the net work rate produced
d) Calculate both energy and exergy efficiencies of the cycle.

Solution:
In the thermodynamic analysis, it is first critical to draw the schematic diagram of the system studied. Also, for a thermodynamic cycle, its T-s diagram should be drawn along with it. The schematic diagram of the Rankine cycle and its T-s diagram are shown in Figure 3.23.

Next, in order to reduce the complexity of the problem, the conceptually correct assumptions can be done. In the present example, the following assumptions are applied to the problem.

- The pressure losses in the boiler and condenser are negligible.
- The changes in kinetic and potential energies are negligible.
- The isentropic efficiency of the turbine is 90%.
- The reference temperature and pressure are 25 °C and 100 kPa.

a. The following thermodynamic balance equations are provided for each component.

Turbine:

MBE: $\dot{m}_1 = \dot{m}_2$
EBE: $\dot{m}_1 h_1 = \dot{m}_2 h_2 + \dot{W}_{turbine}$
EnBE: $\dot{m}_1 s_1 + \dot{S}_{gen,turbine} = \dot{m}_2 s_2$

ExBE: $\dot{m}_1 ex_1 = \dot{ex}_2 h_2 + \dot{W}_{turbine} + \dot{Ex}_{dest,\,turbine}$

Condenser:

MBE: $\dot{m}_2 = \dot{m}_3$
EBE: $\dot{m}_2 h_2 = \dot{m}_3 h_3 + \dot{W}_{turbine}$

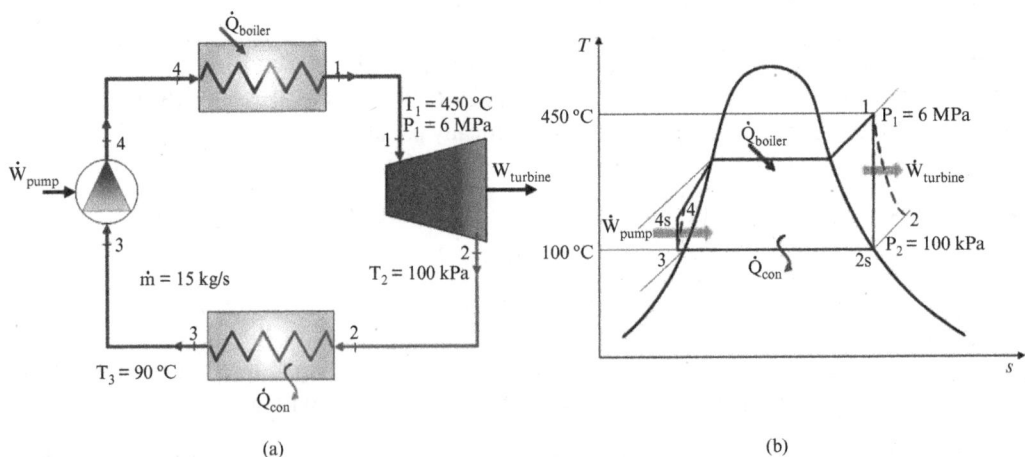

(a)   (b)

Figure 3.23 A Rankine cycle: (a) a system illustration with the components and (b) its T-s diagram.

EnBE: $\dot{m}_2 s_2 + \dot{S}_{gen,con} = \dot{m}_3 s_3 + \dfrac{\dot{Q}_{con}}{T_b}$

ExBE: $\dot{m}_2 ex_2 = \dot{ex}_3 h_4 + \dot{Ex}_{\dot{Q}_{con}} + \dot{Ex}_{dest,con}$

Pump:

MBE: $\dot{m}_3 = \dot{m}_4$ EBE: $\dot{m}_3 h_3 + \dot{W}_{pump} = \dot{m}_4 h_4$

EnBE: $\dot{m}_3 s_3 + \dot{S}_{gen,pump} = \dot{m}_4 s_4$

ExBE: $\dot{m}_3 ex_3 + \dot{W}_{pump} = \dot{ex}_4 h_4 + \dot{Ex}_{dest,pump}$

Boiler:

MBE: $\dot{m}_4 = \dot{m}_1$

EBE: $\dot{m}_4 h_4 + \dot{Q}_{boiler} = \dot{m}_1 h_1$

EnBE: $\dot{m}_4 s_4 + \dfrac{\dot{Q}_{boiler}}{T_s} + \dot{S}_{gen,boiler} = \dot{m}_1 s_1$

ExBE: $\dot{m}_4 ex_4 + \dot{Ex}_{\dot{Q}_{boiler}} = \dot{ex}_1 h_1 + \dot{Ex}_{dest,\,boiler}$

b. In order to determine the heat given to the boiler, we need to use the EBE for the boiler. Here, we need to know that inlet and outlet states of the water circulated in the cycle. We already know the outlet state as it is turbine inlet. It is required to find the inlet state of boiler. The enthalpy of the stream entering the boiler is the same with the exiting from the pump.

$$w_{pump} = v_3 \left(P_4 - P_3 \right) = 0.001010 \; x \; \left(6000 - 100 \right) = 5.959 \; \left(kJ \, / \, kg \right)$$

where v_3 is taken as the specific volume of the saturated liquid at T_1 and P_1.

$$w_{pump} = h_4 - h_3 \rightarrow 5.959 = \left(h_3 - 209.4 \right) \rightarrow h_4 = 215.4$$

For the EBE of the boiler:

$$\dot{m}_4 h_4 + \dot{Q}_{boiler} = \dot{m}_1 h_1 \; \rightarrow \dot{Q}_{boiler} = 53883 \; kW$$

c. In the cycle, while the steam turbine produces power, the pump consumes the power. The power generation in the turbine can be calculated as by applying EBE.

$$\dot{m}_1 h_1 = \dot{m}_2 h_2 + \dot{W}_{turbine}$$

In order to calculate the turbine work, first, the isentropic work of the turbine should be calculated:

$$\dot{W}_{turbine,is} = \dot{m}(h_1 - h_{2s}) = 15 \; x \; (3411 - 2675) = 736 \; kW$$

Actual turbine work can be calculated:

$$\eta_{is} = \dfrac{\dot{W}_{turbine}}{\dot{W}_{turbine,is}} \rightarrow 0.90 = \dfrac{\dot{W}_{turbine,net}}{736} \rightarrow \dot{W}_{turbine} = 15520.32 \; kW$$

d. The energy and exergy efficiencies of the Rankine cycle:
The overall energy efficiency of the Rankine cycle is obtained as:

$$\eta_{en} = \dfrac{\left(\dot{W}_{turbine,net} - \dot{W}_{pump} \right)}{\dot{Q}_{boiler}} = 0.37$$

The exergy efficiency of the Rankine cycle is obtained as:

$$\eta_{ex} = \frac{\left(\dot{W}_{turbine,net} - \dot{W}_{pump}\right)}{\dot{Ex}_{\dot{Q}_{boiler}}} = 0.18$$

3.9 Closing Remarks

This chapter provides the basic principles, fundamental information, essential concepts, and definitions of thermodynamic laws. Thermodynamic efficiencies are a unique tool to evaluate systems, and it further discusses the performance assessment criteria, including energy and exergy efficiencies for heat engines as well as energetic and exergetic coefficients of performance particularly for heat pumps and refrigerators. There are various approaches presented about how to better understand thermodynamics and its concepts and follow a logical pattern in every action. The thermodynamics laws, including zeroth, first, second and third, are discussed with examples and linkages to some practical matters. More importantly, the balance equations are introduced for mass, energy, entropy, and exergy for various types of systems, including closed and open, as well as some types of steady-state steady-flow and uniform-state uniform-flow processes. There are numerous examples given to illustrate how to write the balance equations and make the necessary calculations.

Reference

1 Dincer, I. (2020). *Thermodynamics: A Smart Approach*, 1e. John Wiley & Sons, Ltd.

Questions/Problems

Questions

1 Discuss the importance of thermodynamic for energy, economy, environment and sustainable development.
2 Describe the thermodynamic laws of ZLT, FLT, SLT, and TLT with an example and link to the practical applications.
3 Explain how thermodynamic laws work together in harmony.
4 Illustrate the six steps in thermodynamic analysis and discuss each step with an example.
5 Discuss the seven steps in better understanding of thermodynamics and its principles.
6 Discuss the types of thermodynamic systems and illustrate each with an example.
7 Explain the differences between SSSF and USUF processes.
8 Describe Carnot heat engine, Carnot heat pump, and Carnot refrigerator, and illustrate the thermodynamic limits for each accordingly.
9 Someone claims to develop a new power plant operating with an efficiency of 71% at a heat source temperature of 750°C and condenser temperature of 50°C. Is this claim possible? Prove mathematically and discuss thermodynamically.
10 An engineer claims that there is a heat pump operating with a COP of 12 in the summer months for air conditioning purposes at an evaporator temperature of 18°C and a condenser temperature of 35°C. Is this claim possible? Prove and discuss.

Problems

1 Consider a rigid tank filled with argon that is heated by a heat source at a temperature of 100°C. The rigid tank has a volume of 1 m³ and contains 1 kg of a real gas helium. (a) Write mass, energy and entropy, and exergy balance equations for the system, (b) calculate both pressure and temperature of helium in the rigid tank after a total of 10 kJ of heat is transferred to the system while knowing that the system initially is at ambient temperature of 25°C, (c) find the entropy generation, and (d) calculate the exergy destruction values based on the entropy generation and the exergy balance equation and compare the results.

2 A piston–cylinder device with an initial volume of 0.1 m³, as shown in the figure below, contains saturated water vapor at a temperature of 100°C. The water is compressed so that the final volume is 80% of the initial volume. 40 kJ of heat is lost to the environment through the compression process. Write all balance equations and calculate the electrical power supplied, entropy generation, and the exergy destruction values based on the entropy generation and by using the exergy balance equation and compare the results.

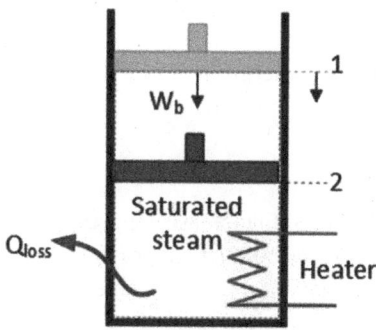

3 Consider a turbine in which hot combustion gases enter at state 1 and exit at state 2. The temperature at the inlet is measured to be 400°C and pressure is set at 1 MPa. The exit pressure is set at 100 kPa and the exit temperature is measured to be 130.1°C. For a mass flow rate of 1.5 kg/s, considering the combustion gases to entail ideal gas properties for air:
a Write all mass, energy, entropy and exergy balance equations.
b Calculate the work output from the turbine if the rate of heat loss is known to be 100 kW.
c Find the entropy generation rate.
d Calculate the exergy destruction rate.

4 An air compressor is used to compress air from atmospheric pressure and a temperature of 280 K to a discharge pressure and a temperature of 600 kPa and 500 K, respectively. The mass flow rate of the air entering the compressor is 1 kg/s.
a Write the balance equations for mass, energy, entropy and exergy.
b Find the work rate consumed by the compressor.
c Find the entropy generation rate.
d Find the exergy destruction rate.

5 Water at ambient pressure and a temperature of 99°C is used to maintain the temperature of a building at 30°C and exits the building heating system at a temperature of 35°C. The mass flow rate of water is 2.8 kg/s.

   a Write the balance equations for mass, energy, entropy and exergy.
   b Find the amount of thermal energy released from the hot water to the building per unit time.
   c Find the entropy generation rate.
   d Find the exergy destruction rate.

6 A 32 kg/s of water flow is heated at a constant pressure by 4.8×10^6 kg/h of exhausts gases, as shown in the figure. Assume the exhaust gases consist only of air where you can use ideal gas equations.

$c_{p,air} = 1\,005$ kJ/kg K.

   a Write the balance equations for mass, energy, entropy and exergy.
   b Calculate the exit temperature of the water.
   c Calculate the exit temperature of the exhaust.
   d Calculate the amount of heat rate transferred to the water.

7 Consider a water pump, as shown in Figure 3.19, which is used to increase the pressure of the water from 100 kPa to 800 kPa at the flowrate of 8 kg/s. Take the inlet water temperature as 15°C and reference temperature as 25°C. Write the balance equations for mass, energy, entropy and exergy and calculate the pump work rate and exergy destruction rate.

8 Reconsider Example 3.4, write the balance equations for mass, energy, entropy and exergy, and repeat the calculations for a flowrate of 10 kg/s.

9 Reconsider Example 3.8, write the balance equations for mass, energy, entropy and exergy, and repeat the calculations for a flowrate of 10 kg/s and isentropic efficiencies of 85%.

10 Reconsider Example 3.8, write the balance equations for mass, energy, entropy and exergy, and repeat the calculations for a flowrate of 15 kg/s and isentropic efficiencies of 80%.

4

Fuels and Combustion

4.1 Introduction

Fuels and combustion are considered oldest subjects, probably as old as humanity. Wood was the first fuel for humanity due to its easiness to obtain and easiness to burn features. It is the first bio-mass-based fuel to meet the energy demand for heating, cooking, and other purposes for a long time. Figure 4.1 shows the average annual energy consumption levels with the historical perspectives and the ages of developments along with the humanity's developments. As it is clearly seen here that the era started with Industrial Revolution which depended heavily upon fossil fuels (in particular coals) and resulted in a *peak oil* during the recent years. When you look at the history overall and what we see the fossil fuels era technically started after the Industrial Revolution. One has to note that wood, charcoal, and coal were the main fuel for the steam engines. As indicated in Figure 4.1, it is expected that the utilization of fossil fuels will end in a few decades due to the depletion of fossil fuels and environmental concerns. So, these years are called the fossil fuel age. However, COVID-19 pandemic has shown that humanity needs urgent action to change the way of the of the globe. Prof. Dincer has stated these in a perspective article that COVID-19 coronavirus pandemic is closing carbon age, but opening hydrogen age. This is clear in Figure 4.1 that the year of 2020 was a turning point for many countries that declared their commitment for hydrogen economy and prepared their road maps and strategic plans. So, we are now in hydrogen era; we need hydrogen-based energy and fuel solutions.

One may need to note that fossil fuels were going back as they were formed many hundreds of millions of years ago before the time of the dinosaurs. The age they were formed is called the carboniferous period (from about 360 to 286 million years ago). As illustrated above, their heavy use began with the Industrial Revolution, particularly between 1760 and 1840.

Figure 4.2 demonstrates the energy consumption in the world by source for a narrow timespan. The use of fossil fuels has increased significantly after 1800s with the deployment of the steam engines. Steam engines have improved the industrial activities and transportation in the world. That's why it is called the Industrial Revolution (1760–1840). It is also named as the industry 1.0 as the industrialization has just started. From ancient times to mid-1800s, the traditional biomass, mostly wood, was the main fuel. After the mid-1800s, coal entered the game as a main player as it was only effective fuel for steam engines and steel production. Oil has been used since the late 1800s and early 1900s. In the second quarter of 1900, natural gas started to be used. With the petrol crisis in 1973, the use of renewables has taken some attention, especially for the countries that do not have fossil fuel source. Forming up the fossil fuels requires thousands of years. That is really a very long period going back to dinosaurs. Therefore, fossil fuels will be over in the future.

Introduction to Energy Systems, First Edition. Ibrahim Dincer and Dogan Erdemir.
Companion Website: www.wiley.com/go/Dincer/Introduction_to_Energy_Systems

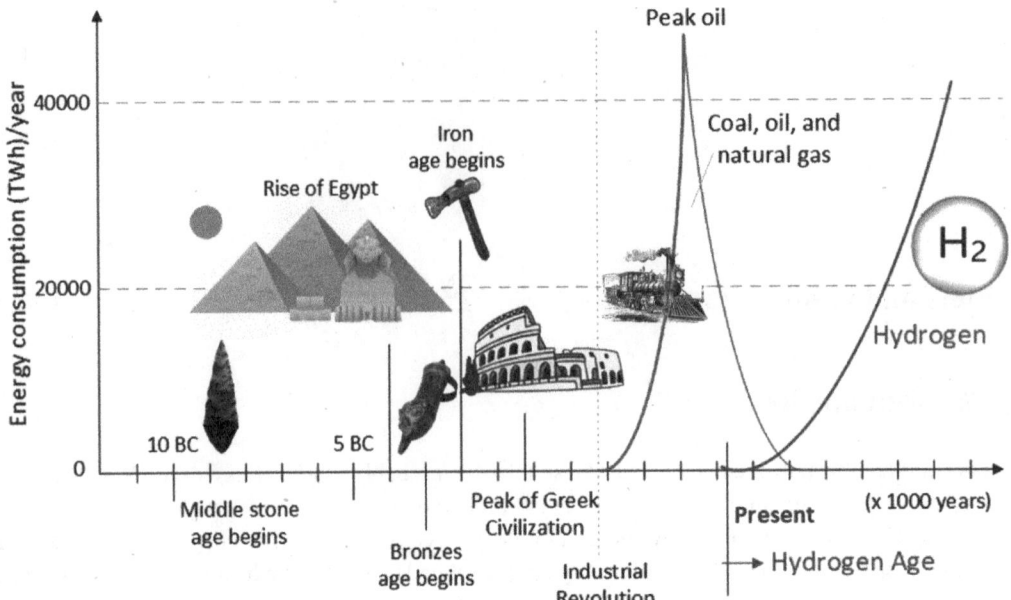

Figure 4.1 A historical perspective of fossil fuel utilization in the world.

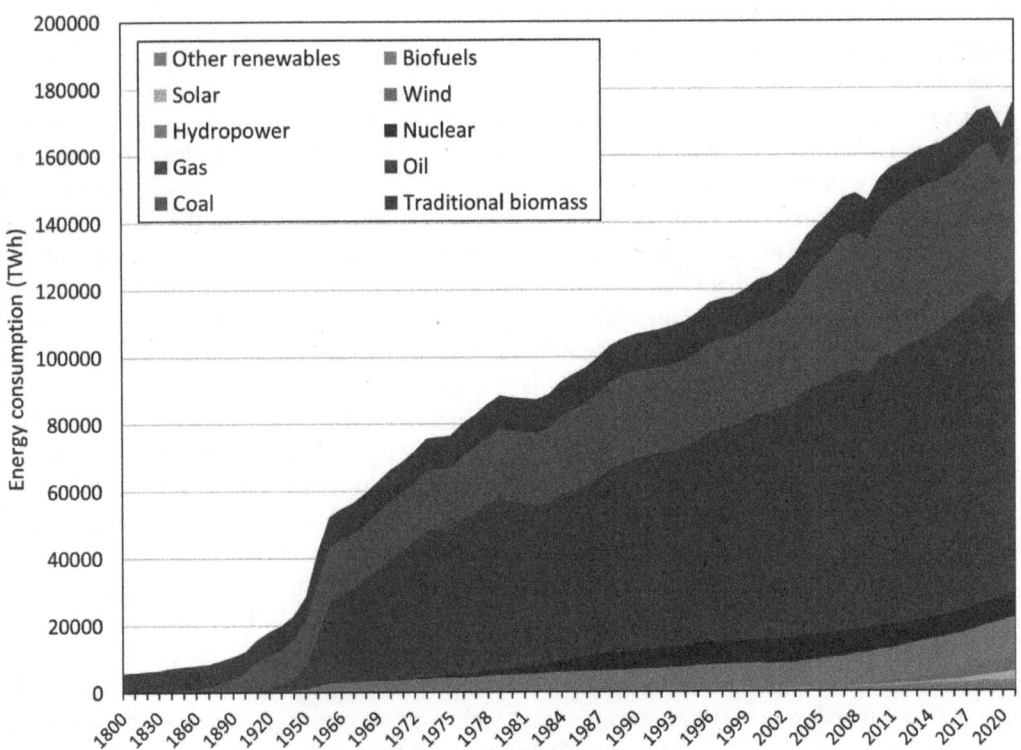

Figure 4.2 Global primary energy consumption by source (data from [1]).

The increase in the amount of fossil fuels has caused many issues such as global warming and air pollution. Although they cause many issues, the most known one by people is carbon emissions. Table 4.1 shows the common fossil fuels, their lower heating values, and carbon content. It is not surprising that coal has the lowest heating value and the highest carbon content. They have been used for many years and are relatively easy to obtain and process. Among fossil fuels, natural gas has the lowest carbon content. Because of its relatively lower carbon ratio, the natural gas systems called as the environmentally friendly enough. Especially, compressed, or liquid natural gas-powered busses advertised as the clean solution. Of course, comparing their competitors which are diesel and gasoline fueled ones, they release less carbon dioxide due to the lower carbon content.

This chapter aims to begin with some introductory information and types of fossil fuels which is followed by the impacts of utilizing fossil fuels. In addition, the combustion of fossil fuels is presented in detail, along with thermodynamic analysis of combustion processes through energy and exergy analyses for closed and open types of systems and their performance assessments through energy and exergy efficiencies.

4.2 Fossil Fuels

Fossil fuels, generally called hydrocarbons, contain materials of biological origin occurring within Earth's crust formed millions of years ago. They have been heavily used as a source of energy since ancient times. Fossil fuels are made from decomposing organic materials such as plants, animals, etc. These fuels are found in the Earth's crust. The composition of fossil fuels is carbon and hydrogen. Although fossil fuels include coal, petroleum, natural gas, oil shales, bitumen, tar sands, and heavy oils, today's fuel consumptions are considered, there are three main ones which are coal, oil, and natural gas. Table 4.2 compares various types of fuels in terms of their densities, higher heating values and costs.

The coal is usually found in sedimentary rock deposits where dead plants and animals are piled up in layers. Fossilized plants must make up more than half of the weight of coal. The oil was originally found as a solid material between sedimentary rocks, such as shale. Fuel is made from this material by heating it to make thick oil. Oil deposits usually have pockets of natural gas above them. Sedimentary rocks without oil can also contain it. Methane is the main component of natural gas.

Table 4.1 A comparison of lower heating values and carbon contents of some common fossil fuels.

Type	Lower Heating Value (kJ)	Carbon Content (%)
Raw	20,908	25.8
Coke	28,435	29.2
Washed coal	25,344	26.2
Crude oil	41,816	20.0
Gasoline	43,070	18.9
Kerosene	43,070	19.6
Diesel	42,652	20.2
Natural gas	38,931	15.3

Source: Ref. [2].

Table 4.2 Comparison of various types of fuels.

Fuel/Storage Option	P (bar)	Density (kg/m^3)	HHV (MJ/kg)	Energy Density (GJ/m^3)	Specific Volumetric Cost ($ m^3)	Specific Energetic Cost ($/GJ)
Gasoline (C$_8$H$_{18}$)/ liquid tank	1	736	46.7	34.4	1000	29.1
CNG (CH$_4$)/integrated storage system	250	188	55.5	10.4	400	38.3
LPG (C$_3$H$_8$)/ pressurized tank	14	388	48.9	19.0	542	28.5
Methanol (CH$_3$OH)/ liquid tank	1	749	15.2	11.4	114	10
Hydrogen (H$_2$)/metal hydrides	14	25	142	3.6	125	35.2
Ammonia (NH$_3$)/ pressurized tank	10	603	22.5	13.6	181	13.3
Ammonia (NH$_3$)/metal amines	1	610	17.1	10.4	183	17.5

Source: Ref. [3].

4.2.1 Oil

Crude oil (so-called petroleum) is recognized as a mixture of aliphatic hydrocarbons which is illustrated by the alkanes of chemical formulas of C$_n$H2$_{n+2}$ along with several other compounds and impurities, including sulfur being one of most significant quality criteria to grade it as sweet crude oil with low sulfur content and sour crude oil with high sulfur content. Following oil discovery had developed an oil-based economy and brought an important scientific development through refineries where crude oil is refined and processed into various other types of fuels (such as gasoline, diesel, fuel oil, jet fuel, kerosene, liquefied gases, still gas, petroleum coke, asphalt, etc.) and many other chemicals as well through various additional processes in refineries. It is important to note that the final products and their quantities depend essentially on the composition of crude oil. One may need to keep in mind that the primary process is distillation in these refineries where crude oil is heated up to 360°C where the constituents are separated based on their boiling point temperatures under atmospheric pressure (so-called atmospheric distillation) or vacuum pressures (so-called vacuum distillation).

For example, in a crude oil distillation process in refineries, heat is important and required specifically for distillation. The temperature levels determine what will come out after the distillation, and here are some example products: bitumen after 350°C, lubricating oils between 300°C and 350°C, fuel oil between 250°C and 300°C, diesel between 220°C and 250°C, paraffin between 180°C and 220°C, naphtha between 60°C and 180°C, gasoline between 25°C and 60°C, and liquefied petroleum gases below 25°C.

Crude oil-based fuels, chemicals, and other products have been an essential part of our life. Of course, more and more they have played a more critical role in transportation through transportation fuels, including gasoline, diesel, kerosene, jet fuel, etc. Even if the trend turns into electric and hydrogen vehicles in transportation, the oil is the main fuel for the transportation sector so far, which is responsible for the one-third of the total energy consumption. Figure 4.3 shows the distribution of the crude oil reserves in the world. It is clear from Figure 4.3 that crude oil reserves are uniform. While some countries have fossil fuel resources, some on them do not have. Their

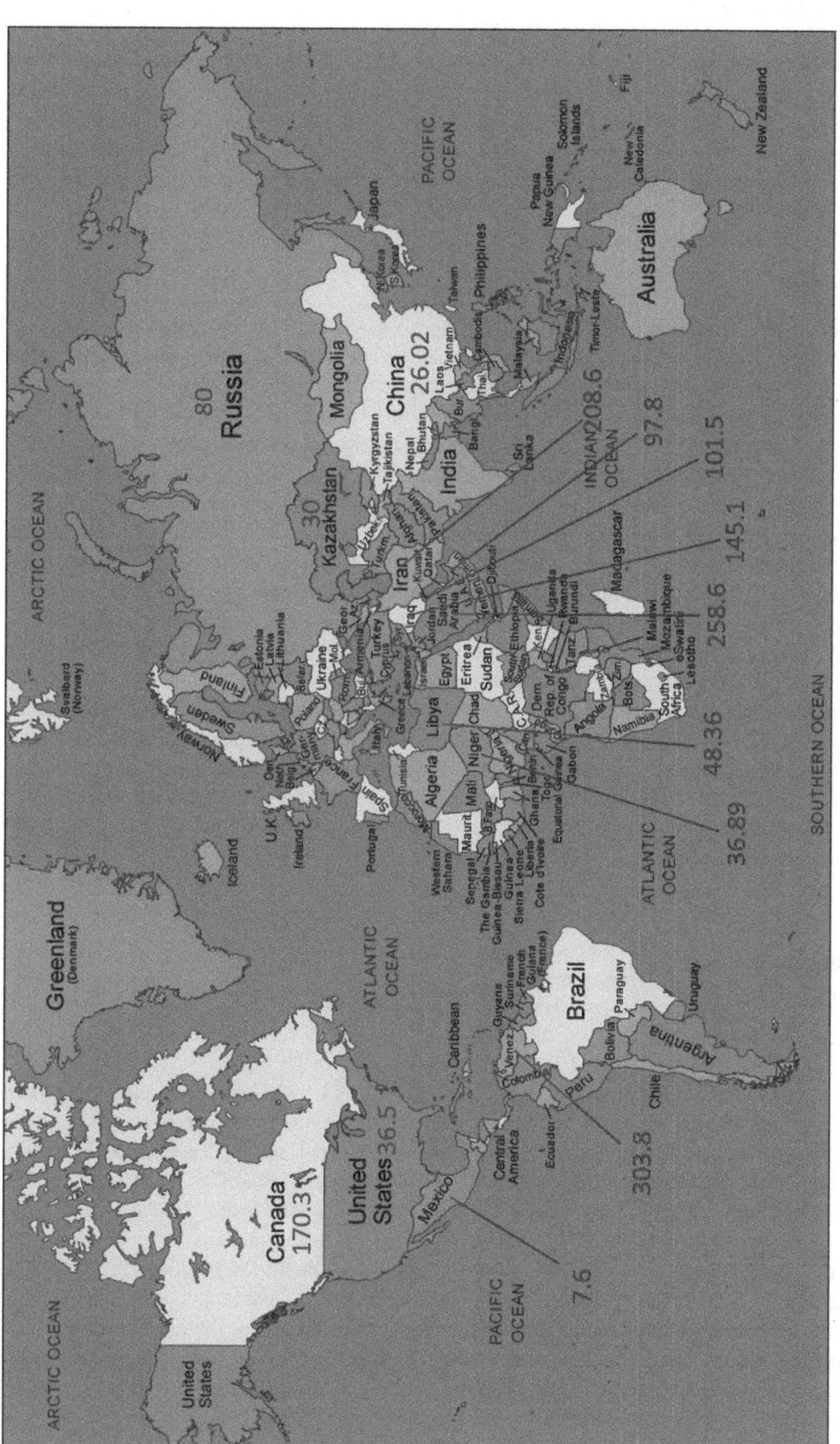

Figure 4.3 Distribution of proven crude oil reserves (in billion barrels) for 2021 in the world (data from [4]).

capacity is also changing by countries. Venezuela, Saudi Arabia, Canada, Iran, and Iraq are the top five countries to hold the largest crude oil reserves. Venezuela has the highest crude oil reserves in the world with the capacity of 303.8 billion barrel. Saudi Arabia is the second oil dominant country in the world with its 258.6 billion barrels of capacity. This is followed by Iran (208 billion barrels). Canada has about 170 billion barrels of crude oil reserves, which is the fourth highest rank in the world. Iraq has the fifth largest capacity in the terms of crude oil in the world.

4.2.2 Coal

Coals (chemically represented by carbon (C)) exist in a large variety of geologic forms in various parts of the world and at varying depths, and its level of quality is primarily measured by the specific heating value. Coals may generally be characterized or categorized in a continuous range of ranks, beginning with peat, then lignites (known as brown coal), then bituminous (known as soft coal), and finally anthracite (known as hard coal) and graphite. Looking at the world's coal resources, it is important to mention that low-rank coals accounting for about 47%, which may be categorized into about lignite coals taking 1/3 of it and used particularly for power generation and subbituminous taking 2/3 of it and used also for power generation, cement, glass, and steel manufacturing. In regards to hard coals accounting for about 53% which are essentially bituminous with only about a small percentage for anthracite, they are commonly utilized for domestic, industrial, and power generation purposes.

Coals can be assumed as the first industrial fuel as they were used in steam engines and heating purposes for many years. It is also the first fuel for thermal power plants for electricity generation, which are still used by the world. It is also very critical for the steel industry. It can also be used for synthetic gas production such as hydrogen, methane, etc. The United States has the world's largest coal reserves, and coal is the world's most abundant fossil fuel. Coals, also for all fossil fuels, are formed from the remains of ancient organisms which make it a nonrenewable resource. It took millions of years for the coal to be formed. Coals occur in underground formations which are recognized as coal seams. The largest coal reserves are located in the United States, Russia, China, Australia, and India.

The common types of coals are listed below:

- peat
- lignite
- subbituminous coal
- bituminous coal
- anthracite
- graphite

There are two different ways to obtain coal from the earth, which are surface mining or underground mining. This classification is done according to the deepness of the coal source. If coal is obtained from less than 61 meters underground, it is called surface mining. Deeper than 61 meters, it is called the underground mining. As the surface mining requires less effort and cost, it is preferable; however, most of the coal reserves in the world are obtained from deep underground.

The immediate environmental impact of underground mining appears less dramatic than surface mining. There is little overburden, but underground mining operations leave significant tailings. Tailings are the often-toxic residue left over from the process of separating coal from gangue, or economically unimportant minerals. Toxic coal tailings can pollute local water supplies.

Figure 4.4 demonstrates the distribution of coal reserves in the world. The top counties which have the highest coal reserves in the world. The USA has the highest number. This is followed by Russia, Australia, China, and India. These five counties share the 33.2%, 15.5%, 14%, 13.1%, and 9.5%, respectively [5].

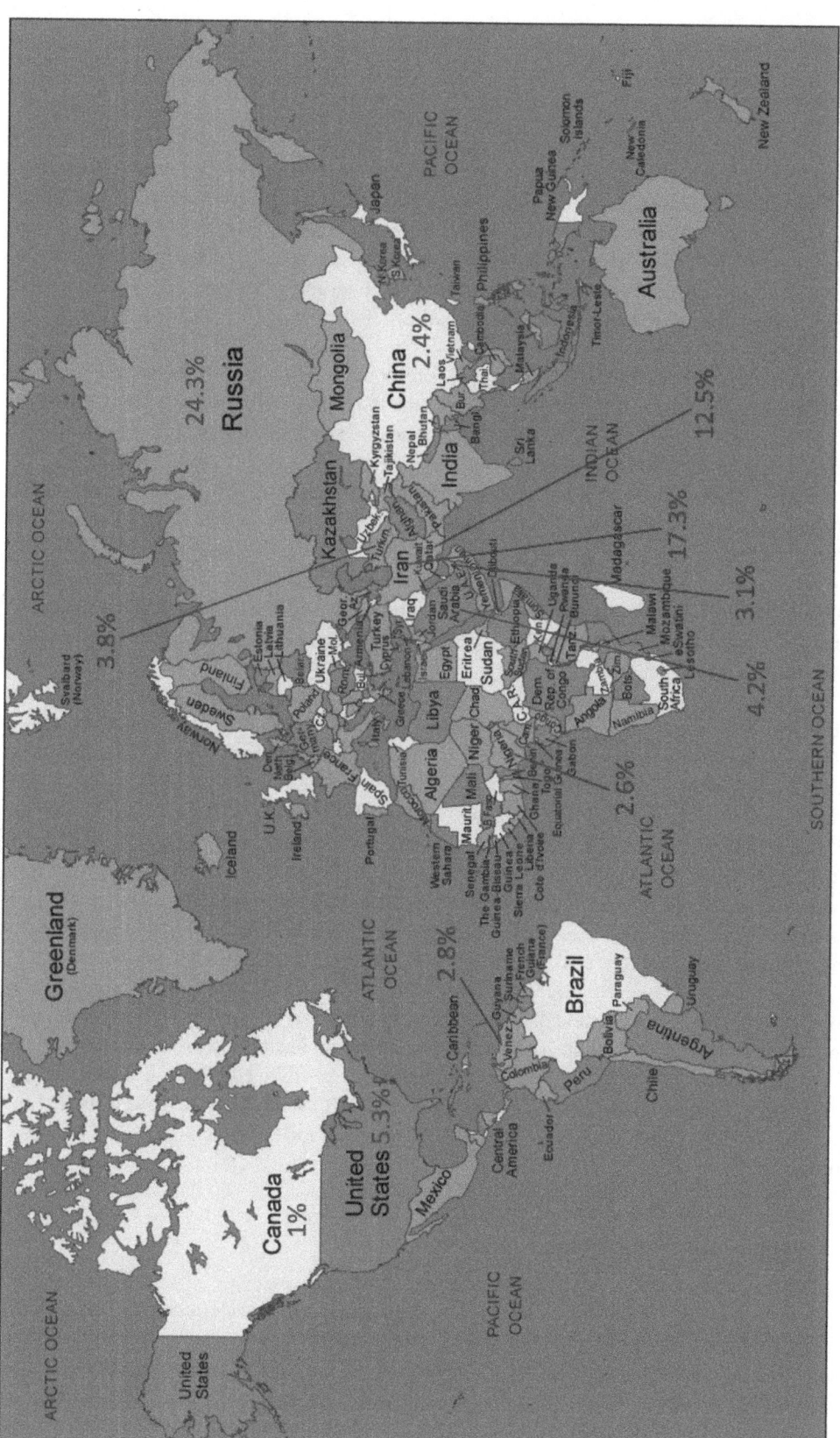

Figure 4.4 Shares of proven coal reserves for the top ten countries in 2021 (data from [4]).

4.2.3 Natural Gas

Natural gas mainly consists of methane (chemically represented by CH_4) which is mainly found in sedimentary rocks, generally in the presence of crude oil as "wet" gas (associated gas) or separately as "dry" gas (nonassociated gas). One may further note that wet gas generally contains several other hydrocarbon gases, and several impurity gases which need to be removed before delivery. Dry gas generally does not contain many other gases besides methane. Natural gas is classified as "sour gas" with high H_2S. The composition varies depending upon the type and deposit, particularly typical constituents. Moreover, the heating value of natural gas also depends on the chemical composition.

Natural gas (also called fossil gas or simply gas) is recognized as a naturally occurring mixture of gaseous hydrocarbons consisting primarily of methane in addition to various smaller amounts of other higher alkanes. Usually, low levels of trace gases like carbon dioxide, nitrogen, hydrogen sulfide, and helium are also present.

Natural gas is colorless and odorless, so odorizes such as mercaptan, which smells like sulfur or rotten eggs, are commonly added to natural gas supplies for safety so that leaks can be readily detected. Natural gas is a fossil fuel and nonrenewable resource that is formed when layers of organic matter decompose under anaerobic conditions and are subjected to intense heat and pressure underground over millions of years. The energy that the decayed organisms originally obtained from the sun via photosynthesis is stored as chemical energy within the molecules of methane and other hydrocarbons.

Natural gas can be burned for heating, cooking, and electricity generation. It is also used as a chemical feedstock in the manufacture of plastics and other commercially important organic chemicals and is less commonly used as a fuel for vehicles. It is also used for hydrogen production in chemical sectors like ammonia production.

Natural gas is usually found in pockets above oil deposits. It can also be found in sedimentary rock layers that don't contain oil. Natural gas is primarily made up of methane. According to the National Academies of Sciences, 81% of the total energy used in the United States comes from coal, oil, and natural gas. This is the energy that is used to heat and provide electricity to homes and businesses and to run cars and factories. Unfortunately, fossil fuels are a nonrenewable resource and waiting millions of years for new coal, oil, and natural gas deposits to form is not a realistic solution. Fossil fuels are also responsible for almost three-fourths of the emissions from human activities in the last 20 years. Now, scientists and engineers have been looking for ways to reduce our dependence on fossil fuels and to make burning these fuels cleaner and healthier for the environment.

At the start of 2021, proved gas reserves were dominated by three countries: Russia, Iran, and Qatar dominate the natural gas production in the world as illustrated Figure 4.5. Today, natural gas is mainly used for power generation, hydrogen production for chemical production, and heating purposes.

4.3 Impacts of Fossil Fuels

When fossil fuels are burned, they release large amounts of carbon dioxide, a greenhouse gas, into the air. Greenhouse gases trap heat in our atmosphere, causing global warming. Already the average global temperature has increased by 1°C. Warming above 1.5°C risks further sea level rise, extreme weather, biodiversity loss, and species extinction, as well as food scarcity, worsening health, and poverty for millions of people worldwide. Avoiding the worst impacts of climate change will require aggressive action to reduce the greenhouse gas emissions that are causing Earth to

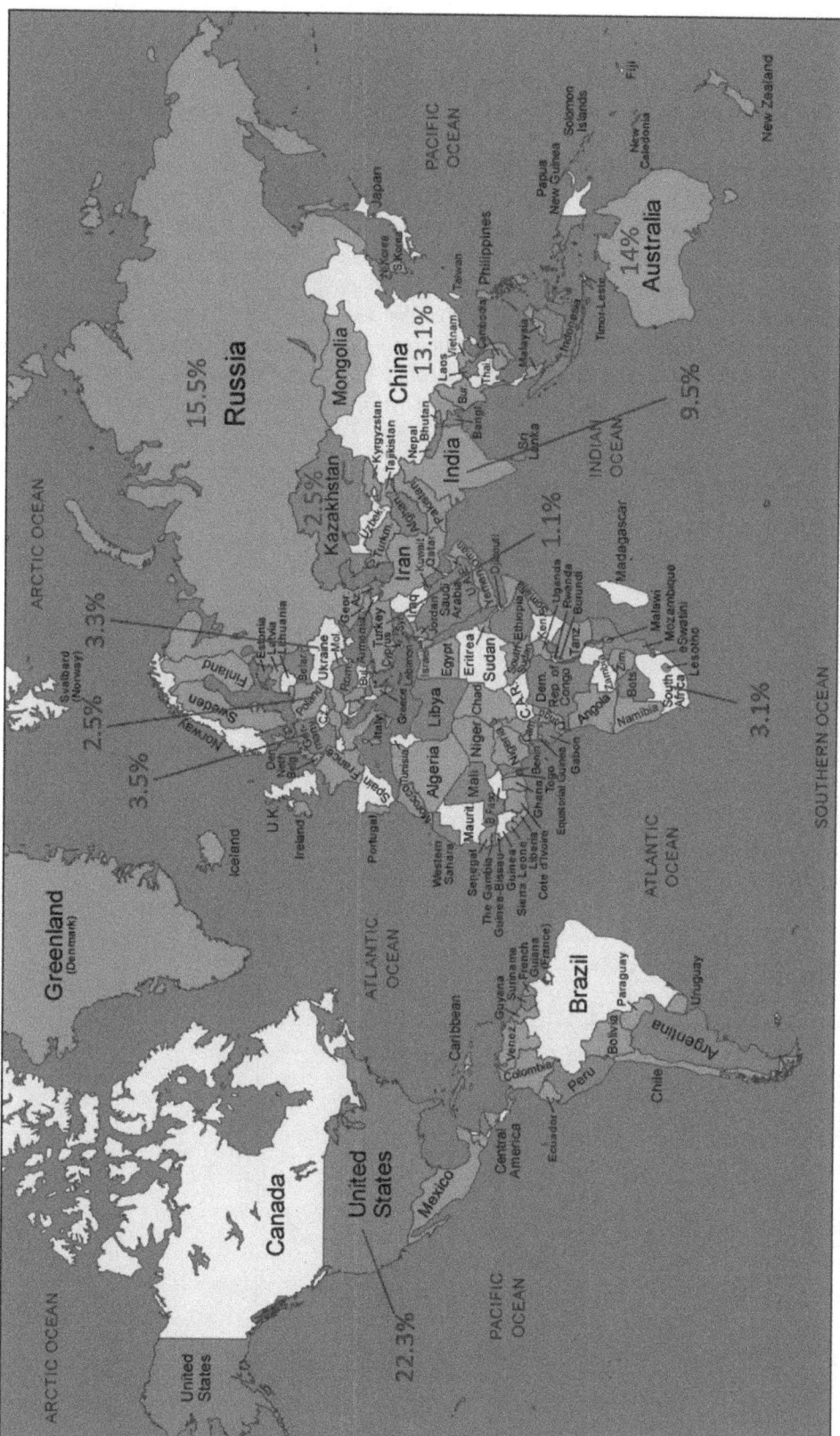

Figure 4.5 Shares of the proven natural gas reserves for top ten counties in 2021(data from [4]).

warm. A number of expert reports from the National Academies have assessed the latest in climate science, technology options, and socioeconomic dimensions related to the goal of reaching net-zero emissions by the year 2050.

The Intergovernmental Panel on Climate Change (IPCC) has found that emissions from fossil fuels are the dominant cause of global warming. In 2018, 89% of global CO_2 emissions came from fossil fuels and industry. Coal is a fossil fuel and is the dirtiest of them all, responsible for over 0.3°C of the 1°C increase in global average temperatures. This makes it the single largest source of global temperature rise. Oil releases a huge amount of carbon when burned – approximately a third of the world's total carbon emissions. There have also been a number of oil spills in recent years that have had a devastating impact on our ocean's ecosystem. Natural gas is often promoted as a cleaner energy source than coal and oil. However, natural gas is still a fossil fuel and accounts for a fifth of the world's total carbon emissions. Figure 4.6 depicts the annual total CO_2 emissions in the world and by some counties. As

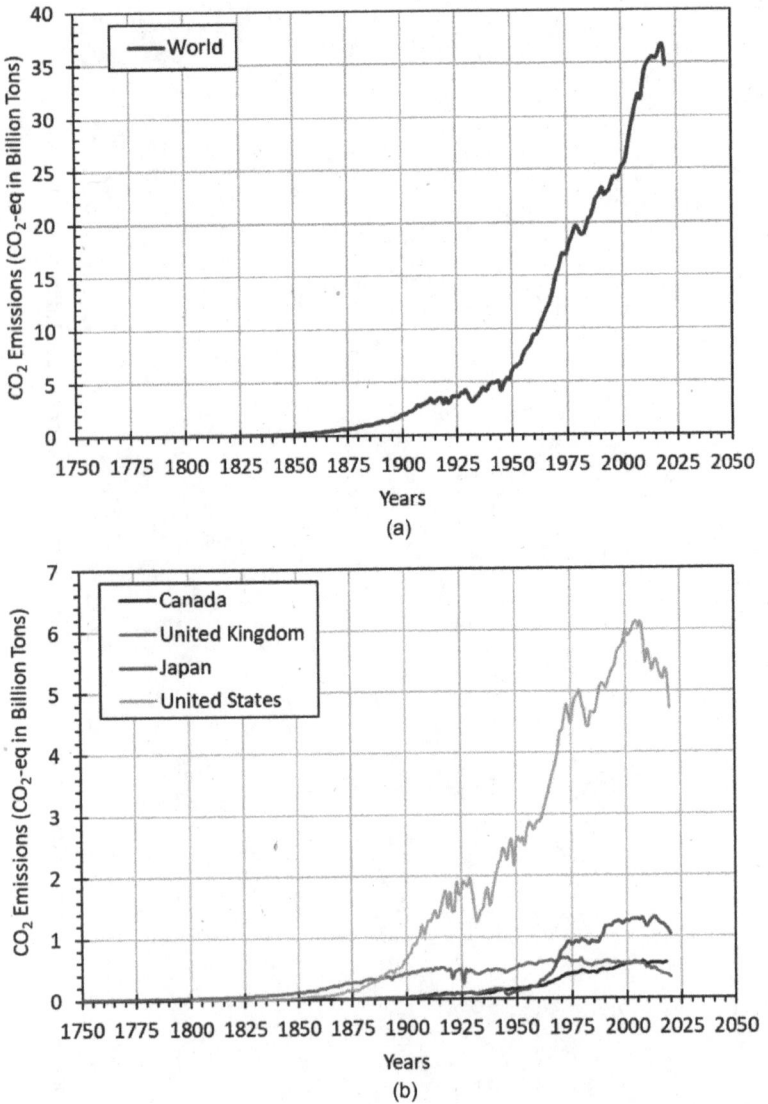

Figure 4.6 Total CO_2 emissions for (a) the world and (b) some countries (data from [6]).

illustrated in Figure 4.6a, there is a certain increase in CO_2 emissions just after the Industrial Revolution. Today, it came up to the level of 35 billion tons. Also, CO_2 emission level is checked from Figure 4.6b, the same growing trend is seen with the world. Despite there being timely decreases in CO_2 emissions, its cumulative value tended to increase. Because of population and industrialization in the USA, CO_2 emission is higher than the other countries presented in Figure 4.6b. After 2010, with the impact of the carbon neutralization actions, emission's values have started to decrease.

Figure 4.7 shows the rise in energy-related CO_2 emissions pushed overall greenhouse gas emissions from energy to their highest ever level in 2021. Total greenhouse gas emissions reached 40.8 Gt of CO_2-eq in 2021 when using a 100-year global warming potential time, above the previous all-time high in 2019. CO_2 emissions from energy combustion and industrial process account for close to 89% of energy sector greenhouse gas emissions in 2021. CO_2 emissions from gas flaring accounted for another 0.7%. Beyond CO_2, fugitive and combustion-related methane emissions represented 10% of the total, and combustion-related emissions of nitrous oxide 0.7%. Methane emissions from the energy sector rose by just under 5% in 2021 but remain below their 2019 level.

Figure 4.8 shows projection of global greenhouse gas emissions with different strategies. Current policies presently in place around the world are projected to result between 2.5°C and 3°C warming above preindustrial levels. When binding short-term or net-zero targets are included warming would be limited to about 2°C above. Warming estimates for the pledges and targets scenario has fallen by 0.3°C compared to last assessment due, primarily, to the inclusion of the USA and China's net-zero targets, now that both countries have submitted their long-term strategies to the United Nations. The optimistic targets scenario shows the effect of net-zero emissions targets of over 140 countries that are adopted or under discussion. Under the optimistic assumption that governments will achieve these targets, the median warming estimate is 1.7°C. On the other hand, the net-zero target aims to neutralize carbon emissions until 2050.

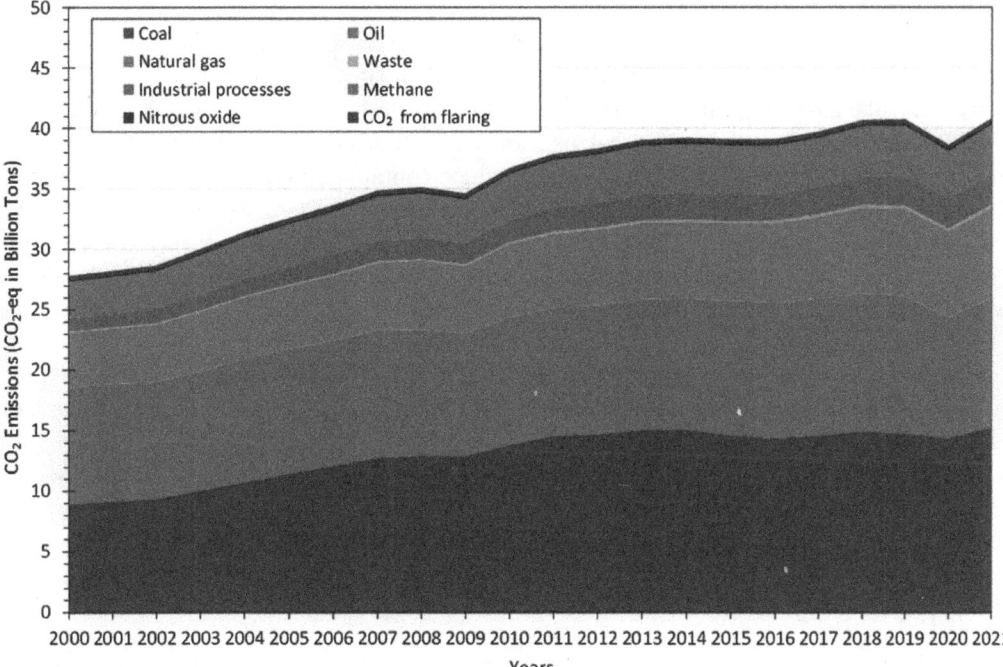

Figure 4.7 CO_2 emissions change with years by fossil fuels (data from [4]).

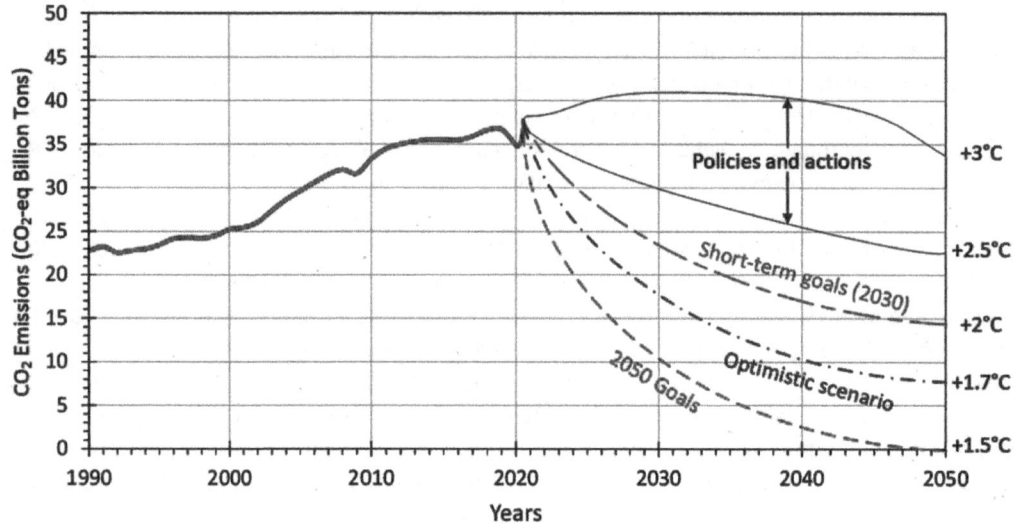

Figure 4.8 The actual CO_2 emissions from 1990 to 2022 and the projections variations of global greenhouse gas emissions with different scenarios (data from [12]).

4.4 Combustion of Fuels

Any material that can be burned to release thermal energy is called a fuel. Combustion is a chemical process in which a substance reacts rapidly with oxygen and gives off heat. The original substance is called the fuel, and the source of oxygen is called the oxidizer. The fuel can be a solid, liquid, or gas, although for airplane propulsion the fuel is usually a liquid. The oxidizer, likewise, could be a solid, liquid, or gas. During combustion, new chemical substances are created from the fuel and the oxidizer. These substances are called exhaust. Most of the exhaust comes from chemical combinations of the fuel and oxygen. When a hydrogen-carbon-based fuel (like gasoline) burns, the exhaust includes water (hydrogen + oxygen) and carbon dioxide (carbon + oxygen). But the exhaust can also include chemical combinations from the oxidizer alone. If the gasoline is burned in air, which contains 21% oxygen and 78% nitrogen, the exhaust can also include nitrous oxides (NOX, nitrogen + oxygen). The temperature of the exhaust is high because of the heat that is transferred to the exhaust during combustion. Because of the high temperatures, exhaust usually occurs as a gas, but there can be liquid or solid exhaust products as well. Soot, for example, is a form of solid exhaust that occurs in some combustion processes.

All fossil fuels can be burned in air or with oxygen derived from air to provide heat. This heat may be employed directly, as in the case of home furnaces, or used to produce steam to drive generators that can supply electricity. In still other cases – for example, gas turbines used in jet aircraft – the heat yielded by burning fossil fuel serves to increase both the pressure and the temperature of the combustion products to furnish motive power.

In Table 4.3, the common fuels used in internal combustion engines are given with their combustion properties.

Table 4.3 Comparison of common fuels used in internal combustion engines.

Properties	Units	Gasoline	Diesel	LPG	CNG	Gaseous Hydrogen	Liquid Hydrogen	Ammonia
Formula		C_8H_{18}	$C_{12}H_{23}$	C_3H_8	CH_4	H_2	H_2	NH_3
Lower heating value	MJ/kg	44.5	43.5	45.7	38.1	120.1	120.1	18.8
Flammability limits, gas in air	Vol. %	1.4–7.6	0.6–5.5	1.81–8.86	5.0–15.0	4–75	4–75	16.25
Flame speed	m/s	0.58	0.87	0.83	8.45	3.51	3.51	0.15
Autoignition temperature	°C	300	230	470	450	571	571	651
Minimum ignition energy	MJ	0.14	N/A	N/A	N/A	0.018	N/A	8
Flash point	°C	−42.7	73.8	−87.7	−184.4	N/A	N/A	−33.4
Octane		90–98	N/A	112	107	>130	>130	110
Fuel density	kg/m³	698.3	838.8	1898	187.2	17.5	71.1	602.8
Energy density	MJ/m³	31,074	36,403	86,487	7132	2101	8539	11,333
Latent heat of vaporization	kJ/kg	71.78	47.86	44.4	104.8	0	N/A	1369
Storage method		Liquid	Liquid	Comp. liquid	Comp. gas	Comp. gas	Comp. liquid	Comp. liquid
Storage temperature	°C	25	25	25	25	25	−253	25
Storage pressure	kPa	101.3	101.3	850	24,821	24,821	102	1030
Cost*	$/liter	0.58	0.65	0.72	0.57	0.14	0.18	0.24

*Cost data for April 2020. Source: Data from [3,7–12].

A chemical reaction during which a fuel is oxidized, and a large quantity of energy is released is called combustion. For oxidization, air is usually used. Dry air is approximated as 21% oxygen and 79% nitrogen by mole numbers. Each mole of oxygen entering a combustion chamber is accompanied by 0.79/0.21 = 3.76 mol of nitrogen. During a combustion reaction, the entering components are called reactants and the exiting components after the reaction are called products, as illustrated in Figure 4.9. This figure illustrates the mass conservation and allows to write the mass balance

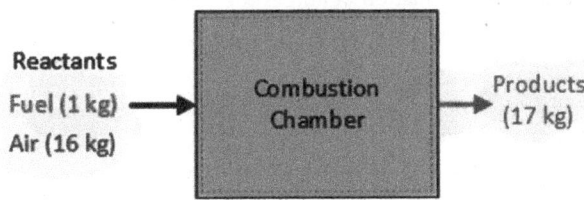

Figure 4.9 An illustration of a steady-flow combustion process.

accordingly. Here, the sum of the fuel and air, which are so-called reactants, should be equal to the sum of the products. As illustrated, 1 kg of fuel and 16 kg of air are taken into a combustion chamber to burn to generate heat exothermically and deliver products with a total mass of 17 kg. If one wants to calculate the air-fuel ratio, which is symbolized as AF or AFR. Here we will use AFR which is defined as the ratio of mass of air to the mass of fuel as follows:

$$AF \ or \ AFR = \frac{m_{air}}{m_{fuel}} \tag{4.1}$$

Going back to Figure 4.9 to calculate the AFR for this illustrated combustion, it becomes 16 which means that 16 kg of air is needed to burn 1 kg of fuel.

One may need to remember that the mass, m, is $m = NM$, where M is the molar mass. N is the number of moles. The total number of moles is not conserved during a chemical reaction.

Further to illustrate the combustion of hydrogen as follows: For example, the formation of the water is combustion, as hydrogen oxidized with oxygen. For 2 kg of hydrogen, 16 kg of oxygen is required. Thus, 18 kg of water is formed.

$$H_2(2 \ kg \ hydrogen) + \frac{1}{2}O_2(16 \ kg \ oxygen) \rightarrow H_2O(18 \ kg \ of \ water)$$

The critical step for analyzing the combustion processes is to write the combustion reaction and balance it accordingly. Various illustrative examples are presented here to better show how to set up the chemical reactions and balance them accordingly.

Illustrative Example 4.1   Consider one kmol of gasoline (octane: C_8H_{18}) is burned with 25 kmol of air (consisting of O_2 and N_2). It is assumed that the products contain CO_2, H_2O, O_2, and N_2. Determine the mole number of each gas in the products. Calculate the AFR for this combustion reaction.

Solution:

A specific illustration of this combustion process is shown in Figure 4.10. There are two reactants, namely gasoline and air and four products in terms of CO_2, H_2O, O_2, and N_2. For the chemical reaction, the coefficients of a, b, c, and d are to be found first which can easily be determined by balancing the chemical reaction accordingly.

$$C_8H_{18} + e(O_2 + 3.76N_2) \rightarrow aCO_2 + bH_2O + cO_2 + dN_2$$

Note that air is considered, as stated before, 1 kmol of O_2 and 3.76 kmol of N_2. Since the problem provides that 25 kmol of dry air is supplied to the combustion chamber for burning, we can take $e = 25$. So, the reaction becomes

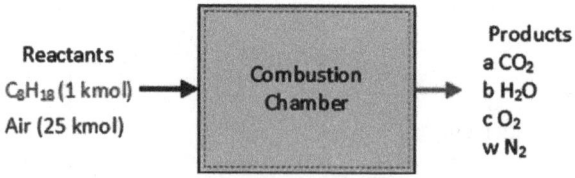

Figure 4.10   Illustration of the combustion process in the Illustrative Example 4.1.

$$C_8H_{18} + 25(O_2 + 3.76N_2) \rightarrow aCO_2 + bH_2O + cO_2 + dN_2$$

Then, we first balance the carbon and find it as 8 ($a = 8$):

$$C_8H_{18} + 25(O_2 + 3.76N_2) \rightarrow 8CO_2 + bH_2O + cO_2 + dN_2$$

Next, we repeat it for diatomic hydrogen (H_2) and find it as 9 ($b = 9$):

$$C_8H_{18} + 25(O_2 + 3.76N_2) \rightarrow 8CO_2 + 9H_2O + cO_2 + dN_2$$

For monoatomic oxygen (O), we calculate it through $25 \times 2 = 8 \times 2 + 9 \times 1 + c \times 2$ and find it as 12.5 ($c = 12.5$):

$$C_8H_{18} + 25(O_2 + 3.76N_2) \rightarrow 8CO_2 + 9H_2O + 12.5O_2 + dN_2$$

In addition, we now repeat it for diatomic nitrogen as $25 \times 3.76 = d$ and calculate it as $d = 94$. Substituting all coefficients calculated above, we obtain the final chemical reaction as

$$C_8H_{18} + 25(O_2 + 3.76N_2) \rightarrow 8CO_2 + 9H_2O + 12.5O_2 + 94N_2$$

Here, in order to calculate the mass of air, we need to consider the masses of O_2 and N_2 summed which requires multiplying the number of moles of each by its molecular weight. Similarly, one needs to repeat it for fuel as well. The molecular weights are taken from Table A1 as 12 kg/kmol for C, 16 kg/kmol for O, 14 kg/kmol for N, and 1 kg/kmol for H. In conjunction with this, the air fuel ratio is obtained through Eq. 4.1 as follows:

$$AFR = \frac{m_{air}}{m_{fuel}} = \frac{(25\,kmol \times 32\ kg\,/\,kmol) + (94\,kmol \times 28\,kg\,/\,kmol)}{(8\,kmol \times 12\ kg\,/\,kmol) + (18\,kmol \times 1\ kg\,/\,kmol)} = \frac{3432}{114} = 30.1$$

The result indicates that a total of 30.1 kg of air is necessary to burn each kg of the fuel.

Discussion: AFR is a significant parameter for the combustion process. It directly affects the performance of the combustion. The ideal air and fuel combustion mixture is called to be the stoichiometric ratio. When the AFR is ideal, the mixture burns completely during combustion. The stoichiometric ratio of gasoline is 14.7. When an air/fuel mixture has too much fuel, namely, AFR is higher than the stoichiometric ratio it is *rich*. When there is not enough fuel in the mixture, it is called *lean*. In summary,

- If AFR is higher than the stoichiometric ratio: lean
- If AFR is lower than the stoichiometric ratio: rich

Illustrative Example 4.2 Consider the fuel of methane (CH_4) being combusted with air. Write the combustion chemical reaction for stoichiometric conditions and calculate the AFR.

Solution:
For methane, the general combustion reaction can be written as follows:

$$CH_4 + a(O_2 + 3.76N_2) \rightarrow bCO_2 + cH_2O + dN_2$$

In order to balance the chemical reaction, let's first begin with carbon balance and find $b = 1$:

$$CH_4 + a(O_2 + 3.76N_2) \rightarrow CO_2 + cH_2O + dN_2$$

For diatomic hydrogen (H_2) and find c as 2 ($c = 2$). Substituting this into the chemical reaction will be

$$CH_4 + a(O_2 + 3.76N_2) \rightarrow 1CO_2 + 2H_2O + dN_2$$

For monoatomic oxygen (O), it can be calculated as: $a \times 2 = 2 + 2 \times 1 \Rightarrow a = 2$. The chemical reaction that results in

$$CH_4 + 2(O_2 + 3.76N_2) \rightarrow CO_2 + 2H_2O + dN_2$$

Finally, for the diatomic nitrogen: $2 \times 3.76 = d \Rightarrow d = 7.52$.
    Substituting all coefficients calculated above, we obtain the final chemical reaction as

$$CH_4 + 2(O_2 + 3.76N_2) \rightarrow CO_2 + 2H_2O + 7.52N_2$$

In order to calculate the AFR, the molar weights of the elements are taken from Table A1. They are 12 kg/kmol for carbon, 1 kg/kmol for hydrogen, 16 kg/kmol for oxygen, and 14 kg/kmol for nitrogen.

$$AFR = \frac{m_{air}}{m_{fuel}} = \frac{(2\,kmol \times 32\ kg\,/\,kmol) + (7.52\,kmol \times 28\ kg\,/\,kmol)}{(1\,kmol \times 12\ kg\,/\,kmol) + (2\,kmol \times 2\ kg\,/\,kmol)} = 25.8$$

The result indicates that a total of 25.8 kg of air is necessary to burn each kg of methane fuel.

Illustrative Example 4.3 Ammonia (NH_3) is considered as a potentially carbon-free fuel. Write the chemical reaction for its combustion under stoichiometric conditions and calculate the AFR accordingly.

Solution:

The ideal combustion equation for ammonia is written as

$$NH_3 + b(O_2 + 3.76N_2) \rightarrow cH_2O + dN_2$$

Let's first begin with hydrogen to balance it as $3 = c \times 2$ and calculate it as $c = 1.5$.

$$NH_3 + b(O_2 + 3.76N_2) \rightarrow 1.5H_2O + dN_2$$

Through balancing this time for monoatomic oxygen, $b \times 2 = 1.5$, and we find $b = 0.75$.

$$NH_3 + 0.75(O_2 + 3.76N_2) \rightarrow 1.5H_2O + dN_2$$

Further to balance it for monoatomic nitrogen, $1 + 0.75 \times 3.76 = d$, and we obtain $d = 3.82$.
    The final chemical reaction will be written as

$$NH_3 + 0.75(O_2 + 3.76N_2) \rightarrow 1.5H_2O + 3.82N_2$$

In order to calculate the AFR, first, we need to obtain the molar weights of the elements from Table A1 as 16 kg/kmol for O, 14 kg/kmol for N, and 1 kg/kmol for H, respectively. The AFR for stoichiometric combustion is determined as

$$AFR = \frac{m_{air}}{m_{fuel}} = \frac{(0.75\,kmol \times 32\ kg/kmol) + (2.82\,kmol \times 28\ kg/kmol)}{(1\,kmol \times 7\ kg/kmol) + (3\,kmol \times 1\ kg/kmol)} = 10.35$$

The resultant indicates that a total of 10.35 kg of air is necessary to burn each kg of ammonia.

4.4.1 Theoretical and Actual Combustion Processes

All the combustion compounds of a fuel are burned to the completion during a complete combustion process. The earlier Illustrative Examples (4.1 through 4.3) presented in this chapter considered the complete combustion process or complete chemical reaction or ideal combustion. A complete combustion is illustrated in Figure 4.11a, which is an example of a complete chemical reaction where all the carbon atoms in the fuel (for the case when the fuel is carbon based) are converted into CO_2, all the hydrogen atoms in the fuel are converted into H_2O, and the sulfur atoms if there is sulfur in the fuel are converted into SO_2. In general, a complete combustion is when all of the fuel is completely burned and there is no unreacted oxygen left. On the other hand, an incomplete combustion is illustrated in Figure 4.11b where the fuel is not completely burned or converted to products, which might be due to many different reasons, is mainly due to not having enough oxidants, which in many cases is the oxygen in the air, to oxidize the fuel. Other reasons for not having a complete combustion are not having enough contact time between the oxidant molecules and the fuel molecules, or that there is not appropriate mixing between the fuel and the oxidant. For example, all combustion processes given in the previous illustrative examples are complete combustion. All carbon contents turn into CO_2 and hydrogen contents are converted into H_2O. If there is unburned fuel and free oxygen in products, that means the combustion is incomplete.

The ideal air and fuel combustion mixture is called to be the stoichiometric ratio. When the AFR is ideal, the mixture burns completely during combustion. The stoichiometric ratio of gasoline is 14.7. When an air/fuel mixture has too much fuel, namely, AFR is higher than the stoichiometric ratio it is *rich*. When there is not enough fuel in the mixture, it is called *lean*. In summary,

- Lean mixture: If AFR is higher than the stoichiometric ratio.
- Rich mixture: If AFR is lower than the stoichiometric ratio.

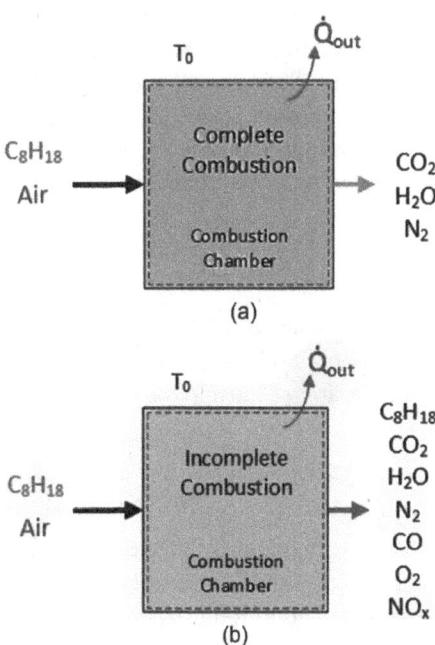

Figure 4.11 Type of combustion (a) complete combustion (b) incomplete combustion.

It is also possible to define combustion processes in terms of the amount of air. The amount of air in excess of the stoichiometric amount is called excess air. The amount of excess air is usually expressed in terms of the stoichiometric air as percent excess air or percent theoretical air. For example, 50% excess air is equivalent to 150% theoretical air and 200% excess air is equivalent to 300% theoretical air. Anything less than the stoichiometric amount is so-called deficiency of air (or deficient air) and is often expressed as percent deficiency of air. The ratio of the reduction in volume to the original volume is the volume fraction of the CO_2, which is equivalent to the mole fraction if ideal-gas behavior is assumed.

4.4.2 Enthalpy of Formation and Enthalpy of Combustion

A process that involves chemical reactions involves changes in the chemical energies, which must be accounted for in an energy balance.

By assuming no nuclear reactions and disregarding changes in kinetic and potential energies. The energy change of a system during a chemical reaction due to a change in state and a change in chemical composition:

$$\Delta E_{sys} = \Delta E_{state} + \Delta E_{chem} \tag{4.2}$$

Considering the reference state is 25°C and 1 atm, which is known as the standard reference state, consider the formation of CO_2 from its elements, carbon and oxygen, during a steady-flow combustion process.

The reactants are carbon 1 kmol of C and 1 kmol of O_2 at the reference conditions; CO_2 is the product at the reference conditions. The combustion of carbon is an exothermic reaction:

$$Q = H_{prod} - H_{react} = -393.520 kJ \,/\, kmol \tag{4.3}$$

As both reactants and product are at the reference conditions, the enthalpy change in this combustion process is only due to the changes in the chemical composition of the system. This enthalpy change is different for different reactions. It is a property to represent the changes in chemical energy during a chemical reaction. It is called enthalpy of reaction, h_r.

For combustion processes, the enthalpy of reaction is generally defined as the enthalpy of combustion, h_c. The enthalpy of combustion denotes the amount of heat released during a steady-flow combustion process when 1 kmol of fuel is burned completely at a specific temperature and pressure. Thus, Eq. 4.3 becomes

$$h_r = h_c = H_{prod} - H_{react} \tag{4.4}$$

For the combustion of the fuels, the heating value of the fuel is commonly used. The heating value is the amount of heat released when a fuel is burned completely in a steady-flow process and the products are returned to the state of the reactants.

$$Heating\ value = |h_c| \left(\frac{kJ}{kg} \right)$$

The heating value is called the higher heating value (HHV) of the fuel when the H_2O in the products is in the liquid form, and it is called the lower heating value (LHW) of the fuel when the H_2O in the products is in the vapor form. So, the relation between HHV and LHW can be written as:

$$HHV = LHV + (mh_{fg})_{H_2O} \tag{4.5}$$

where m is the mass of water in the products per unit mass of fuel and h_{fg} is the enthalpy of vaporization of water at the specified temperature. Table 4.4 illustrates the properties of the common fuels.

4.5 Thermodynamic Analysis of Combustion

In order to perform a complete thermodynamic analysis, the laws of thermodynamics should be considered. Here, we first begin with the first law analysis of the reacting systems.

The enthalpy of a component on a unit mole basis as

$$Entalpy = \bar{h}_f^o + \bar{h} - \bar{h}^o \left(\frac{kJ}{kmol} \right) \tag{4.6}$$

The steady-flow energy balance relation $\dot{E}_{in} = \dot{E}_{out}$ can be expressed for a chemically reacting steady-flow system more explicitly as

$$\dot{Q}_{in} + \dot{W}_{in} + \sum \dot{n}_r \left(\bar{h}_f^o + \bar{h} - \bar{h}^o \right)_r = \dot{Q}_{out} + \dot{W}_{out} + \sum \dot{n}_p \left(\bar{h}_f^o + \bar{h} - \bar{h}^o \right)_p \tag{4.7}$$

where the left-hand side of Eq. 4.7 is the rate of net energy transfer in by heat, work, and mass and the right-hand size of Eq. 4.7 is the rate of net energy transfer in by heat, work, and mass. $\dot{n}_r$ and $\dot{n}_p$ are the molar flow rates of the reactants and products.

Equation 4.7 can be written for per mole of fuel as follows:

$$Q_{in} + W_{in} + \sum N_r (\bar{h}_f^o + \bar{h} - \bar{h}^o)_r = Q_{out} + W_{out} + \Sigma N_p (\bar{h}_f^o + \bar{h} - \bar{h}^o)_p \tag{4.8}$$

Here, N_r and N_p are the number of moles of the reactants and products per mole of fuel.

Table 4.4 Properties of various hydrocarbons and fuels.

Substance	Formula	Molar Mass (kg/kmol)	Enthalpy of Vaporization (kJ/kg)	Specific Heat (kJ/kg K)	Higher Heating Value (kJ/kg)	Lower Heating Value (kJ/kg)
Acetylene	C_2H_2	26.04		1.695	49,960	48,271
Benzene	C_6H_6	78.11	433.6	1.049	42,263	40,573
Butane	C_4H_{10}	58.12	360.8	1.686	49,132	45,348
Carbon monoxide	CO	28.01	119.6	1.037	10,102	10,102
Diesel nr.2		233		1.934	45,579	42,601
Ethane	C_2H_6	30.07		1.76	51,896	47,508
Hexane	C_6H_{14}	86.17		1.655	48,316	44,742
Methane	CH_4	16.04	510.8	2.239	55,516	50,023
Octane	C_8H_{18}	114.2		1.644	47,896	44,430
Propane	C_3H_8	44.1	335.3	1.657	50,321	46,330

Also, the energy balance relation in Eq. 4.8 can be written as follows:

$$Q - W = H_{prod} - H_{react} \tag{4.9}$$

where H_{prod} and H_{react} are written as:

$$H_{prod} = \sum N_p \left(\bar{h}_f^o + \bar{h} - \bar{h}^o \right)_p \tag{4.10}$$

and

$$H_{react} = \sum N_r \left(\bar{h}_f^o + \bar{h} - \bar{h}^o \right)_r \tag{4.11}$$

The steady-flow energy equation per mole of fuel is written as follows:

$$Q - W = \bar{h}_c^o + \sum N_r \left(\bar{h} - \bar{h}^o \right)_r - \sum N_p \left(\bar{h} - \bar{h}^o \right)_p \tag{4.12}$$

Finally, the energy balance for a typical steady-flow combustion process becomes

$$Q_{out} = \sum N_r \left(\bar{h}_f^o + \bar{h} - \bar{h}^o \right)_r - \sum N_p \left(\bar{h}_f^o + \bar{h} - \bar{h}^o \right)_p \tag{4.13}$$

4.5.1 Combustion in a Closed System

Combustion chambers can be considered as reactors and can be analyzed either on the concept of closed systems. For closed combustion chamber analysis, we usually use the concept of the heating value since for this concept we are concerned with the amount of fuel burned only and how much energy it will release; we still do not focus on the outputs of the reaction other than the heat energy.

A general energy balance equation for a closed combustion chamber is written as follows:

$$Q_{in} + W_{in} + U_r = Q_{out} + W_{out} + U_p \tag{4.14}$$

where U_r and U_p are the internal energy of the reactants and products, respectively.
The internal energy is related to enthalpy as follows:

$$\bar{u} = \bar{h} - \overline{Pv} \tag{4.15}$$

In the terms of chemical formation properties, as given in Eq. 4.6,

$$\bar{u}_f^o + \bar{u} - \bar{u}^o = \bar{h}_f^o + \bar{h} - \bar{h}^o - P\bar{v} \tag{4.16}$$

When we write the final enthalpy equation in the general energy balance equation given in Eq. 4.14. The following equations are obtained:

$$Q - W = \bar{h}_c^o + \sum N_r \left(\bar{h} - \bar{h}^o \right)_r - \sum N_p \left(\bar{h} - \bar{h}^o \right)_p \tag{4.17}$$

$$Q_{out} = \sum N_r \left(\bar{h}_f^o + \bar{h} - \bar{h}^o \right)_r - \sum N_p \left(\bar{h}_f^o + \bar{h} - \bar{h}^o \right)_p \tag{4.18}$$

In order to perform an SLT analysis for a closed combustion chamber, we need to write the entropy and exergy balance equations for the combustion process. The general entropy balance equation is written as follows:

$$S_{in} - S_{out} + S_{gen} = \Delta S_{system} \qquad (4.19)$$

Here, S_{in} is the entropies enter with the reactants. S_{out} is the entropies exit from the system with products. The entropy balance equation can be expressed more explicitly for a closed or steady-flow reacting system as

$$\sum \frac{Q_k}{T_k} + S_{gen} = S_{prod} - S_{react} \qquad (4.20)$$

For adiabatic flame temperature condition, $Q=0$. Thus, the entropy balance equation results in

$$S_{gen,adiabatic} = S_{prod} - S_{react} \geq 0$$

As the adiabatic flame temperature condition defines the theoretical limit of the combustion.
The absolute entropy is written as follows:

$$\bar{s}(T,P_i) = \bar{s}_i^0(T,P_0) - R_u ln \frac{P}{P_0} \qquad (4.21)$$

For component i of an ideal-gas mixture:

$$\bar{s}(T,P_i) = \bar{s}_i^o(T,P_0) - R_u ln \frac{y_i P_m}{P_0} \left(\frac{kJ}{kmolK} \right) \qquad (4.22)$$

The exergy destroyed, Ex_d, associated with a chemical reaction can be determined from:

$$Ex_d = T_0 S_{gen} \qquad (4.23)$$

When analyzing reaction systems, we are more concerned with the changes in the exergy of reaction systems than with the values of exergy at various states (therefore $Ex = W_{rev}$):

$$W_{rev} = \sum N_r \left(\bar{h}_f^o + \bar{h} - \bar{h}^o - T_0 \bar{s} \right)_r - \sum N_p \left(\bar{h}_f^o + \bar{h} - \bar{h}^o - T_0 \bar{s} \right)_p \qquad (4.24)$$

Like in all thermodynamic systems, the performance of a combustion process can be evaluated through the thermodynamic efficiencies which are energy and exergy efficiencies.
The energy efficiency for a closed combustion chamber can be defined as the ratio of the heat output from combustion to energy input via fuel. It is therefore written as follows:

$$\eta_{en,c} = \frac{Q_{out}}{m_{fuel} LHV_{fuel}} \qquad (4.25)$$

The exergy efficiency of a closed combustion chamber is defined as the ratio of the exergy content of heat output to the exergy input with fuel. It is then written as:

$$\eta_{en,c} = \frac{Ex^{Q_{out}}}{m_{fuel} ex_{fuel}} \qquad (4.26)$$

where $Ex^{Q_{out}}$ can be written as follows:

$$Ex^{Q_{out}} = \left(1 - \frac{T_0}{T_c}\right)Q_{out}$$ (4.27)

4.5.2 Combustion in an Open System

Combustion chambers may technically be treated as reactors where the reactants enter to the combustion chamber and the products exit from it continuously. In the combustion chamber, there may be heat and work coming in and leaving out. The illustration of the open combustion chamber is shown in Figure 4.12. The mass balance equation is written for the open combustion chamber shown in Figure 4.12 as follows:

$$\dot{m}_{r1} + \dot{m}_{r2} = \dot{m}_p$$ (4.28)

where r_1 and r_2 denote the reactants 1 and 2 which are fuel and air, respectively.

The energy balance equation for the open combustion chamber:

$$\dot{Q}_{in} + \dot{W}_{in} + \dot{n}_{r1}\left(\overline{h}_f^o + \overline{h} - \overline{h}^o\right)_{r1} + \dot{n}_{r2}\left(\overline{h}_f^o + \overline{h} - \overline{h}^o\right)_{r2}$$
$$= \dot{Q}_{out} + \dot{W}_{out} + \dot{n}_{p1}\left(\overline{h}_f^o + \overline{h} - \overline{h}^o\right)_{p1} + \ldots + \dot{n}_{pn}\left(\overline{h}_f^o + \overline{h} - \overline{h}^o\right)_{pn}$$ (4.29)

or in terms of mass flow rates of reactants and products:

$$\dot{Q}_{in} + \dot{W}_{in} + \dot{m}_{r1}\left(\overline{h}_f^o + \overline{h} - \overline{h}^o\right)_{r1} + \dot{m}_{r2}\left(\overline{h}_f^o + \overline{h} - \overline{h}^o\right)_{r2}$$
$$= \dot{Q}_{out} + \dot{W}_{out} + \dot{m}_{p1}\left(\overline{h}_f^o + \overline{h} - \overline{h}^o\right)_{p1} + \ldots + \dot{m}_{pn}\left(\overline{h}_f^o + \overline{h} - \overline{h}^o\right)_{pn}$$ (4.30)

In order to perform an SLT analysis for an open combustion chamber, we need to write the entropy and exergy balance equations for the combustion process. The general entropy balance equation is written as follows:

$$\dot{S}_{in} - \dot{S}_{out} + \dot{S}_{gen} = \Delta S_{system}$$ (4.31)

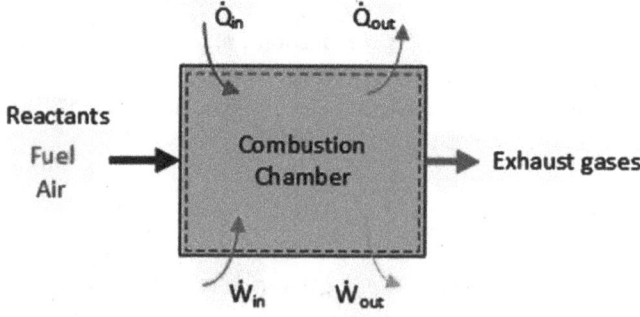

Figure 4.12 An illustration for an open combustion chamber.

Here, S_{in} is the entropies enter with the reactants. S_{out} is the entropies exit from the system with products. The entropy balance equation can be expressed more explicitly for a closed or steady-flow reacting system as

$$\sum \frac{\dot{Q}_k}{T_k} + \dot{S}_{gen} = \dot{S}_{prod} - \dot{S}_{react} \tag{4.32}$$

For adiabatic flame temperature condition, $Q=0$. Thus, the entropy balance equation becomes

$$\dot{S}_{gen,adiabatic} = \dot{S}_{prod} - \dot{S}_{react} \geq 0$$

As the adiabatic flame temperature condition defines the theoretical limit of the combustion. The absolute entropy is then written as

$$\bar{s}(T, P_i) = \bar{s}_i^0 (T, P_0) - R_u ln \frac{P}{P_0} \tag{4.33}$$

For component i of an ideal-gas mixture:

$$\bar{s}(T, P_i) = \bar{s}_i^o (T, P_0) - R_u ln \frac{y_i P_m}{P_0} \left(\frac{kJ}{kmol\, K} \right) \tag{4.34}$$

The exergy destroyed, $\dot{E}x_d$, associated with a chemical reaction can be determined from:

$$\dot{E}x_d = T_0 \dot{S}_{gen} \tag{4.35}$$

When analyzing reaction systems, we are more concerned with the changes in the exergy of reaction systems than with the values of exergy at various states (therefore $\dot{E}x = \dot{W}_{rev}$):

$$\dot{W}_{rev} = \sum N_r \left(\bar{h}_f^o + \bar{h} - \bar{h}^o - T_0 \bar{s} \right)_r - \sum N_p \left(\bar{h}_f^o + \bar{h} - \bar{h}^o - T_0 \bar{s} \right)_p \tag{4.36}$$

The energy efficiency for an open combustion chamber can be defined as the ratio of the heat output rate from combustion to energy input via fuel flow through the combustion chamber which is written as follows:

$$\eta_{en,c} = \frac{\dot{Q}_{out}}{\dot{m}_{fuel} LHV_{fuel}} \tag{4.37}$$

The exergy efficiency of a closed combustion chamber is defined as the ratio of the exergy content of heat output to the exergy input with fuel which is then written as:

$$\eta_{ex,c} = \frac{\dot{E}x^{\dot{Q}_{out}}}{\dot{m}_{fuel} ex_{fuel}} \tag{4.38}$$

where $\dot{E}x^{\dot{Q}_{out}}$ can be written as follows:

$$\dot{E}x^{\dot{Q}_{out}} = (1 - \frac{T_0}{T_c}) \dot{Q}_{out} \tag{4.39}$$

Illustrative Example 4.4 Consider a steady-state combustion chamber, as shown in Figure 4.13. The combustion occurs at a constant pressure of 101.3 kPa and temperature of 25°C. In the combustion chamber, hydrogen is supplied as the fuel and combusted with pure oxygen as the reactants. The only product is water.

a) Write the mass and energy balance equations for the combustion chamber.
b) Find the molar flow rate of the product, H_2O for 2 kmol/s of H_2 and 1 kmol/s of O_2.
c) Determine the heat released by the chamber.
d) Calculate the energy and exergy efficiency of the combustion process.

Solution:

For the inputs, it is assumed that they are fed to the combustion chamber at the pressure and temperature of the combustion chamber. Also, the following assumptions are applied to the system:

- The combustion of H_2 and O_2 occurs in a steady-state process.
- The product, H_2O, exits from the combustion chamber in a liquid form at 25°C and 101.3 kPa.
- The chemical reaction occurs in the stoichiometric chemical reaction.
- The volume of the combustion chamber is constant.

a) The mass balance equation for the combustion system:

$$\dot{m}_{H_2} + \dot{m}_{O_2} = \dot{m}_{H_2O}$$

The energy balance equation is written

$$\dot{n}_{H_2}\left(\bar{h}_f^o + \bar{h} - \bar{h}^o\right)_{H_2} + \dot{n}_{O_2}\left(\bar{h}_f^o + \bar{h} - \bar{h}^o\right)_{O_2} = \dot{Q}_{out} + \dot{n}_{H_2O}\left(\bar{h}_f^o + \bar{h} - \bar{h}^o\right)_{H_2O}$$

$$\dot{m}_{H_2} + \dot{m}_{O_2} = \dot{m}_{H_2O}$$

$$\dot{n}_{H_2}\bar{h}_{H_2} + \dot{n}_{O_2}\bar{h}_{O_2} = \dot{n}_{H_2O}\bar{h}_{H_2O}$$

Using the data from Table A1 the molecular weights are taken as 32 kg/kmol for O_2, 2 kg/kmol for H_2, and 18 kg/kmol for H_2O.

$$\left(2\frac{kmol\,H_2}{s} \times 2.0\frac{g}{kmol\,H_2}\right) + \left(1\frac{kmol\,O_2}{s} \times 32\frac{g}{kmol\,O_2}\right) = \dot{n}_{H_2O}\left(18\frac{g}{kmol\,H_2O}\right)$$

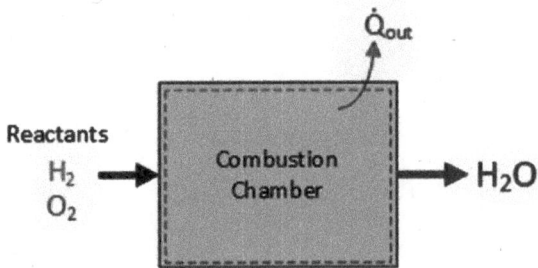

Figure 4.13 An illustration for an open combustion chamber in the Illustrative Example 4.4

$$\dot{n}_{H_2O} = 2mol/s$$

$$\dot{n}_{H_2}\left(\bar{h}_f^o + \bar{h} - \bar{h}^o\right)_{H_2} + \dot{n}_{O_2}\left(\bar{h}_f^o + \bar{h} - \bar{h}^o\right)_{O_2} = \dot{Q}_{out} + \dot{n}_{p1}\left(\bar{h}_f^o + \bar{h} - \bar{h}^o\right)_{H_2O}$$

By eliminating the null terms in the energy balance equation and accounting for the fact that the reactants and the products are at ambient pressure and temperature, the energy equation reduces to:

$$\dot{n}_{H_2}\left(\bar{h}_f^o\right)_{H_2} + \dot{n}_{O_2}\left(\bar{h}_f^o\right)_{O_2} = \dot{Q}_{out} + \dot{n}_{p1}\left(\bar{h}_f^o\right)_{H_2O}$$

The enthalpy of formation for H_2O can be obtained from Table A1 as $-285,830\left(kJ\middle/ kmol\, H_2O\right)$.

$$\left(2\frac{kmol\, H_2}{s} \times 0\frac{kJ}{kmol\, H_2}\right) + \left(1\frac{kmol\, O_2}{s} \times 0\frac{kJ}{kmol\, O_2}\right) = \dot{Q}_{out} + \left(2\frac{kmol\, H_2O}{s} \times -285,830\frac{kJ}{kmol\, H_2O}\right)$$

$$\dot{Q}_{out} = 285,830 kW$$

b) The energy efficiency of the combustion can be found by using Eq. 4.38:

$$\eta_{en,c} = \frac{\dot{Q}_{out}}{\dot{m}_{H_2} LHV_{H_2}}$$

where LHV_{H_2O} is taken from Table A1 as 1,200,000 kJ/kg.

$$\eta_{en,c} = \frac{285,830\,(kW)}{2\left(\frac{kmol}{s}\right) \times (2 \times 10^{-3})\left(\frac{kg}{kmol}\right) \times 120,000\left(\frac{kJ}{kg}\right)} = 0.5980$$

Also, the exergy efficiency is obtained from Eq. 4.39 as

$$\eta_{ex,c} = \frac{\dot{Ex}^{\dot{Q}_{out}}}{\dot{m}_{H_2} ex_{H_2}}$$

with

$$\dot{Ex}^{\dot{Q}_{out}} = \left(1 - \frac{T_0}{T_c}\right)\dot{Q}_{out}$$

Here, T_c is the combustion temperature. In this example, T_c is assumed to be 1000 K.

$$\dot{Ex}^{\dot{Q}_{out}} = \left(1 - \frac{T_0}{T_c}\right)\dot{Q}_{out} = \left(1 - \frac{T_0}{T_c}\right)285,830 = 181,698.6 kW$$

Also, the exergy efficiency is calculated as follows with the specific chemical exergy of hydrogen which is 117,200 kJ/kg from Table A1:

$$\eta_{ex,c} = \frac{\dot{Ex}^{\dot{Q}_{out}}}{\dot{m}_{fuel} LHV} = \frac{181,698.6\,(kW)}{2\frac{kmol}{s} \times (2)\frac{kg}{kmol} \times 117,200\frac{kJ}{kg}} = 0.3875$$

4.5.3 Adiabatic Flame Temperature

The adiabatic flame temperature, also called adiabatic combustion temperature, is defined as the maximum temperature of products coming out of a combustion process where the system is adiabatic with no heat transfer in or out in the absence of work and kinetic and potential energies. Adiabatic flame temperature is considered the ideal condition. However, in actual combustion, the temperature of the products (exhaust gases) will be significantly lower due to heat transfer, incomplete combustion, and dissociation of combustion gases.

In adiabatic flame conditions, there will be no heat transfer crossing the chamber's boundary which will result in $\dot{Q}_{out} = 0$. So, the energy balance equation for the adiabatic process becomes:

$$\dot{n}_{r_1}\left(\bar{h}_f^o + \bar{h} - \bar{h}^o\right)_{r_1} + \dot{n}_{r_2}\left(\bar{h}_f^o + \bar{h} - \bar{h}^o\right)_{r_2} = \dot{n}_{p1}\left(\bar{h}_f^o + \bar{h} - \bar{h}^o\right)_{p1} \tag{4.33}$$

Illustrative Example 4.5 Consider the combustion process for CH_3OH with 200% excess air, as shown in Figure 4.14. The relevant data and information given in Table 4.5 and Table 4.6 are known to make the necessary calculations:

a) Calculate the heat output from the combustion.
b) Calculate the entropy generation.
c) Find the exergy destruction.

Solution:
a) The chemical reaction for the combustion is written as follows:

$$CH_3OH + 4.5\,(O_2 + 3.76N_2) \rightarrow CO_2 + 2\,H_2O + 3\,O_2 + 16.92\,N_2$$

The energy balance equation is written first, and the heat output is then calculated using the data taken from the tables as follows:

$$\dot{n}_r\left(\bar{h}_f^o + \bar{h} - \bar{h}^o\right)_r + \dot{n}_{O_2}\left(\bar{h}_f^o + \bar{h} - \bar{h}^o\right)_{O_2} = \dot{Q}_{out} + \dot{n}_p\left(\bar{h}_f^o + \bar{h} - \bar{h}^o\right)_p$$

$$\dot{Q}_{out} = \left[-(1)(-393520) + 11351 - 9364\right] - \left[16.92(0 + 10180 + 8669)\right] +$$
$$1(-2000670 - 9904) - 3(0 + 10213 - 8682) - 16.92(0 + 10180 - 8669) + 1(-20067)$$

$$\dot{Q}_{out} = 663{,}550 \frac{kJ}{kmol} \quad \text{of fuel which is needed for this combustion.}$$

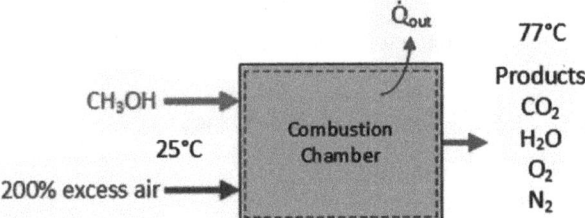

Figure 4.14 The schematic view of the combustion chamber for Illustrative Example 4.5

Table 4.5 Features for the reactants and products.

	N_i	y_i	$\bar{s}_i^o(T, 1\ atm)$	$R_u \ln(y_i P_m)$	$N_i \bar{s}_i$
CH_3OH	1	–	239.70	–	239.70
O_2	4.5	0.21	205.04	−12.98	981.09
N_2	16.92	0.79	191.61	−1.960	3275.20
					S_R=4496 kJ/K
CO_2	1	0.0436	219.831	−26.05	245.88
H_2O	2	0.0873	194.125	−20.27	428.79
O_2	3	0.1309	209.765	−16.91	680.03
N_2	16.92	0.7382	196.173	−2.52	3361.89
					S_P=4717 kJ/K

Table 4.6 Properties for the reactants and products.

Substance	$\bar{h}_f^o$ (kJ/kmol)	$\bar{s}_i^o(T, 1\ atm)$ (kJ/kmol)	$\bar{h}_f^o$ (kJ/kmol) (kJ/kmol)
CH_3OH	−200,670	–	–
O_2	0	8682	10,213
N_2	0	8669	10,180
H_2O	−241,820	9904	11,652
CO_2	−393,510	9364	11,351

b) For entropy balance equation, one can write the following:

$$\sum N\bar{s}_R + S_{gen} = \sum N\bar{s}_P + \frac{Q_{out}}{T_0}$$

$$S_i = N_i \bar{s}_i(T, P_0) - R\ln(y_i, P_m)$$

The entropy generation is also calculated as

$$S_{gen} = S_P - S_R + \frac{Q_{out}}{T_0} = 4717 - 4496 + \frac{663,550}{298}$$

$$S_{gen} = 2448\frac{kJ}{kmol}$$

c) The exergy destruction is then calculated accordingly as follows:

$$Ex_d = T_0 S_{gen} = 298 \times 2448 = 729,400\frac{kJ}{kmol}$$

or

$$Ex_d = Ex_R - Ex_P = \sum N_r (\bar{h}_f^o + \bar{h} - \bar{h}^o - T_0 \bar{s})_r - \sum N_p (\bar{h}_f^o + \bar{h} - \bar{h}^o - T_0 \bar{s})_p$$

$$Ex_d = 729,400 \frac{kJ}{kmol}$$

4.6 Closing Remarks

This chapter is in general about fuels and their combustion. In this regard, the basic aspects of fuels and combustion processes are considered and discussed through various important parameters, including air-fuel ratio, heating values, percent theoretical air, adiabatic flame temperature, etc. The fossil fuels are discussed from various perspectives including their classifications, processing, and utilization for heat generation. For the combustion processes, both closed and open systems are considered through first- and second-law of thermodynamics. Furthermore, both energy and exergy efficiencies are introduced. Moreover, there are numerous illustrative examples presented to show how to analyze combustion reactions thermodynamically and make the necessary performance evaluations through energy and exergy efficiencies.

References

1 Ritchie, H., Roser, M., and Rosado, P. (2022). Energy production and consumption. Available online https://ourworldindata.org/energy-production-consumption (accessed 9 November 2022).

2 Luo, G., Zhang, J., Rao, Y. et al. (2017). Coal supply chains: a whole-process-based measurement of carbon emissions in a mining city of China. *Energies* 10 (11): 1855. doi: 10.3390/en10111855.

3 Zamfirescu, C. and Dincer, I. (2008). Using ammonia as a sustainable fuel. *Journal of Power Sources* 185: 459–465. doi: 10.1016/j.jpowsour.2008.02.097.

4 Oil reserves – country rankings. (2022). Available online https://www.theglobaleconomy.com/rankings/oil_reserves/#:~:text=The%20average%20for%202021%20based,available%20from%201980%20to%202021 (accessed 9 November 2022).

5 U.S. Energy Information Administration, Coal Explained. Available online https://www.eia.gov/energyexplained/coal/how-much-coal-is-left.php (accessed 9 November 2022).

6 Ritchie, H., Roser, M., and Rosado, P. (2020). CO_2 and greenhouse gas emissions Available online https://ourworldindata.org/co2-and-other-greenhouse-gas-emissions (accessed 9 November 2022).

7 Mishra, D.P. and Rahman, A. (2003). An experimental study of flammability limits of LPG/air mixtures. *Fuel* 82: 863–866. doi: 10.1016/S0016-2361(02)00325-3.

8 Chong, C.T. and Hochgreb, S. (2011). Measurements of laminar flame speeds of liquid fuels: Jet-A1, diesel, palm methyl esters and blends using particle imaging velocimetry (PIV). *Proceedings of the Combustion Institute* 33: 979–986. doi: 10.1016/j.proci.2010.05.106.

9 Ramasamy, D., Kadirgama, K., Rahman, M.M., and Zainal, Z.A. (2015). Analysis of compressed natural gas burn rate and flame propagation on a sub-compact vehicle engine. *International Journal of Automotive and Mechanical Engineering* 11: 2405–2416. doi: 10.15282/ijame.11.2015.21.0202.

10 Alternative Fuels Data Center, U. S. Department of Energy (n.d.). https://afdc.energy.gov/fuels/properties (accessed 21 October 2020).

11 Dincer, I. and Acar, C. (2015). Review and evaluation of hydrogen production methods for better sustainability. *International Journal of Hydrogen Energy* 40: 11094–11111.

12 Climate Action Tracker (2021). *2100* Warming Projections: emissions and expected warming based on pledges and current policies. November 2021. Available online https://climateactiontracker. org/global/temperatures.

13 Dincer, I. (2020). *Thermodynamics: A Smart Approach*, 1st ed., John Wiley & Sons, Ltd.

14 Statistical review of world energy 2021. 70e. Available online https://www.bp.com/content/dam/ bp/business-sites/en/global/corporate/pdfs/energy-economics/statistical-review/bp-stats-review-2021-oil.pdf (accessed 9 November 2022).

Questions/Problems

Questions

1 Classify fossil fuels and discuss their specific advantages and disadvantages.
2 Discuss the environmental impacts of fossil fuels use.
3 List the types of coals and explain what makes them different.
4 What differentiates sweet and sour crude oil?
5 Explain what process is critical in obtaining useful fuels and chemicals.
6 Describe both atmospheric and vacuum distillation processes and why they are used for.
7 Define the heating value for fuels and explain its types for specific combustion processes.
8 What is air-fuel ratio? Write its equation and highlight the practical importance.
9 Compare both complete and incomplete combustion processes and illustrate the differences in terms of reactants and products.
10 Define the adiabatic flame temperature and explain its role in combustion.

Problems

1 Consider C_8H_{18} as the fuel being combusted with air at the stoichiometric ratio. Construct the chemical reaction taking place during this combustion process and balance it correctly. Calculate the AFR accordingly.
2 Consider the C_8H_{18} as given in Problem 1, now write and balance the chemical reaction for the combustion of C_8H_{18} in the case of 100% excess air conditions. Determine the AFR for this combustion.
3 Consider diesel ($C_{12}H_{23}$) as the fuel.
   a Write and balance the chemical reaction for the combustion under the stoichiometric conditions and calculate the AFR accordingly.
   b Consider the case of 50% excess air this time and repeat the above calculations for comparison.
4 Consider the combustion of C_8H_{18} with 50% excess air under complete combustion conditions, as illustrated in figure below.
   a Write the chemical reaction equation for the process and balance it accordingly.
   b Calculate the heat extracted heat from the combustion process.
   c Calculate the entropy generation for the combustion process.
   d Find the exergy destruction for the combustion process.
   e Calculate both energy and exergy efficiencies.

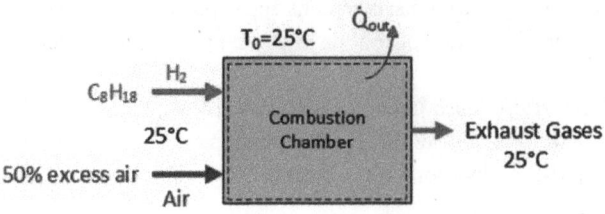

The following data and information are known for this combustion:
The specific enthalpy of formation values for the reactants and products:

Substance	$\bar{h}_f^o$ (kJ/kmol)
C_8H_{18}	−249.950
O_2	0
N_2	0
H_2O	−285,830
CO_2	−393,510

The mole numbers, fractions and entropy of formation values for the reactants and products:

	N_i	y_i	$\bar{s}_i^o\,(T,1\,atm)$	$R_u\,\ln(y_iP_m)$	$N_i\bar{s}_i$
C_8H_{18}	1	1	360.79	–	360.79
O_2	18.75	0.21	205.14	−12.98	4089.75
N_2	70.50	0.79	191.61	−1.960	13,646.69
					S_R=18,097.23 kJ/K
CO_2	1	0.0436	213.80	−19.62	1867.3
H_2O	2	0.0873	69.92	–	629.3
O_2	3	0.1309	205.04	−21.68	1417.6
N_2	16.92	0.7382	191.61	−1.53	13,616.3
					S_P=17,531 kJ/K

5 Liquid ethyl alcohol (C_2H_5OH) at 25°C is burned in a steady-flow combustion process with 50% excess air that also enters in the chamber at 25°C. The products leave the combustion chamber at 600 K. At 25°C, the density of liquid ethyl alcohol is 790 kg/m³. The combustion equation is written as follows:

$$C_2H_5OH + A[O_2 + 3.76N_2] \rightarrow BCO_2 + DH_2O + 1.5O_2 + FN_2$$

a Balance the combustion reaction given and find the values of the coefficients a_{th}, B, D, and F.
b Determine the air fuel ratio and required heat of combustion.
c Determine the entropy generation rate considering the combustion temperature as 600 K.
d Determine the exergy destruction rate if the reference temperature is taken as 25°C.
e Calculate both energy and exergy efficiencies.

Substance	$\bar{h}_f^0$ (kJ/ kmol)	$\bar{h}_{600K}$ (kJ/ kmol)	$\bar{h}^0$(kJ/ kmol)	$s^0{}_{600K}$(kJ/ kmol K)	$R_u ln\left(\frac{y_i P_m}{P_0}\right)$	Molecular Mass (kg/kmol)
CO_2	−393,520	22,280	9364	243.2	−20.5	44
H_2O	−241,820	20,402	9904	212.9	−17.1	18
N_2	0	17,563	8669	212.1	−2.7	28
O_2	0	17,929	8682	226.3	−22.8	32
C_2H_5OH	−277,650	–	–	340.6	–	
O_2	0	–	–	205.04	−12.9	
N_2	0	–	–	191.61	−1.9	

Substance	$\bar{h}_f^0$(kJ/kmol)	$s^0{}_{600K}$(kJ/kmol K)	$R_u ln\left(\frac{y_i P_m}{P_0}\right)$	Molecular Mass (kg/kmol)
C_2H_5OH	−277,650	340.6	–	46
O_2	0	205.04	−12.9	32
N_2	0	191.61	−1.9	28

6 The Illinois type coal is burned with 40% excess air according to the combustion reaction:

$$0.63C + 0.29H_2 + 0.053O_2 + 0.0058N_2 + 0.0083S$$
$$+ 1.4a_{th}(O_2 + 3.76N_2) \rightarrow xCO_2 + yH_2O + zSO_2 + kN_2 + 0.29O_2$$

The molar masses and thermophysical properties of the reactants and products are listed in the table below. All products and reactants are considered at 1 atm pressure and 25°C.

Substance	$\bar{h}_f^0$(kJ/kmol)	$\bar{s}^0$(kJ/kmol K)	$R_u ln\left(\frac{y_i P_m}{P_0}\right)$	Molecular Mass (kg/kmol)
Products				
CO_2	−393,520	213.80	−17.39	44
H_2O	−241,820	188.83	−23.62	18
SO_2	−396,000	248.21	−53.41	64
N_2	0	191.61	−2.29	28
O_2	0	205.04	−23.73	32
Reactants				
Carbon	0	5.74	–	12
H_2	0	30.74	–	2
Sulfur	0	32.05	–	32
O_2 in coal	0	31.04	–	32
N_2 in coal	0	28.61	–	28
O_2 in air	0	205.04	−12.98	32
N_2 in air	0	191.61	−1.96	28

a   Write the balanced combustion chemical reaction.
b   Find the air-fuel ratio and % theoretical air (considering 29 kg/kmol as the molecular mass of air).
c   Find the heat of combustion.
d   Find the entropy generation.
e   Find the exergy destruction if the reference temperature is taken as 25°C.
f   Calculate both energy and exergy efficiencies.

7   A cogeneration cycle powered by solar energy and natural gas is used for power and heat generations. The total heat input rate required by the cycle is 50 MW. Solar collectors provide 40% of this requirement while natural gas provides 60%. The reactants and products of the combustion reaction are maintained at 25°C. Natural gas is burned in the combustion chamber with 50% excess air according to the following reaction:

$$CH_4 + 3(O_2 + 3.76N_2) \rightarrow CO_2 + XH_2O + YO_2 + ZN_2$$

The required properties are provided in the table below.

Substance	N_i	$\bar{h}_f^0$ (kJ/kmol)	$\bar{s}^0$ (kJ/kmol K)	$R_u ln\left(\dfrac{y_i P_m}{P_0}\right)$	$N_i \bar{s}_i$	Molecular Mass (kg/kmol)
Products						
CO_2	1	−393,520	213.8	−21.7	231.2	44
H_2O	TBO	−241,820	188.8	−28.8	TBO	18
N_2	TBO	0	191.6	−1.6	TBO	28
O_2	1	0	205.0	−21.7	226.7	32
Reactants						
CH_4	1	−74,850	186.2	Not applicable	186.2	16
O_2	3	0	205.0	−12.9	653.7	32
N_2	11.28	0	191.6	−1.9	2182.7	28

a   Find **X** and **Y**, and calculate the amount of heat released during the combustion process for every 1 kmol of natural gas burned.
b   Determine the mass flow rate of natural gas required in the combustion chamber (take molar mass of CH_4 as 16 kg/kmol).
c   Find the exergy destruction rate in the combustion chamber.
d   Calculate the overall energy and exergy efficiencies (take T_0=298 K, T_c =1110 K, lower heating value of CH_4= 50,000 kJ/kg, specific exergy of CH_4=51,950 kJ/kg).

8   Consider problem 7 for the case of 200% excess air.
a   Find **X, Y,** and **Z**, and calculate the amount of heat released during the combustion process for every 1 kmol of natural gas burned.
b   Determine the mass flow rate of natural gas required in the combustion chamber (take molar mass of CH_4 as 16 kg/kmol).
c   Find the exergy destruction rate in the combustion chamber.
d   Calculate the overall energy and exergy efficiencies (Take T_0=298 K, T_c =1110 K, lower heating value of CH_4= 50,000 kJ/kg, specific exergy of CH_4=51,950 kJ/kg).

9   Consider Illustrative Example 4.4 and repeat the calculations for a different fuel, namely dimethyl ether (DME) C_2H_6O.

10 Consider Illustrative Example 4.5 and repeat the calculations this time for a diesel fuel, $C_{12}H_{23}$.

5

Nuclear Energy

5.1 Introduction

Energy plays a vital role in all sectoral activities as well as the development of countries as it is considered a major commodity needed for almost anything and every application. Things are, thus, not possible without energy. In conjunction with this, it is significant for a country to be self-sufficient and independent in meeting its energy needs. Today, it is possible to see various types of applications, ranging from community-based district energy systems to large plants with mega-level capacities. As mentioned earlier, energy demands have drastically been increasing since the Industrial Revolution as started with coal which is part of fossil fuels family. The demand has further increased even exponentially with the spread of electrification. In order to meet the gross electricity demands, there was an urgent need to develop to play a critical role for baseload coverage, which brought nuclear power into the energy picture as the second key option.

Nuclear energy has been considered a significant option due to its high energy content and capacity factor. Figure 5.1 demonstrates the capacity factor of various energy sources. Nuclear energy has by far the highest capacity factor of any other energy source. This basically means nuclear power plants are producing maximum power more than 92% of the time during the year. This is almost doubling what is obtained by natural gas units and is almost three times or more reliable than wind and solar plants. On the other hand, one uranium pellet has as much energy available as 3 barrels of oil or 1 ton of coal or 481 m^3 of natural gas. Uranium has essentially no other essential application, except for nuclear power production.

Since nuclear power plants require less maintenance and last longer before refueling (typically every 1.5 or 2 years), they are typically used more often. As a result of routine maintenance and/or refueling, natural gas and coal capacity factors are generally lower. It is generally important to recognize that some of the renewable energy sources are only available in intermittent nature or become quite variable sources, and they are often limited by a lack of resource (e.g., wind, sun, or water). The plants need a backup power source, such as large-scale energy storage or a reliable baseload power source, such as nuclear power.

In order to observe and evaluate the current potential of nuclear energy, one may benefit from the online databases which are called "grid watch." Figure 5.2 shows the spontaneous power generation capacities obtained for December 9, 2022, at 12:00 am. As seen from Figure 5.2a, almost half of the electricity demand of the Ontario, Canada, is met by nuclear energy. Natural gas-powered power generation only covers 8.5% of the total electricity production. It is clear that 90% of the

Introduction to Energy Systems, First Edition. Ibrahim Dincer and Dogan Erdemir.
© 2023 John Wiley & Sons, Inc. Published 2023 by John Wiley & Sons, Inc.
Companion Website: www.wiley.com/go/Dincer/Introduction_to_Energy_Systems

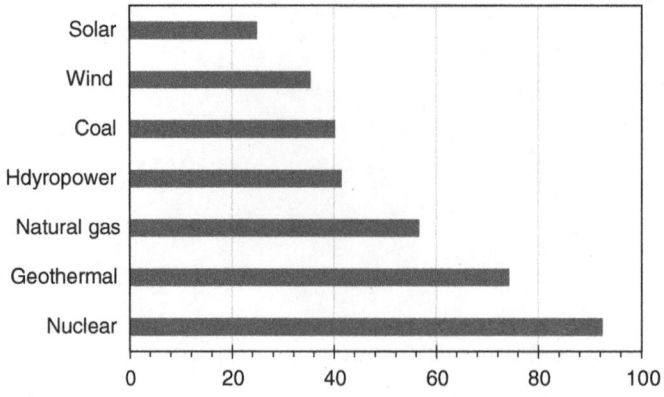

Figure 5.1 Capacity factors by energy source (data from [1]).

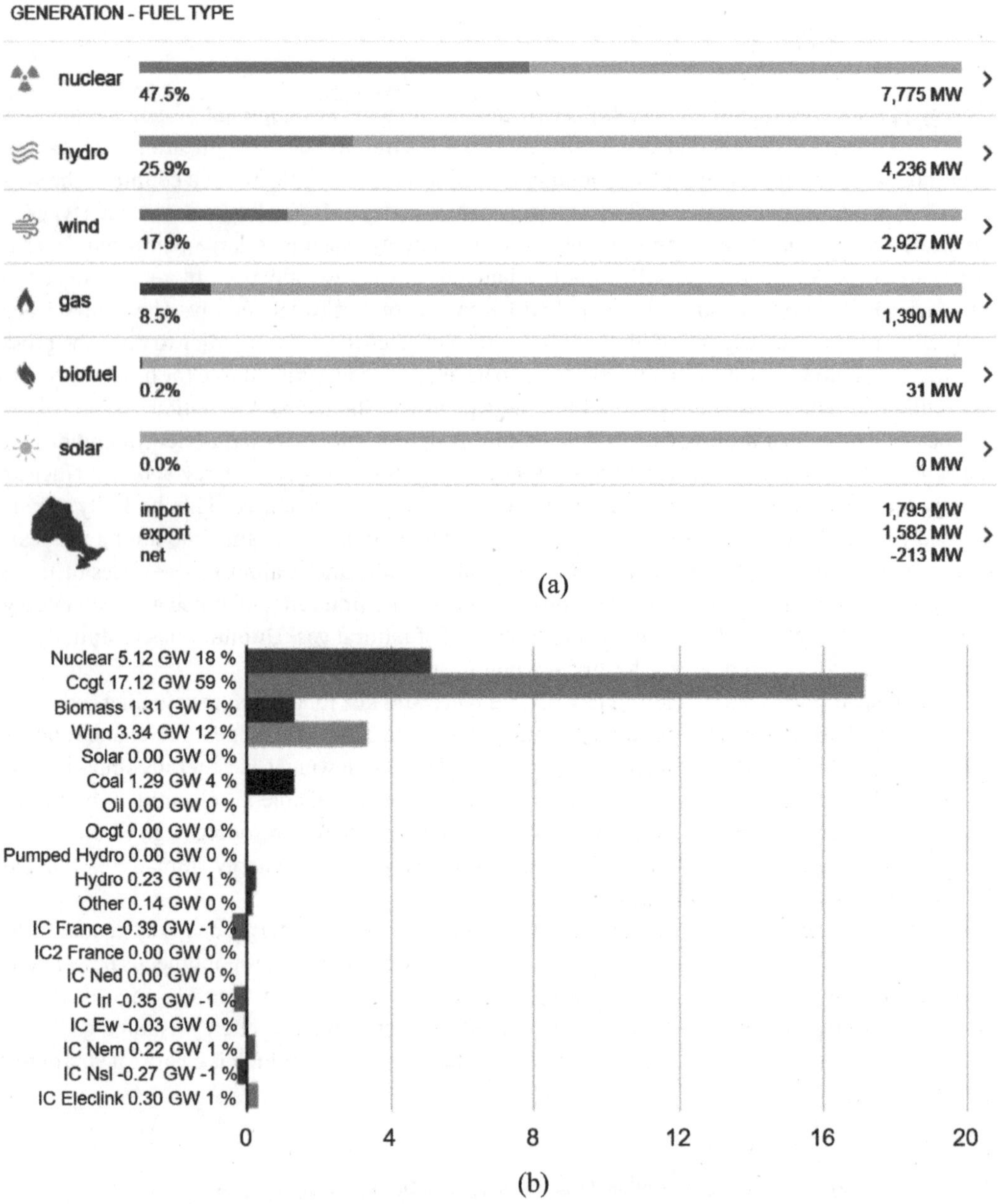

GENERATION - FUEL TYPE

☢ nuclear	47.5%		7,775 MW >
≋ hydro	25.9%		4,236 MW >
☸ wind	17.9%		2,927 MW >
♦ gas	8.5%		1,390 MW >
◗ biofuel	0.2%		31 MW >
☀ solar	0.0%		0 MW >

	import	1,795 MW
	export	1,582 MW >
	net	-213 MW

(a)

Nuclear 5.12 GW 18 %
Ccgt 17.12 GW 59 %
Biomass 1.31 GW 5 %
Wind 3.34 GW 12 %
Solar 0.00 GW 0 %
Coal 1.29 GW 4 %
Oil 0.00 GW 0 %
Ocgt 0.00 GW 0 %
Pumped Hydro 0.00 GW 0 %
Hydro 0.23 GW 1 %
Other 0.14 GW 0 %
IC France -0.39 GW -1 %
IC2 France 0.00 GW 0 %
IC Ned 0.00 GW 0 %
IC Irl -0.35 GW -1 %
IC Ew -0.03 GW 0 %
IC Nem 0.22 GW 1 %
IC Nsl -0.27 GW -1 %
IC Eleclink 0.30 GW 1 %

0   4   8   12   16   20

(b)

Figure 5.2 The energy production capacities for (a) Ontario, Canada [2] and (b) the UK [3].

total energy production is met by carbon-free sources. For the UK, a total of 18% of electricity generation comes from nuclear energy.

The primary objective of this chapter is to provide the fundamentals of nuclear energy and its importance. We begin with the historical perspective of nuclear energy. Next, the types of nuclear energy are presented with nuclear fuels and their production methods. Then, nuclear power generation and the type of nuclear reactors are discussed. Small modular reactors are presented after nuclear cogeneration techniques. Furthermore, nuclear hydrogen production is briefly discussed. Finally, integrated and/or hybridized nuclear energy systems, particularly with renewable energy sources, for communities are presented to provide a potential spectrum for the readers.

5.2 Historical Perspectives

According to 2021 data, almost 5% of the global power generation capacity is done by nuclear energy. Also, nuclear power generation covers more than 70% of the total power generation capacity in some countries. Reaching that much of level wasn't easy for nuclear energy. A brief information on some historical perspectives in nuclear power and related developments along with some landmark events is presented in Figure 5.3. At first glance, we can start with the discovery of the Uranium which is main source for nuclear energy by Martin Klaproth in 1789. Another milestone for the nuclear sector was the discovery of the Ionian radiation by Wilhelm Rontgen.

Nuclear power was first proved by Enrico Fermi by discovering neutrons could split atoms in 1934. Right after this development, a team led by Fermi achieved the first nuclear chain reaction 1942 in Chicago. At the same year, the Manhattan project, which covers the development of nuclear weapon, was officially declared the USA. The first plutonium bomb, code-named Trinity, was denoted on July 16, 1945, in New Mexico. In just less than a month after this development, on August 6, 1945, the first uranium-based bomb was used in Hiroshima, Japan. Three days later a plutonium bomb was dropped on Nagasaki, Japan. After the serious effects of the use of nuclear energy as a weapon, peaceful use of nuclear power for civilian purposes was declared, parallel to the attempts to develop nuclear weapons in the world. Then, the Atomic Energy Commission was founded by the USA.

The first electricity produced from nuclear power is performed in Idaho's Experimental Breeder Reactor I in 1951. Nuclear energy met all lighting purposes of the facility. In 1954, the first nuclear power plant was started to generate electricity in the former Soviet Union. The first commercial nuclear power plant was found in Shippingport, Pennsylvania, in 1957. After the mid-1950s, nuclear energy became popular many countries started to build their nuclear power plants. In 1959, the first nuclear power plant without using any government funding was created.

After use of nuclear power as the weapon, another disaster regarding the nuclear energy is the accidents in the nuclear power plants, which happed in Harrisburg, Pennsylvania, USA, in 1979 and in Chernobyl, Russia, in 1986.

The use of nuclear energy did not stay limited to electricity generation. The USA and Russia started to build nuclear reactors for submarines and ships. Here, one may ask the question. Why only submarines and ships? The answer is clear, due to the huge need for the cooling in the power cycle. Submarines and ships are floating on the cooling sources. Also, here, readers can understand why nuclear power plants are located near water sources like lakes, seas, oceans, etc.

Today, nuclear energy is proven technology for power generation. Also, there are serious attempts and developments on small modular reactors (SMRs) which are more reliable and feasible. Additionally, there are significant developments in fusion reactors. The Joint European Torus in Oxford, UK, reached 11 MW of power with nuclear fusion; this is almost double the previous record of 1997.

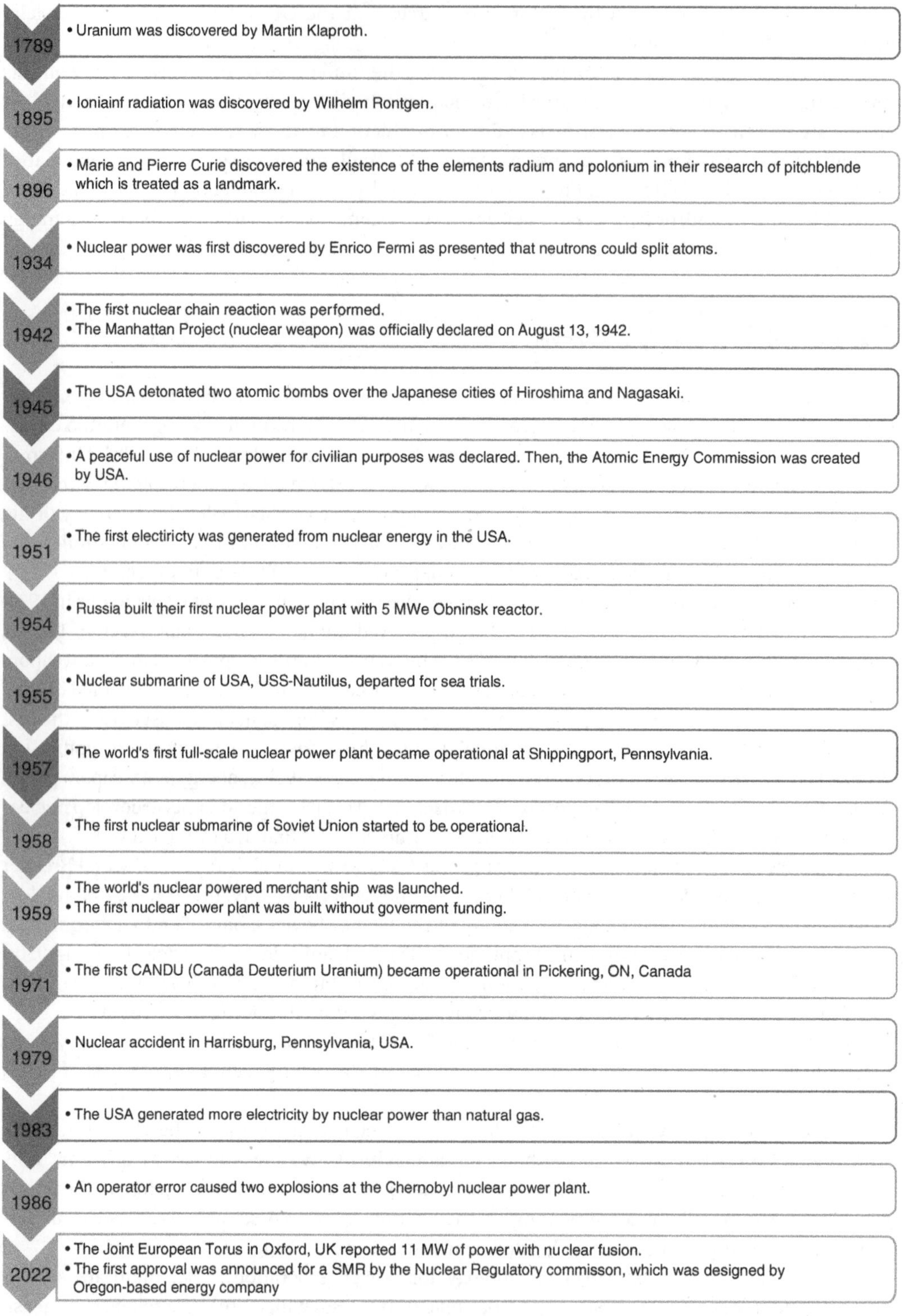

1789 • Uranium was discovered by Martin Klaproth.

1895 • Ioniainf radiation was discovered by Wilhelm Rontgen.

1896 • Marie and Pierre Curie discovered the existence of the elements radium and polonium in their research of pitchblende which is treated as a landmark.

1934 • Nuclear power was first discovered by Enrico Fermi as presented that neutrons could split atoms.

1942 • The first nuclear chain reaction was performed.
• The Manhattan Project (nuclear weapon) was officially declared on August 13, 1942.

1945 • The USA detonated two atomic bombs over the Japanese cities of Hiroshima and Nagasaki.

1946 • A peaceful use of nuclear power for civilian purposes was declared. Then, the Atomic Energy Commission was created by USA.

1951 • The first electiricty was generated from nuclear energy in the USA.

1954 • Russia built their first nuclear power plant with 5 MWe Obninsk reactor.

1955 • Nuclear submarine of USA, USS-Nautilus, departed for sea trials.

1957 • The world's first full-scale nuclear power plant became operational at Shippingport, Pennsylvania.

1958 • The first nuclear submarine of Soviet Union started to be operational.

1959 • The world's nuclear powered merchant ship was launched.
• The first nuclear power plant was built without goverment funding.

1971 • The first CANDU (Canada Deuterium Uranium) became operational in Pickering, ON, Canada

1979 • Nuclear accident in Harrisburg, Pennsylvania, USA.

1983 • The USA generated more electricity by nuclear power than natural gas.

1986 • An operator error caused two explosions at the Chernobyl nuclear power plant.

2022 • The Joint European Torus in Oxford, UK reported 11 MW of power with nuclear fusion.
• The first approval was announced for a SMR by the Nuclear Regulatory commisson, which was designed by Oregon-based energy company

Figure 5.3 Some historical perspective in nuclear energy.

5.3   Types of Nuclear Energy

As nuclear energy is a form of energy released from the nucleus, which is core of atom, before starting the type of nuclear energy, we need to remember the basic atomic structure which is illustrated in Figure 5.4. Ernest Rutherford discovered that every atom has a nucleus at the core in 1911. The nucleus consists of electrically positive protons and electrically neutral neutrons. Protons and neutral in the nucleus are held together by the strongest known fundamental force, called the strong force. Most of the atomic nucleus are spherical or ellipsoidal. It is also possible to see some different shapes. Nucleus can vibrate and rotate around itself when attacked by other particles.

As a result of this strong force, the nucleus covers less than 0.01% of the atom's volume, while containing 99.9% of its mass. An atom's chemical properties are determined by its negatively charged electrons. It is common for the number of electrons in the nucleus to match the number of protons. Electrons move around the nucleus.

The nucleus is normally stable. It is possible for nuclei to be unstable and undergo radioactive decay, eventually reaching a stable state through photon emission (gamma decay), electron or positron emission or capture (beta decay), or helium nuclei emission (alpha decay), or a combination of these processes.

Nuclear energy is a form of energy, which is heat, released from the nucleus, which consists of protons and neutrons. Nuclear energy can be generated in two methods: fission and fusion. While nuclei of atoms split into several parts in fission, nuclei fuse together in fusion. In the next subsections, nuclear fission and nuclear fusion will be discussed.

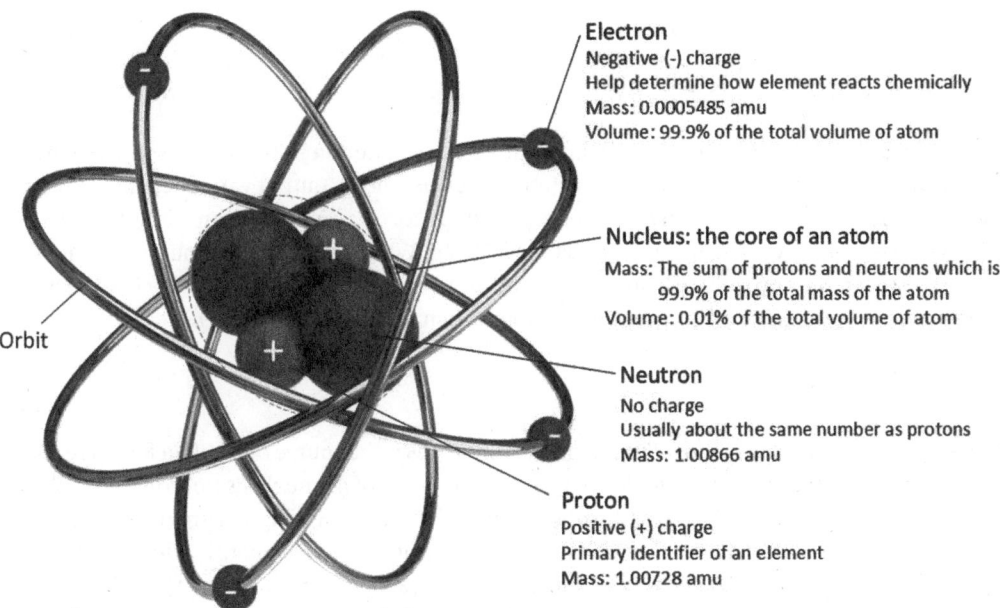

Figure 5.4   The illustration of a basic atomic structure.

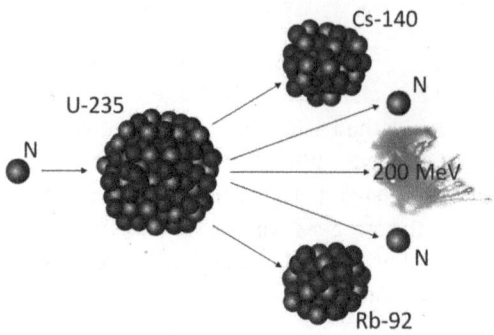

Figure 5.5 The illustration of the fission reaction of uranium-235.

5.3.1 Nuclear Fission

The fission of an atom takes place when its nucleus splits into two or smaller nuclei while releasing energy. Let's consider uranium-235, which has 92 protons and 143 neutrons (a total of 235). The particles within uranium-235 are unstable, and the nucleus may disintegrate when attacked by an external source. When a neutron hits a uranium-235 atom, the nucleus of the uranium-235 atom splits into two nuclei. As a result of this process, uranium-235 releases two or three neutrons, a barium nucleus, a krypton nucleus, etc., as illustrated in Figure 5.5. Hence, the possibility exists for creating a chain reaction by these extra neutrons hitting other surrounding uranium-235 atoms. This creates a multiplying effect which will also split and generate additional neutrons while releasing heat and radiation. The heat released can be used for power generation and other useful purposes.

The number of neutrons and other fission products such as barium, krypton, etc. are not stable and constant reactions. They are governed by statistical probability. However, the total number of nucleons and energy are conserved due to conversation laws. The number of nucleons, which are protons + neutrons, is conserved. The results of the fission reaction of uranium-235 may be barium (Ba), krypton (Kr), strontium (St), cesium (Cs), iodine (I), and xenon (Xe). Their atomic masses vary from 95 to 135. Some typical reactions are written as follows:

$$U235 + n \rightarrow Ba144 + Kr90 + 2n + \sim 200 \text{ MeV}$$

$$U235 + n \rightarrow Ba141 + Kr92 + 3n + \sim 170 \text{ MeV}$$

$$U235 + n \rightarrow Zr94 + Te139 + 3n + \sim 197 \text{ MeV}$$

Figure 5.5 shows an illustration of the nuclear fission of uranium-235. Because of the changing characteristics of the fission reaction of an atomic nucleus, energy output varies depending on how chemical reactions occur. The average energy output for uranium-135 can be assumed to be 3.2×10^{-11} J (200 MeV) which corresponds to 82 TJ/kg. When the energy output potential of the nuclear reaction is compared with an ideal combustion reaction, the nuclear energy is quite higher energy output than an ideal combustion reaction, which is 6.5×10^{-19} J (4 eV).

5.3.2 Nuclear Fusion

Nuclear fusion is the power of the Sun; in other words, this is the source of life. In a fusion reaction, two light nuclei merge to form a single heavier nucleus. Mass of the nuclei is not conserved. So, the total mass of the resulting nucleus of the nuclear fusion reaction is lighter than the sum of the reactants of the nuclear fusion. That's why nuclear fusion releases energy. The leftover mass becomes energy. This can be defined by Einstein's discovery, which is $(E = mc^2)$, which declares that mass and energy can be converted into each other. Nuclear fusion has a significant potential to meet energy demands.

There are serious attempts to develop a working nuclear fusion. Nuclear fusion is already an achieved technique; however, today's technology does not allow us to use it practically. The reason behind this is that there are uncertainties in the processes. It is not controlled properly. Possible accidents and damage are unpredictable. Therefore, the controlled nuclear fusion is at the developmental stage and expected that it will successfully be achieved in the future. It is accepted as one of the keys in solving the energy issues in the world. Researchers are working on fusion energy technologies, specifically in the deuterium-tritium (DT) fusion reaction. Figure 5.6 demonstrates the nuclear fusion of deuterium and tritium. DT fusion produces a neutron and a helium nucleus. Nuclear fusion can involve many different elements. Nuclear fusion reaction can release much more energy than most fusion reactions.

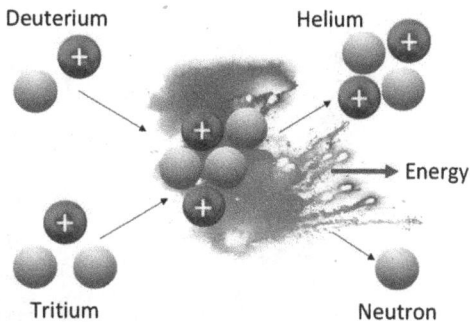

Figure 5.6 The illustration of the fusion reaction of deuterium and tritium.

Thermonuclear Fusion:

Potentially, one can duplicate on earth the way in which energy is produced in the sun, through thermonuclear fusion. It can be defined over Boltzmann equation:

$$E = k \times T$$

where k = 8.62×10^{-5} eV/K.

The fusion reaction produced on the sun is introduced as follows:

$$2H + {}^2H \rightarrow {}^4He + Be(\gamma)$$

where ^{2}H is deuterium and Be(γ) is the binding energy of helium atom retrieved in for emitted gamma radiation. The uncontrolled fusion reaction proved on earth through hydrogen bomb which is:

$$2H + {}^3H \rightarrow {}^4He + {}^1n + Be(\gamma) + Q$$

Here, the energy output, Q, is 17.6 MeV. This reaction lasts less than 1 μs.

In order to control thermonuclear fusion has to maintain extremely high energies into a "container" which appears to be possible by laser or magnetic fusion but is not yet technologically achieved.

5.4 Types of Nuclear Radiation and Potential Effects

Atoms decay to reach stable states when they decay into radiation. In chemistry, a half-life describes how long it takes for a sample of an element to decay half-way. Despite the fact that natural radiation exists everywhere, including rocks, water, and sunlight, manmade radiation is much more potent. A total of 37 radioactive elements are currently on the periodic table, 26 of which are manmade and include plutonium and americium (used in smoke detectors).

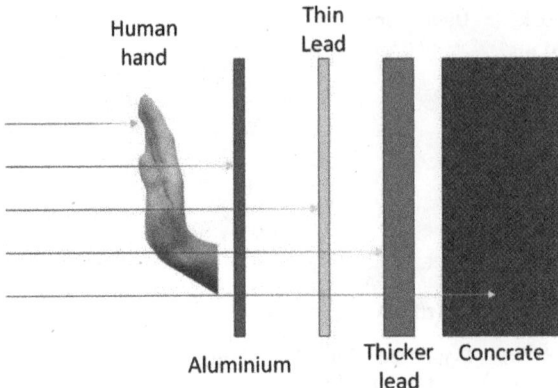

Figure 5.7 Penetration of radioactive particles.

There are four types of radiation: alpha radiation, beta radiation, gamma rays, and neutron emission. During the fission of heavier elements, alpha radiation is released as two protons and two neutrons. Figure 5.7 shows the penetration of radioactive particles. Despite being heavy, alpha particles cannot penetrate human skin. If consumed, however, alpha particles can cause serious health problems. Neutrons are turned into protons by beta radiation or vice versa. This change releases beta radiation. The skin can be penetrated by beta particles but not by light metals. Electromagnetic radiation such as X-rays, light, radio waves, and microwaves is called gamma rays after alpha and beta are released. There are several severe effects that can occur, including burns, vomiting, hemorrhage, blood changes, hair loss, increased susceptibility to infection, and even death. Lower exposure levels result in cancers (such as thyroid, leukemia, breast, and skin cancers), as well as cataracts in the eyes.

5.5 Nuclear Fuels and Production

In nuclear reactors, nuclear fuel is used as a means of withstanding the chain reaction caused by nuclear reactions. All processes involved in obtaining, refining, and utilizing this fuel make up a cycle known as the nuclear fuel cycle. Uranium-235 and plutonium-239 are the most widely known nuclear fuels. Depending on the reactor type, uranium-235 can be used at different concentrations. A CANDU reactor, for instance, uses natural uranium with a concentration of 0.7%. There are some other reactors that require enriched uranium. The concentration of the uranium varies from 3% to 5% in the enriched-uranium reactors. There is a significant amount of uranium-238 in fast breeder reactors, which use plutonium-239 as fuel. Further, thorium-232 can be used as a nuclear fuel.

The uranium resources in the world are presented in Table 5.1. Kazakhstan, Canada, and Australia account for more than two-thirds of the total world production of uranium taken from mines. There is an increase in the amount of uranium that is produced through in situ leaching, which reaches over 60% at present. On the other hand, the International Atomic Energy Agency (IAEA) reported that Kazakhstan produced 45% of the world's uranium supply from mines in 2021. This is followed by Namibia, with a share of 12% of the global supply. Canada ranks third with a 10% share of the uranium supply market.

In some nuclear reactors, enriched uranium is used. During uranium enrichment, there is an increase in the amount of uranium-235 in comparison to uranium-238. The enrichment process in domestic power plants increases uranium content by 3–5%. The most common fuel for domestic reactors is uranium dioxide. The metal form of highly enriched uranium up to 90%, which can be used for nuclear weapons, research facilities, or naval reactors, is generally used in nuclear weapons. In addition to being radioactive and chemically toxic, uranium hexafluoride is also very dangerous during the conversion and enrichment process of uranium-235. Additionally, uranium hexafluoride will release hydrofluoric acid if it comes into contact with moisture.

Table 5.2 gives the top companies producing uranium as nuclear fuel in the world according to 2021 data. Kazatomprom from Astana, Kazakhstan, meets the one-fourth of the total global

Table 5.1 Uranium resources of the world by countries in 2019.

	Tons Uranium	Percentage (%)
Australia	1,692,700	28%
Kazakhstan	906,800	15%
Canada	564,900	9%
Russia	486,000	8%
Namibia	448,300	7%
South Africa	320,900	5%
Brazil	276,800	5%
Niger	276,400	4%
China	248,900	4%
Mongolia	143,500	2%
Uzbekistan	132,300	2%
Ukraine	108,700	2%
Botswana	87,200	1%
Tanzania	58,200	1%
Jordan	52,500	1%
USA	47,900	1%
Other	295,800	5%
World total	6,147,800	100%

Source: Ref. [4]

Table 5.2 The companies producing uranium nuclear fuels in 2021.

Company	Tons U	% Share of World Total
Kazatomprom	11,858	25
Orano	4541	9
Uranium One	4514	9
Cameco	4397	9
CGN	4112	9
Navoi Mining	3500	7
CNNC	3562	7
ARMZ	2635	5
General Atomics/Quasar	2241	5
BHP	1922	4
Energy Asia	900	2
Sopamin	809	2
VostGok	455	1
Other	2886	6
Total	48,332	100

Source: Ref. [4]

production. Orano (Canada), Uranium One (Canada), Cameco (Canada), and CGN (China) meet more than one-third of the global nuclear fuel demand.

The nuclear fuel cycle is a collection of all the steps taken to procure, refine, and use this fuel. The uranium nuclear fuel cycle is shown in Figure 5.8. The front-end and back-end cycles of the nuclear fuel cycle are its two sub-cycles. The front end of the cycle refers to the steps taken before to the nuclear reactor's power generation, while the back end of the cycle refers to the steps taken to safely handle, prepare, and dispose of used nuclear fuel.

Nuclear fuel is produced by extracting uranium ore from mines and exploring for uranium. Mining the ore using the following techniques is the next step in the fuel cycle after locating ore deposits that are economically recoverable:

- underground mining
- open pit mining
- in-place (in-situ) solution mining
- heap leaching

After uranium ore is extracted from a source (i.e., open pit or underground mine), it is refined into uranium concentrate at a uranium mill. The fine powder is created by crushing, pulverizing, and grinding the ore. Uranium is separated from other minerals using chemicals added to the fine powder. For uranium extraction and concentration, groundwater from solution mining operations circulates through a resin bed.

In the next step, at a converter facility, yellowcake is converted into uranium hexafluoride gas. U-234, U-235, and U-238 are the three forms of uranium found in nature. To operate efficiently, U-235 isotope concentrations (enrichment) need to be higher in U-235 reactor designs. As a matter of fact, the concentrations of uranium isotopes in the gas are unchanged; it is called natural UF6. A uranium enrichment plant separates the individual uranium isotopes from the converted UF6 gas in order to produce enriched UF6 with a 3–5% U-235 concentration.

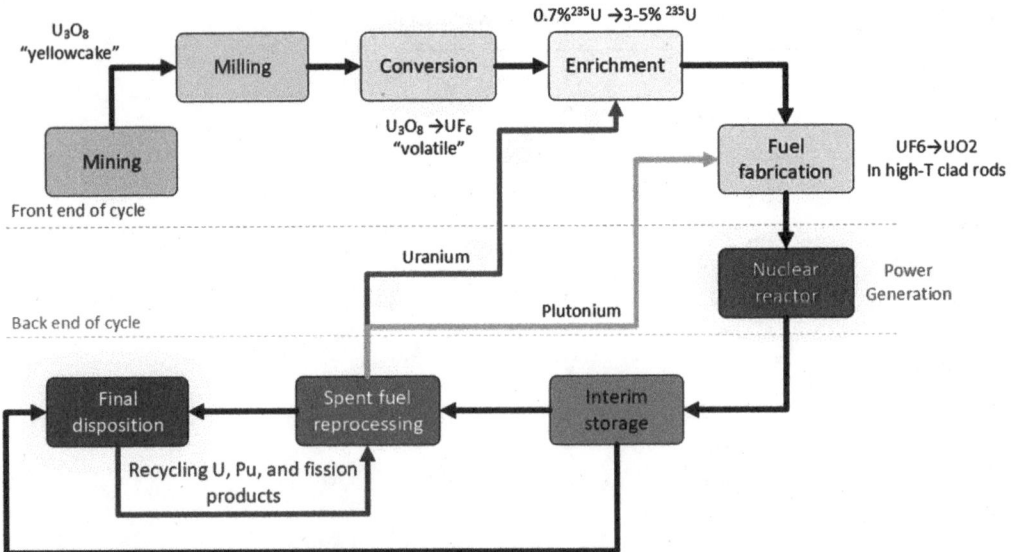

Figure 5.8 An illustration of the nuclear fuel cycle.

Uranium can be converted into nuclear fuel once it has been enriched. Nuclear fuel fabrication facilities process UF6 into uranium dioxide (UO_2) powder by heating it to gaseous form and then chemically processing it. A small ceramic fuel pellet is formed from the powder after it has been compressed and formed. Fuel rods are formed by stacking and sealing pellets into 1-cm diameter metal tubes. A fuel assembly is then formed by bundling the fuel rods together. Every fuel assembly contains between 150 and 300 fuel rods, depending on the reactor type. A typical reactor core holds 100–200 fuel assemblies.

Once the fuel assemblies are fabricated, they are transported to the reactor site. Fuel assemblies are stored onsite in fresh fuel storage bins until they are needed by reactor operators. Radiation is primarily contained within the metal tubes at this stage, and the uranium is only mildly radioactive. Every 12–24 months, reactor operators typically replace about one-third of the reactor core. Figure 5.9 demonstrates the view of a nuclear fuel assembly.

At the back end of the nuclear fuel cycle shown in Figure 5.8, after using the fuel rods in the reactor, as fuel assemblies become highly radioactive, they must be

Figure 5.9 A nuclear fuel assembly (Adapted from [5]).

removed and stored underwater at the reactor site for several years. The spent fuel continues to emit heat even after the fission reaction has stopped because the radioactive elements created when the uranium atoms were split apart continue to decay. As well as cooling the fuel, the water in the pool blocks radiation from escaping. From 1968 through 2018, a total of 276,879 nuclear fuel assemblies were disassembled and stored at the sites. They are either closed or operating nuclear sites in the USA.

After cooling in the pool, the spent fuel is moved to a dry cask storage container at the power plant. In many reactors, older spent fuel is stored in these air-cooled concrete or steel containers. In the nuclear fuel cycle, spent fuel assemblies are collected from interim storage sites and disposed of in a permanent underground repository.

Another major sector of nuclear products is medical isotopes. A fundamental aspect of nuclear medicine is the use of medical isotopes. Many know that medical sciences use radioactive sources, atoms and molecules to diagnose, characterize, and treat diseases. In the production of medical isotopes, two technologies are used: nuclear reactors and particle accelerators (namely linear accelerators, cyclotrons). Note that radiation therapy typically uses cobalt-60 as its main isotope. Sterilizing medical equipment, such as gowns, gloves, masks, syringes and implants, is also done with colbalt-60 in hospitals and clinics. Each year, more than 40 million medical procedures use isotopes worldwide, including 36 million diagnostic nuclear medicine procedures and four million radiation therapy procedures [6]. The use of iodine-125 in medicine includes diagnostic procedures (nuclear medicine imaging, biological assays) and brachytherapy for treating certain types of cancer. Further details are available in CNA [6].

5.6 Types of Nuclear Reactors

Nuclear reactors generate electricity by using the heat energy released from nuclear fuel rods through nuclear reactions. The market is flooded with a variety of nuclear reactors. Existing reactors are mostly power reactors, which generate electricity by generating heat for turbines to turn. There are also numerous research reactors. Submarines and surface ships powered by nuclear reactors are used by some navies throughout the world. Several types of power reactors exist, but the light-water reactor is the most widely used.

A historical evolution of the nuclear power plants is shown in Figure 5.10. The sectoral activities in Canada started with the heavy water moderated reached reactors in the 1940s. In the mid-1900s, the works in CANDU started, and the first prototype was built, which is called Gen-I. Then, large CANDU commercial plants were deployed with the CANDU-based reactors called Gen-I+. This is followed by Gen-II CANDU reactors which are multi-unit CANDU plants. While Gen-II+ reactors that CANDU-6 in the 1980s were developed, CANDU-9 was the Gen-III reactor in the 1990s. ACR reactor is the Gen-III+ reactor in the early 2000s. In the 2010s, the first SCWR reactors were announced.

The types of the nuclear reactors used globally are listed below with some brief details, particularly about some operations aspects:

- **Pressurized Water Reactor (PRW)**: Water at high pressure and temperature removes heat from the core of the PWR and is transported to a steam generator via a pressurizer as illustrated in Figure 5.11. The heat from the primary loop is transferred to a lower-pressure secondary loop which also contains water. During the secondary loop, water enters the steam generator at a pressure and temperature slightly below those required to initiate boiling. In spite of this,

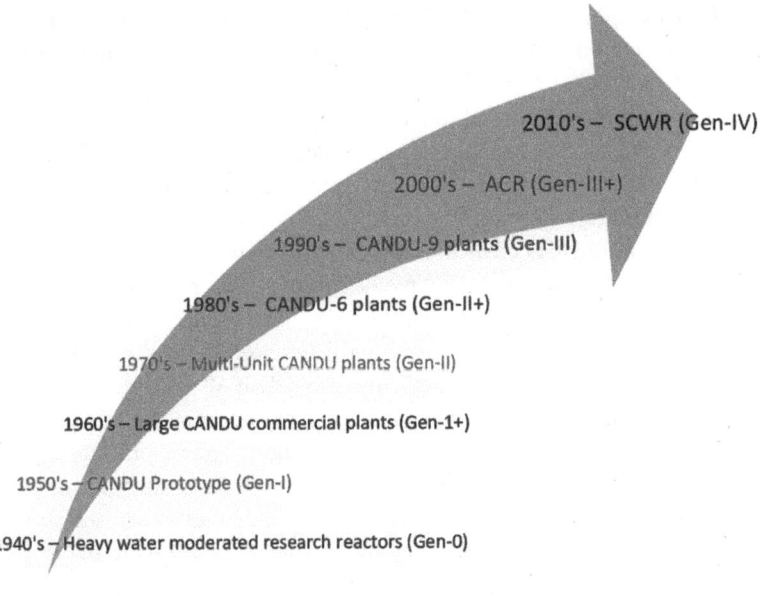

Figure 5.10 Evolution of the nuclear reactors/plants in Canada.

it becomes slightly superheated after absorbing heat from the primary loop. As a result of this process, steam is generated and ultimately used as the operating fluid for steam turbines.

- **Boiling Water Reactor (BWR)**: Electricity is generated by boiling water reactors (BWRs), a type of light water nuclear reactor. Figure 5.12 illustrates how the boiling water reactor (BWR) power generation system is laid out. A BWR uses a direct power cycle to generate electricity. At an intermediate pressure level, water passing through the core boils directly. In a BWR, the reactor core heats water, which becomes steam and drives a steam turbine. Within the reactor vessel, a series of separators and dryers promote a superheated state for the saturated steam that exits the core. A steam turbine is then powered by superheated water vapor. Argonne National Laboratory and General Electric (GE) developed the BWR in the mid-1950s. Note that the PRW and BRW reactors are known as light-water reactors (LWRs). There are many types of nuclear reactors, but these are the most common for power generation. They cover about 60% of the total nuclear reactors in the world.

- **Graphite Boiling Water Reactor (GBWR)**: It is quite similar to BWR, as shown in Figure 5.13. Different from the BWR, it uses graphite and oxygen in the control rods. A version of GBWR with moderated pressure tubes is a water-cooled power reactor with dual-use potential developed by the Soviet Union based on their graphite-moderated plutonium production military reactors (so called: RBMK reactors). It is one of the earliest and most widely deployed commercial reactor designs from the Gen-II era. It was the reactor used in Chernobyl. After the Chernobyl disaster, it has been retrofitted with a number of safety updates.

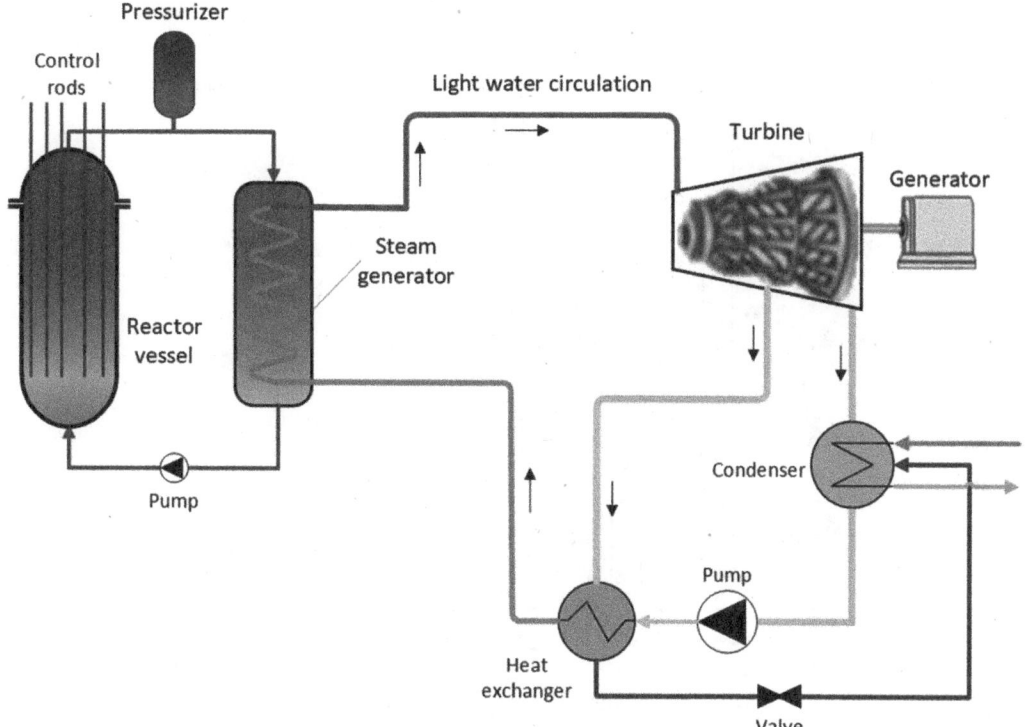

Figure 5.11   Schematic view of the pressurized water reactor (PRW).

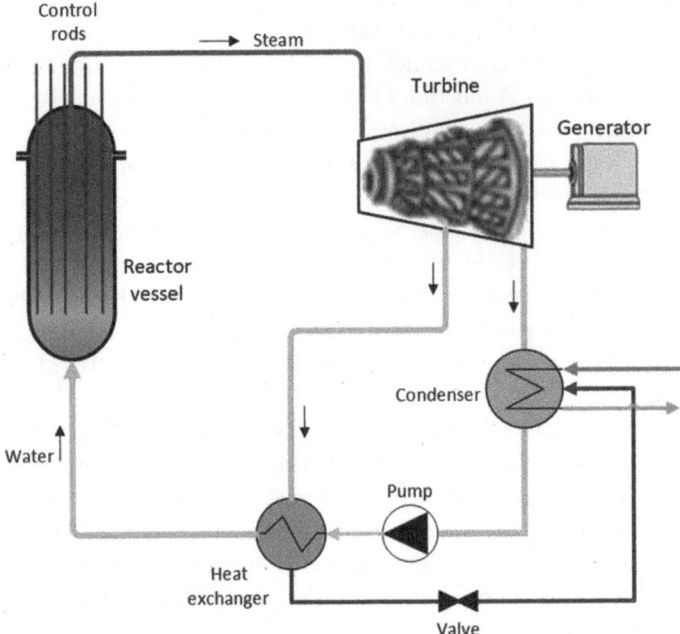

Figure 5.12 Schematic view of the boiling water reactor (BWR).

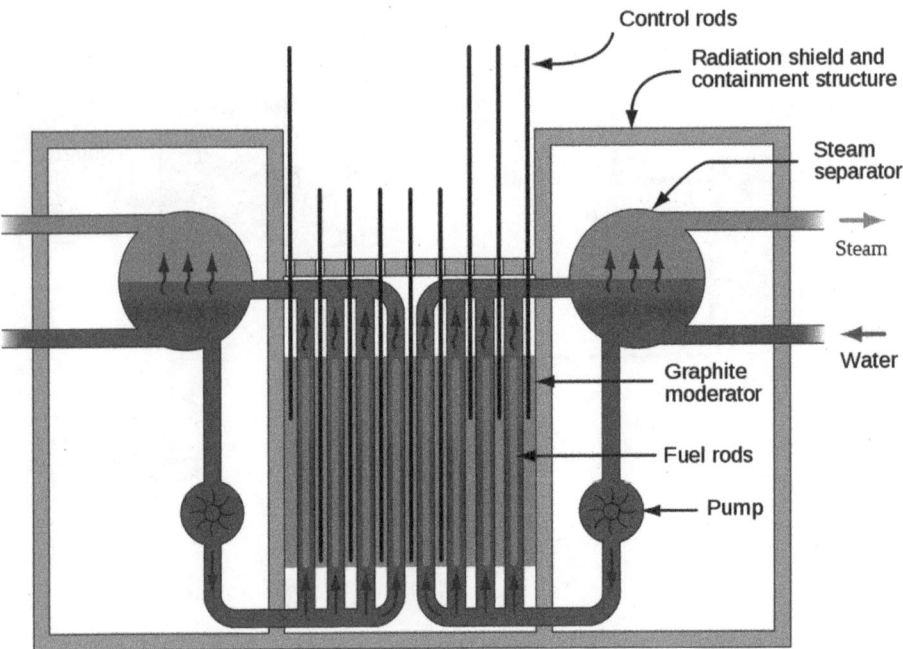

Figure 5.13 Schematic illustration of the RBMK reactor (modified from [7]).

- **Canadian Deuterium Uranium (CANDU)**: The reactor does not use enriched uranium. Natural uranium is used. In a CANDU reactor, the fuel assembly consists of short zirconium alloy-clad tubes containing natural uranium dioxide pellets, which can be changed while the reactor is running. In a pressure tube, a new assembly is pushed into one end, and spent fuel is extruded from the opposite end. This feature provides higher capacity factors for CANDU than other reactor types. More details about CANDU reactor and power generation from it are given in Section 5.7.

- **Gas-cooled Reactor (Magnox)**: Magnox reactors run on natural uranium and use graphite as a moderator and carbon dioxide gas as a heat exchange coolant, as shown in Figure 5.14. It is cooled with carbon dioxide or helium and uses natural uranium. It is generally used in the UK and France. During reactor construction, magnesium-aluminum alloys were used to cover the fuel rods. In addition to producing electrical power, the Magnox was designed to produce plutonium-239 for the British nascent nuclear weapons program.

- **Advanced Gas-cooled Reactor (AGR)**: A gas-cooled reactor utilizing graphite as the neutron moderator and carbon dioxide as the coolant, the Advanced Gas-cooled Reactor (AGR) is a second generation of British gas-cooled reactors. Figure 5.15 demonstrates the AGR. This reactor uses uranium dioxide pellets enriched to 2.5–3.5%, which are housed in stainless steel tubes. Like CANDU reactors, AGRs can be refueled without being shut down. It is cooled with carbon dioxide or helium like in Magnox reactors. Different from Magnox, it uses enriched uranium as fuel.

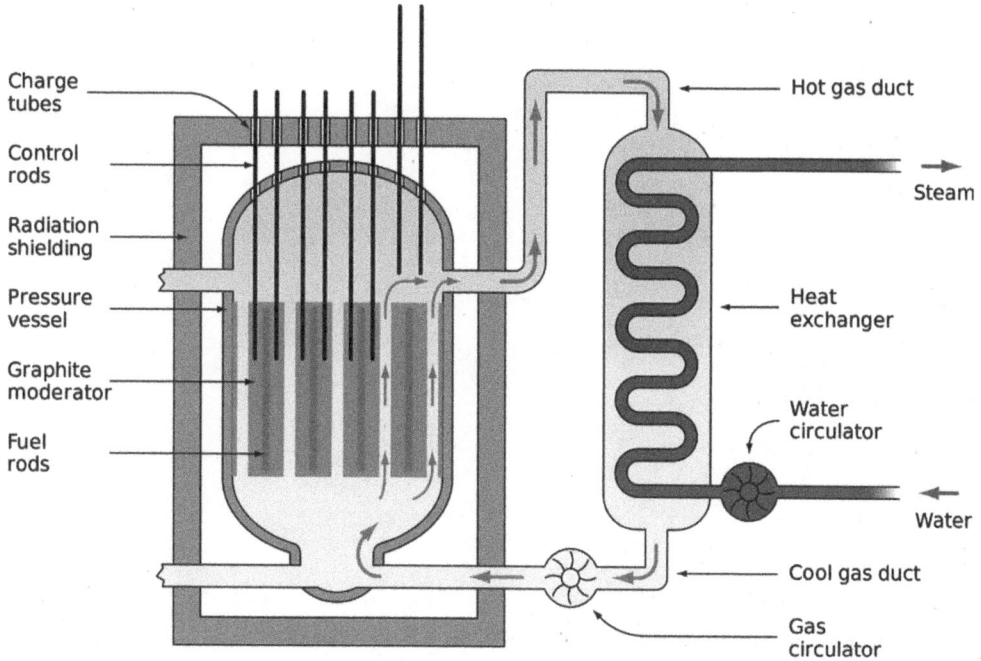

Figure 5.14 Schematic illustration of the RBMK reactor (modified from [7]).

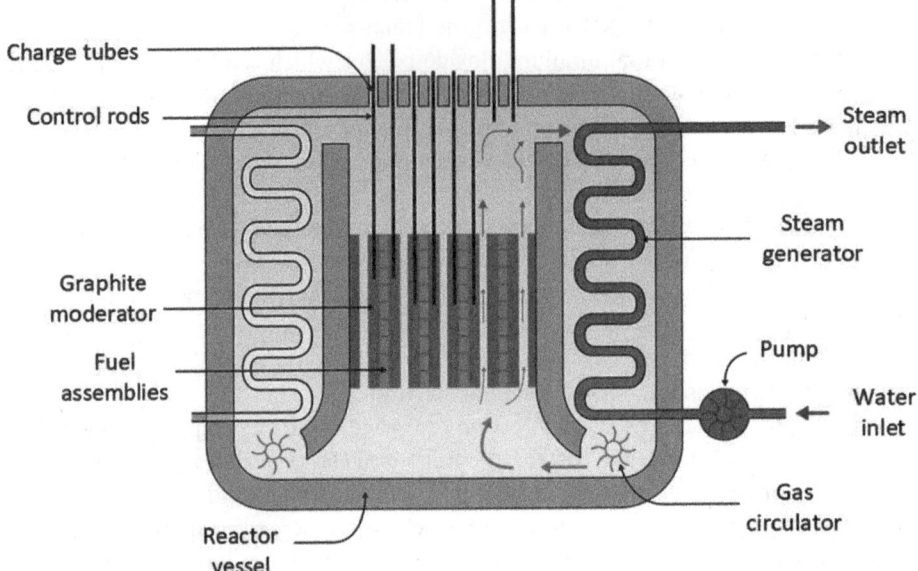

Figure 5.15 Schematic illustration of the advanced gas-cooled reactor (AGR) (modified from [7]).

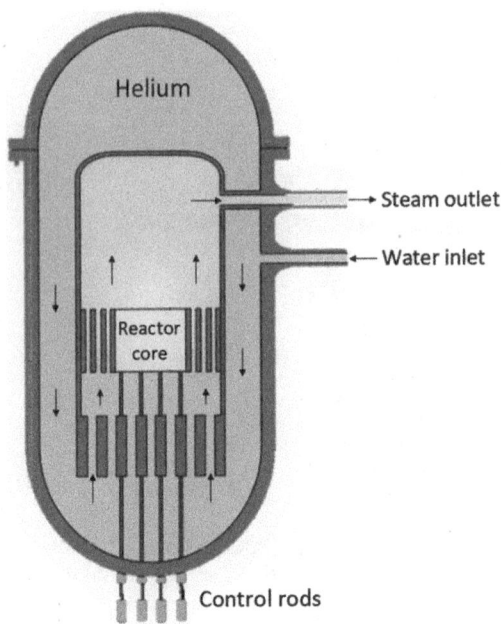

Figure 5.16 Schematic illustration of the high temperature gas reactor (fast breeder) (modified from [7]).

- **High-temperature Gas Reactor (Fast Breeder)**: It uses uranium-235 and uraninum-238 and plutonium-239. As it uses liquid sodium in the main cycle, it is quite dangerous to use since liquid sodium is inflammable with air and explosive with water. The reference reactor design is a helium-cooled system operating with an outlet temperature of 850 °C. It is a nuclear reactor design which is currently in development, which is among the Gen-IV reactors. The schematic view of the fast breeder reactor is shown in Figure 5.16.

The details of those nuclear power cycles using those reactors will be given in the following section in detail. In addition to these reactors, there are some advanced reactors which are called Generation IV (Gen-IV). These are then listed as follows:

- Gas-cooled fast reactor (GFR)
- Lead-cooled fast reactor (LFR)
- Molten salt reactor (MSR)
- Sodium-Cooled fast reactor (SFR)
- Supercritical water-cooled reactor (SCWR)
- Very high temperature reactor (VHTR)

5.7 Nuclear Power Production

In order to generate electricity from nuclear energy, the working principle is actually simple. Figure 5.17 shows the system layout for a basic nuclear power plant. The heat released by the nuclear reaction in the reactor is used for producing superheated steam via a working fluid circulating between reactor and heat exchanger. Then, the steam produced is used for the power generation in steam turbine(s). This principle is the same for almost all nuclear power generation which use different nuclear reactors, which are mentioned in the previous section. Differences come from the operating and design conditions such as type of the working fluid, temperature level, cooling type, etc.

For example, in a CANDU reactor-based power generation, natural uranium is used in the reactor, so there is no enrichment. Figure 5.18 shows the system layout for a nuclear power generation cycle which has a CANDU reactor. The reactor is cooled down with heavy water. The heavy water is Deuterium oxide (D_2O), which is the form of water that contains two atoms of the ^2H, or D, isotope. The term heavy water is also used for water in which ^2H atoms replace only some of the ^1H atoms.

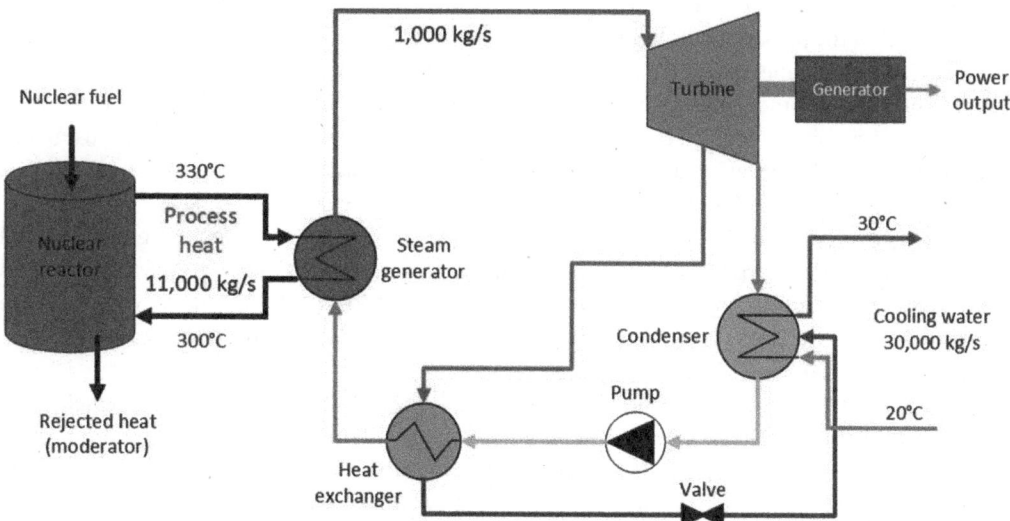

Figure 5.17 The system layout of a simple nuclear power plant.

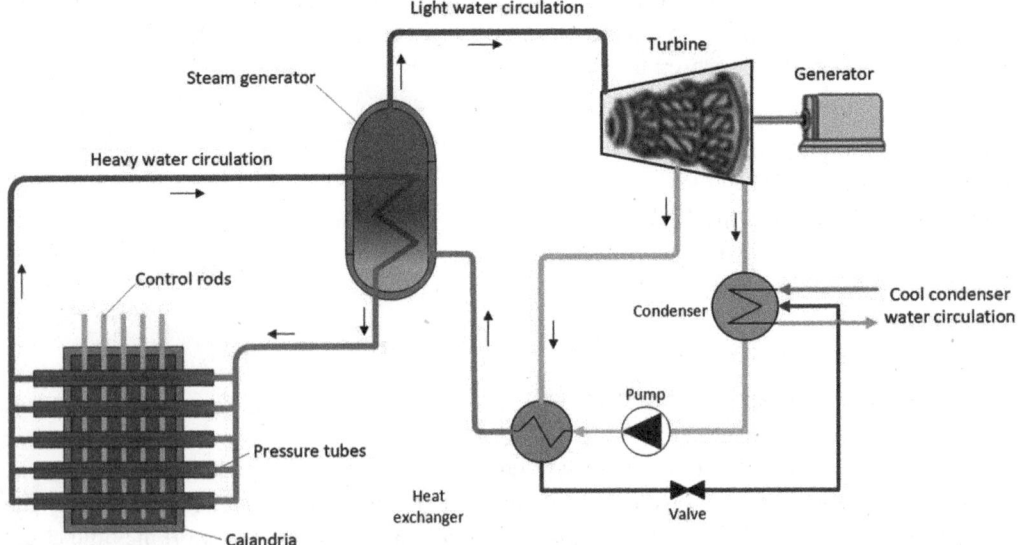

Figure 5.18 The system layout of a nuclear power plant using a CANDU reactor.

A CANDU reactor consists of a tank, or calandria vessel, containing a cold heavy water moderator at atmospheric pressure. There are zirconium alloy tubes piercing the calandria, through which heavy water coolant circulates and the natural uranium fuel is placed. A CANDU reactor vessel is aligned horizontally instead of vertically as is typical for LWR reactor vessels. A steam generator generates electricity by transferring the heat of fission into a heavy-water coolant. In the secondary loop, light-water steam exits the steam generator and is transported through a conventional turbine cycle.

In a CANDU reactor, the fuel assembly is made up of zirconium alloy-clad tubes containing natural uranium dioxide pellets, which can be changed while the reactor is running. The most important feature that sets it apart from other reactors is this. As spent fuel is extruded from the opposite end of the pressure tube, a new assembly replaces the old one. As a result of this feature, the CANDU has a higher capacity factor than other reactor types.

It is important to understand what capacity factor is here. Capacity factor refers to the ratio between the time a reactor is operating at full power during a given period versus the total amount of time available. The capacity factor of 1.0 indicates that the reactor operated continuously 24 h a day for the entire period of time.

As another nuclear power cycle, let's consider a nuclear power generation system which is a liquid-metal reactor (**fast breeder**). In a liquid-metal reactor-based power generation, liquid sodium is used for reactor cooling, as shown in Figure 5.19. Most LMRs are fueled with uranium dioxide or mixed uranium-plutonium dioxide. In the United States, however, the greatest success has been with metal fuels.

In some LMRs, heat exchangers and pumps are located outside the primary reactor vessel, while in others, the primary sodium is contained in a pool along with primary pumps and a primary-to-secondary heat exchanger. Regardless of type, primary sodium extracts heat from the core and transfers it to a secondary, nonradioactive sodium loop, which powers a steam generator that heats water in a tertiary loop to power a turbine. Since the large volume of primary sodium heats up

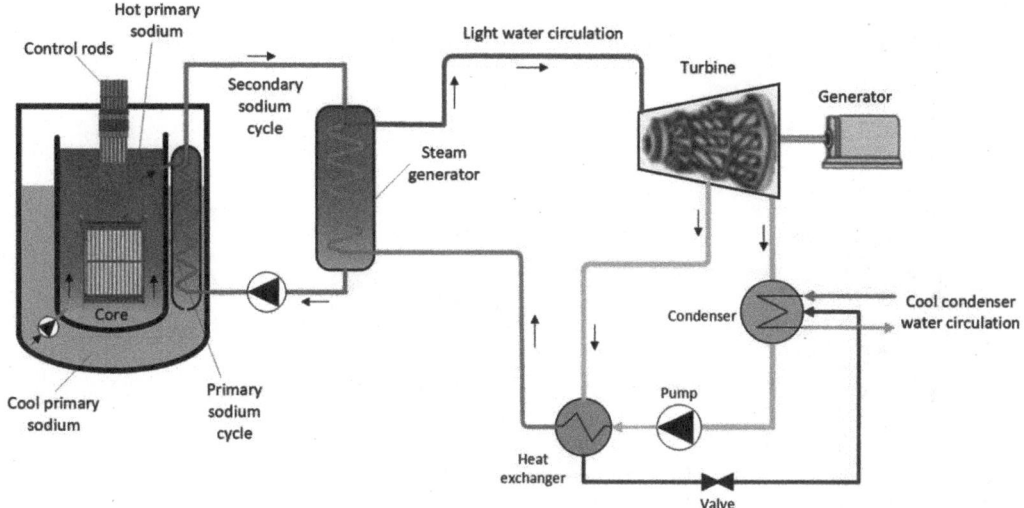

Figure 5.19 Schematic illustration of a nuclear power plant using a pool-type sodium-cooled liquid-metal reactor.

slowly even if no power is extracted, the reactor is effectively isolated from upsets in the balance of the plant. All such systems have a tightly packed core of fuel and steel cladding through which sodium coolant flows. In general, LMRs are breeders or are capable of breeding, which means they produce more fissile material than they consume.

Supercritical water-cooled reactor (SCWR) among the GenIV reactors, one of the common nuclear reactors. Figure 5.20 shows the schematic illustration of a nuclear power plant using a SCWR reactor. SCWRs are high temperature, high-pressure, light-water-cooled reactors that operate above the thermodynamic critical point of water (such as 374 °C, 22.1 MPa). The reactor core may have a thermal or a fast-neutron spectrum, depending on the core design. Unlike current water-cooled reactors, the coolant will experience a significantly higher enthalpy rise in the core, which reduces the core mass flow for a given thermal power and increases the core outlet enthalpy to superheated conditions. For both pressure vessel and pressure-tube designs, a once through steam cycle has been envisaged, omitting any coolant recirculation inside the reactor.

A supercritical water-cooled reactor (SCWR) is one of the most common nuclear reactors among GenIV reactors. A schematic illustration of a nuclear power plant with an SCWR reactor can be seen in Figure 5.20. These reactors operate above the thermodynamic critical point of water, which is above 374 °C, 22.1 MPa. Reactor cores can either have thermal or fast-neutron spectra. Coolant in the core will experience a significantly higher enthalpy rise than in water-cooled reactors, reducing the core mass flow for a given thermal power and increasing the core outlet enthalpy. There is no coolant recirculation in pressure vessels or pressure tubes, allowing a once-through steam cycle to be used.

Some key advantages of the SCWR-based nuclear power plant may be listed as follows:

- Economically viable
- High level of safety
- Flexible fuel cycle and system(s)
- Potentially direct and indirect production of hydrogen

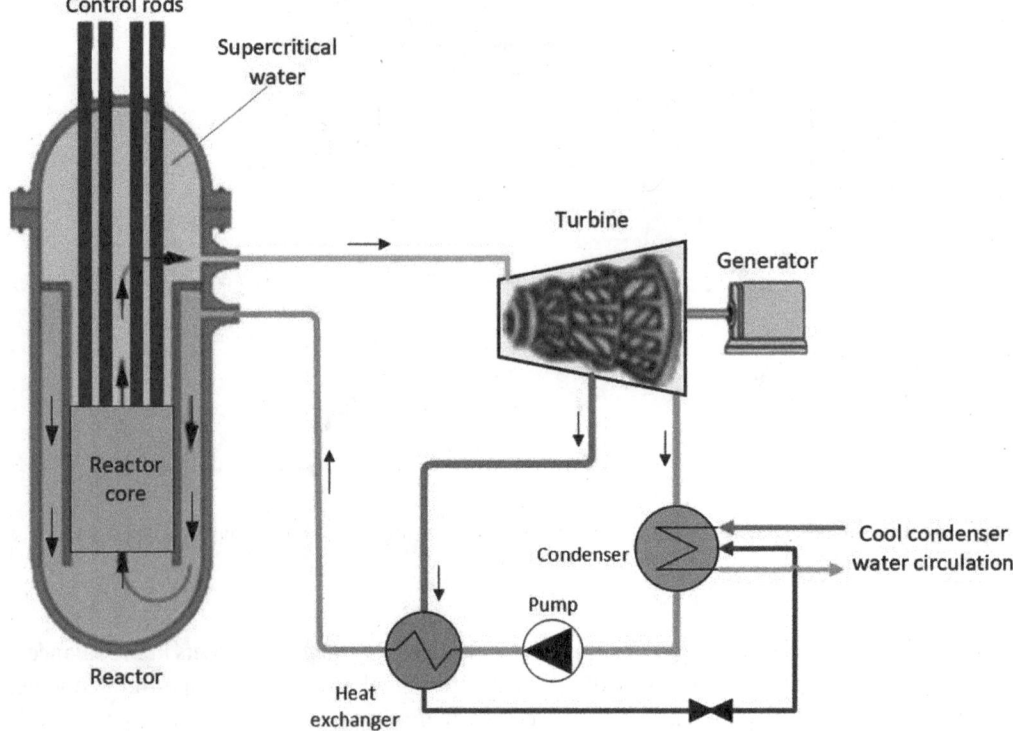

Figure 5.20 The system layout of a nuclear power plant using a SCWR reactor.

- Direct SCW cycle (with no steam generators)
- Higher efficiency (higher than 40%)
- Availabilities of efficient and effective turbine technologies

Here, the critical point is to analyze the nuclear power plant thermodynamically in order to assess the system, optimize, and further develop it. In order to perform a complete thermodynamic analysis, mass, energy, entropy, and exergy balances should be written for each component in the nuclear power plant. One may check and study the following illustrative example to further get into the thermodynamic analysis of nuclear power plants.

Illustrative Example 5.1 Let's consider a nuclear power plant which deploys a CANDU reactor, as shown in Figure 5.21. In the reactor, there are 380 fuel channels, each heats heavy water at 6704.21 kW heating rate. The temperature of the reactor is $T_{reactor} = 670$ K. The following parameters are known about the system as illustrated in Figure 5.21. Further consider the following data and make the necessary calculations:

$v_{f@10kPA} = 0.001010 \text{ m}^3/\text{kg}$, $h_1 = 3001$ kJ/kg, $h_2 = 2800$ kJ/kg, $h_3 = 2584$ kJ/kg, $P_4 = 0.1$ bar, $P_5 = 65$ bar. $h_7 = 453.4$ kJ/kg, $h_8 = 1728.9$ kJ/kg, $h_9 = 1578$ kJ/kg, $s_1 = 6.233$ kJ/kg K, $s_2 = 6.343$ kJ/kg K, $s_3 = 8.15$ kJ/kg K, $ex_1 = 1158.9$ kJ/kg, $ex_2 = 925.4$ kJ/kg, $ex_3 = 174.2$ kJ/kg.

a) Find mass flow rates of the heavy water and light water.
b) Determine the power output of the nuclear power plant, $\dot{W}_{net}$.
c) Find energy and exergy efficiencies of the nuclear power plant.

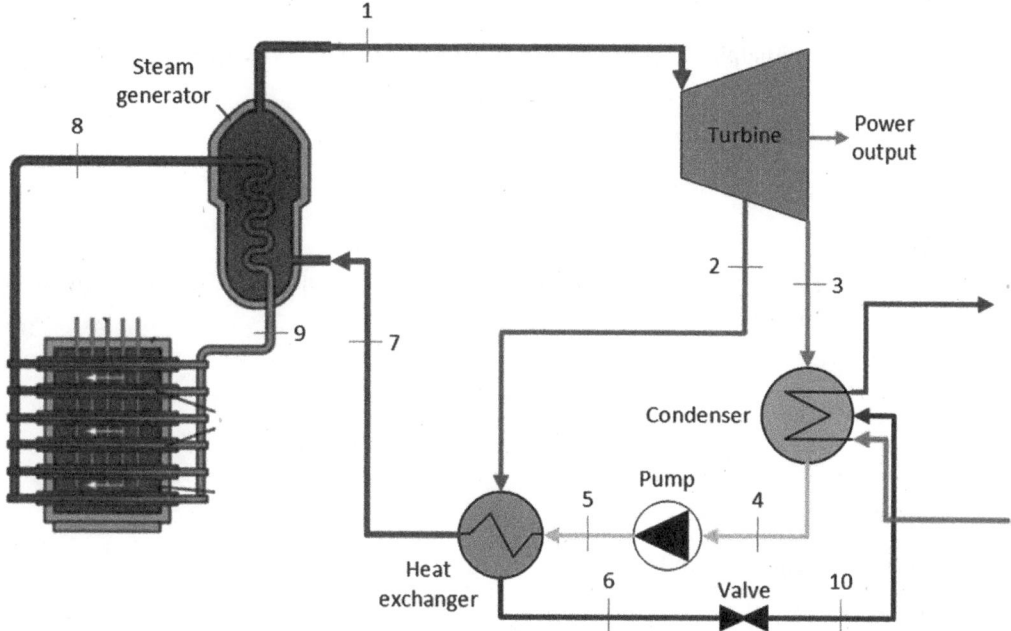

Figure 5.21 The system layout of the nuclear power plant in Illustrative Example 5.1.

Solution:

a) In order to determine the mass flow rate of the heavy water, we need to write the energy balance equation for the steam generator. The EBE for the steam generator is written as follows:

$$\dot{m}_8 h_8 + \dot{Q}_{\text{total,fc}} = \dot{m}_9 h_9$$

Here, h_8 and h_9 are already known as given. From the mass balance equation,

$$\dot{m}_8 = \dot{m}_9$$

We need to calculate $\dot{Q}_{\text{total,fc}}$. As given in the problem, there are 380 fuel channels and each release 6704.21 kW of heat. So, the total heat rate is calculated as

$$\dot{Q}_{\text{total,fc}} = 380 \times 6704.2 = 2{,}547{,}600 \text{ kW}$$

From the EBE for CANDU:

$$\dot{m}_8 (h_8 - h_9) = \dot{Q}_{\text{total,fc}}$$

$$\dot{m}_8 (1728 - 1578) = 2{,}547{,}600$$

Thus, the flow rate of the heavy water is found to be:

$$\dot{m}_8 = 16882.7 \text{ kg/s}$$

b) In order to determine the flow rate of the light water, it is required to write the energy balance equation for the steam generator. It is written as follows:

$$\dot{m}_8 h_8 + \dot{m}_7 h_7 = \dot{m}_9 h_9 + \dot{m}_1 h_1$$

$$\dot{m}_8(h_8 - h_9) = \dot{m}_1(h_1 - h_7)$$

Since we know all enthalpies in the equation above, we proceed to calculate mass flow rate as follows:

$$16882.7 \times (1728 - 1578) = \dot{m}_1(3001 - 453.4)$$

So, the mass flow rate of the light water is found to be

$$\dot{m}_1 = 1000 \text{ kg/s}$$

c) In order to determine the power output of the turbine, it is required to write the energy balance equation for the turbine. It can be written as follows:

$$\dot{m}_1 h_1 = \dot{W}_{out} + \dot{m}_2 h_2 + \dot{m}_3 h_3$$

From the mass balance equation for the turbine:

$$\dot{m}_1 = \dot{m}_2 + \dot{m}_3$$

$$1000 \times 3001 = \dot{W}_{out} + 100 \times 2800 + 900 \times 2584$$

The turbine work output is found to be

$$\dot{W}_{out} = 395,400 \text{ kW}$$

d) Both energy and exergy efficiencies for the nuclear power plant can be written as follows:

$$\eta_{en} = \frac{\dot{W}_{net}}{\dot{Q}_{total,fc}}$$

and

$$\eta_{ex} = \frac{\dot{W}_{net}}{\dot{Ex}_{\dot{Q}_{total,fc}}}$$

where

$$\dot{W}_{net} = \dot{W}_{out} - \dot{W}_{pump}$$

So, in order to calculate the energy efficiency of the nuclear power plant, we need to calculate the pumping power.

The pump power can be found via the following equation, based on the information provided in the problem:

$$\dot{W}_{pump} = \dot{m}_4 v_4 (P_5 - P_4)$$

$$\dot{W}_{pump} = 1000 \times 0.001010 \times (6500 - 10)$$

$$\dot{W}_{pump} = 6554 \text{ kW}$$

Thus, the energy efficiency of the system is found to be

$$\eta_{en} = \frac{\dot{W}_{out} - \dot{W}_{pump}}{\dot{Q}_{total,fc}} = \frac{395,400 - 6554}{2,547,600} = 0.1526$$

For the exergy efficiency, we need to calculate the exergy content of the heat released from the nuclear reactor.

$$\dot{Ex}_{\dot{Q}_{\text{total,fc}}} = \dot{Q}_{\text{total,fc}}\left(1 - \frac{T_0}{T_{\text{reactor}}}\right)$$

$$\dot{Ex}_{\dot{Q}_{\text{total,fc}}} = 395{,}400 \times \left(1 - \frac{293}{670}\right) = 1{,}452{,}132 \text{ kW}$$

Thus, the exergy efficiency of the system is calculated as

$$\eta_{\text{ex}} = \frac{\dot{W}_{\text{net}}}{\dot{Ex}_{\dot{Q}_{\text{total,fc}}}} = \frac{395{,}400 - 6554}{1{,}452{,}132} = 0.2677$$

Illustrative Example 5.2   The schematic of a nuclear power plant is considered as an example as shown in Figure 5.22. The power plant is to be designed to produce 1 MW of electrical power. The efficiency of the electrical generator is known to be 95%. The thermodynamic properties of different state points are provided in Table 5.3 Also, a total of 0.7 MW of thermal energy can be obtained from the nuclear reactor at a fuel consumption rate of 0.042 g/h.

a) Write all mass, energy, entropy, and exergy balance equations for all system components.
b) Determine the actual fuel consumption rate in the designed power plant.
c) Find the missing data and fill in Table 5.3 accordingly.
d) Find the overall system energy and exergy efficiencies.

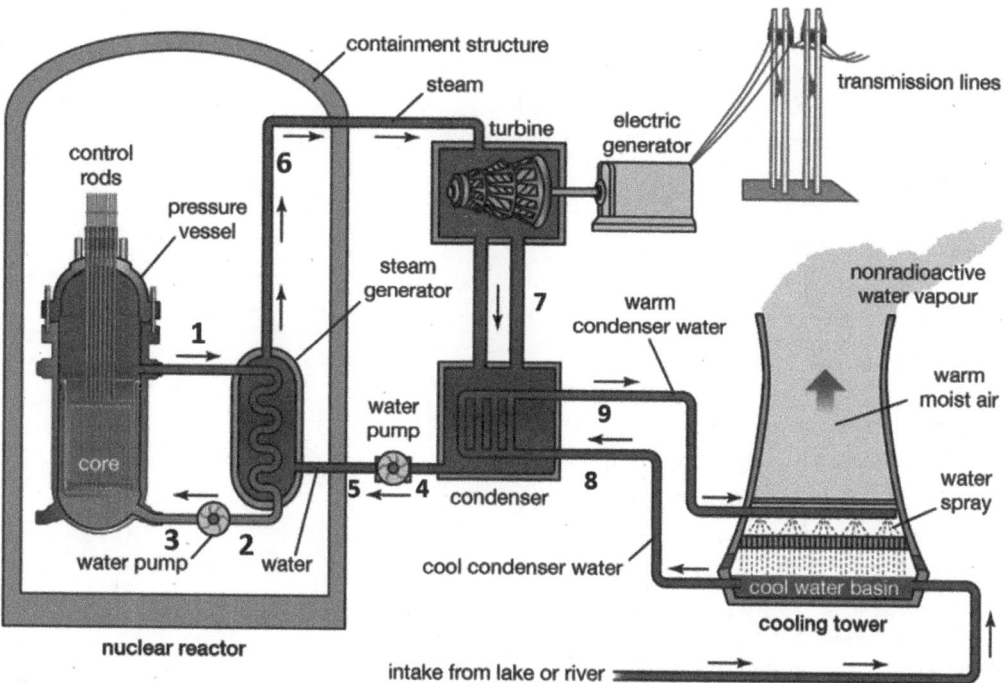

Figure 5.22   A schematic illustration of the nuclear power plant considered in Illustrative Example 5.2.

Table 5.3 The state point table for Illustrative Example 5.2.

State	$\dot{m}$ (kg/s)	Temperature (°C)	Pressure (kPa)	Specific Enthalpy (kJ/kg)
0		25	101.3	104.2
1	?	800	30,000	4016
2	?	600	29,950	3443
3	?		30,000	3450
5	10	50	30,000	1200
6	10		29,950	?
7	?		10	1305

Solution:

a) Let's write thermodynamic balance equations for each component of the nuclear power plant, as summarized below:

For nuclear reactor:

MBE: $\dot{m}_3 = \dot{m}_1$

EBE: $\dot{m}_3 h_3 + \dot{Q}_{NR} = \dot{m}_1 h_1$

EnBE: $\dot{m}_3 s_3 + \dfrac{\dot{Q}_{NR}}{T_s} + \dot{S}_{gen,NR} = \dot{m}_1 s_1$

ExBE: $\dot{m}_3 ex_3 + \dot{Q}_{in}(1 - \dfrac{T_0}{T}) = \dot{m}_1 ex_1 + \dot{Ex}_{dest,NR}$

For steam generator:

MBE: $\dot{m}_1 = \dot{m}_2$ and $\dot{m}_5 = \dot{m}_6$

EBE: $\dot{m}_1 h_1 + \dot{m}_5 h_5 = \dot{m}_2 h_2 + \dot{m}_6 h_6$

EnBE: $\dot{m}_1 s_1 + \dot{m}_5 s_5 + \dot{S}_{gen,SG} = \dot{m}_2 s_2 + \dot{m}_6 s_6$

ExBE: $\dot{m}_1 ex_1 + \dot{m}_5 ex_5 = \dot{m}_2 ex_2 + \dot{m}_6 ex_6 + \dot{Ex}_{dest,SG}$

For turbine:

MBE: $\dot{m}_6 = \dot{m}_7$

EBE: $\dot{m}_6 h_6 = \dot{m}_7 h_7 + \dot{W}_T$

EnBE: $\dot{m}_6 s_6 + \dot{S}_{gen,T} = \dot{m}_7 s_7$

ExBE: $\dot{m}_6 ex_6 = \dot{m}_7 ex_7 + \dot{W}_T + \dot{Ex}_{dest,T}$

For condenser:

MBE: $\dot{m}_7 = \dot{m}_4$ and $\dot{m}_8 = \dot{m}_9$

EBE: $\dot{m}_7 h_7 + \dot{m}_8 h_8 = \dot{m}_4 h_4 + \dot{m}_9 h_9$

EnBE: $\dot{m}_7 s_7 + \dot{m}_8 s_8 + \dot{S}_{gen,con} = \dot{m}_4 s_4 + \dot{m}_9 s_9$

ExBE: $\dot{m}_7 ex_7 + \dot{m}_8 ex_8 = \dot{m}_4 ex_4 + \dot{m}_9 ex_9 + \dot{Ex}_{dest,con}$

For pump:

MBE: $\dot{m}_4 = \dot{m}_5$

EBE: $\dot{m}_4 h_4 + \dot{W}_P = \dot{m}_5 h_5$

EnBE: $\dot{m}_4 s_4 + \dot{S}_{gen,P} = \dot{m}_5 s_5$

ExBE: $\dot{m}_4 ex_4 + \dot{W}_P = \dot{m}_5 ex_5 + \dot{Ex}_{dest,P}$

b) Here, one may calculate the actual fuel consumption as follows:

$$1000 \text{ kW} = \dot{W}_T \eta_{\text{gen}}$$

$$1000 \text{ kW} = (\dot{m}_6 h_6 - \dot{m}_7 h_7)(0.95)$$

Substitute the values of $\dot{m}_6$, h_6, $\dot{m}_7$, and h_7 in the above equation (from the properties given in the question) to find $h_6 = 1410 \text{ kJ/kg}$. Thus, we determined h_6 requested in Table 5.3.

Next, applying energy balance on the steam generator:

$$\dot{m}_1 h_1 + \dot{m}_5 h_5 = \dot{m}_2 h_2 + \dot{m}_6 h_6$$

Substituting the known values in the above equation to find:

$$\dot{m}_1 = \dot{m}_2 = 3.7 \text{ kg/s}$$

Now, we have $\dot{m}_1$ and $\dot{m}_2$ which are requested in Table 5.3. Also, applying energy balance equation in the nuclear reactor aims to obtain heat input as follows:

$$\dot{m}_3 h_3 + \dot{Q}_{\text{in}} = \dot{m}_1 h_1$$

Substituting the known values in the above equation, one finds

$$\dot{Q}_{\text{in}} = 2094.2 \text{ kW}$$

Hence, to find actual fuel consumption rate, let's proceed further:

$$\frac{2094.2 \text{ kW}}{700 \text{ kW}} = \frac{X}{0.042 \text{ kg/h}}$$

Solving the above equation to get, one may find the fuel consumption rate:

$$X = 0.125 \text{ kg/h}$$

c) Some of the requested values in Table 5.3 are found previously.

$$h_6 = 1410 \text{ kJ/kg}$$

$$\dot{m}_1 = \dot{m}_2 = 3.7 \text{ kg/s}$$

Only missing item in Table 5.3 is $\dot{m}_3$. $\dot{m}_3$ can be calculated from the mass balance equation for the pump. The EBE for pump is written as follows:

$$\dot{m}_2 = \dot{m}_3 = 3.7 \text{ kg/s}$$

Thus, we have everything required for Table 5.3.

d) In order to calculate the overall energy and exergy efficiencies, one may write both efficiencies and substitute the values obtained above as follows:

$$\eta_{\text{en}} = \frac{\dot{W}_{\text{out,electrical}}}{\dot{Q}_{\text{in,(from nuclear)}}} = \frac{1000 \text{ kW}}{2094.2 \text{ kW}} = 47.8\%$$

and

$$\eta_{\text{ex}} = \frac{\dot{W}_{\text{out,electrical}}}{\dot{Q}_{\text{in,(from nuclear)}}\left(1 - \dfrac{T_0}{T_{\text{NR}}}\right)} = \frac{1000 \text{ kW}}{2094.2\left(1 - \dfrac{298}{3000}\right) \text{ kW}} = 53.1\%$$

5.8 Small Modular Reactors and Their Utilization

Worldwide, nuclear reactors of various sizes and power outputs are used for a variety of purposes, including research, materials testing, medical treatment, shaft power for ships and submarines, and electrical power generation. Recent advances in reactor technology have made it possible to provide power to isolated or off-grid locations. A small modular reactor (SMR) is the common name for this revolutionary technology. Although their sizes vary, nuclear power plants are typically larger than SMRs.

An SMR is a nuclear power plant that typically consists of a smaller reactor and uses a variety of innovative technologies, including passive safety features and factory-built modules. A modular reactor is one of the most commonly used terms to describe advanced reactor technologies internationally.

Comparison of the nuclear power plants are demonstrated in Figure 5.23. The conventional nuclear power plants are huge power plants, and it is required to be located neat water sources. They are able to meet the power demands of the large cities and communities which consist of all sectors. Small modular reactors are relatively smaller than conventional nuclear power plants. They can meet up to 300 MWe. Therefore, it can be used for small cities, industrial zones, campuses. Finally, microreactors, which are still under development, quite small nuclear power plants. Their capacity may reach 10 MWe as conceptually designed. Therefore, they may potentially be used in factories, campuses, hospitals, and small communities.

SMRs may vary significantly in size, design features, and cooling types. Examples of different SMR technologies include:

- integral pressurized water reactors
- molten salt reactors
- high-temperature gas reactors
- liquid metal-cooled reactors
- solid state or heat pipe reactors

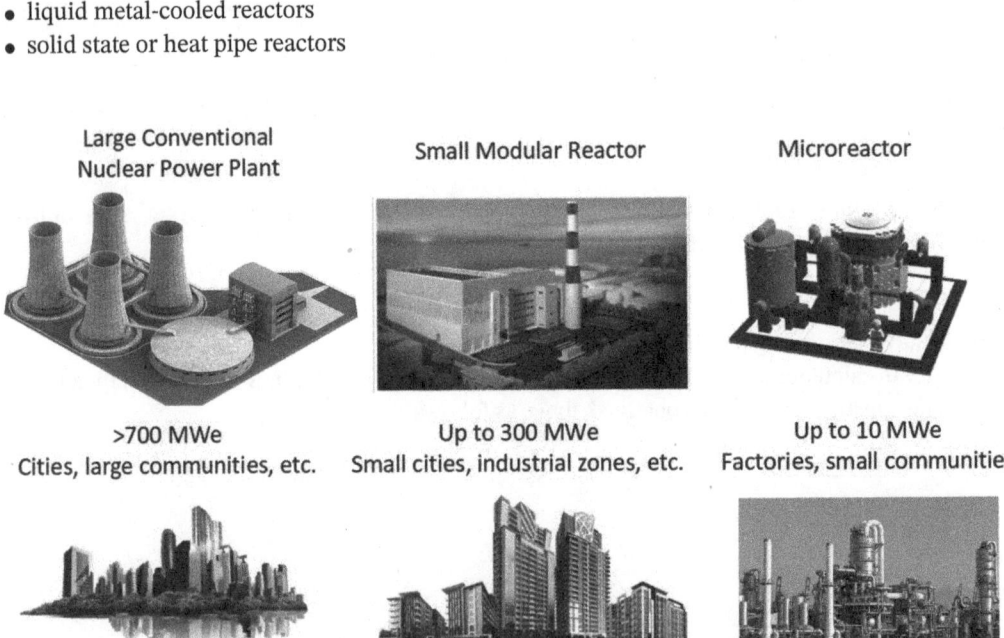

Figure 5.23 Three nuclear power plants at various capacities. *Source:* Mamag / cleanpng.com/ Averiguao/ Dasratkaka/Milton/Ralphh/Tomungr.

Unlike traditional nuclear power plants, SMRs can also be located on different sites. It is possible to establish SMRs on small grids where power generation needs are usually less than 300 megawatts electric (MWe) per facility, or at edge-of-grid or off-grid locations where power needs are small.

Globally, electrical utilities, industry groups, and governments are exploring alternative uses for SMRs besides electricity generation. These include producing steam supplies for industrial applications and district heating systems, hydrogen production, and desalinated drinking water.

The small and modular nature of SMRs contributes to many of their advantages. Due to their smaller footprint, SMRs can be located in locations not suitable for larger nuclear power plants. Unlike large power reactors, which are often custom-built for a particular location, SMRs are prefabricated and shipped and installed on-site, which makes the construction process more affordable. Cost and construction time can be reduced with SMRs, and they can be deployed incrementally to meet increasing energy demands.

Some challenges to access to energy in rural areas can be listed as follow:

- infrastructure
- limited grid coverage
- cost of grid connection

It is recommended that no more than 10% of the total installed grid capacity be accounted for by a single power plant. Because of their smaller electrical output, SMRs can be installed into existing grids or remotely off-grid in areas lacking sufficient transmission lines and grid capacity, providing low-carbon power for industry and the population. Microreactors, which are a subset of SMRs that typically generate electrical power up to 10 MW(e), are particularly relevant in this context. Microreactors have smaller footprints than other SMRs and will be better suited to areas without access to clean, reliable, and affordable energy. In rural communities or remote businesses, microreactors could also replace diesel-fueled power generators that are often used in emergency situations.

Some proposed SMR designs as conceptually developed in the literature are generally simpler than existing reactors, and the safety concept often relies more on passive systems and the inherent safety characteristics of the reactor, such as low power and operating pressure. Due to physical phenomena such as natural circulation, convection, gravity, and self-pressurization, passive systems do not require human intervention or external power to shut down. Increasing safety margins can, in some cases, eliminate or significantly reduce the potential for radioactive release to the environment and the public in the event of an accident.

Fuel requirements have been reduced with SMRs. In contrast to conventional plants, SMRs may require less frequent refueling, every 3–7 years, instead of every 1–2 years. SMRs can operate without refueling for up to 30 years.

5.9 Nuclear Cogeneration

Cogeneration is a way to use a single source of energy efficiently to produce power, useful thermal energy, and other commodities such as hydrogen, cooling, freshwater, etc. The Carnot cycle tells us that a heat engine needs to work between a hot source of energy and a cold environment (sink) to produce a specific amount of power. Therefore, there is always a considerable amount of heat to be released to the cold source. The idea of cogeneration is to use this rejected energy to obtain more useful output. Thus, this makes it possible to increase the efficiency of the system further.

Nuclear cogeneration concept is demonstrated in Figure 5.24. In a conventional nuclear power plant, nuclear fuel (source) is used for generating electricity (service) via a nuclear reactor (system). So, the system undergoes the 3S concept as we use a source to meet a demand item with a system. As mentioned often through this book, we need to benefit from a source as much as possible in a multigenerational manner. Namely, we need to increase the number of useful outputs. Additional useful outputs will help to reduce energy consumption and emissions because it will not require external systems to meet those demand items.

The temperature of the exhaust flow (or working fluid), which transfers heat to the surrounding environment, is usually high in power generating cycles. Thus, recovering heat for useful purposes is really essential which will help improve the efficiency and make the project at large more feasible. Some key questions here may be:

(i) where to use this heat?, (ii) how to design a heat recovery system?, (iii) how to incorporate the designed heat recovery system into the existing system?, (iv) how to achieve thermal management in en effective manner? and (v) what control and monitoring tools to use for managing effective operation?

During the past decade the subject of nuclear cogeneration has become important where there is another useful commodity is simultaneously produced in addition to the electricity generated. The other useful commodity is primarily about recovering heat and using it for useful purposes, district heating, hot water, cooling, thermal desalination, hydrogen production, etc. Producing another useful commodity will increase the number of useful outputs to two and allow capacities increased which will result in efficiency improvement. Such an efficiency improvement may go up to 10%–20% energetically and/or exergetically. This will bring another benefit to reduce the environmental impact and hence the life cycle emissions at tangible quantities.

Among the considered power generation options, nuclear power plants may have slightly higher efficiencies and much less life cycle emissions (and hence drastically reduced environmental impacts compared to thermal power plants driven by fossil fuels. The main benefits of nuclear cogeneration include the following, but not limited to:

- Reduced life cycle GHG emissions,
- More effective and resilient operation,
- Higher electricity output,
- Reduced area and space requirement per MW electricity generated,
- Longer plant lifetime,
- More efficient operation.

Although nuclear-based cogeneration systems have several advantages, there may be some challenges, such as increased cost, risks, operational complexity, management and safety requirements.

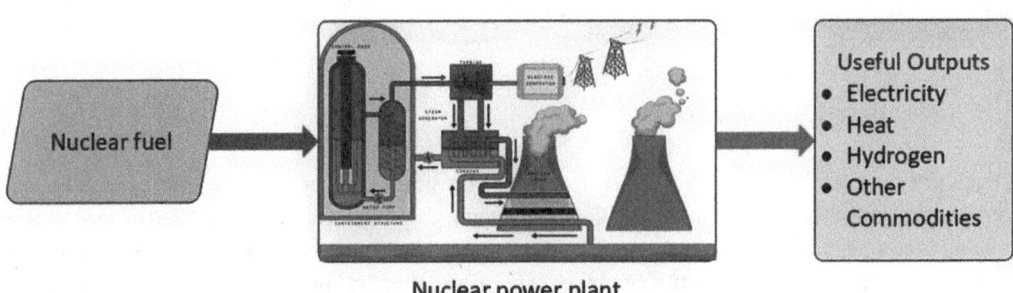

Nuclear power plant

Figure 5.24 Nuclear cogeneration concept, covering the production of at least two useful outputs.

- Figure 5.25 demonstrates various heat application processes of interest for cogeneration in nuclear power plant. Applications, such as water electrolysis for hydrogen production and reverse osmosis (RO) for desalination are not included in this section because they do not necessarily require a nuclear power plant. The temperature requirement for heat source varies widely according to the process. In the low temperature range are district heating of water or steam (80–150 °C) and desalination of sea water (65–120 °C) by thermal processes, such as multistage flash (MSF) and MED.

A large number of heat applications exists in the medium temperature range, although only a few major processes are identified as presented in Figure 5.25. In the past, a significant application for nuclear cogeneration was the intensive supply of steam for bitumen extraction from oil sands in Alberta, Canada, where oil reserves are the world's second largest after Saudi Arabia. To extract bitumen from deep sand deposits, significant amounts of high-pressure steam (400 °C) are required,

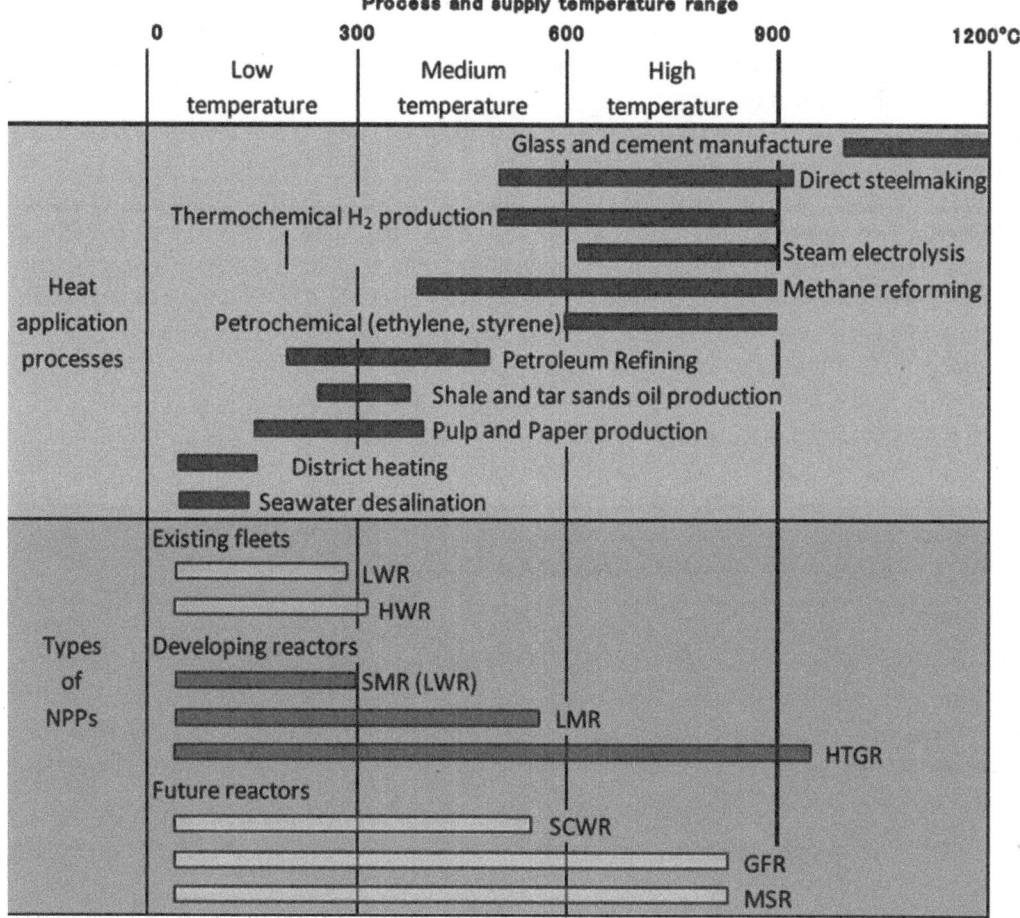

Figure 5.25 Temperature ranges of heat application processes and types of nuclear power plant (adapted from [8]). (GFR: gas-cooled fast reactor, HTGR: high temperature gas reactor, HWR: heavy water reactor, LMR: liquid metal reactor, LWR: light water reactor, MSR: molten salt reactor, NPP: nuclear power plant, SCWR: supercritical water reactor, SMR: small modular reactor.)

using by in situ methods, such as steam-assisted gravity drainage. Further energy is required to upgrade bitumen to synthetic crude oil as ready feedstock for refineries. Currently, the industry uses natural gas as the prime energy source to produce crude oil from sand oil.

The high temperature heat is typically demanded in petrochemical, steel, and hydrogen production. The hydrogen produced has current and future applications in the following:

- Ammonia synthesis
- Methanol synthesis
- Heavy crude oil upgrade by hydrocracking
- Fischer-Tropsch hydrocarbon synthesis
- Methanation in long distance energy transport
- Hydrogasification
- Fuel for power generators and fuel cell vehicles

Nuclear power plants support cogeneration in the processes identified above by meeting part or all of the consumption for heat and additional power. In comparison to single purpose generation, cogeneration with nuclear power provides major benefits, such as: reduced or zero fossil fuel energy uses; reduced or zero CO_2 emission; greater overall thermal efficiency through the utilization of low grade or waste heat from the nuclear power plants; and lower combined generation costs and environmental impacts.

More details about nuclear-based multigenerational system are given in the following sections with some case studies.

Illustrative Example 5.3 Consider a CANDU reactor-based nuclear power plant as illustrated in Illustrative Example 5.1. This is a conventional power plant as it produces only electricity. An engineering team wants to upgrade the nuclear power plant by adding additional useful outputs which are heat and hydrogen. The production potential of each commodity is given as follows:

(a) Heat: 50 MW of thermal energy will be extracted as a useful output.

(b) Hydrogen: 50 kg of hydrogen will be produced in an hour.

By assuming the heat recovery is done at the temperature of 150 °C, and the lower heating value of hydrogen is 120 MJ/kg along with that the chemical exergy of hydrogen is 138 MJ/kg. Calculate the overall energy and exergy efficiencies of this trigeneration nuclear plant and compare with the efficiencies obtained for a conventional single generation plant.

Solution:

As presented earlier in Illustrative Example 5.1, a total of 2,547,600 kW of heat generated by nuclear fuel was able to produce 395,400 kW of electricity. In the present system, there are two more useful outputs (such as heat and hydrogen by keeping the input identical. So, it is required to redefine the energy and exergy efficiency equations for the trigenerational system. The overall energy efficiency is then written as:

$$\eta_{en} = \frac{\left(\dot{W}_{out} - \dot{W}_{pump}\right) + \dot{Q}_{heat} + \dot{m}_{H_2} \times LHV_{H_2}}{\dot{Q}_{total,fc}}$$

Remember, here, $\left(\dot{W}_{out} - \dot{W}_{pump}\right)$ is the net power output obtained from the power plant. $\dot{Q}_{heat}$ is the useful heat output and $(\dot{m}_{H_2} \times LHV_{H_2})$ is the energy content of the hydrogen output.

The exergy efficiency for the trigenerational power plant is written as:

$$\eta_{ex} = \frac{\left(\dot{W}_{out} - \dot{W}_{pump}\right) + \dot{Ex}_{\dot{Q}_{total,fc}} + \dot{m}_{H_2} \times ex_{H_2}}{\dot{Ex}_{\dot{Q}_{total,fc}}}$$

where $\dot{Ex}_{\dot{Q}_{total,fc}}$ is the exergy content of the useful heat output, and $(\dot{m}_{H_2} \times ex_{H_2})$ is the exergy output of the hydrogen.

Let's start with calculating energy efficiency. First, we need to calculate the hydrogen flow rate. As the hourly hydrogen production rate is 50 kg.

$$\dot{m}_{H_2} = \frac{\dot{m}_{H_2}\left(\dfrac{kg}{h}\right)}{3600\left(\dfrac{h}{s}\right)} = \frac{50}{3600} = 0.014 \frac{kg}{s}$$

So, we know everything to calculate the overall energy efficiency as follows:

$$\eta_{en} = \frac{\left(\dot{W}_{out} - \dot{W}_{pump}\right) + \dot{Q}_{heat} + \dot{m}_{H_2} \times LHV_{H_2}}{\dot{Q}_{total,fc}} =$$

$$\frac{(395,400 - 6554) + 50,000 + (0.014 \times 120,000)}{2,547,600}$$

Thus, the overall energy efficiency is found to be $\eta_{en} = 0.1729$.

In order to calculate the overall exergy efficiency, we need to find the exergy content of the useful heat output.

$$\dot{Ex}_{\dot{Q}_{total,fc}} = \dot{Q}_{heat}\left(1 - \frac{T_0}{T_s}\right) = 50,000 \times \left(1 - \frac{273}{423}\right) = 17,730 \; kW$$

Now, we proceed to calculate the overall exergy efficiency as follows:

$$\eta_{ex} = \frac{(395,400 - 6554) + 17,730 + (0.014 \times 138,000)}{1,452,132}$$

Thus, the exergy efficiency is found to be $\eta_{ex} = 0.2813$.

Without cogeneration the energy and exergy efficiencies were found to be $\eta_{en} = 0.1526$ and $\eta_{ex} = 0.2677$, respectively. An integration of heat output and hydrogen production increases the energy and exergy efficiencies to, for example: $\eta_{en} = 0.1730$ and $\eta_{ex} = 0.2813$.

5.10 Nuclear Hydrogen Production

The use of hydrogen as an energy carrier is expected to become more common in the future. Almost any application in which fossil fuels are currently used can be fuelled by hydrogen. Unlike fossil fuels, its combustion produces no harmful emissions, except for NO_x emissions, which can be controlled. Hydrogen is also converted more efficiently into useful forms of energy than fossil fuels. Even though hydrogen is frequently perceived as a dangerous fuel, it is as safe as any other fuel.

It is therefore necessary to produce hydrogen from a variety of sources, including water, which is the most abundant source of hydrogen. Hydrogen can be produced from water by splitting it, but this process requires more energy than hydrogen can provide. Hydrogen is therefore considered an energy carrier for a suitable form of energy, like electricity. It is widely accepted that hydrogen is one of the promising energy carriers and that the demand for it will greatly increase in the near future, since it can be utilized as a clean fuel in diverse energy end-use sectors, including converting it into electricity without releasing any carbon dioxide.

Figure 5.26 shows the nuclear-based hydrogen production concept. The thermal efficiency of this production process is considered as the ratio of the lower heating value of the produced hydrogen to the energy input into the system from all sources. The production process could vary from electrolysis to steam reforming. However, there are mainly two methods depending on what energy input is used for hydrogen production, which are heat or electricity. Figure 5.26a depicts hydrogen production using the process or waste heat obtained from the nuclear reactor where there are, in this regard, some thermochemical cycles, such as Cu-Cl, Ca-Br, etc. to consider the heat for hydrogen production. On the other hand, electricity generated via nuclear energy can directly be used in water electrolysis. While thermochemical cycles have a significant advantage to use waste heat or nuclear heat directly for hydrogen production. Cogeneration for hydrogen production using electrolysis is considered a candidate for integration with nuclear energy because it may be combined with either existing nuclear electrical generating plants or with new, highly efficient nuclear plants. Furthermore, Figure 5.26b illustrates a scenario where nuclear process or waste heat is used to first generate electricity using some systems, for example organic Rankine cycles, to first generate electricity (in addition to the primarily generated electricity) and use this electricity to do water electrolysis for producing hydrogen and oxygen. These two options illustrated in Figures 5.26a and 5.26b are considered under nuclear cogeneration options.

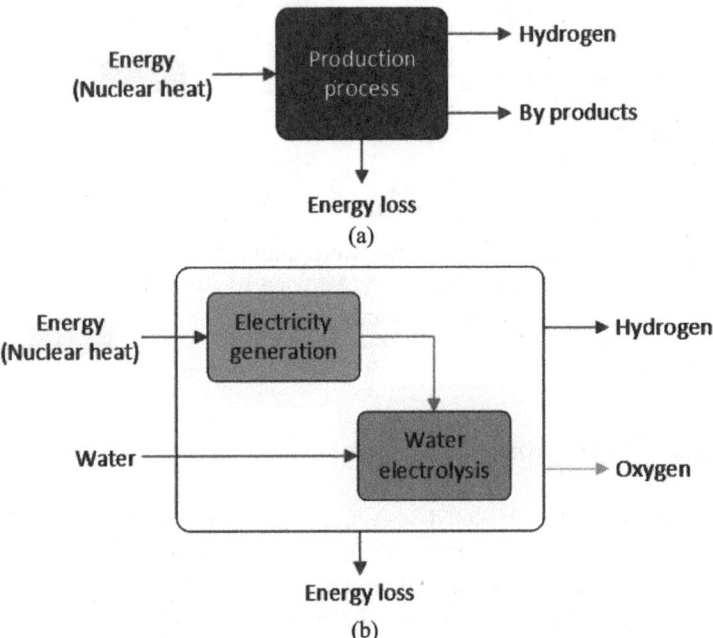

Figure 5.26 Nuclear cogeneration-based hydrogen production (a) general concept and (b) water electrolysis-based hydrogen production.

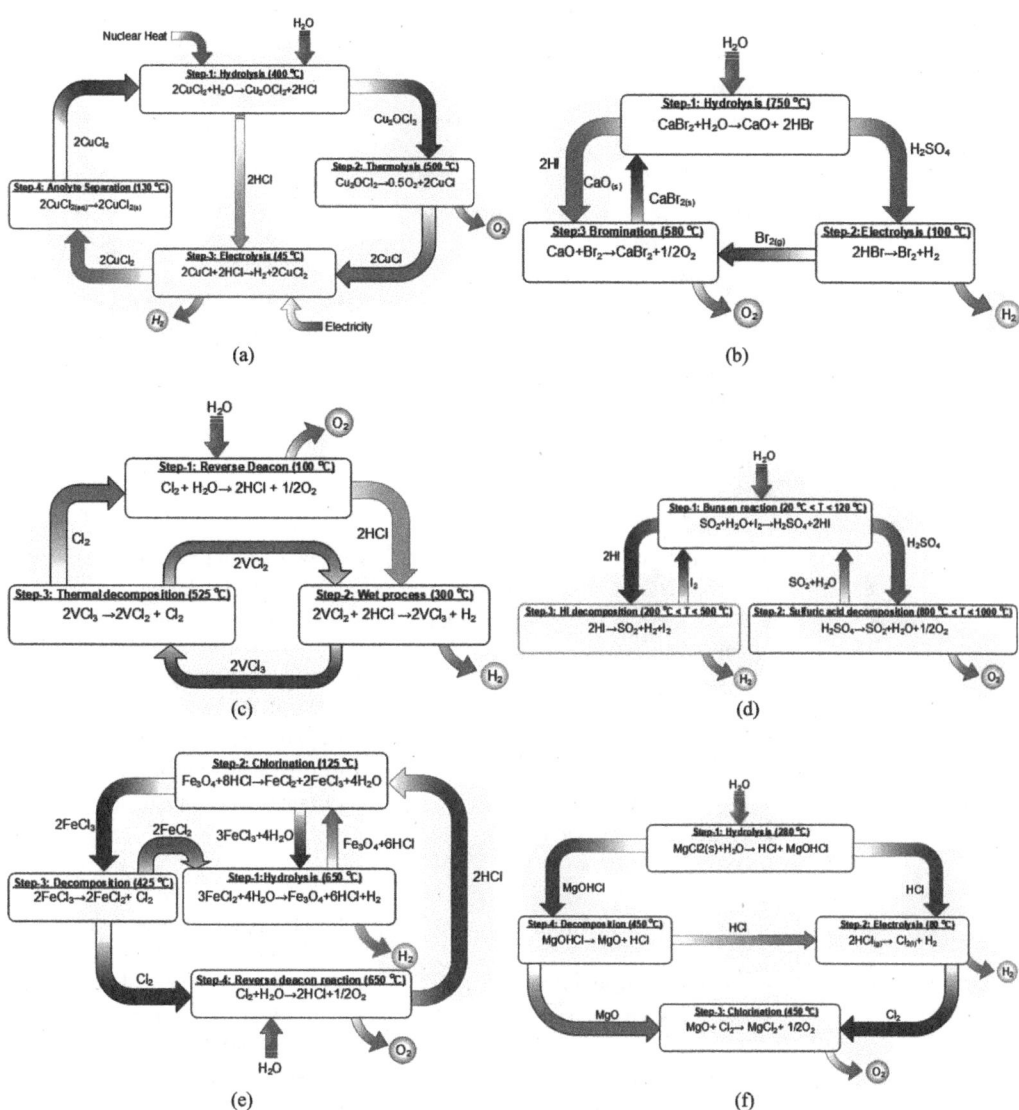

Figure 5.27 Nuclear-based thermochemical cycles for hydrogen production (a) four-step Cu-Cl thermochemical cycle, (b) three-step Ca-Br thermochemical cycle, (c) three-step V-Cl thermochemical cycle, (d) three-step S-I thermochemical cycle, (e) four-step Fe-Cl thermochemical cycle, and (f) four-step Mg-Cl thermochemical cycle.

Figure 5.27 demonstrates the common thermochemical cycles using heat and a set of chemical reactions to produce hydrogen. During thermochemical cycles, hydrogen and oxygen are produced through chemical reactions and heat. Water splitting can be achieved at desired cycle temperatures (usually below 1000 °C) than the direct one-step thermal water decomposition which may up to much higher temperatures (beyond 2000 K). Furthermore, these systems recycle the chemicals internally without emitting waste to the environment, making them more environmentally friendly than fossil fuel-based hydrogen production systems. The Cu–Cl thermochemical cycle (see Figure 5.27a), which splits water into hydrogen and oxygen through intermediate copper and chlorine compounds, operates at lower temperatures than most thermochemical cycles. This cycle can be integrated with various nuclear power plants due to its low-temperature requirements. Those cycles bring pros and cons with them. Table 5.4 gives the

Table 5.4 A basic comparison of some thermochemical cycles.

Thermochemical Cycle	Opportunities	Challenges
Cu-Cl	• Low operating temperature • Reactions demonstrated at lab scale • H_2 production at low pressures • Low environmental impacts • H_2 production at reasonable costs • High energy and exergy efficiencies • H_2 production at large scales	• Corrosive working agent • Handling issues with solids among processes • Issues associated with scale-up at multistep reactions
Ca-Br	• Reasonably low operating temperatures • Feasible for integration to Gen-IV nuclear reactor concepts • Electrolysis step with reasonable overpotentials	• CaO deposition at bromination inlet • No existing commercial application to evaluate feasibility and real-life performance
V-Cl	• High thermal efficiency expected • Low reaction temperature • Use of thermal energy from various sources including solar and nuclear.	• Slow reaction kinetics and difficult separation process • R&D required to improve the performance of reverse Deacon cycle • Need for feasibility and production cost evaluation
S-I	• High thermal efficiency • Low environmental impacts • Use of low-cost materials • Products in gas or liquid form reducing energetic cost of transportation	• High operational temperatures • Corrosive material use • Complex reaction kinetics
Fe-Cl	• Mature and well-known chemical process • Low operating temperatures • Use of low-cost materials	• Low energetic and exergetic efficiencies • Dimerization from $FeCl_3$ to Fe_2Cl_6 • R&D required to enhance yield in reverse Deacon reaction
Mg-Cl	• Low operating temperatures • Feasible to operate with thermal solar and nuclear power systems • High energetic and exergetic efficiencies • Reasonably low environmental impacts • Easy-to-handle feasible reactions	• Hydrolysis reaction to be optimized • Complex electrochemical reactions • MgO chlorination issues • Formation of azeotrope and $MgCl_2$ hydrates

opportunities and challenges in the four thermochemical cycles given in Figure 5.27. Additionally, Cu-Cl, Mg-Cl, and V-Cl cycles appear to be suitable choices for integration into almost all nuclear reactor types as long as the cycle temperature requirements are fulfilled.

For example, the Pickering Nuclear Generating Station (PNGS) has been in operation since 1971 and is located in Ontario, Canada. The plant is operated by Ontario Power Generation, a provincial electricity company. The net electrical output of the plant is approximately 500 MW and is a good representative of recent nuclear technologies. For detailed information on the PNGS system description and analyses, Dincer and Rosen [9] highlight the importance of transforming a nuclear power plant into cogeneration and multigenerational systems.

Figure 5.28a represents a nuclear-based advanced integrated system, as taken from Al-Zareer et al. [10] to give an example case study, which is using a super-critical water-cooled reactor as an energy generating source. A four-step Cu-Cl cycle produces hydrogen through the decomposition of water. In a hybrid thermochemical water decomposition cycle and hydrogen compression system, Rankine cycles generate power, part of which is used for electrolysis. Hydrolysis and oxygen production reactors are the only ones that receive thermal energy from the nuclear reactor in the proposed four-step thermochemical and electrical water decomposition cycle. A supercritical fluid delivers nuclear thermal energy to the integrated system. It is determined that the energy and exergy efficiencies of the proposed system are 31.6% and 56.2%, respectively. This means that the integrated system is able to produce 2.02 kg/s of highly compressed hydrogen and 553 MW of electrical power. Further details about this particular system are available in Al-Zareer et al. [10].

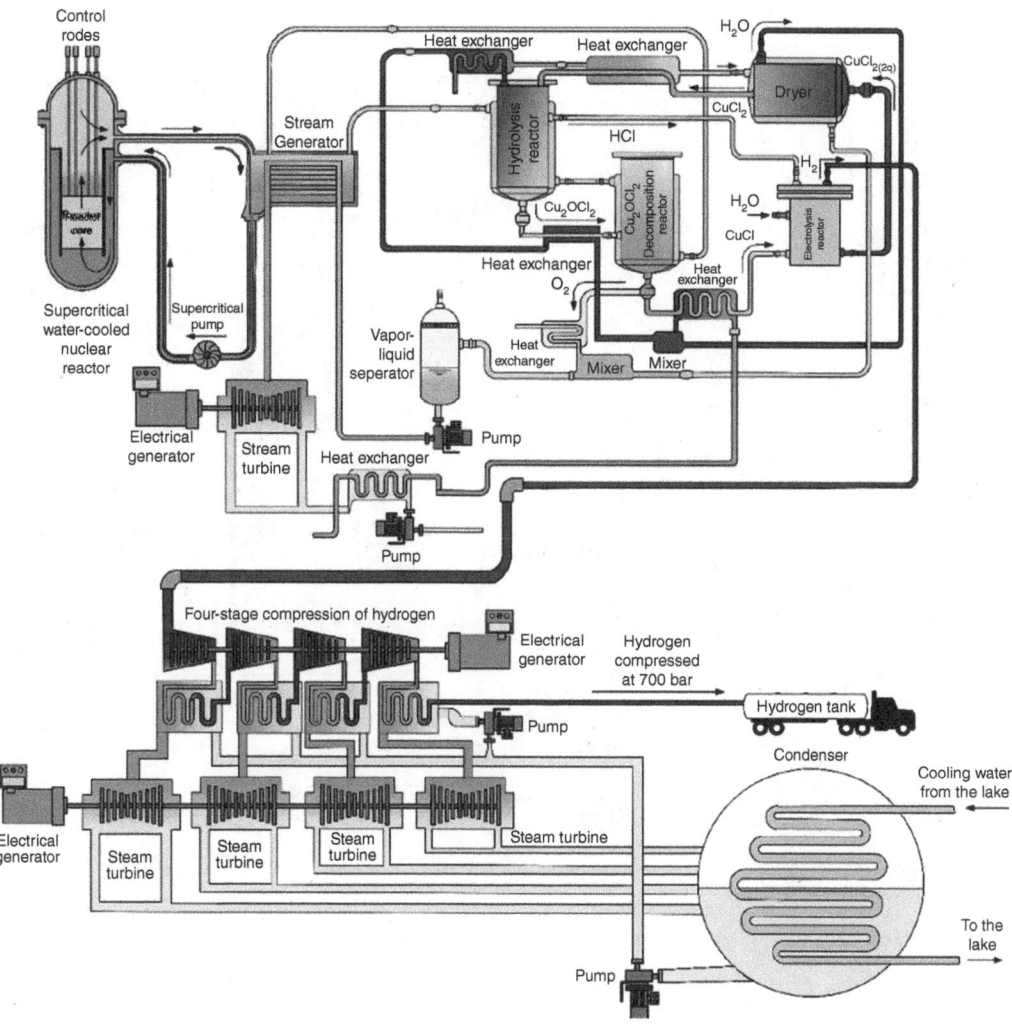

Figure 5.28a System layout for a nuclear power plant integrated with hydrogen production with Cu-Cl cycle (adapted from [10]).

Figure 5.28b depicts an integrated system for compressed hydrogen and electrical power production based on a Generation IV nuclear reactor (a lead-cooled reactor) as taken from Al-Zareer et al. [11] to give an example case study. An integrated system produces hydrogen by decomposing water thermochemically and electrically. Here, the copper and chlorine compounds decompose water through four main steps in the water decomposition cycle. As well as generating electrical power, the Rankine cycle also contributes to cooling compressed hydrogen between compression stages and provides the electrical power required for the electrolysis step in water decomposition. This system incorporates a heat recovery network into the water decomposition cycle, so that only the hydrogen production and oxygen production reactors receive thermal energy from lead-cooled nuclear reactors. According to energy and exergy analyses, the integrated system has an energy efficiency of 25.4% and an exergy efficiency of 40.6%. It produces 467.2 kW of electricity and 3.45 g/s of compressed hydrogen. Further details about this particular system are available in Al-Zareer et al. [11].

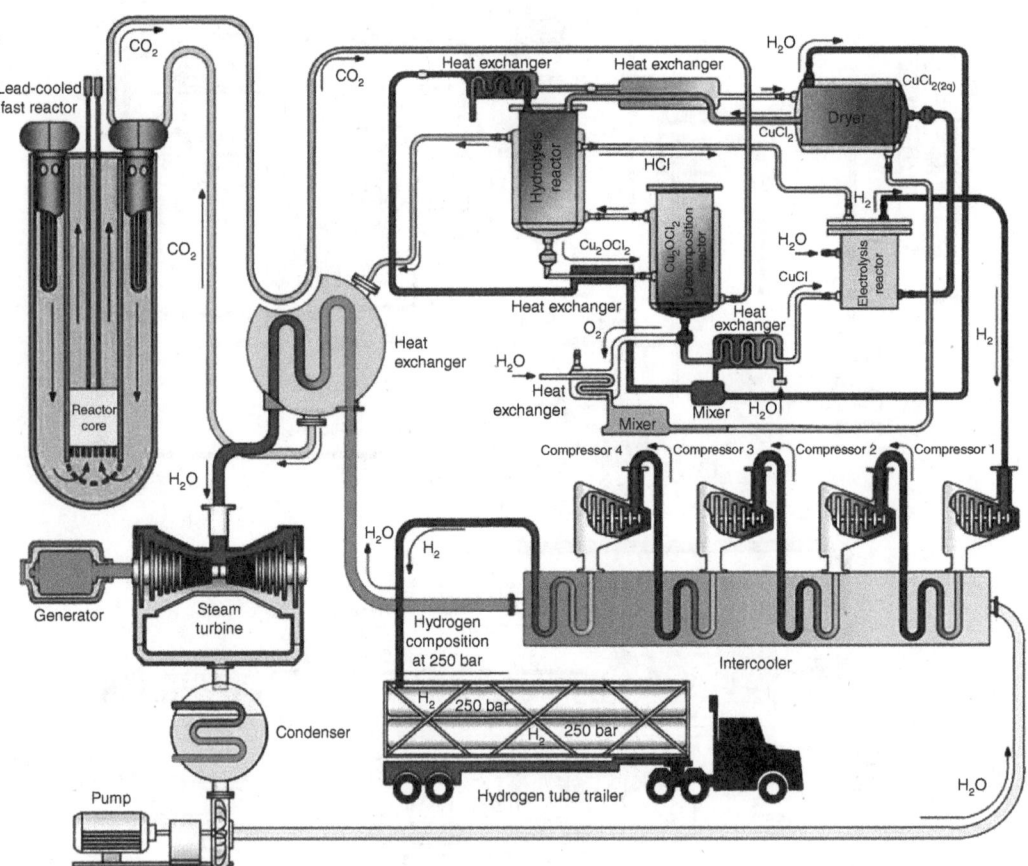

Figure 5.28b Schematic illustration of the nuclear electricity and hydrogen production system (Adapted from [11]).

5.11 Integrated Nuclear Energy Systems for Communities

Integrated systems, which consist of renewable and nuclear energy sources for multigenerational manner with energy storage options, are the key for the today's and future's energy-related issues. They also help to reduce the environmental problem. Integrated system ensures meeting more than two demands from a source. In a conventional nuclear energy plant, there is the only useful output which is electricity. Later, in order to benefit from the heat released in the nuclear power plant, nuclear cogeneration power plants have been deployed. In those systems, the heat is another useful output along with the electricity. However, today's technology offers more than electricity and heat as useful outputs. It is possible to obtain hydrogen, freshwater, cooling, etc. in addition to nuclear power plant cogeneration. They also include energy storage systems to better management of energy supply and demand profiles. This concept is illustrated in Figure 5.24 previously. As mentioned earlier, as the system is able to produce more useful commodities while keeping the input the same, it really tends to increase the system efficiency.

In nuclear power generation systems, there is a significant potential to use it as an energy source of integrated system to meet multiple demand items, especially for the communities. Communities have already defined demands, such as electricity, heat, cooling, ventilation, freshwater, food, etc. Nuclear energy, especially SMRs, can be used as only energy source, or it can be combined with other clean energy sources. Energy storage, of course, will be the part of the integrated system to balance and manage energy supply and demand profiles.

For example, a CANDU PHWR is located near Saugeen First Nation in Southampton, Ontario, Canada. In order to support the community, CANDU operator collaborating with Saugeen First Nation on various development projects and programs. The Saugeen Community is an indigenous community with a total of 810 inhabitants. Currently, food and freshwater projects are undertaken among the Saugeen First Nation. A multigenerational system was conceptually developed to address multiple needs of Saugeen Community, such as power, heat, freshwater, food, fuel. Besides nuclear energy, sun energy is used as another low carbon energy source by Temiz and Dincer [12]. Their proposed system has a parabolic trough concentrated solar plant along with the CANDU pressurized water reactor, in order to reach high temperatures for the Cu-Cl cycle. The summary of this example system is given in Table 5.5, which is illustrated in Figure 5.29 along with all system details, including all inputs and outputs.

Table 5.5 Summary of the system represented in Figure 5.29.

Location	Saugeen Community, Southampton, Ontario
Main goals	Food security, clean water, clean energy
Resources and fuels	Solar, nuclear, Lake Huron
Electricity generation methods	Steam turbines, PV plant
Heat consumption reasons	Food drying, greenhouse heating, district heating, fish farm heating
Heat generation methods	Nuclear based, solar based
Hydrogen generation method	Thermochemical Cu-Cl cycle
Freshwater generation	Multi-effect water desalination
Nuclear technology	CANDU PHWR
Solar technologies	Parabolic trough CSP, Bifacial PV

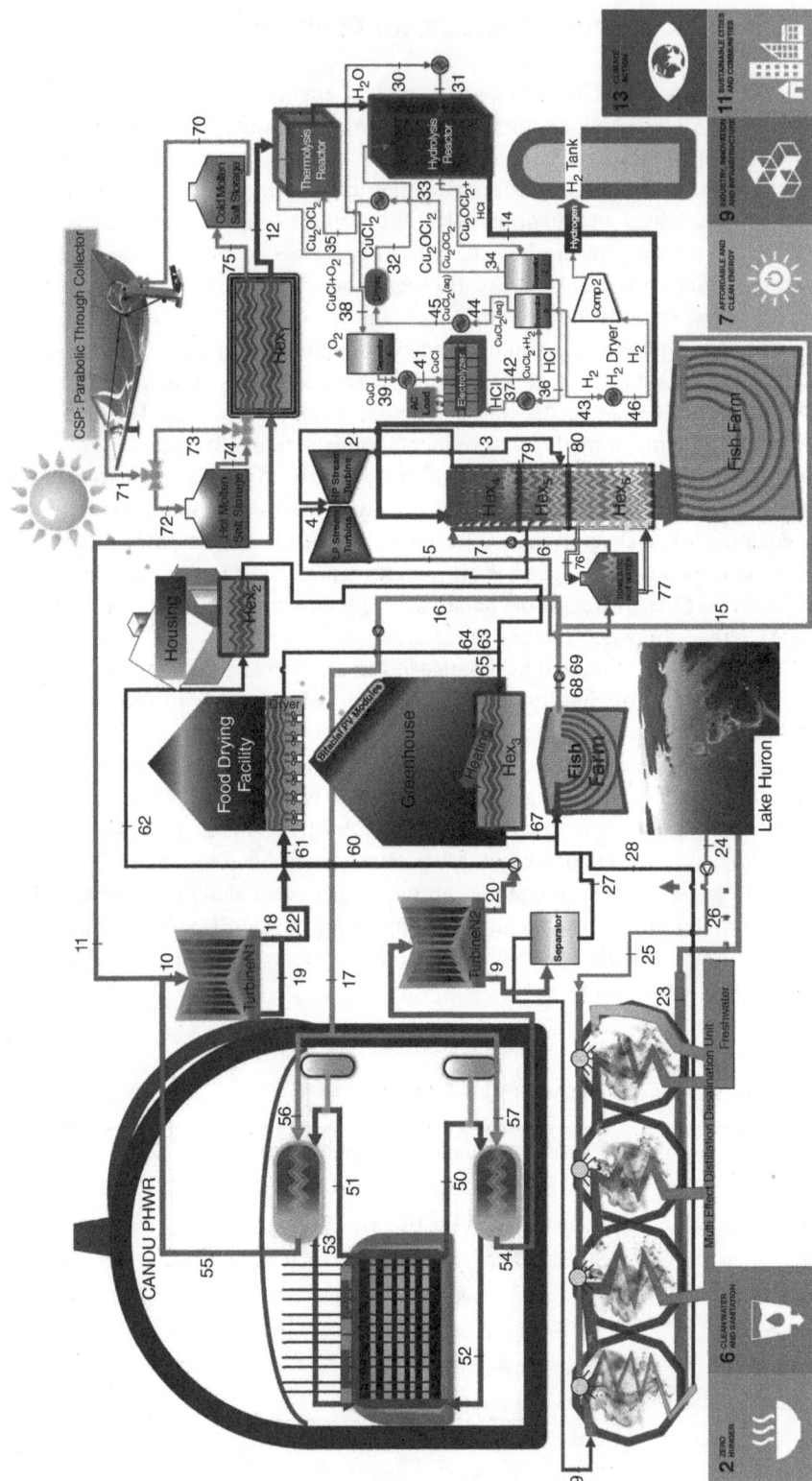

Figure 5.29 The schematic illustration of the nuclear-driven integrated system (adapted from [12]).

The capacities of each component are designed by considering the Saugeen Community's demand profile. A molten salt storage system is added within the concentrated solar plant, in order to operate the system with less fluctuations. Thermal energy storage systems with nuclear energy possess various advantages such as low cost, less pollution, better reliability, less complexity, maturity, and fewer safety issues, compared to the other energy storage methods. However, thermal energy storage has demerits such as higher energy losses overtime, less storage density, and corrosion.

The nuclear reactor unit in this study is a CANDU pressurized heavy reactor which generates 830 MWe electricity. The heat output from the nuclear reactor is enhanced via parabolic trough concentrated solar system with 86.2 GWh of additional heat. A flow chart is given in Figure 5.29 to illustrate the system operation. The rejected heat from the heavy water reactor is collected by two circuits. One of the circuits is used to generate power, and the other one is used in the thermochemical Cu-Cl cycle via heat upgrading by concentrated solar plant. The excess heat is recovered via various processes in order to produce useful outputs such as freshwater, food, space heating, domestic hot water, and dry food. The recovered heat was exploited to produce 1420 tons of salmonid fish and 41.7 tons of vegetables, 124,416 tons of freshwater in a typical meteorological year

The overall system efficiency of the nuclear plant is significantly improved. In the average ambient conditions, the overall energy and exergy efficiencies are found to be 65.8% and 40.1% (see Figures 5.30). The hydrogen is generated at 126.04 mol/s rate by the thermochemical Cu-Cl cycle. Table 5.6 summarizes technical details of CANDU PHWR case study for Saugeen Community.

The proposed system also addresses United Nation's six of the sustainable development goals, namely, (2) zero hunger, (6) clean water and sanitation, (7) affordable and clean energy, (9) industry, innovation, and infrastructure, (11) sustainable cities and communities, (13) climate action.

Here, we now consider another example case study with some specific system related details (Table 5.7) and energy and exergy results (Table 5.8) as illustrated in Figure 5.31 for another remote indigenous community. Nunavut Territory has a massive land area which is enough to make it world's 13th largest country. However, only 38,780 people live in the territory. Part of Nunavut is located in the Arctic. Weather conditions and massive land make logistics very challenging. The current energy infrastructure is diesel based and it requires fuel transportation. Distributed energy systems near communities would be more feasible, rather than a central energy system in Nunavut. Solar and nuclear-based energy systems can replace the current energy systems as proposed by Temiz and Dincer [13]. Moreover, their proposed system can address power and heating requirements with freshwater, food, and fuel needs of Nunavut Territory.

A high-temperature gas-cooled reactor type of SMR used for indigenous communities among the Nunavut. Their proposed system consists of steam Rankine cycle, closed-cycle gas circuit with a gas turbine and a thermochemical Cu-Cl cycle, multi-effect desalination unit, bifacial PV plant, greenhouse, residential, and fish farm heating systems and a domestic water heater. Helium gas cools the high-temperature SMR and exchanges its heat with carbon dioxide and steam. Carbon dioxide expands in the gas turbine and then supplies heat to the thermochemical Cu-Cl cycle. The pebble bed reactor design uses spherical fuel elements at 6 cm diameter, where 27,000 fuel elements can be contained by the core, which is 1.8 m in diameter for a 10 MWth HTGR-type SMR.

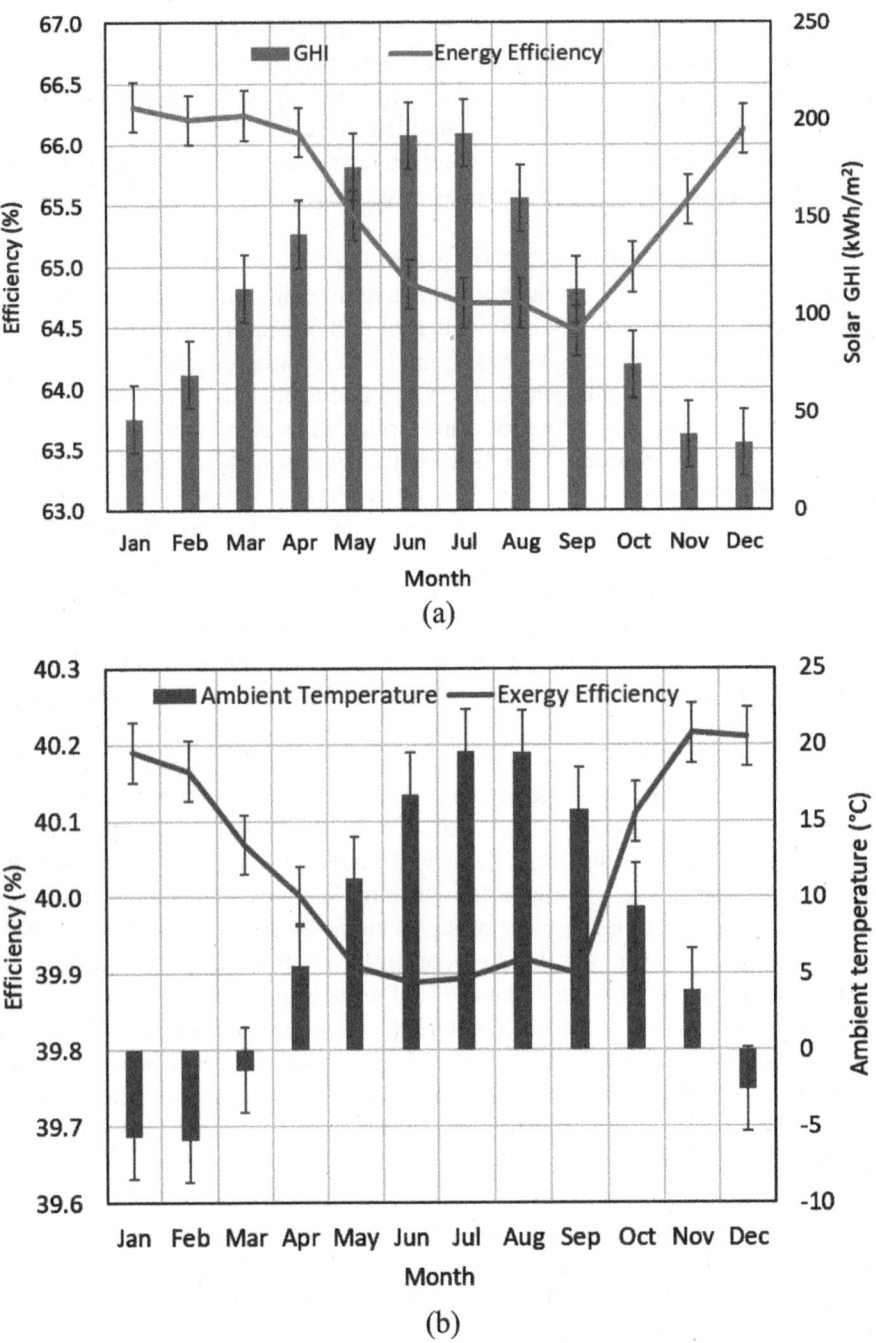

Figure 5.30 Energy and exergy efficiencies variation of the overall system for average monthly conditions: (a) with respect to solar radiation intensity and (b) ambient temperature (data from [12]).

Table 5.6 Technical Summary of the case study for the Saugeen Community.

Details	Amount	Unit
H_2 production rate	126.04	mol/s
Hydrogen storage pressure	365	bar
Hydrogen temperature	15	°C
Total exergy destruction rate	1645.00	MW
Net electricity production rate	586.00	MW
H_2 production LHV content rate	30,248.92	kW
Cu-Cl Q input	63,904.36	kW
Electrolyzer W input	18,655.80	kW
H_2 conversion efficiency	36.64	%
Molten salt storage capacity	651.00	MW_{th} h
Freshwater production	124,416.00	ton/year
Fish production	1420.00	ton/year
Agricultural production	41.74	ton/year
CSP plant power capacity	50.10	MWth
BiPV plant power capacity	280	kWp
Nuclear plant power capacity	565.60	MWp
Annual solar radiation GHI	1354.90	kWh/m^2 year
Global horizontal irradiance	1354.90	kWh/m^2 year
CSP: Reflective aperture area	178,880.00	m^2
BiPV: Module area	1405.00	m^2
CSP heat generation	86,157.97	MWh/year
BiPV electricity generation	430.60	MWh/year
CSP energy conversion efficiency	33.5	%
BiPV energy conversion efficiency	22.6	%
Overall energy efficiency	65.8	%
Overall exergy efficiency	40.1	%

Source: Temiz and Dincer [12].

A pilot system with 10 MWth HTGR-type SMR with 1.3 MWp bifacial PV plant in the ambient conditions of Iqaluit, Nunavut, can produce 97 tons of hydrogen, 2.1 tons of food, 12,003 tons of freshwater, 22.6 GWh of electricity. The overall energy and exergy efficiencies of their developed system for the average ambient conditions were found to be 62.64% and 68.91%. Their conceptually developed system's specific summary details and results are available in Tables 5.7 and 5.8.

Table 5.7 Details of the pilot system.

Pilot system location	Iqaluit, Nunavut Territory
Resources and fuels	Solar, enriched Uranium, sea water
Useful outputs	Electricity, hydrogen, food, freshwater, and space heating
Electricity generation methods	Rankine cycle, PV conversion
Hydrogen generation method	Cu-Cl thermochemical hydrogen production cycle
Freshwater generation	Multi-effect desalination
Nuclear technology	HTR-type pebble bed reactor
Solar power technology	Bifacial PV
Solar intensity (GHI)	980.6 kWh/m^2
Pilot system nuclear capacity	10 MW$_t$ HTR-type pebble bed reactor plant
Pilot system solar capacity	1.35 MWp Bifacial PV plant

Source: Temiz and Dincer [13].

Table 5.8 Result summary.

Details	Amount	Unit
H$_2$ production rate*	1.54	mol/s
Hydrogen storage pressure	36,000	kPa
Global exergy destruction rate*	2023.40	kW
Power production rate from turbines*	2354.31	MW
Cu-Cl Q input*	768	kW
Electrolyzer W input*	226	kW
The Cu-Cl cycle energy efficiency*	37.19	%
Freshwater production	120,003	ton/year
Solar radiation GHI	980.60	kWh/m^2 year
Total solar module (cell) area	7135 (6279)	m^2
Solar radiation GHI on array	6996.6	MWh/year
Converted solar energy	2368.6	MWh/year
Overall energy efficiency*	62.64	%
Overall exergy efficiency*	68.91	%

*: During steady-state operation at the average ambient conditions.

Source: Temiz and Dincer [13].

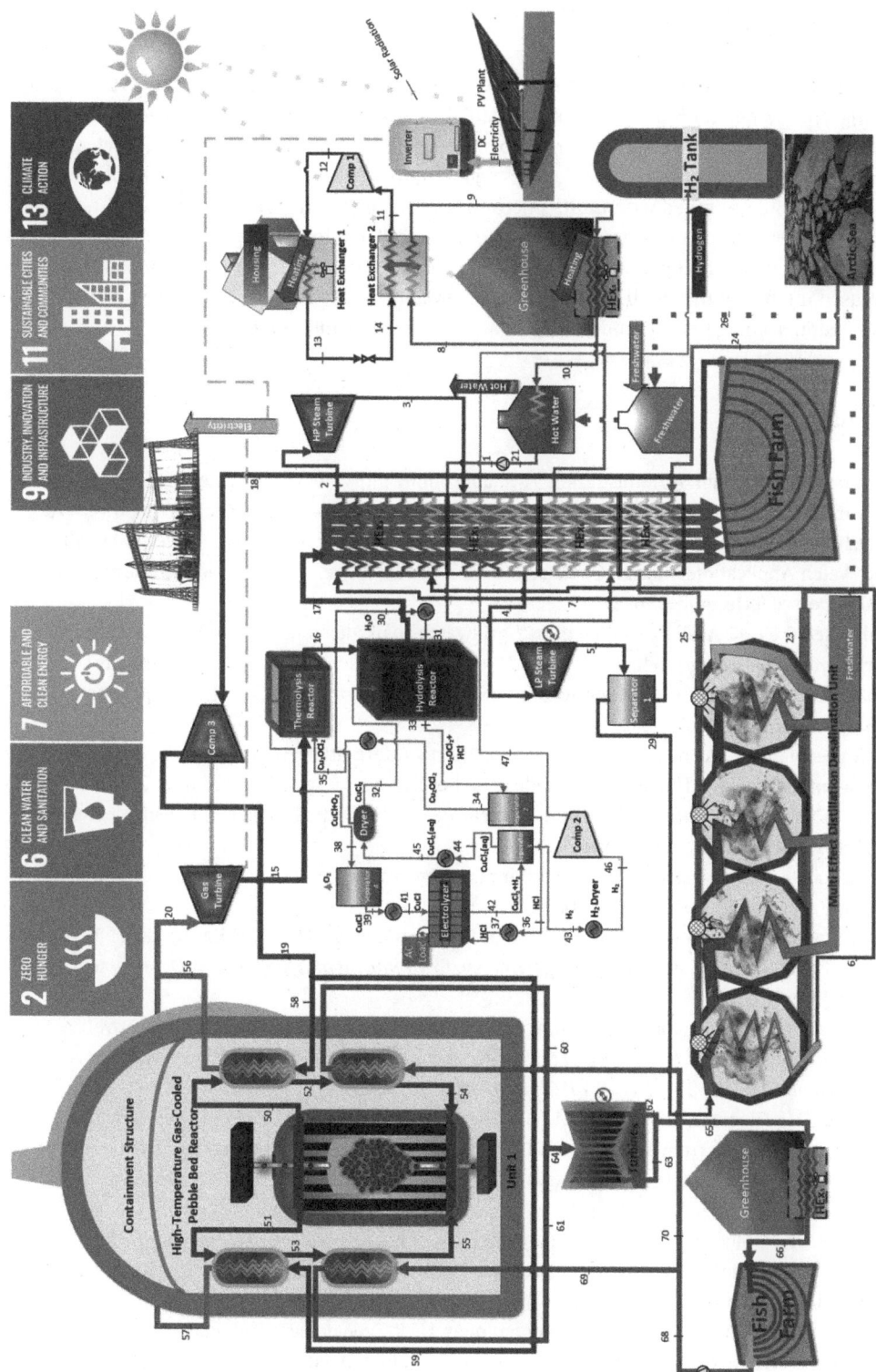

Figure 5.31 The schematic illustration of the nuclear-based integrated system for Nunavut community (adapted from [13]).

5.12 Closing Remarks

Nuclear energy has been one of the key sources for electricity production since 1951. Due to increasing energy demands in the world, it still has significant potential to meet the energy demands. There are many research attempts to better performance and more reliable nuclear reactors. Today, nuclear energy is not for electricity generation. It can be integrated with other clean energy sources and improved with multigenerational systems. In the nuclear power plant, we can potentially use excess power or process or waste heat available for hydrogen production. Gen IV reactors provide more efficient, effective sustainable operation for nuclear power plants. There is an increased interest in the small modular reactors system. Hybridization with renewables is very critical. Fission is going to be a kind of future target to develop and implement.

References

1 U.S. Energy Information Administration. Available online https://www.energy.gov/ne/articles/ nuclear-power-most-reliable-energy-source-and-its-not-even-close#:~:text=Nuclear%20power%20 plants%20are%20typically,or%20refueling%20at%20these%20facilities (accessed 10 December 2022).

2 Grid Watch Application Ontario, Canada. Available online https://live.gridwatch.ca/home-page. html (accessed 9 December 2022).

3 Grid Watch the U.K. Available online https://gridwatch.co.uk/demand (accessed 9 December 2022).

4 World Uranium Mining Production, World Nuclear Association. https://world-nuclear.org/ information-library/nuclear-fuel-cycle/mining-of-uranium/world-uranium-mining-production. aspx. (accessed 8 December 2022).

5 Nuclear explained, The nuclear fuel cycle. U.S. Energy Information Administration. https://www. eia.gov/energyexplained/nuclear/the-nuclear-fuel-cycle.php. (accessed 9 December 2022).

6 CNA (2023). Medical Isotopes, Canadian Nuclear Association. https://cna.ca/nuclear-medicine/ medical-isotopes/ (accessed 07 June 2023).

7 Encyclopedia Britannica. https://www.britannica.com/technology/nuclear-reactor/Liquid-metal-reactors (accessed 10 December 2022).

8 IAEA Nuclear Energy Series, Opportunities for Cogeneration with Nuclear Energy, No. NP-T-4.1.

9 Dincer, I. and Rosen, M.A. (2020). *Exergy: Energy, Environment and Sustainable Development*, 3e. Oxford: Elsevier.

10 Al-Zareer, M., Dincer, I., and Rosen, M.A. (2017a). Development and assessment of a novel integrated nuclear plant for electricity and hydrogen production. *Energy Conversion and Management* 134: 221–234. doi: 10.1016/j.enconman.2016.12.004.

11 Al-Zareer, M., Dincer, I., and Rosen, M.A. (2017b). Assessment and analysis of hydrogen and electricity production from a Generation IV lead-cooled nuclear reactor integrated with a copper-chlorine thermochemical cycle. *International Journal of Energy Research* 42: 91–103. doi: 10.1002/ er.3819.

12 Temiz, M. and Dincer, I. (2021). Design and analysis of nuclear and solar-based energy, food, fuel, and water production system for an indigenous community. *Journal of Cleaner Production* 314: 127890. doi: 10.1016/j.jclepro.2021.127890.

13 Temiz, M. and Dincer, I. (2021). Development of an HTR-Type nuclear and bifacial PV solar based integrated system to meet the needs of energy, food and fuel for sustainable indigenous cities. *Sustainable Cities and Society* 74: 103198. doi: 10.1016/j.scs.2021.103198.

Questions/Problems

Questions

1 Discuss the capacity factors for various energy sources.
2 Define the common types of nuclear energy.
3 Define the mechanism of nuclear fusion.
4 Describe the mechanism of nuclear fission.
5 Briefly explain what thermonuclear fusion is.
6 What are nuclear fuels?
7 Explain the steps of nuclear fuel production.
8 Classify the nuclear reactors based on the generational perspectives.
9 Define how a CANDU reactor works.
10 Discuss small modular reactors and their importance.
11 Explain working principles of a nuclear power plant.
12 Why nuclear cogeneration is important? Explain advantages and disadvantages with examples.
13 Explain the potential methods to produce hydrogen from a nuclear source.
14 How can nuclear energy be used for communities to meet multiple demand items?
15 Prepare a short essay about public acceptance of nuclear energy.

Problems

1 Consider a nuclear power plant which is to be designed to produce 10 MW of electrical power, as shown in the figure below. The efficiency of the electrical generator is known to be 95%. The thermodynamic properties of different state points are provided in the table below. Also, a total of 0.9 MW of thermal energy is obtained from the nuclear fuel at a fuel consumption rate of 0.052 g/h.
   a Write all mass, energy, entropy and exergy balance equations for all system components.
   b Calculate the actual fuel consumption rate in the designed power plant.
   c Calculate/obtain the missing data in the following table.
   d Find the overall energy and exergy efficiencies.

State	$\dot{m}$ (kg/s)	Temperature	Pressure	Specific Enthalpy (kJ/kg)
0		25	101.3	104.2
1	?	800	30,000	4016
2	?	600	29,950	3443
3	?		30,000	3450
5	10	50	30,000	1200
6	10		29,950	?
7	?		10	1305

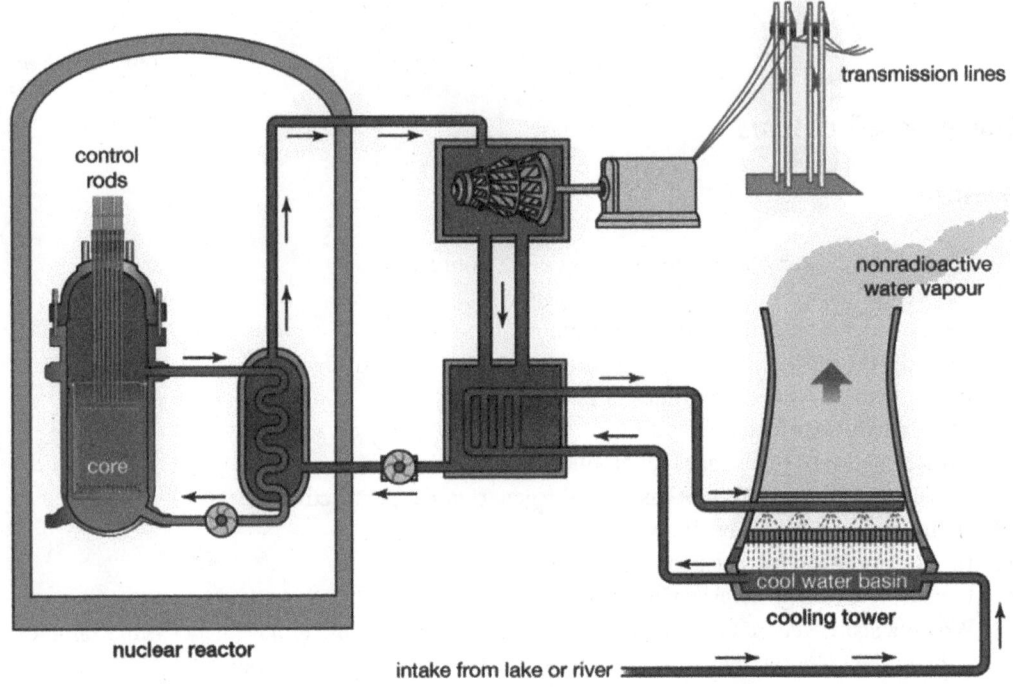

Adapted from [14].

2 A nuclear power plant consists of CANDU reactors, shown in the figure below. In the reactor, there 400 fuel channels, each heats heavy water at 7104.21 kW heating rate. The temperature of the steam in the generator is $T_{generator} = 670$ K. The following specific properties and parameters are known about the system.

$v_{f@10kPA} = 0.001010$ m^3/kg, $h_1 = 3001$ kJ/kg, $h_2 = 2800$ kJ/kg, $h_3 = 2584$ kJ/kg, $P_4 = 0.1$ bar, $P_5 = 65$ bar. $h_7 = 453.4$ kJ/kg, $h_8 = 1728.9$ kJ/kg, $h_9 = 1578$ kJ/kg, $s_1 = 6.233$ kJ/kg K, $s_2 = 6.343$ kJ/kg K, $s_3 = 8.15$ kJ/kg K, $ex_1 = 1158.9$ kJ/kg, $ex_2 = 925.4$ kJ/kg, $ex_3 = 174.2$ kJ/kg.

a Find mass flow rates of the heavy water and light water.
b Determine the power output of the nuclear power plant, $\dot{W}_{net}$.
c Find energy and exergy efficiencies of the nuclear power plant.

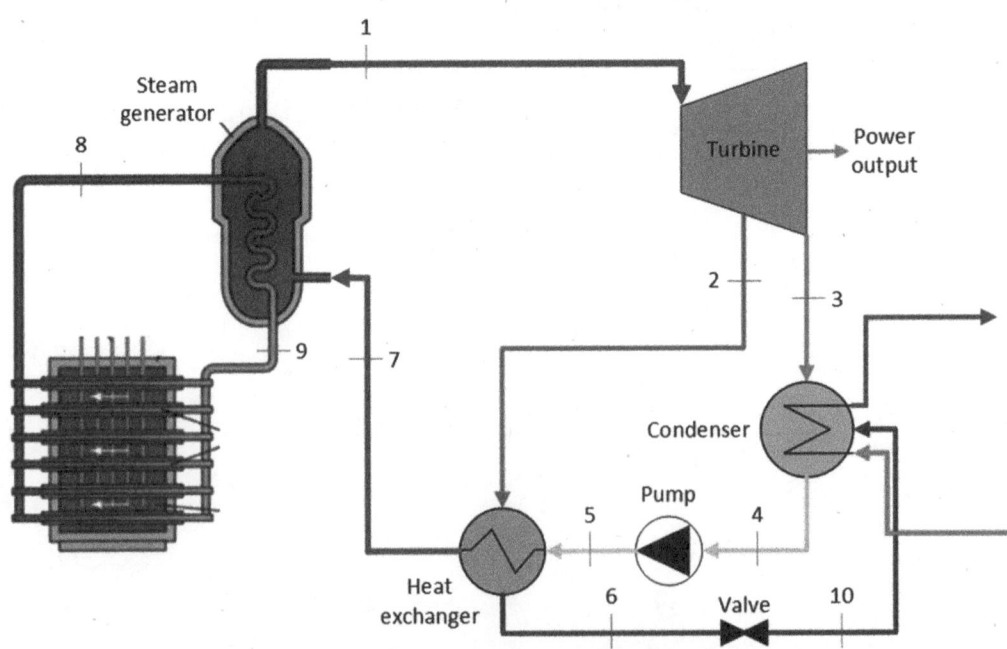

3 Consider the nuclear power plant in problem 1 which is a single-generation power plant as it produces only electricity. An engineering team wants to upgrade the power plant by adding additional useful outputs which are heat and hydrogen. The production capacities of each commodity are given as follows:

Heat: a total of 75 MW of thermal energy which is extracted as a useful output.
Hydrogen: the mass flowrate of hydrogen produced is 100 kg/h.

Note that the heat recovery is done at the temperature of 150 °C and lower heating value of hydrogen is 120 MJ/kg and the chemical exergy of the hydrogen is 138 MJ/kg, calculate the energy and exergy efficiencies for the nuclear-based multigenerational system. Compare the results with the efficiencies obtained for the conventional single-generation plant.

4 In order to improve efficiency of the nuclear power plant given in problem 2, it is requested to add hydrogen, heat, and freshwater as the useful outputs. A total of 2000 kg of hydrogen will be produced daily. The heat output will be 20 MW, and 10,000 kg of freshwater will be produced daily. Calculate both energy and exergy efficiencies for the system.

5 Design an integrated energy system where both nuclear and renewable energy sources are considered to cover the needs of a community in terms of power, heat, hot water, cooling, freshwater and hydrogen. See the system illustrated in Figure 5.31 to get some ideas.

6

Solar Energy Systems

6.1 Introduction

Solar energy, which is recognized as the prime energy source of universe, is the first energy source as humanity begun with. It can be assumed as the first renewable energy source used by people as it has been used for thousands of years by people for heating purposes for space and water, and drying purposes for vegetables, fruits, and meats. Today, solar energy technologies are recognized as one of the major and most mature renewable energy technologies. Interest in solar energy and its applications has significantly risen since the petrol crisis in 1973, especially in using solar heating and hot water systems. After the petrol crisis, the countries which did not have fossil fuel resources have begun deploying solar domestic hot water and space heating systems, such as Turkey, Greece, Spain, Croatia, Italy, etc. So, the industrial sector for solar energy system began with the solar domestic hot water and space heating systems.

Power generation by converting sunlight into electricity has primarily started directly using photovoltaics (PV) and concentrated solar power (CSP). The first PV-based megawatt-scale power plant has been built in California, USA, in 1982. On the other hand, the first commercial CSP, also called solar thermal power plants, were built in Sevilla, Spain, in 2007; the capacity of the power plant is 11 MW. The development of the solar energy sector is shown in Figure 6.1, based on some selected achievements and events. As indicated in this figure, there are many milestones in the progress of solar energy, but not limited to what are listed there. Today, the solar power plants meet a total of 2% of the global energy demand which corresponds to the energy of 2702 TWh, as depicted in Figure 6.2. Today, solar energy systems are really reasonable energy systems in terms of cost and performance. Therefore, it is expected that the deployment of the solar energy systems will grow significantly due to environmental concerns and carbon neutral targets comes with the Paris Agreement.

Worldwide usage of solar energy varies greatly by country, with the top 10 countries representing approximately 74% of the photovoltaic market. As of 2021, China has the largest solar energy capacity in the world at 306,973 MW, which produces roughly 4.8–6% of the country's total energy consumption. It is followed by the United States at 95,209 MW and Japan at 74,191 MW.

The basics of solar energy systems are similar to any energy systems, which covers use of energy source, systems for conversion into the useful output(s) and their utilization. This concept has been defined as the 3S concept in Chapter 1. Figure 6.3 now illustrates the 3S concept for solar energy being used as a source. In solar energy systems, the sun is the only energy source if it is not combined with the other energy sources like wind, fossil fuels, grid electricity, etc. In order to

Introduction to Energy Systems, First Edition. Ibrahim Dincer and Dogan Erdemir.
© 2023 John Wiley & Sons, Inc. Published 2023 by John Wiley & Sons, Inc.
Companion Website: www.wiley.com/go/Dincer/Introduction_to_Energy_Systems

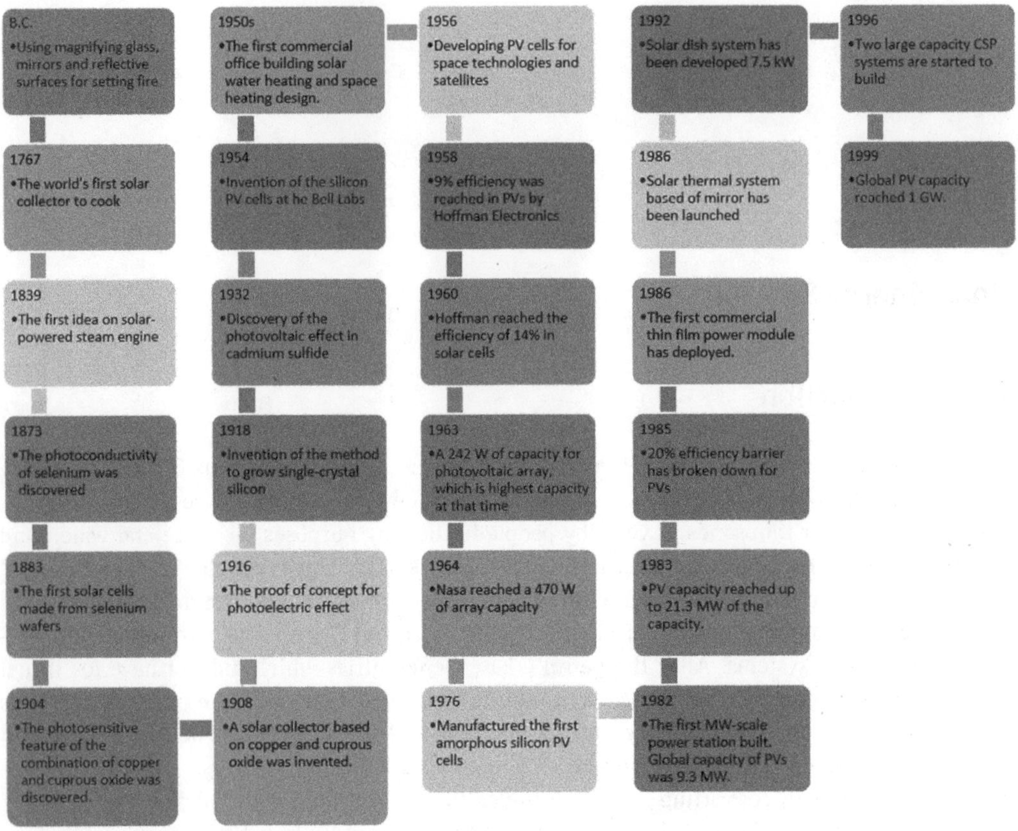

B.C.
•Using magnifying glass, mirrors and reflective surfaces for setting fire
1767
•The world's first solar collector to cook
1839
•The first idea on solar-powered steam engine
1873
•The photoconductivity of selenium was discovered
1883
•The first solar cells made from selenium wafers
1904
•The photosensitive feature of the combination of copper and cuprous oxide was discovered.
1908
•A solar collector based on copper and cuprous oxide was invented.
1916
•The proof of concept for photoelectric effect
1918
•Invention of the method to grow single-crystal silicon
1932
•Discovery of the photovoltaic effect in cadmium sulfide
1954
•Invention of the silicon PV cells at he Bell Labs
1950s
•The first commercial office building solar water heating and space heating design.
1956
•Developing PV cells for space technologies and satellites
1958
•9% efficiency was reached in PVs by Hoffman Electronics
1960
•Hoffman reached the efficiency of 14% in solar cells
1963
•A 242 W of capacity for photovoltaic array, which is highest capacity at that time
1964
•Nasa reached a 470 W of array capacity
1976
•Manufactured the first amorphous silicon PV cells
1982
•The first MW-scale power station built. Global capacity of PVs was 9.3 MW.
1983
•PV capacity reached up to 21.3 MW of the capacity.
1985
•20% efficiency barrier has broken down for PVs
1986
•The first commercial thin film power module has deployed.
1986
•Solar thermal system based of mirror has been launched
1992
•Solar dish system has been developed 7.5 kW
1996
•Two large capacity CSP systems are started to build
1999
•Global PV capacity reached 1 GW.

Figure 6.1 The historical milestones in the development of solar energy sector.

Biofuels, 1140, 1%
Solar, 2702, 2%
Wind, 4872, 3%
Other renewables, 2373,1%
Hydropower 11183, 6%
Traditional biomass, 11111, 6%
Nuclear, 7031, 4%
Coal, 44473, 25%
Natural gas, 40375, 23%
Oil, 51170, 29%

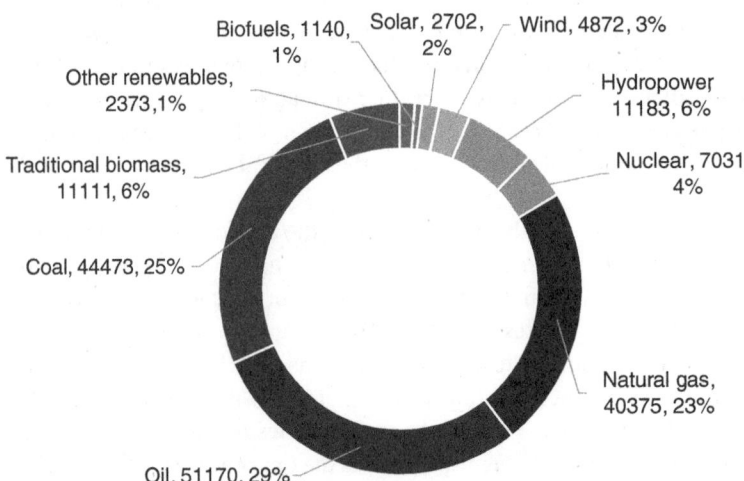

Figure 6.2 The global power generation capacity for 2021 (data from [1]).

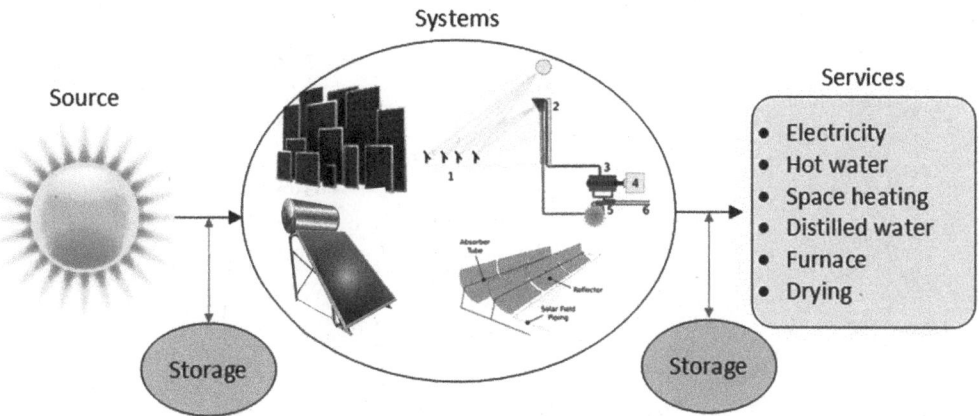

Figure 6.3 3S concept for solar energy systems.

convert solar energy into useful energy, solar collectors, PV panels, and solar towers are used. It is converted into either heat or electricity, or both. While it is possible to use these outputs directly to meet the demands, it can be converted into any form of energy. For example, solar thermal power plant, solar energy is firstly converted into heat, and heat generated is used for steam production in the Rankine cycle for power generation. The services to be able to be obtained by solar energy includes but are limited to electricity, hot water, space heating, distilled water, heat for cooking purposes, and heat for agricultural drying processes.

As illustrated in Figure 6.3, energy storage is the essential part of most solar energy applications due to intermittent and fluctuating availability profile of solar energy. Energy storage may be part of solar energy systems either to compensate for the power output when solar radiation is not adequate or to continue meeting the energy demand during the nighttime. Because it is not available all day and it has fluctuating available thought a year. This is forced to integrate the energy storage options to continue benefiting from the solar energy when it is adequate for available. While heat storage techniques such as hot water tanks, latent heat storage mediums are used for solar energy in the form of heat, battery technologies, hydrogen and mechanical energy storage methods are used for storing solar energy especially for electricity outputs. It is also possible to use energy storage techniques for long-term application. For instance, solar energy can be stored as heat in various heat storage mediums such as sand, rock beds, phase change materials, etc. for winter. In order to continue to use solar energy during the wintertime, it should be stored during the summertime when solar radiation has reached its highest values.

The applications of the solar energy systems, especially its photovoltaic applications, have increased in the last decade. Of course, the environmental concerns and the technical, geographical and political challenges and uncertainties regarding fossil fuels have played a significant role in decreasing the interest; the other main reason is substantially reducing installment cost of solar energy systems. Figure 6.4 demonstrates the changes in solar energy costs for various countries between 2010 and 2021. For all countries considered, there is a significant decrease in the solar PV investment costs. The solid black line shows the average of the 15 counties given in Figure 6.4. When a regression analysis is performed for it, it is clearly seen that the reduction in solar energy cost has occurred exponentially. The cost changes in solar PV systems are correlated as Average Cost = $4{\times}10^{15}{\times}e^{-0.168(\text{Year})}$. Such a reduction trend in solar energy costs is almost

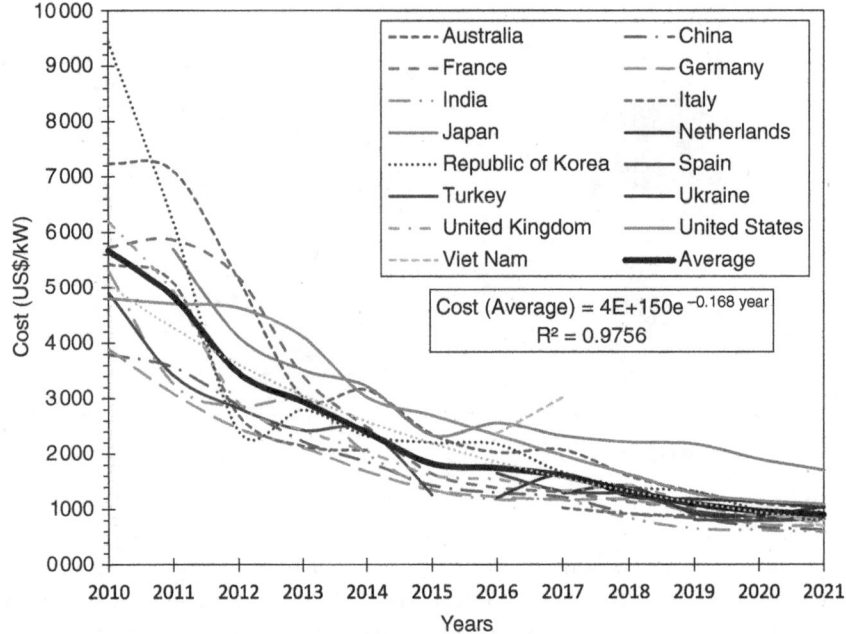

Figure 6.4 Utility-scale PV total installed cost trends for selected countries between 2010 and 2021 (data from IRENA [2]).

identical for all counties. It is expected that the reduction trend in the solar energy installment costs will continue. So, solar energy deployment all around the world will be feasible.

This chapter particularly aims to provide the basics on solar energy systems and applications. Firstly, solar radiation aspects are presented along with key measurement devices and techniques. Then, solar energy applications are introduced with practical examples. This is followed by the introductions and analyses of the solar energy thermal and PV systems. Finally, photovoltaic thermal (PVT) systems are discussed. There are numerous examples and applications presented to better reflect the content and provide a true understanding.

6.2 Atmospheric and Direct Solar Radiation

Before entering solar energy applications, we first need to learn about solar radiation and its measurement techniques, as both subjects are quite important for solar energy systems. For solar energy systems, there are two common types of radiation which are solar and atmospheric radiation. When solar radiation hits a surface, while a part of it is absorbed by the surface, the rest of it is emitted.

There are also two types of radiation which are atmospheric and direct solar radiation. Direct solar radiation is the radiation coming from the sun directly with an angle. At the first glance, we can say that the solar collectors or receivers should get solar radiation in a direct form with an angle. Atmospheric radiation is about the radiation coming from the sky and clouds, which happens due to reflected solar radiation from earth surface and clouds.

In order to understand what atmospheric radiation and direct radiation are, one needs to check Figure 6.5. As seen from this figure, a direct solar radiation comes to the surface directly with a

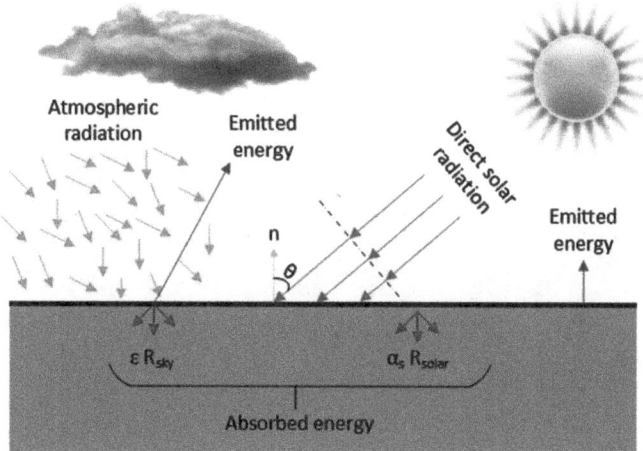

Figure 6.5 Illustration of direct solar and atmospheric radiation.

specific angle, while atmospheric radiation occurs due to the direct solar radiation reflected from the surfaces and solar radiation emitted by the clouds and atmospheric gases.

The total solar energy (direct and atmospheric) incident on the unit area of a horizontal surface on the ground is shown in Figure 6.5. It can be formulated as follows:

$$R_{solar} = R_D \cos\theta + R_d \tag{6.1}$$

where R_D is the direct solar radiation, R_d is the diffuse solar radiation, and θ angle of incidence of direct solar radiation.

The radiation emission from the atmosphere to the earth's surface is calculated as follows:

$$R_{sky} = \sigma T_{sky}^4 \ \left(W/m^2\right) \tag{6.2}$$

where σ is the Stephan-Boltzmann constant and $\sigma = 5.67 \times 10^{-8}$ W/m²K⁴. T_{sky} is the effective sky temperature ranges from 230 K for cold, clear-sky conditions to about 285 K for warm, cloudy-sky conditions.

A part of the solar radiation is absorbed by the sky. The sky radiation absorbed by a surface is calculated as follows:

$$E_{sky,absorbed} = \sigma R_{sky} = \alpha \sigma T_{sky}^4 = \varepsilon \sigma T_{sky}^4 \ \left(W/m^2\right) \tag{6.3}$$

The net radiation heat transfer (energy transfer) rate to a surface exposed to solar and atmospheric radiation (as illustrated in Figure 6.5) is written as

$$\begin{aligned}
\dot{E}_{net} = \dot{R}_{net,rad} &= \sum E_{absorbed} - \sum E_{emitted} \\
&= E_{solar,absorbed} + E_{sky,absorbed} - E_{emitted} \\
&= \alpha_s G_{solar} + \varepsilon \sigma T_{sky}^4 - \varepsilon \sigma T_s^4 \\
&= \alpha_s G_{solar} + \varepsilon \sigma \left(T_{sky}^4 - T_s^4\right)
\end{aligned} \tag{6.4}$$

where α is absorptivity of the surface and ε is the emissivity of the surface.

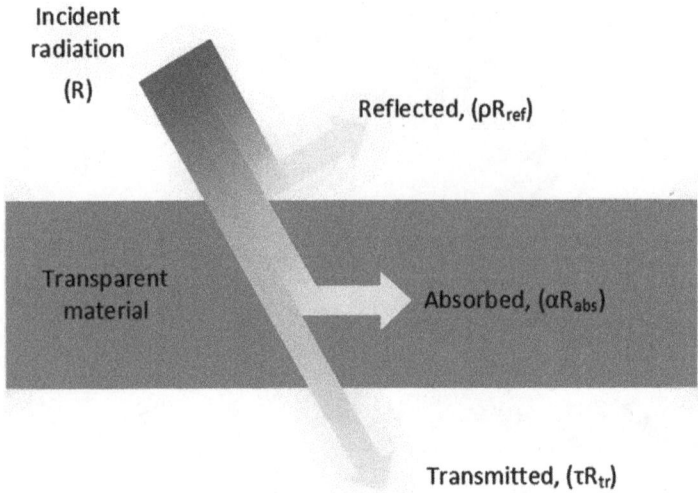

Figure 6.6 Radiation transmitted through a transparent material.

Many solar collectors and PV panels have a transparent cover on it. Therefore, solar radiation transmission should be considered. Figure 6.6 shows the adsorption, reflection, and transmission of incident radiation modes. When solar radiation comes to a transparent material, a part of it is absorbed by the material and its temperature increases. A part of it is reflected by the surface and the rest is transmitted. Absorption, reflection, and transmission are called the surface features of the material. Each can be written as follows:

$$\alpha = \frac{Absorbed\ radiation}{Total\ radiation\ input} = \frac{G_{abs}}{G}, \qquad 0 \leq \alpha \leq 1 \tag{6.5}$$

$$\rho = \frac{Reflected\ radiation}{Total\ radiation\ input} = \frac{G_{ref}}{G}, \qquad 0 \leq \rho \leq 1 \tag{66}$$

$$\tau = \frac{Transmitted\ radiation}{Total\ radiation\ input} = \frac{G_{tr}}{G}, \qquad 0 \leq \tau \leq 1 \tag{6.7}$$

From an energy balance for an object which is subjected to the solar radiation:

$$R = R_{abs} + R_{ref} + R_{tr} \tag{6.8}$$

and

$$\alpha + \rho + \tau = 1 \tag{6.9}$$

Here, it should be noted about black body behavior in the solar radiation. A black body (or black-body) is considered an ideal body that absorbs all incident radiation at any angle and frequency. It is named as "black body" because it absorbs all colors of light, nothing emitted from the body or transmitted. So, for a black body, α will be 1, ρ and τ will be 0 ($\alpha = 1$, $\rho = 0$ and $\tau = 0$).

Illustrative Example 6.1   Consider a solar thermal collector which is covered with a glass cover. The surface properties are specified as $\alpha = 0.1$ and $\rho = 0.2$. The surface area of the solar panel is 1 m². Calculate the transmitted, absorbed, and reflected energies in the case of 1000 W/m² of solar radiation.

Solution:
The surface properties are provided to be $\alpha = 0.1$ and $\rho = 0.2$. Here, we can find the transmissivity coefficient from Eq. 5.9.

$$\alpha + \rho + \tau = 1$$
$$0.1 + 0.2 + \tau = 1$$

so, the transmittivity is found as $\tau = 0.7$.
Also, the absorbed radiation is found from Eq. 5.5 as follows:

$$\alpha = \frac{G_{abs}}{G} \rightarrow 0.1 = \frac{G_{abs}}{1000}$$

So, the absorbed energy can be found as:

$$G_{abs} = 100 \ W/m^2$$

The reflected energy can be calculated as:

$$\rho = \frac{G_{ref}}{G} \rightarrow 0.2 = \frac{G_{abs}}{1000} \rightarrow G_{abs} = 200 \ W/m^2$$

The transmitted energy can be calculated from Eq. 5.8.

$$G = G_{abs} + G_{ref} + G_{tr}$$
$$1000 = 100 + 200 + G_{tr}$$

So,

$$G_{tr} = 700 \ W/m^2$$

6.2.1   Solar Radiation Measurements

After learning about solar radiation, the key question is how solar radiation can be measured. As solar radiation is the input parameter for the solar energy systems, it is required to measure it accurately to calculate and optimize the subject matter solar energy systems. The solar radiation can be specified by measuring the solar radiation (power per area at an instant, W/m²) and via solar insolation (energy delivered per area). Pyrheliometers, net radiometers, and pyranometers are some common devices used for measuring solar radiation.

In practical applications, pyranometers are widely used for measuring both direct and diffuse solar radiation. Figure 6.7 shows different pyranometers which are used in practical applications for both commercial and research purposes. When sunlight falls on a pyranometer, the thermopile sensor produces a proportional response typically in 30 s or less: the more sunlight, the hotter the sensor gets and the greater the electric current it eventually generates. The thermopile is designed to be precisely linear (so a doubling of solar radiation produces twice as much current) and also

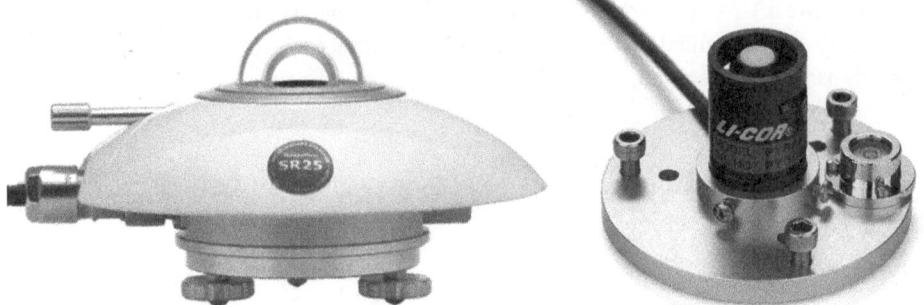

Figure 6.7 Photos for the common pyranometers used in the practical applications.

has a directional response: it produces maximum output when the Sun is directly overhead (at midday) and zero output when the Sun is on the horizon (at dawn or dusk). This is so-called a cosine response (or cosine correction) because the electrical signal from the pyranometer varies with the cosine of the angle between the sun's rays and the vertical surface. Pyranometers generally generates a μA output with the changing solar radiation like how PV panels convert solar energy into electricity. The current generated is converted to a voltage via a shunt resistor. That voltage information is converted to the solar radiation data via data acquisition system.

Pyrheliometer (as shown in Figure 6.8) is used for instantaneous measurement of direct solar irradiance from 300 to 4000 nm at normal incidence. Such devices are capable of making measurements with high accuracy and high stability. When used with colored glass broadband pass filter we get spectral distribution of direct solar irradiance. The unit of irradiance is watts per square meter (W/m^2). A net radiometer measures the solar radiation by differentiating the incoming and outgoing short-wave radiations via two thermopile pyranometers.

As mentioned earlier, the incident angle of solar radiation is significant for solar systems. In order to increase the effectiveness of solar energy, or increase solar energy input, solar collectors and receivers should be positioned as possible vertically to the solar incoming radiation. Figure 6.9 shows the comparison of global horizontal radiation and direct normal radiation in a generic form for a selected location.

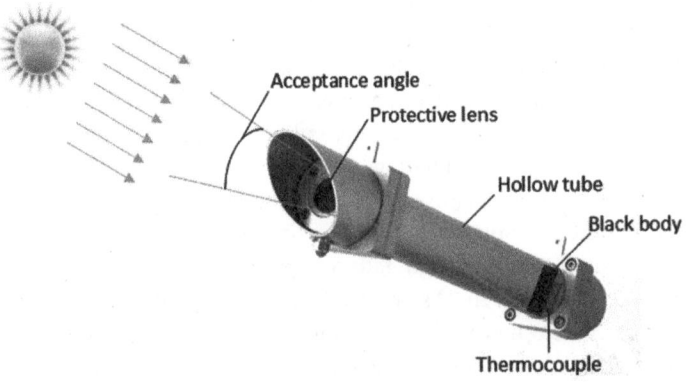

Figure 6.8 The view of the pyrheliometer.

Figure 6.9 Yearly average of the global horizontal radiation and direct normal radiation for a selected location.

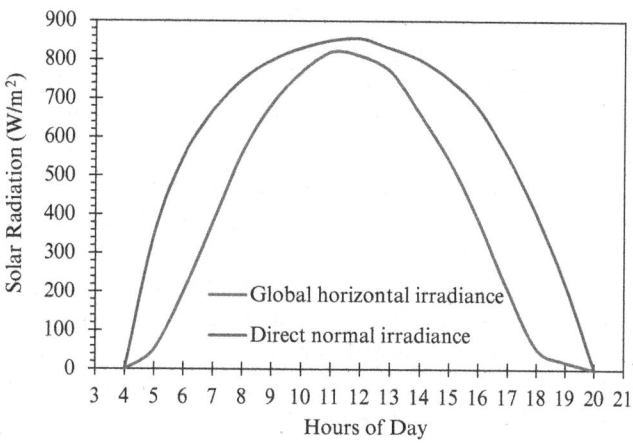

6.2.2 Solar Radiation Potential

Regarding fossil-fueled energy systems, energy and the environment are two critical subjects and sit on opposite sides of the same coin. In other words, energy generation means environmental problems, specifically CO_2 emissions and global warming potential. Global issues have forced people to use renewable energy sources. Solar energy was one of the significant candidates as a renewable energy source, and today it has ever-increasing capacity in the world. In order to demonstrate the potential of solar energy, the illustration provided in Figure 6.10 is prepared for a common use. Thus, Figure 6.10 shows the potential of solar energy compared to other common renewable and nonrenewable energy sources. It is clear that the potential of solar energy is the highest to meet all

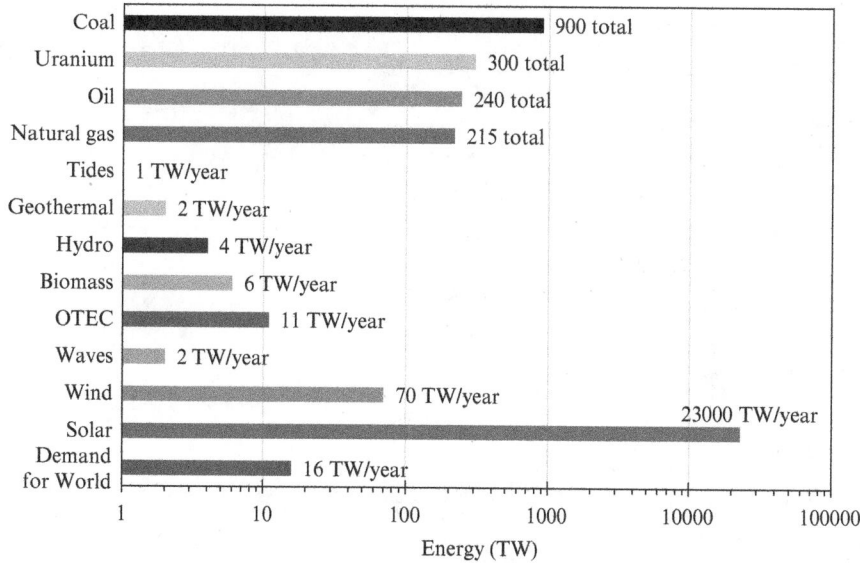

Figure 6.10 The potential of various energy sources to meet the global energy demands (data from [3]).

energy demand of the world today and the future, as solar radiation, also called solar input is almost 10,000 times higher than the world's energy demand. Also, it is quite high compared to other renewable energy sources and fossil fuel sources. However, there are certain restrictions on benefiting from solar energy, such as the efficiency of the solar collectors and PV panels, environmental and geographical conditions, intermittent availability, and also some side effects etc. Although it has big potential to meet the energy demands, it is not the only solution as a clean energy source. Therefore, all clean energy sources should be considered together in a multigenerational manner with energy storage options, as demonstrated in Figure 6.10.

Solar radiation input depends on the geographical conditions. Therefore, each location in the world takes different solar radiation. Figure 6.11 demonstrates solar irradiation in the world. Around the world, solar thermal collectors are widely used in the zones where are between 25° and 45° of latitude that spans both the Northern and Southern Hemispheres. This zone is called the solar belt zone. However, PV panels can be used for more extended areas. In the past a few decades, the counties which have fossil fuel sources did not pay serious attention to solar energy, such as such as Iran, Saudi Arabia, Kuwait, Mexico, etc. However, today, the entire world is trying to benefit from solar energy and other renewables due to lowering initial costs, increasing performance, and longer lifetimes.

Solar energy maps are very useful on any scale from the world to a specific location to see the potential of solar energy for possible investments. However, only the direct or horizontal solar energy maps are not enough to see the potential of the solar energy systems. In order to clarify it, solar exergy maps which are specific to the different applications should be prepared accordingly. The exergy maps provide more details for the solar energy system and better selection criteria as they are specific to application and considering to exergy destructions. Joshi et al. [4] have developed novel solar exergy maps for photovoltaic thermal (PVT) systems for various locations by considering the solar radiation potential available on the selected locations and their conversion into useful outputs, such as electricity and heat, through the respective efficiencies.

Figure 6.12 demonstrates performance of PVT system in terms of exergy efficiency for different American climatic conditions for different month. In Figure 6.12, the solar exergetic values for

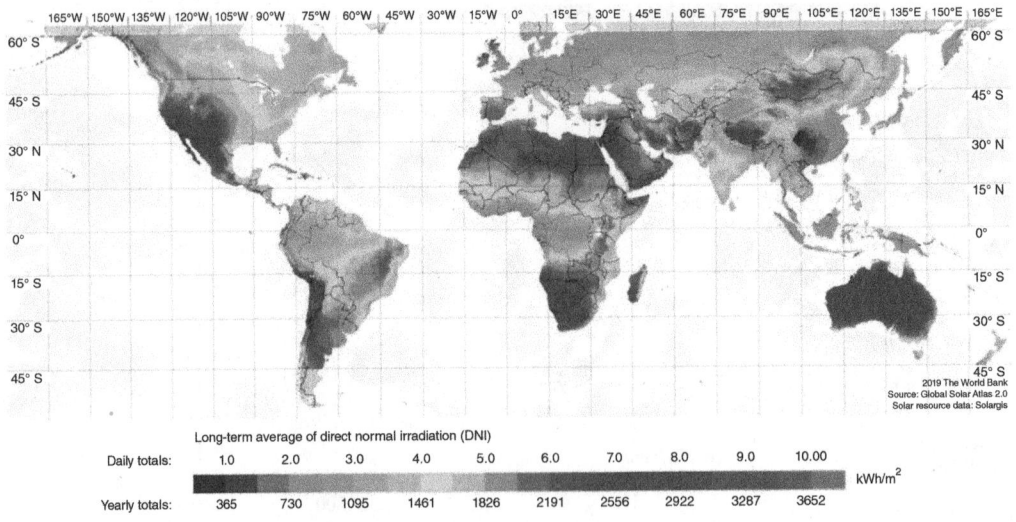

Figure 6.11 Direct normal solar radiation in the world (adapted from [4]).

Chicago, Las Vegas, Miami, New York, Portland, San Antonio, San Francisco, Tucson, and Tulsa for different months of January, April, June, and October. In order to obtain those maps, the average total solar radiation and ambient temperature data obtained from The National Renewable Energy Laboratory (NREL). In Figure 6.12, a negative sign convention is adopted for the west longitude for different USA cities. It is clear from the same figure that the PVT system gives best performance in terms of exergy efficiency in Las Vegas (with 32%) and Tucson (with 32.5–31.5%) in April and June as compared to other above-mentioned USA cities as the average total solar radiation received in both cities is more in respective months. A lower value of exergy efficiency in Las Vegas (with 21.5%) and Tucson (with 23%) is seen in January due to less solar radiation received on PV/T surface and a reasonable value of exergy efficiency in Las Vegas (with 25.2%) and Tucson (with 26%) is observed in October.

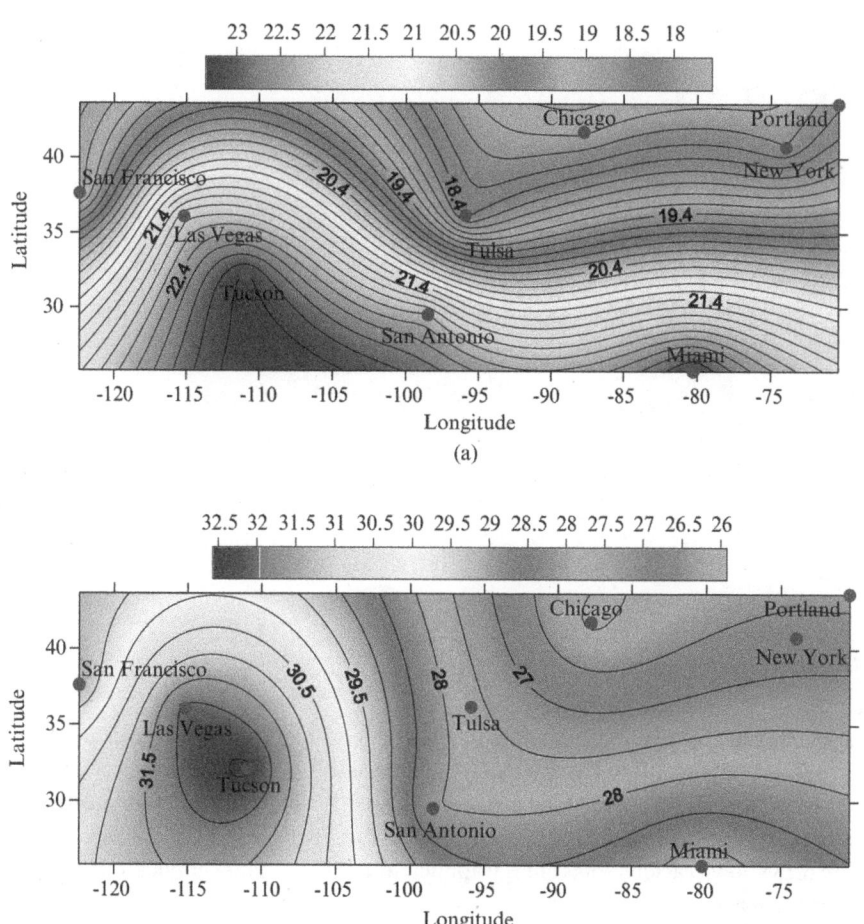

Figure 6.12 Solar exergy maps for PVT applications for different American climatic conditions for different month: (a) January, (b) April, (c) June, and (d) October (adapted from [5]).

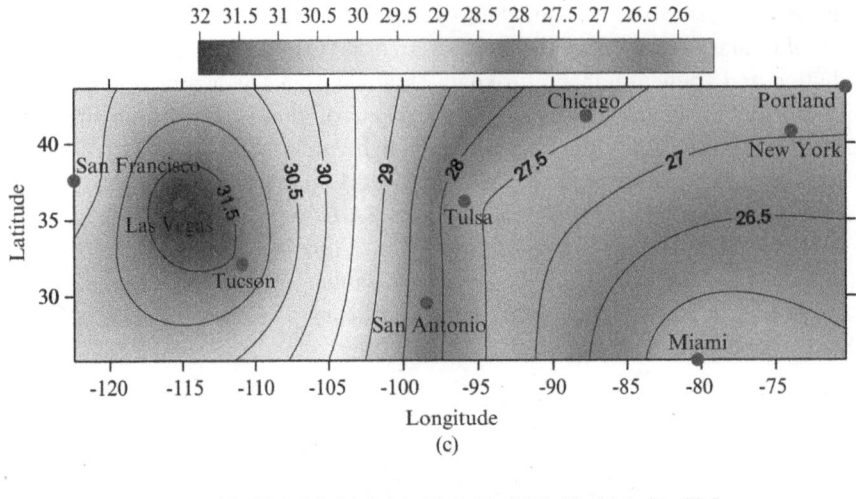

(c)

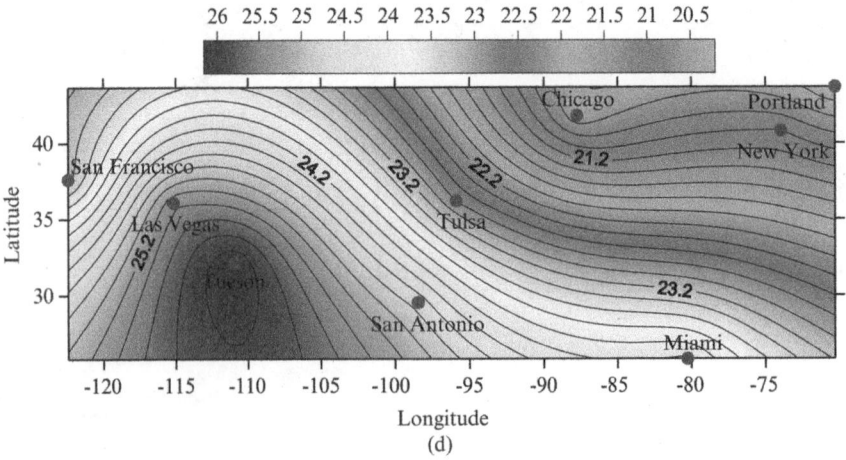

(d)

Figure 6.12 (Cont'd)

6.3 Solar Energy Applications

Today, solar energy is a prime source of energy with advanced and mature technologies used for converting solar radiation into useful outputs which is an end result of continuous use for many years. Solar energy is an auspicious technology as it can be used for both heating and electricity production purposes. It can be deployed in any size and capacity. There are many applications of solar energy. Figure 6.13 shows various solar energy applications. Solar energy applications can be classified under four main branches as follows:

- Photovoltaic (PV) applications
- Thermal applications
- Photochemical and biochemical applications
- Lighting

When we consider the total capacity of solar energy systems, PV and thermal applications are more practically applied ones. There are also serious attempts in photochemical and biochemical applications of solar energy. As the sun is the main source of photosynthesis, which is the oxygen

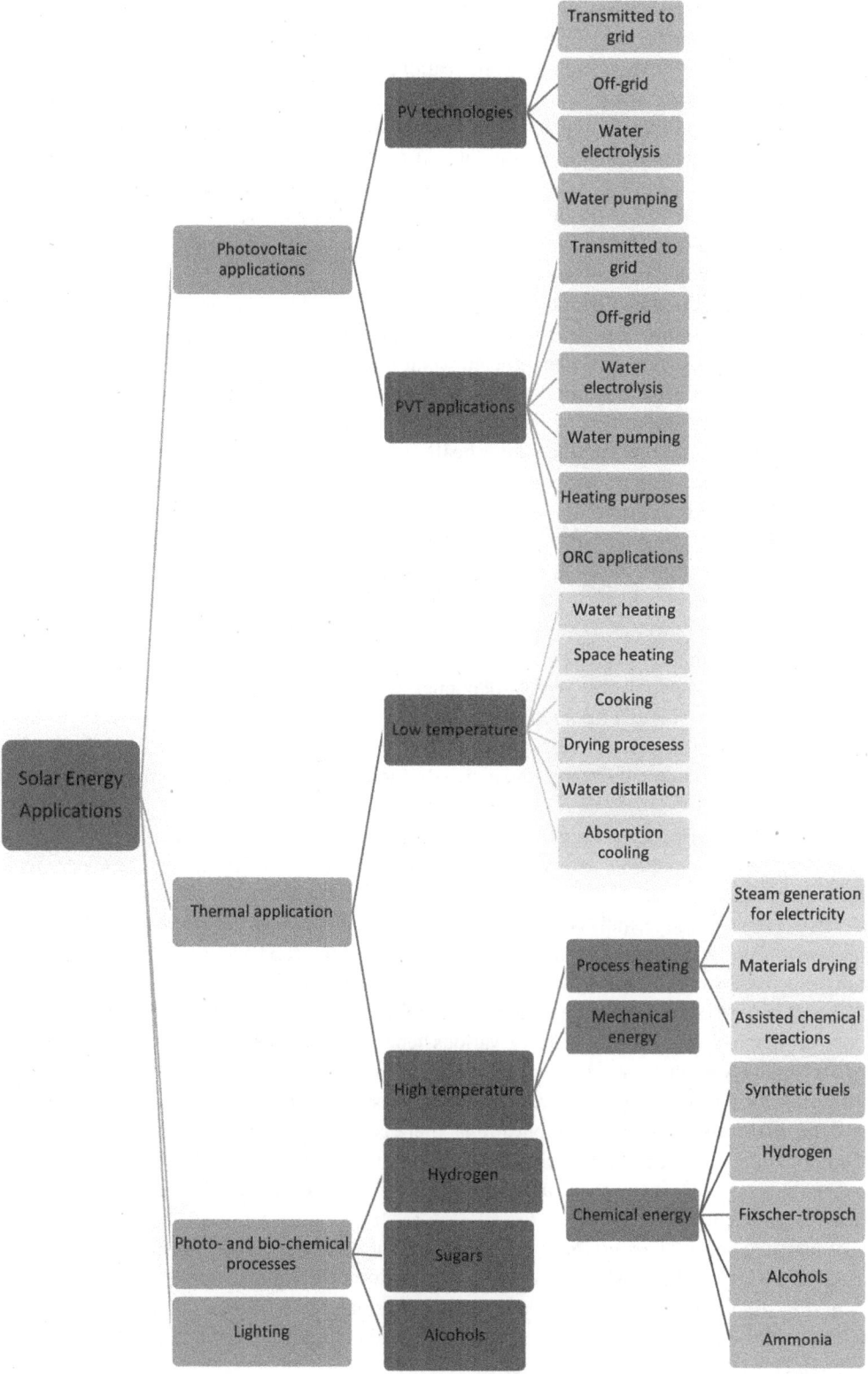

Figure 6.13 Classification of the solar energy applications.

source of life, it is crucial for human beings. On the other hand, photochemical reactors have a significant potential for hydrogen production and wastewater treatments. The sun is also a natural light source for the world. There are many applications to carry via fiber optic cables the sunlight to underground and under water to provide natural light.

Application of solar energy thermally is performed based on basic heating purposes and power generation with concentrated solar power plants. Also, it can be used for water desalination, drying for agricultural products, cooking purposes. On the other hand, photovoltaic application of solar energy is essentially used for power generation. It is also possible to harness heat from PV panels with PVT systems where liquid or gas circulation is used to extract the heat. Solar energy applications are presented in brief below.

Solar Heating Applications: These are the first industrial and sectoral activities of solar energy applications. There are many perks to living a solar lifestyle. Using any of the following solar energy applications, one may create an efficient solar energy system right at home. For example, solar water heating is becoming an eco-friendly alternative to traditional water heaters. Active water systems in solar water heating include direct circulation systems and indirect circulation systems. Passive water systems in solar water heating may become another option involving integral collector-storage and thermosiphon systems. Additionally, solar water heaters are commonly used in hotels, hospitals, guest houses, and more. There are also options available where houses or buildings are heated through active or passive solar heating systems.

Solar Distillation Applications: The solar-distillation method requires ample sunlight to transform saline water into distilled water. Once the solar radiation turns into heat, it creates purified water for cooling purposes. Distilled water is normally expensive in other electrical avenues, but solar distillation makes this type of electrical energy more cost-effective.

Solar Drying for Agriculture: Using solar energy to dry agricultural and animal products improves airflow and fruit quality, protecting sensitive agricultural products from harsh sunlight and preventing low moisture. Solar drying may be achieved in two ways, namely passive drying without using any mechanical systems and active drying where mechanical systems are employed in the drying system.

Solar Furnace Applications: Solar furnaces must operate at extremely high temperatures. In this method, solar radiation requires slanted, rotating mirrors to generate high heat. As the energy market faces continued risks with fuel supply, solar cooking is becoming more necessary. Unfortunately, fuels, such as coal, kerosene, and cooking gas are quite scarce. The solar cooker reduces heat loss in convection through an airtight box. While low maintenance costs are a benefit of solar cooking, foods cannot be cooked in a solar cooking system in unpredictable conditions or at night.

PV Applications: Solar electric generation is important where photovoltaic (PV) cells generate electricity through direct sunlight. There are various benefits to using solar electric power generation, such as reliability, low maintenance costs, durability, and eco-friendly. Solar electric power generation is most beneficial for irrigation, commercial-grid power systems, public transportation, and more.

PVT Applications: In solar PVT applications, there is useful heat output along with electricity generation. Therefore, the overall efficiency for PVTs is higher than the regular PV panels. Also, as the working fluid used in the thermal output purposes cools down the PV panels, it is able to increase the efficiency of the PV panel, too.

Concentrated Solar Thermal Power Generation Applications: In these systems, photons are focused on the certain point where heat transfer fluid flows through it. Thus, the temperature of the working fluid has reached higher levels. Solar thermal power production is a method that transforms solar energy into electricity. It stores thermal energy by heating fluids through a turbine, producing steam to generate electricity.

6.4 Solar Thermal Systems

Solar thermal systems generally aim to convert solar energy to useful heat which can be used for either heating, drying, and distillation purposes directly or electricity production after steam generation in power generating systems, such as Rankine cycle. It is also possible to combine those two outputs in the solar thermal power generation power plants, so-called solar cogeneration power plants. There are various methods in solar thermal systems. In this section, solar thermal systems are presented with practical applications and illustrative examples.

Before starting the thermal applications of solar energy, we need to glance at the solar collectors which are used in the solar thermal systems to convert solar energy to heat. Figure 6.14 demonstrates a classification of the solar thermal collectors. Generally, solar thermal collectors are classified into two including non-concentrating and concentrating. In the non-concentration ones, solar

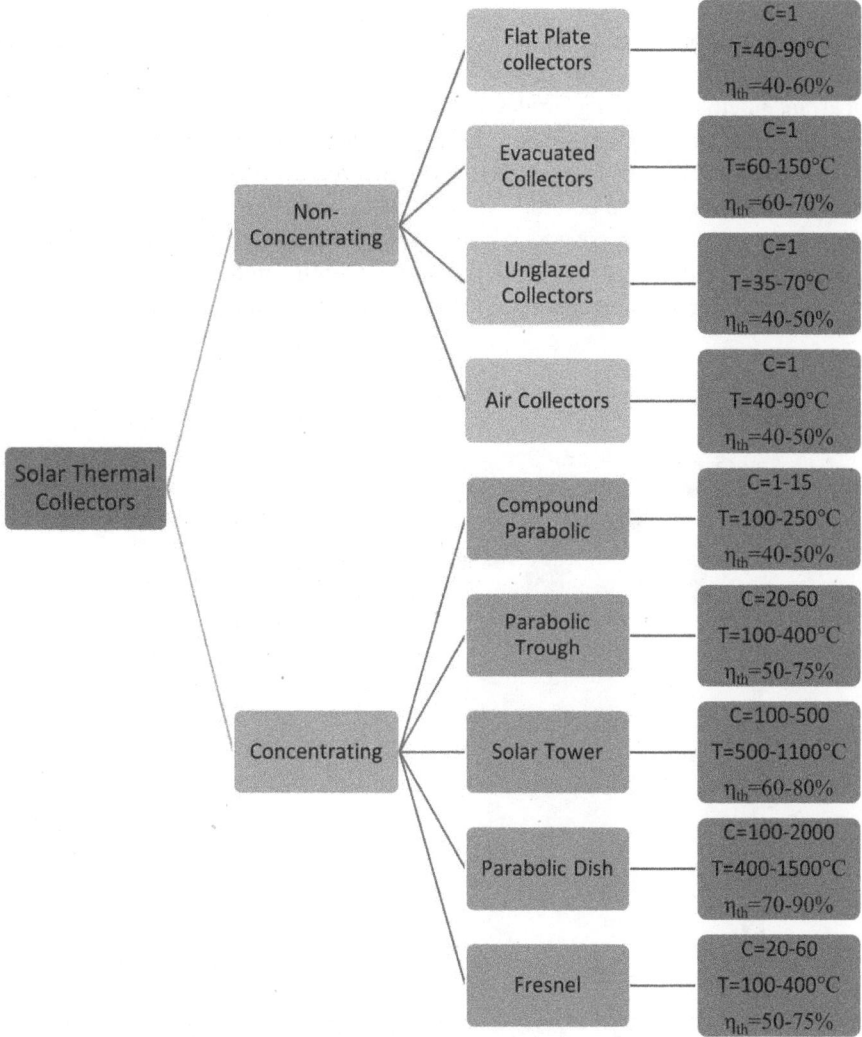

Figure 6.14 Classification of solar thermal collectors.

radiation is directly converted to heat. In concentrating solar thermal collectors, solar radiation is focused on a point or a line. "C" denotes the concentration rate of the solar system given in Figure 6.14. Non-concentrating solar collectors are used for basic heating purposes such as water and space heating and drying. The output temperature of non-concentrating solar collectors is relatively lower. Generally, it may not exceed 90 °C. Therefore, they are usually used in low-temperature applications given in Figure 6.13. On the other hand, concentrating solar collectors are used in high-temperature applications which can be varied from electricity generation to chemical processes. Each of these solar collector systems may be included in various applications in the following sections with practical examples.

Note that solar thermal power systems may also have energy storage units that allow us to benefit from solar energy when it is not available or adequate due to its intermittent and fluctuating behavior. These energy storage techniques will be introduced in Chapter 10.

6.4.1 Passive Solar Heating

Solar passive heating is a method of using solar energy through designs that use few or no mechanical systems (i.e., no pumps, or fans, or solar collectors). Passive solar energy system mainly involves special arrangements of building facades and windows. Passive solar technology also may include solar stills, greenhouses, or solar ponds. For example, there is a new trend in creating fully closed spaces with glasses, called a winter yard. In the winter yards, on sunny winter days it is possible to keep the spaces at a certain temperature without any external heater.

Figure 6.15 shows an example of a passive solar heating system. In this system, solar energy transmits inside the closed space and hit surfaces. Thus, the surfaces' temperature increases. Due to convection heat transfer between the absorber surface, the ambient temperature increases. For passive

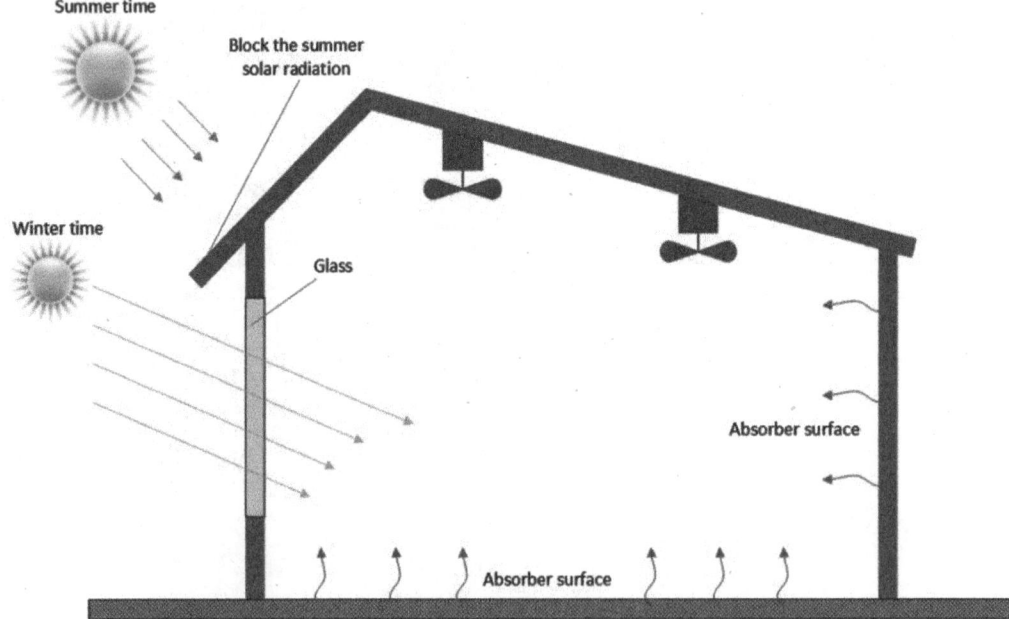

Figure 6.15 An illustration for the passive solar heating.

solar heating systems, heat storage mediums can be included in the absorber surfaces for later heating when solar energy is not available. Also, for summer season, in order to block to excess solar energy, special geometries can be used to block the solar radiation to reach inside the closed space.

Simple arrangements like south-facing widows allow a good transmittance of photonic energy inside the building; the light is then reflected and partially absorbed by the walls and eventually it reaches windows from the inside where it is partially reflected back and partly refracted toward the outside. Thus, there is a positive accumulation of solar energy inside the building, during the daytime. Construction materials, such as internal walls, the building shell, and so on, store the thermal energy accumulated during the day and enable a temporal phase shift between solar radiation availability and heat utilization.

6.4.2 Solar Hot Water Systems

Around the world, utilization of solar collectors for thermal applications is well-suited to countries located in what is called the solar belt, a zone between 25 and 45° of latitude that spans both the Northern and Southern Hemispheres. Some countries in the solar belt area which have abundant fossil fuel reserves – such as Iran, Saudi Arabia, Kuwait, Mexico, among others – don't use solar energy. Turkey located in solar belt. So, it takes high solar radiation. There are three different solar collector types for solar water heating collectors. These are flat-plate, vacuum tube, and unglazed solar water heating collectors. Solar water heating collectors are seen in Figure 6.16.

While flat-plate and vacuum tube collectors are used for the domestic hot water applications, unglazed ones are used generally for the swimming pool heating applications. In recent years, there is increasing attention in vacuum tube collectors due to lower weightiness and costs compared to the flat-plate collectors. In the manufacturing of solar collectors, in terms of production capacity and utilization of China and Turkey are known as two leading countries in the sector. Unglazed solar collectors are mainly used in USA, Australia, Canada.

Figure 6.17 demonstrates the solar domestic hot water system. In the solar domestic hot water systems, the water at the main line temperature generally around 10 °C–25 °C is heated up to 90 °C. So, there is a significant change in the density of the water. In other words, the water expands while being heated. In order to protect the system from volume and pressure changes, an expansion tank is integrated into the system. Also, there is a check valve attached to the hot water storage tank to release some hot water in case pressure inside the tank increases.

In the solar domestic hot water systems, the water circulation may be done either with a circulation pump or without it. In systems without pump, water is circulated thanks to the density difference due to temperature difference. In this case, the hot water storage tank should be placed at a

(a)   (b)   (c)

Figure 6.16 Solar water heating collectors a) flat-plate, b) vacuum tube, and c) unglazed.

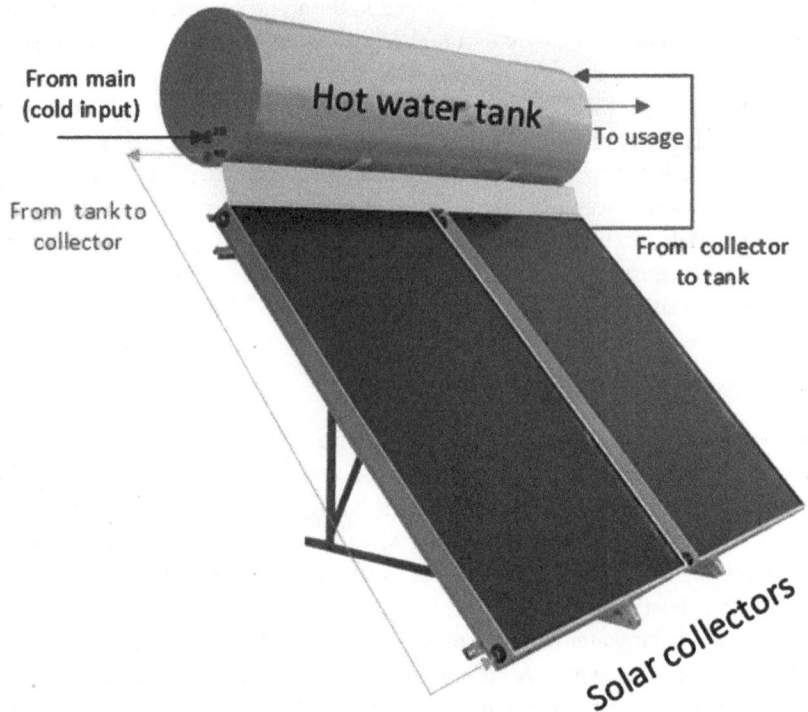

Figure 6.17 An illustration for solar domestic water heating system.

higher level from the collector. A pump is generally used when storage tank is placed at a lower level than collector such as attic or basement, and high-capacity systems.

In order to calculate the outlet temperature of water from the collector, first, we need to calculate heat losses from the collector as all solar radiation converted into heat in the collector does not transfer into the water. For a flat-plate collector, heat losses occur on all surfaces of the solar collector. The loss factor can be calculated in general with the following equation [6].

$$K = K_{sides} + K_{bottom} + K_{top} \tag{6.10}$$

As a practical approach, K_{sides} may be calculated as follows:

$$K_{sides} = \frac{K_{top}}{2} \tag{6.11}$$

Here, K_{top} can be calculated by the following equation presented in [6]:

$$K_{\ddot{u}st} = \left[\frac{N}{\frac{C}{T_y}\left[\frac{T_y - T_{\varsigma ev}}{N+f}\right]^{0,33}} + \frac{1}{h_{td}} \right]^{-1} + \left[\frac{\sigma\left(T_y + T_{\varsigma ev}\right)\left(T_y^4 + T_{\varsigma ev}^4\right)}{\left[\varepsilon_y + 0,05N\left(1 - \varepsilon_s\right)\right]^{-1} + \frac{2N+f-1}{\varepsilon_s} - N} \right] \tag{6.12}$$

where

$$h_{td} = 5.7 + 3.8\,V_r \tag{6.13}$$

$$f = \left(1 - 0.04 h_{td} + 0.0005 h_{td}^2\right)\left(1 + 0.091 N\right) \tag{6.14}$$

$$C = 250\left[1 - 0.0044\left(\beta - 90\right)\right] \tag{6.15}$$

The amount of heat transferred in the water is called the useful heat (Q_f). Q_f can be calculated as:

$$\dot{Q}_f = \dot{m}.c_p.\left(T_{out} - T_{in}\right) \tag{6.16}$$

and

$$\dot{Q}_f = A_c F_t\left[S - K\left(T_{in} - T_{env}\right)\right] \tag{6.17}$$

In order to calculate the outlet temperature of the water (T_{out}), the first Eq. 5.17 is used and Q_f is determined. Then, from Eq. 5.16, the outlet temperature of the water can be calculated. F_t in Eq. 5.17 is calculated as follows:

$$F_t = \frac{\dot{m}\,c_p}{A_c K}\left[1 - exp\left(-\frac{A_c\,K\,F_v}{\dot{m}\,c_p}\right)\right] \tag{6.18}$$

where

$$F_t = \frac{\dot{m}\,c_p}{A_c K}\left[1 - exp\left(-\frac{A_c\,K\,F_v}{\dot{m}\,c_p}\right)\right] \tag{6.19}$$

Illustrative Example 6.2   A hotel management wants to install a solar domestic hot water system for proving the hot water without using any external source. Daily hot water usage can be considered 50 L per person. The capacity of the hotel is 200 people. The average solar radiation at the location is 500 W/m^2. The thermal efficiency of solar collectors is 0.50. It is requested that the hot water required at the hotel should be prepared in 3 h. Calculate the total solar collector area required. Find the total number of solar collectors if a solar collector is size 2 m^2. The initial temperature of the water is 20 °C, and it is heated up to 70 °C .

Solution:
First, we need to calculate the total hot water demand:

$$V_{total\,hot\,water} = v \times N = 50\frac{L}{person} \times 200\left(people\right)$$

Thus,

$$V_{total\,hot\,water} = 10,000\,L$$

A total of 10,000 L of hot water is required at the hotel.
   The heat demand of the hot water can be calculated as follows:

$$Q_{demand} = m_{water} \times c_p \times \left(T_{initial} - T_{final}\right) = V_{water} \times \rho \times c_p \times \left(T_{final} - T_{initial}\right)$$

$$Q_{demand} = 10,000(L) \times \frac{1(m^3)}{1000\ (L)} \times 998 \left(\frac{kg}{m^3} \right) \times 4.18 \left(\frac{kJ}{kgK} \right) \times (70 - 20)$$

and

$$Q_{demand} = 2,085,820\ kJ$$

For 3 h of charging period, the heat transfer rate can be determined as:

$$\dot{Q}_{demand} = \frac{Q_{demand}}{\Delta t} = \frac{2,085,820\ kJ}{3(h) \times 3600 \left(\dfrac{h}{s} \right)} = 193\ kW$$

The energy balance of solar collector is written as:

$$\dot{Q}_{demand} = R_{solar} \times A_{coll} \times \eta_{th}$$

From this equation, one may proceed for calculation further:

$$193\ (kW) = 500 \left(\frac{W}{m^2} \right) \times \frac{1\ (kW)}{1000(W)} \times A_{coll}\ (m^2) \times 0.50$$

Thus, the total collector area is found to be

$$A_{coll} = 772\ m^2$$

The number of solar collectors is the determined to be

$$N_{coll} = \frac{A_{coll}\ (m^2)}{2(m^2)} = \frac{772}{2} = 386\ \text{collectors}$$

6.4.3 Solar Space Heating Systems with Air Collectors

Another common use of solar thermal systems is solar space heating systems which are used for building heating purposes. In such systems, despite the fact that it is possible to use solar water collectors, solar air collectors, depending on the system and capacity. Figure 6.18 shows a system layout for a solar space heating system that includes solar air collector integrated with heat storage unit. The system mainly consists of a solar air collector, fans, heat storage tank, and heat storage unit. During the daytime, the air is circulated through the solar collector and heated it up. In order to continue to meet the heating demand of the building when solar energy is not adequate or available, it is stored in a heat storage unit. The working principle of the solar space heating system is similar to the solar domestic hot water system. A practical example is presented in Chapter 10 where Energy Storage is discussed.

6.4.4 Concentrated Solar Power

Concentrated solar power (CSP) systems, also called concentrating solar power, concentrated solar thermal systems are thermal applications of solar energy to generate electricity. In order to reach higher temperatures, sunlights are focused on a certain zone with mirrors and lenses. Solar energy in a large area is collected in a quite small area. The temperature of a working fluid increases

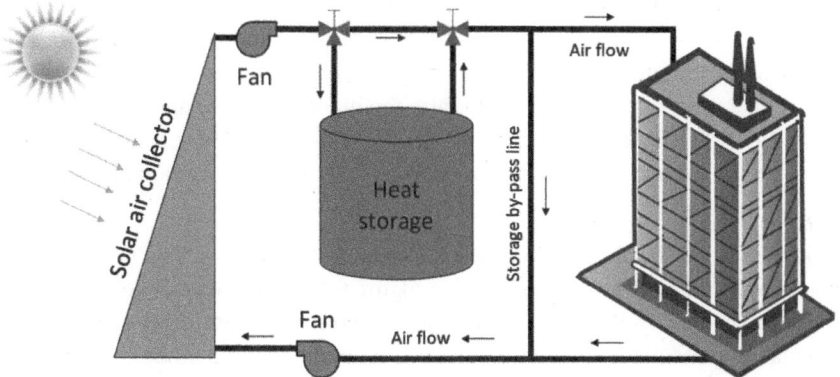

Figure 6.18 The schematic view of the sensible heat storage integrated solar heating system.

significantly at the location where solar photons are directed and it turns into a superheated vapor. This superheated vapor is then used for power generation in a turbine, usually a steam turbine.

Global capacity of the CSP systems is reached 6000 MW in 2021, while it was 354 MW in the very early 2000s. Spain has the highest CSP capacity in the world, which covers almost one-third of the total CSP capacity. This is followed by the USA which has approximately 1.8 GW of capacity. The largest CSP project in the world is the Ouarzazate Solar Power Station in Morocco. It has a total of 510 MW capacity and integrated with molted-salt heat storage option for power generation at night. The Ivanpah Solar Power Facility (392 MW) in the United States is the second power plant in the world with the capacity of 392 MW. There is no heat storage option at their Ivanpah Solar Power Facility.

There are a couple of applications of CSPs which are parabolic through, solar power tower, Fresnel reflectors, and dish Stirling. Each of them is presented below with practical applications.

As a thermal energy generating power station, CSP has more in common with thermal power stations such as coal, gas, or geothermal. A CSP plant can incorporate thermal energy storage, which stores energy either in the form of sensible heat or as latent heat (for example, using molten salt), which enables these plants to continue to generate electricity whenever it is needed, day or night. This makes CSP a dispatchable form of solar. Dispatchable renewable energy is particularly valuable in places where there is already a high penetration of PVs, such as California, because demand for electric power peaks near sunset just as PV capacity ramps down.

CSP is often compared to PV since they both use solar energy. While solar PV experienced huge growth in recent years due to falling prices, Solar CSP growth is slow due to technical difficulties, complexity of the system, and high costs. In 2017, CSP represented less than 2% of worldwide installed capacity of solar electricity plants. However, CSP can more easily store energy during the night, making it more competitive with dispatchable generators and baseload plants. Consequently, the CSP systems still have significant advantages when solar will be stored for later use.

Parabolic trough:

Parabolic trough CSP systems are the most widely used one in the practical applications. In the parabolic trough CSP power plant, there are linear parabolic reflectors that concentrate light onto a receiver positioned along the reflector's focal line, as shown in Figure 6.19. Vacuum tube solar collectors are placed at the focal point of reflectors. Vacuum tube solar collectors have a selective glass surface feature to reduce reflection from the tubes to increase effectiveness of the collector. Also, the reflectors are able to rotate to track the sun.

The largest capacity for a parabolic trough CSP, it is also the highest capacity among the CSPs, is the Ouarzazate Solar Power Station in Morocco which has a total of 510 MW capacity. A molten salt-based heat storage is then integrated to continue when solar energy is not available.

Solar power tower:

A solar power consists of a set of mirrors, also called heliostats, that concentrate sunlight on a central receiver placed at top of a tower, as illustrated in Figure 6.20. A heat exchanger is included in the system to transfer the heat in the working fluid to the water to generate steam. Generally, the water is not circulated in the receiver. A working fluid, generally thermal oil or molten salt, is

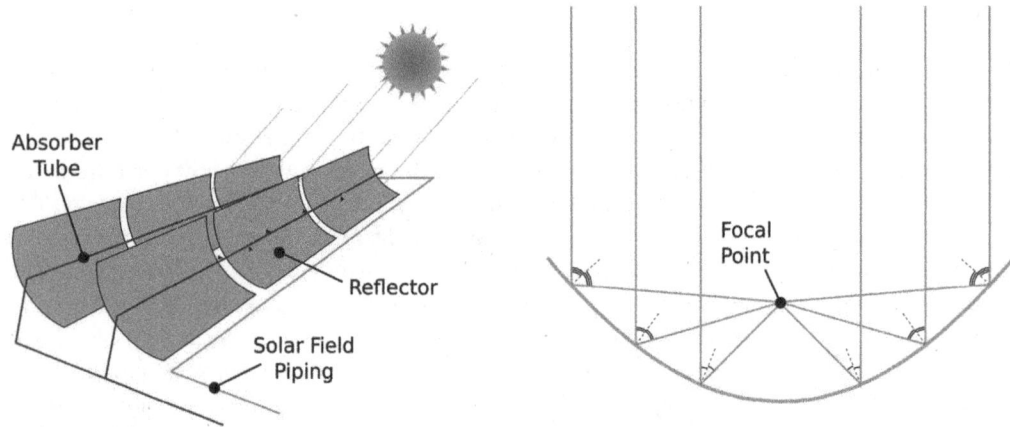

Figure 6.19 The schematic illustration of the parabolic trough collectors.

Figure 6.20 The view of the concentrated solar tower power plant.

used between receiver and heat exchanger. The working fluid in the receiver is heated to 500 °C–1200 °C. Water flows on the counter side of the heat exchanger and steam is generated. Steam is used for power generation in a steam turbine.

The mirrors used in the solar tower systems have the capability to track the sun. The biggest advantage of the solar tower system is not to circulate the working fluid like in parabolic trough system. The working fluid is circulated in a limited area between the receiver and heat exchanger. In order to improve the system performance, some additives are used in the working fluid to enhance its heat transfer properties.

A cost/performance comparison between power tower and parabolic trough concentrators was made by the NREL which estimated that by 2020 electricity could be produced from power towers for 5.47 ¢/kWh and for 6.21 ¢/kWh from parabolic troughs. The capacity factor for power towers was estimated to be 73% and 56% for parabolic troughs. Further details are available in Ref. [4].

The first commercial tower power plant was PS10 in Spain with a capacity of 11 MW, completed in 2007. Since then, a number of plants have been proposed, and several of them have been built in a number of countries (Spain, Germany, US, Turkey, China, India) but several proposed plants were cancelled as photovoltaic solar prices plummeted. A solar power tower went online in South Africa in 2016. Ivanpah Solar Power Facility in California generates 392 MW of electricity from three towers, making it the largest solar power tower plant when it came online in late 2013. Further details are available in Ref. [4].

Fresnel reflectors:

A linear Fresnel reflector power plant uses a series of long, narrow, shallow-curvature (or even flat) mirrors to focus light onto one or more linear receivers positioned above the mirrors, as illustrated in Figure 6.21. On top of the receiver a small parabolic mirror can be attached for further focusing the light. These systems aim to offer lower overall costs by sharing a receiver between several mirrors (as compared with trough and dish concepts), while still using the simple line-focus geometry with one axis for tracking. This is similar to the trough design (and different from central towers and dishes with dual axis). The receiver is stationary, so fluid couplings are not required (as in troughs and dishes). The mirrors also do not need to support the receiver, so they are structurally simpler.

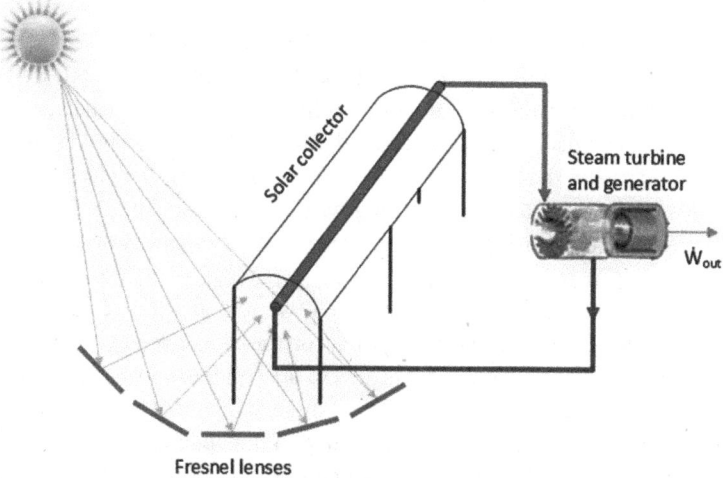

Figure 6.21 The schematic illustration of the CSP with Fresnel reflectors.

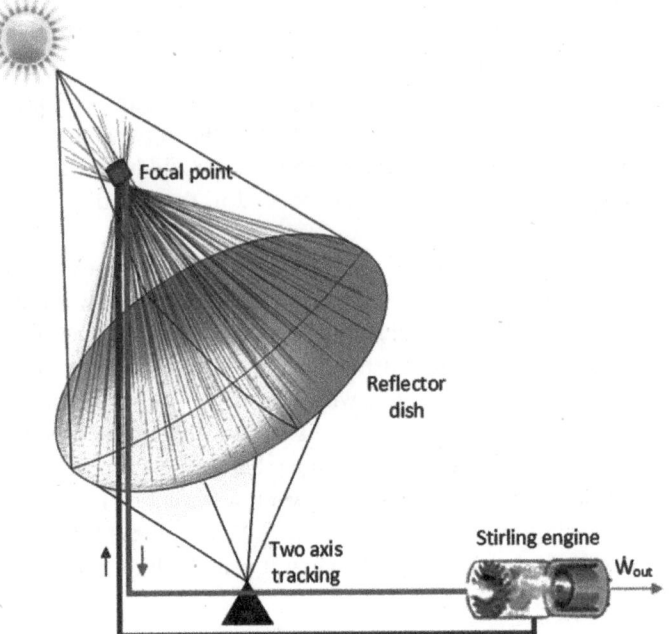

Figure 6.22 A solar dish Stirling power generation system.

When suitable aiming strategies are used (mirrors aimed at different receivers at different times of day), this can allow a denser packing of mirrors on available land area.

Rival single axis tracking technologies include the relatively new linear Fresnel reflector (LFR) and compact-LFR (CLFR) technologies. The LFR differs from that of the parabolic trough in that the absorber is fixed in space above the mirror field. Also, the reflector is composed of many low row segments, which focus collectively on an elevated long tower receiver running parallel to the reflector rotational axis.

Dish Stirling:

A dish Stirling system uses a large, reflective, parabolic dish (similar in shape to a satellite television dish), as depicted in Figure 6.22. It focuses all the sunlight that strikes the dish up onto a single point above the dish, where a receiver captures the heat and transforms it into a useful form. Typically, the dish is coupled with a Stirling engine in a dish Stirling system, but also sometimes a steam engine is used. These create rotational kinetic energy that can be converted to electricity using an electric generator. Rispasso Energy, a Swedish firm, in 2015 its Dish Sterling system being tested in the Kalahari Desert in South Africa where they obtained satisfactory performance with an efficiency of 34%.

Illustrative Example 6.3 Consider a solar dish power plant as illustrated in Figure 6.23. Steam enters the turbine at 10 MPa and exits from turbine at 1.2 MPa for the closed feedwater heater (FWD). It exits at 0.6 MPa for open FWD. The feed water is heated to the condensation temperature of the extracted steam leaves the closed feedwater heater. The extracted steam leaves the closed feedwater heater as saturated water, which is subsequently throttled to the open feedwater heater. The total work output of the turbine is 400 MW. Calculate both energy and exergy efficiencies for the system. The pumping power of the pump 3 which is used for circulation of the working between solar dish to turbine is negligible.

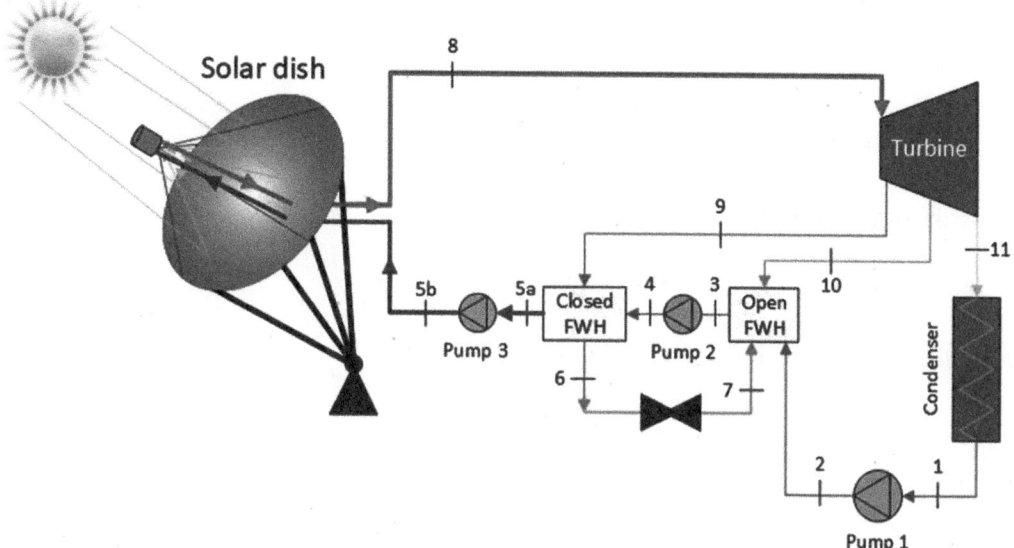

Figure 6.23 The schematic illustration of the solar dish CSP system in Illustrative Example 6.3.

Solution:
We can start to solve the problem from the pump which is between state points 1 and 2. For pump 1, the mass balance equation can be written for under steady-state conditions as follows:

$$\dot{m}_1 = \dot{m}_2 = \dot{m}$$

EBE for the pump is

$$\dot{m}_1 h_1 + \dot{W}_{p1} = \dot{m}_2 h_2$$

As it is $\dot{m}_1 = \dot{m}_2$ and $\dot{W}_{p1} = \dot{m} \times \dot{w}_{p1}$, we can write the EBE as follows:

$$h_1 + \dot{w}_{p1} = h_2$$

where

$$\dot{w}_{p1} = \nu_1 \left(P_2 - P_1 \right)$$

Here, ν_1 and h_1 can be obtained from Table A1 for $P_1 = 10$ kPa and $x_1 = 0$:

$$h_1 = 191.81 \ kJ/kg$$

$$\nu_1 = 0.001001 \ m^3/kg$$

So,

$$\dot{w}_{p1} = 0.00101 \times 590 = 0.6 \frac{kJ}{kg}$$

Thus, from the energy balance equation, h_2 is found to be:

$$h_2 = 192.4 \frac{kJ}{kg}$$

For the second pump in the system, inlet conditions are $P_3 = 600$ kPa and $x = 0$, so, from Table A1

$$h_3 = 670.38 \frac{kJ}{kg}$$

$$v_3 = 0.001101 \frac{m^3}{kg}$$

Thus, the specific pumping power is found as follows

$$\dot{w}_{p2} = v_3 (P_4 - P_4) = 0.001101 \times (10{,}000 - 6000) = 10.35 \frac{kj}{kg}$$

From the EBE equation for the pump 2

$$h_3 + \dot{w}_{p2} = h_4$$

$$670.38 + 10.35 = h_4$$

$$h_4 = 680.73 \frac{kJ}{kg}$$

For the closed feedwater, $P_9 = P_6 = 1.2$ MPa and state 6 is the fluid. Thus,

$$h_6 = h_{f@1.2MPa} = 798.33 \frac{kJ}{kg}$$

As h_6 and P_6 are known, we can find T_6 from Table A1.

$$T_6 = 188°C$$

In order to find the values for state 9, it is required to analyze the turbine first. State 8 is already known with

$$P_8 = 10 \ MPa \ and \ T_8 \ = \ 600°C$$

So, from Table A1, the inlet properties for the turbine inlet are

$$h_8 = 3625.8 \ kJ/kg$$

$$s_8 = 6.9045 \frac{kJ}{kgK}$$

We can find the outlet conditions of the turbine by assuming first the expansion the turbine is isentropic:

$$\eta_{isentropic} = \frac{h_8 - h_9}{h_8 - h_{9s}}$$

Under isentropic condition, s_{9s}, s_{10s}, and s_{11s} will become equal to s_8. As we know all pressure values for each outlet state, we can find the actual enthalpy values for turbine outlet.

$$\left. \begin{array}{l} P_9 = 1.2 \ MPa \\ s_{9s} = s_8 = 6.9045 \frac{kJ}{kg} s \end{array} \right| h_{9s} = 2974 \frac{kJ}{kg}$$

$$P_{10} = 0.6 \; MPa$$
$$s_{10s} = s_8 = 6.9045 \frac{kJ}{kg} \cdot s$$
$$\left. \right\} h_{10s} = 2810 \frac{kJ}{kg}$$

$$P_{11} = 10 \; kPa$$
$$s_{11s} = s_8 = 6.9045 \frac{kJ}{kg} \cdot s$$
$$\left. \right\} h_{11s} = 2187 \frac{kJ}{kg}$$

Assuming an isentropic efficiency of the turbine as 0.85, we obtain

$$\eta_{isentropic} = \frac{h_8 - h_9}{h_8 - h_{9s}} \rightarrow 0.85 = \frac{3625 - h_9}{3625 - 2974} \rightarrow h_9 = 3072 \frac{kJ}{kg}$$

$$\eta_{isentropic} = \frac{h_8 - h_{10}}{h_8 - h_{10s}} \rightarrow 0.85 = \frac{3625 - h_{10}}{3625 - 2018} \rightarrow h_9 = 2941 \frac{kJ}{kg}$$

$$\eta_{isentropic} = \frac{h_8 - h_{11}}{h_8 - h_{11s}} \rightarrow 0.85 = \frac{3625 - h_{11}}{3625 - 2187} \rightarrow h_9 = 2402 \frac{kJ}{kg}$$

The energy balance equation for the closed feedwater heater is written as:

$$\dot{m}_9 h_9 + \dot{m}_4 h_4 = \dot{m}_6 h_6 + \dot{m}_5 h_5$$

By applying mass balance equation $\dot{m}_4 = \dot{m}_5$ and $\dot{m}_9 = \dot{m}_6$, we can write it as follows:

$$\dot{m}_9 (h_9 - h_6) = \dot{m}_5 (h_5 - h_4)$$

From this equation, we can find the pressure ratios (x, y, and z) as illustrated in Figure 6.23:

$$y = \frac{\dot{m}_9}{\dot{m}_5} = \frac{h_9 - h_6}{h_5 - h_4} = 0.05404$$

In order to find z, we need to write the energy balance equation for the open feedwater heater as:

$$\dot{m}_7 h_7 + \dot{m}_2 h_2 + \dot{m}_{10} h_{10} = \dot{m}_3 h_3$$

$$z = \frac{\dot{m}_{10}}{\dot{m}_3} = \frac{(h_3 - h_2) - y(h_7 - h_2)}{h_{10} - h_2} = 0.1624$$

Thus, x can be found as follows:

$$x = 1 - y - z = 0.7858$$

The power output of the turbine was supposed to be 400,000 kW in the question. From the energy balance equation for the turbine:

$$\dot{m}_8 h_8 = \dot{W}_t + (\dot{m}_9 h_9 + \dot{m}_{10} h_{10} + \dot{m}_{11} h_{11})$$

We now proceed with x, y, and z values instead of the mass flow rates,

$$\dot{m}_8 h_8 = \dot{W}_t + (y h_9 + z h_{10} + x h_{11})$$

where $\dot{W}_t$ can be written as

$$\dot{W}_t = w_t \times \dot{m}_8$$

One can proceed with x, y, and z values instead of the mass flow rates. Thus, the EBE for the turbine

$$w_t = y(h_8 - h_9) + z(h_8 - h_{10}) + x(h_8 - h_{11})$$

$$w_t = 961.044 \ (kJ/kg)$$

$$\dot{W}_t = w_t \times \dot{m}_8 \rightarrow 400{,}000(kW) = 961.044\left(\frac{kJ}{kg}\right) \times \dot{m}_8$$

$$\dot{m}_8 = 416.21\left(\frac{kg}{s}\right)$$

In order to calculate the overall energy efficiency, one may proceed

$$\eta_{en} = \frac{\dot{W}_{net}}{\dot{Q}_{in}}$$

where $\dot{W}_{net}$ can be calculated to be

$$\dot{W}_{net} = \dot{W}_t - (\dot{W}_{p1} + \dot{W}_{p2}) = 400{,}000 - (196.3 + 4307) = 395{,}500 \ kW$$

and $\dot{Q}_{in}$ is determined as

$$\dot{Q}_{in} = \dot{m}_8(h_8 - h_5) = 1176845.56 \ kW$$

Thus, the overall energy efficiency is calculated through

$$\eta_{en} = \frac{\dot{W}_{net}}{\dot{Q}_{in}} = \frac{400{,}000}{1{,}176{,}845} = 33.6\%$$

For the overall exergy efficiency, it can be calculated as follows:

$$\eta_{ex} = \frac{\dot{W}_{net}}{\dot{Ex}_{\dot{Q}_{in}}}$$

where $\dot{Ex}_{\dot{Q}_{in}}$ is determined from

$$\dot{Ex}_{\dot{Q}_{in}} = \dot{Q}_{in}\left(1 - \frac{T_0}{T_s}\right) = 1{,}176{,}845\left(1 - \frac{298}{5780}\right) = 35.43\%.$$

Illustrative Example 6.4 Consider a solar tower based combined power plant as shown in Figure 6.24. There are two cycles (Brayton and Rankine cycles) in the system. The net work output is 500 MW. Calculate overall energy and exergy efficiencies of the combined system. The following are known about to the system:

T_3	1000 °C
P_3	100 MPa
P_4	2692 kPa
Isentropic efficiency	0.85
x_1	0
x_5	0
T_7	300 °C
P_7	100 kPa
P_8	8.226 kPa
Thermal efficiency of solar collector	0.6

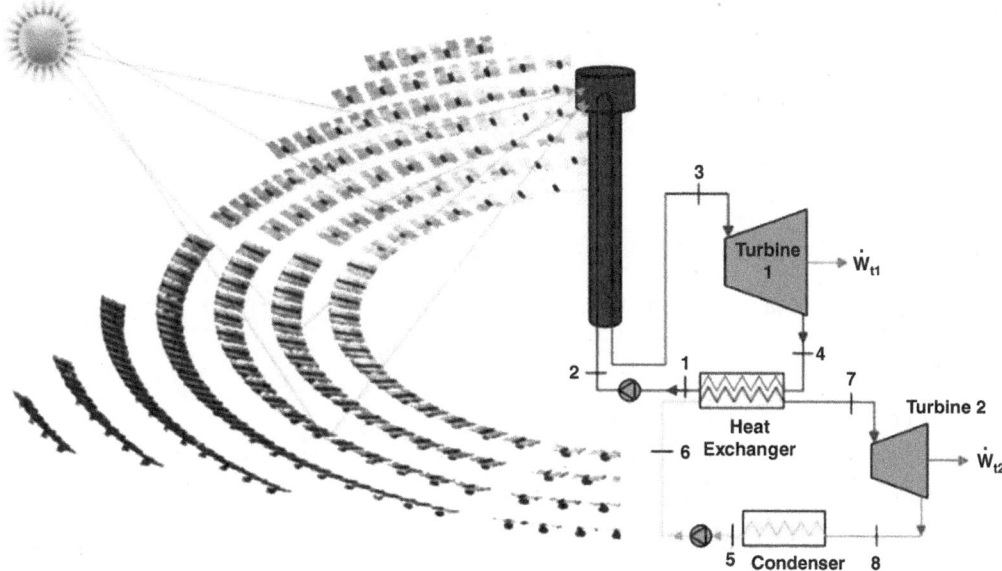

Figure 6.24   The schematic illustration of the solar tower power plant in Illustrative Example 6.4.

Solution:

For the state point 3, $T_3 = 1000\,°C$ and $P=100$ MPa from Table A1:

$$\left.\begin{array}{l} T_3 = 1000°C \\ P_3 = 100\,MPa \end{array}\right| \begin{array}{l} h_3 = 4373\ kJ/kg \\ s_3 = 6.604\ kJ/kgK \end{array}$$

For turbine 1, first, assuming isentropic conditions where we obtain

$$s_{4s} = s_3$$

$$\left.\begin{array}{l} P_4 = 2692\ kPa \\ s_{4s} = 6.904\dfrac{kJ}{kgK} \end{array}\right\} h_{4s} = 3004\dfrac{kJ}{kg}$$

From the isentropic efficiency definition of the turbine, we find

$$0.85 = \frac{h_3 - h_4}{h_3 - h_{4s}} = \frac{4373 - h_4}{4373 - 3004} \rightarrow h_4 = 3209.3\frac{kJ}{kg}$$

From the EBE of the turbine, one may write

$$h_3 = w_{t1} + h_4 \rightarrow w_{t1} = 1163.6\ kJ/kg$$

For the state point 1:

$$\left.\begin{array}{l} x_1 = 0 \\ P_1 = 2692\,MPa \end{array}\right| \begin{array}{l} h_1 = 976.1\ kJ/kg \\ v_1 = 0.001203\ m^3/kg \end{array}$$

In the system $P_2 = P_3$,

$$w_{p1} = v_1\left(P_1 - P_2\right) = 0.001203\left(2692 - 100\right) = 137.72 kJ/kg$$

For turbine 2:

$$T_7 = 300°C \mid h_7 = 3075\ kJ/kg$$
$$P_7 = 100 kPa \mid s_7 = 8.217\ kJ/kgK$$

Under the isentropic conditions, $s_{8s} = s_7$

$$0.85 = \frac{h_7 - h_8}{h_7 - h_{8s}} = \frac{3075 - h_8}{3075 - 2577} \rightarrow h_4 = 2651.7\frac{kJ}{kg}$$

The energy consumed in pump 2 can be calculated as

$$w_{p2} = v_5\left(P_6 - P_5\right) = 0.001009\left(100 - 8.226\right) = 0.1089\ kJ/kg$$

For the overall energy efficiency one can write

$$\eta_{en} = \frac{\dot{W}_{net}}{\dot{Q}_{solar}}$$

where

$$\eta_{th} = \frac{\dot{m}_2\left(h_3 - h_2\right)}{\dot{Q}_{solar}} \rightarrow 0.60 = \frac{412.0055\times\left(4373 - 976.1\right)}{\dot{Q}_{solar}}$$
$$\dot{Q}_{solar} = 1,724,173\ kW$$

Thus, the overall energy efficiency can be found as

$$\eta_{en} = 0.2899$$

The exergy content of the solar input can be calculated as follows:

$$\dot{Ex}_{\dot{Q}_{solar}} = \dot{Q}_{solar}\times\left(1 - \frac{T_0}{T_{sun}}\right) = 1,724,173\times\left(1 - \frac{298}{5780}\right) = 1,035,279$$

In closing, the overall exergy efficiency is found to be

$$\eta_{ex} = \frac{\dot{W}_{net}}{\dot{Ex}_{\dot{Q}_{solar}}} = 0.2899$$

6.5 Solar PV Systems

Photovoltaics (PV) is the conversion of light into electricity using semiconducting materials that exhibit the photovoltaic effect, a phenomenon studied in physics, photochemistry, and electrochemistry. The photovoltaic effect is commercially used for electricity generation and as photosensors. A photovoltaic system employs solar modules, each comprising a number of solar cells, which generate electrical power. PV installations may be ground-mounted, rooftop-mounted, wall-mounted, or floating. The mount may be fixed or use a solar tracker to follow the sun across the sky.

Photovoltaic applications can be beneficial wherever electrical energy is needed. Photovoltaic technologies have always had an upper edge over other conventional technologies as it is pollution free, and it uses solar energy that is freely and immensely available. Another advantage with the PV technology is that it does not emit any greenhouse gases during the operation and hence environmentally friendly. Also, here, it should be noted that there are still many attempts to reduce the environmental impact of solar PV panel production since it is pretty much fossil fuel dependent.

The intermittent of solar radiation can be a limitation to the technology as it can't supply electricity continuously during the off-sunshine periods, but this problem can be encountered by using energy storage. However, there is a need to understand the application of this technology to make it feasible for its users. For example, to water a field, a farmer can use a solar water pumping system during daytime and get benefited by the technology as he doesn't have to worry about the unwanted load shedding or power failure or to pay bills for the electrical power consumptions. Another example could be solar street lighting; the electricity converted by PV panels during sunshine hours can be stored in a battery and can be utilized to power the streetlights in the off-sunshine periods. Here, in this section, we are discussing various photovoltaic applications based on their performance in terms of efficiency.

Photovoltaic technology helps to mitigate climate change because it emits much less carbon dioxide than fossil fuels. Solar PV has specific advantages as an energy source: once installed, its operation generates no pollution and no greenhouse gas emissions, it shows scalability in respect of power needs and silicon has large availability in the Earth's crust, although other materials required in PV system manufacture such as silver may constrain further growth in the technology. Other major constraints identified are competition for land use. The use of PV as a main source requires energy storage systems or global distribution by high-voltage direct current power lines causing additional costs, and also has a number of other specific disadvantages such as variable power generation which have to be balanced. Production and installation do cause some pollution and greenhouse gas emissions, though only a fraction of the emissions caused by fossil fuels.

Photovoltaic systems have long been used in specialized applications as space technologies, stand-alone installations and grid-connected PV systems have been in use since the 1990s. Mass production of PV panels was first performed in 2000 in Germany. In those years, due to its high cost, it was used in space technologies and satellite applications. Parallel to the development in PV materials and production technologies, their costs have started to drop, as indicated in Figure 6.4. Thus, PV has become a significant and cost-effective energy contributor in global energy networks.

In 2021, worldwide installed PV capacity has reached to more than 940 GW covering approximately 3% of global electricity demand. Approximately 170 GW of capacity has been included in the energy network in 2021. The solar PV is the third highest renewable energy source used in the world, after hydro and wind power options. It is expected that the global capacity growth in the PV will increase exponentially with reducing PV costs and global issues which are pandemics, wars, environmental issues.

6.5.1 Solar Cells

Today, photovoltaics is best known method to generate electricity by using solar cells to convert solar energy from the sun. Figure 6.25 shows a view of the solar cell which is the basic part of the PV panels and its working principles. Solar cells produce direct current electricity from sunlight which can be used to power equipment or to recharge batteries. The first practical application of photovoltaics was to power orbiting satellites and other spacecraft, but today the majority of photovoltaic modules are used for grid-connected systems for power generation. In this case an inverter is required to convert the DC to AC. There is still a smaller market for standalone systems for remote dwellings, boats, recreational vehicles, electric cars, roadside emergency telephones, remote sensing, and cathodic protection of pipelines.

Photovoltaic power generation employs solar modules composed of a number of solar cells containing a semiconductor material. Copper solar cables connect modules (module cable), arrays (array cable), and subfields. As illustrated in Figure 6.25a, when the sun photons hit the solar cell, the electrons in atoms outer shell will be lost, and there will be abundance of free electrons in each cell. If conductors are connected to the negative and positive side of the solar cell, an electrical circuit will be formed, and consequently an electrical current will follow if the circuit is closed (electricity generation).

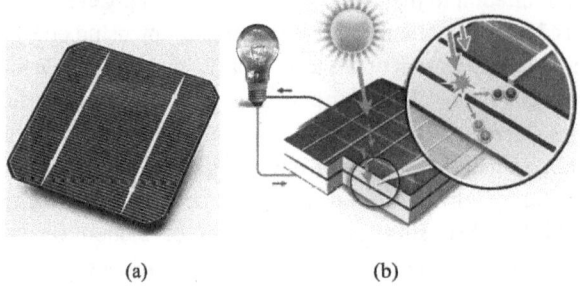

(a)   (b)

Figure 6.25 View of solar cell generated electricity directly from sunlight (a) view of cell and (b) working principle of a PV cell.

As mentioned, that number of solar cells connected to each other will form the solar panel power inverter (called module as well). And also, connecting a number of solar panels in series with each other will form solar string, then connecting number of solar strings in parallel will form the solar array which represents final solar panels system. So, the more solar panels we use, the larger solar array we get and of course, the higher electrical energy we can obtain from the solar system.

Photovoltaic module power is measured under standard test conditions in watts peak (Wp). The actual power output at a particular place may be less than or greater than this rated value, depending on geographical location, time of day, weather conditions, and other factors. Solar photovoltaic array capacity factors are typically under 25%, which is lower than many other industrial sources of electricity.

Illustrative Examples 6.5   Consider a PV-powered pumping system as shown in Figure 6.26. In the system, during the spring season, an agricultural irrigation tank at 50 m of height is filled with 50,000 tons of water by using a solar PV-driven pumping system. Considering pump efficiency is 85%, the PV module efficiency is 20%, the temperature of sun is 5778 K, and ambient temperature

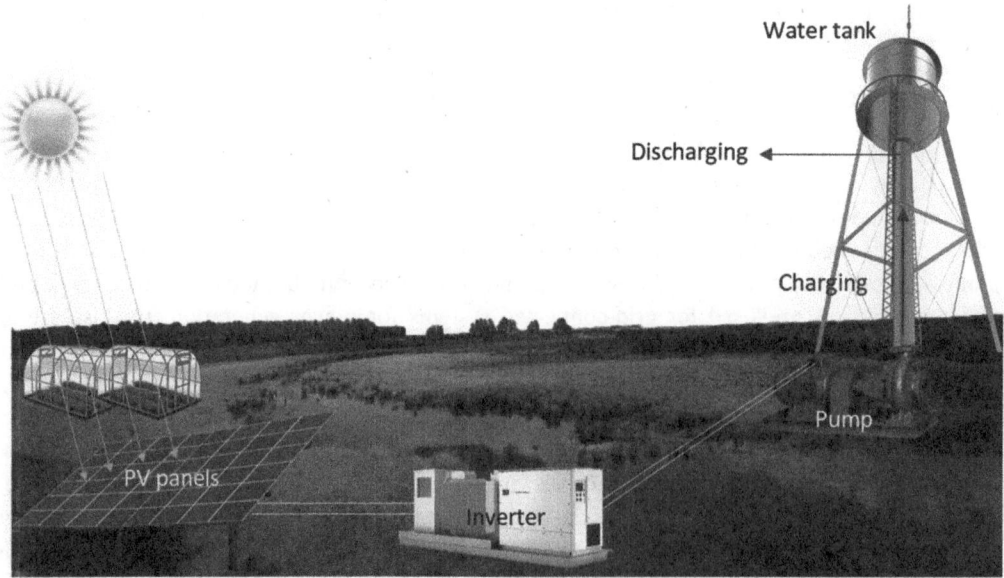

Figure 6.26 The schematic view of PV-powered pumping station.

is 25 °C. Calculate the total surface area of the PV panels. Calculate the overall energy and exergy efficiencies for the system. The monthly solar radiation data as given in the following table are provided to consider for calculations:

Months	Global Horizontal Solar Radiation (kWh/m²)
January	42.5
February	59.4
March	108.8
April	128.8
May	167.8
June	176.0
July	181.2
August	163.7
September	120.3
October	85.1
November	42.2
December	34.9

Solution:

In the system, the pump power is utilized to pump up to a height of 50 m. So, the pumping power is found as

$$W_{pump} = mgh = 50,000,000\,(kg) \times 9.81\left(\frac{m}{s^2}\right) \times 50\,(m) = 24,525,000\ kgm^2/s^2$$

$$W_{pump} = 6812.5\ kWh$$

The electricity consumption of the pump is calculated by

$$\eta_{elect} = \frac{W_{pump}}{W_{elect}} \rightarrow 0.85 = \frac{6812.5}{W_{elect}}$$

and

$$W_{elect} = 8014.71\ kWh$$

The efficiency of the PV panels is given to be 0.20. The electricity input of the pump needs to be met via solar energy. In order to find the solar energy output from PV panel can be calculated as follows:

$$W_{solar,output} = I_{solar} \times A_{total\,PV} \times \eta_{PV}$$

$W_{solar,output}$ should be equal to W_{elect} which is energy demand of the pump. Thus,

$$W_{solar,output} = W_{elect} = I_{solar} \times A_{total\,PV} \times \eta_{PV}$$
$$6812.5 = (108.8 + 128.8 + 167.8) \times A_{total\,PV} \times \eta_{PV}$$

So, the total area of the PV panel is found to be

$$A_{totalPV} = 98.85\ m^2$$

6.6 Photovoltaic Thermal Hybrid Solar Panels (PVTs)

Photovoltaic thermal collectors, typically abbreviated as PVT collectors and also known as hybrid solar collectors, photovoltaic thermal solar collectors, PVT collectors or solar cogeneration systems, are power generation technologies that convert solar radiation into usable thermal and electrical energy. PVT collectors combine photovoltaic solar cells (often arranged in solar panels), which convert sunlight into electricity, with a solar thermal collector, which transfers the otherwise unused waste heat from the PV module to a heat transfer fluid. By combining electricity and heat generation within the same component, these technologies can reach a higher overall efficiency than solar photovoltaic (PV) or solar thermal (T) alone. Figure 6.27 demonstrates a view of the PVT.

Significant research has gone into developing a diverse range of PVT technologies since the 1970s. The different PVT collector technologies differ substantially in their collector design and heat transfer fluid and address different applications ranging from low temperature heat below ambient up to high temperature heat above 100 °C.

There are a multitude of technical possibilities to combine PV cells and solar thermal collectors. A number of PVT collectors are available as commercial products, which can be divided into the following categories according to their basic design and heat transfer fluid:

- PVT liquid collector
- PVT air collector

While water or any heat transfer liquid flows through in the PVT liquid collector, air is used in the PVT air collector.

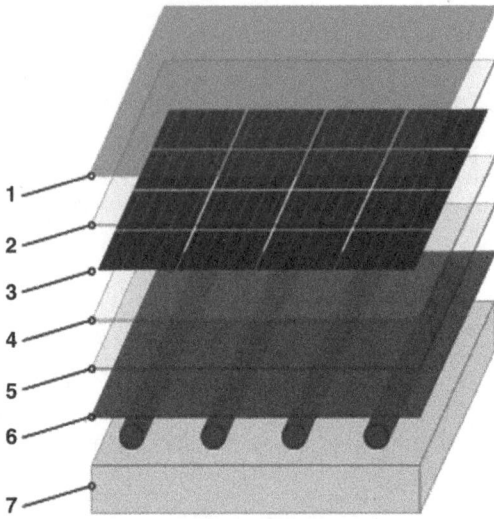

Figure 6.27 Schematic cross section of an uncovered PVT collector with sheet-and-tube-type heat exchanger and rear insulation 1. Antireflective glass, 2. Encapsulant, 3. Solar PV cells. 4. Encapsulant, 5. Back sheet, 6. Heat exchanger, and 7. Thermal insulation.

Illustrative Example 6.6 A greenhouse dryer uses PVT collector to produce electricity and dry crops. The schematic illustration of the greenhouse is shown in Figure 6.28. For a total of 500 hours of operation in year, PVT collector heats air with

$$\dot{m}_{air} = 10 \ kg/s$$

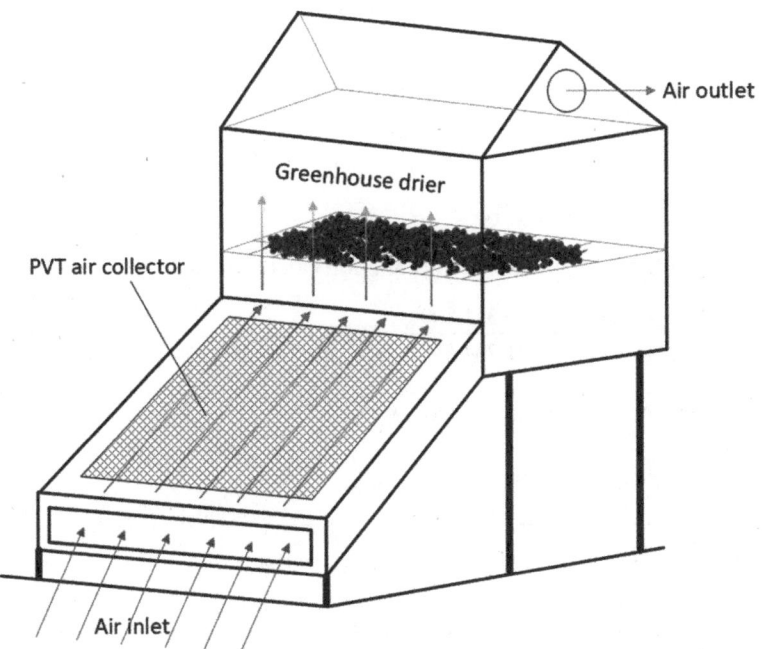

Figure 6.28 The schematic illustration of the PVT air heater coupled with a greenhouse dryer.

The incoming and outgoing air temperatures are given as follows:

$$T_{in} = 24.85°C$$

and

$$T_{out} = 41.85°C$$

If the annual electricity generation of PVT is three times higher as much as heat generation of PVT, find the required PV module area. The electricity efficiency of the PV panel is 0.20. Calculate overall energy and exergy efficiencies. The ambient temperature will be assumed as 5.14 °C. The total yearly solar radiation of the location is 1474.7 kW/m².

Solution:
The heat output from PVT panels can be calculated as follows

$$Q_{PVT} = \dot{m}(h_{out} - h_{in}) \times t$$

From Table A1, the enthalpy values are obtained as:

$$h_{in} = 298.18 \ kJ/kg$$

$$h_{out} = 315.27 \ kJ/kg$$

Thus,

$$Q_{PVT} = \dot{m}(h_{out} - h_{in}) \times t = 10 \times (315.27 - 298.18) \times 500 = 85450 \ kWh$$

For electricity generation, it is given that it is three times higher than the heat output.

$$W_{PV} = 85,450 \times 3 = 256,350 \ kWh$$

$$W_{PV} = I_{solar} \times A_{PV} \times \eta_{PV}$$

$$256,350 = 1474.7 \times A_{PV} \times 0.20$$

and the total PV area becomes

$$A_{PV} = 869 \; m^2$$

In order to calculate the energy efficiency, the following equation can be written

$$\eta_{en} = \frac{Q_{PVT} + W_{PV}}{I_{solar} \times A_{PV}} = \frac{85,450 + 256,350}{1474.7 \times 869} = \frac{341,800}{1,284,514.3} = 0.26$$

The exergy efficiency for the system can be written as follows:

$$\eta_{ex} = \frac{Ex_{Q_{PVT}} + W_{PV}}{Ex_{I_{solar}} \times A_{PV}} = \frac{\left[\left(1 - \frac{T_0}{T_s}\right)Q_{PVT}\right] + W_{PV}}{\left[\left(1 - \frac{T_0}{T_s}\right)I_{solar}\right] \times A_{PV}} = \frac{\left[\left(1 - \frac{T_0}{T_s}\right)Q_{PVT}\right] + W_{PV}}{\left[\left(1 - \frac{T_0}{T_s}\right)I_{solar}\right] \times A_{PV}} = 0.29$$

Illustrative Example 6.7 In Oshawa, Ontario, Canada, a PVT system is designed for a greenhouse, to generate both electricity and heat by using PVT modules which have 150 m² of total area. Air with 0.2 kg/s mass flow rate is heated from $T_{in} = 7\,°C$ to $T_{out} = 25\,°C$. If PVT manufacturer's data show $V_{oc} = 740$ C, $S_{sc} = 22.6$ A, FF $= 0.72$.

Consider $h_{air,in} = 280.13$ kJ/kg, $h_{air,out} = 298.18$ kJ/kg, $T_{cell} = 50\,°C$, and solar intensity as 400 W/m²

a) Calculate the useful heat generation rate of PVT.
b) Calculate the electricity generation rate of PVT.
c) Calculate the energy and exergy efficiencies of the PVT system.

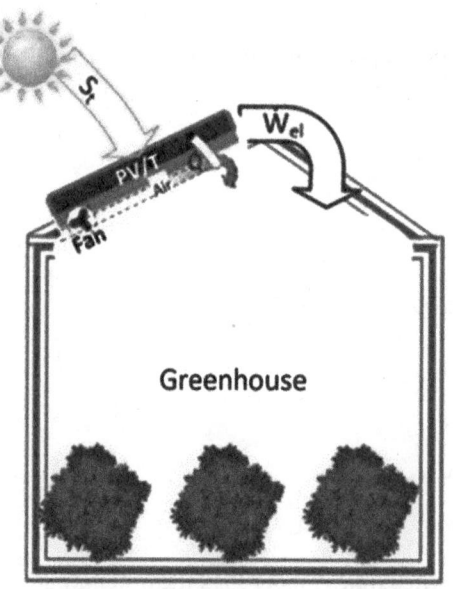

Greenhouse

Solution:

For the thermal process of the PVT system, the energy balance is written as:

$$\dot{m}_{air} h_{air,in} + \dot{Q}_{heating} = \dot{m}_{air} h_{air,out}$$

where $\dot{Q}_{heating}$ can be calculated as follows:

$$\dot{Q}_{heating} = \dot{m}_{air} \left(h_{air,out} - h_{air,in} \right)$$

$$\dot{Q}_{heating} = 0.2 \times (298.18 - 280.13)$$

and

$$\dot{Q}_{heating} = 3.61 \ kW$$

For the electricity generation, we find

$$P_{act} = FF \times V_{oc} \times I_{sc}$$

$$P_{act} = 0.72 \times 740 \times 22.6$$

Thus, electricity output is found as follows:

$$\dot{W}_{elect} = P_{act} = 12.04 \ kW$$

The energy efficiency of the PVT system can be found from

$$\eta_{en} = \frac{\dot{W}_{elect} + \dot{Q}_{heating}}{R_{solar} \times A_{coll}}$$

$$\eta_{en} = \frac{12.04 + 3.61}{0.4 \times 150}$$

So, the energy efficiency is found to be

$$\eta_{en} = 26.09\%$$

The exergy efficiency of the PVT system can be calculated from,

$$\eta_{ex} = \frac{\dot{W}_{elect} + \dot{Ex}_{\dot{Q}_{heating}}}{R_{solar} \times A_{coll} \times \left(1 - \dfrac{T_0}{T_{sun}} \right)}$$

where $\dot{Ex}_{\dot{Q}_{heating}}$ is calculated from

$$\dot{Ex}_{\dot{Q}_{heating}} = \dot{Q}_{heating} \times \left(1 - \frac{T_0}{T_s} \right) = 3.61 \times \left(1 - \frac{T_0}{T_{cell}} \right)$$

Thus, the exergy efficiency is found to be

$$\eta_{ex} = 21.78\%$$

6.7 Closing Remarks

This chapter introduces solar energy systems and their practical applications, and discusses numerous topics related to solar energy, ranging from radiation aspects to fundamental concepts and from system operational aspects to system energy and exergy efficiencies. Due to wide range of availability and technical maturity, solar energy has a significant potential to solve the energy problem in the world. Like in each renewable energy source, it has a few challenges like intermittent availability and fluctuating power generation. Therefore, often solar energy systems are integrated with energy storage methods for later use when solar is not available or adequate to meet the demand. Today's costs and technological readiness make solar energy as logical energy solution. This chapter also presents numerous examples to clearly highlight how to analyze solar energy systems for various practical applications and define their efficiencies accordingly for performance assessment.

References

1 Ritchie, H., Roser, M., and Rosado, P. (2022). Energy production and consumption. Available online https://ourworldindata.org/energy-production-consumption.

2 International Renewable Energy Agency (IRENA). (2022). Renewable power generation costs in 2021. Available online https://www.irena.org/publications/2022/Jul/Renewable-Power-Generation-Costs-in-2021 (accessed 25 November 2022).

3 Perez, R. and Perez, M.J.R. (2015). A fundamental look at energy reserves for the planet. For publication in the IEA/SHC Solar Update. Available online http://asrc.albany.edu/people/faculty/perez/Kit/pdf/a-fundamental-look-at%20the-planetary-energy-reserves.pdf.

4 Joshi, A.S., Dincer, I., and Reddy, B.V. (2009). Development of new solar exergy maps. *International Journal of Energy Research* 33: 709–718. doi: 10.1002/er.1506.

5 Global Horizontal Irradiation, Solar Resource Map. Available online https://globalsolaratlas.info/download/world.

6 Duffie, J.A. and Beckman, W.A. (1991). *Solar Engineering of Thermal Processes*. Wiley-Interscience.

Questions/Problems

Questions

1 Discuss the potential benefits of solar energy to meet global energy demand by comparing to other conventional energy sources.

2 How the 3S concept is developed for solar energy systems. Explain with practical examples.

3 Discuss the importance of energy storage in solar storage systems.

4 How are solar energy systems classified?

5 Briefly discuss solar thermal energy systems with practical examples.

6 Explain the differences between PV and PVT systems through pros and cons.

7 Classify and discuss solar thermal collectors.

8 Describe each of absorption, reflection and transmission concepts in radiation heat transfer.

9 Classify the commonly used devices for measuring solar radiation and explain their operational aspects.

10 Discuss how a solar space heating system works.

11 Define working principles of a PV cell.

12 What is the functionalities of using mirrors and lenses in CSP systems.

13 Discuss the use of solar energy systems in agricultural sector.
14 Is it really possible to supply the electricity and heat required for greenhouses in Ontario, Canada by employing solar PV systems?
15 Discuss the differences between energy and exergy efficiencies for PV and PVT systems.

Problems

1 Consider a glass used at the top of the greenhouse. The surface properties are specified as $\alpha=0.3$ and $\rho=0.3$. The total surface area of the glass roof is 100 m². Calculate how much energy will be transmitted inside the greenhouse when solar radiation is 750 W/m².
2 A hotel management wants to install a solar domestic hot water system for proving the hot water without using any external source. The daily hot water usage is considered 50 L per person. The capacity of the hotel is 1000 people. The solar radiation intensity at the location is 750 W/m². The thermal efficiency of solar collectors is 0.60. It is requested that the hot water required at the hotel should be prepared in 2 h. Calculate the total solar collector area. Find the number of solar collectors if each one is 2.5 m². The initial temperature of water is 20 °C, and it is then heated up to 80 °C.
3 It is requested to heat up a swimming pool with solar energy. The total volume of the swimming pool is 500 m³. It is desired to keep the temperature of the pool is 35 °C. The initial temperature of water in the pool is 10 °C. The heating should be completed in 2 h under 250 W/m² of solar radiation. For heating, unglazed solar collectors, which have the efficiency of 0.40, will be used. Calculate the total surface of the unglazed solar collectors required for this pool.
4 A cogeneration cycle powered by solar energy is shown in the figure below. The total heat input rate required by the cycle is 50 MW. The solar radiation intensity is measured to be 1.1 kW/m² and the efficiency of the solar collectors is 80%. The turbine has an isentropic efficiency of 85%.

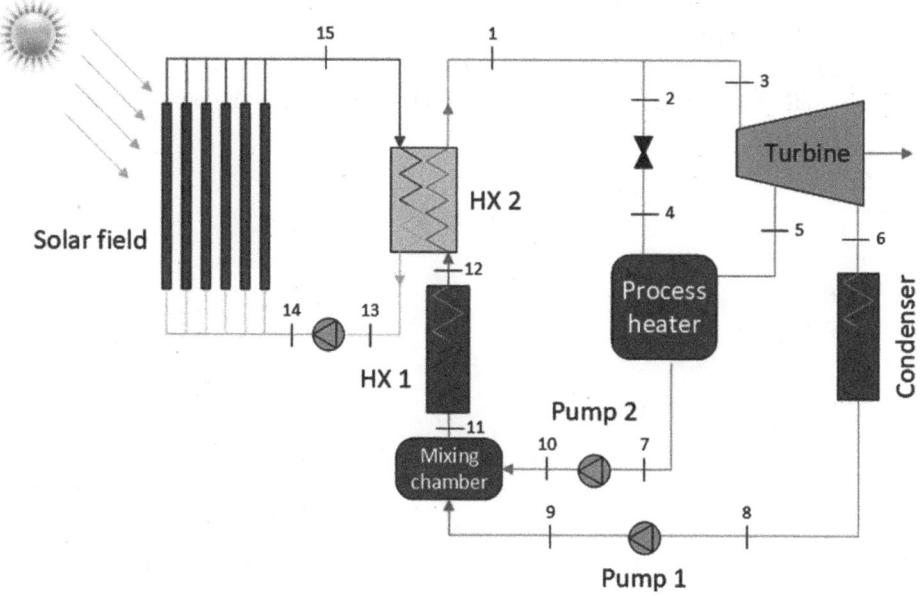

a Write the mass, energy, entropy and exergy balance equations for all system components.
b Calculate the total area of solar collectors required.
c Determine the turbine work output.

 d Find the total useful heat output from the process heater and condenser.
 e Calculate the overall energy and exergy efficiencies (Take $T_0 = 298$ K, $T_{sun} = 5777$ K, $T_{heater} = 523$ K).

5 A solar-driven reheat and regenerative steam Rankine cycle is shown in the figure below. The solar collectors absorb the incoming solar energy and transfer it to the power generation cycle with an efficiency of 70%. On a particular day, the solar radiation intensity is measured to be 900 W/m². The area of the solar collectors is 60,299 m². It is known that the ratio of the steam mass flow rate passed directly to the open feedwater heater (FWH) and the total mass flow rate $\dot{m}_8 / \dot{m}_7$) is 0.3 and the total steam mass flow rate entering the high-pressure turbine is 10 kg/s. The ambient temperature is 25 °C and the sun temperature is 5777 °C. If the known thermophysical properties at different state points is given in the table below:

State	Temperature (°C)	Pressure (kPa)	Specific enthalpy (kJ/kg)
1	45.82	10	191.8
4	46.32	15,000	207
5	600	15,000	3581
6	415.5	4000	?
7	600	4000	3674
8	439.6	1200	3346
9	45.82	10	241.2

Calculate the following:
 a The rate of energy transfer to the power generation Rankine cycle from the sun.
 b The enthalpy at state point 6.
 c The work outputs of the high- and low-pressure turbines.
 d The rate of heat rejected from the condenser.
 e The overall system energy and exergy efficiencies while neglecting pump work.

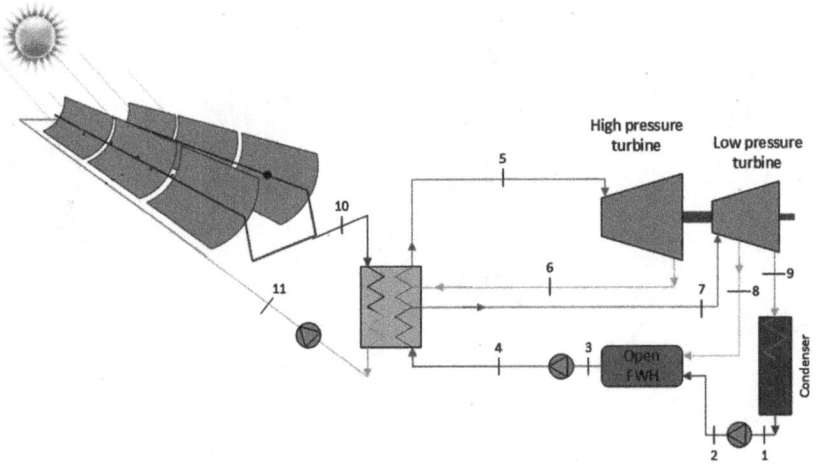

6 A solar collector-based steam Rankine cycle is shown in the figure below. The information given in the table below are known about the system.

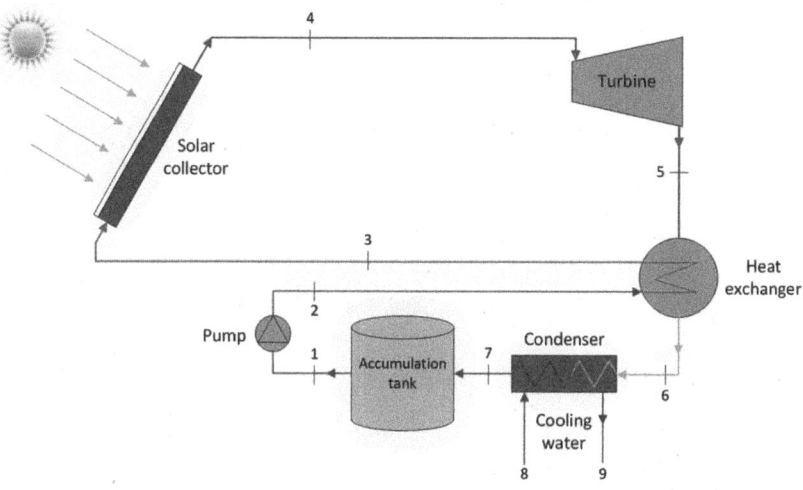

State Point	Temperature (°C)	Pressure (kPa)	Specific Enthalpy (kJ/kg)	Specific Entropy (kJ/kgK)	Specific Exergy (kJ/kg)
0	25	101	104.9	0.367	0
1	53.8	15	225.9	0.7549	5.5
2	53.9	1000	225.8	0.7544	6.5
3	60	1000	252	0.831	?
4	501.3	1000	3482	7.777	?
5	88.3	15	2664	8.199	?
6	75.1	15	2639	8.128	221.5

a Write the mass, energy, entropy and exergy balance equations for all system components.
b For a power output of 3.67 MW, determine the total area of the solar collector required considering a solar radiation intensity of 0.85 kW/m^2.
c Determine the specific exergy at states 3, 4, and 5 if the ambient temperature is 25 °C.
d Determine the exergy destruction rates for turbine and solar collector.
e Calculate the energy and exergy efficiencies of the plant.

7 A space heating application for the house where multiple PV modules are installed on the ground and coupled with a living space is presented in the figure. The system is an example of forced mode PVT module as it uses a small fan to circulate the room air to come in contact with the collector and get heated where the cell temperature T is 45.7 °C.

   Single monocrystalline silicon photovoltaic module has 20% efficiency. The global solar radiation is 700 W/m^2 and PV module area A_{cell} of 1.65 m^2. The mass flow of the circulating air is 0.5 kg/s and the inlet temperature of the air to the DC fan equals to the ambient temperature $T_0 = T_{in} = 15$°C. The convective heat transfer coefficient of the circulating air is 25.41 W/m^{2}K.

a Write the mass, energy, entropy and exergy balance equations.
b Find both system energy and exergy efficiencies.

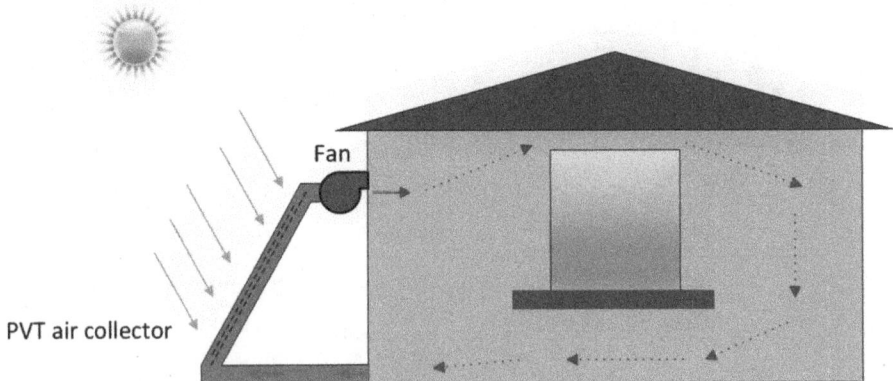

8 Consider a solar-driven ideal steam regenerative Rankine cycle with two feedwater heaters (one closed and one open as mixing chamber) as shown in the figure below. The thermal oil is circulated between the solar collector and boiler. The solar collector efficiency is 60%. The work rate of circulation pump is 1.2 kW. The steam enters the steam turbine at 10 MPa and 600 °C and exits the condenser at 1.2 MPa for the closed feedwater heater and at 0.6 MPa for the open one. The feedwater is heated to the condensation temperature of the extracted steam in the closed feedwater heater. The extracted steam leaves the closed feedwater heater as saturated liquid, which is subsequently throttled to the open feedwater heater. Take the ambient temperature T_0 as 25 °C.

a Write all mass, energy, entropy and exergy balance equations for each component of the system.
b Calculate the net work output (in rate form).
c Calculate both energy and exergy efficiencies of the overall system.

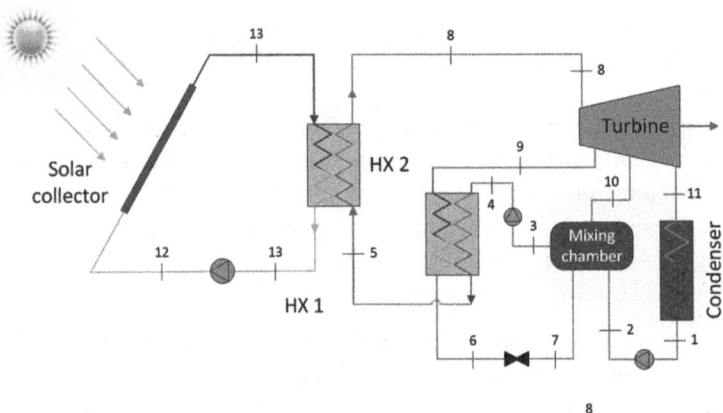

9 A greenhouse dryer uses PVT collector to produce electricity and dry crops. The schematic illustration of the greenhouse is shown below. For 500 h operation in year, the PVT collector heats air to following conditions:

$$\dot{m}_{air} = 15 \ kg/s$$

$$T_{in} = 22°C$$

$$T_{out} = 45°C$$

If the annual electricity generation of PVT is three times higher as much as heat generation of PVT, find the required PV module area. The electricity efficiency of the PV panel is 0.22. Calculate the overall energy and exergy efficiencies. The ambient temperature is assumed as 5 °C. The total yearly solar radiation of the location is 1250 kW/m².

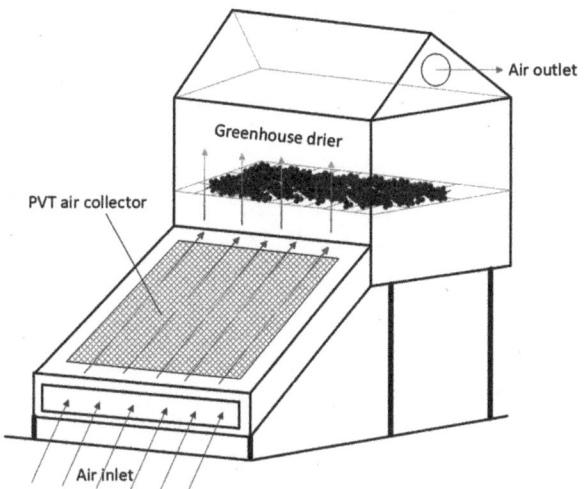

10 A PVT system is designed for a greenhouse, to generate electricity and heat by using PVT modules with 150 m² of total area. Air with 0.2 kg/s mass flow rate is heated from $T_{in} = 10$ °C to $T_{out} = 35$ °C. If PVT manufacturer's data shows $V_{oc} = 720$ V, $S_{sc} = 20.6$ A, FF=0.68.

Consider $h_{air,in} = 280.13$ kJ/kg, $h_{air,out} = 298.18$ kJ/kg, $T_{cell} = 50$ °C, and solar intensity as 400 W/m²

a Calculate the useful heat generation rate of PVT.
b Calculate the electricity generation rate of PVT.
c Calculate the energy and exergy efficiencies of the PVT system.

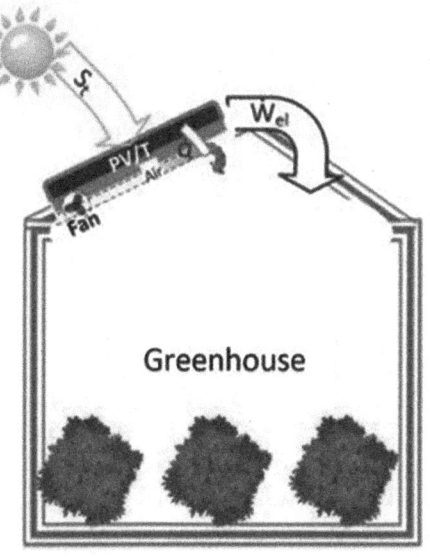

7

Wind Energy

7.1 Introduction

Renewable energy is generally known as the energy coming from a natural source in a renewable manner which is not depleted when used, such as wind. So, this means that we will never run out of wind source as long as the universe survives. Wind energy is not a new source of energy for humanity. Since ancient times, it has been used to meet specific needs, including milling, lifting water, elevating objects, etc. In such systems, the kinetic energy of wind (air) is converted into mechanical power, which is then used for useful purposes. As far as wind energy systems are concerned, the principle remains the same. Today, when wind energy is considered, electricity comes to mind.

In light of today's energy sources and potential choices, some may wonder, why wind power is considered? It is, of course, a renewable resource that should be the first response. It is also environmentally benign and sustainable as it exists naturally. Furthermore, it is abundant; it produces no greenhouse gas emissions or other toxic emissions during power generation; and the source is free. It has relatively higher efficiency than other renewables. It is a commercially viable and cost-effective energy conversion system. It has many types of diverse applications. It can be applied to all sectors. Almost all of them are reliable systems. Today, their manufacturer's warranties have reached up to 20–30 years.

Today, wind energy is a proven and mature technology for the energy sector due to its practical advantages. Its contribution to the world energy supply has been ever-increasing, and Figure 7.1, in this regard, shows how the global wind power capacity has changed over time, between 1995 and 2021 which shows an exponential increase. Its recorded data for the wind power was about 4 GW. Wind power capacity increased significantly after 2005, and the total capacity reached 872 GW alone in 2021.

The benefits of wind energy explain why it is the fastest-growing energy source in the world. On the other hand, researchers and technology developers are trying to address technical and socio-economic challenges in order to expand wind energy's capabilities and community benefits.

This chapter starts with the historical perspective of wind energy. This is followed by the wind effects and global patterns of wind energy. Then, wind power is defined. The types of wind turbines are categorized and presented. Next, it is explained how thermodynamic analysis is applied to the specific wind turbines. Finally, some illustrative examples and case studies are presented.

Figure 7.2 shows the installed wind energy capacity in the world and countries which have a capacity higher than 10 GW. In other words, the countries shown in Figure 7.2 are the top 12

Introduction to Energy Systems, First Edition. Ibrahim Dincer and Dogan Erdemir.
© 2023 John Wiley & Sons, Inc. Published 2023 by John Wiley & Sons, Inc.
Companion Website: www.wiley.com/go/Dincer/Introduction_to_Energy_Systems

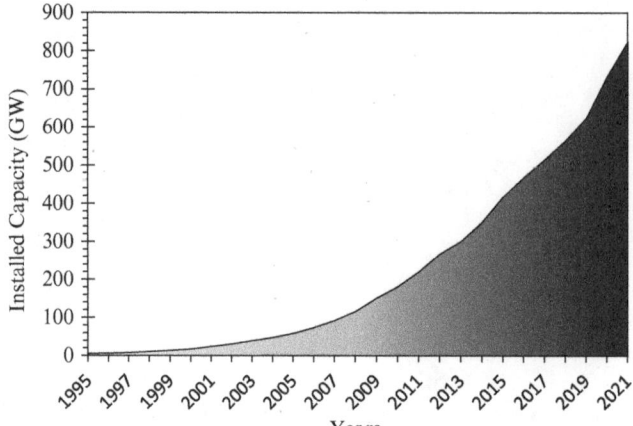

Figure 7.1 The global installed wind power capacity by years (Adapted from [1]).

countries in addition to the world's total installed capacity. China covers almost 40% of the total wind capacity in the world. The US is the second rank with a capacity of 132.7 GW, which covers 16.1% of the global capacity. The installed capacity in Germany is 63.8 GW which is 7.7% of the global capacity. While India has the share of 4.9% in the total capacity, Spain and the UK have 3.3%. Twelve counties given in Figure 7.2 generate almost 85% of the global wind capacity. The capacity of the wind power will increase exponentially with reducing initial and operation costs.

Figure 7.3 demonstrates the global weighted average total installed costs, capacity factors, and levelized cost of electricity for solar PV, offshore wind and onshore wind powers. Solar photovoltaics (PV), onshore and offshore wind power project costs globally fell in 2021. In spite of rising

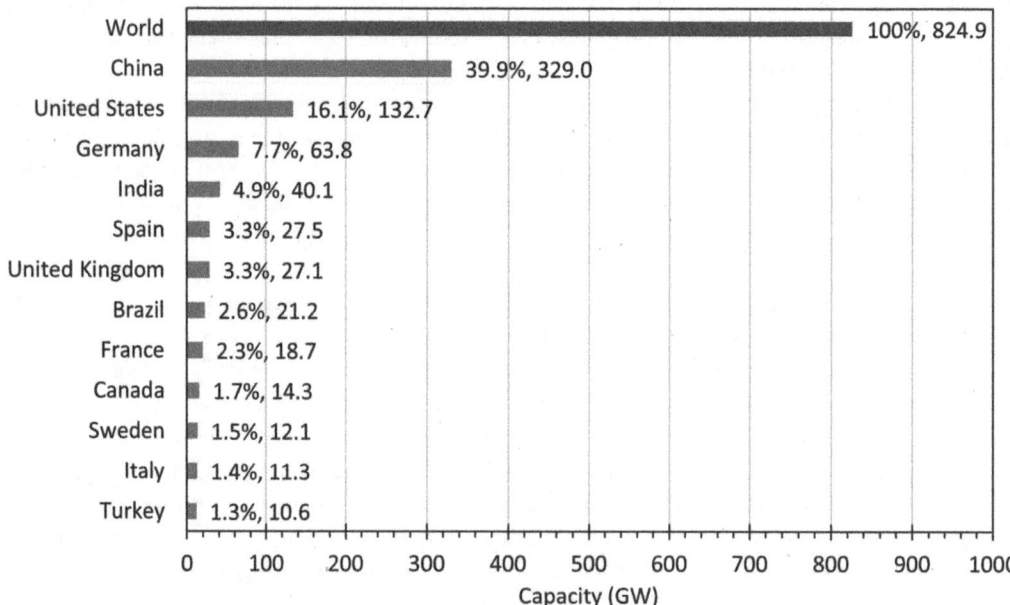

Figure 7.2 The installed wind energy capacities of top 12 countries in 2021 (Adapted from [1]).

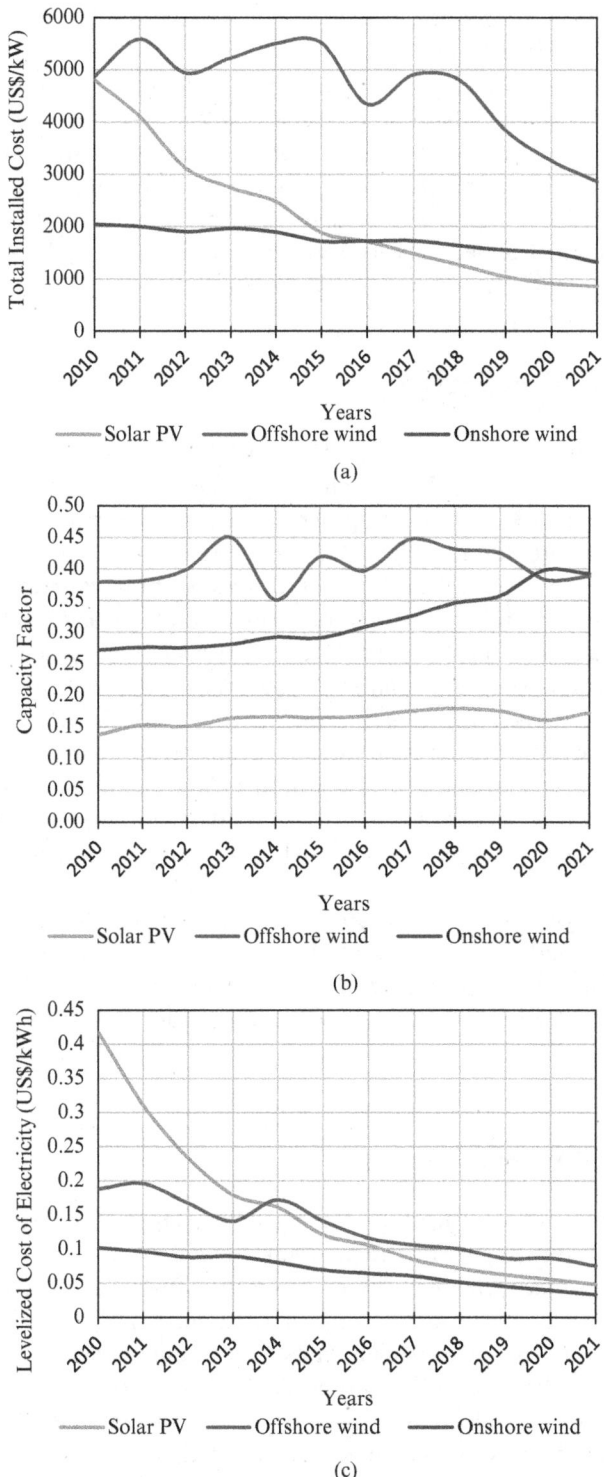

Figure 7.3 (a) Variations of total installed costs, (b) capacity factors and (c) levelized costs of electricity for solar PV, offshore and onshore wind energy systems (Adapted from [2]).

commodity and renewable equipment prices in 2021, the performance of onshore wind projects improved significantly in 2021, which raised capacity factors.

From 0.039 US$/kWh in 2020 to 0.033 US$/kWh in 2021, the global weighted average levelized cost of electricity (LCOE) of new onshore wind projects dropped by 15%. In 2021, China continued to dominate new onshore wind capacity additions, and its wind turbine prices fell against the rest of the world. For new onshore wind projects excluding China, electricity costs fell by a more modest 12% year-on-year to 0.037 US$/kWh. During 2021, the offshore wind market saw unprecedented growth (21.5 GW), driven by China's increased capacity additions and a drop in the global weighted average cost of electricity from 0.086 US$/kWh to 0.075 US$/kWh.

7.2 Historical Development of Wind Energy

Using wind energy for useful purposes is not a new technology. It is one of the oldest technologies historically, and its use started with sailboats in 5000 BC. This was followed by windmills. Today, wind energy is used to be a address for electricity as it is mainly used for generating electricity. Some historical developments and events of wind energy are shown in Figure 7.4. Some are summarized here in brief. The first electricity from wind energy was generated in 1888 by Charles F. Brush by using a windmill. In 1908, 72 wind power systems were alone deployed in Denmark. The vertical-axis wind turbine is patented by Georges Jean Marie Darrieus in 1931. The capacity of the wind power systems was less than one megawatt. The first megawatt-size wind turbine was connected to the grid in 1941.

After the 1940s, wind energy systems received great attention, and the NASA launched a program to develop utility-scale wind turbines. The Danish wind turbine manufacturer, Vestas, currently covers one-fifth of the total wind power production; produced their first wind turbine in 1978. At the same year, the first US wind farm became operational to meet the power demand of the 4149 residential units. In 1982, the first fruit of the NASA's wind turbine program was harvested as a 3.2 MW wind turbine.

All turbines up to this point required high wind speed. In 1999, Vestas started to develop their new generation wind turbine to be able to work under low-speed wind speeds. In 2010, the cost of electricity generation from wind was reduced to 0.08 US$/kWh and 0.05 US$/kWh in 2017. Today, wind power has reached 906 GW of installed capacity. It is expected that this capacity will further continue growing exponentially.

7.3 Wind Effect and Global Wind Patterns

The wind is basically formed up according to the world's movement and the temperature difference between day and night. Figure 7.5 illustrates the wind patters around the globe. Today's meteorologists define a prevailing wind as a surface wind blowing predominantly from a particular direction in a particular region. Over a particular point on the Earth's surface, dominant winds are the winds with the highest speeds. Global patterns of movement in the Earth's atmosphere determine a region's prevailing and dominant winds. Easterly winds generally characterize low latitudes. The strength of westerly winds is heavily influenced by polar cyclones at mid-latitudes. In areas with light winds, the sea breeze/land breeze cycle dominates the wind pattern; in areas with variable terrain, mountain and valley breezes dominate the wind pattern. The wind flow can be enhanced by highly elevated surfaces, which induce a thermal low.

It is possible to display the direction of the prevailing wind using wind roses. It is possible to develop wind erosion prevention strategies for agricultural lands, such as those across the Great

2022
- The installed capacity of wind power has reached 906 GW globally.

2010
- The cost of electricity generation from wind was reduced down to $0.08/kWh.

2009
- The first large-capacity floating wind turbine in the world started operation on the coast of Norway.

2007
- The UK announced plans to install thousands of offshore wind turbines, enough to provide electricity for every home in Britain by 2020.

2003
- The UK's first offshore wind farm was opened in north Wales. It included 30 wind turbines, each with a power capacity of 2 megawatts.

1999
- Vestas launched a wind turbine with "OptiSpeed," which was able to make it suitable for low-wind sites.

1995
- The wind farm in Cadiz became operational with 90 turbines.

1987
- A 3.2-megawatt wind turbine was developed by the NASA wind turbine program.

1980
- The world's first wind farm including 20 turbines was operational.

1978
- Danish wind turbine manufacturer, Vestas produced their first wind turbine.
- The first US wind farm is put online, producing enough power for up to 4149 homes.

1975
- A NASA wind turbine program was launched to develop utility-scale wind turbines.

1941
- The first megawatt-size wind turbine was connected to a local electrical distribution grid.

1931
- A vertical-axis wind turbine design called the Darrieus wind turbine was patented by Georges Jean Marie Darrieus.

1908
- 72 electricity-generating wind power systems became operational across Denmark. The windmills range from 5 kW to 25 kW in size.

1888
- Charles F. Brush achieved generating electricity by using a windmill.

1880
- Early experimental wind power generating works were done by Aermotor Company.

1300
- The first horizontal-axis windmills were deployed in Europe.

500-900
- First windmills were developed in Persia.

5000 BC
- Sailboats were used on the Nile indicate the power of wind.

Figure 7.4 Historical perspectives of wind energy.

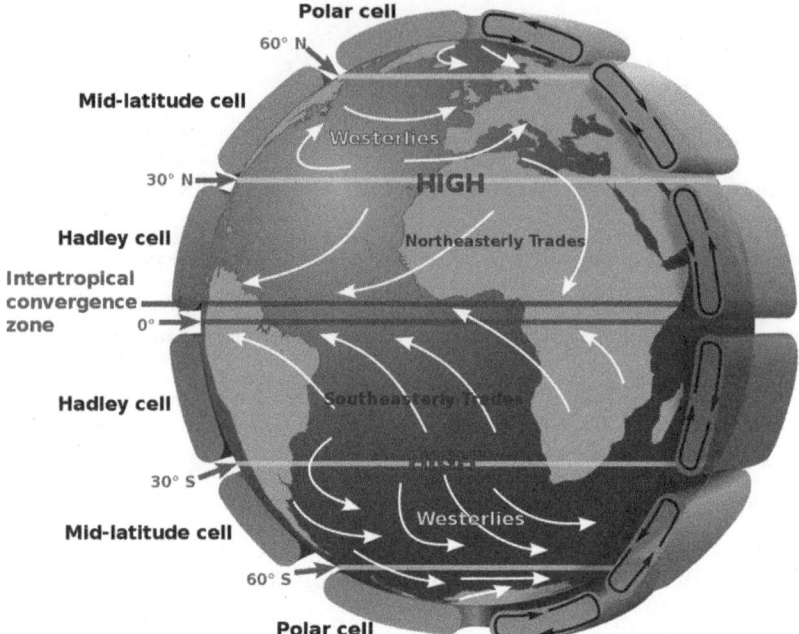

Figure 7.5 An illustration of the global wind patterns as part of Earth's atmospheric circulation (Adapted from [3]).

Plains, based on knowledge of the prevailing wind. Coastal and desert sand dunes can align themselves perpendicular to prevailing winds. The flight of insects follows the prevailing wind, but birds are less affected by it. It is common for mountain locations to experience significant gradients of rainfall, ranging from wet to desert-like conditions along slopes facing windward. Due to the uneven heating of the Earth, prevailing winds can vary.

Before starting the wind energy systems, we need to glance at the mechanism of how the wind is formed. The primary mechanism of wind forming is the temperature changes due to solar energy. The sun shines on land and water. The land heats up faster than the water. The temperature of the air in contact with the land increases, and it rises due to buoyancy forces depending on density changes. The density of air increases with increasing temperature. The cold air over the water takes the air volume rising place. This kind of mechanism is illustrated in Figure 7.6.

7.4 Wind Power

The characteristics of wind affect the design of systems to exploit its power. The friction slows wind as it flows over the ground and vegetation, resulting in little wind at ground level. Major landforms can accelerate wind, resulting in some regions being very windy while others remain relatively calm. By converting wind power into electricity, wind power can be transported over long distances, serving the needs of urban centers with large populations.

One of the fastest-growing renewable energy sources in the world is wind energy. Technology developments, environmental concerns, and the continuous increase in conventional energy use have led to a reduction in relative wind energy costs in many locations to economically acceptable

Figure 7.6 Wind effect from water sources to the mainland.

levels as a result of concerns over fossil fuel demand. As a result, many jurisdictions are considering wind energy farms as an alternative energy source because they have been installed and operating for more than 25 years.

Electricity is produced by wind turbines by converting the kinetic energy of the wind to shaft power. The shaft power is transmitted to the generator by transmission. Figure 7.7 demonstrates the wind power generation mechanism. Modern large-scale wind turbines convert wind kinetic energy into rotational motion by mounting a rotor on which the device to catch the wind is mounted. Wind turbines usually have a three-bladed assembly at the front, but other geometries and types are also available. Wind turbines have a rotor that spins a shaft, which transfers motion to the nacelle. A gearbox inside the nacelle increases the rotational speed of the slowly rotating shaft. Several hundred volts of electricity are generated by converting the rotational motion of the output shaft into electricity at a medium voltage. A transformer (a few thousand volts) increases the voltage of the electric power to a level more appropriate for distribution (a few thousand volts) by passing it through heavy electric cables inside the tower. By using higher voltage electricity, fewer power losses and less heat will be generated through electric lines as a result of fewer resistances. In order to combine the power from other turbines, the distribution-voltage power flows through underground cables or other lines. In many cases, the electricity is distributed for use to nearby farms, residences, and towns. Instead, distribution-voltage power is sent to a substation, where the voltage is dramatically raised to transmission-voltage levels (a few hundred thousand volts) and transmitted through transmission lines many kilometers away.

One may write the kinetic energy of wind as calculated by the following equation:

$$E_{wind} = \frac{1}{2} \dot{m}_{air} V^2 \tag{7.1}$$

where $\dot{m}_{air}$ is the flow rate of air and V^2 is the velocity of the air. It is clear from Eq. 7.1 that the energy content of the wind is primarily dependent on the wind velocity. Higher velocity means higher wind energy. Mass flow rate is calculated depending on cross-sectional area of the wind flow varied from wind turbine. As the kinetic energy equals its exergy content, both are equal to each other.

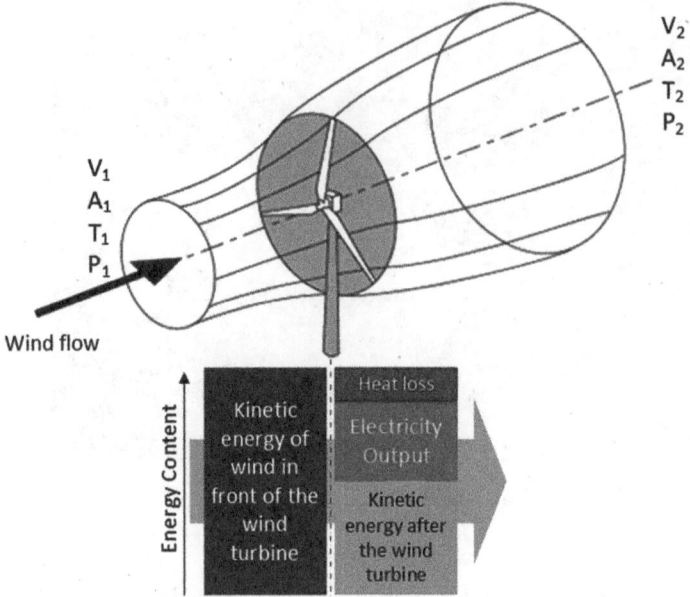

Figure 7.7 Illustration of conversion of wind energy to the power.

Illustrative Example 7.1 Consider a wind power generating system where the flow rate of the air is measured to be 1000 m³/h and wind speed is observed to be 15 m/s. Calculate the energy and exergy contents of the wind flow. The density of air can be assumed to be 1.10 kg/m³.

Solution:
In order to calculate the energy content of the wind, Eq. 7.1. should be used. First, we need to calculate the mass flow rate:

$$\dot{m}_{air} = \rho_{air}\dot{V}_{air} = 1.10 \times 100,000 \times \frac{1}{3600} = 30.5 \; kg/s$$

Thus, the wind energy is then obtained as:

$$E_{wind} = \frac{1}{2} \times 30.5 \times 15^2$$

Finally, the wind energy is found to be E_{wind} = 3431.3 W.

As the energy and exergy contents of mechanical works or electrical powers are identical, the exergy content of wind becomes the same as Ex_{wind} = 3431.3 W.

7.5 Classification of Wind Turbines

When wind turbines are considered, 3-blade wind turbines come to people's mind first as they are commonly deployed all around the world as seen by many almost everywhere; however, basically, they are classified as the horizontal and vertical axis, depending on the direction of the air flow. Figure 7.8 shows the basic classification of the wind turbines. A wind turbine is often described by the position of the

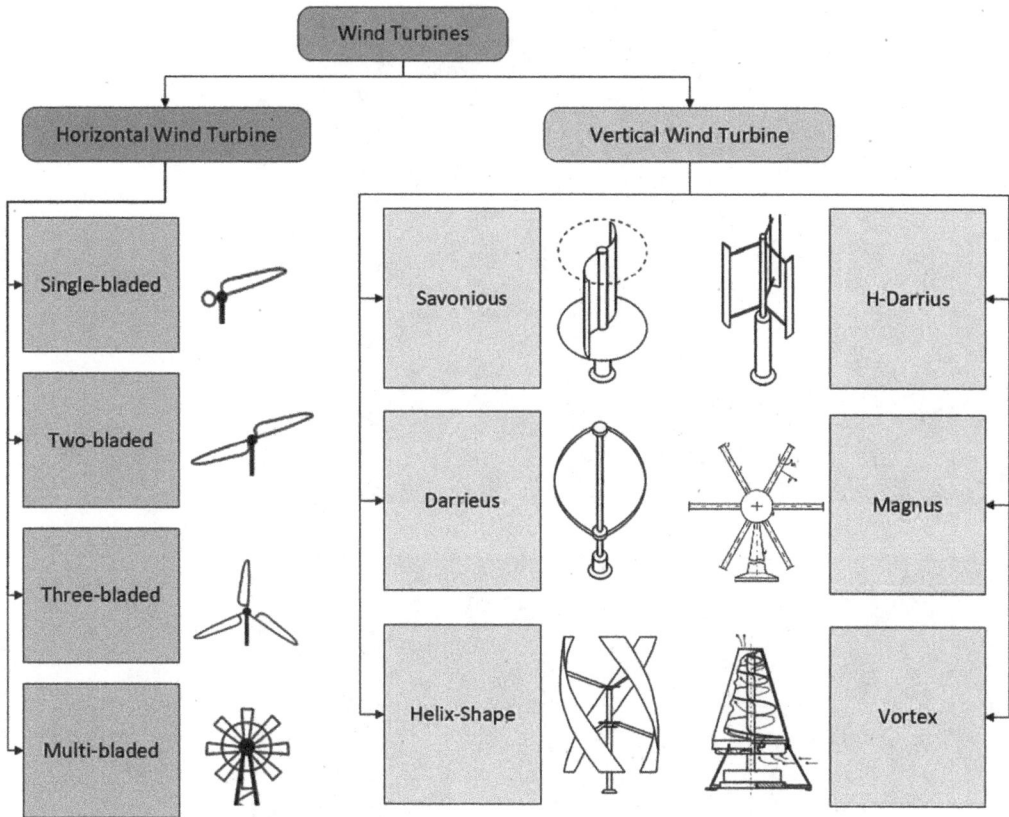

Figure 7.8 Classification of the wind turbines.

main rotating shaft with respect to the earth's surface. Therefore, a horizontal-axis wind turbine has a shaft that is parallel to the earth's surface. The rotation of the rotor is perpendicular to the airflow across the earth's surface. In a vertical-axis wind turbine, the shaft is perpendicular to the earth's surface. In other words, the rotation of the rotor is parallel to the airflow. Both horizontal- and vertical-axis wind turbines have advantages and disadvantages according to each other. All wind turbines will be discussed in the following sections. Also, they can also include blades or be bladeless.

7.6 Horizontal-Axis Wind Turbines

The horizontal-axis wind turbines are classified according to the number of blades which are single-bladed, double-bladed, triple-bladed, and multibladed, or direction of the air through wind turbine which is upwind and downwind. Figure 7.9 shows the types of horizontal wind turbines. Wind turbines come in a variety of sizes. Wind turbines generate electricity based on the length of their blades. Wind turbines with a generating capacity of 10 kW can easily power a single home. Electricity generating capacities of the largest wind turbines in operation are around 15 MW, and larger turbines are being developed. Wind power plants, or wind farms, are often created by grouping large turbines together.

Figure 7.9 Views of the horizontal-axis wind turbines [4] NASA / Wikimedia Commons / Public Domain/ Dirtsc / Wikimedia Commons / CC BY-SA 3.0.

In addition to the number of blade, it is possible to classify the wind turbine according to the incidence angle of wind which are upwind and downwind wind turbines, as illustrated in Figure 7.10. Horizontal-axis wind turbines are typically designed as upwind configurations. It is crucial to position the rotor blades far enough from the turbine tower and design them as inflexible as possible in order to avoid blade strikes. A complicated yaw control system is also required to keep the turbine rotor facing the incoming wind. To overcome these disadvantages, the downwind configuration of the turbine is proposed. Because rotor blades can be designed more flexibly since there is no danger of blade strikes. Also, if the nacelle is designed appropriately to passively follow the incoming wind direction, a yaw control system can be eliminated.

In general, horizontal-axis turbines have three blades, like airplane propellers. Vertical-axis turbines can reach heights of 20–25 stories and have blades that measure more than 40 m in length.

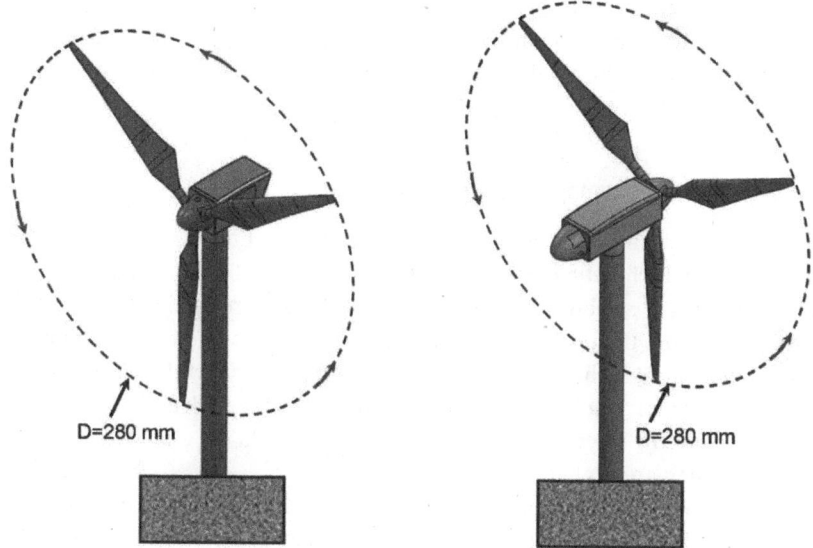

Figure 7.10 Types of horizontal wind turbines according to the incidence angle of wind (a) upwind and (b) downwind [5] / with permission of Elsevier.

Electricity is generated more efficiently by taller turbines with longer blades. Most wind turbines in use today are horizontal-axis turbines. A Chinese wind turbine manufacturer currently holds the record for the biggest wind turbine in the world which is the MySE 16.0-242, an offshore hybrid drive wind turbine. Its diameter is 242 m long, its blades are 118 m long, and the turbine has a 46,000 m^2 swept area.

A large number of wind power in the world today is generated by horizontal-axis wind turbines with three blades upwind of the tower. The main rotor shaft and electrical generator are at the top of the tower, and they must be pointed into the wind. Main parts of three-bladed upwind wind turbine is shown in Figure 7.11.

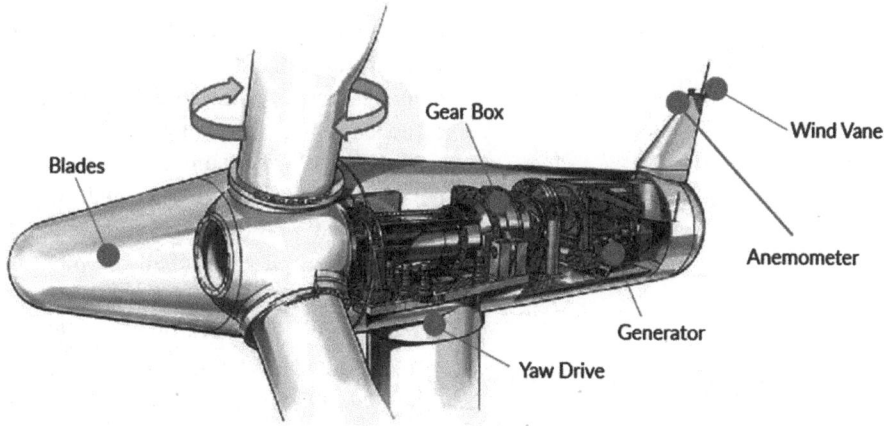

Figure 7.11 Main parts of a common three-bladed upwind wind turbine.

The direction of the wind turbine according to the wind direction is adjusted in small turbines by a simple wind vane and in large turbines with a wind sensor coupled with a yaw system. A gearbox transforms the slow rotation of the blades into a faster rotation that can drive an electrical generator.

7.6.1 Betz Limit, Coefficient of Performance, and Tip-Speed Ratio

Before starting the analysis of the wind turbine, we need to explain Betz limit. Betz limit can be defined as the Carnot efficiency which defines the limit of the heat and cooling machine working between hot and cold mediums. It is exhibited by a German physicist, Albert Betz. Betz limit tells us no wind turbine can convert more than 59.3% of the kinetic energy of the wind into mechanical energy, that is shaft power. Betz limit is the theoretical limit for wind turbine, which is able to be achieved. In reality, turbines cannot reach the Betz limit, and common efficiencies are in the 35–45% range. In order to account the electricity generation, we need to add the efficiencies of the transmission and generator, which are explained above.

For example, the rate of electricity conversion is 70% for a wind turbine, the maximum wind power generating efficiency can be achieved with this wind turbine will 0.75×0.593 (which is the Betz limit) $= 0.4447$. In other words, the electricity conversion ratio will be 0.4447. In other words, a maximum of 44.47% of the kinetic energy of wind can be converted to electricity. This parameter is called as the coefficient of performance. The coefficient of performance, C_p is the ratio of the electricity generated to the kinetic energy of the wind. It can be written as follows:

$$C_p = \frac{Electricity\ produced\ by\ wind\ turbine}{Total\ wind\ energy\ input} \tag{7.2}$$

where the total energy available in the wind can be calculated by using Eq. 7.1.

Another significant parameter for the wind turbine is the tip-speed ratio (TSR or λ). It is the ratio between the tangential speed of the tip of a blade and the actual speed of the wind. It can be formulated to be:

$$TSR = \lambda = \frac{Tip-speed\ of\ blade}{Wind\ speed} = \frac{wR}{V} \tag{7.3}$$

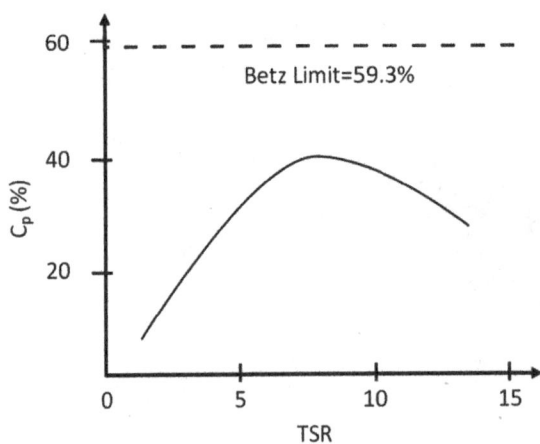

where w is the rotational speed of the rotor which turbine blades are attached. R is the diameter of the rotor, in other words, the blade length. V is the wind speed.

In general, it is assumed to depend on both the tip-speed ratio and pitch angle. Here is a plot of the power coefficient versus the tip-speed ratio when the pitch is held constant, in Figure 7.12. The performance of the wind turbine should be below the Betz limit. It is clear from Figure 7.12 that not only wind speed but also design of the wind turbine is critical for the performance of wind turbine. A value between 7 and 8 is optimal. Additionally,

Figure 7.12 Tip-speed ratio versus coefficient of performance for a wind turbine.

it should be noted that higher tip-speeds cause higher noise levels and require stronger blades due to larger centrifugal forces.

7.6.2 Analysis of Horizontal-Axis Wind Turbines

A wind turbine is a device for extracting kinetic energy from the wind. In other words, wind speed should decrease after the wind turbine, as it loses its kinetic energy. Assuming that the affected mass of air remains separate from the air which does not pass through the rotor disc and does not slow down, a boundary surface can be drawn containing the affected air mass and this boundary can be extended upstream as well as downstream forming a long stream-tube of circular cross section. Figure 7.13 shows such a concept. The air flowing along the stream-tube does not cross the boundary, so its mass flow rate will be the same at all stream-wise positions. As the air inside the stream tube slows down, but does not compress, its cross-sectional area must expand to accommodate the slower movement. It works like a gas turbine that expands gas flow through it.

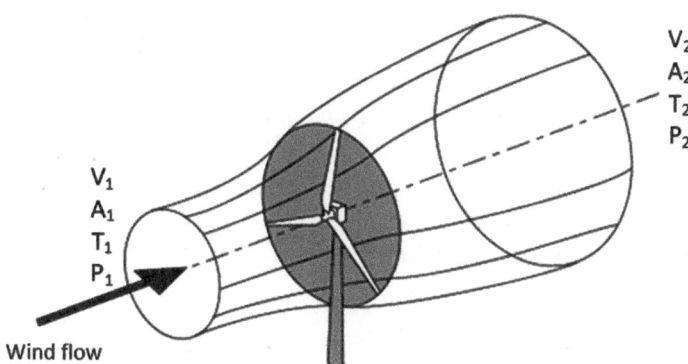

Figure 7.13 Working mechanism of the horizontal wind turbines.

Like in every energy system, thermodynamic analysis is considered a unique tool to design and assess wind turbines. This section describes a complete thermodynamic approach for wind energy systems. First, we need to define the potential of wind power. Let's consider a horizontal-axis wind turbine with the swept areas and diameter indicated as illustrated in Figure 7.14.

In practice, the energy which can be obtained from the wind, as a function of kinetic energy, can be defined as:

$$E_{wind} = \frac{1}{2}\rho A V^3 \tag{7.4}$$

where ρ is the air density, A is the swept area, and V is the wind speed. The swept area is calculated as follows:

$$A = \pi R^2 \tag{7.5}$$

The kinetic energy of wind is calculated by the following equation:

$$E_k = \frac{1}{2}\dot{m}_{air} V^2 \tag{7.6}$$

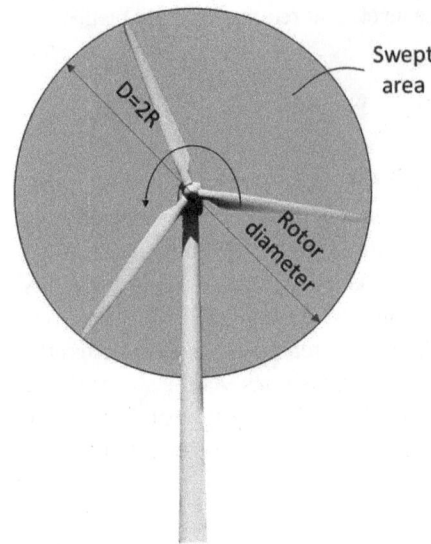

Swept area

Rotor diameter

Figure 7.14 Illustration of the analysis of wind turbine.

where $\dot{m}_{air}$ is the mass flow rate of the air. As we know the cross-sectional area of the wind flow through the wind turbine:

$$\dot{m}_{air} = \rho AV \tag{7.7}$$

Thus, the wind power can be found as:

$$P_{wind} = \frac{1}{2}\rho\pi R^2 V^3 \tag{7.8}$$

which is defined as the wind power potential. However, in order to perform a complete thermodynamic analysis, it is required to calculate the electricity output from the turbine. In a wind turbine, the rotor is attached to a transmission and generator. The maximum electricity to be able to be achieved from wind turbine is calculated as follows:

$$P_{electricity,max} = C_B \eta_g \eta_t P_{wind} \tag{7.9}$$

where C_B is the Betz limit. η_g and η_t are efficiency values for generator and transmission, which can be obtained from the manufacturers. Generally, $\eta_g = 0.75$ and $\eta_t = 0.95$.

The actual electricity production can be calculated as follows:

$$P_{electricity,actual} = C_p \eta_g \eta_t P_{wind} \tag{7.10}$$

where C_p is the coefficient of performance for wind turbine.

Also, the exergy content of the wind can be calculated by:

$$Ex_{wind} = E_{generated} + \dot{m}C_p(T_2 - T_1) + \dot{m}T_0\left[C_p \ln\left(\frac{T_2}{T_1}\right) - R \ln\left(\frac{P_2}{P_1}\right) - \frac{Q_{loss}}{T_0}\right] \tag{7.11}$$

Illustrative Example 7.2   Consider a wind turbine which has a 50 m long blade. The average wind speed is measured to be 12 m/s. The site of the installation is about 500 m above sea level. Calculate the maximum electricity generation rate to be achieved from the wind turbine. Assume the following: $\eta_g = 0.75$ and $\eta_t = 0.95$. Air density will be taken as 1.16 kg/m^3. The heat losses are neglected.

Solution:

First, we need to calculate the wind power by using Eq. 7.8:

$$P_{wind} = \frac{1}{2}\rho\pi R^2 V^3$$

$$P_{wind} = \frac{1}{2} \times 1.16 \times \pi \times 50^2 \times 12^3$$

Thus, the wind power is found as $P_{wind} = 7,867,585$ W.

The maximum electricity that can be generated is calculated via Eq. 7.9.

$$P_{electricity,max} = C_P \eta_g \eta_t P_{wind}$$

$$P_{electricity,max} = 0.593 \times 0.75 \times 0.95 \times 7,867,585$$

Thus, the electricity generated from the wind turbine is found to be $P_{electricity,max} = 3,324,153$ W or 3.32 MW.

Illustrative Example 7.3   Consider the wind turbine given in the above example. The manufacturer declares that the coefficient of performance for the wind turbine is 0.40. Considering C_p provided by the manufacturer. Determine the actual electricity to be achieved.

Solution:
The wind power is found to be $P_{wind} = 7,867,585$ W in the previous example. The actual electricity can be found from Eq. 7.10.

$$P_{electricity,actual} = C_p \eta_g \eta_t P_{wind}$$

$$P_{electricity,actual} = 0.40 \times 0.75 \times 0.95 \times 7,867,585$$

Thus, the electricity generated from the wind turbine is found to be $P_{electricity,actual} = 2,242,261.4$ W or 2.24 MW.

7.6.3   Thermodynamic Analysis of Horizontal-Axis Wind Turbines

In order to perform a complete thermodynamic analysis, first, we need to define system boundaries. Figure 7.13 demonstrates the thermodynamic system boundaries. As seen from Figure 7.13, the wind turbine behaves as a gas turbine which expands the air.

In order to perform a complete thermodynamic analysis, four balance equations, which are the mass, energy, entropy, and exergy balance equations, are required to be written for the system shown in Figure 7.13 accordingly.

The mass balance equation for the control volume of the wind turbine is written as follows:

$$\dot{m}_1 = \dot{m}_2 \tag{7.12}$$

or

$$\rho_1 A_1 V_1 = \rho_2 A_2 V_2 \tag{7.13}$$

The energy balance equation is written as follows:

$$\dot{m}_1 (h_1 + ke_1 + pe_1) = \dot{m}_2 (h_2 + ke_2 + pe_2) + \dot{Q}_{loss} + \dot{W}_{wt} \tag{7.14}$$

Here, $(h_1 + ke_1 + pe_1)$ and $(h_2 + ke_2 + pe_2)$ are the total energy of air at the inlet and outlet of the control volume. For a wind turbine, potential energy changes can be neglected.

The entropy balance equation can be written for the control volume, where some heat losses are considered from the turbine as whole, as follows:

$$\dot{m}_1 (s_1) + \dot{S}_{gen} = \dot{m}_2 (s_2) + \frac{\dot{Q}_{loss}}{T_0} \tag{7.15}$$

Finally, the exergy balance equation can be written as:

$$\dot{m}_1\left(ex_1 + ke_1 + pe_1\right) = \dot{m}_2\left(ex_2 + ke_2 + pe_2\right) + \dot{Ex}_{\dot{Q}_{loss}} + \dot{W}_{wt} + \dot{Ex}_{dest} \tag{7.16}$$

Also, the relation between exergy destruction and entropy generation can be written as follows:

$$\dot{Ex}_{dest} = T_0 \dot{S}_{gen} \tag{7.17}$$

The energy and exergy efficiencies for a horizontal wind turbine can be written as follows:

$$\eta_{en} = \frac{\dot{W}_{electricity}}{\dot{E}_{wind}} \tag{7.18}$$

and

$$\eta_{ex} = \frac{\dot{W}_{electricity}}{\dot{Ex}_{wind}} \tag{7.19}$$

Now, let's use these balance equations in the following illustrative example.

Illustrative Example 7.4 Consider a wind farm which has 50 wind turbines. The blade length of the wind turbine is 50 m. The wind speed is measured to be 9.44 m/s before wind turbine and the same for each turbine. The density of the air is 1.1839 kg/m³. The outlet velocity is measured as 5.28 m/s. There is 3.5 kW of heat loss in the turbine. Calculate the total electricity output from this wind farm. Further find the energy efficiency of this farm.

Solution:
In order to calculate the electricity generated with wind turbines, it is required to calculate the mass flow rate. From Eqs. 7.10 and 7.11:

$$\dot{m}_1 = \rho_1\, A_1\, V_1 = 0.1839 \times \pi \times 50 \wedge 2 \times 9.44^2 = 128,645.9 \; kg/m^3$$

The energy balance equation for the control volume is

$$\dot{m}_1(h_1 + ke_1) = \dot{m}_2(h_2 + ke_2) + \dot{W}_{wt} + \dot{Q}_{loss}$$

Assuming there is no enthalpy and entropy changes in the air,

$$128,645.9 \times \frac{9.44^2}{2} = 128,645.9 \times \frac{5.28^2}{2} + \dot{W}_{wt} + 3500$$

$$5,734,949.4 = 1,796,720.9 + \dot{W}_{wt}$$

Thus, the turbine output is found to be $\dot{W}_{wt} = 3,938,228.5$ W which corresponds to 3.9 MW.

In the farm, there are 50 wind turbines. Thus, the total wind power to be obtained from the farm is 195 MW.

In order to calculate the energy and exergy efficiencies, we need to calculate the energy and exergy content of the wind. The wind energy can be calculated by using Eq. 7.4.

$$E_{wind} = \frac{1}{2}\dot{m}_1 V^2 = \frac{1}{2} \times 128,645.9 \times 9.44^2$$

Thus, the kinetic energy of the wind is found to be 5,732,049.6 W or 5.73 GW.

So, energy efficiency of the wind farm can be calculated as follows:

$$\eta_{en} = \frac{\dot{W}_{electricity}}{\dot{E}_{wind}} = \frac{1950MW}{5732MW}$$

Thus, energy efficiency of the farm is found to be $\eta_{en} = 34.01\%$.

The size and number of the turbine blades have a great impact on the system performance and capacity. Although three-bladed wind turbines are widely used in the practical applications, the following question may arise that how number of blades are affecting the performance of the wind turbines. Figure 7.15 shows us the effect of the number of blades on the wind turbine performance over TSR and C_p relation. TSR can be found with Eq. 7.3. In order to write the equation in an open form:

$$TSR = \lambda = \frac{Tip - speed \ of \ blade}{Wind \ speed} = \frac{wR}{V} = \frac{2\pi n}{60V} \tag{7.20}$$

When the value of TSR>1 means more lift force in part of the blades, but if TSR<1 more drag in part of the blades. TSR maximum or TSR_{max} can be determine related with number of blades as,

$$N_b = \frac{4\pi}{TSR_{max}} \tag{7.21}$$

where N_b is number of blades ($N_b = 2,3$, or 4).

The number of blades will influence the performance of the wind turbine, conforming with the solidity of the turbine. Solidity of wind turbines is the ratio between the area of blades and rotor sweeping, determined by

$$\sigma = \frac{N_b C}{r} \tag{7.22}$$

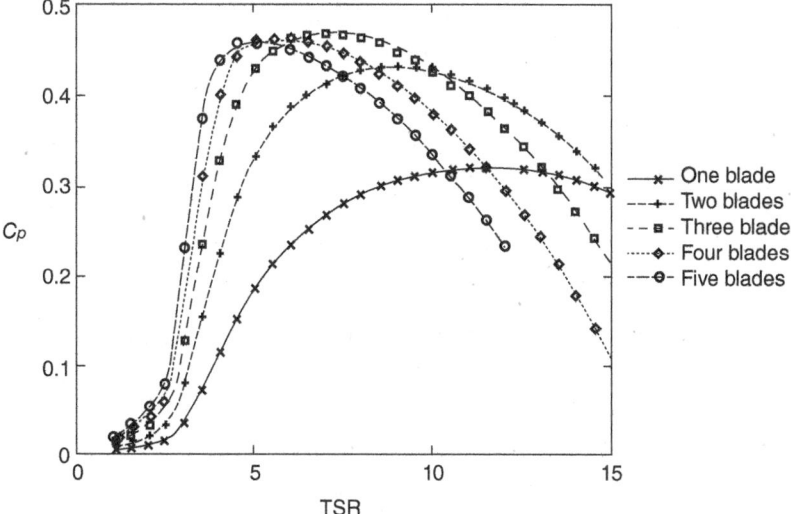

Figure 7.15 Curve of power coefficient C_p and tip speed ratio TSR of wind turbine [5] / with permission of Elsevier.

where C is the length of blade chord (m) and r is radius of rotor (m).

It is clear from Figure 7.15 that three blades are ideal for the horizontal-axis wind turbine for energy output. Also, three-blade can provide better balance while rotating and cost effectiveness. On the other hand, four and five blades wind turbines can show better performance under lower wind speed conditions. They provide higher C_p than three-blades one up to TSR of 5.

Illustrative Example 7.5   A small-scale wind farm which has a 100 kW of capacity is desired to be built for an apple orchard. The average wind speed is measured to be 15 m/s. The blade length for each turbine is 20 m. The rotational speed of the rotor is 5 rad/s. Calculate the C_p values for from one to five blades. Discuss which one has to be preferred.

Solution:

In order to obtain the C_p for different number of blades, we need to use Figure 7.15. First, we need to calculate TSR value for the wind turbines. TSR can be calculated by using Eq. 7.18.

$$TSR = \frac{wR}{V} = \frac{5 \times 20}{15} = 6.66$$

So, we know the TSR for the wind turbines. To use Figure 7.15, we need to draw a vertical line (red) from 6.6 of TSR as shown below. Horizontal lines (blue) are drawn from the intersections for each of the number of blade curve.

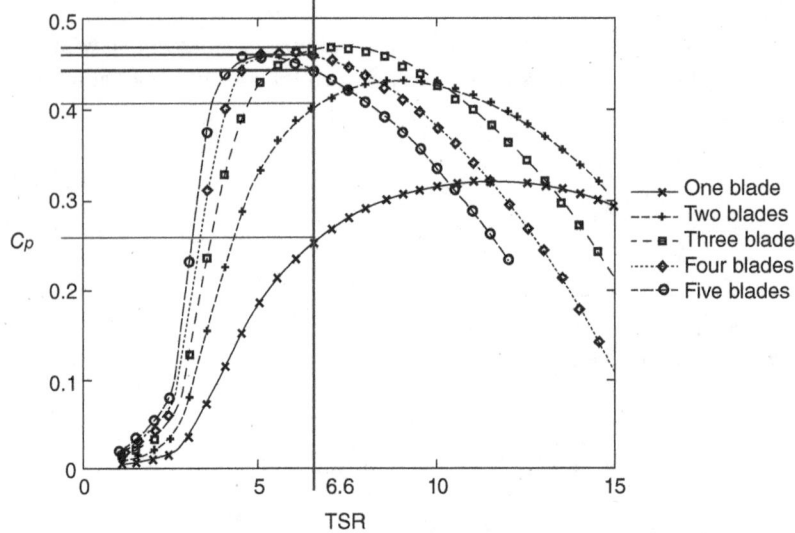

C_p values for each wind turbine which has different number of blades can be found as:

$$C_{p,\text{one-bladed}} = 0.262$$

$$C_{p,\text{two-bladed}} = 0.405$$

$$C_{p,\text{three-bladed}} = 0.484$$

$$C_{p,\text{four-bladed}} = 0.468$$

$$C_{p,\text{five-bladed}} = 0.437$$

It is clear from these values that the three-bladed wind turbines gives the best performance and provides the best performance than others under the same conditions considered.

7.7 Vertical-Axis Wind Turbines

Wind turbines with vertical rotor shafts have their main rotor shafts vertically oriented. An advantage of this arrangement is that the turbine does not need to be pointed into the wind to be effective, which is useful on sites where wind direction is highly variable. In addition, when the turbine is integrated into a building, it is inherently less steerable. A direct drive from the rotor assembly to the ground-based gearbox allows the generator and gearbox to be placed near the ground, making maintenance easier. Over time, these designs produce much less energy, which is a major disadvantage.

Standard horizontal turbine designs are much more efficient than vertical turbine designs. One of the main disadvantages is the relatively low rotational speed, resulting in a higher torque and, therefore, a higher cost of the drive train. It has an inherently lower power coefficient. Due to the highly dynamic loading on the blade, there may be significant fatigue stress-related issues with mechanical vibrations.

The subtypes of the vertical-axis designs are summarized in the following sections:

Darrieus Turbine:
Darrieus turbines, also called eggbeater turbines, are named after the French inventor Georges Darrieus. Figure 7.16 shows a Darrieus turbine. It tends to be more efficient, but it produces cyclical stress on the tower, which contributes to poor reliability. Due to their very low starting torque, they also require an external power source or an additional Savonius rotor to turn. Three or more blades reduce torque ripples, resulting in a more solid rotor. By dividing the blade area by the rotor area, solidity is measured. Unlike older Darrieus turbines, the newer ones have an external superstructure connected to the top bearing instead of guywires.

Savonius Turbine:
Savonius wind turbines are vertical-axis wind turbines that convert wind force into torque on rotating shafts. Figure 7.17 demonstrates the view of a Savonius turbine. The turbine consists of several aero foils mounted vertically on a rotating shaft or framework, either on the ground or airborne. Savonius wind turbine was invented by the Finnish engineer Sigurd Johannes Savonius in 1922 and patented in 1926. The Savonius turbine is one of the simplest turbines. In terms of aerodynamics, it is a drag-type device with two or three scoops. When moving against the wind, the scoops experience less drag because of their

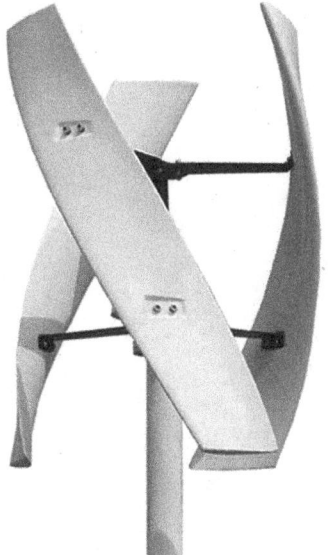

Figure 7.16 The view of the Darrieus turbine.

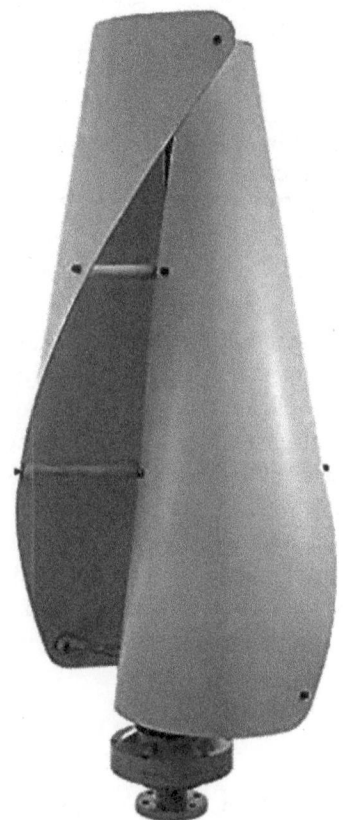

Figure 7.17 Illustration of a Savonius turbine.

Figure 7.18 The view of the magnus turbine (Adapted from [8] Jinbo et.al., 2015 / Hindawi / CC BY 3.0).

curvature. The advantages of vertical-axis wind turbines include low noise levels, the ability to operate with low wind speeds, and relative independence from wind direction.

Giromill Turbine:

A giromill is a Darrieus-type vertical-axis wind turbine. Unlike the more common Savonius turbine, the giromill wind turbine uses lift forces generated by vertical airfoils to convert wind energy into rotational mechanical energy. The giromill turbine is included in Darrieus's 1927 patent for vertical aero foil-powered vertical-axis wind turbines, but new giromill are now under development, utilizing modern ultra-strong and light materials in order to produce turbine blades that are strong enough to withstand the stresses they face. Two or three vertical airfoils are mounted on a central mast by horizontal supports and drive the giromill. It is cheaper and easier to build than a Darrieus turbine, but is less efficient, and requires strong winds (or a motor) to start. It is also possible for them to have trouble maintaining a steady rotational speed. Giromills, however, can be used in turbulent wind conditions and are an affordable alternative to horizontal-axis windmills.

Magnus Turbine:

The Magnus wind turbine is an invention that uses rotating cylinders as blades to extract energy from the wind. This invention overcomes the limitation of operating a wind turbine at low wind speed conditions. Figure 7.18 depicts the illustration of Magnus effect wind turbine. Further to note that wind turbines can be driven by the Magnus effect since it was described in 1852. Figure 7.18 illustrates how a horizontal-axis wind turbine's structure is not difficult to understand. It must, however, be implemented as a vertical-axis wind turbine so that the rotor rotation directions on the leeward and windward sides are opposite, so that the torques cannot cancel each other out and drive the turbine. Originally, it was made using gears and connecting rods, which were difficult to implement because of the complicated structure. The rapid advancement of semiconductor technology, micro-controllers, and motor drive technology has made it possible to realize this concept more easily in recent years.

Vortex Turbine:

Vortex wind turbine is a bladeless one, as illustrated in Figure 7.19. The bladeless turbines stand at 3 m high, a curve-topped cylinder fixed vertically with an elastic rod. To the untrained eye it appears to waggle back and forth, not unlike a car dashboard toy. In reality, it is designed to oscillate within

Figure 7.19   View of the vortex wind turbine. *Source:* UTS ePRESS.

the wind range and generate electricity from the vibration. The turbine uses the force created by natural wind vortices to generate electricity. Basically, if you put an object in the path of wind, it will create an undulating vortex behind the barrier. Structural engineers have understood this for some time and have designed tall-standing buildings and structures to either withstand or channel them.

Illustrative Example 7.6   Wind blows over a vertical wind turbine at a velocity of 25 km/h. The wind turbine has a diameter of 20 m, and the wind exit velocity is measured to be 15.8 km/h. Also, the overall heat loss from the turbine is provided to be 2 kW.

a) Write all mass, energy, entropy, and exergy balance equations for the turbine.
b) Calculate the power generated by the turbine.
c) Calculate both energy and exergy efficiencies.

Note: The air inlet and outlet temperature is measured to be 30°C. Treat air as an ideal gas and assume air temperature remains constant and uniform at inlet and exit. The density of air is 1.18 kg/m^3 and the specific heat capacity (c_p) is 1.005 kJ/kgK.

Solution:

a) Four balance equation can be written as follows for the vertical wind turbine:

MBE: $\dot{m}_{in} = \dot{m}_{out}$

EBE: $\dot{m}_{in}\left(h_{in} + \dfrac{V_{in}^2}{2} \right) = \dot{m}_{out}\left(h_{out} + \dfrac{V_{out}^2}{2} \right) + \dot{W}_t + \dot{Q}_{loss}$

EnBE: $\dot{m}_{in} s_{in} + \dot{S}_{gen} = \dot{m}_{out} s_{out} + \dfrac{\dot{Q}_{loss}}{T_{surf}}$

ExBE: $\dot{m}_{in}\left(ex_{in} + \dfrac{V_{in}^2}{2} \right) = \dot{m}_{out}\left(ex_{out} + \dfrac{V_{out}^2}{2} \right) + \dot{W}_t + \dot{Q}_{loss}\left(1 - \dfrac{T_0}{T_{surf}} \right)$

b) In order to calculate the power output from the wind turbine, we need to write the energy balance equation for the wind turbine considered. The energy balance equation can be written as follows:

$$\dot{m}_{in}\left(h_{in}+\frac{V_{in}^2}{2}\right)=\dot{m}_{out}\left(h_{out}+\frac{V_{out}^2}{2}\right)+\dot{W}_t+\dot{Q}_{loss}$$

Substituting the given and known parameters, one obtains

$$\dot{m}_{in}=\rho\dot{V}=\rho Av=(1.18)\left(\frac{\pi\left(20^2\right)}{4}\right)\left(\frac{25{,}000}{3600}\right)=2574.7\,kg\,/\,s$$

Here, $h_{in}=h_{out}$ as they are at the same temperature. Thus,

$$(2574.7)\frac{\left(\frac{25{,}000}{3600}\right)^2}{2}=(2574.7)\frac{\left(\frac{15{,}800}{3600}\right)^2}{2}+2000+\dot{W}_t$$

Solving the above equation to get: $\dot{W}_t=35286.3W$

c) The energy and exergy efficiencies are the identical by neglecting the thermal effects through flow enthalpies and obtained as follows:

$$\eta_{en}=\eta_{ex}=\frac{\dot{W}_t}{\frac{1}{2}\rho AV_1^3}=\frac{35{,}286.3W}{0.5(1.18)\left(\frac{\pi\left(20^2\right)}{4}\right)\left(\frac{25{,}000}{3600}\right)^3}=56\%$$

7.8 Offshore and Onshore Types of Wind Energy

Wind energy systems may also be classified according to the locations they are erected/installed, which are offshore and onshore wind systems. Wind turbines are placed on the earth's surface in onshore wind systems; offshore wind systems are placed over the water surfaces. The technology of wind turbines is already well known. Basically, the wind turns blades around a rotor, generating electricity. There is a difference in the wind speed between offshore wind and onshore wind. Offshore wind speeds are much faster. And faster wind speeds mean exponentially more energy output. For example, the power generation is almost double with a 20 km/h wind than a 15 km/h

(a)   (b)

Figure 7.20 Illustration of (a) offshore and (b) onshore wind farms. *Source:* Sergiy Serdyuk / Adobe Stock.

wind. Also, offshore winds are steadier. Therefore, they are higher capacity factor than onshore wind power systems. Additionally, large components can easily be transported by sea, making it easier to build larger turbines. The blades can be longer, but the towers can be shorter, because open water does not have the vegetation and topography that cause wind shear and turbulence. Figure 7.20 illustrates the offshore and onshore wind farms.

7.9 Case Studies

In this section, the case studies related to wind energy systems will be presented. First, a couple of concept wind farm calculations will be presented over illustrative examples. Then, energy and exergy maps for wind power systems will be given.

Illustrative Example 7.7 Wind blows steadily over an offshore horizontal wind turbine at a velocity of 36 km/h, having a diameter of 20 m. The wind exit velocity is assumed to be 28.8 km/h. The inlet air temperature is 30°C, and the outlet temperature is 29.992°C. The heat loss from the turbine is 3.5 kW. For a reference temperature of 25°C and pressure of 100 kPa.

a) Write all mass, energy, entropy and exergy balance equations for the turbine.
b) Calculate the power generated by the turbine.
c) Calculate the energy and exergy efficiencies.

Note: Take $h_1 = 303.6 \frac{kJ}{kg}$, $s_1 = 5.713 \frac{kJ}{kgK}$, $h_0 = 301.6 \frac{kJ}{kg}$, $s_0 = 5.707 \frac{kJ}{kg} K$.

Consider the ambient temperature as 25°C and density of air as 1.1839 kg/m³. $c_p = 1.005$ kJ/Kg K. Treat air as an ideal gas and assume air temperature remains constant at inlet and exit.

Solution:

a) Four balance equation can be written as follows for the vertical wind turbine:

MBE: $\dot{m}_{in} = \dot{m}_{out}$

EBE: $\dot{m}_{in}\left(h_{in} + \frac{V_{in}^2}{2}\right) = \dot{m}_{out}\left(h_{out} + \frac{V_{out}^2}{2}\right) + \dot{W}_t + \dot{Q}_{loss}$

EnBE: $\dot{m}_{in}s_{in} + \dot{S}_{gen} = \dot{m}_{out}s_{out} + \frac{\dot{Q}_{loss}}{T_{surf}}$

ExBE: $\dot{m}_{in}\left(ex_{in} + \frac{V_{in}^2}{2}\right) = \dot{m}_{out}\left(ex_{out} + \frac{V_{out}^2}{2}\right) + \dot{W}_t + \dot{Q}_{loss}\left(1 - \frac{T_0}{T_{surf}}\right)$

b) In order to calculate the power output by using the energy balance equation, we need to determine the mass flow rate of the air. The mass flow rate can be found to be:

$$\dot{m}_{air} = \rho AV = 1.1839 \times \frac{\pi}{4} \times 20^2 \times 10$$

Thus, the mass flow rate of the air is found to be $\dot{m}_{air} = 3719.3$ kg/s

Rearranging the energy balance equation, one may find

$$\dot{m}_{air}\left(h_1 - h_2 + \frac{V_1^2}{2} - \frac{V_2^2}{2}\right) - \dot{Q}_{loss} = \dot{W}_{turbine}$$

This equation can be written as follows since air is assumed to be the ideal gas.

$$\dot{m}_{air}\left[c_p\left(T_1 - T_2\right) + \frac{V_1^2}{2} - \frac{V_2^2}{2}\right] - \dot{Q}_{loss} = \dot{W}_{turbine}$$

Substituting all information provided, the power output from the wind turbine is found to be $\dot{W}_{turbine} = 93.4$ kW.

c) In order to calculate the energy and exergy efficiencies, first, we need to calculate the energy content of the wind.

$$E_{wind} = \frac{1}{2}\rho\pi R^2 V^3 = 0.5 \times 1.1839 \times \pi \times 10^3 \times 10^2$$

Thus, the wind energy input is found to be $E_{wind} = 185.8$ kW.
The wind exergy can be calculated as follows:

$$Ex_{wind} = \dot{m}_{air}\left(ex_1 + \frac{V_1^2}{2}\right) = 3719.3 \times \left(0.03839 + \frac{10^2}{2}\right) = 186.1 kW$$

Thus, the energy efficiency is calculated to be

$$\eta_{en} = \frac{\dot{W}}{\dot{E}_{in}} = \frac{93.4}{185.8} = 0.5026$$

and the exergy efficiency is also found to be

$$\eta_{ex} = \frac{\dot{W}}{\dot{E}_{in}} = \frac{93.4}{186.1} = 0.5018$$

As seen from this illustrative example, the energy and exergy efficiencies of the wind turbine are almost the same as the energy and exergy contents of the wind is the almost same.

Illustrative Example 7.8 Consider a wind farm used for residential units, as shown in Figure 7.21. The blade length of the wind turbine is 10 m. Air density is 1.1839 kg/m³. For seasonal operation, in winter, the wind speed is measured before and after the wind turbine to be 9.44 m/s and 6.39 m/s. The heat loss from wind turbine is 5700 W in the winter. During the summer, the wind speeds before and after wind turbine are 6.94 m/s and 5.28 m/s. The heat loss from wind turbine is 1700 W in the summer. The electricity consumption of the residential area is 20 kW in winter and 15 kW for summer. The electricity sell back rate is 1.33 US$/kWh. Determine the entropy generation, exergy destruction, and energy and exergy efficiencies of the wind power plant. Calculate the winter and summer revenues.

Solution:

First, let's write the thermodynamic balance equations for the wind turbine as follows:

MBE: $\dot{m}_1 = \dot{m}_2$

EBE: $\dot{m}_1(h_1 + ke_1) = \dot{m}_2(h_2 + ke_2) + \dot{Q}_{loss} + \dot{W}_{wt}$

EnBE: $\dot{m}_1 s_1 + \dot{S}_{gen} = \dot{m}_2 s_2 + \frac{\dot{Q}_{loss}}{T_0}$

ExBE: $\dot{m}_1(ex_1 + ke_1) = \dot{m}_2(ex_2 + ke_2) + \dot{Ex}_{\dot{Q}_{loss}} + \dot{W}_{wt} + \dot{Ex}_{dest}$

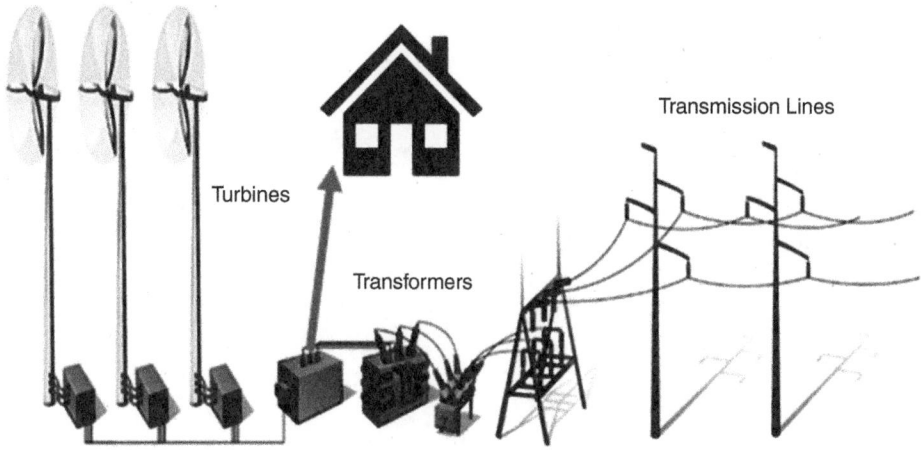

Figure 7.21 Illustration of the wind farm used for residential units.

In order to use balance equations, it is required to calculate the mass flow rate of the air for both winter and summer.

$$\dot{m}_{1,winter} = \rho_1 A_1 V_1 = 0.1839 \times \pi \times 10^2 \times 9.44 = 545.4 \; kg/m^3$$

$$\dot{m}_{1,summer} = \rho_1 A_1 V_1 = 0.1839 \times \pi \times 10^2 \times 6.94 = 400.9 \; kg/m^3$$

By assuming inlet and outlet conditions are the same, one obtains

$$h_1 = h_2 \text{ and } s_1 = s_2$$

The EBE for winter is written as:

$$\dot{m}_1(ke_1) = \dot{m}_2(ke_2) + \dot{Q}_{loss,winter} + \dot{W}_{wt,winter}$$

$$545.4 \times \frac{9.44^2}{2} = 545.4 \times \frac{6.39^2}{2} + 5,700 + \dot{W}_{wt,winter}$$

Thus, the wind power in the winter is found to be $\dot{W}_{wt,winter} = 3411.1$ W.
The EBE for summer is written as:

$$\dot{m}_1(ke_1) = \dot{m}_2(ke_2) + \dot{Q}_{loss,summer} + \dot{W}_{wt,summer}$$

$$400.9 \times \frac{6.94^2}{2} = 400.9 \times \frac{5.28^2}{2} + 1700 + \dot{W}_{wt,winter}$$

Thus, the wind power in the summer is found to be $\dot{W}_{wt,summer} = 2366.1$ W.
The entropy balance equation can be written as follow as $s_1 = s_2$.

$$\dot{S}_{gen} = \frac{\dot{Q}_{loss}}{T_0}$$

Thus,

$$\dot{S}_{gen,winter} = \frac{5700}{6+273} = 20.43 \ W$$

$$\dot{S}_{gen,summer} = \frac{1700}{25+273} = 5.70 \ W$$

The exergy destruction can be calculated as:

$$\dot{Ex}_{dest,winter} = T_0\dot{S}_{gen,winter} = 298 \times 20.43 = 6088.1\frac{W}{K}$$

and

$$\dot{Ex}_{dest,winter} = T_0\dot{S}_{gen,winter} = 298 \times 5.70 = 1698.6\frac{W}{K}$$

The energy and exergy efficiencies can be calculated as:

$$\eta_{en} = 40.15\%$$

and

$$\eta_{ex} = 42.54\%$$

The electricity to be sent the grid is the difference between the power generated by wind and the energy demand of the building.

$$\dot{E}_{grid} = \dot{W}_{wt}N - \dot{E}_{demand}$$

Here, N is the number of wind turbines. For the winter:

$$\dot{E}_{grid,winter} = 3.4 \ (kW) \times 10 - 20(kW)$$

$$\dot{E}_{grid,winter} = 14kW$$

For the summer:

$$\dot{E}_{grid,summer} = 2.3 \ (kW) \times 10 - 15(kW)$$

$$\dot{E}_{grid,summer} = 8kW$$

In order to perform the cost analysis, we need to find the amount of energy. Assuming 3 months is 2160 hours.

$$E_{grid,winter} = \dot{E}_{grid,winter}t = 14 \times 2160 = 30,240 \ kWh$$

and

$$E_{grid,summer} = \dot{E}_{grid,summer}t = 8 \times 2160 = 17,280 \ kWh$$

The cost benefits from the wind turbines are

$$Winter = 30,240 \times 1.33 = US\$40,219.2$$

and

$$Summer = 17{,}280 \times 1.33 = US\$2{,}298{,}242$$

Illustrative Example 7.9   Wind blows over a wind turbine at a velocity of 34 km/h and exits the wind turbine at a velocity of 24 km/h during winter. During summer, wind blows over a wind turbine at a velocity of 25 km/h and exits the wind turbine at a velocity of 18 km/h. Wind turbine has a diameter of 20 m and air density is $\rho_{air} = 1.1839$ kg/m³. The average ambient temperature can be considered 6°C during winter and 12°C during summer. The heat losses are given as 15.7 kW for winter and 3.5 kW for summer.

a) Write the mass, energy, entropy and exergy balance equations for the wind turbine.
b) Find the windchill temperatures for inlet and exit of the wind turbine both for winter and summer conditions.
c) Calculate the power output from the wind turbine.
d) Calculate the entropy generation and exergy destruction rates.
e) Calculate the energy efficiencies for winter and summer operations.

Windchill temperature can be found from the following equation. In this equation, wind speed is in mph and temperature is in F.

$$T_{windchill} = 35.74 + 0.6215 T_{air} - 35.75 V^{0.16} + 0.4274 T_{air} V^{0.16}$$

Solution:

a) Let's first write all thermodynamic balance equations for the wind turbine as follows:
   MBE: $\dot{m}_1 = \dot{m}_2$
   EBE: $\dot{m}_1 (h_1 + ke_1) = \dot{m}_2 (h_2 + ke_2) + \dot{Q}_{loss} + \dot{W}_{wt}$
   EnBE: $\dot{m}_1 s_1 + \dot{S}_{gen} = \dot{m}_2 s_2 + \dfrac{\dot{Q}_{loss}}{T_0}$
   ExBE: $\dot{m}_1 (ex_1 + ke_1) = \dot{m}_2 (ex_2 + ke_2) + \dot{Ex}_{\dot{Q}_{loss}} + \dot{W}_{wt} + \dot{Ex}_{dest}$

b) In order to calculate the windchill temperature, the equation given in the problem can be used.

$$T_{windchill} = 35.74 + 0.6215 T_{air} - 35.75 V^{0.16} + 0.4274 T_{air} V^{0.16}$$

The wind speeds are given as follows:

$V_{1,winter} = 34$ km/h $= 9.44$ m/s $= 21.13$ mph

$V_{2,winter} = 24$ km/h $= 6.67$ m/s $= 14.91$ mph

$V_{1,summer} = 25$ km/h $= 6.94$ m/s $= 15.53$ mph

$V_{2,summer} = 18$ km/h $= 5$ m/s $= 11.19$ mph

$T_{air,winter} = 6°C = 42.8$ F

$T_{air,summer} = 12°C = 53.6$ F

Substituting the values above for the winter and summer:

$$T_{windchill,1,winter} = 35.74 + 0.6215 \times 42.8 - 35.75 \times 21.13^{0.16} + 0.4274 \times 42.8 \times 21.13^{0.16}$$

$$T_{windchill,1,winter} = 33.9F = 1.05°C$$

$$T_{windchill,2,winter} = 35.74 + 0.6215 \times 42.8 - 35.75 \times 14.91^{0.16} + 0.4274 \times 42.8 \times 14.91^{0.16}$$

$$T_{windchill,2,winter} = 35.44F = 1.91°C$$

$$T_{windchill,1,summer} = 35.74 + 0.6215 \times 53.6 - 35.75 \times 15.53^{0.16} + 0.4274 \times 53.6 \times 15.53^{0.16}$$

$$T_{windchill,1,summer} = 44.14F = 9.52°C$$

$$T_{windchill,2,summer} = 35.74 + 0.6215 \times 53.6 - 35.75 \times 11.19^{0.16} + 0.4274 \times 53.6 \times 11.19^{0.16}$$

$$T_{windchill,2,summer} = 50.16F = 10.09°C$$

c) In order to calculate the power output from the wind turbine, the energy balance equation should be used accordingly. After rearranging the energy balance equation by assuming $h_1 = h_2$ and $s_1 = s_2$.

$$\dot{W}_{wt} = \dot{m}_1 \left(\frac{V_1^2}{2} - \frac{V_2^2}{2} \right) - \dot{Q}_{loss}$$

In order to calculate the wind power for the winter and summer, we need to calculate the mass flow rates for winter and summer separately.

$$\dot{m}_{winter} = \rho A V_{winter} = 1.1839 \times \pi \times 10^2 \times 9.44 = 3512.7 \ kg/s$$

and

$$\dot{m}_{summer} = \rho A V_{summer} = 1.1839 \times \pi \times 10^2 \times 6.94 = 2582.9 \ kg/s$$

Now, we can calculate the wind turbine powers for winter and summer operations:

$$\dot{W}_{wt,winter} = 3512.7 \times \left(\frac{9.44^2}{2} - \frac{6.67^2}{2} \right) = 26.49 \ kW$$

and

$$\dot{W}_{wt,summer} = 2582.9 \times \left(\frac{6.94^2}{2} - \frac{5^2}{2} \right) = 62.90 \ kW$$

d) The entropy generation rates can be calculated as follows:

$$\dot{S}_{gen,winter} = \frac{\dot{Q}_{loss,winter}}{T_{windchill,winter}} = \frac{15.7}{\frac{(1.05+1.91)}{2}+273} = 0.0573 \frac{kW}{K}$$

$$\dot{S}_{gen,summer} = \frac{\dot{Q}_{loss,summer}}{T_{windchill,summer}} = \frac{3.5}{\frac{(9.92+10.09)}{2}+273} = 0.0555 \frac{kW}{K}$$

Thus, the exergy destructions are found to be

$$\dot{Ex}_{d,winter} = T_0 \dot{S}_{gen,winter} = 298 \times 0.0573 = 15.70 \ kW$$

and

$$\dot{Ex}_{d,summer} = T_0 \dot{S}_{gen,summer} = 283 \times 0.0555 = 15.71 \ kW$$

d) The energy efficiencies for winter and summer operations can be calculated as follows:

$$\eta_{en} = \frac{\dot{W}_{wt,winter}}{\frac{1}{2}\rho A V^3} = \frac{26.49}{\frac{1}{2} \times 1.1839 \times \pi \times 10^2 \times 9.44^3} = 40.15\%$$

and

$$\eta_{en} = \frac{\dot{W}_{wt,summer}}{\frac{1}{2}\rho A V^3} = \frac{62.90}{\frac{1}{2} \times 1.1839 \times \pi \times 10^2 \times 6.94^3} = 42.54\%$$

7.10 Energy and Exergy Maps for Wind Energy Systems

The characteristics and properties of the atmosphere vary primarily by location. Thus, wind engineering studies require spatial modeling of wind. Spatial variation is generally studied separately from temporal variation. In meteorology and wind engineering literature, spatial modeling of wind is achieved through mapping and objective analysis methods. It is possible to interpolate data from measurement stations to any desired point using various methods.

As wind power is directly dependent on the location, energy, and exergy maps can be critical decision-makers. Here, as a case study, energy and exergy maps for wind power in Ontario, Canada, are presented based on the data taken from 21 stations, based on the average values of the last 30 years. The capacity factor is considered 45% for wind speeds of 8–11 m/s. The location of the stations is illustrated in Table 7.1. Wind speed values are interpolated from 10 to 30 m. While creating the maps, a 100 kW wind turbine with a 30 m hub height is specifically selected to minimize wind speed interpolation errors. It should be noted here that the selected area is a lake area, so interactions between water and land surfaces are very high. This water and land interaction causes a continuous high wind speed in the location.

Wind energy maps which are covering wind speed, energy efficiency, exergy efficiency, and % difference between energy and exergy efficiencies are shown in Figures 7.22–7.25. Here are developed and discussed geostatistical spatiotemporal maps for January, April, July, and October. Exergy analysis is used to display the differences between the energy and exergy efficiencies of a particular wind turbine system at the same conditions. Besides providing more information, this analysis provides a more detailed description of how wind energy is used, how much energy is lost, and where these losses and inefficiencies are located. Each month is representative of the season that belongs to it. There is no climatological data for Lake Ontario at the bottom right of this map, so this area is not discussed. Low wind speeds in the east and north characterize January in Ontario. In this month, Atikokan's average wind speed is below the typical turbine cut-in wind speed, so no electricity is generated. A monthly maximum average wind speed of 9–10 m/s has been observed in southwestern side of the area studied, as illustrated in Figure 7.22a.

Table 7.1 Details of selected meteorological stations used for creating wind maps.

Station	Latitude (°N)	Longitude (°W)	Altitude (m)
Atikokan	48.45	91.37	395
Big Trout Lake	53.50	89.52	220
Dryden Airport	49.50	92.45	413
Kapuskasing	49.25	82.28	227
Kenora	49.47	94.22	407
Kingston	44.13	76.36	93
London	43.02	81.09	278
Moosonee	51.16	80.39	10
North Bay	46.21	79.26	358
Ottawa	45.19	75.40	116
Red Lake	51.04	93.48	375
Simcoe	46.29	84.30	187
Sault Ste Marie	42.51	80.16	241
Sioux Lookout	50.07	91.54	398
Sudbury	46.37	80.48	348
Thunder Bay	48.22	89.19	199
Timmins	48.34	81.22	295
Toronto Pearson Airport	43.40	79.38	173
Trenton	44.07	77.32	85
Wiarton	44.45	81.06	222
Windsor	42.16	82.58	190

Figure 7.22b shows an estimate of energy efficiency maps for January. At low wind speeds, although energy efficiencies seem high, this does not mean that the wind turbine is more efficient than rated for that wind speed. In other words, at these wind speeds, there is a low potential for wind energy as well as a low generation of electricity. Therefore, the ratio of generated electricity to potential energy is high (Figure 7.22b). For all regions, the contours for energy efficiency are lower than those for exergy efficiency. There is an average value of 40% for exergy efficiency. By using this exergy map, parameters in regions where measurements are lacking can be estimated using interpolation. Therefore, this type of map in Figure 7.22c can be used for practical engineering applications. Figure 7.22d shows the relative differences between energy and energy efficiency. Low wind speeds result in large relative differences in energy-efficiency values. At high wind speeds, however, the relative differences between energy and exergy efficiencies are smaller. At all stations, these values are higher than 10%. In energy management and planning, these differences should not be ignored.

In April (Figure 7.23), electricity can be generated at all stations as wind speed is higher than 10 m/s. Similar to the January map, the highest wind speed values are observed in southwestern Ontario in April (Figure 7.23a). South-to-north efficiencies increase sequentially in the energy-efficiency map for April. In this region's northern parts, wind speeds are low, resulting in energy efficiency of approximately 50%. There are also three clusters in this efficiency map (Figure 7.23b).

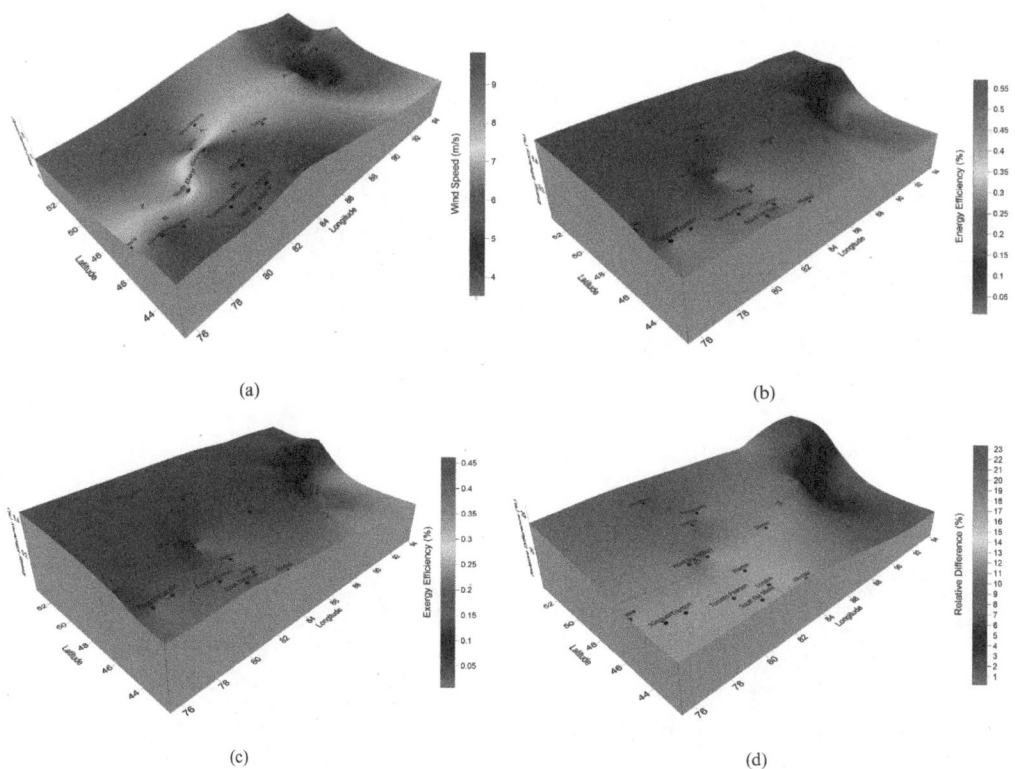

Figure 7.22 Maps of wind energy for January (a) wind speed, (b) energy efficiency, (c) exergy efficiency, and (d) relative differences (in %) (adapted from [7]).

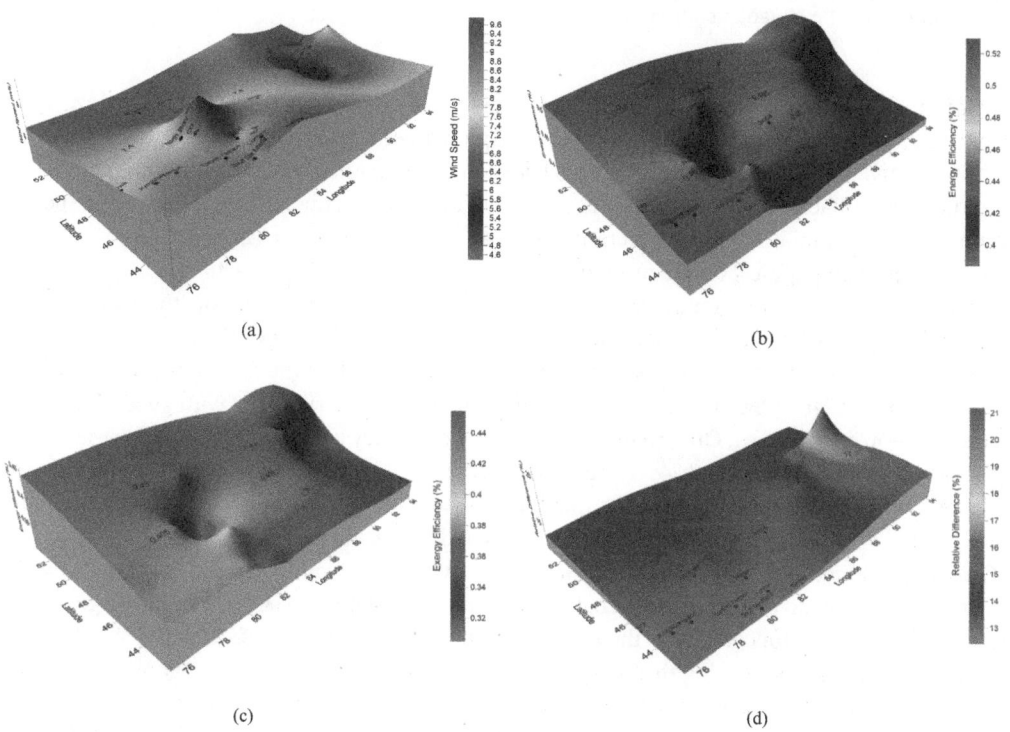

Figure 7.23 Maps of wind energy for April (a) wind speed, (b) energy efficiency, (c) exergy efficiency, and (d) relative differences (in %) (adapted from [7]).

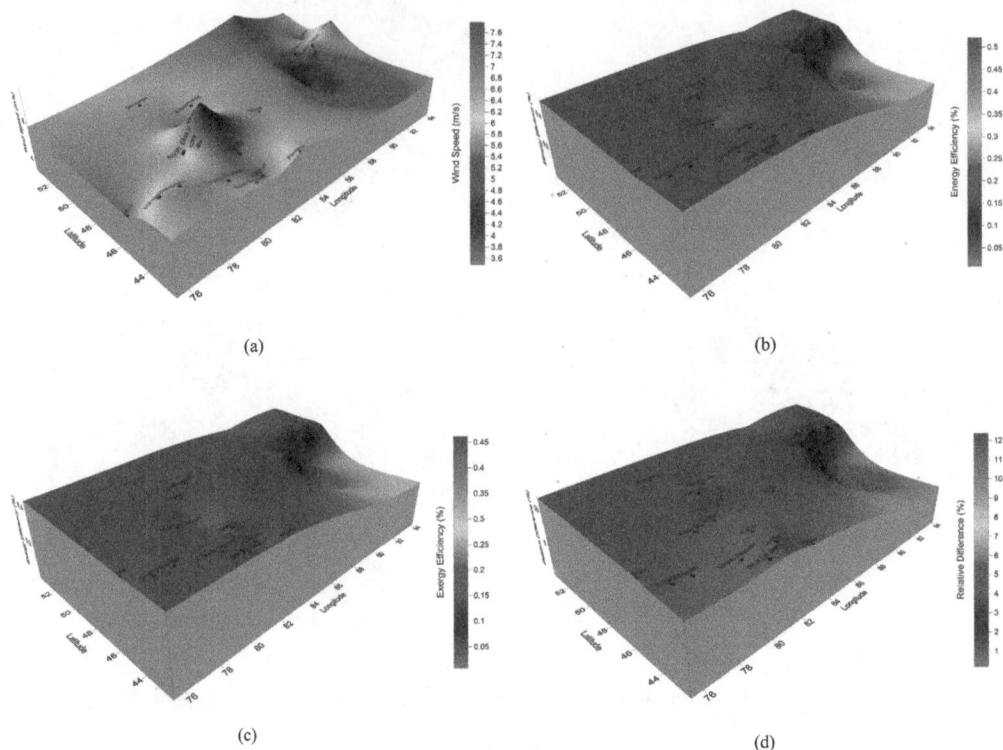

Figure 7.24 Maps of wind energy for July (a) wind speed, (b) energy efficiency, (c) exergy efficiency, and (d) relative differences (in %) (adapted from [7]).

According to alternate energy efficiency definitions, wind energy efficiencies decreased in April (Figure 7.23c). Lines of constant latitude tend to align with energy and energy efficiency contours in April. Conversely, the relative differences between the two efficiencies are roughly parallel to longitude lines. There is a range of 14–22% relative difference. Among the lowest wind speeds observed in Atikokan, the relative difference between the two efficiencies is 22% (Figure 7.23d).

In summer, topographical effects result in different clusters of wind speeds for July. As a result of the high heating during this month, surface conditions are unstable. In one station, the average wind speed is lower than the cut-in value, resulting in zero energy and energy efficiency. Figure 7.24a shows that the highest wind speed for this month is the lowest of all the maximums for the other months. Figure 7.24b shows three clusters of energy efficiency distributions with general contour values of 40–50%. Northwest Ontario has a high energy efficiency zone, but it has the lowest exergy efficiency. As shown in Figure 7.24c, the dominant efficiency in July is approximately 40%. Energy and energy efficiencies in July are similar, and the relative differences between these efficiencies are small (Figure 7.24d). It is important to note that the windchill is not a significant factor in July.

October is selected as a representative month for fall. As seen in Figure 7.25a, there are three wind speed clusters and wind power systems generate electricity in all stations. There are two main groups of energy efficiency. The topography at these stations causes some localized effects in October (Figure 7.25b). Exergy efficiencies during this month are lower than energy efficiencies. Figure 7.25c shows that one of the highest energy-efficiency areas, located in western Ontario, is less significant on the basis of exergy. During October in most parts of Ontario, the relative differences between these efficiencies are low without summer topographical heating, but windchill becomes more noticeable (Figure 7.25d).

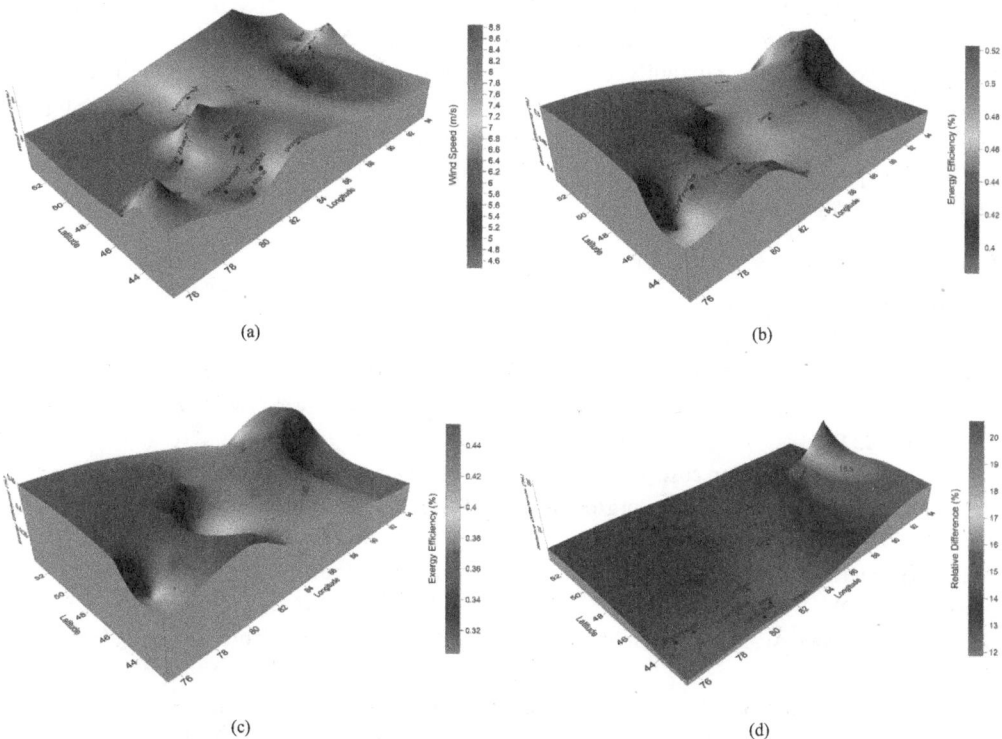

Figure 7.25 Maps of wind energy for October (a) wind speed, (b) energy efficiency, (c) exergy efficiency, and (d) relative differences (in %) (adapted from [7]).

Wind energy can be described energetically and exergetically using spatiotemporal energy and exergy maps. In the form of geostatistical maps, seasonal energy efficiency and exergy are presented. Wind energy sources can be viewed from a new perspective with the application of exergy analysis to each system, along with the ensuing point-by-point map analysis. Wind turbine exergy maps provide valuable information regarding efficiency, losses, and performance. Furthermore, the approach simplifies analyses and facilitates practical applications and analyses. One may draw a few important conclusions. First, the relative differences between energy and energy efficiency are highest in winter and lowest in summer. Second, the exergy efficiencies are lower than energy efficiencies for each station for every month considered. For wind energy systems, the exergy approach provides useful results, and the tools presented here for approximating wind energy efficiencies are widely applicable. Using such tools can increase the use of wind systems, optimize designs, and identify appropriate applications and optimal system arrangements.

7.11 Closing Remarks

This chapter deals with wind energy as part of renewable energy family and discusses wind energy concepts, systems, and applications. Due to some particular practical advantages of wind power systems, they are one of the significant contributors in achieving the carbon-free future. The global capacity of the wind power systems has been increasing exponentially and providing a unique opportunity to help overcome power generation challenges in an environmentally-friendly and sustainable manner. Although there are various wind turbines, the three-bladed upwind wind turbines are the most commonly deployed ones due to their particular advantages, efficiency, effectiveness, cost, viability, reliability, etc. On the other hand, in order to determine the potential of wind energy, energy and exergy maps are useful. Offshore wind power systems have higher potential than onshore systems due to higher and more reliable wind. Furthermore, wind energy and exergy maps are presented to highlight the need for more efficient and effective applications based on the exergy efficiency.

References

1 Installed wind energy capacity. *Our World in Data*. Available online https://ourworldindata.org/grapher/cumulative-installed-wind-energy-capacity-gigawatts?country=~OWID_WRL (accessed 10 December 2022).

2 Renewable power generation costs in 2021. *International Renewable Energy Agency (IRENA)*. Available online https://www.irena.org/publications/2022/Jul/Renewable-Power-Generation-Costs-in-2021 (accessed 10 December 2022).

3 *Prevailing winds*. Available online https://en.wikipedia.org/wiki/Prevailing_winds (accessed 10 December 2022).

4 *Wind turbine design*. Available online https://en.wikipedia.org/wiki/Wind_turbine_design (accessed 12 December 2022).

5 Wang, Z., Tian, W., and Hu, H. (2018). A comparative study on the aeromechanic performances of upwind and downwind horizontal-axis wind turbines. *Energy Conversion and Management* 163: 100–110. doi: 10.1016/j.enconman.2018.02.038.

6 Jinbo, M., Farret, F.A., Junior, G.C. et al. (2015). MPPT of magnus wind system with DC servo drive for the cylinders and boost converter. *Journal of Wind Energy* 2015: 148680. doi: 10.1155/2015/148680.

7 Dincer, I. and Rosen, M.A. (2020). *Exergy – Energy, Environment, and Sustainable Development*, 3e. Elsevier. ISBN: 978-0-08-097089-9.

Questions/Problems

Questions

1 Explain and illustrate how winds occur around the world.

2 Explain the global wind patterns and their role locally.

3 What is wind power? Briefly describe and illustrate.

4 Classify wind turbines and explain their pros and cons briefly.

5 Describe offshore and onshore wind power plants and discuss their advantages and disadvantages.

6 Describe vertical-axis wind turbines with an example to illustrate their basic components.

7 Which parameters are critical for wind power systems? List them and explain each.

8 Write mass, energy, entropy and exergy balance equations for a basic horizontal-axis wind turbine with a system boundary and all inputs and outputs indicated.

9 Explain how the wind turbine is affected by varying windchill temperature.

10 Explain the role of energy and exergy maps and their potential utilization for practical system design.

Problems

1 Consider a wind turbine which has a 30 m long blade. The wind speed is measured to be 20 m/s. The site of the installation is about 100 m above sea level. Assume $C_b = 0.35$, $\eta_g = 0.70$, and $\eta_t = 0.95$. Air density is taken as 1.09 kg/m³.

2 Consider a wind turbine used in an offshore application which has a 80 m long blade. The wind speed is measured to be 25 m/s. Assume $C_b = 0.35$, $\eta_g = 0.75$, and $\eta_t = 0.95$. Air density is taken as 1.1 kg/m³.

3 Consider a wind farm with 75 wind turbines. The blade length of the wind turbine is 30 m. The wind speed is measured to be 12 m/s before wind turbine and the same for each turbine. The density of the air is 1.1839 kg/m³. The outlet velocity is measured as 7.28 m/s. There is a total of 5.1 kW of heat loss in the turbine. Calculate the total electricity generation from this wind farm.

4 Repeat Illustrative Example 7.6 by changing the average inlet and exit velocities of wind to 8 m/s and 5 m/s, respectively.

5 Repeat Illustrative Example 7.6 with a parametric study where one may consider average inlet velocity changing from 4 m/s to 12 m/s and the exit velocity 20% less than the corresponding inlet velocity.

6 The power output capacity of a wind farm is 100 MW. The blade length of the wind turbine is 40 m. The wind speed is measured to be 12 m/s before wind turbine and the same for each turbine. The outlet velocity is measured as 7.28 m/s. There is a total of 4000 W of heat loss in the turbine. Calculate the number of wind turbines required to obtain desired power output. The density of the air is 1.1839 kg/m³.

7 For an offshore wind power plant, it is expected to reach a capacity of 300 GW. The test values indicate that the wind speeds are measured to be 20 m/s and 11 m/s for inlet and outlet of the control volume. The density of air can be assumed to be 1.10 kg/m³. The heat loss is 5 kW. Find the number of wind turbines required to obtain desired power output from offshore wind farm.

8 Consider a wind farm used for commercial units. The blade length of the wind turbine is 25 m. Air density is 1.1839 kg/m³. For seasonal operation, in winter, the wind speed is measured before and after the wind turbine to be 12 m/s and 7 m/s. The heat loss from wind turbine is 3200 W in the winter. During the summer, the wind speeds before and after wind turbine are 9.5 m/s and 5.2 m/s. The heat loss from wind turbine is 1700 W in the summer. The electricity consumption of the residential area is 25 kW in winter and 20 kW for summer. The electricity sell back rate is 1.5 US$/kWh. Calculate the entropy generation, exergy destruction, and energy and exergy efficiencies of the wind power plant. Calculate winter and summer revenues.

9 Wind blows steadily over a vertical wind turbine at a velocity of 36 km/h, having a diameter of 20 m. The wind exit velocity is assumed to be 28.8 km/h. The heat loss from the turbine is 3.5 kW. For an outside temperature of 25°C and pressure of 100 kPa.
   a Write all mass, energy, entropy and exergy balance equations for the turbine.
   b Calculate the actual and maximum power generated by the turbine.
   c Calculate both energy and exergy efficiencies.
   Note: Take temperature as 30°C and density of air as 1.1839 kg/m³. $C_p = 1.005$ kJ/kg K. Treat air as an ideal gas and assume air temperature remains constant at inlet and exit.

10 Wind blows over a wind turbine at a velocity of 34 km/h and exits the wind turbine at a velocity of 24 km/h, during winter. During summer, wind blows over a wind turbine at a velocity of 25 km/h and exits the wind turbine at a velocity of 18 km/h. Wind turbine has a diameter of

20 m and air density is $\rho_{air} = 1.1839$ kg/m^3. The ambient temperature can be considered 0°C during winter and 25°C during summer. The heat losses are given as 11 kW for winter and 3 kW for summer.

a Write the mass, energy, entropy and exergy balance equations for the wind turbine.
b Find the windchill temperatures for inlet and exit of the wind turbine both for winter and summer conditions.
c Calculate the power output from the wind turbine.
d Calculate the entropy generation and exergy destruction rates.
e Calculate the energy efficiency.

The windchill temperature can be found from the following equation. Here in this equation, wind speed is in mph and temperature is in F.

$$T_{windchill} = 35.74 + 0.6215 T_{air} - 35.75 V^{0.16} + 0.4274 T_{air} V^{0.16}$$

8

Geothermal Energy

8.1 Introduction

There is no doubt that geothermal energy has existed since the world's formation going back to billions of years ago. It is fair enough to consider it as one of the human history's oldest energy sources. Moreover, geothermal energy is by nature recognized as one of renewable energy sources and reveals its potential for better environment and better sustainable development. Archeological research indicates that the first use of geothermal energy coincides with the Paleolithic age. Geothermal energy was used to meet their basic heating demands for daily personal needs during this time in addition to using such sources for curing skin diseases.

The name of the geothermal comes from "geo" that means earth and "thermal" that means heat; so geothermal represents the earth's heat. Essentially, it is a heat source coming from the core of the earth. It is a renewable, reliable, and efficient energy source. It can be used directly or indirectly. Today, the cost of geothermal electricity varies 5–8 cents per kWh, depending various aspects, location, capacity, system, operation, etc. Drilling through the earth core is necessary to reach the heat source in geothermal applications. The temperature of the rock increases about 3°C for every 100 m below ground. Deep beneath the surface, water sometimes makes its way close to the hot rock and turns into boiling water or steam.

Although the hot water reservoirs may reach temperatures of more than 150°C, which are more suitable for power generation, low hot water sources are used in hot springs and spas which have been very popular around the world for thousands of years. Hot water sources with low sulfur content are used for this purpose. Other applications of geothermal hot water involve the use of reservoirs for agricultural production, aquaculture, industrial applications, and heating.

Besides the hot water applications of geothermal sources that hot springs, spas and heating, it is a reliable, mature and advanced technology for power generation. As of 2021, the capacity of geothermal electricity reached a capacity of 15,854 MW. Figure 8.1 demonstrates the installed geothermal electricity capacity for the top ten countries and the world. These top ten countries cover about 93% of the total capacity. The USA is leading the geothermal capacity with 3722 MW of installed capacity. Indonesia, Philippines, Turkey and New Zealand are following it. These five countries are called the GIga Watt (GW) club as they have an installed capacity higher than 1 GW of capacity. There are many forthcoming geothermal power plants and additional capacities all around the world.

Figure 8.2 depicts the largest ten geothermal power plants in the world. The largest geothermal power plant is the Geysers power plants in San Francisco, USA. There are more than 350 wells and 22 power plants. Larderello in Italy is following it with a capacity of 769 MW. Turkey and New

Introduction to Energy Systems, First Edition. Ibrahim Dincer and Dogan Erdemir.
© 2023 John Wiley & Sons, Inc. Published 2023 by John Wiley & Sons, Inc.
Companion Website: www.wiley.com/go/Dincer/Introduction_to_Energy_Systems

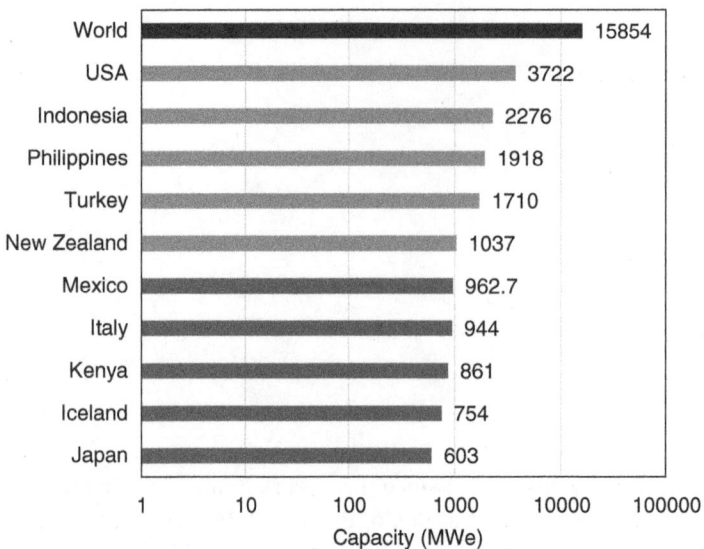

Figure 8.1 Installed capacities of geothermal power plants for the top ten countries (data from [1]).

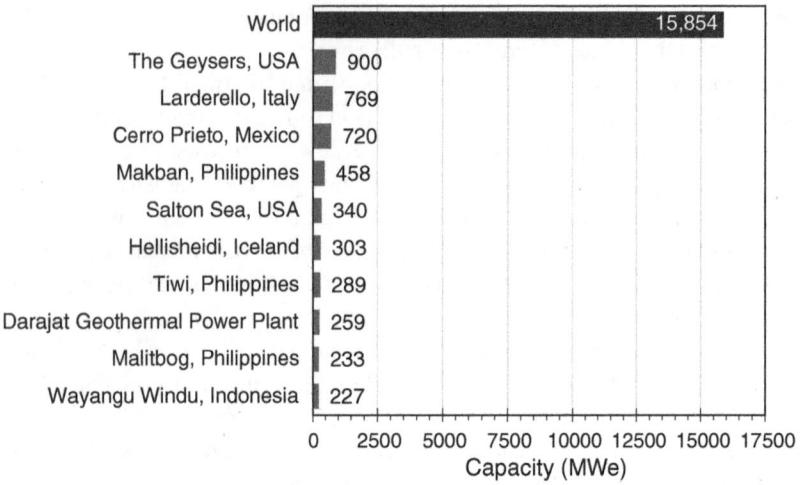

Figure 8.2 The top ten specific geothermal power plants in various locations of the world (data from [2]).

Zealand are active in geothermal energy production and utilization, but do not have a large-scale geothermal power plant.

Although geothermal has a big potential in generating electricity and heating demand, the biggest challenge is drilling the wells. This requires significant amount of cost and effort. A geothermal energy source is created by underground hot fluid sources, and its utilization varies depending on its temperature value. The use of geothermal energy will be greater if its temperature is high and technology allows it to be utilized, according to the Carnot heat engine. As deeper wells provide higher temperature, the depth of the geothermal wells will increase parallel to the developing technologies.

Today's technology reaches a depth of 10,000 m into the Earth's crust, and technologies utilizing geothermal energy at this depth are further developed. As mentioned earlier, the temperature increases 3°C from the Earth's surface every 100 m. While the temperature on the Earth's surface is approximately 14°C, the temperature within the Earth's crust may reach from 1000°C to 3500°C. The reason for the temperature increases with depth is the heat source in the center of the world. So, we have a significant heat source in the core of the earth.

The historical development of geothermal energy is presented with some example events and activities in Figure 8.3. Geothermal energy is one of the oldest renewable energy sources used by

2000 onward • Next generation and integrated geothermal energy systems have received increased attention.

1989 • The first geopressure-geothermal power plant was founded.

1987 • The geothermal fluids were used for gold recovery processes.

1972 • The Geothermal Energy Association was founded.

1960 • The world's largest geothermal power plant became operational, in Geysers, San Francisco, USA.

1948 • The first ground source heat pump was developed.

1942 • The capacity of geothermal electricity was reached 128 MW.

1921 • A well was drilled for the purpose of electricity generation in the USA.

1919 • The first geothermal well was drilled in Japan.

1913 • The site of the world's first commercial geothermal power plant was activated.

1904 • The first geothermal power plant was built in the USA.

1892 • The world's first district heating system was built as water is piped from hot springs to town buildings.

1860 • The geothermal sources were used in large scale hotel for heating and wellness.

8000 BC • The first reported geothermal use was in the North America.

Figure 8.3 Historical development of geothermal energy along with some events and activities.

humans. Technically, geothermal power was first used in North America at least 10,000 years ago, when indigenous peoples were attracted to hot springs for spiritual and practical reasons. Furthermore, the Romans used geothermal energy as a heating source for their buildings.

The first effort to harness geothermal energy for an industrial purpose occurred in 1818 in Italy. It was used for extracting boric acid from hot springs by François Jacques de Larderel, instead of using burning fossil fuel. The use of geothermal for electricity generation was achieved by Piero Ginori Conto in 1904. He was able to power a few light bulbs. In 1913, the site of the world's first commercial geothermal power plant was established with a capacity of 250 kW.

In the mid-1900s, the USA became a major user of geothermal energy sources. As of 1942, geothermal-based electricity generation capacity had reached almost 128 MW. Many countries developed geothermal energy systems between 1950 and 1960, especially in the USA. The foundations of today's highest geothermal capacity facility were laid in San Francisco in 1960. Today, it is still operational and reached a capacity of 900 MW by using more than 350 wells and 22 power plants.

This chapter deals with geothermal energy systems and applications, ranging from fundamentals to system analysis and assessment. First, geothermal resources and applications are classified and evaluated from various perspectives. Second, geothermal power generation techniques are presented where a unique feature of flashing process is defined and discussed. Next, the analysis and assessment studies on geothermal energy systems are presented through energy and exergy methods. Furthermore, geothermal heat pump and district heating applications are given separately with practical examples. Moreover, some illustrative examples are provided to help readers for better understanding the systems and their analyses and performance evaluation.

8.2 Geothermal Resources

In geothermal energy studies, geothermal energy can be defined and classified differently than other sources. The common method is to classify them according to the enthalpy of geothermal fluid or its temperature. Figure 8.4 shows a classification of geothermal sources. It is further known that enthalpy is a function of temperature; so, higher temperature means higher enthalpy. According to the temperature values, it is possible to classify into four branches which are low temperature, medium temperature, high temperature, and very high temperature geothermal sources. Low temperature geothermal sources are used for basic heating and hot water demands, such as direct heating, spas, hot springs, drying, etc. Its temperature varies from 50°C to 100°C and it is almost all in liquid form. Medium temperature geothermal source covers the source temperature ranges from 100 to 150°C. It is generally used for high-demand heating loads and power generation with organic Rankine cycle. Geothermal fluid, which has temperatures between 150°C and 200°C, is called the high-temperature geothermal source. Higher temperatures than 200°C is called a very high temperature geothermal energy source. High and very high temperature geothermal sources are generally used for power generation. Also, geothermal power generation is often combined with heating applications.

The enthalpy of the geothermal source can also be used for classification. As enthalpy is also a function of pressure, the classification of the geothermal source in terms of enthalpy is very critical. Pressure of the geothermal fluid directly affects the performance of geothermal power generation.

Whichever temperature or enthalpy is chosen here, this phenomenon is all about the thermal energy of geothermal fluid. Both indicate the energy level of a geothermal source. Geothermal energy sources are classified based on their enthalpies to express the thermal energy levels of geothermal fluids.

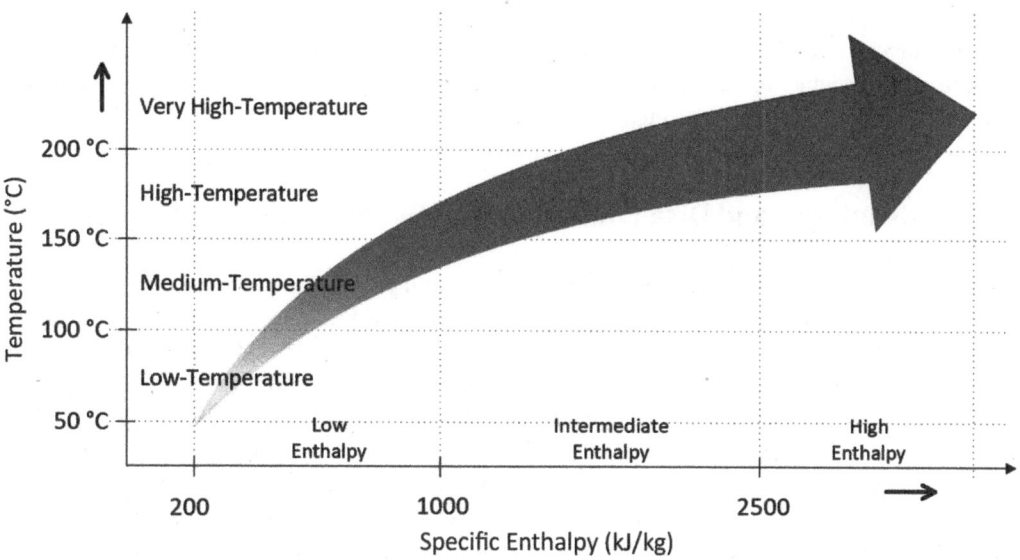

Figure 8.4 Classification of geothermal sources according to their temperature and enthalpy values.

Geothermal water provides thermal energy for producing useful outputs such as heating, electricity, distilled water, etc. It can basically be used as a heat source. While it may directly be utilized to meet the demands, it can be transferred into another medium (secondary cycle) for further process. Technically, the temperature of the geothermal sources is parallel to the depth of the source. However, the temperature value may not be the same for the same depths in different locations. Despite the depth is the dominant parameter for the temperature, it depends on various parameters as well, like location, type of the materials in the earth. From a geothermal reservoir, the following may be obtained:

- Liquid water
- Steam-liquid mixture
- Dry steam

Depending on the content of geothermal source listed above, the type of the geothermal power generation system is determined. High-temperature vapor is used for power generation with flashing chamber. Steam/liquid mixture is separated in the separator. While steam is used in the steam turbine for power generation, the liquid separated can be evaluated in the flash chamber or heating purposes. Dry steam can be directly used in power generation applications. Depending on the temperature of the source, it can be used for various applications.

- High temperature: T>150°C, it is primarily used for electricity production.
- Medium temperature: 100°C < T < 150°C, it is generally used for direct heating applications or combined with other sources for electricity production.
- Low temperature: T<100°C, it is used for direct heating purposes or water distillation.

Geothermal resources may be used either for power generation or direct consumption. Also, it is possible to combine it with other sources for heat upgrade. For example, medium- or low-temperature geothermal reservoirs can be combined with solar thermal systems for power generation. Most common uses of shallower reservoirs of lower temperature are for bathing and spas, agriculture, aquaculture, industrial use, and district heating. Some industrial uses of the geothermal resources are drying purposes, wool washing, cotton treatments, cloth dying, paper manufacture, milk pasteurization, melting ice on the roads, runways, sidewalks by piping under them, industrial evaporation processes, etc.

8.3 Advantages and Disadvantageous of Geothermal Energy Systems

Geothermal energy sources offer many advantages for both people and the environment. It is commonly accepted that its most important benefit for our world is that it is an environmentally-friendly energy source. However, it has some disadvantages as well like all other energy sources. The advantages and disadvantages of geothermal energy sources are summarized in Figure 8.5.

Without any combustion cycle, this geothermal energy source can be used directly to produce useful outputs. Compared to combustion-based energy generation systems, geothermal power generation systems emit rather very low emission values. Geothermal energy also provides the community where it is used with reliability since it is a consistent source of energy. Therefore, geothermal energy sources will be able to provide continuous electricity or useful output to the country where they are located, regardless of weather conditions. An energy generation system can also integrate geothermal energy with other renewable energy sources.

Despite its significant advantages, it brings some complications. For example, geothermal energy sources may emit harmful gases from underground to atmosphere. As it is really depended on the earth's body, surface instability and earthquakes are the main risk factor for geothermal energy systems. All these disadvantages can be solved by technological developments readily available. However, the main challenge in the geothermal energy systems is still the high installation costs which mainly come from the well drilling.

Advantages	Disadvantages
• Environmentally friendly	• High initial costs
• Fossil fuel-free operation	• Surface instability
• Wide-range operational capacity	• Effects of earthquakes
• Low operational cost	• Environmental impact due to some
• No complicated systems or components	harmful emissions
• Stability and reliablity	
• Widespread availability	
• Easy-integration with other renewable energy systems	

Figure 8.5 Advantages and disadvantages of geothermal energy sources.

8.4 Geothermal Applications

Geothermal energy sources are used in various applications to meet different demands items. The main objective is to use geothermal (heat) source for beneficial purposes. Figure 8.6 demonstrates the applications of geothermal sources. Geothermal energy sources are mainly used for power

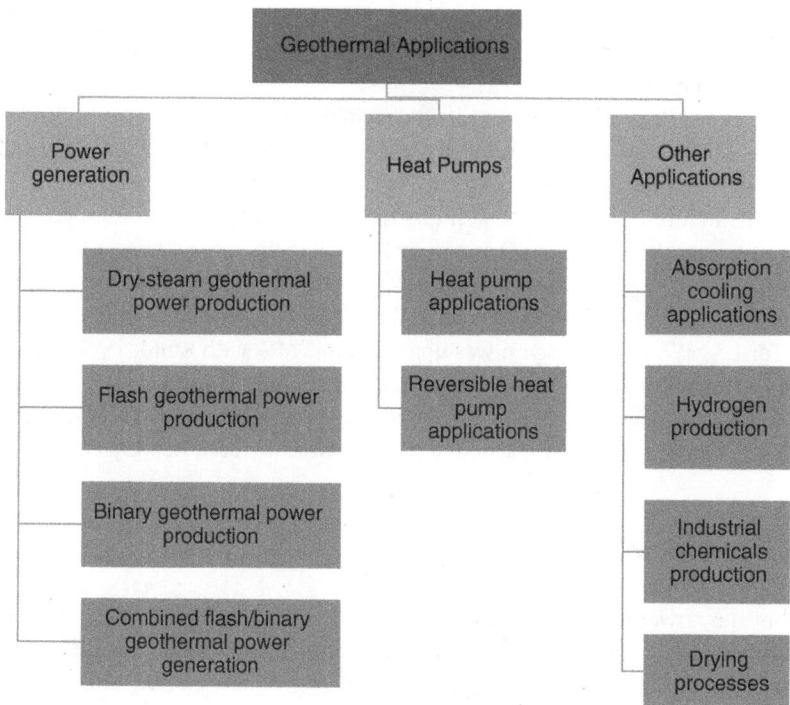

Figure 8.6 Classification of geothermal energy applications.

generation and heat pumps applications for heating and cooling purposes. Also, it can be used for green hydrogen production, absorption cooling, chemical production, and drying processes.

The heating application of geothermal can easily be performed with direct use, heat pump applications, and dying processes. In direct use, the heat extracted from geothermal source is used directly for heating purposes. It may also be performed individually or with district heating applications. While geothermal water sources with low and medium temperature can be directly used, they can be used for a heat exchanger connected to the secondary loop heat distribution. Ground-based heat pump and underground water-based heat pumps are also used for heating purposes, too. They use geothermal sources as the low-temperature energy source of the refrigeration cycle used in the heat pump. It is also possible to use geothermal energy for heating up the air that is used in the drying processes.

Geothermal energy sources find utilization possibility in absorption cooling systems, as the heat source of the cooling cycle. Also, the low temperature-ground can be used for direct cooling purposes for low-capacity cooling applications. It can be used in reversible heat pump for high demand cooling purposes.

In today's world geothermal sources are more preferred for power generation as long as the temperature level is suitable. There are various techniques to produce electricity from geothermal sources. The main focus of geothermal power generation is to use the heat in the geothermal source to run a steam cycle directly or a binary cycle using steam or another working fluid. The techniques given in Figure 8.6 will be discussed in detail in the following section.

In addition to the main use of geothermal sources like heating and power generation, there are many attempts to use it for green hydrogen production and production of some industrial chemical and synthetic fuels. Technically, geothermal energy systems can be the part of the any system requiring heat or electricity that comes from the geothermal. Integrated geothermal energy systems will be discussed in detail in Chapter 11.

It is also possible to use geothermal energy in an integrated manner in order to meet multiple demands at the same time. An interesting example for the multiuse of geothermal energy is shown in Figure 8.7, from Húsavík, Iceland. As can be seen from Figure 8.7, geothermal energy is used in power generation, water distillation, district heating, industrial purposes, spa, hot springs, greenhouse heating, snow melting, etc. The location, Húsavík, is known as the second largest town in the northeast region of Iceland. The population is around 2500. The main livelihood in the community is fishing. The power generation with geothermal is first made by a binary organic Rankine cycle using Isopentane for 1.5 MW of electricity. It was upgraded to 2 MW with Kalina cycle using ammonia and water. The total cost of the Húsavík project is approximately 12 million euro. While 4 million (one-third) of the total cost is for power generation, the rest (two-third) is for district heating system. One may read more about this particular community project in Ref. [3].

8.5 Geothermal Power Generation

The working principle of the power generation from geothermal sources is based on the power production from heat. Depending on the type of geothermal power plant, the use of heat differs. It works based on Rankine cycle which is using geothermal energy as the energy source. Figure 8.8

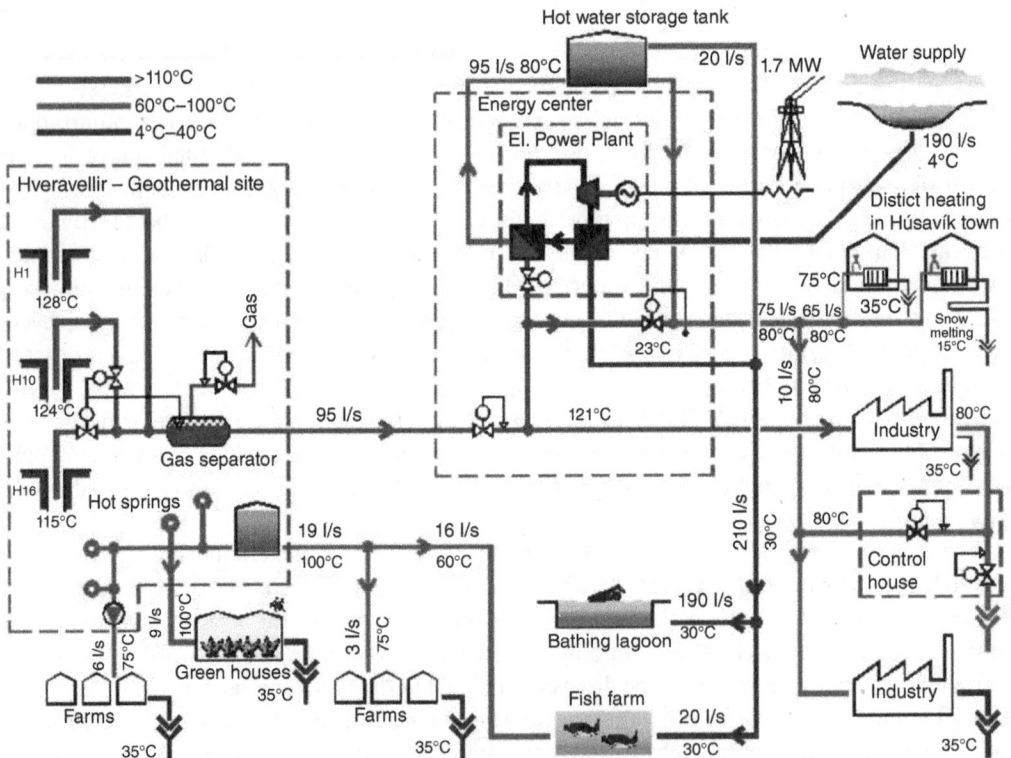

Figure 8.7 A unique illustration of multiple use of geothermal energy – Húsavík Energy, Iceland (adapted from [3]).

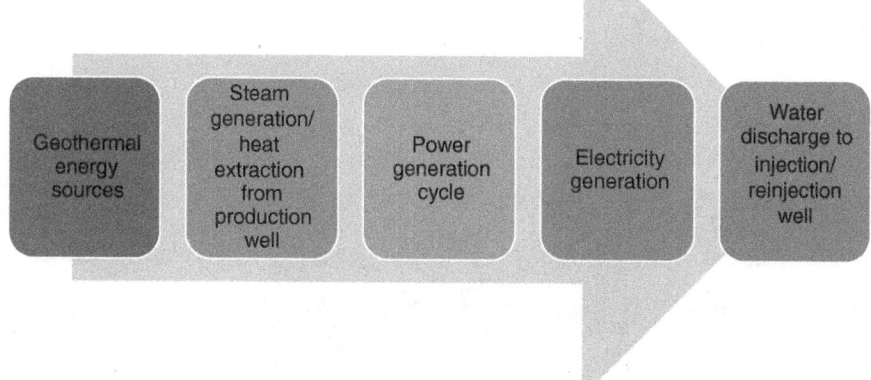

Figure 8.8 The flow diagram for geothermal power generation.

demonstrates the basics of the power generation from geothermal sources. A working fluid harvests heat from the geothermal well. The heat is transferred to the water for steam production. Steam generated expands in the turbine for power generation. A generator is included in the system for electricity generation. The steam is cooled down in the condenser and pumped back to the heat exchanger. The cold working fluid is then reinjected into the well.

There are four common methods of geothermal power generation. These are essentially listed as follows:

- Direct steam (dry steam) power generation
- Flash-steam power generation
- Binary cycle power generation
- Combined flash/binary power generation

Note that each of these techniques will be discussed in detail with various practical examples in the following sections.

8.5.1 Direct Steam (Dry Steam) Power Generation

This is recognized as the most basic system for power generation from geothermal sources. As far as simplicity and economics are concerned, these systems are the most basic ones in terms cycle/system. The temperature of the incoming fluid should be high enough in vapor phase to benefit directly from geothermal energy. It is possible to integrate these systems into a variety of system configurations since the underground fluid consists only of steam. It is also important to take into account changes to the mathematical model of the system when integrating.

Figure 8.9 demonstrates the schematic illustration of the direct steam power generation. Explaining the working principle of the system, in general, is useful for understanding how the system works. Direct steam or dry steam geothermal power generation is the most basic technique to produce electricity. Geothermal sources are used for steam generation. Then, the steam generated is used for power generation in a turbine. Figure 8.9a shows the basic dry steam geothermal power generation system.

On the other hand, in actual dry steam systems, to eliminate particles in the fluid from the geothermal energy source, the fluid is sent to the particle separator. Also, since the pressure in the

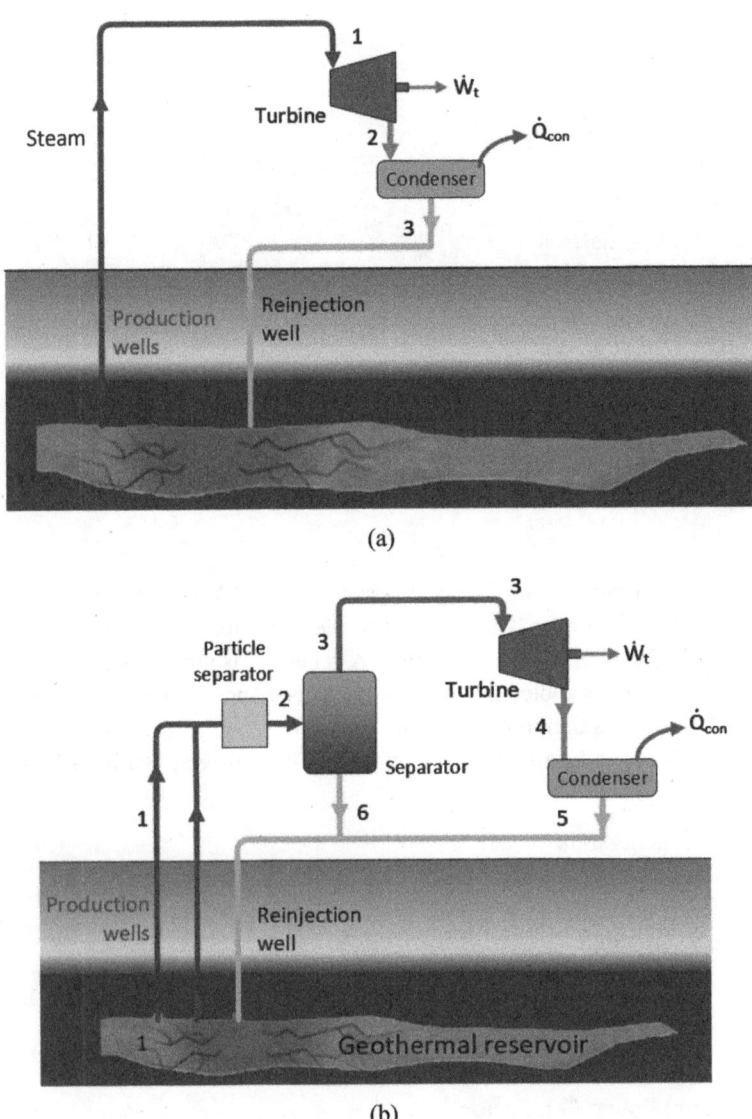

Figure 8.9 System layout of a direct steam geothermal power systems (a) basic dry steam geothermal power generation and (b) actual dry steam geothermal power generation with particle separator and moisture separator.

geothermal energy source is higher than where the particle lifter is located, it is transmitted to the particle lifter after pressure adjustment. After the particles in the fluid are removed, the fluid coming out of the particle extractor is sent to the moisture separator to remove the moisture in the geothermal fluid, as illustrated in Figure 8.9b. Here, the moisture released from the fluid is sent to the geothermal reservoir. After moisture separation, the dry steam is sent to the turbine. With the generation of power, the fluid coming out of the turbine is sent to the condenser to be condensed. The fluid coming out of the condenser is sent again to the reinjection well.

In order to assess the system's performance, thermodynamic analysis is then applied to the system, based on four thermodynamic balance equations, which are presented in Chapter 3.

Illustrative Examples 8.1 Consider a dry steam geothermal power plant. There are three production wells in the geothermal power plant facility. The temperature and pressure of the geothermal fluid is 243°C and 2315 kPa, respectively. The flow rate of the working fluid from each well is 50 kg/s. The steam that comes from the geothermal source is expanded to 100 kPa in a single stage turbine. The outlet temperature is found to be 102°C. A generator attached to the turbine is used for electricity generation. The efficiency of the generator is 0.85. Calculate the power generation capacity of the power plant.

Solution:

As provided in the question, the temperature and pressure of the geothermal fluid are 243°C and 2315 kPa. At this condition the fluid is superheated vapor. There is no need for separator in this case as the working fluid is fully vapor. The thermophysical properties of the geothermal fluid are

$h_1 = 2870$ kJ/kg

$s_1 = 6.422$ kJ/kg K

There are three wells in the system; the total flow rate at the inlet of the turbine is

$\dot{m}_{total} = 3 \times 50 = 150\,kg/s$

The power generation is found from the EBE of the turbine. So,

$\dot{m}_{total}h_{in} = \dot{m}_{total}h_{out} + \dot{W}_t$

The outlet state is found from Table A1 as:

$h_2 = 2680$ kJ/kg

$s_2 = 7.372$ kJ/kg K

Substituting these values in the EBE, one obtains,

$\dot{W}_t = 150 \times (2870 - 2680)$

The turbine work is found to be $\dot{W}_t$=28,500 kW. As the generator efficiency is 0.85, the power output is found as

$\dot{W}_g = 28750 \times 0.85 = 24,437.5\,kW$

Illustrative Examples 8.2 A direct steam geothermal power plant, as shown in Figure 8.10, is utilized for power generation. The geothermal working fluid enters the particle separator at the temperature and pressure of 250°C and 2350 kPa, respectively, and a mass flow rate of 50 kg/s. The mass flow rate of unwanted particles ($\dot{m}_{particles}$) is equal to 10% of the input geothermal water mass flow rate. Then, the clean geothermal working fluid enters the moisture separator and is separated in a saturated form in this plant; the fraction of saturated water at flow 5 is 0.4. The exit pressure of the steam from the separator is 780 kPa, and then the dry steam is utilized to produce power in a turbine. Finally, the discharged geothermal working fluid from the steam turbine is at 25 kPa with a quality of 0.84, which is then condensed and reinjected to the injection well

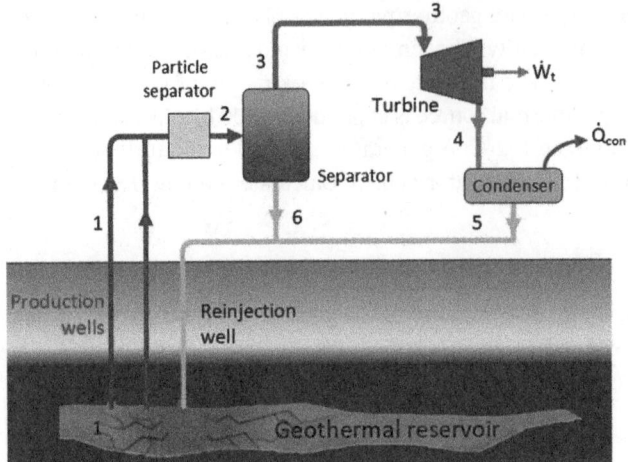

Figure 8.10 Schematic illustration of the dry-steam geothermal power generation given in the Illustrative Example 8.2.

a) Write mass, energy, entropy and exergy balance equations for each system component.
b) Find the work rate of the turbine.
c) Calculate both energy and exergy efficiencies of the system.
d) Recalculate the energy and exergy efficiencies in the case of 50% of the heat discharged in the condenser are recovered for useful purposes.

Solution:

a) Let's write the balance equations for each component of the system as illustrated in Figure 8.10:

Particle separator:

MBE: $\dot{m}_1 = \dot{m}_2$

EBE: $\dot{m}_1 h_1 = \dot{m}_2 h_2$

EnBE: $\dot{m}_1 s_1 + \dot{S}_{gen,separator} = \dot{m}_2 s_2$

ExBE: $\dot{m}_1 ex_1 = \dot{m}_2 ex_2 + \dot{Ex}_{dest,ps}$

Separator:

MBE: $\dot{m}_2 = \dot{m}_3 + \dot{m}_4$

EBE: $\dot{m}_2 h_2 = \dot{m}_3 h_3 + \dot{m}_4 h_4$

EnBE: $\dot{m}_2 s_2 + \dot{S}_{gen,separator} = \dot{m}_3 s_3 + \dot{m}_4 s_4$

ExBE: $\dot{m}_2 ex_2 = \dot{m}_4 ex_3 + \dot{m}_4 ex_4 + \dot{Ex}_{dest,ms}$

Turbine:

MBE: $\dot{m}_5 = \dot{m}_6$

EBE: $\dot{m}_5 h_5 = \dot{m}_6 h_6 + \dot{W}_{tur}$

EnBE: $\dot{m}_5 s_5 + \dot{S}_{gen,tur} = \dot{m}_6 s_6$

ExBE: $\dot{m}_5 ex_5 = \dot{m}_6 ex_6 + \dot{W}_{tur} + \dot{Ex}_{dest,t}$

Condenser:

MBE: $\dot{m}_6 = \dot{m}_7$

EBE: $\dot{m}_6 h_6 = \dot{m}_7 h_7 + \dot{Q}_{con}$

EnBE: $\dot{m}_6 s_6 + \dot{S}_{gen,con} = \dot{m}_7 s_7 + \dfrac{\dot{Q}_{loss}}{T_0}$

ExBE: $\dot{m}_6 ex_6 = \dot{m}_7 ex_7 + \dot{Ex}_{\dot{Q}_{loss}} + \dot{Ex}_{dest,con}$

b) In order to determine the turbine output, we first need to determine flow rates accordingly. As 10% of the mixture coming from production well goes to the particle, $\dot{m}_2$ can be calculated as follows:

$$\dot{m}_2 = \dot{m}_1 \times 0.9 = 50 \times 0.9 = 45 \; kg/s$$

As the fraction of saturated water at stream 5 is 0.4. $\dot{m}_6$ can be found as

$$\dot{m}_6 = \dot{m}_2 \times 0.4 = 45 \times 0.4 = 18 \; kg/s$$

From mass balance equation for the moisture separator:

$$\dot{m}_2 = \dot{m}_3 + \dot{m}_6$$

$$45 = 18 + \dot{m}_4$$

Thus, the steam flow rate is found to be $\dot{m}_4 = 27$ kg/s.

It is given in the problem that $T_1 = 250°$ and $P_1 = 2350$ kPa. Thus, from Table A1, the enthalpy and entropy values of the geothermal source is taken as follows:

$$h_1 = 2801 \; kJ \, / \, kg$$

$$s_1 = 6.285 \; \frac{kJ}{kgK}$$

From the energy balance equations for the particle and moisture separators, the enthalpy values are the same in these streams. So, $h_1 = h_2 = h_3 = 2810$ kJ/kg. The pressure of stream 3 is given as $P_3 = 780$ kPa. From Table A1, by using the enthalpy and pressure values, the temperature is found to be 185°C. $s_3 = 7.74$ kJ/kg K.

For the turbine, outlet conditions are provided in the problem as $x_4 = 0.84$, $P_4 = 25$ kPa. As we know two properties for this stream, we can find the rest of it by using Table A1. The enthalpy and entropy values for the stream 4 is obtained as

$$T_4 = 67°C$$

$$h_4 = 2245 \; kJ/kg$$

$$s_4 = 6.75 \; kJ/kg \; K$$

Thus, we calculate the turbine output by using the energy balance equation for turbine as follows:

$$\dot{m}_3 h_3 = \dot{m}_4 h_4 + \dot{W}_{tur}$$

$$27 \times 2810 = 27 \times 2245 + \dot{W}_{tur}$$

So, the useful work of the turbine is found to be $\dot{W}_{tur} = 15{,}255$ kW.

c) The energy and exergy efficiencies of the system can be defined as follows:

$$\eta_{en} = \frac{\dot{W}_{tur}}{\dot{m}_1 h_1 - \left[\dot{m}_5 h_5 + \dot{m}_6 h_6 \right]}$$

By substituting the values in this equation:

$$\eta_{en} = \frac{15{,}255}{2801 \times 50 - \left[45 \times 272 + 18 \times 941.5 \right]} = \frac{15{,}225}{110{,}863} = 0.1373$$

$$\eta_{ex} = \frac{\dot{W}_{tur}}{\dot{m}_1 ex_1 - \left[\dot{m}_5 ex_5 + \dot{m}_6 ex_6 \right]}$$

where ex_1, ex_5, and ex_6 should be calculated.

$$ex_1 = (h_1 - h_0) - T_0(s_1 - s_0)$$

The specific enthalpy and entropy values at the reference conditions are

$h_0 = 104.0$ kJ/kg

$s_0 = 0.3672$ kJ/kg K

$ex_1 = 1037.7$ kJ/kg

$ex_5 = 96.3$ kJ/kg

$ex_6 = 136.8$ kJ/kg

Thus, the exergy efficiency is found to be

$$\eta_{en} = \frac{15{,}225}{1027.7 \times 50 - \left[45 \times 96.3 + 18 \times 136.8\right]} = \frac{15{,}225}{44{,}590} = 0.3339$$

Finally, the energy and exergy efficiencies are obtained as 13.73% and 33.39%, respectively.

d) The heat released in the condenser is found from the energy balance equation for the condenser.

$$\dot{m}_4 h_4 = \dot{m}_5 h_5 + \dot{Q}_{con}$$

$$27 \times 2245 = 27 \times 272 + \dot{Q}_{con}$$

Thus, $\dot{Q}_{con}$ is found to be 53,271 kW. If 50% of this heat is recovered,

$$\dot{Q}_{recovered} = \dot{Q}_{con} \times 0.50 = 26{,}635.5 \ kW$$

The energy efficiency equation can be written by considering heat output as well:

$$\eta_{en} = \frac{\dot{W}_{tur} + \dot{Q}_{recovered}}{\dot{m}_1 h_1 - \left[\dot{m}_5 h_5 + \dot{m}_6 h_6\right]} = \frac{15{,}255 + 26{,}635.5}{2801 \times 50 - \left[45 \times 272 + 18 \times 941.5\right]}$$

$$\eta_{en} = 0.83$$

The exergy efficiency can be written as

$$\eta_{ex} = \frac{\dot{W}_{tur} + \dot{Ex}_{\dot{W}_{tur}}}{\dot{m}_1 ex_1 - \left[\dot{m}_5 ex_5 + \dot{m}_6 ex_6\right]} = \frac{\dot{W}_{tur} + \left[\dot{Q}_{con}(1 - \frac{T_0}{T_{ave}})\right]}{\dot{m}_1 ex_1 - \left[\dot{m}_5 ex_5 + \dot{m}_6 ex_6\right]}$$

$$\eta_{ex} = \frac{15{,}255 + 13{,}325}{2801 \times 50 - \left[45 \times 272 + 18 \times 941.5\right]} 0.6470$$

In the case of 50% heat recovery option in the condenser, the energy and exergy efficiencies are calculated as 83% and 64.7%, respectively. The heat recovery (cogeneration) has a significant potential to increase the system efficiency.

8.5.2 Flash-Steam Power Generation

Flashing process is very important in geothermal power generating systems and confirms how much vapor can be extracted from the incoming saturated geothermal liquid. Figure 8.11 demonstrates the flashing process on a P-h diagram. Normally, high-temperature saturated liquid, such as point (1) in the figure comes from the geothermal reservoir. In order to generate power, the working fluid should be turned fully into vapor to use in the steam turbine. For this process, flash

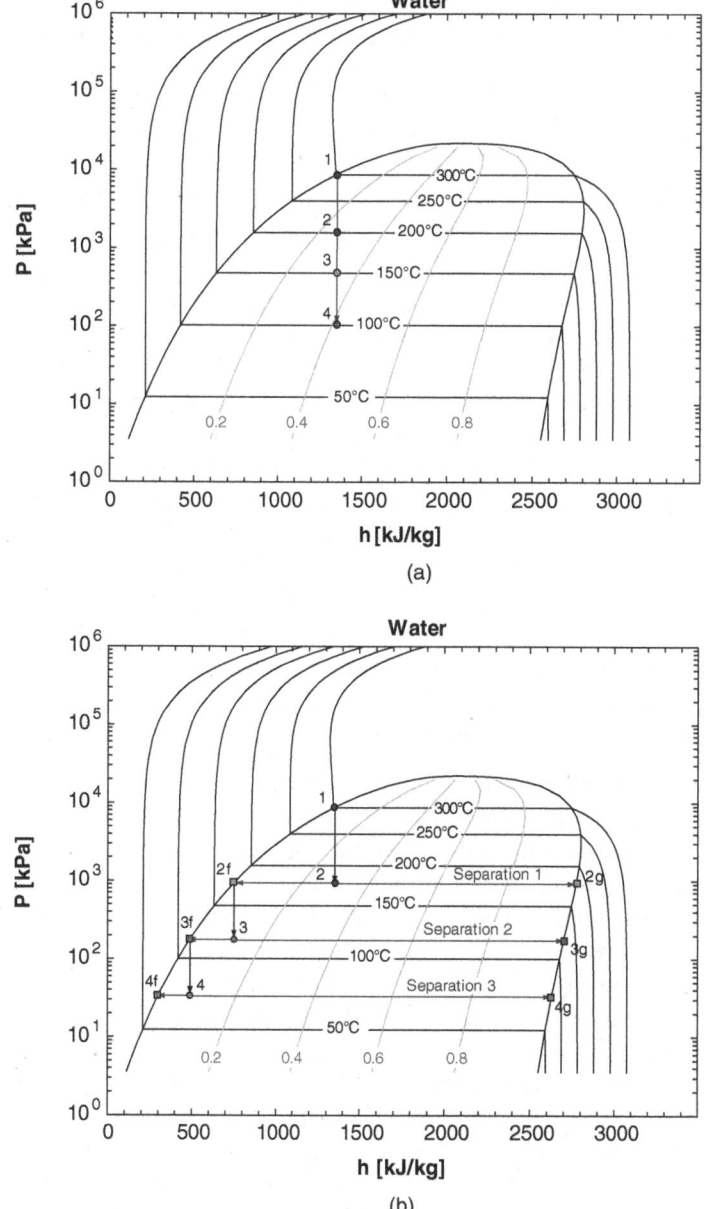

Figure 8.11 Illustration of flashing process on P-h diagram (a) the effect of pressure drops in the flashing process and (b) the effect of the number of the flashing process.

(or flashing) chambers are used. In the flash chamber, high-temperature and high-pressure saturated water turns into the mixture of vapor and liquid water while reducing the pressure. The vapor is the separated and sent to the turbine.

The main parameter in the flashing process is the pressure drop which will determine how much vapor can be achieved. Figure 8.11a demonstrates the effect of the pressure drop in a flashing process. At the point (1) in Figure 8.11a, there is a geothermal fluid as the saturated fluid at 300°C. At this condition, the pressure, P_1, is 8771 kPa. The specific enthalpy and entropy are 1345 kJ/kg and 3.254 kJ/kgK, respectively. Considering the reference conditions as $T_0 = 25°C$ and $P_0 = 100$ kPa, the specific exergy is calculated to be 1168 kJ/kg. In the flashing chamber, it is assumed that the pressure drop occurs isenthalpically. In other words, while the pressure drops, the enthalpy stays constant. Therefore, the phase of the geothermal fluid changes. The quality of the geothermal fluid changes. As illustrated in Figure 8.11a, while the pressure of the geothermal fluid drops, the quality of the geothermal fluid increases. In other words, the vapor content of the mixture increases. On the other hand, the temperature of the mixture reduces. A higher vapor content (quality) means a higher flow rate for the steam which increases the turbine output. In contrast, the lower pressure and temperature tend to reduce the turbine output. Therefore, the flashing processes should be optimized carefully.

Further to note that the phase change is changed from 1 to 2 in Figure 8.11b by assuming an isenthalpic process. The geothermal fluid is the saturated fluid state at 1. It is converted to water and vapor mixture at 2. Then, the mixture goes to the separator. It is separated into steam and water. This process is illustrated as separation 1 in Figure 8.11b. It is possible to perform more than one flashing process. After the first separation, the pressure of the saturated water is reduced again in another flashing chamber. Again, the saturated water turns into the water and steam mixture. The mixture goes to the second separator to get vapor from the mixture for power generation. It is possible to increase the number of flash processes. Like in the case of pressure drop, as given in Figure 8.11a, the number of the flashing process should be optimized carefully to maximize the turbine output.

Flash-steam or flashing geothermal power plants are named according to the number of flash chambers in the system. If there is one flash chamber, it is called a single-flash steam geothermal power plant. If there are two of them, it is named as double-flash steam geothermal power plant. Let's start with the single-flash steam geothermal power plants.

Single-Flash Steam Geothermal Power Generation:
In the single-flash geothermal power plants, the fluid that comes from the geothermal source is the mixture of steam and liquid. Figure 8.12 demonstrates the system layout for a single-flash steam geothermal power generation. Here, we should first define what flash is. The flashing is a process which is used for obtaining pressurized liquid and vapor from the liquid with this saturation pressure by decreasing the pressure of the underground pressurized liquid.

The high-pressure (HP) geothermal fluid is sent to the valve, where the flashing takes place with flow number 1. A mixture of liquid and vapor is transmitted to the separator with flow number 2 as the fluid whose pressure is reduced. The steam and brine are separated from the incoming fluid in the separator. Flow number 4 is used to increase the performance of the fluid by sending steam from the separator to the purifier. Flow number 5 removes fouling from the fluid by removing it from the system. Flow number 6 is used to send steam fluid with improved quality to the turbine, which produces electricity. Condensation occurs when steam comes out of the turbine and goes to the condenser with flow number 7. Research on the selection of the condenser used for condensation in the literature or in the application. The heat is generated by the condenser, which is a useful output. Furthermore, flow number 8 is used to send the fluid from the condenser to the ground.

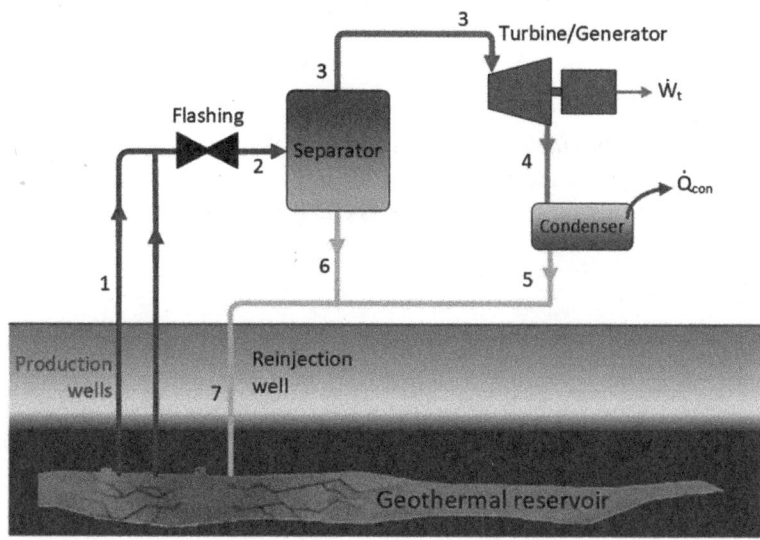

Figure 8.12 Schematic illustration of the single-flash geothermal power generation system.

In order to analyze the single-flash steam geothermal power generation system, four main thermodynamic balance equations should be written accordingly, as presented below:

The mass, energy, entropy and exergy balance equations for the flash chamber are written as follows:

$$\text{MBE: } \dot{m}_1 = \dot{m}_2 \tag{8.1}$$

$$\text{EBE: } \dot{m}_1 h_1 = \dot{m}_2 h_2 \tag{8.2}$$

$$\text{EnBE: } \dot{m}_1 s_1 + \dot{S}_{gen,fc} = \dot{m}_2 s_2 \tag{8.3}$$

$$\text{ExBE: } \dot{m}_1 ex_1 = \dot{m}_2 ex_2 + \dot{Ex}_{dest,fc} \tag{8.4}$$

For separator:

$$\text{MBE: } \dot{m}_2 = \dot{m}_3 + \dot{m}_4 \tag{8.5}$$

$$\text{EBE: } \dot{m}_2 h_2 = \dot{m}_3 h_3 + \dot{m}_4 h_4 \tag{8.6}$$

$$\text{EnBE: } \dot{m}_2 s_2 + \dot{S}_{gen,separator} = \dot{m}_3 s_3 + \dot{m}_4 s_4 \tag{8.7}$$

$$\text{ExBE: } \dot{m}_2 ex_2 = \dot{m}_4 ex_3 + \dot{m}_4 ex_4 + \dot{Ex}_{dest,separator} \tag{8.8}$$

For turbine:

$$\text{MBE: } \dot{m}_4 = \dot{m}_5 \tag{8.9}$$

$$\text{EBE: } \dot{m}_4 h_4 = \dot{m}_5 h_5 + \dot{W}_{tur} \tag{8.10}$$

$$\text{EnBE: } \dot{m}_4 s_4 + \dot{S}_{gen,t} = \dot{m}_5 s_5 \tag{8.11}$$

$$\text{ExBE: } \dot{m}_4 ex_4 = \dot{m}_5 ex_5 + \dot{W}_{tur} + \dot{Ex}_{dest,t} \tag{8.12}$$

For condenser:

$$\text{MBE: } \dot{m}_5 = \dot{m}_6 \tag{8.13}$$

$$\text{EBE: } \dot{m}_5 h_5 = \dot{m}_6 h_6 + \dot{Q}_{con} \tag{8.14}$$

$$\text{EnBE: } \dot{m}_5 s_5 + \dot{S}_{gen,con} = \dot{m}_6 s_6 + \frac{\dot{Q}_{con}}{T_0} \tag{8.15}$$

$$\text{ExBE: } \dot{m}_5 ex_5 = \dot{m}_6 ex_6 + \dot{Ex}_{\dot{Q}_{loss}} + \dot{Ex}_{dest,con} \tag{8.16}$$

The energy and exergy efficiencies for the single-flash steam geothermal power generation system may be defined according to the cases with and without reinjection.

If there is no reinjection in the system, the energy and exergy efficiencies are written as:

$$\eta_{en} = \frac{\dot{W}_t}{\dot{m}_1 h_1} \tag{8.17}$$

and

$$\eta_{ex} = \frac{\dot{W}_t}{\dot{m}_1 ex_1} \tag{8.18}$$

In the case of reinjection, the energy and exergy efficiencies are defined as:

$$\eta_{en} = \frac{\dot{W}_t}{\dot{m}_1 h_1 - \dot{m}_8 h_8} \tag{8.19}$$

and

$$\eta_{ex} = \frac{\dot{W}_t}{\dot{m}_1 ex_1 - \dot{m}_8 ex_8} \tag{8.20}$$

It should be noted here that most common steam plant is the single-flash steam plant where it generates electricity from hot and high-pressure liquid-dominated reservoirs by flashing the entering liquid into steam by reducing the pressure. The steam is then piped directly to a steam turbine.

Illustrative Example 8.3   A single-flash geothermal power plant (as illustrated in Figure 8.13) uses hot geothermal water at 210°C and 2500 kPa as the heat source. The flow rate of geothermal fluid is 150 kg/s. The geothermal fluid isenthalpically flashed in a flash chamber at a pressure ratio of 3 and separated in the saturated form in a vapor-liquid separator. The fraction of vapor at the flash part outlet is 0.15. Then the produced vapor is used to generate power in a steam turbine. Finally, the discharged geothermal working fluid from the steam turbine is at 11 kPa with a quality of 0.84, which is then condensed and discharged to the injection well.

a) Calculate the work generation in the steam turbine.
b) Find both energy and exergy efficiencies of the single-flash geothermal power plant.

Solution:
The thermodynamic balance equation for a single-flash geothermal power plant is already given earlier from Eqs. 8.1 to 8.16.

a) In order to determine the turbine output, we need to write the energy balance equation for the turbine. From Eq. 8.10, we obtain

$$\dot{m}_3 h_3 = \dot{m}_4 h_4 + \dot{W}_{tur}$$

In order to this equation, we need to determine the mass flow rates and enthalpy values. Let's begin with 1, for $T_1 = 210°C$ and $P_1 = 2500$ kPa. The specific enthalpy and entropy values are obtained from Table A1.

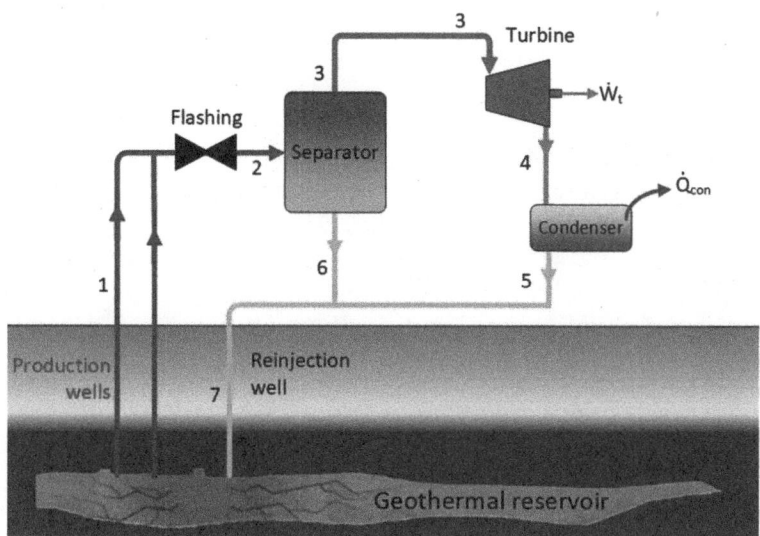

Figure 8.13 Schematic view of the single-flash geothermal power plant in Illustrative Example 8.2.

$h_1 = 897.8$ kJ/kg

$s_1 = 2.424$ kJ/kg K

For the stream 2, $X_2 = 0.15$ and pressure ratio is 1/3. Thus, $P_2 = P_1/3$ and $P_2 = 833.3$ kPa. Thus, for the stream 2, one may find

$T_2 = 173°C$

$h_2 = 719.8$ kJ/kg

$s_2 = 2.046$ kJ/kg K

For the separator, it is required to determine the flow rates:

$$\dot{m}_3 = 0.11 \times \dot{m}_2 = 0.11 \times 150 = 16.5 \ kg \ / \ s$$

From the mass balance equation of the separator (Eq. 8.5):

$$\dot{m}_2 = \dot{m}_3 + \dot{m}_4$$

$$150 = 16.5 + \dot{m}_4$$

Thus, $\dot{m}_4$ is found to be 133.5 kg/s.

For stream 3, $P_2 = P_3$ and $X_3 = 1$ (saturated vapor only). Thus, the temperature, and specific enthalpy and entropy values of the stream are obtained as:

$T_3 = 173°C$

$h_3 = 2791$ kJ/kg

$s_3 = 6.683$ kJ/kg K

So, the turbine output can be found as

$$\dot{m}_3 h_3 = \dot{m}_4 h_4 + \dot{W}_{tur}$$

$$124.6 \times 2791 - 124.6 \times 2205 = \dot{W}_{tur}$$

Finally, the turbine output is found to be $\dot{W}_{tur} = 73{,}015.6$ kW.

b) The energy and exergy efficiencies can be calculated from Eq. 8.19 and 8.20, respectively.

$$\eta_{en} = \frac{\dot{W}_t}{\dot{m}_1 h_1 - \dot{m}_5 h_5 - \dot{m}_6 h_6}$$

$$\eta_{en} = \frac{73{,}015.6}{150 \times 897.8 - 16.5 \times 272 - 133.5 \times 941.5} = 0.224 \; or \; 22.4\%$$

and

$$\eta_{ex} = \frac{\dot{W}_t}{\dot{m}_1 ex_1 - \dot{m}_5 ex_5 - \dot{m}_6 ex_6}$$

$$\eta_{en} = \frac{73{,}015.6}{150 \times 426.6 - 16.5 \times 97.5 - 133.5 \times 175.3} = 0.669 \; or \; 66.9\%$$

Double-Flash Steam Geothermal Power Generation:

A double-flash steam geothermal power plant is a result of upgrading the performance of a single-flash one in order to recover the energy that comes from the geothermal reservoir by adding one extra flash chamber. Although it is quite similar to the single-flash steam geothermal power plant, the double-flash steam geothermal system may require extra maintenance and cost requirements compared to the single-flash system due to its little more complex operational matters and requirements. An overview of the double-flash steam geothermal power plant is shown in Figure 8.14. In the double-flash system, the working fluid that comes from the geothermal reservoir enters the flash chamber twice. Thus, it aims to get more energy from the geothermal source and more power generation.

In order to perform a complete thermodynamic analysis, the mass, energy, entropy and exergy balance equations can be written as follows:

For flash chamber 1:

$$\text{MBE: } \dot{m}_1 = \dot{m}_2 \tag{8.21}$$

$$\text{EBE: } \dot{m}_1 h_1 = \dot{m}_2 h_2 \tag{8.22}$$

$$\text{EnBE: } \dot{m}_1 s_1 + \dot{S}_{gen,fc} = \dot{m}_2 s_2 \tag{8.23}$$

$$\text{ExBE: } \dot{m}_1 ex_1 = \dot{m}_2 ex_2 + \dot{Ex}_{dest,fc} \tag{8.24}$$

For separator 1:

$$\text{MBE: } \dot{m}_2 = \dot{m}_3 + \dot{m}_6 \tag{8.25}$$

$$\text{EBE: } \dot{m}_2 h_2 = \dot{m}_3 h_3 + \dot{m}_6 h_6 \tag{8.26}$$

$$\text{EnBE: } \dot{m}_2 s_2 + \dot{S}_{gen,separator,1} = \dot{m}_3 s_3 + \dot{m}_6 s_6 \tag{8.27}$$

$$\text{ExBE: } \dot{m}_2 ex_2 = \dot{m}_4 ex_3 + \dot{m}_6 ex_6 + \dot{Ex}_{dest,separator,1} \tag{8.28}$$

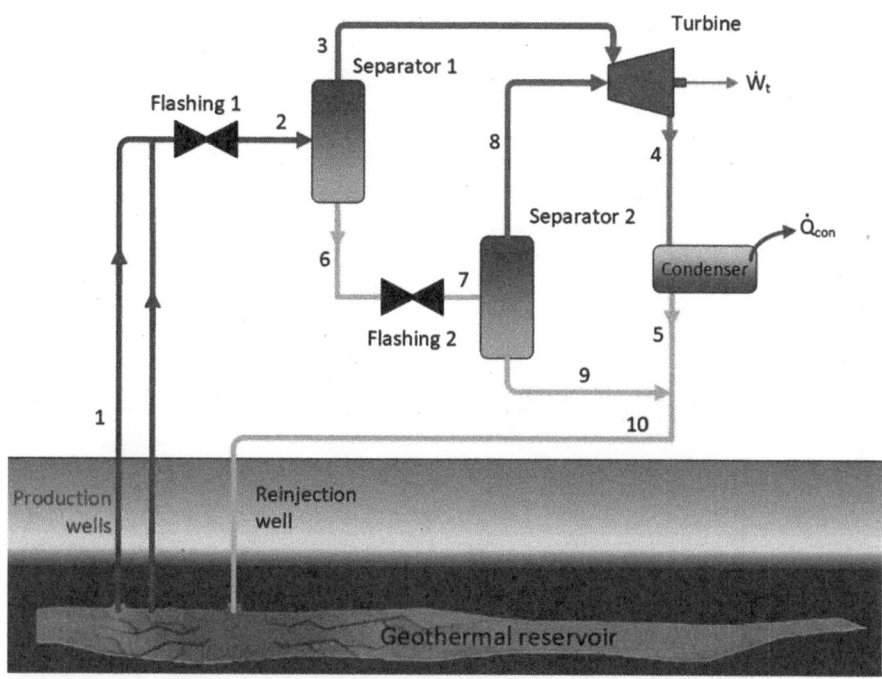

Figure 8.14 Schematic view of a double-flash steam geothermal power plant.

For flash chamber 2:

$$\text{MBE: } \dot{m}_6 = \dot{m}_7 \tag{8.29}$$

$$\text{EBE: } \dot{m}_6 h_6 = \dot{m}_7 h_7 \tag{8.30}$$

$$\text{EnBE: } \dot{m}_6 s_6 + \dot{S}_{gen,fc,2} = \dot{m}_7 s_7 \tag{8.31}$$

$$\text{ExBE: } \dot{m}_6 ex_6 = \dot{m}_7 ex_7 + \dot{Ex}_{dest,fc,2} \tag{8.32}$$

For separator 2:

$$\text{MBE: } \dot{m}_7 = \dot{m}_8 + \dot{m}_9 \tag{8.33}$$

$$\text{EBE: } \dot{m}_7 h_7 = \dot{m}_8 h_8 + \dot{m}_9 h_9 \tag{8.34}$$

$$\text{EnBE: } \dot{m}_7 s_7 + \dot{S}_{gen,separator,2} = \dot{m}_8 s_8 + \dot{m}_9 s_9 \tag{8.35}$$

$$\text{ExBE: } \dot{m}_7 ex_7 = \dot{m}_8 ex_8 + \dot{m}_9 ex_9 + \dot{Ex}_{dest,separator,2} \tag{8.36}$$

For turbine:

$$\text{MBE: } \dot{m}_3 + \dot{m}_8 = +\dot{m}_4 \tag{8.37}$$

$$\text{EBE: } \dot{m}_3 h_3 + \dot{m}_8 h_8 = \dot{m}_4 h_4 + \dot{W}_t \tag{8.38}$$

$$\text{EnBE: } \dot{m}_3 s_3 + \dot{m}_8 s_8 + \dot{S}_{gen,t} = \dot{m}_4 s_4 \tag{8.39}$$

$$\text{ExBE: } \dot{m}_3 ex_3 + \dot{m}_8 ex_8 = \dot{m}_4 ex_4 + \dot{W}_t + \dot{Ex}_{dest,t} \tag{8.40}$$

For condenser:

$$\text{MBE: } \dot{m}_4 = \dot{m}_5 \tag{8.41}$$

$$\text{EBE: } \dot{m}_4 h_4 = \dot{m}_5 h_5 + \dot{Q}_{con} \tag{8.42}$$

$$\text{EnBE: } \dot{m}_4 s_4 + \dot{S}_{gen,con} = \dot{m}_5 s_5 + \frac{\dot{Q}_{con}}{T_0} \tag{8.43}$$

$$\text{ExBE: } \dot{m}_4 ex_4 = \dot{m}_7 ex_7 + \dot{Ex}_{\dot{Q}_{con}} + \dot{Ex}_{dest,con} \tag{8.44}$$

The energy and exergy efficiencies are calculated in the case of with reinjection as follows:

$$\eta_{en} = \frac{\dot{W}_t}{\dot{m}_1 h_1 - \dot{m}_{10} h_{10}} \tag{8.45}$$

and

$$\eta_{ex} = \frac{\dot{W}_t}{\dot{m}_1 ex_1 - \dot{m}_{10} ex_{10}} \tag{8.46}$$

For the case of without reinjection, they are listed as:

$$\eta_{en} = \frac{\dot{W}_t}{\dot{m}_1 h_1} \tag{8.47}$$

and

$$\eta_{ex} = \frac{\dot{W}_t}{\dot{m}_1 ex_1} \tag{8.48}$$

Illustrative Example 8.4 Consider a double-flash geothermal power plant where geothermal resources exist as saturated liquid at 230°C. The schematic view of the geothermal power plant is shown in Figure 8.15, The geothermal liquid is withdrawn from the production well at a rate of 230

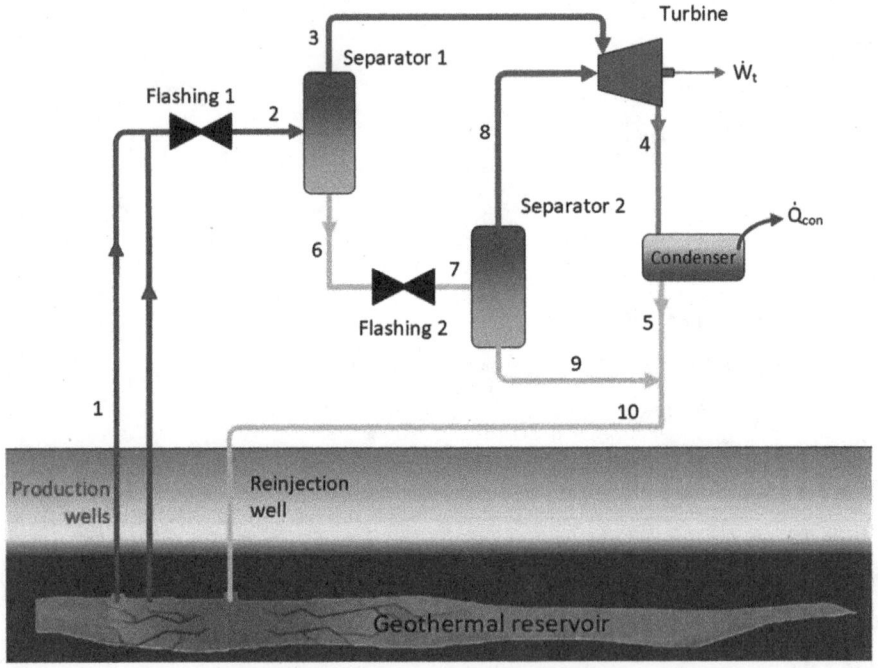

Figure 8.15 Schematic illustration of the double-flash geothermal power plant.

Table 8.1 State points for the Illustrative Example 8.3.

State	h (kJ/kg)	s (kJ/kg K)	X (vapor quality)
Ambient (0)	104.9	0.37	0
1	990.14	2.61	0
2	h_2	s_2	0.166
3	2748.1	6.82	1
4	2345	7.39	1
5	191.8	0.65	0
7	h_7	s_7	0.078
8	2693	7.223	1
9	467.1	1.43	0

kg/s and is flashed to a pressure of 500 kPa by an essentially isenthalpic flashing process where the resulting vapor is separated from the liquid in a separator. It is proposed that the liquid water coming out of the separator route through another flash chamber be maintained at 150 kPa, and the steam produced be directed to the turbine (see the figure). Both streams of steam leave the turbine at the same state (state 4). The required properties of various state points are given in Table 8.1:

a) Calculate the work generation rate in the turbine.
b) Find the missing properties in Table 1 and fill it up.
c) Calculate the exergy destruction rate in the steam turbine.
d) Determine the energy and exergy efficiencies for the power plant with reinjection.
e) Determine the energy and exergy efficiencies for the power plant without reinjection.
f) Calculate the heat recovery potential in the condenser.

Solution:

a) In order to determine the power generation rate in the turbine, we need to use the energy balance equation for the turbine. One may write it using Eq. 8.38:

$$\dot{m}_3 h_3 + \dot{m}_8 h_8 = \dot{m}_4 h_4 + \dot{W}_t$$

In order to use this equation, we need to find the mass flow rates. $\dot{m}_3$ is then found as:

$$\dot{m}_3 = \dot{m}_2 x_2 = 230(0.166) = 38.18 \frac{kg}{s}$$

Then, for $\dot{m}_6$:

$$\dot{m}_6 = \dot{m}_7 = \dot{m}_2 - \dot{m}_3 = 191.82 \frac{kg}{s}$$

$\dot{m}_9$ is found to be

$$\dot{m}_9 = \dot{m}_7 - \dot{m}_8 = 176.92 \frac{kg}{s}$$

and finally,

$$\dot{m}_4 = \dot{m}_8 + \dot{m}_3 = 53.1 \, kg/s$$

Now, we know each parameter in the energy balance equation of the turbine:

$$38.18 \times 2748.1 + 14.9 \times 2693 = 53.1 \times 2345 + \dot{W}_t$$

Thus, the work rate of the turbine is found as $\dot{W}_t = 20{,}575$ kW.

b) In order to determine the missing parameters in Table 8.1, let's start with state point 2 which is the outlet of the first flashing. As flashing process occurs isenthalpically, we write

$$h_1 = h_2 = 990.14 \text{ kJ/kg}$$

s_2 is found from Table A1 as:

$$s_2 = 2.684 \text{ kJ/kg K}$$

For state point 7, as flashing occurs isenthalpically, we obtain

$$h_6 = h_7 = 191.8 \text{ kJ/kg}$$

s_2 is found from Table A1 as:

$$s_7 = 0.7043 \text{ kJ/kg K}$$

c) In order to find the exergy destruction, we need to use the exergy balance equation as given in Eq. 8.40:

$$\dot{m}_3 ex_3 + \dot{m}_8 ex_8 = \dot{m}_4 ex_4 + \dot{W}_t + \dot{E}x_{dest,t}$$

Using this equation, we need to find the specific exergy values. The specific exergy for each stream in the system can be found as follows:

$$ex_1 = h_1 - h_0 - T_0(s_1 - s_0) = 217.7 \text{ kJ / kg}$$
$$ex_3 = h_3 - h_0 - T_0(s_3 - s_0) = 721.1 \text{ kJ / kg}$$
$$ex_4 = h_4 - h_0 - T_0(s_4 - s_0) = 148.1 \text{ kJ / kg}$$
$$ex_5 = h_5 - h_0 - T_0(s_5 - s_0) = 3.46 \text{ kJ/kg}$$
$$ex_8 = h_8 - h_0 - T_0(s_8 - s_0) = 548.9 \text{ kJ / kg}$$
$$ex_9 = h_9 - h_0 - T_0(s_9 - s_0) = 46.32 \text{ kJ / kg}$$

Thus, the exergy destruction rate of the turbine is found as:

$$\dot{E}x_{dest,t} = \dot{m}_3 ex_3 + \dot{m}_8 ex_8 - \dot{m}_4 ex_4 - \dot{W}_T$$
$$\dot{E}x_{dest} = 7271.1 \text{ kW}$$

d) The energy and exergy efficiencies for the case of with reinjection can be found to be:

$$\eta_{en,withRI} = \frac{\dot{W}_T}{\dot{m}_1(h_1) - \dot{m}_5 h_5 - \dot{m}_9 h_9} = \frac{20575}{(230)(990.14) - (53.1)(191.8) - (176.92)(467.1)} = 15.3\%$$

and

$$\eta_{ex,withRI} = \frac{\dot{W}_T}{\dot{m}_1(ex_1) - \dot{m}_5 ex_5 - \dot{m}_9 ex_9}$$
$$= \frac{20575}{(230)(217.7) - (53.1)(3.46) - (176.92)(46.3)} = 49.4\%$$

The energy and exergy efficiencies for the case of without reinjection is found to be:

$$\eta_{en,withoutRI} = \frac{\dot{W}_T}{\dot{m}_1\left(h_1 - h_0\right)} = \frac{20{,}575}{(230)(990.14 - 104.9)} = 10.1\%$$

$$\eta_{ex,withoutRI} = \frac{\dot{W}_T}{\dot{m}_1(h_1 - h_0 - T_0\left(s_1 - s_0\right))} = \frac{20{,}575}{(230)(217.7)} = 41.1\%$$

e) In order to calculate the heat recovery potential in the condenser, it is required to write the balance equation for condenser. The EBE for the condenser:

$$\dot{m}_4 h_4 = \dot{m}_5 h_5 + \dot{Q}_{con}$$

$$53.1 \times 2345 = 53.1 \times 191.8 + \dot{Q}_{con}$$

The heat to able to be recovered in the condenser, $\dot{Q}_{con}$ is found to be 114,334.9 kW, which is almost 5 times higher than the power output in the turbine.

8.5.3 Binary Cycle Power Generation

Binary cycle geothermal power plants operate like the systems that generate power from any energy sources, such as fossil fuel, nuclear, solar, etc. The energy contained by geothermal source is transferred to another working fluid for power generation. This essentially helps avoid any specific issues of direct use of geothermal water. The heat is extracted from the geothermal fluid via a heat exchanger and transferred to a working fluid. Figure 8.16 illustrates a binary cycle geothermal power plant. Geothermal energy is used for power generation via power cycle. The binary cycle geothermal power plants can be classified into two branches:

- Organic Rankine cycle (ORC)
  - Single stage organic Rankine cycle single turbine geothermal power plant
  - Single stage organic Rankine cycle with double turbine geothermal power plant
  - Double stage organic Rankine cycle geothermal power plant
- Kalina cycle geothermal power plant

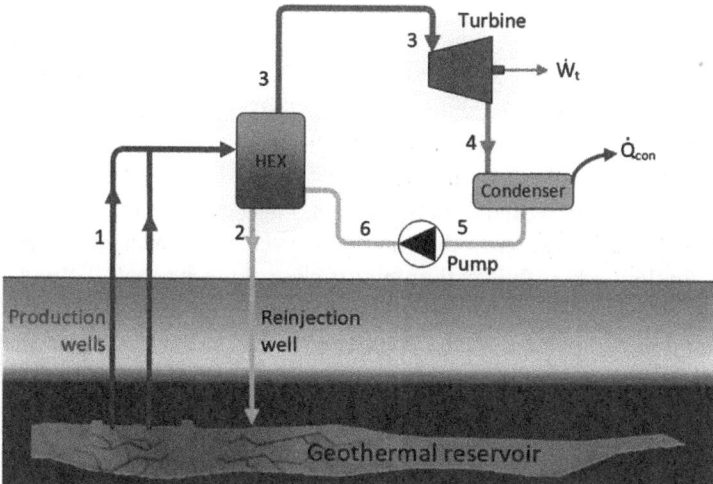

Figure 8.16 Schematic illustration of the binary cycle geothermal power plant.

The thermodynamic balance equation for the binary cycle geothermal power plant can be written as follows:

For heat exchanger:

$$\text{MBE: } \dot{m}_1 = \dot{m}_2 \text{ and } \dot{m}_6 = \dot{m}_3 \tag{8.49}$$

$$\text{EBE: } \dot{m}_1 h_1 + \dot{m}_6 h_6 = \dot{m}_2 h_2 + \dot{m}_3 h_3 \tag{8.50}$$

$$\text{EnBE: } \dot{m}_1 s_1 + \dot{m}_6 s_6 + \dot{S}_{gen,hex} = \dot{m}_2 s_2 + \dot{m}_3 s_3 \tag{8.51}$$

$$\text{ExBE: } \dot{m}_1 ex_1 + \dot{m}_6 ex_6 = \dot{m}_2 ex_2 + \dot{m}_3 ex_3 + \dot{Ex}_{dest,hex} \tag{8.52}$$

For turbine:

$$\text{MBE: } \dot{m}_3 = \dot{m}_4 \tag{8.53}$$

$$\text{EBE: } \dot{m}_3 h_3 = \dot{m}_4 h_4 + \dot{W}_t \tag{8.54}$$

$$\text{EnBE: } \dot{m}_3 s_3 + \dot{S}_{gen,t} = \dot{m}_4 s_4 \tag{8.55}$$

$$\text{ExBE: } \dot{m}_3 ex_3 = \dot{m}_4 ex_4 + \dot{W}_t + \dot{Ex}_{dest,t} \tag{8.56}$$

For condenser:

$$\text{MBE: } \dot{m}_4 = \dot{m}_5 \tag{8.57}$$

$$\text{EBE: } \dot{m}_4 h_4 = \dot{m}_5 h_5 + \dot{Q}_{con} \tag{8.58}$$

$$\text{EnBE: } \dot{m}_4 s_4 + \dot{S}_{gen,con} = \dot{m}_5 s_5 + \frac{\dot{Q}_{con}}{T_b} \tag{8.59}$$

$$\text{ExBE: } \dot{m}_4 ex_4 = \dot{m}_5 ex_5 + \dot{Ex}_{\dot{Q}_{con}} + \dot{Ex}_{dest,con} \tag{8.60}$$

For pump:

$$\text{MBE: } \dot{m}_5 = \dot{m}_6 \tag{8.61}$$

$$\text{EBE: } \dot{m}_5 h_5 + \dot{W}_p = \dot{m}_6 h_6 \tag{8.62}$$

$$\text{EnBE: } \dot{m}_5 s_5 + \dot{S}_{gen,p} = \dot{m}_6 s_6 \tag{8.63}$$

$$\text{ExBE: } \dot{m}_5 ex_5 + \dot{W}_p = \dot{m}_6 ex_6 + \dot{Ex}_{dest,p} \tag{8.64}$$

The energy and exergy efficiencies are then calculated for the binary cycle power generation as follow:

$$\eta_{en} = \frac{\dot{W}_t}{\dot{m}_1 \left(h_1 - h_2 \right)} \tag{8.65}$$

$$\eta_{ex} = \frac{\dot{W}_t}{\dot{m}_1 \left(ex_1 - ex_2 \right)} \tag{8.66}$$

Illustrative Example 8.5 In a binary geothermal power plant, the geothermal fluid is taken at the condition of $T_1 = 280°C$ and $P_1 = 2800$ kPa. It is reinjected back to the reservoir at $T_2 = 100°C$ and $P_2 = 1150$ kPa. The flow rate of the geothermal fluid is 150 kg/s. The heat of geothermal fluid is transferred to the working fluid which is water in the binary cycle via a plate heat exchanger. The inlet and outlet conditions for the working fluid in the binary cycle are known to be $T_6 = 105°C$, $P_6 = 116$ kPa, $T_3 = 250°C$, and $P_3 = 3500$ kPa. The working fluid exits from the turbine at $T_4 = 125°C$ and $P_4 = 150$ kPa. The working fluid that exits from the condenser is the saturated water at the pressure of turbine outlet.

a) Write the mass, energy, entropy and exergy balance equations for each component of the system.
b) Calculate the flow rate of water in the binary cycle.
c) Determine the turbine work rate.
d) Calculate the potential heat rate from the condenser.
e) Calculate the energy and exergy efficiencies for the system considering there is no heat recovery.
f) Calculate the energy and exergy efficiencies for the system including the heat recovered.

Solution:

a) The mass, energy, entropy and exergy balance equations for the heat exchanger are written as follows:

MBE: $\dot{m}_1 = \dot{m}_2$ and $\dot{m}_6 = \dot{m}_3$

EBE: $\dot{m}_1 h_1 + \dot{m}_6 h_6 = \dot{m}_2 h_2 + \dot{m}_3 h_3$

EnBE: $\dot{m}_1 s_1 + \dot{m}_6 s_6 + \dot{S}_{gen,hex} = \dot{m}_2 s_2 + \dot{m}_3 s_3$

ExBE: $\dot{m}_1 ex_1 + \dot{m}_6 ex_6 = \dot{m}_2 ex_2 + \dot{m}_3 ex_3 + \dot{Ex}_{dest,hex}$

The thermodynamic balance equations for the turbine are written as:

MBE: $\dot{m}_3 = \dot{m}_4$

EBE: $\dot{m}_3 h_3 = \dot{m}_4 h_4 + \dot{W}_t$ EnBE: $\dot{m}_3 s_3 + \dot{S}_{gen,t} = \dot{m}_4 s_4$

ExBE: $\dot{m}_3 ex_3 = \dot{m}_4 ex_4 + \dot{W}_t + \dot{Ex}_{dest,t}$

The thermodynamic balance equations for the condenser are written as:

MBE: $\dot{m}_4 = \dot{m}_5$

EBE: $\dot{m}_4 h_4 = \dot{m}_5 h_5 + \dot{Q}_{con}$

EnBE: $\dot{m}_4 s_4 + \dot{S}_{gen,con} = \dot{m}_5 s_5 + \dfrac{\dot{Q}_{con}}{T_b}$

ExBE: $\dot{m}_4 ex_4 = \dot{m}_5 ex_5 + \dot{Ex}_{\dot{Q}_{con}} + \dot{Ex}_{dest,con}$

The thermodynamic balance equations for the pump are written as:

MBE: $\dot{m}_5 = \dot{m}_6$

EBE: $\dot{m}_5 h_5 + \dot{W}_p = \dot{m}_6 h_6$

EnBE: $\dot{m}_5 s_5 + \dot{S}_{gen,p} = \dot{m}_6 s_6$

ExBE: $\dot{m}_5 ex_5 + \dot{W}_p = \dot{m}_6 ex_6 + \dot{Ex}_{dest,p}$

b) In order to calculate the flow rate of the water in the binary cycle, it is required to write the EBE for the heat exchanger.

$$\dot{m}_1 h_1 + \dot{m}_6 h_6 = \dot{m}_2 h_2 + \dot{m}_3 h_3$$

As given in the problem, $\dot{m}_1 = \dot{m}_2 = 150 \ kg/s$. From Table A1, the properties of the working fluids are found as follows:

$h_1 = 2949$ kJ/kg

$s_1 = 6.491$ kJ/kg K

$h_2 = 525.7$ kJ/kg

$s_2 = 1.581$ kJ/kg K

$h_3 = 2830$ kJ/kg

$s_3 = 6.176$ kJ/kg K

$h_6 = 398.1$ kJ/kg

$s_6 = 1.25$ kJ/kg K

By substituting these values into the EBE of the heat exchanger:

$150 \times 2949 + \dot{m}_6 \times 398.1 = 150 \times 525.7 + \dot{m}_3 \times 2830$

As $\dot{m}_6 = \dot{m}_3$ from the EBE, the flow rate of the water in the binary cycle is found to be $\dot{m}_6 = \dot{m}_3 = 149.5$ kg/s.

c) The turbine work are found from the EBE for the turbine.

$\dot{m}_3 h_3 = \dot{m}_4 h_4 + \dot{W}_t$

From the provided information in the problem:

$h_4 = 2722$ kJ/kg

$s_4 = 7.296$ kJ/kg K

$149.5 \times 2949 = 149.5 \times 2722 + \dot{W}_t$

The turbine work rate is found as $\dot{W}_t = 16{,}134$ kW.

d) The heat released from the condenser can be found from the EBE. As the working fluid exits from the condenser at $T_5 = 95°C$ and $P_5 = 100$ kPa. From Table A1,

$h_5 = 398.1$ kJ/kg

$s_5 = 1.25$ kJ/kg K

By substituting the parameters known into the EBE for the condenser:

$149.5 \times 2722 = 149.5 \times 398.1 + \dot{Q}_{con}$

The heat recovered from condenser is then found to be $\dot{Q}_{con} = 347{,}723$ kW.

e) The energy efficiency of the system is calculated from

$$\eta_{en} = \frac{\dot{W}_t}{\dot{m}_1 (h_1 - h_2)} = \frac{16{,}134}{150 \times (2949 - 525.7)} = \frac{16{,}134}{404{,}448} = 0.039 \ or \ 3.9\%$$

The exergy efficiency of the system is found as:

$$\eta_{ex} = \frac{\dot{W}_t}{\dot{m}_1 (ex_1 - ex_2)}$$

In order to use this equation, it is required to calculate ex_1 and ex_2 values. They are obtained as follows:

$ex_1 = (h_1 - h_0) - T_0(s_1 - s_0) = (2949 - 104.9) - 298 \times (6.491 - 0.3672) = 1020 \ kJ/kg$

$ex_2 = (h_2 - h_0) - T_0(s_2 - s_0) = (525.7 - 104.9) - 298 \times (1.581 - 0.3672) = 59.17 \ kJ/kg$

By substituting these values to the exergy efficiency equation, we obtain

$$\eta_{ex} = \frac{16{,}134}{150 \times (1020 - 59.17)} = \frac{16{,}134}{144{,}124.5} = 0.1119 \ or \ 11.19\%$$

f) In order to include the potential heat recovery, it is required to include the heat output in the efficiency equation as an additional useful output:

$$\eta_{en} = \frac{\dot{W}_t + \dot{Q}_{con}}{\dot{m}_1(h_1 - h_2)} = \frac{16,134 + 347,723}{404,448} = 0.8996 \, or \, 89.96\%$$

The exergy efficiency in the case of heat recovered can be calculated as follows:

$$\eta_{ex} = \frac{\dot{W}_t + \dot{E}x_{\dot{Q}_{con}}}{\dot{m}_1(ex_1 - ex_2)}$$

Here, it is required to calculate $\dot{E}x_{\dot{Q}_{con}}$.

$$\dot{E}x_{\dot{Q}_{con}} = \dot{Q}_{con}\left(1 - \frac{T_0}{T_b}\right)$$

In this case, the boundary or surface temperature is not given. So, we need to use the average temperature instead of it as follows:

$$T_{ave} = \frac{T_4 + T_5}{2} = \frac{125 + 95}{2} = 110°C \, or \, 383K$$

Thus,

$$\dot{E}x_{\dot{Q}_{con}} = 347,723 \times \left(1 - \frac{298}{383}\right) = 77,194.5 \, kW$$

Finally, the exergy efficiency is found as

$$\eta_{ex} = \frac{16,134 + 77,194.5}{150 \times (1020 - 59.17)} = \frac{93,328.5}{144,124.5} = 0.6475 \, or \, 64.75\%$$

8.5.4 Combined Flash-Binary Geothermal Power Generation

Combined flash-binary geothermal power plants are the power plants where flash and binary cycle power plants are combined for more power generation and hence better performance. Figure 8.17 shows the system layout of a flash-binary geothermal power plant. Like in the double-flash geothermal power generation system, there are two turbines. Here, the main difference is that there is another power cycle for the second turbine. In this kind of geothermal power plant, the geothermal fluid is extracted from the production well and goes to the flashing chamber. Then it goes to the separator. After the separator, while the steam goes to the steam turbine, the liquid goes to the heat exchanger working as an evaporator for the secondary power cycle.

Illustrative Example 8.6   A flash-binary geothermal power plant schematic is provided in Figure 8.18. The temperature and pressure of the geothermal water as it enters the system are 230°C and 2700 kPa. The associated mass flow rate is determined to be 100 kg/s at point 1 and 5 kg/s at point 8. The fluid (water) in point 8 is saturated vapor. The ambient temperature and pressure are at 25°C and 1 atm. The isentropic efficiencies of the turbines are 85%. The vapor quality of steam at 2 is 0.2. The specific enthalpy and entropy values at state 12 are found to be 500 kJ/kg and 0.4 kJ/kg K, respectively.

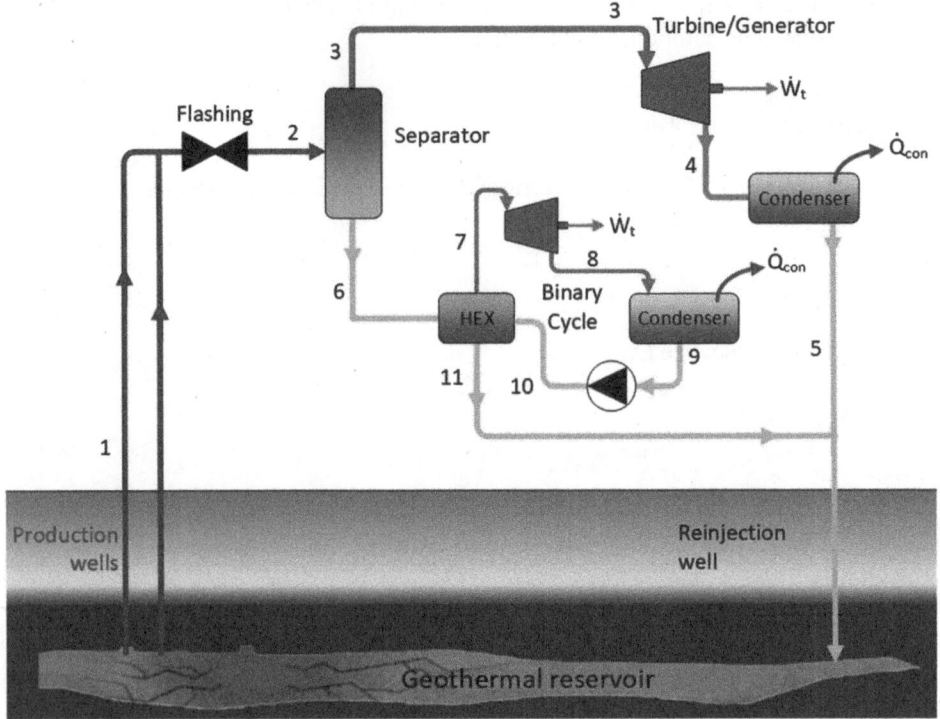

Figure 8.17 Schematic illustration of the flash-binary power plant.

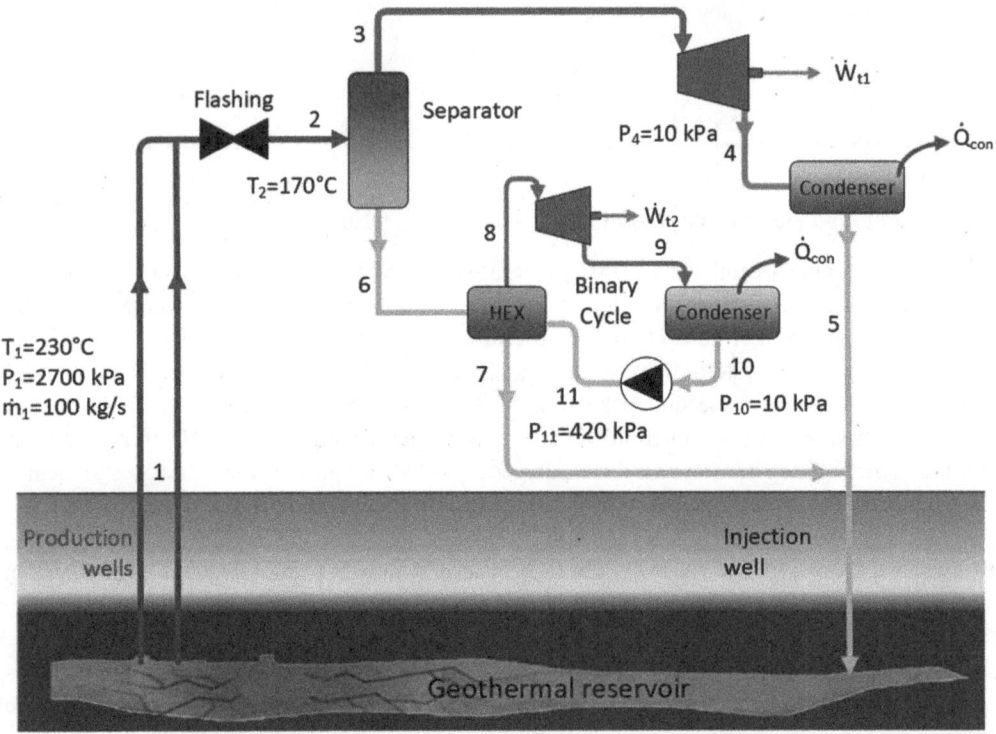

Figure 8.18 A schematic view of the flash-binary geothermal power plant in the Illustrative Example 8.4.

a) Write the thermodynamic balance equations associated with each component.
b) Determine the energy and exergy efficiencies of the system with and without reinjection.

Solution:

a) The thermodynamic balance equations associated with each component are written below accordingly:

Throttle valve:

MBE: $\dot{m}_1 = \dot{m}_2$

EBE: $\dot{m}_1 h_1 = \dot{m}_2 h_2$

EnBE: $\dot{m}_1 s_1 + \dot{S}_{gen,TV} = \dot{m}_2 s_2$

ExBE: $\dot{m}_1 ex_1 = \dot{m}_2 ex_2 + \dot{Ex}_{dest,TV}$

Flash chamber:

MBE: $\dot{m}_2 = \dot{m}_3 + \dot{m}_6$

EBE: $\dot{m}_2 h_2 = \dot{m}_3 h_3 + \dot{m}_6 h_6$

EnBE: $\dot{m}_2 s_2 + \dot{S}_{gen,FC} = \dot{m}_3 s_3 + \dot{m}_6 s_6$

ExBE: $\dot{m}_2 ex_2 = \dot{m}_3 ex_3 + \dot{m}_6 ex_6 + \dot{Ex}_{dest,FC}$

Turbine 1:

MBE: $\dot{m}_3 = \dot{m}_4$

EBE: $\dot{m}_3 h_3 = \dot{m}_4 h_4 + \dot{W}_{T1}$

EnBE: $\dot{m}_3 s_3 + \dot{S}_{gen,T1} = \dot{m}_4 s_4$

ExBE: $\dot{m}_3 ex_3 = \dot{m}_4 ex_4 + \dot{W}_{T1} + \dot{Ex}_{dest,T1}$

Condenser 1:

MBE: $\dot{m}_4 = \dot{m}_5$

EBE: $\dot{m}_4 h_4 = \dot{m}_5 h_5 + \dot{Q}_{con}$

EnBE: $\dot{m}_4 s_4 + \dot{S}_{gen,con} = \dot{m}_5 s_5 + \dfrac{\dot{Q}_{con}}{T}$

ExBE: $\dot{m}_4 ex_4 = \dot{m}_5 ex_5 + \dot{Q}_{con}\left(1 - \dfrac{T_0}{T}\right) + \dot{Ex}_{dest,con}$

Evaporator:

MBE: $\dot{m}_6 = \dot{m}_7$ and $\dot{m}_{11} = \dot{m}_8$

EBE: $\dot{m}_6 h_6 + \dot{m}_{11} h_{11} = \dot{m}_7 h_7 + \dot{m}_8 h_8$

EnBE: $\dot{m}_6 s_6 + \dot{m}_{11} s_{11} + \dot{S}_{gen,eva} = \dot{m}_7 s_7 + \dot{m}_8 s_8$

ExBE: $\dot{m}_6 ex_6 + \dot{m}_{11} ex_{11} = \dot{m}_7 ex_7 + \dot{m}_8 ex_8$
$$+ \dot{Ex}_{dest,eva}$$

Turbine 2:

MBE: $\dot{m}_8 = \dot{m}_9$

EBE: $\dot{m}_8 h_8 = \dot{m}_9 h_9 + \dot{W}_{T2}$

EnBE: $\dot{m}_8 s_8 + \dot{S}_{gen,T2} = \dot{m}_9 s_9$

ExBE: $\dot{m}_8 ex_8 = \dot{m}_9 ex_9 + \dot{W}_{T2} + \dot{Ex}_{dest,T2}$

Condenser 2:

MBE: $\dot{m}_9 = \dot{m}_{10}$

EBE: $\dot{m}_9 h_9 = \dot{m}_{10} h_{10} + \dot{Q}_{con2}$

EnBE: $\dot{m}_9 s_9 + \dot{S}_{gen,con2} = \dot{m}_{10} s_{10} + \dfrac{\dot{Q}_{con2}}{T}$

ExBE: $\dot{m}_9 ex_9 = \dot{m}_{10} ex_{10} + \dot{Q}_{con}\left(1 - \dfrac{T_0}{T}\right)$
$$+ \dot{Ex}_{dest,con2}$$

Pump:

MBE: $\dot{m}_{10} = \dot{m}_{11}$

EBE: $\dot{m}_{10} h_{10} + \dot{W}_P = \dot{m}_{11} h_{11}$

EnBE: $\dot{m}_{10} s_{10} + \dot{S}_{gen,P} = \dot{m}_{11} s_{11}$

ExBE: $\dot{m}_{10} ex_{10} + \dot{W}_P = \dot{m}_{11} ex_{11} + \dot{Ex}_{dest,T1}$

b) The energy and exergy efficiencies without reinjection are expressed as:

$$\eta_{en} = \frac{\dot{W}_{T1} + \dot{W}_{T2}}{\dot{m}_1 \left(h_1 - h_0\right)}$$

and

$$\eta_{ex} = \frac{\dot{W}_{T1} + \dot{W}_{T2}}{\dot{m}_1 \left[h_1 - h_0 - T_0 \left(s_1 - s_0\right)\right]}$$

The energy and exergy efficiencies with reinjection are then expressed as:

$$\eta_{en} = \frac{\dot{W}_{T1} + \dot{W}_{T2}}{\dot{m}_1 \left(h_1 - h_{12}\right)}$$

and

$$\eta_{ex} = \frac{\dot{W}_{T1} + \dot{W}_{T2}}{\dot{m}_1 \left[h_1 - h_{12} - T_0 \left(s_1 - s_{12} \right) \right]}$$

As provided in the problem, $T_1 = 230°C$ and $P_1 = 2700$ kPa. So, the enthalpy and entropy values can be found from Table A1. $h_1 = 2809$ kJ/kg and $s_1 = 6.24$ kJ/kg K.

The flow rate can be determined according to the quality of the mixture in the separator.

$$\dot{m}_3 = x_2 \dot{m}_2 = 0.2 \left(100 \frac{kg}{s} \right) = 20 \ kg/s$$

$$h_3 = h_{g@T=170°C} = 2768 \ kJ/kg$$

$$s_3 = s_{g@T=170°C} = 6.67 \ kJ/kg \ K$$

The turbine output is found as follows:

$$\eta_{is,T1} = \frac{h_3 - h_4}{h_3 - h_{4,is}} = 0.85$$

$s_{4is} = s_3 = 6.67 \ kJ/kgK$ and $P_4 = 10 \ kPa \ h_{4,is} = 2111 \ kJ/kg$

Thus, rearrange the equation $\eta_{is,T1} = \dfrac{h_3 - h_4}{h_3 - h_{4,is}}$ and solve for h_4 to find $h_4 = 2209$ kJ/kg.

Hence, $\dot{W}_{T1} = \dot{m}_3 (h_3 - h_4) = 20 \dfrac{kg}{s} \left(2768 \dfrac{kJ}{kg} - 2209 \dfrac{kJ}{kg} \right) = 11,180 \ kW$

$P_8 = 420 \ kPa$ and $X_8 = 1 \ h_8 = 2740 \ kJ/kg$ and $s_8 = 6.88 \ kJ/kgK$

$$\eta_{is,T2} = \frac{h_8 - h_9}{h_8 - h_{9,is}} = 0.85$$

$s_{9,is} = s_8 = 6.88 \ kJ/kgK$ and $P_9 = 10 \ kPa \ h_{9,is} = 2179 \ kJ/kg$

Thus, rearrange the equation $\eta_{is,T2} = \dfrac{h_8 - h_9}{h_8 - h_{9,is}}$ and solve for h_9 to find $h_9 = 2263$ kJ/kg.

Hence, $\dot{W}_{T2} = \dot{m}_8 (h_8 - h_9) = 5 \dfrac{kg}{s} \left(2740 \dfrac{kJ}{kg} - 2263 \dfrac{kJ}{kg} \right) = 2385 \ kW$

$T_0 = 25°C$ and $P_0 = 101 \ kPa \ h_0 = 104.9$ kJ/kg and $s_0 = 0.367 \dfrac{kJ}{kg \ K}$

Hence, the energy and exergy efficiencies without reinjection are:

$$\eta_{en} = \frac{\dot{W}_{T1} + \dot{W}_{T2}}{\dot{m}_1 (h_1 - h_0)} = \frac{11,180 \ kW + 2385 \ kW}{100 \dfrac{kg}{s} \left(2809 \dfrac{kJ}{kg} - 104.9 \dfrac{kJ}{kg} \right)} = 5.0\%$$

and

$$\eta_{ex} = \frac{\dot{W}_{T1} + \dot{W}_{T2}}{\dot{m}_1 \left[h_1 - h_0 - T_0 \left(s_1 - s_0 \right) \right]} = \frac{11,180 + 2385}{100 \times (2809 - 104.9 - 298 (6.24 - 0.367))} = 0.129 \ or \ 12.9\%$$

In addition, the respective energy and exergy efficiencies with reinjection are found to be:

$$\eta_{en} = \frac{\dot{W}_{T1} + \dot{W}_{T2}}{\dot{m}_1\left(h_1 - h_{12}\right)} = \frac{11{,}180 \, kW + 2385 \, kW}{100 \, \frac{kg}{s}\left(2809 \, \frac{kJ}{kg} - 500 \, \frac{kJ}{kg}\right)} = 0.059 \; or \; 5.9\%$$

and

$$\eta_{ex} = \frac{\dot{W}_{T1} + \dot{W}_{T2}}{\dot{m}_1\left[h_1 - h_{12} - T_0\left(s_1 - s_{12}\right)\right]} = \frac{11{,}180 + 2385}{100 \, \frac{kg}{s}\left[2809 - 500 - 298(6.24 - 0.4)\right]} = 0.239 \; or \; 23.9\%$$

8.6 Geothermal Heat Pumps

A geothermal or ground source heat pump system is operated at the constant temperature under the ground to heat or cool a building or drying crops. Their utilization helps reduce carbon-based source consumption and contribute to electrification by using geothermal energy for heating applications. Space heating with geothermal energy requires certain considerations and studies. Note that numerous climate, population, building type, technological aspect, and economic aspect can all be considered. Instead of using reference air, which is unstable, ground source heat pumps use the stable temperature level of the planet to exchange heat. Based on these phenomena, a geothermal-sourced heat pump leads to a higher temperature variation between higher and lower working point temperatures, bringing a higher Carnot effectiveness. Geothermal-sourced heat pumps are also examined in this subsection in terms of indicators affecting their performance and running. There are three main components that make up a geothermal heating and cooling system: a heat pump, an underground heat exchanger, and a distribution system like air ducts.

Figure 8.19 illustrates ground source heat pump processes with horizontal circulation loops and vertical circulation loops for heating and cooling applications. The main difference between vertical and horizontal loops is the effective area of the loops. Horizontal loops require a larger area in the building than vertical one for the same capacity. If there is an available area, horizontal loop will be the cost-effective solution. Because vertical loop requires drilling which makes it costly. Also, vertical loops are lying down the deeper levels than the horizontal loops, it will be more thermally effective.

In general, heat pumps operate between a high-temperature point (heat sink at T_H) and a low-temperature point (heat source at T_L). Heat pumps that use the ground as a heat sink are called geothermal source heat pumps or ground heat pumps. Using ground heat exchangers, heat energy is exchanged between the ground and the heat pump system. By using heating coils, heat is exchanged between the heat pump cycle and space for long periods of time.

As mentioned in Chapter 3, the performance of the heat pumps is assessed over the COP. In order to analyze the heat pump systems, the thermodynamic balance equations should be written accordingly. As a reminder for the COP values, the theoretical limit of a heating and cooling purposes can be defined as follows, respectively:

$$COP_{Carnot,heating} = \frac{Q_H}{Q_H - Q_L} \tag{8.67}$$

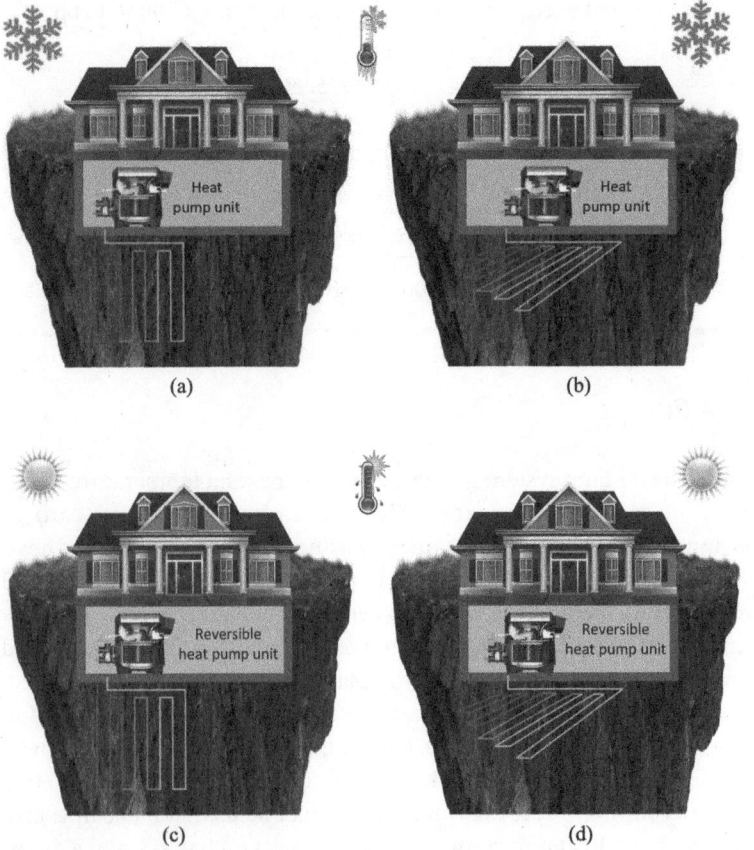

Figure 8.19 Ground source heat pumps (a) heating application with vertical circulation loops, (b) heating applications with vertical circulation loops, (c) cooling application with vertical loops, and (d) cooling application with horizontal loop.

and

$$COP_{Carnot,cooling} = \frac{Q_L}{Q_H - Q_L} \tag{8.68}$$

From the temperature equivalence, the Carnot COPs for heating and cooling can be written as follows:

$$COP_{Carnot,heating} = \frac{T_H}{T_H - T_L} \tag{8.69}$$

and

$$COP_{Carnot,cooling} = \frac{T_L}{T_H - T_L} \tag{8.70}$$

The actual COPs are calculated as follows for both heating and cooling applications, respectively.

$$COP_{heating} = \frac{\dot{Q}_{heating}}{\dot{W}_{comp}} \tag{8.71}$$

and

$$COP_{cooling} = \frac{\dot{Q}_{cooling}}{\dot{W}_{comp}} \tag{8.72}$$

The exergetic COP values are then written for heating and cooling as follows:

$$COP_{heating} = \frac{\dot{Ex}_{\dot{Q}_{heating}}}{\dot{W}_{comp}} \tag{8.73}$$

and

$$COP_{cooling} = \frac{\dot{Ex}_{\dot{Q}_{cooling}}}{\dot{W}_{comp}} \tag{8.74}$$

Illustrative Example 8.7   A ground source heat pump system, consisting of a compressor, a condenser, an expansion valve and an evaporator, works with R-134a working fluid in two configurations for winter and summer conditions. So, it is then employed for heating and cooling applications. The temperature of the soil stays constant at 13.5°C throughout the year. The heat pump system provides 50 kW heat to the residential area during winter and collects 50 kW heat from the residential area in order to provide cooling effect during summer.

During winter, the outside temperature is considered as −10°C. The compressor inlet and outlet conditions are given as $T_{comp,in} = 13°C$, $P_{comp,in} = 458.1\,kPa$, $h_{comp,in} = 257.8\,kJ/kg$, $T_{comp,out} = 64.6°C$, $P_{comp,out} = 1683$ kPa, $h_{comp,out} = 287.6$ kJ/kg. The working fluid condensed and cooled down to 60°C at the condenser outlet with $h_{cond,out} = 139.4$ kJ/kg enthalpy value.

During summer, the outside temperature is considered as 35°C. The compressor inlet and outlet conditions are given as $T_{comp,in} = 2°C$, $h_{comp,in} = 251.6$ kJ/kg, $T_{comp,out} = 14.55°C$, $P_{comp,out} = 1683$ kPa, $h_{comp,out} = 260.2$ kJ/kg. The working fluid enters the evaporator as saturated liquid at 2°C with $h_{evap,in} = 139.4$ kJ/kg and exits as saturated vapor at 2°C.

a) Write mass, energy, entropy and exergy balance equations for the system and its components.
b) Calculate both COP_{en} and COP_{ex} values.

Solution:
a) The mass, energy, entropy and exergy balance equations for the system components in the heat pump cycle can be written as follows:

For expansion valve:
MBE: $\dot{m}_i = \dot{m}_e$
EBE: $\dot{m}_i h_i = \dot{m}_e h_e$
EnBE: $\dot{m}_i s_i + \dot{S}_{gen,XV} = \dot{m}_e s_e$
ExBE: $\dot{m}_i ex_i = \dot{m}_e ex_e + \dot{Ex}_{dest,XV}$

For evaporator:
MBE: $\dot{m}_i = \dot{m}_e$ and $\dot{m}_{gf,i} = \dot{m}_{gf,e}$
EBE: $\dot{m}_i h_i + \dot{m}_{gf,i} h_{gf,i} = \dot{m}_e h_e + \dot{m}_{gf,e} h_{gf,e}$
EnBE: $\dot{m}_i s_i + \dot{m}_{gf,i} s_{gf,i} + \dot{S}_{gen,eva} = \dot{m}_7 s_7 + \dot{m}_{gf,e} s_{gf,e}$
ExBE: $\dot{m}_i ex_i + \dot{m}_{gf,i} ex_{gf,i} = \dot{m}_e ex_e + \dot{m}_{gf,e} ex_{gf,e} + \dot{Ex}_{dest,eva}$

For condenser:
MBE: $\dot{m}_i = \dot{m}_e$

EBE: $\dot{m}_i h_i = \dot{m}_e h_e + \dot{Q}_{con}$

EnBE: $\dot{m}_i s_i + \dot{S}_{gen,con} = \dot{m}_e s_e + \dfrac{\dot{Q}_{con}}{T_0}$

ExBE: $\dot{m}_i ex_i = \dot{m}_e ex_e + \dot{Q}_{con}\left(1 - \dfrac{T_0}{T_c}\right) + \dot{Ex}_{dest,con}$

For compressor:

MBE: $\dot{m}_i = \dot{m}_e$

EBE: $\dot{m}_i h_i + \dot{W}_{Comp} = \dot{m}_e h_e$

EnBE: $\dot{m}_i s_i + \dot{S}_{gen,Comp} = \dot{m}_e s_e$

ExBE: $\dot{m}_i ex_i + \dot{W}_{Comp} = \dot{m}_e ex_e + \dot{Ex}_{dest,Comp}$

b) In order to determine the COP values, we first need to calculate the compressor work rate for both winter and summer operations. To find the compressor work, it is required to find the mass flow rate of R134a. From the EBE for condenser:

$$\dot{m}_{con,in} h_{con,in} = \dot{m}_{con,out} h_{con,out} + \dot{Q}_{con,out}$$

$\dot{Q}_{con,out}$ is given to be 50 kW. Substituting this into the above given equation, one obtains

$$\dot{Q}_{con,out} = \dot{m}_{con,in}\left(h_{con,in} - h_{con,out}\right)$$

$$50 = \dot{m}_{con,in}\left(287.6 - 139.4\right)$$

Thus, the mass flow rate of R134a is found to be $\dot{m}_{con,in} = \dot{m}_{R134a} = 0.337\ \text{kg/s}$.
Finally, the compressor work rate is found by using the EBE:

$$\dot{m}_{R134a} h_{comp,in} + \dot{W}_{comp} = \dot{m}_{R134a} h_{comp,out}$$

Here, $h_{comp,in}$ and $h_{comp,out}$ are given as 257.8 and 287.6 kJ/kg. By writing the enthalpy values and flow rate into the equation above:

$$\dot{W}_{comp} = \dot{m}_{R134a}\left(h_{comp,out} - h_{comp,in}\right)$$

$$\dot{W}_{comp} = 0.337 \times \left(287.6 - 257.8\right)$$

Thus, the compressor work rate is found as $\dot{W}_{comp} = 10.044\ \text{kW}$.
Therefore, COP_{en} of the geothermal-sourced heat pump is calculated via Eq. 8.51.

$$COP_{heating,en} = \dfrac{\dot{Q}_{heating}}{\dot{W}_{comp}} = \dfrac{50}{4.977} = 4.977$$

In addition, COP_{ex} can be calculated by using the following equation:

$$COP_{heating,ex} = \dfrac{\dot{Ex}_{\dot{Q}_{heating}}}{\dot{W}_{comp}}$$

where

$$\dot{Ex}_{\dot{Q}_{heating}} = \dot{Ex}_{\dot{Q}_{con}} = \dot{Q}_{con}\left(1 - \frac{T_0}{T_{gt}}\right) = 50 \times \left(1 - \frac{263}{335}\right) = 11 \, kW$$

Thus, $COP_{heating,ex}$ can be found as

$$COP_{heating,ex} = \frac{11}{10.044} = 1.095$$

For summer, it is required to recalculate the mass flow rates and COP values. The flow rate of the R134a in the summer:

$$\dot{m}_{con,in} h_{con,eva} = \dot{m}_{con,out} h_{con,eva} + \dot{Q}_{eva,out}$$

$$50 = \dot{m}_{con,in}\left(251.6 - 139.4\right)$$

Thus, the mass flow rate of R134a in the summer is found to be $\dot{m}_{con,in} = \dot{m}_{R134a} = 0.275$ kg/s. The compressor work rate can be determined from the EBE for the compressor:

$$\dot{W}_{comp} = \dot{m}_{R134a}\left(h_{comp,out} - h_{comp,in}\right)$$

$$\dot{W}_{comp} = 0.275 \times \left(286.4 - 251.4\right)$$

Thus, $\dot{W}_{comp}$ is found to be $\dot{W}_{comp} = 2.35$ kW.

The energetic COP for the system in the cooling mode is found as:

$$COP_{cooling,en} = \frac{\dot{Q}_{cooling}}{\dot{W}_{comp}} = \frac{50}{2.35} = 21.27$$

The exergetic COP for the system can be determined as:

$$COP_{cooling,ex} = \frac{\dot{Ex}_{\dot{Q}_{cooling}}}{\dot{W}_{comp}}$$

$$COP_{cooling,ex} = \frac{\dot{Ex}_{\dot{Q}_{cooling}}}{\dot{W}_{comp}} = \frac{\dot{Q}_{cooling}\left(1 - \frac{T_0}{T_s}\right)}{\dot{W}_{comp}}$$

where $T_0 = 35°C$ and $T_s = 2°C$. Thus,

$$COP_{cooling,ex} = 2.16$$

8.7 Geothermal District Heating

In addition to power generation, another useful purpose of the geothermal sources is to manage district heating. It is a really significant option for meeting the heating demand of the buildings. Figure 8.20 shows the schematic illustration of the geothermal source district heating system. The

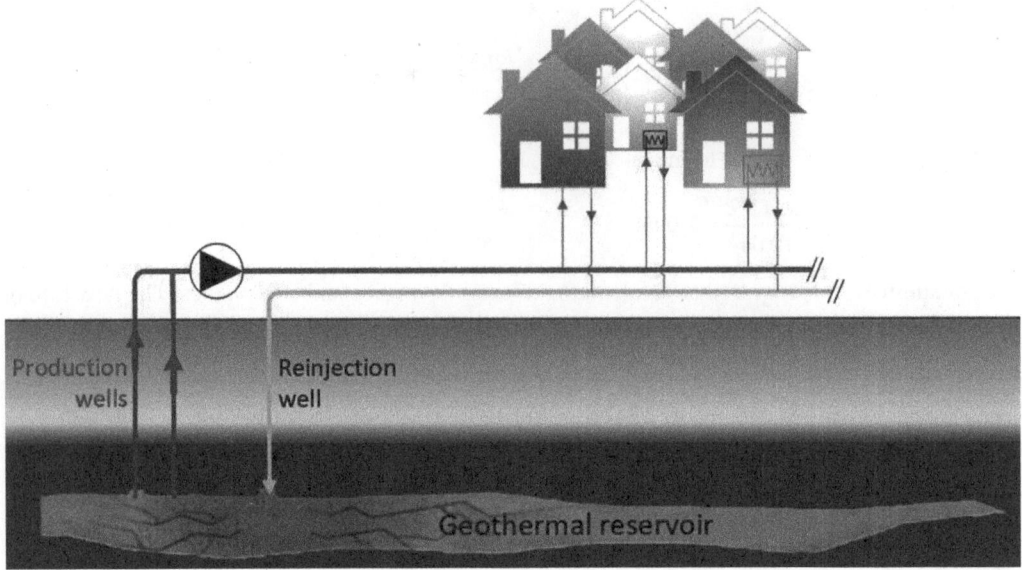

Figure 8.20 Geothermal-based district heating system.

use of geothermal source for direct heating purposes is quite useful when the temperature of the geothermal fluid is around 80°C to 120°C which are it is really hard to produce steam for power generation. The geothermal fluid obtained from the production well is directly pumped to the buildings. Every building has a heat exchanger, generally a plate heat exchanger. The geothermal heat is transferred to the building cycle via this heat exchanger. Namely, in geothermal district heating system, heat exchangers take place the furnaces which are used for meeting heating demand.

Thanks to the geothermal district heating systems, since no fossil fuel or electricity will be consumed for heating in buildings, it will not cause any environmental issues. Also, it is not complicated to apply. The only need is the piping infrastructure. Also, geothermal district heating can be part of any geothermal power plant. A part of the heat thrown out in the condenser in geothermal power plant can be recovered for district heating. Also, in the flashing geothermal power plants, the exiting geothermal fluid in the separator again can be used for district heating, especially when its temperature is not enough to run a binary cycle or additional steam cycle.

Illustrative Example 8.8   In a region where there are 1000 residential and 250 commercial units, a district heating energy system with geothermal source is considered. The information shows that there is geothermal water at 90°C in the region. The average heating demand is 25 kW for the residential units and 40 kW for the commercial units. In order to meet the heating demand with geothermal source, calculate the flow rate of the geothermal fluid. Assume that heat losses in the piping line are neglected. Also, due to the regional regulation, the temperature of the geothermal fluid to be injected is expected to be a minimum of 20°C lower than one obtained from production well.

Solution:
First, it is required to determine the total heating demand. There are 1000 residential and 250 commercial units in the region. So, the total heating load is calculated as:

$$\dot{Q}_{totalheating} = 1000 \times 25 + 250 \times 40$$

Thus, the total heating demand is found to be $\dot{Q}_{totalheating} = 35{,}000 \text{ kW}$.

This heating demand is met by geothermal source. It is given in the problem that the temperature difference should be 20°C. Basically,

$$\dot{Q}_{totalheating} = \dot{m}_{gt} c_p \Delta T$$

Substituting the given parameters and assuming c_p of the geothermal water is 4.18 kJ/kg K,

$$35{,}000 = \dot{m}_{gt} \times 4.18 \times (90 - 70)$$

Finally, the flow rate of the geothermal fluid is found to be $\dot{m}_{gt} = 418.66 \text{ kg/s}$.

8.8 Other Applications of Geothermal Energy

Geothermal energy is potentially used as an energy source in any process or application requiring heat. Therefore, in addition to the applications of power generation, heat pumps, and direct heating, it can also be used in absorption cooling, hydrogen production, industrial chemical and synthetic fuel production, and drying processes. It can be combined with other renewable energy sources and used in an integrated manner with a multigenerational manner to meet multiple demand items.

8.8.1 Absorption Cooling Applications

Cooling is one of the useful outputs of geothermal energy where a geothermally driven absorption cooling system is employed. Geothermal energy generation systems generally use two fluids to produce cooling output. While refrigeration cycles using a single fluid are known as vapor compression cycles. Absorption refrigeration cycles involve the use of two fluids, such as refrigerant and absorbent for cooling. Thus, a geothermal energy source with two fluids and a system input can be used to produce cooling. There is a difference in the pressure parameter between these two cooling cycles. There is also a difference between these two cycles regarding the fluid used to perform the cooling operation. In fact, these two differences occur interconnectedly in cycles.

Figure 8.21 shows the geothermal-sourced single-effect absorption cooling plant. Ammonia and water are used as working fluids in the cycle where ammonia is refrigerant and water is used as absorbent. Three basic principles govern this absorption cooling cycle:

- A liquid boil (vaporizes) when heated, and a gas condenses when cooled,
- Lowering the pressure above a liquid lowers its boiling point, and
- Heat flows from warmer to cooler surfaces.

The process of absorption cooling is dependent on a thermal process using heat used instead of a compressor in the vapor compression refrigeration cycle using work input (i.e., electricity). This process uses two different fluids: a refrigerant and an absorbent. Geothermal fluid is used as the heat source in the generator of the absorption cooling cycle.

8.8.2 Hydrogen Production

Using geothermal energy for hydrogen production has great potential for increasing performance and producing a new energy source. The production of hydrogen with geothermal energy is

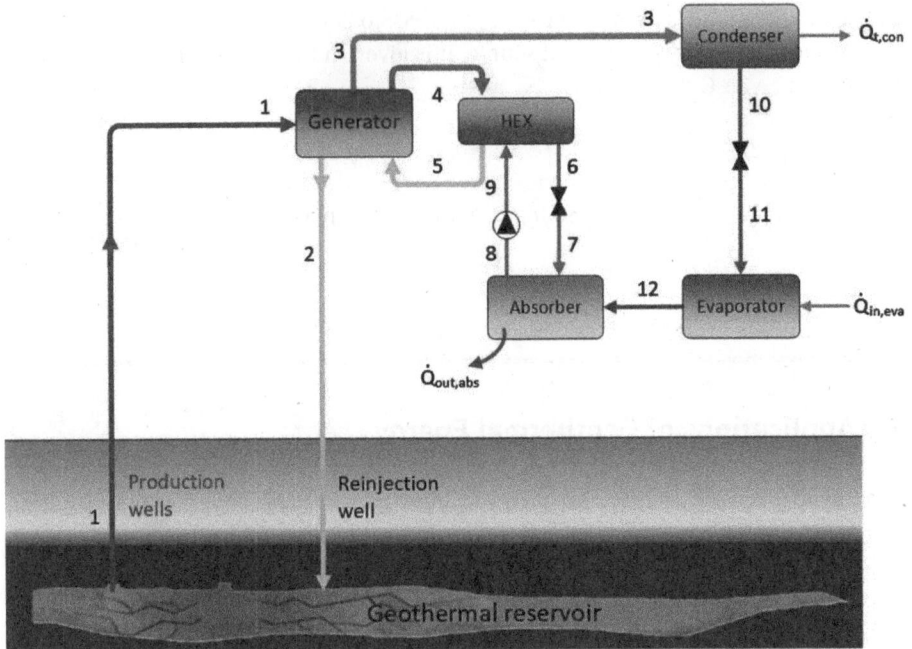

Figure 8.21 Geothermal-based absorption cooling system.

therefore of great interest for researchers and technology developers. By involving the hydrogen economy all around the world, the use of geothermal energy for hydrogen generation is expected to significantly increase for regions with abundant geothermal reservoirs. Geothermal energy may used for hydrogen production in mainly four ways:

- direct generation,
- water electrolysis-based generation,
- thermal energy-based thermochemical generation, and
- hybrid thermochemical generation.

First, geothermal steam can be used to produce hydrogen. Some regions of the world release geothermal steam with hydrogen; that is, along with geothermal gases, hydrogen gas is also released into the atmosphere. The use of hydrogen requires the cleaning of other gases. Thus, hydrogen continues to benefit a wide variety of technologies. Hydrogen can also be produced using thermal energy, which is divided into two groups. Heat is used as thermal energy in the first group of systems, and electricity is used as thermal energy in the second group. Electrolysis is a third method of generating hydrogen. Different technologies can be used to electrolyze water to produce hydrogen with geothermal electricity. The disadvantage of water electrolysis is that electricity is considered an expensive fuel. However, it is great potential to store off-peak electricity generation. The fourth technique is a hybrid thermochemical generation process that combines geothermal power with hydrogen generation. According to thermodynamic analysis of thermochemical hydrogen generation reactions, some steps may require both electric and thermal energy to provide the required power. Also, for high-temperature geothermal sources, hydrogen can be produced by thermochemical cycles such as Cu-Cl, Mg-Cl, etc. For these cycles, the temperature should be higher than 500°C. Further information on hydrogen production by geothermal energy systems are available elsewhere [4, 5].

8.8.3 Industrial Chemical and Synthetic Gas Production

As mentioned earlier, geothermal sources take place in almost all processes and systems requiring heat. First, the first chemical, which is geothermal may be used in the production procedure as the heat source, is ammonia. Ammonia is one of the mostly produced chemical in the world, which is mainly used in agricultural sector as a fertilizer. It has a great potential to use as hydrogen carrier and carbon-free fuel. Ammonia is generally produced with the so-called Haber-Bosch process. For the Haber-Bosch, high pressure and high temperature are required. This high-temperature and high-pressure demands can be met by geothermal sources. On the other hand, ammonia production requires a high amount of hydrogen, too. So, combining the ammonia production with green hydrogen that comes from geothermal sources can be critical for sustainability of the ammonia production.

Synthetic fuel can be produced with geothermal energy sources. Synthetic fuels could replace conventional fuels if they can be produced more economically. In power generation systems, synthetic fuels are produced using a variety of energy sources. Geothermal energy sources could be one of them. It is necessary to establish energy generation systems in regions where geothermal energy resources are available in order to produce synthetic fuel efficiently. Various systems for synthesizing fuel using geothermal energy have been proposed in the literature and are still being proposed. There are several technologies that can use geothermal energy. In the appropriate thermochemical cycle, thermal energy is used to generate electricity and, later, synthetic fuels such as ethanol, methanol, butanol, propane, ammonia, and nonfossil methane are produced. It is possible to store these synthetic fuels in liquid form, and then transport them to their final destinations. Furthermore, these fuels can be used in fuel cells and as chemicals for transportation, space heating, and power generation.

It is possible for energy production systems to release some carbon-containing tail gases and pollutants. One example of such an energy production system is the production of liquid fuel. These systems can produce tail gases, pollutants, and additional synthesis gases. Along with these synthesis gases, liquid fuel and hydrocarbon can be produced. The liquid fuel production cycle is another way to utilize these tail gases and pollutants. It is possible to improve the conversion of carbon to liquid fuel or hydrocarbons by using these cycles.

8.8.4 Other Industrial and Agricultural Use of Geothermal Energy

Geothermal energy sources can readily be available in areas to be used for many purposes. Geothermal fluid responds to different needs around the world because of its temperature differences. Due to the fact that the supply of fluid is costless, industrial systems have great potential. Agriculture is another sector that uses geothermal fluid. The temperature of geothermal fluid is preferred to be high in industrial applications, but low-temperature geothermal fluid can be used in agricultural activities. This sector is attracted to geothermal fluid due to some of its features.

First, it is a cost-effective energy source. Due to the fact that the energy source itself comes from nature, this source is cost-effective for these two sectors despite the initial investment. Another feature is the quality of the energy. The energy quality varies from region to region, but it is generally accepted as suitable for these two industries. Reliability is the third feature. After a system has been operational for a long time, reliability becomes apparent. Different sectors use geothermal sources at different temperatures, proving the reliability of geothermal power generation.

Furthermore, the methods of pre-drying and postproduction drying differ based on the type of dried product, such as grains, vegetables, and fruits, as well as the desired results, such as moisture, shape, and further processing. On the basis of these results, geothermal energy resources should be combined with general drying lines for grains and fruits/vegetables.

8.9 Closing Remarks

Geothermal sources are recognized as one of the most significant renewable energy options for effective power generation and suitable for many applications where heat is required, such as heating, hot water, drying and cooling, and where both electricity and heat will potentially be utilized for applications, such as hydrogen production. Geothermal sources are treated as a renewable and environmentally-friendly energy source. Although higher initial, hence capital, cost may in some cases appear to be a kind of disadvantage for implementing geothermal energy systems, their lower operating costs make them significantly appealing. One may need to note that more than half of the initial/capital cost is essentially for drilling of the production and injection wells. The rest is for the power plant built and other expenses. Depending on the ground structure, the initial cost may vary from 2000 to 4000 US$ per kW. Furthermore, this chapter discusses the geothermal energy sources and their categories with numerous examples and applications. It also discusses main types of geothermal power generation techniques, namely dry steam, flash, binary, and combined flash/binary. Geothermal (or ground source) heat pumps stand out with its high performance compared to other conventional systems. Geothermal district heating systems are widely preferred to meet the heating demand almost for free compared to fossil fuel- or electricity-based heating systems. Moreover, geothermal energy systems are capable of integrating with other renewable energy systems for multigeneration purposes. Some examples of the integrated geothermal energy systems are additionally presented in Chapter 11.

References

1 ThinkGeoEnergy Research 2022. Top 10 Geothermal Countries 2021 – installed power generation capacity (MWe). Available online https://www.thinkgeoenergy.com/thinkgeoenergys-top-10-geothermal-countries-2021-installed-power-generation-capacity-mwe/#:~:text=Global%20geothermal%20power%20generation%20capacity,was%20added%20in%20several%20countries (accessed 18 December 2022).

2 The world's 10 biggest geothermal energy plants. Available online https://www.climatecare.com/blog/the-worlds-10-biggest-geothermal-energy-plants (accessed 18 December 2022).

3 Húsavík Energy, Orkuveita Húsavíkur. Multiple-use of geothermal energy in Húsavík. Available online https://www.oh.is/static/files/Skyrslur_og_greinar/Enskar/Multiple_use_of_geothermal_energy_in_Husavik.pdf (accessed 27 December 2022).

4 Dincer, I. and Zamfirescu, C., *Sustainable Hydrogen Production*, Elsevier Science, Ltd., Oxford, 479 p., 2016.

5 Dincer, I. and Ozturk, M., *Geothermal Energy Systems*, Elsevier Science, Ltd., Oxford, 513 p., 2021.

Questions/Problems

Questions

1 What is geothermal energy? Briefly describe it with some practical applications.
2 Classify the geothermal energy sources based on temperature and enthalpy and identify suitable applications every temperature level.

3 How does temperature change from Earth's surface to deeper levels?

4 Classify the geothermal power generation systems and give an example for each.

5 Define the direct steam geothermal power generation system.

6 Explain the flash geothermal power generation systems.

7 Explain why binary geothermal power generation systems are preferred over flashing systems.

8 Define the combined flash-binary geothermal power generation system.

9 How does geothermal heat work? Briefly explain.

10 What are the advantages of geothermal heat pumps?

11 Explain why geothermal district energy systems are preferred.

12 Discuss potential use of geothermal sources for hydrogen generation and make a report.

13 Discuss potential use of geothermal sources for clean fuels and chemicals and make a report.

14 Explain how to determine a typical use of geothermal use for greenhouse applications.

15 Write an essay about potential use of geothermal energy sources.

Problems

1 Consider a dry steam geothermal power plant where there are three wells in the geothermal power plant facility. The temperature and pressure of the geothermal fluid are 256°C and 2405 kPa, respectively. The flow rate of the working fluid from each well is 45 kg/s. The steam that comes from the geothermal source is expanded to 100 kPa in a single stage turbine. The outlet temperature is found to be 102°C. A generator attached to the turbine is used for electricity generation. The efficiency of the generator is 0.85. Calculate the power generation capacity of the subject matter power plant.

2 A direct steam geothermal power plant where there are four wells in the geothermal power plant facility. The temperature and pressure of the geothermal fluid are 257°C and 2345 kPa, respectively. The flow rate of the working fluid from each well is 40 kg/s. The steam that comes from the geothermal source is expanded to 100 kPa in a single stage turbine. The outlet temperature is found to be 102°C. A generator attached to the turbine is used for electricity generation. The efficiency of the generator is 0.85. Calculate the power generation capacity of the power plant.

3 Recalculate the power generation capacities in Problem 2 considering the geothermal water temperatures are at 150, 200, 250, 300°C, respectively.

4 A direct steam geothermal power plant, as shown in the figure below, is utilized for power generation. The geothermal working fluid enters the particle separator at the temperature and pressure of 350°C and 2100 kPa, respectively, and a mass flow rate of 75 kg/s. There is no unwanted particles in geothermal water. Geothermal working fluid enters the moisture separator and is separated in a saturated form in this plant; the fraction of saturated water at flow 3 is 0.25. The exit pressure from the valve is 800 kPa, and then the dry steam is utilized to produce power in a turbine. Finally, the discharged geothermal working fluid from the steam turbine is at 20 kPa with a quality of 0.80, which is then condensed and discharged into the injection well.

   a Write thermodynamic balance equations for each system component.

   b Find the power production rate of the turbine.

   c Calculate the energy and exergy efficiency of the system.

   d Recalculate the energy and exergy efficiencies in the case of 50% of the heat released in the condenser are recovered for useful purposes.

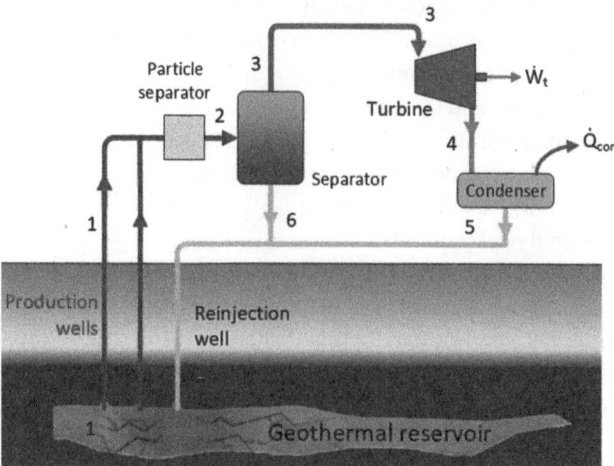

5 Resolve Problem 4 this time by considering that the geothermal working fluid enters the particle separator at the temperature and pressure of 400°C and 3 MPa, respectively, and a mass flow rate of 85 kg/s. The rest of the information will remain the same.

6 A single-flash geothermal power plant is considered, as shown in the figure below, where it uses hot geothermal water at 200°C and 2250 kPa as the heat source. The flow rate of geothermal fluid is 125 kg/s. The geothermal fluid isenthalpically flashed in a flash chamber at a pressure ratio of 4 and separated in the saturated form in a vapor-liquid separator. The fraction of vapor at the flash part outlet is 0.10. Then the produced vapor is used to generate power in a steam turbine. Finally, the discharged geothermal working fluid from the steam turbine is at 20 kPa with a quality of 0.80, which is then condensed and discharged into the injection well.

a Calculate the work production in the steam turbine.

b Find both energy and exergy efficiencies of the single-flash geothermal power plant.

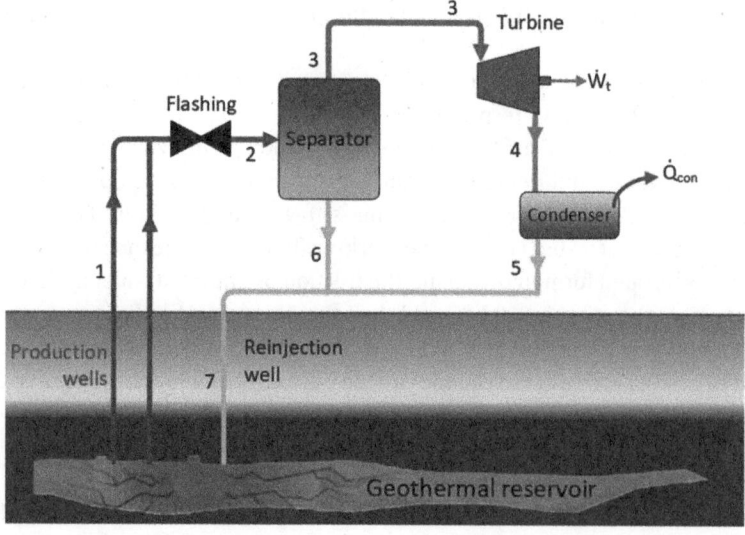

7 Resolve Problem 6 this time by considering the hot geothermal water temperature at 300°C and pressure 2750 kPa with a flow rate of 135 kg/s. The rest of the problem will remain the same.

8 Consider a geothermal system where the geothermal water comes in as saturated liquid at 210°C. The schematic view of the geothermal power plant is shown below. The geothermal liquid is withdrawn from the production well at a rate of 230 kg/s and is flashed to a pressure of 500 kPa by an essentially isenthalpic flashing process where the resulting vapor is separated from the liquid in a separator. It is proposed that the liquid water coming out of the separator route through another flash chamber be maintained at 150 kPa, and the steam produced be directed to the turbine (see figure). Both streams of steam leave the turbine at the same state (state 4). The required properties of various state points are given in table below:
  a Calculate the power generation rate in the turbine.
  b Calculate the missing properties and fill the table given below.
  c Find the exergy destruction rate in the steam turbine.
  d Determine the energy and exergy efficiencies for the power plant with reinjection.
  e Determine the energy and exergy efficiencies for the power plant without reinjection.
  f Calculate the heat recovery potential in the condenser.

State	h (kJ/kg)	s (kJ/kg K)	X (vapor quality)
Ambient (0)	104.9	0.37	0
1	994.34	2.68	0
2	h_2	s_2	0.178
3	2798.4	6.91	1
4	2390	7.43	1
5	190.1	0.62	0
7	h_7	s_7	0.104
8	2693	7.223	1
9	467.1	1.43	

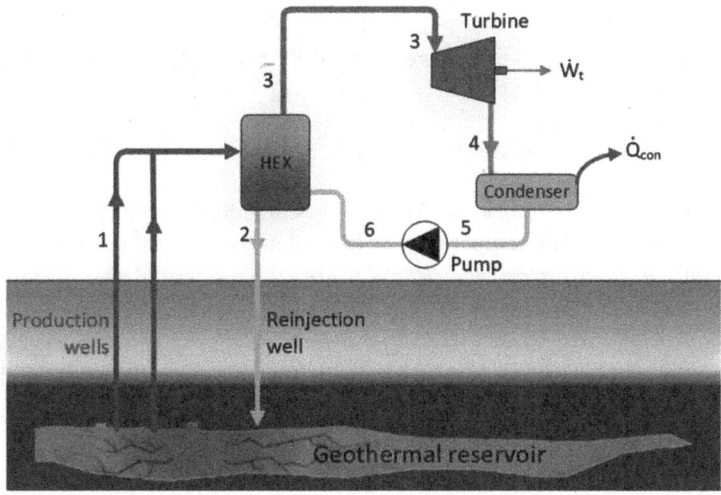

9 In a binary geothermal power plant given below, the geothermal fluid is taken at the condition of $T_1=310°C$ and $P_1=2950$ kPa. It is reinjected back to the reservoir at $T_2=100°C$ and $P_2=1150$ kPa. The flow rate of the geothermal fluid is 120 kg/s. The heat of geothermal fluid is transferred to the working fluid which is water in the binary cycle via a plate heat exchanger. The inlet and outlet conditions for the working fluid in the binary cycle are known to be $T_6=105°C$, $P_6=116$ kPa, $T_3=250°C$ and $P_3=3500$ kPa. The working fluid exits from the turbine is $T_4=125°C$ and $P_4=150$ kPa. The working fluid that exits from the condenser is the saturated water at the pressure of turbine outlet.

  a Write the thermodynamic balance equations for each component of the system.
  b Calculate the flow rate of the water in the binary cycle.
  c Determine the turbine work output.
  d Calculate the heat released from the condenser.
  e Calculate the energy and exergy efficiencies for the system considering that there is no heat recovery.
  f Calculate the energy and exergy efficiencies for the system including the heat recovered.

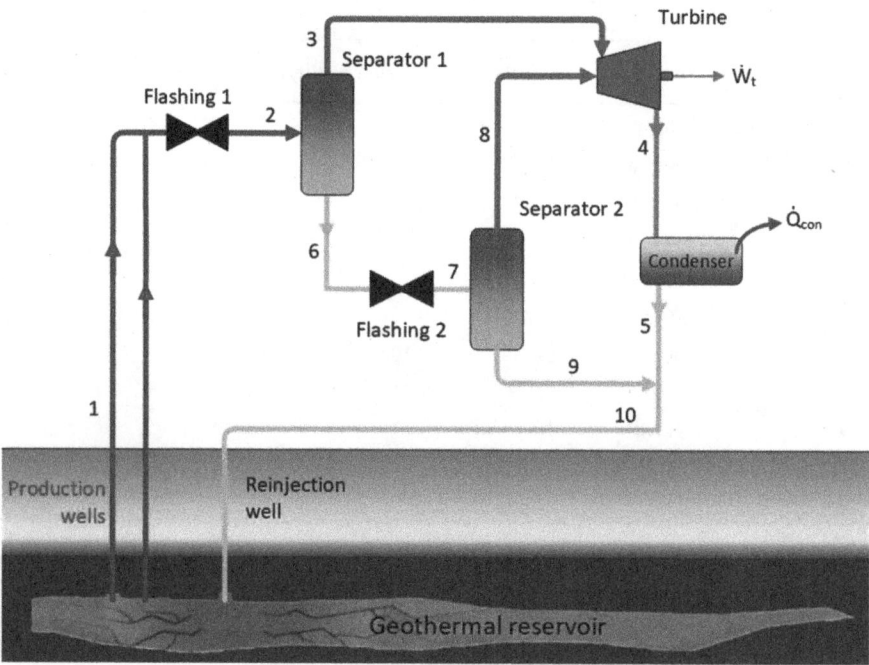

10 A ground source heat pump system works with R-410a working fluid in two configurations for winter and summer conditions. The temperature of the soil stays constant at 13.5°C throughout the year. The heat pump system provides 50 kW heat to the residential area during winter and collects 50 kW heat from the residential area in order to provide cooling effect during summer.

During winter, the outside temperature is considered at −10°C. The compressor inlet and outlet conditions are given as $T_{comp,in}=13°C$, $P_{comp,in}=458.1$ kPa. The working fluid is then condensed and cooled down to 60°C at the condenser outlet.

During summer, the outside temperature is considered at 35°C. The compressor inlet and outlet conditions are given as $T_{comp,in}=2°C$, $T_{comp,out}=14.55°C$, and $P_{comp,out}=1683$ kPa. The working fluid enters the evaporator as saturated liquid at 2°C and exits as saturated vapor at 2°C.

  a Write thermodynamic balance equations for system components.

  b Calculate COP_{en} and COP_{ex} values.

11 At a region where there are 1500 residential and 400 commercial units, District heating with geothermal source is considered. There is a report indicating that there is geothermal water at 90°C in the region. The average heating demand is 35 kW for the residential units and 50 kW for the commercial units. In order to meet the heating demand with geothermal source, calculate the flow rate of the geothermal fluid. The heat losses in the piping line will be neglected. Also, due to the regional regulation, the temperature of the geothermal fluid to be injected can be a minimum of 25°C lower than one obtained from production well.

12 At a region where there are 1000 residential and 250 commercial units, District heating with geothermal source is considered. The measurement shows that there is geothermal water at 100°C in the region. The average heating demand is 25 kW for the residential units and 40 kW for the commercial units. The heat losses in the system are negligible. Also, due to the regional regulation, the temperature of the geothermal fluid to be injected can be a minimum of 20°C lower than one obtained from production well. In order to meet the heating demand with geothermal source, calculate the flow rate of the geothermal fluid. If the maximum flow rate for a production well is 100 kg/s, determine how many drills are required to meet the heating demand.

13 Rework on Problem 10 by considering an open ended problem with 3 different refrigerants, namely ammonia, R32 and R-1234yf (HFO-1234yf) and make the same calculations and compare accordingly.

14 Rework on Problem 11 by considering an open ended problems by considering different regions with different number of residential and commercial units and make the necessary calculations accordingly.

15 Repeat Problem 12 as an open problem for a campus community and make the necessary calculations for different number of units and capacities.

9

Biofuels and Biomass Energy Systems

9.1 Introduction

Biofuels are recognized as one of the oldest energy sources and fuels on the Earth where wood was the only fuel for humanity for many years. It was used for meeting the heating demands and cooking purposes. So, biomass has started with the wood. Today, although we still have wood as a fuel especially for the undeveloped countries and indigenous and remote communities, biomass and biofuel are really in different stage. Biomass originates from the photosynthesis portion of the solar energy distribution and includes all plants, life (terrestrial and marine), all subsequent species in the food chain, and eventually all organic wastes. Sources of the biomass are demonstrated in Figure 9.1. The common biomass sources include municipal solid wastes, sewage, animal residues, agricultural crops and residues, industrial residues, and forestry crops and residues. Actually, all residues mentioned in Figure 9.1 are generally called wastes. Those six are the key domains for biomass as the source. So, biomass resources come in a large variety of wood forms, crop forms, and waste forms. The basic characteristic of biomass is its chemical composition in such forms as sugar, starch, cellulose, hemicellulose, lignin, resins, and tannins. Biomass is combustible. When combusted heat is released, this heat can be used for useful purposes such as commercial heat, electricity, transportation fuel applications. The biggest advantages of using biomass are threefold as follows:

- carbon neutrality due to the carbon cycle,
- containing almost no sulfur which causes almost no SO_2 emissions, and
- having significantly less nitrogen than other fossil fuels, which cause much less NO_x emissions.

These three features make it a significant option as a fuel. Biomass energy (or bioenergy) can be used for many things, including commercial heat, electricity and transportation fuel production.

Figure 9.2 illustrates the historical development of biomass energy utilization in the world. The utilization of biomass started with the existence of humanity. Wood was used for meeting daily energy demands. Wood was the first and only fuel source up until industrial revolution where coal started being used as the prime fuel. The first methane production from biowaste was achieved from cattle manure in 1808. The first commercial biomass gasifier became operational in France in 1840. The use of wood has drastically been reduced by using coal and natural gas. However, biogas and biofuel became quite popular with increasing environmental impacts and hence concerns. In this regard, the capacity of biomass-based energy production increased up to 1084 TWh in 2021.

Introduction to Energy Systems, First Edition. Ibrahim Dincer and Dogan Erdemir.
© 2023 John Wiley & Sons, Inc. Published 2023 by John Wiley & Sons, Inc.
Companion Website: www.wiley.com/go/Dincer/Introduction_to_Energy_Systems

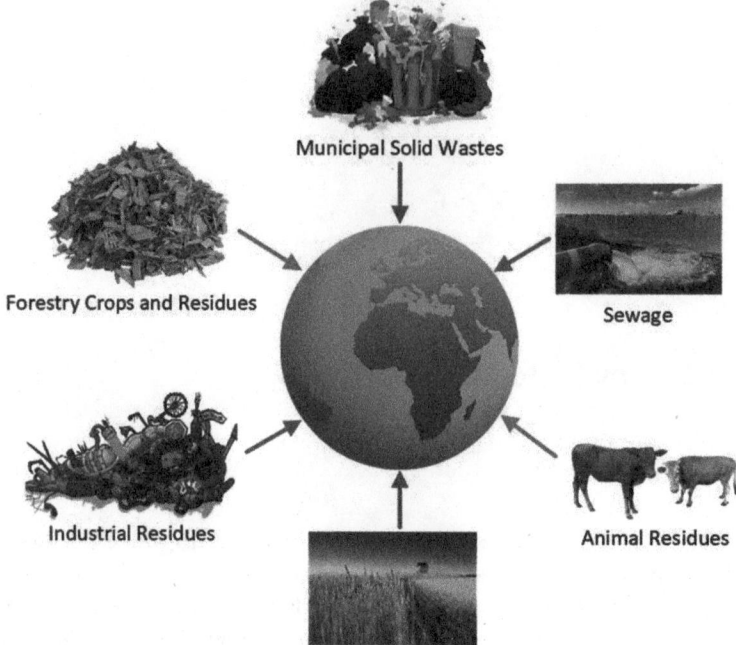

Figure 9.1 Common biomass sources.

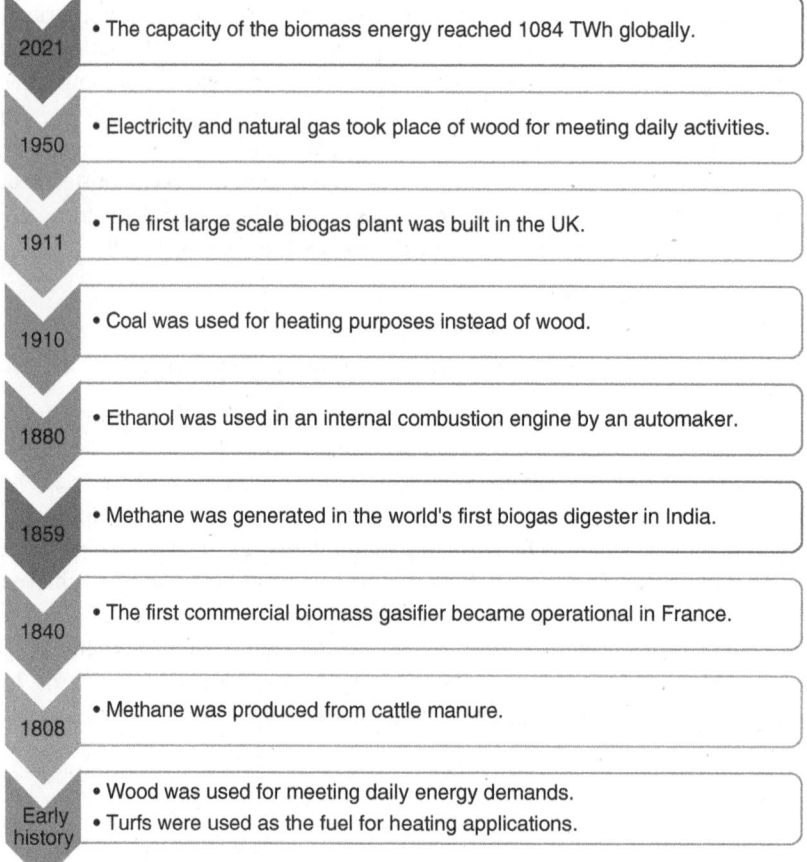

Figure 9.2 Historical development of biomass energy.

The variations of biomass in several countries and world between 19915 and 2021 are shown cumulatively in Figure 9.3. The global capacity of biomass energy has reached 1084 TWh in 2021. In 1990, Brazil was the first and the USA was the second in terms of biomass energy capacity. The picture later changed in a way by the end of 2005 where the USA became the first and Brazil second. Indonesia has recently put a significant attention on the biomass energy. While there were no significant biomass energy activities in the country, they are in third place in the biomass production capacity. This situation is the same for many countries. After 2005, many counties started to include biomass energy systems in the power network. France and Germany are other two counties that have started to use biomass energy before 2000.

Figure 9.4 shows the top ten countries in terms of biomass energy production capacity according to the data provided in Ref. [1] for 2021. The USA is leading the biomass energy production and Brazil is following it. Although Indonesia started biomass energy production later than many counties, they are in the third place. This is the same for China as well. Despite China started the biomass energy production in 2002, it is now in the fourth rank. The top ten countries cover almost 88% of the total biomass energy production. The remaining 12% is produced by the other countries.

This chapter presents the basics of biomass and biofuel technologies. First, CO_2 balance is presented to express how biomass and biofuels are environmentally friendly sources. Then, combustion, gasification, and pyrolysis processes are introduced. This is followed by biofuel and biogas processes. Next, biodigestion processes along with biodigesters are then introduced. Micro-gas turbines which use biofuels as fuels are given to illustrate the enhanced utilization for stationary power generation. Finally, a couple of case studies and illustrative examples are provided to show the importance and potential of the biomass energy system for better sustainability.

9.2 CO₂ Balance

Many people question about biomass energy systems whether or not biomasses, biofuels and biogases contain carbon. If they do contain, are they as undesirable as fossil fuels? If not, what makes them different? Of course, the answer simply lies on carbon cycle (in other terms, CO_2

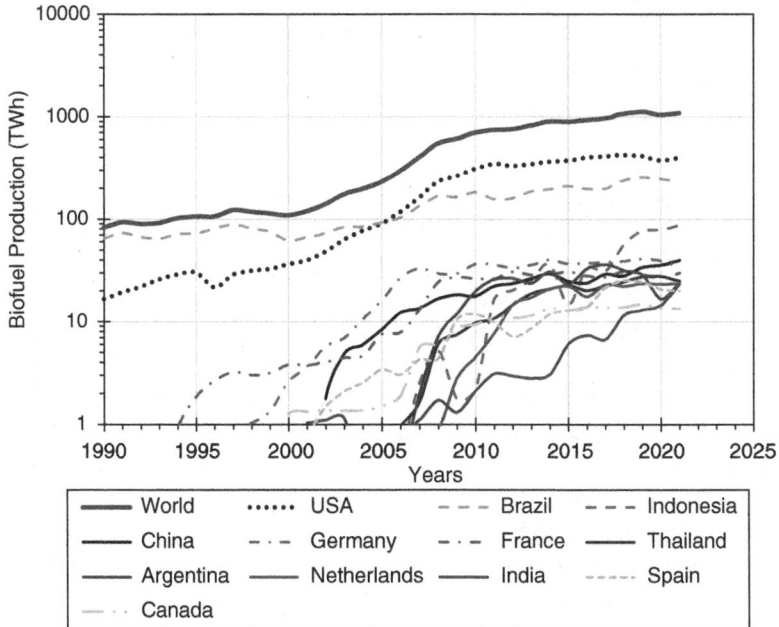

Figure 9.3 Biofuel production change by years for the world and top ten countries (data from [1]).

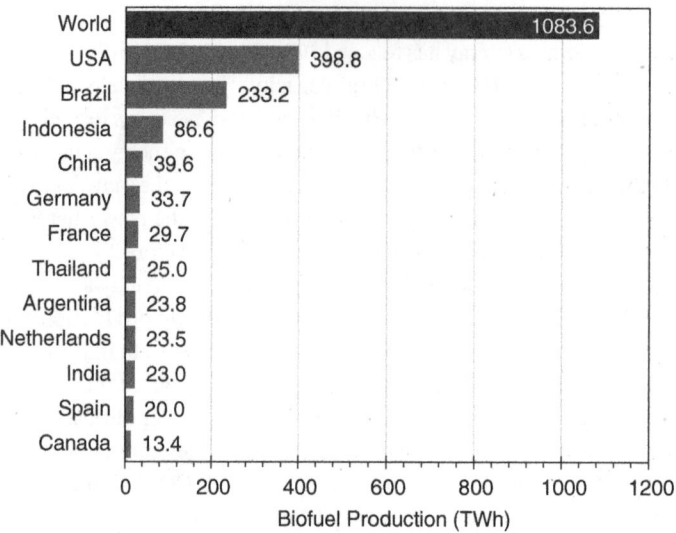

Figure 9.4 Biofuel production capacity for the world and the top ten countries in 2021 (data from [1]).

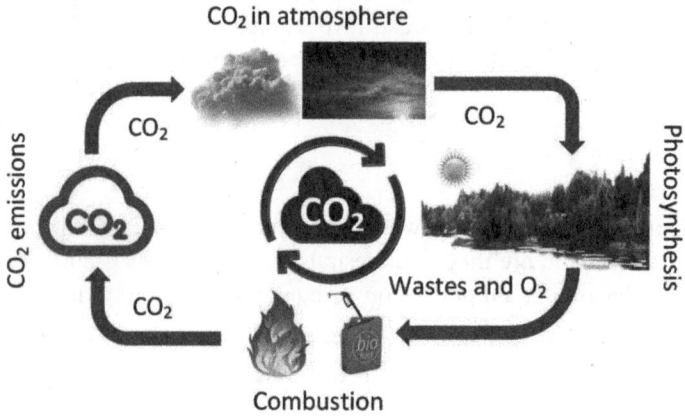

Figure 9.5 Illustration of the CO_2 balance in the globe.

balance). Figure 9.5 shows how CO_2 balance, which is known as carbon cycle, works in the world. This makes biomass sources carbon neutral sources. Biomass generates about the same amount of CO_2 as do fossil fuels (when burned); but from a chemical balance point of view, every time a new plant grows, CO_2 is actually removed from the atmosphere. The net emission of CO_2 will be zero as long as plants continue to be replenished for biomass energy purposes. When we check the carbon balance in the world, if wood is used for meeting the energy demands, CO_2 is emitted to the atmosphere after it is combusted. CO_2 in the atmosphere is converted into O_2 by trees and plants through the photosynthesis process. Carbon content is neutralized inside cycle which makes it carbon cycle. So, it seems there is balance in nature in terms of CO_2. However absolution there is a CO_2 accumulation in the world. This imbalance comes from the combustion of other fossil fuels and reducing trees and plants all over the world.

Today, the only solution to maintain the CO_2 balance in the world is to reduce CO_2 emissions, increasing trees, forest, plants etc. and CO_2 capturing from atmosphere. If the biomass is converted through gasification or pyrolysis, the net balance may even result in removal of CO_2. Energy crops, such as fast-growing trees and grasses are called biomass feedstocks. The use of biomass feedstocks is expected to help increase profits for the agricultural industry

9.3 Biomass

Biomass sources, including bio-wastes, are extensively available around the world, and there are three common methods, as illustrated in Figure 9.6, to covert these biomass sources into useful outputs as follows:

- physical conversion,
- thermochemical conversion, and
- biological conversion.

Physical conversion is one of the oldest engineered techniques to benefit from the plant biomasses. The technique is simple: squeeze the plant-based biomasses to extract the oil from them and use the oil extracted in a combustion process. The squeezing process can be done with and without heat input. Solar heating can also be used to heat up the biomass mixture for more bio-oil gathering. This process is similar to olive oil or sunflower production. However, it should be reminded that olive oil is not combustible.

Biological conversions occur in two methods, either fermentation or anaerobic digestion. The common one is anaerobic digestion. In order to achieve anaerobic digestion, digesters are required for the process. Fermentation is the process to produce plant-based alcohols. These alcohols can be burned or upgraded to further use such as fuel. It can also be used as a burner gas. That's why it is called biogas.

In thermochemical conversion, a chemical process or reaction occurs to get the bio-useful output. Let's start with the combustion. Every plant-based biomass has carbon content which makes it combustible. Thus, basically it is possible to combustion them directly in a combustion chamber to use as

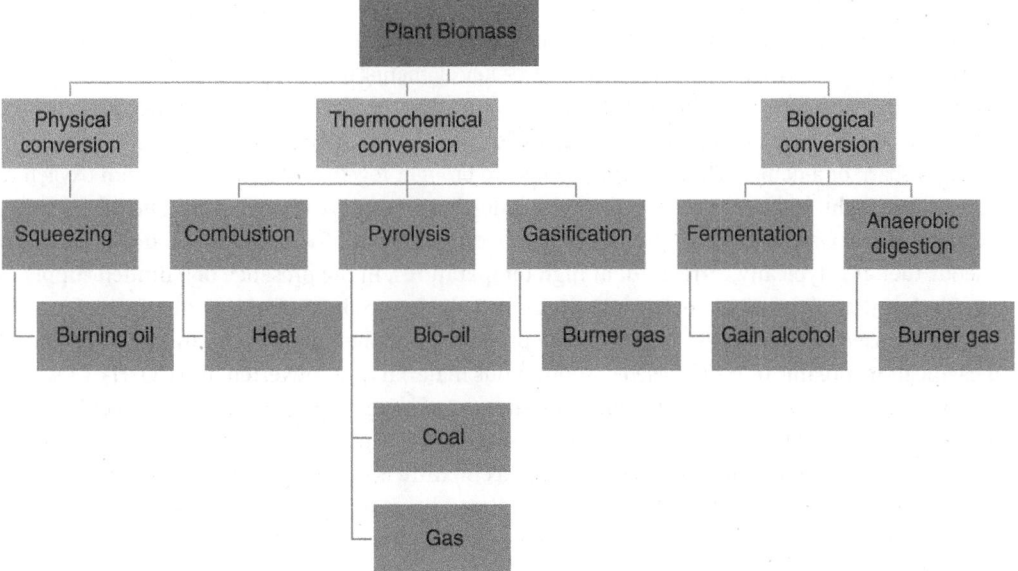

Figure 9.6 A classification of the biomass formation.

fuel. Biomasses can be blended with natural gas, especially in power production facilities. Biomass here can be biomass, biomass wastes, municipal wastes, industrial wastes, or agricultural wastes. They burn to supply heat. So, large volumes of waste are burned, and heat is supplied for useful purposes along with a lower volume of ash. When biomasses are used in this way, CO_2 is emitted. That's why this method is still under question in terms of environment despite CO_2 balance.

In pyrolysis, there are three methods including bio-oil, biogas, and coal (charcoal). So, depending on the biomass, it is possible to obtain bio-oil, biogas, or charcoal. All of these biomass products can be used for various applications. The output form of the conversion can be solid, liquid, or gas.

Gasification is a process that converts biomass- or fossil fuel-based carbonaceous materials into gases without combustion. Combustion is an oxidation process, and its performance depends on oxygen or air contents. In gasification, the chemical reactions occur with almost no oxygen conditions. The products are nitrogen (N_2), carbon monoxide (CO), hydrogen (H_2), and carbon dioxide (CO_2). The output form of the conversion is gas with a really small volume of ash. The gas obtained from gasification can be used for syngas production. For gasification, higher temperatures are required going up to 2000°C. In order to achieve gasification in a sustainable manner, it is often combined with renewable energy sources such as solar thermal, geothermal, etc.

Note that combustion, pyrolysis, and gasification are briefly discussed in the next section.

9.4 Combustion, Gasification and Pyrolysis

Combustion, gasification and pyrolysis are all thermochemical processes and widely utilized for biomass sources. Let's start discussing them in brief here:

Combustion: A burning process, including the sequence of exothermic chemical reactions between a fuel and an oxidant accompanied by the production of heat and conversion of chemical species. In a combustion, fuel and air (consisting essentially of oxygen and nitrogen) are considered reactants, while the chemical species coming out of the reaction are so called products. Combustion reaches a maximum temperature when the air fuel ratio permits all of the hydrogen and carbon in the fuel to be burned to H_2O and CO_2 (stoichiometric combustion). One has to remember that combustion process requires excess air more than the stoichiometric air to practically achieve the combustion. Combustion reactions are characterized by the presence of three key elements: fuel, oxygen, and an ignition source. The fuel is the substance that is burned, and the oxygen is the source of oxygen that is necessary for the reaction to occur. The ignition source is a source of energy that is used to initiate combustion reactions, such as a spark or a flame. When these elements are brought together, the fuel reacts with oxygen to produce heat, light, and a variety of products, including water vapor, carbon dioxide, and other gases.

Gasification: Gasification is a process that converts organic or fossil fuel-based materials into a gaseous fuel. It is typically carried out at high temperatures, in the presence of a limited supply of oxygen (less than the stoichiometric air), and results in the production of a synthetic gas (syngas) that can be used as a fuel or feedstock for the production of chemicals, fuels, and other products. In gasification, organic or fossil-based carbonaceous materials are converted into CO, H_2, CO_2, and CH_4 without combustion by controlling the amount of oxygen and/or steam. It is achieved by reacting the material at high temperatures (>700°C), without combustion, with a controlled amount of oxygen and/or steam. The resulting gas mixture is called syngas (from synthesis gas or synthetic gas) or producer gas and is itself a fuel. In addition to biomass gasification, plasma gasification occurs more than 2000°C. With plasma gasification, any waste is converted into a gas mixture and a small volume of ash. It is a critical solution for waste-to-energy phenomena. There are several types of gasification processes, including steam gasification, air gasification, and oxygen-blown gasification. The specific process used depends on the feedstock being used and the desired products. Gasification can be used to convert a variety of feedstocks, including coal, biomass, and waste materials, into a range of products, including hydrogen, methane, and other hydrocarbons. It is often used as an alternative to traditional fossil fuel-based processes, as it can help to reduce emissions and increase the efficiency of energy production.

Pyrolysis: Pyrolysis is a thermochemical decomposition process that occurs at high temperatures varied from 400 to 800°C, typically in the absence of oxygen. It is used to convert organic materials into a variety of products, including fuels, chemicals, and materials. In pyrolysis, the organic material is heated to a high temperature in a reactor, and the resulting chemical reactions break down the organic molecules into simpler compounds. The products of pyrolysis depend on the feedstock being used and the conditions under which the pyrolysis is carried out. Pyrolysis has a variety of applications, including waste management, energy production, and chemical synthesis.

Further to note on the above listed processes, each of them is critical in dealing with biomass and biomass sources to convert into useful fuel and energy. As mentioned earlier, air or oxygen amount is the critical parameter for combustion, gasification, and pyrolysis processes. Figure 9.7 demonstrates the temperature level and air (also oxygen) content of these three types of processes. The air content of these processes can be defined over air/fuel ratio (AFR) and equivalence ratio (λ). They are formulated as follows:

$$AFR = \frac{m_{air}}{m_{fuel}} \tag{9.1}$$

and

$$\lambda = \frac{Actual\ air\ /\ fuel\ ratio}{Stoichiometric\ air\ /\ fuel\ ratio} = \frac{AFR_{actual}}{AFR_{stoichiometric}} \tag{9.2}$$

Table 9.1 shows the stoichiometric AFR values for the common fuels. AFR tends to increase with increasing carbon content in the fuel. Air/fuel equivalence ratio (λ) compares the air content of the

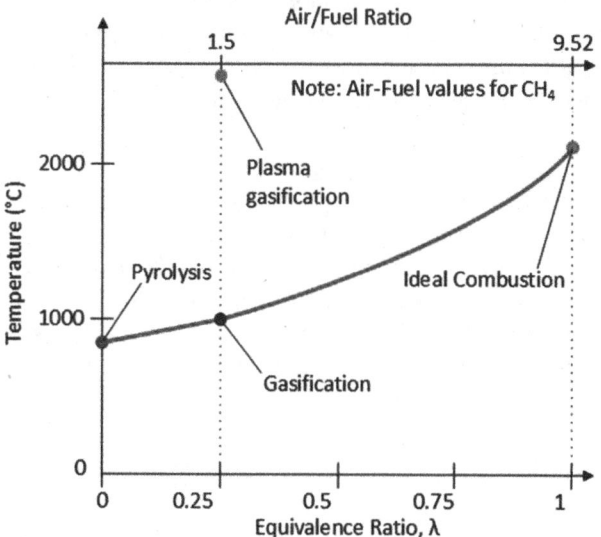

Figure 9.7 The equivalence ratio for combustion, pyrolysis, and gasification.

Table 9.1 Stoichiometric AFR values for various fuels.

Fuel Type	AFR$_{stoichiometric}$
Gasoline (C_8H_{18})	14.7
Methanol (CH_3OH)	6.5
Methane (CH_4)	9.52
Diesel ($C_{12}H_{24}$)	18
Ethanol (C_2H_6O)	16.66
Propane (C_3H_8)	15.64

thermochemical process with the ideal ratio ($\lambda=1$). A lean air-fuel ratio refers to a mixture that has more air than the stoichiometric ratio ($\lambda<1$), while a rich air-fuel ratio refers to a mixture that has less air than the stoichiometric ratio ($\lambda>1$). The stoichiometric ratio is the ideal air-fuel ratio at which all of the fuel is burned, and it is determined by the chemical properties of the fuel and the oxygen in the air. In general, a lean air-fuel ratio helps improve fuel efficiency, but may lead to incomplete combustion, while a rich air-fuel ratio definitely improves power output, but may lead to increased emissions.

As it was previously mentioned in Chapter 4, in order to get the highest possible temperature and useful output from a combustion process, the fuel should be combusted ideally. Namely, the products will be CO_2, H_2O, and N_2. It is called the ideal combustion. In Figure 9.7, it is indicated with yellow at the intersection of $\lambda=1$ and the highest temperature. On the other hand, combustion requires the highest air/fuel ratio. While 100% air is required for the stoichiometric combustion (ideal combustion), in actual combustion excess air is required to achieve a complete combustion.

For gasification, the oxygen content should be lower than a combustion process. It generally occurs when equivalence is around 0.25 and the temperature is around 1000°C, which is lower than combustion. Low air content is provided not to combust the mixture. The resulting gas mixture can be separated for useful purposes such as hydrogen production. The biggest advantage of gasification is the low volume and mass of ash. A large portion of mass the biomass input is gasified.

In pyrolysis, there is no need for air. It results in a lower temperature compared to both gasification and combustion. One has to note that it requires specific process and reactor type, depending on biomass source, as follows: vacuum, flash, etc. in various types of reactors ranging from rotating cone to fluidized bed reactors.

Illustrative Example 9.1 Municipal solid wastes are processed by pyrolysis, gasifier, and combustion in a facility. For pyrolysis, the process is performed underground. Air is not fed to the pyrolysis chamber, and there is no air gap after waste charging. For gasifier, 500 kg of air is provided for gasification. In the combustion, 15,000 kg of air is fed to the combustion chamber. The stoichiometric AFRs of municipal waste for pyrolysis, gasifier and combustion are known to be 0, 1.5, and 13.1, respectively. Calculate the actual AFR value and λ for each process.

Solution:

The AFR values are calculated by using Eq. 9.1.

$$AFR_{pyrolysis} = \frac{m_{air}}{m_{waste}} = \frac{0}{1000} = 0$$

$$AFR_{gasification} = \frac{m_{air}}{m_{waste}} = \frac{500}{1000} = 0.5$$

$$AFR_{combustion} = \frac{m_{air}}{m_{waste}} = \frac{15000}{1000} = 15$$

The air/fuel equivalence (λ) can be found by using Eq. 9.2. $\lambda_{pyrolysis} = 0$, as there is no air in the pyrolysis process. For the gasification,

$$\lambda_{gasification} = \frac{AFR_{actual}}{AFR_{stoichiometric}} = \frac{0.5}{1.5} = 0.33$$

For the combustion,

$$\lambda_{combustion} = \frac{AFR_{actual}}{AFR_{stoichiometric}} = \frac{15}{13.1} = 1.145$$

Further to note that it is always helpful to categorize and illustrate processes. In this regard, Figure 9.8 is given to better show the production and their utilization of gasification, pyrolysis and combustion processes. The only useful output of the combustion is heat. Heat may directly be used for heating demands, such as boilers and stoves. It is also possible to generate electricity with integration of a Rankine power cycle. As gasification is able to produce gaseous products, the products of gasification process can be used in the synthetic gas production processes. This syngas can be used in gas turbines for power generation, internal combustion engines in transportation and power generators, boilers. The products of pyrolysis can be solid, liquid, and gas. While gaseous products can be used in syngas production, charcoal, which is solid product, is used for barbeque and metallurgical industry. The liquid product of pyrolysis, which is bio-oil, is used as lubricant and fuel after the upgrading process.

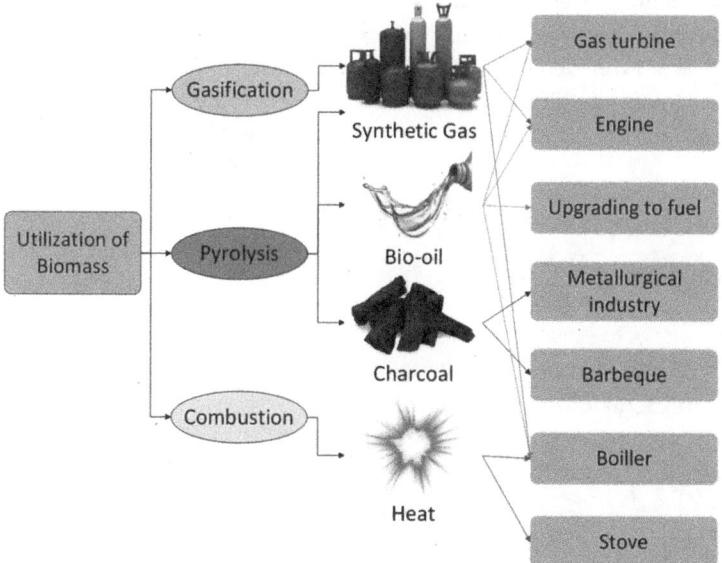

Figure 9.8 Utilization of various biomass sources to generate useful outputs through combustion, gasification, and pyrolysis.

9.5 Biofuels

Biofuel may be in the form of a gas or liquid fuel obtained from plant biomass including wood, wood waste, wood liquors, wood sludge, peat, used sulfite liquors, agricultural wastes, municipal wastes, landfill gases, and ethanol blended gasoline, waste alcohol, etc. Biofuels are often classified according to its generation which are the first, second, third, and fourth generation. Figure 9.9 demonstrates the generational classification of the biofuels. The first generation biofuels are the food-related biofuels, which are obtained from the food waste along with some agricultural wastes. The second generation biofuels cover the agricultural and animal wastes, that's why it is called non-food wastes. The third generation is coming down to the algae. Algae is becoming popular due to its high production capacities. The fourth generation covers using industrial and chemical wastes to obtain biofuels. The fourth generation of biofuels is generally called synthetic fuels. Biofuels are produced and utilized based on the circulation given in Figure 9.5 which is CO_2 balance.

As illustrated in Figure 9.10, there are various biomass sources and conversion processes to obtain many types of biofuels. Biomass-to-biofuel processes work based on the source, system, and service (3S) approach. Biomass sources are converted into biofuels and useful outputs via various systems. Biomass sources are generally classified as agricultural wastes (non-lignin and lignin vegetable), organic residues (food and animal wastes), and biofluids (waste oils). These biomass sources can be processed with various techniques such as fermentation, liquefaction, combustion, mechanical processes, gasification, biodigestion, cracking, esterification, etc. Bioethanol is one of the most popular biofuels which is resulted from the fermentation processes of non-lignin wastes. It is often blended with gasoline to reduce the environmental impact of gasoline combustion, especially in the transportation sector. Up to 15% of it is blended with gasoline.

It is also defined as a fuel produced from dry organic matter or combustible oils produced by plants, including alcohols (from fermented sugar), biodiesel from vegetable oil and wood. Unlike fossil fuels such as petroleum, coal, and natural gas, biofuels can be replenished readily, which makes them renewable sources of energy. Biofuels have become popular due to their cost-effectiveness and environmental benefits over petroleum and other fossil fuels, particularly in light of rising petroleum prices and concerns over greenhouse gas emissions from fossil fuels.

Wood has been used directly as a raw material and burned to produce heat for many years. In a power plant, the heat can be used to run a turbine to generate electricity. There are a number of existing power plants that burn grass, wood or other kinds of biomass. Since liquid biofuels can be used

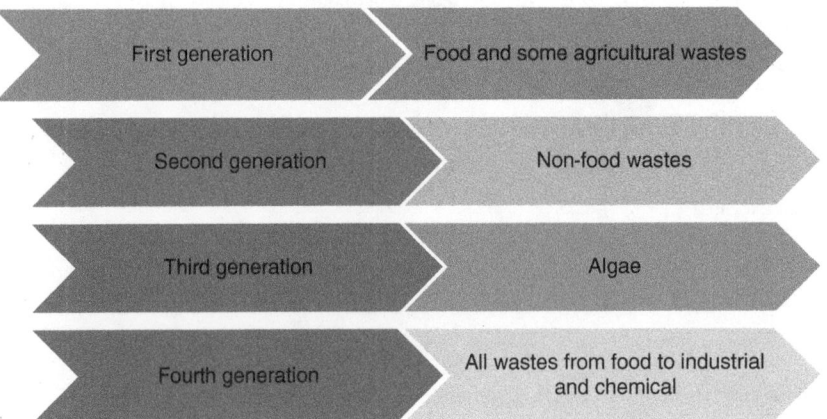

Figure 9.9 Classification of biofuels.

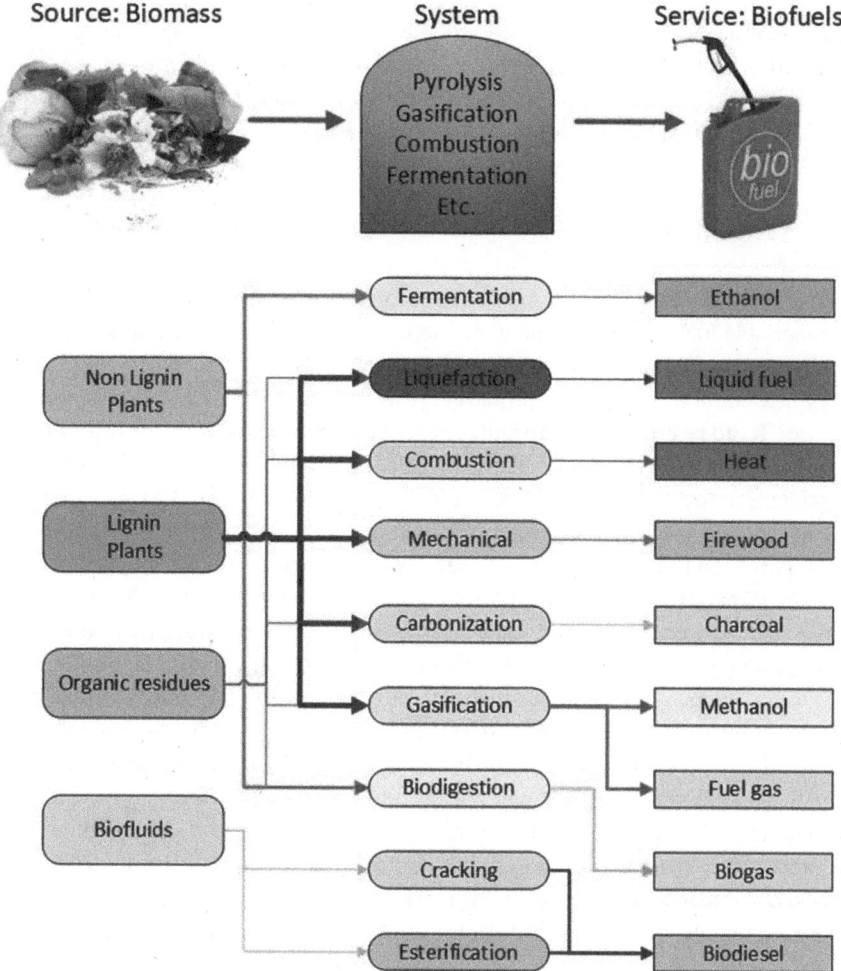

Figure 9.10 Processes of conversion of biomass into biofuels based on 3S (Source-System-Service) approach.

in transportation, they are of particular interest because of the extensive infrastructure already in place. A popular liquid biofuel that is ethanol is generally produced by fermenting starch or sugar. Ethanol is produced primarily in the USA and Brazil, as mentioned earlier. In the USA, ethanol is made primarily from corn grain. A common use for it is to blend it with gasoline to make "gasohol," a 10% ethanol-based fuel. Sugarcane is the primary source of ethanol in Brazil, where it is commonly used as an ethanol fuel or in gasoline blends containing 85% ethanol.

Unlike the ethanol produced from food crops expressed above, cellulosic ethanol is derived from low-value biomass that possesses a high cellulose content, including wood chips, crop residues, and municipal waste. A common source of cellulosic ethanol is sugarcane bagasse, a waste product from sugar processing, or various types of grass that can be grown on low-quality land. Cellulosic ethanol is primarily used as a gasoline additive due to its lower conversion rate than previous ethanol production methods.

Biodiesel is the second most common liquid biofuel, made primarily of oily plants (such as soybeans or oil palm) and to a lesser extent from waste cooking fat from restaurants. Generally, biodiesel is blended with petroleum diesel fuel at various percentages and is used in diesel engines, where it has found the greatest acceptance. There are many attempts to produce biodiesel from algae and cyanobacteria. They are very critical for the decarbonization of the power generation and transportation sectors. Methane gas and biogas can be produced from biomass decomposition without oxygen, as well as methanol, butanol, and dimethyl ether, which are in development.

9.6 Biogas

Biogas is a gas or gas mixture obtained from the breakdown of organic matter by biochemical processes, which makes it a renewable fuel. An environmentally friendly, renewable energy source, biogas is a great alternative to fossil fuels. It is a fuel of high calorific value, resulting from anaerobic fermentation of biomass. The calorific value of biogas depends primarily on the amount of methane in its composition, e.g., 5000–6000 kcal/m^3. It can be used for stove heating, internal combustion engines, gas turbines, etc.

It is mainly produced by a process, so-called: anaerobic digestion. Organic matter is broken down by microorganisms in the absence of oxygen. In order to accomplish this, the waste material must be enclosed in an oxygen-free environment. The production of biogas for use as fuel can occur naturally or as part of an industrial process. Several types of waste materials break down into biogas, including animal manure, municipal rubbish, plant material, food waste, and wastewater. Today, in many regions, organic wastes and yard wastes are collected separately to be used in biogas facilities. It is also possible to separate the organic waste from the bulk waste collection. Methane and carbon dioxide are the main components of biogas. A small amount of hydrogen sulfide, siloxanes, and moisture can also be present. Depending on the type of waste used in biogas production, these quantities vary. Most commonly, biogas is referred to as biomethane. The gas is also known as marsh gas, sewer gas, compost gas, and swamp gas.

Table 9.2 lists the potential biogas production from various organic sources. Although animal waste produces almost seven times higher biogas than agricultural residues, it is difficult to harness, provision, and discharge in biodigesters. The type of biomass source has a great impact on the biogas production rate and its purity, as shown in Table 9.2. It is most appropriate to use

Table 9.2 Biogas production potential of some organic residues (modified from [2]).

Biomass Source	Production of Biogas (m^3/ton)	Methane (%)
Sunflower leaves	300	58
Rice straws	300	Variable
Wheat straws	300	Variable
Bean straws	380	59
Soy straws	300	57
Linen stem	359	59
Grapevine leaves	270	Variable
Potato leaves	270	Variable
Dry leaves of trees	245	58

intermittent biodigesters for this biomass. Due to their lightness, agricultural remains cannot be mixed with water; they emerge only after initial decomposition, when gas is produced. When triturated, they can be used in any type of biodigester, although this requires more energy and labor. Biogas is produced after 60–120 days of fermentation and provisioning. Chlorine prevents methanogenic bacteria from growing on residues containing agricultural chlorinated pesticides.

Today, when biogas is considered, power generation generally comes to mind. Figure 9.11 demonstrates the diagram of a biogas cogeneration power plant. The organic wastes are collected in the forms of silage, slurry, and other organic wastes. They go to a mixer passing through a chopper. Then, they are pumped into fermentation tanks for fermentation. The biogas exposed is collected at the top of the tank. As the gas exposed during the fermentation is not pure methane, it is required to purify before using it. It can also be blended with promoters such as natural gas and hydrogen. There are also some attempts to mix with ammonia and biogas to reach a carbon-free solution.

As mentioned earlier, the composition of the biogas is the critical parameter to obtain the useful output. Table 9.3 gives the variation of the composition of the biogas. Methane (CH_4) content varies from 30% to 65%. Therefore, before using biomethane, it is required to purify it. Depending on the biomass used in the biogas production, carbon dioxide content of the end gas mixture ranges from 20% to 40%. Therefore, there are various attempts for capturing CO_2 in the mixture. When carbon capture is included in the system, the system may even become a negative carbon system. Also, nitrogen, hydrogen and oxygen take place in the biogas process. Also, hydrogen sulfide may emit even if the rates are low. Furthermore, it is important to note that hydrogen sulfide is a hazardous content.

In order to show the potential of the biogas to meet the daily demand of the people, Table 9.4 gives the amount of biogas to be needed to meet the demands. While excluding the fuel demand, the daily demand of the house can be met by $6 \, m^3$ of biogas. These numbers indicate that biogas has a significant potential to meet the daily demands.

9.7 Waste to Energy Power Generation Systems

Waste and/or trash are used as fuel for waste-to-energy plants, just like coal, oil, or natural gas. Water is heated into steam by the burning fuel that comes from biomass, which drives a turbine to

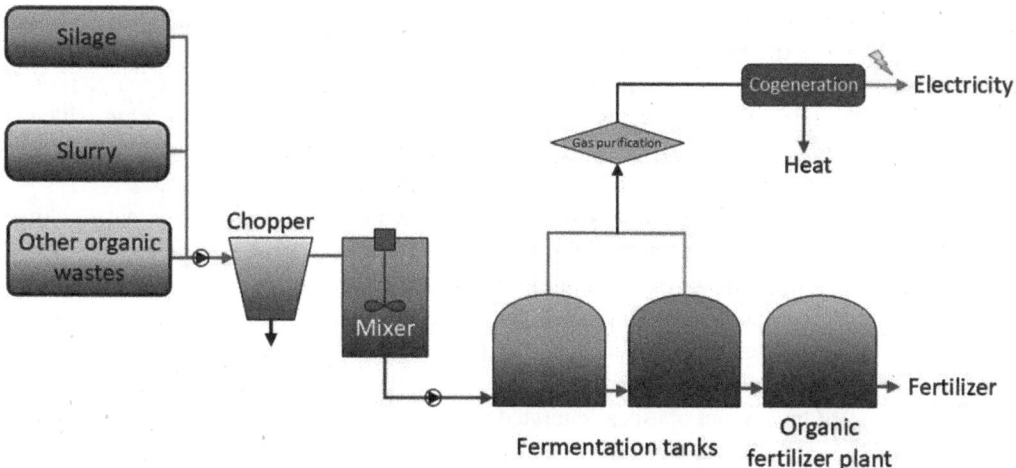

Figure 9.11 System layout of a biogas cogeneration system.

Table 9.3 Variation of the composition of the biogas (modified from [3]).

Gas	Variation Range (%)	Average Content (%)
Methane, CH_4	30–65	45
Carbon dioxide, CO_2	20–40	35
Nitrogen, N_2	5–40	15
Hydrogen, H_2	1–3	1
Oxygen, O_2	0–5	1
Hydrogen sulfide, H_2S	0–0.01	0.003

Table 9.4 Daily domestic biogas utilization by a family of five or six people (modified from [2]).

Demand Item	Amount of Biogas
Kitchen	$1.960 \, m^3$
Hot water	$0.924 \, m^3$
Refrigeration	$2.800 \, m^3$
Lighting	$0.140 \, m^3$
Fuel	$0.5 \, m^3/hp \, h$

generate electricity. A community's landfill volume can be reduced by up to 90% through the process, preventing up to one ton of carbon dioxide from being released for every ton of waste burned. Figure 9.12 demonstrates a schematic illustration for a waste-to-energy power plant. In a waste-to-energy power, the process starts with the waste collection. Depending on the waste type, it may require a significant effort, especially for the municipal wastes. Here, the critical challenge is CO_2 emissions due to waste collection which comes with collection trucks. After collection, the bulk waste is sorted out for adjusting as possible as higher heating values for the mixture. In many counties and regions, the waste is collected separately as organic wastes, yard wastes and recyclables which are primarily plastics and papers.

The next process is the fuel preparation process which covers mixing and drying (dehumidifying). The waste mixture processed is burned in the combustion chamber. The heat released is used for steam generation in a boiler. Steam is used for power generation in the steam turbine, which is attached to the generator. The heat is then recovered just after the steam turbine for heating purposes. The heat recovered can also be used for the preheating of the waste mixture.

In the filter baghouse, an induction fan draws air through fabric bags toward the stack or chimney to remove the finest airborne particulates. This process removes any remaining particulates by 96%. During vibration, particulates caked on the outer and inner surfaces of the bags are shaken loose. Fly ash is often disposed of in landfills. Magnets and eddy current separators remove ferrous (steel and iron) and other metals – such as copper, brass, nickel, and aluminum – from the unburned remains of combustion. For roadbeds and rail embankments, the remaining ash can be used as aggregate. The amount of ashes generated is roughly 10% of the original volume and 30% of the original weight of the waste.

Another waste-to-energy solution is to produce biogas and use it in the internal combustion engine (ICE) and gas turbine. Figure 9.13 shows the schematics for a waste-to-energy system

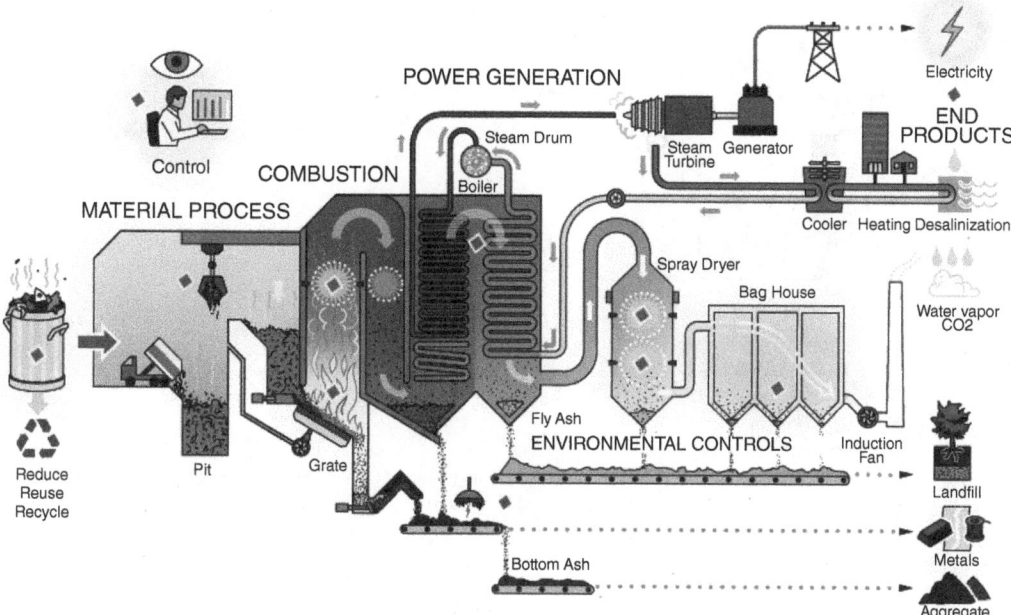

Figure 9.12 Waste-to-energy power plant (adapted from [4]).

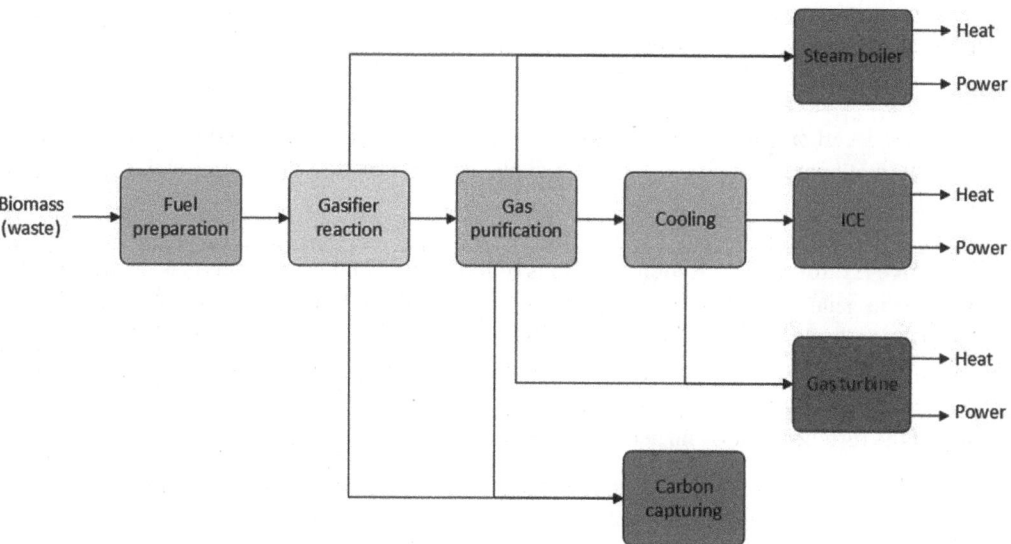

Figure 9.13 Flow diagram for a waste-to-energy-plant.

which is based on gasification. They can also combine with cogeneration systems to use heat released from the system. In order to increase the environmental impact of CO_2 capturing can be included in the system. When carbon capturing is included in the system, it is called negative carbon system due to the carbon balance phenomena. Using biogas in gas turbines and ICEs is a significant option for meeting power and heat demands in remote communities. It is a quite critical

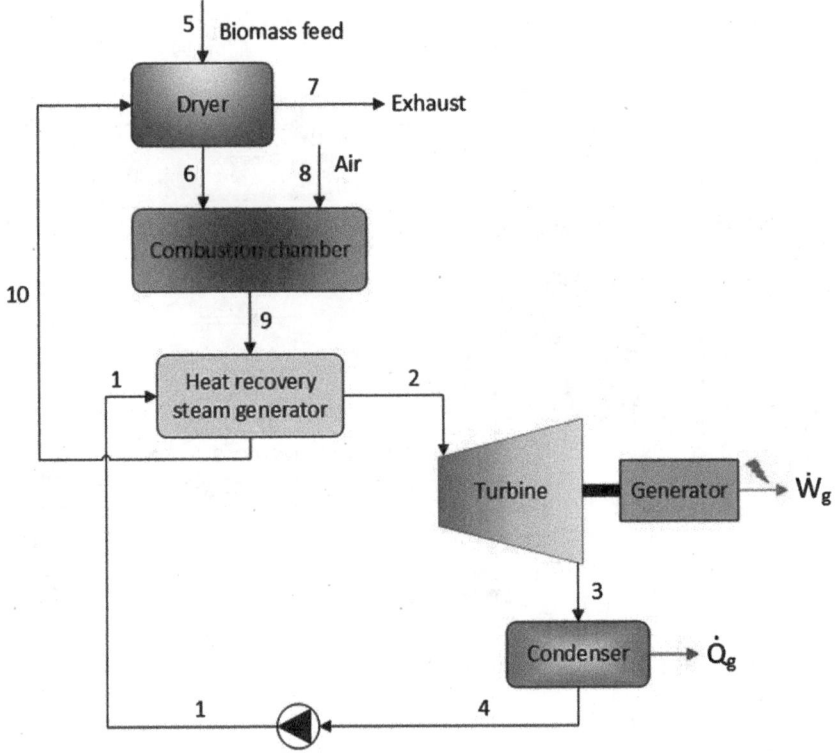

Figure 9.14 System layout for the Illustrative Example 9.2.

solution for decarbonization of the remote communities. Biogas can be used to replace the diesel used in the stationary generators. The main challenge is biogas-fueled systems is the purity of the biomethane.

Illustrative Example 9.2 A total of 10 tons of municipal solid waste (MSW) is burned directly in a boiler to generate saturated steam at 200°C, as illustrated in Figure 9.14. Water enters the boiler at 20°C. The heating value of the MSW is 18 MJ/kg and the chemical exergy is 16 MJ/kg. The boiler efficiency is 75%. The steam produced in the boiler is run through a turbine-generator group to generate electricity. The steam exits from the turbine at 100 kPa of pressure. The isentropic efficiency of the turbine is 0.85. The generator efficiency is 95%.

a) Write the mass, energy, entropy and exergy balance equations for system components.
b) Find the steam generation rate in the boiler.
c) Calculate the amount of electricity generated.
d) Determine the energy and exergy efficiencies of the system.

Solution:
a) The mass, energy, entropy and exergy balance equations for the system are written below accordingly:
   For pump:
   MBE: $\dot{m}_4 = \dot{m}_1$
   EBE: $\dot{m}_4 h_4 + \dot{W}_p = \dot{m}_1 h_1$

EnBE: $\dot{m}_4 s_4 + \dot{S}_{gen,p} = \dot{m}_1 s_1$

ExBE: $\dot{m}_4 ex_4 + \dot{W}_p = \dot{m}_1 ex_1 + \dot{Ex}_{dest,p}$

For steam turbine:

MBE: $\dot{m}_2 = \dot{m}_3$

EBE: $\dot{m}_2 h_2 = \dot{W}_t + \dot{m}_3 h_3$

EnBE: $\dot{m}_2 s_2 + \dot{S}_{gen,t} = \dot{m}_3 s_3$

ExBE: $\dot{m}_2 ex_2 = \dot{W}_t + \dot{m}_3 ex_3 + \dot{Ex}_{dest,t}$

For heat recovery steam generator (HRSG):

MBE: $\dot{m}_9 = \dot{m}_{10}$ and $\dot{m}_1 = \dot{m}_2$

EBE: $\dot{m}_9 h_9 + \dot{m}_1 h_1 = \dot{m}_{10} h_{10} + \dot{m}_2 h_2$

EnBE: $\dot{m}_9 s_9 + \dot{m}_1 s_1 + \dot{S}_{gen,HRSG} = \dot{m}_{10} s_{10} + \dot{m}_2 s_2$

ExBE: $\dot{m}_9 ex_9 + \dot{m}_1 ex_1 = \dot{m}_{10} ex_{10} + \dot{m}_2 ex_2 + \dot{Ex}_{dest,HRSG}$

For condenser:

MBE: $\dot{m}_3 = \dot{m}_4$

EBE: $\dot{m}_9 h_9 + \dot{m}_1 h_1 = \dot{m}_{10} h_{10} + \dot{m}_2 h_2$

EnBE: $\dot{m}_9 s_9 + \dot{m}_1 s_1 + \dot{S}_{gen,HRSG} = \dot{m}_{10} s_{10} + \dot{m}_2 s_2$

ExBE: $\dot{m}_9 ex_9 + \dot{m}_1 ex_1 = \dot{m}_{10} ex_{10} + \dot{m}_2 ex_2 + \dot{Ex}_{dest,HRSG}$

For dryer:

MBE: $\dot{m}_5 = \dot{m}_6$ and $\dot{m}_{10} = \dot{m}_7$

EBE: $\dot{m}_5 h_5 + \dot{m}_6 h_6 = \dot{m}_6 h_6 + \dot{m}_7 h_7$

EnBE: $\dot{m}_5 s_5 + \dot{m}_6 s_6 + \dot{S}_{gen,dryer} = \dot{m}_6 s_6 + \dot{m}_7 s_7$

ExBE: $\dot{m}_5 ex_5 + \dot{m}_6 ex_6 = \dot{m}_6 ex_6 + \dot{m}_7 ex_7 + \dot{Ex}_{dest,dryer}$

For combustion chamber:

MBE: $\dot{m}_6 + \dot{m}_8 = \dot{m}_9$

EBE: $\dot{m}_6 h_6 + \dot{m}_8 h_8 = \dot{m}_9 h_9$

EnBE: $\dot{m}_6 s_6 + \dot{m}_8 s_8 + \dot{S}_{gen,cc} = \dot{m}_9 h_9$

ExBE: $\dot{m}_6 s_6 + \dot{m}_8 s_8 = \dot{m}_9 h_9 + \dot{Ex}_{dest,cc}$

b) In order to find the steam generation rate in the steam generator, it is required to write the energy balance equation for it.

$$\dot{m}_9 h_9 + \dot{m}_1 h_1 = \dot{m}_{10} h_{10} + \dot{m}_2 h_2$$

Here, h_1 and h_2 are taken from Table A1.

$$T_1 = 20°C \atop x_1 = 0 \Bigg\} h_1 = 83.91 kJ / kg$$

$$T_2 = 200°C \atop x_1 = 1 \Bigg\} h_2 = 2792 kJ / kg$$

Also, the efficiency of the boiler is provided to be 0.75 in the problem. Thus,

$$\eta_{boiler} = \frac{Q_{useful}}{Q_{in}} = \frac{m_{steam}(h_2 - h_1)}{m_{fuel} HV_{MSW}}$$

$$0.75 = \frac{m_{steam}\left(2792 - 83.91\right)}{10,000 \times 18,000}$$

Thus, the amount of the steam is determined to be $m_{steam} = 49,850$ kg.

c) In order to determine the amount of electricity generation, it is required to first write the EBE for the turbine.

$$\dot{m}_2 h_2 = \dot{W}_t + \dot{m}_3 h_3$$

The EBE can be modified as follows:

$$\dot{W}_t = \dot{m}_3 h_3 - \dot{m}_2 h_2$$

As the $\dot{m}_2 = \dot{m}_3 = \dot{m}_{steam}$ from the MBE for the turbine, it can be written as:

$$\dot{W}_t = \dot{m}_{steam}\left(h_3 - h_2\right)$$

As we know the total amount of the steam, the energy balance equation of the turbine is then written in its final form as follows:

$$W_t = m_{steam}\left(h_3 - h_2\right)$$

$$\left.\begin{array}{l} P_3 = 100 \; kPa \\ S_3 = S_2 \end{array}\right\} h_{3s} = 2328 \; kJ/kg$$

Since the isentropic efficiency of the turbine is given in the problem, the actual enthalpy at the state point 3 is found from:

$$\eta_T = \frac{h_2 - h_3}{h_2 - h_{3s}} \rightarrow 0.85 = \frac{2792 - h_3}{2792 - 2328.8} \rightarrow h_3 = 2398.3 \; kJ/kg$$

We proceed now by using the final form of the energy balance equation for the turbine by including the efficiency of the generation,

$$W_e = \eta_g W_t = \eta_g m_{steam}\left(h_2 - h_3\right) = 0.95 \times 49,850 \times \left(2792 - 2398.3\right)$$

The electricity generation is found to be 5180 kWh.

d) The energy and exergy efficiencies of the system can be found from the following equations:

$$\eta_{en} = \frac{W_e}{m_{MSW} HV} = \frac{5180}{10,000 \times 18,000} = 0.1036 \; or \; 10.36\%$$

and

$$\eta_{ex} = \frac{W_e}{m_{MSW} ex_{MSW}} = \frac{5180}{10,000 \times 16,000} = 0.1165 \; or \; 11.65\%$$

Illustrative Example 9.3 Consider the system given in Illustrative Examples 9.2. Calculate CO_2 emissions to be emitted from the system for the same amount of electricity generation found in the previous example for different types of fuels. Use Table 9.5 in this regard for the fuel properties.

Solution:

In order to determine the emission values, it is first required to determine the amount of fuel to be used. The amount of fuel is calculated by using the energy efficiency equation for the overall system.

Table 9.5 Details of the fuels used in Illustrative Example 9.3.

Fuel	LHV (kWh/kg)	CO_2 Emissions (kg CO_2/kg fuel)	Fuel to Electricity Efficiency
MSW	5	1.1	0.10
Wood	4.28	1.65	0.05
Coal	7.9	2.69	0.3
Hydrogen	33.33	0	0.56
Natural gas	13.1	1.91	0.34

$$\eta_{en} = \frac{W_e}{m_{fuel}HV}$$

We obtain the efficiency and LHV values for the fuels to be studied. Thus, the amount of the wood used as a fuel is calculated by modifying the energy efficiency equation as follows:

$$m_{wood} = \frac{W_e}{\eta_{en}LHV} = \frac{5180\ kWh}{0.05 \times 4.28} = 24,205\ kg$$

Similarly, for the coal,

$$m_{coal} = \frac{W_e}{\eta_{en}LHV} = \frac{5180\ kWh}{0.3 \times 7.9} = 2,185\ kg$$

For hydrogen,

$$m_{hydrogen} = \frac{W_e}{\eta_{en}LHV} = \frac{5180\ kWh}{0.56 \times 33.33} = 277.53\ kg$$

For natural gas,

$$m_{hydrogen} = \frac{W_e}{\eta_{en}LHV} = \frac{5180\ kWh}{0.34 \times 13.1} = 1,163\ kg$$

The CO_2 emissions for each fuel can be calculated by using the unit emission values for each fuel given in Table 9.5. The results obtained in this example are then summarized in Table 9.6.

Table 9.6 The results of Illustrative Example 9.2.

Fuel	LHV (kWh/kg)	CO_2 Emissions (kg CO_2/kg fuel)	Fuel to Electricity Efficiency	Amount of Fuel (kg)	CO_2 Emissions (kg CO_2)
MSW	5	1.1	0.10	10,000	11,000
Wood	4.28	1.65	0.05	24,205	39,939.3
Coal	7.9	2.69	0.3	2185	5,879.4
Hydrogen	33.33	0	0.56	277	0
Natural gas	13.1	1.91	0.34	1163	2221.3

9.8 Biodigestion and Biodigesters

The anaerobic digestion process involves microorganisms breaking down biodegradable matter without oxygen. It is also called biodigestion. Waste is easily be managed, or fuel is then produced by this process. Both industrial fermentation and home fermentation use anaerobic digestion to produce food and drink products. There is anaerobic digestion occurring naturally in some soils, lake sediments, and oceanic basin sediments, which is called anaerobic activity. Figure 9.15 depicts the workflow for anaerobic digestion process. Hydrolysis of input materials is the first step in digestion. Other bacteria can use soluble derivatives of insoluble organic polymers, such as carbohydrates. The sugars and amino acids are then converted by acidogenic bacteria into carbon dioxide, hydrogen, ammonia, and organic acids. By converting these organic acids into acetic acid, bacteria also produce ammonia, hydrogen, and carbon dioxide. In the end, these products are converted to methane and carbon dioxide by methanogens. In anaerobic wastewater treatment, methanogenic archea play a crucial role. Biological waste and sewage sludge are treated through anaerobic digestion. By reducing landfill gas emissions, anaerobic digestion contributes to integrated waste management systems. The digestion of energy crops, such as maize, can also be achieved with the help of anaerobic digesters.

Anaerobic digestion is assumed as a source of renewable energy. During the process, methane, carbon dioxide, and other contaminant gases such as ammonia, hydrogen, hydrogen sulfur, nitrogen, etc. are produced. It is possible to use this biogas directly as fuel, in combined heat and power gas engines, or to upgrade it into natural gas-quality biomethane. Also produced is nutrient-rich digestate that can be used as fertilizer.

Anaerobic digestion has received increased attention from governments in a number of countries due to its reuse of waste as a resource. Also, new technological developments offer a significant reduction in capital costs.

Materials temperature, acidity, nutrients, and their concentration are the critical key factors for biodigestion. Also, concentration of the solid particle in the mixture affects the biogas production

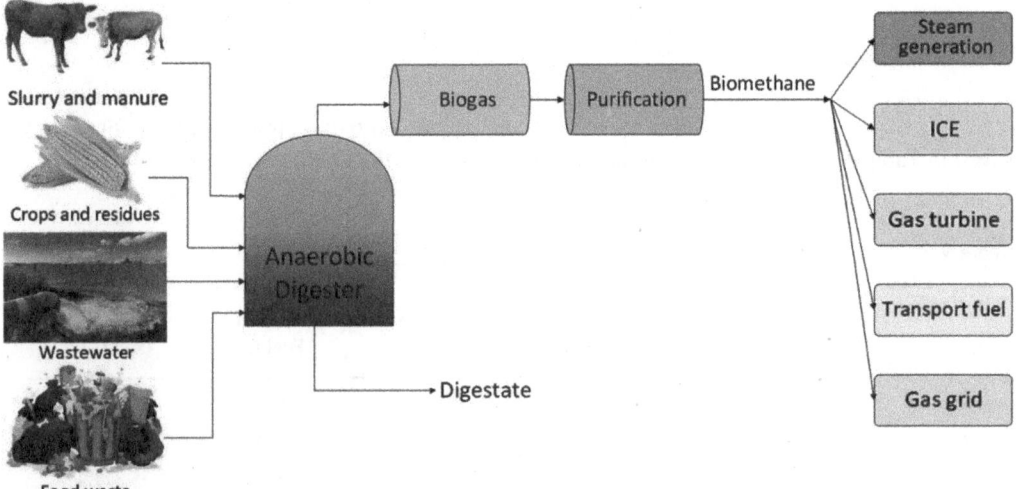

Figure 9.15 Workflow diagram for anaerobic digestion.

rate and potential in per kg of input materials. It is most common for biodigesters to be buried underground due to the higher and more constant temperatures underground. There are several advantages to using biomass and biodigesters in rural applications, such as obtaining biofertilizer from cultural and animal residues (as organic material processed in biodigesters can be used as fertilizer). In addition to providing illumination, heating, and motor power, biomass and biodigesters can also be used to provide energy.

In a biodigester, organic material is biologically digested, either anaerobically (without oxygen) or aerobically (with oxygen). In a biodigester, bacteria and microbes break down organic materials. A biodigester can process most foods, including fats, greases, animal manure, and agricultural wastes.

Due to the closed nature of biodigesters, there will be no odor from the waste; this will improve hygiene by eliminating flies and rodents. The reduction of hauling costs is another benefit of eliminating food waste on-site. Biodigesters can process different amounts of food depending on their size; the larger the digester, the more food it can process. The biodigester is a living system that requires maintenance that brings a significant amount of operating costs. But they are easy-to-use and maintain systems.

Biodigesters are eco-friendly and reduce a facility's carbon footprint significantly, which is one of their greatest advantages. Methane and carbon dioxide are released into the atmosphere by decaying food scraps and other organic materials in landfills. In many countries, food waste is responsible for 30–50% of the total waste. This number indicates that biodigestion and biodigesters are how important for sustainability. Diverting food scraps and other organic materials from landfills, methane and carbon dioxide, can be captured and used efficiently.

Figure 9.16 shows an illustration of a typical biodigester of average. The system consists of a digester and a gasometer. Digesters are underground reservoirs made of bricks or concrete. As part of a biofertilization process, biomass is loaded into a biodigester, which is divided into two

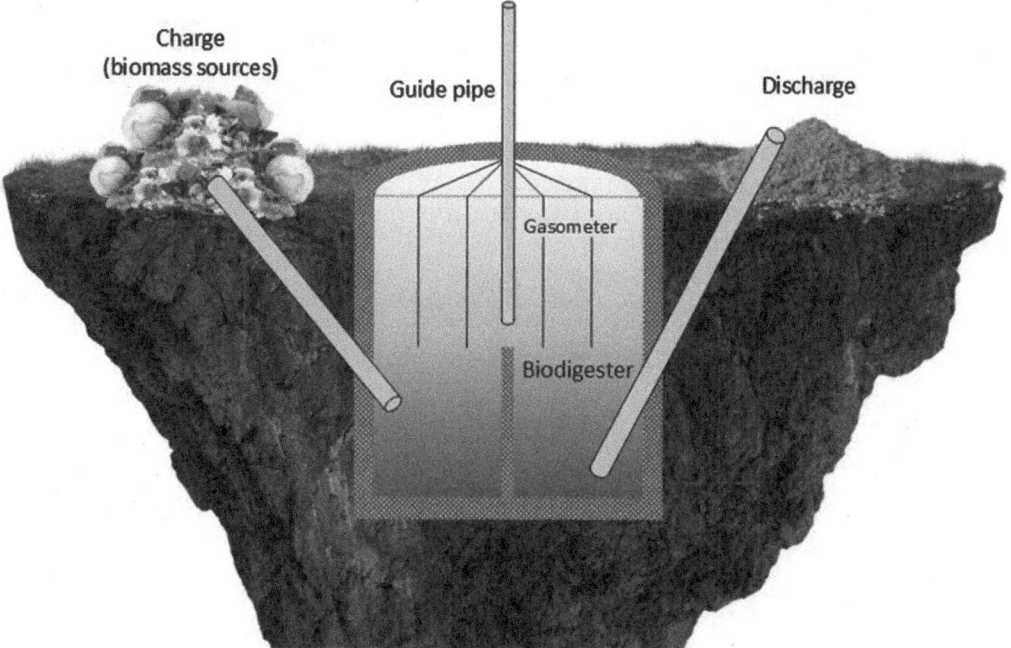

Figure 9.16 Production of biogas in a biodigester. *Source:* Partays / photka / cleanpng.com

semicylindrical parts by a wall. Biodigesters are loaded through charge boxes, which also serve as prefermenters. A pipe connecting the charge box and digester goes down to the bottom. As the biofertilizer leaves the biodigesters through another pipe, the biomass entering the biodigesters is equal to the biomass leaving the biodigesters. In addition, the biofertilizer should be pumped and/ or delivered to the consumer directly from the discharge box, tank, or dam. A biodigester can produce 6 m^3 biogas per day, which corresponds to 8568 kWh of energy when it is charged with 240 L of biomass per day, which is mixture of 120 L is water and 120 L is bovine manure [2]. When the daily demands of the residential unit presented in Table 9.4 are considered, in order to meet the daily demand with biogas, 120 L of biomass is required.

Illustrative Example 9.4 A biodigester used for producing biogas is shown in Figure 9.17. Sludge from wastewater is entered into the digester. The thermodynamic properties of these substances are measured as listed in Table 9.7.

a) Write the mass, energy, entropy and exergy balance equations for the biodigester.
b) Find the mass flow rate of output biogas.
c) Determine the rate of heat input required.
d) Find the exergy destruction rate if the source temperature is 100°C.
e) Find the energy and exergy efficiencies of the bio-digester if the lower heating value of leaving biogas is 10,000 kJ/kg and the lower heating value of sludge is 18,000 kJ/kg.

Solution:

a) The thermodynamic balance equations for the biodigester are written as follows:

MBE: $\dot{m}_{sludge} = \dot{m}_{biogas} + \dot{m}_{digestate}$

EBE: $\dot{m}_{sludge} h_{sludge} + \dot{Q}_{in} = \dot{m}_{biogas} h_{biogas} + \dot{m}_{digestate} h_{digestate}$

EnBE: $\dot{m}_{sludge} s_{sludge} + \dfrac{\dot{Q}_{in}}{T_{source}} + \dot{S}_{gen} = \dot{m}_{biogas} s_{biogas} + \dot{m}_{digestate} s_{digestate}$

ExBE: $\dot{m}_{sludge} ex_{sludge} + \dot{Q}_{in}\left(1 - \dfrac{T_0}{T_{source}}\right) = \dot{m}_{biogas} ex_{biogas} + \dot{m}_{digestate} ex_{digestate} + \dot{Ex}_D$

b) In order to find the mass flow rate of the biogas, it is required to write the mass balance equation for the biodigester. From the MBE:

$$12.06 = \dot{m}_{biogas} + 11.85 = 0.21 \ kg/s$$

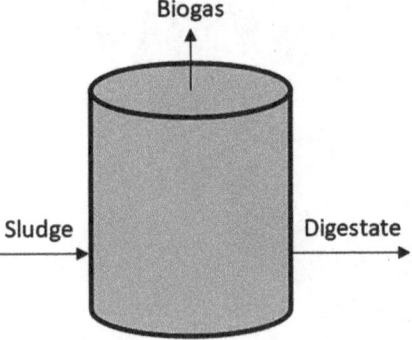

Figure 9.17 Schematic illustration of biodigester in Illustrative Examples 9.4.

Table 9.7 Thermodynamic properties of the sludge, digestate, and biogas.

Substance	h (kJ/kg)	s (kJ/kg K)	ex (kJ/kg)	$\dot{m}$(kg/s)	P (kPa)	T (°C)
Sludge	1308.3	0.37	104.8	?	100	25
Digestate	941.34	0.505	146.7	?	100	25
Biogas	18,570.7	7.282	357.6	15	100	25

c) In order to find the heat required for biodigestion, the EBE should be written accordingly.

$$\dot{m}_{sludge} h_{sludge} + \dot{Q}_{in} = \dot{m}_{biogas} h_{biogas} + \dot{m}_{digestate} h_{digestate}$$

$$12.06 \times 104.8 + \dot{Q}_{in} = 0.21 \times 357.6 + 11.85 \times 146.7$$

Thus, the heat required is found to be $\dot{Q}_{in} = 549.6$ kW.

d) In order to find the exergy destruction in the biodigester, the ExBE is then written. From the ExBE, one obtains

$$\dot{m}_{sludge} ex_{sludge} + \dot{Q}_{in}\left(1 - \frac{T_0}{T_{source}}\right) = \dot{m}_{biogas} ex_{biogas} + \dot{m}_{digestate} ex_{digestate} + \dot{Ex}_D$$

$$12.06 \times 1308.37 + 549.6 \times \left(1 - \frac{298}{373}\right) = 0.21 \times 18570.7 + 11.85 \times 941.3 + \dot{Ex}_D$$

Finally, the exergy destruction is found to be $\dot{Ex}_D = 834.7$ kW.

e) The energy and exergy efficiencies are found as follows:

$$\eta_{en} = \frac{\dot{m}_{biogas} LHV_{biogas}}{\dot{m}_{sludge} LHV_{sludge} + \dot{Q}_{in}}$$

$$\eta_{en} = \frac{0.21 \times 10,000}{12.06 \times 18,000 + 549.6} = 0.009 \; or \; 0.9\%$$

and the exergy efficiency is then calculated as

$$\eta_{ex} = \frac{\dot{m}_{biogas} ex_{biogas}}{\dot{m}_{sludge} ex_{sludge} + \dot{Q}_{in}\left(1 - \frac{T_0}{T_{source}}\right)} = \frac{0.21 \times 18,570.75}{12.06 \times 1308.37 + 549.6 \times \left(1 - \frac{298}{373}\right)}$$

$$\eta_{ex} = 0.245 \; or \; 24.5\%$$

Illustrative Example 9.5 A biomass gasifier is shown in Figure 9.18. Rice husk biomass is entered into the gasifier with separate streams of oxygen and water. The thermodynamic properties of these substances are determined as listed in Table 9.8.

a) Write the mass, energy, entropy and exergy balance equations for the gasifier.
b) Find the mass flow rate of output syngas.
c) Calculate the rate of heat input required.

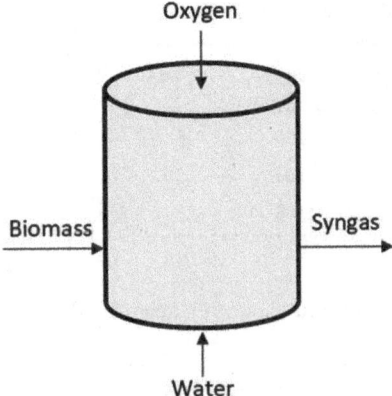

Oxygen

Biomass

Syngas

Water

Figure 9.18 Schematic illustration of biodigester in Illustrative Examples 9.5.

Table 9.8 Thermodynamic properties of the biomass, oxygen, water, syngas.

Substance	h (kJ/kg)	s (kJ/kg K)	ex (kJ/kg)	$\dot{m}$ (kg/s)	P(kPa)	T (°C)
Biomass	1510	3.15	1049.6	1.0	101.325	850
Oxygen	451.3	0.115	428.9	2.76	2431.8	483.8
Water	419.2	1.31	34.13	2.89	111.0	100
Syngas	4950	2.0	3071.6	?	506.6	2056

d) Find the exergy destruction rate if the source temperature is 3000°C.
e) Find the energy and exergy efficiencies of the gasifier if the lower heating value of syngas is 2426 kJ/kg and the lower heating value of biomass is 14,690 kJ/kg.

Solution:

a) The thermodynamic balance equations for the gasifier are written as follows:

MBE: $\dot{m}_{bio} + \dot{m}_w + \dot{m}_o = \dot{m}_{SG}$

EBE: $\dot{m}_{bio}h_{bio} + \dot{m}_w h_w + \dot{m}_o h_o + \dot{Q}_{in} = \dot{m}_{SG}h_{SG}$

EnBE: $\dot{m}_{bio}s_{bio} + \dot{m}_w s_w + \dot{m}_o s_o + \dfrac{\dot{Q}_{in}}{T_{source}} + \dot{S}_{gen} = \dot{m}_{SG}s_{SG}$

ExBE: $\dot{m}_{bio}ex_{bio} + \dot{m}_w ex_w + \dot{m}_o ex_o + \dot{Q}\left(1 - \dfrac{T_0}{T_{source}}\right) = \dot{m}_{SG}ex_{SG} + \dot{Ex}_D$

b) In order to find the mass flow rate of the biogas, it is required to write the mass balance equation for the gasifier. From the MBE:

$$\dot{m}_{bio} + \dot{m}_w + \dot{m}_o = \dot{m}_{SG}$$

$$1 + 2.89 + 2.76 = \dot{m}_{SG}$$

Thus, the biogas is found to be $\dot{m}_{SG} = 6.65$ kg/s.

c) In order to find the heat required for biodigestion, the EBE is written accordingly.

$$\dot{m}_{bio}h_{bio} + \dot{m}_W h_W + \dot{m}_o h_o + \dot{Q}_{in} = \dot{m}_{SG}h_{SG}$$

$$1 \times 1510 + 2.89 \times 419.12 + 2.76 \times 451.3 + \dot{Q}_{in} = 6.65 \times 4950$$

Thus, the heat required is found to be $\dot{Q}_{in} = 28{,}951$ kW.

d) In order to find the exergy destruction in the biodigester, the ExBE is now written. From the ExBE:

$$\dot{m}_{bio}ex_{bio} + \dot{m}_W ex_W + \dot{m}_o ex_o + \dot{Q}\left(1 - \frac{T_0}{T_{source}}\right) = \dot{m}_{SG}ex_{SG} + \dot{Ex}_D$$

$$1 \times 1049.6 + 2.89 \times 34.13 + 2.76 \times 428.9 + 28{,}951 \times \left(1 - \frac{298}{3273}\right) = 6.65 \times 3071.6 + \dot{Ex}_D$$

Finally, the exergy destruction is found to be $\dot{Ex}_D = 7981$ kW.

e) The energy and exergy efficiencies are found as follows:

$$\eta_{en} = \frac{\dot{m}_{SG}LHV_{SG}}{\dot{m}_{bio}LHV_{bio} + \dot{m}_W h_W + \dot{m}_o h_o + \dot{Q}_{in}}$$

$$\eta_{en} = \frac{6.65 \times 2426}{1 \times 14{,}690 + 2.89 \times 419.2 + 2.76 \times 451.3 + 28{,}951} = 0.35 \; or \; 35\%$$

and the exergy efficiency is then calculated as follows:

$$\eta_{ex} = \frac{\dot{m}_{SG}ex_{SG}}{\dot{m}_{bio}ex_{bio} + \dot{m}_W ex_W + \dot{m}_o ex_o + \dot{Q}_{in}\left(1 - T_0 / T_{source}\right)}$$

$$\eta_{ex} = \frac{6.65 \times 3071.6}{1 \times 1049.6 + 2.89 \times 34.13 + 2.76 \times 428.9 + 28{,}951 \times \left(1 - \frac{298}{3273}\right)}$$

$$\eta_{ex} = 0.719 \; or \; 71.9\%$$

9.9 Micro Gas Turbines

Micro gas turbines are not new, but gaining importance for flexible power generation applications. Although they were originally designed for aircraft and helicopters, they have become a critical solution for customer-site electric user applications. Microturbines from 10 to 500 kW are available for small-scale distributed power either for electrical power generation alone, in distributed electrical power generation, or in combined cooling and heat and power systems. They can burn fuels, including natural gas, gasoline, diesel, kerosene, naphtha alcohol, propane, methane and digester gas. They operate on the non-ideal Brayton open cycle with heat recovery. Thermoelectric power generators, battery packs, and fuel cells may be integrated into the micro-gas turbines. Figure 9.19 shows the schematic illustration of a micro-gas turbine.

The most typical applications of the micro gas turbines are

- Peak shaving and base load power (grid parallel)
- Combined heat and power

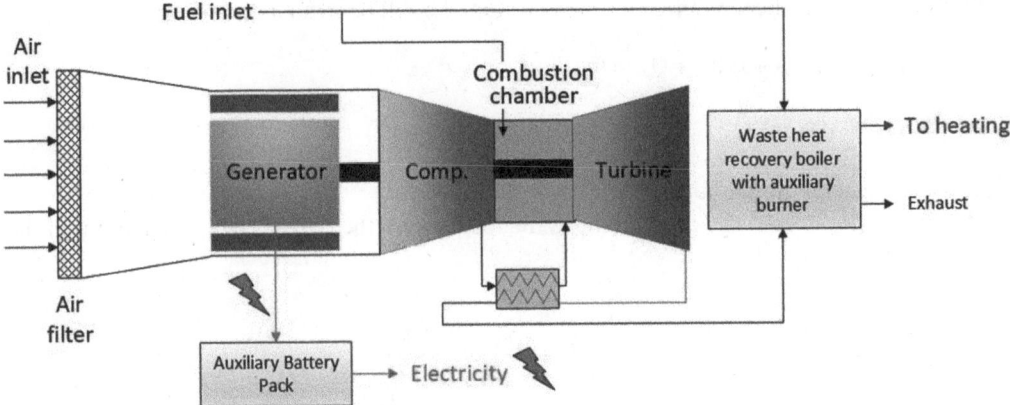

Figure 9.19 Micro gas turbine with heat recovery and battery pack options.

- Stand-alone power
- Backup/standby power
- Ride-through connection
- Primary power with grid as backup
- Microgrid
- Remote communities

In order to operate the micro-gas turbine more environmentally friendly, it is often combined with biogas systems. It may play a key role for remote communities, farms, fish farms, etc.

The working principles of the micro-gas turbines are the same with the large-scale gas turbines, which is Brayton cycle, as mentioned previously and shown in Figure 9.20. As indicated in Figure 9.20, there are mainly two types of gas turbines depending on the air flow:

- Open-type Brayton cycle (Figure 9.20a)
- Closed-type Brayton cycle (Figure 9.20b)

Micro-gas turbines are the open type of Brayton cycle. In the open type of Brayton cycle, the intake air is first compressed in the compressor. The fuel is combusted with compressed air in a combustion chamber. Thus, high-temperature and high-pressure post-combustion gas mixture is converted into the useful shaft work in a turbine. The products of combustion are exhausted after the turbine. Also, it is quite common to benefit from exhaust gases. In such systems, the exhaust gases pass through a heat exchanger to recover the heat for useful purposes. This is called a cogeneration system or combined power and heat system.

In order to analyze the Brayton cycles, it is required to write the mass, energy, entropy, and exergy balance equations accordingly for both open- and closed-type systems. Since micro turbines are an open type of Brayton cycle, here, the thermodynamic aspects for it are presented. The mass, energy, entropy, and exergy balance equation for the open type of Brayton cycle is given in Table 9.9. For more details on the thermodynamic analysis of the Brayton cycles, one may visit Chapters 3 and 4 and Ref. [5].

The energy and exergy efficiencies for an open-type Brayton cycle are written as follows:

$$\eta_{en} = \frac{\dot{W}_{net}}{\dot{m}_{fuel} LHW_{fuel}} \tag{9.3}$$

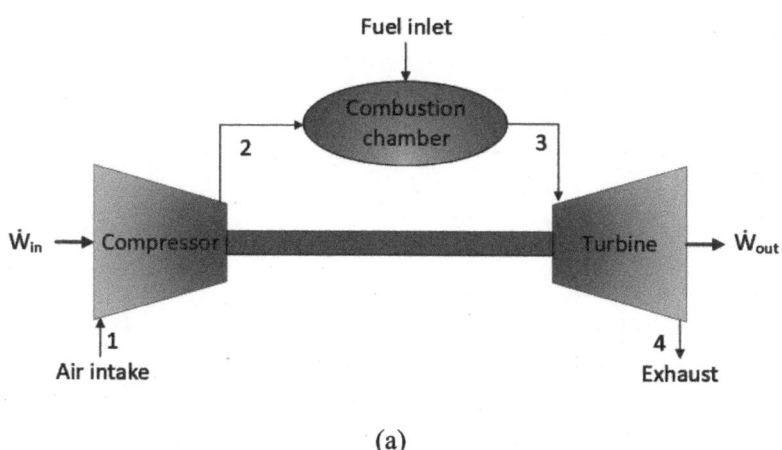

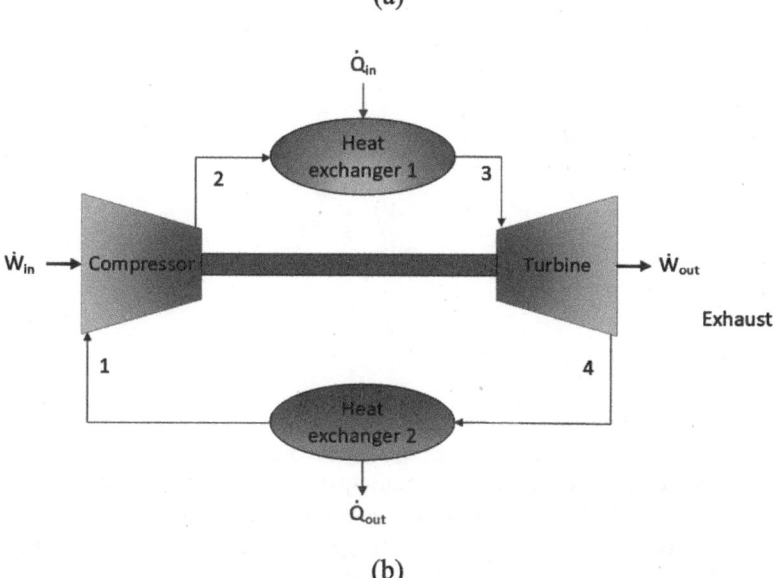

Figure 9.20 Illustration of the Brayton cycle: (a) open type and (b) closed type.

and the exergetic efficiency is

$$\eta_{ex} = \frac{\dot{W}_{net}}{\dot{m}_{fuel}ex_{fuel}} \tag{9.4}$$

where

$$\dot{W}_{net} = \dot{W}_t - \dot{W}_{comp} \tag{9.5}$$

When the heat is recovered in the system and utilized for useful purposes, such as heating, the energy and exergy efficiencies are then revised as follows:

$$\eta_{en} = \frac{\dot{W}_{net} + \dot{Q}_{recovered}}{\dot{m}_{fuel}LHW_{fuel}} \tag{9.6}$$

Table 9.9 The mass, energy, entropy and exergy balance equations for the components of open-type Brayton cycle.

Component		Equation
Compressor	MBE	$\dot{m}_1 = \dot{m}_2$
	EBE	$\dot{m}_1 h_1 + \dot{W}_{comp} = \dot{m}_2 h_2$
	EnBE	$\dot{m}_1 s_1 + \dot{S}_{gen,c} = \dot{m}_2 s_2$
	ExBE	$\dot{m}_1 ex_1 + \dot{W}_{comp} = \dot{m}_2 ex_2 + \dot{Ex}_{dest,comp}$
Combustion chamber	MBE	$\dot{m}_2 + \dot{m}_{fuel} = \dot{m}_3$
	EBE	$\dot{m}_2 h_2 + \dot{m}_{fuel} LHV_{fuel} = \dot{m}_3 h_3$
	EnBE	$\dot{m}_2 s_2 + \dot{m}_{fuel} s_{fuel} + \dot{S}_{gen,cc} = \dot{m}_9 h_9$
	ExBE	$\dot{m}_2 s_2 + \dot{m}_{fuel} ex_{fuel} = \dot{m}_3 h_3 + \dot{Ex}_{dest,cc}$
Turbine	MBE	$\dot{m}_3 = \dot{m}_4$
	EBE	$\dot{m}_3 h_3 = \dot{W}_t + \dot{m}_4 h_4$
	EnBE	$\dot{m}_3 s_3 + \dot{S}_{gen,t} = \dot{m}_4 s_4$
	ExBE	$\dot{m}_3 ex_3 = \dot{W}_t + \dot{m}_4 ex_4 + \dot{Ex}_{dest,t}$

and

$$\eta_{en} = \frac{\dot{W}_{net} + \dot{Ex}_{\dot{Q}_{recovered}}}{\dot{m}_{fuel} ex_{fuel}} \tag{9.7}$$

Here, $\dot{W}_{net}$ is calculated similar with Eq. 9.3.

Illustrative Example 9.6   Biomethane produced from municipal waste collected from residential units is used as fuel in a micro gas turbine unit in a small community, as illustrated in Figure 9.21. Air enters the compressor at the reference conditions which are $T_0 = T_1 = 20°C$ and $P_0 = P_1 = 100$ kPa with mass flow rate 2 kg/s. In the combustion chamber, the compressor compression ratio is 1:12. Air leaves from the compressor at 510°C. Air enters the turbine at 1100°C and exits at 595°C. The combustion temperature is $T_c = 1650°C$. The LHV of biomethane is 40,320 kJ/kg and its chemical exergy is 36,520 kJ/kg.

a) Calculate the turbine output.
b) Determine the compressor work.
c) Determine the flow rate of the fuel.
d) Find the back work ratio.
e) Calculate the energy and exergy efficiencies.

Solution:

a) The thermodynamic properties of the state points are obtained to be:

$$\left. \begin{array}{l} T_1 = 20°C \\ P_1 = 100 \ kPA \end{array} \right\} \rightarrow \begin{array}{l} h_1 = 305.5 \ kJ/kg \\ s_1 = 5.722 \ kJ/kgK \end{array}$$

As the pressure ratio is given 1:12 in the problem, $P_2 = 1200$ kPa, thus,

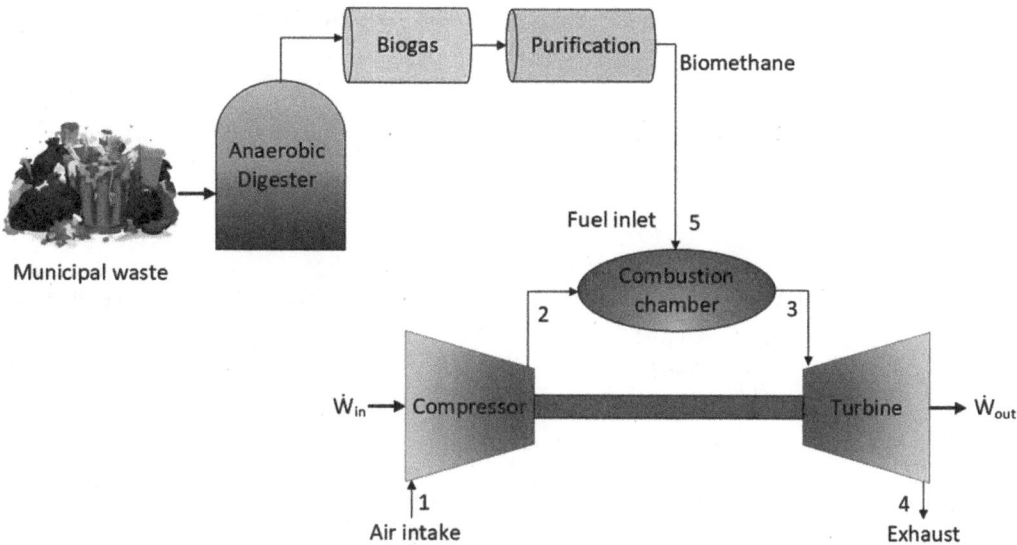

Figure 9.21 The system layout for Illustrative Example 9.6.

$$T_2 = 510°C \atop P_2 = 1200\ kPA \Bigg\} \rightarrow \begin{array}{l} h_2 = 692.1\ kJ/kg \\ s_2 = 5.832\ kJ/kgK \end{array}$$

By assuming the combustion occurs at the constant pressure.

$$T_3 = 1100°C \atop P_3 = 1200\ kPA \Bigg\} \rightarrow \begin{array}{l} h_3 = 1528\ kJ/kg \\ s_3 = 6.66\ kJ/kgK \end{array}$$

As outlet conditions are given as follows:

$$T_4 = 595°C \atop P_4 = 100\ kPA \Bigg\} \rightarrow \begin{array}{l} h_4 = 880.8\ kJ/kg \\ s_4 = 6.792\ kJ/kgK \end{array}$$

In order to calculate the turbine output, it is required to use the energy balance equation for the turbine given in Table 9.9.

$$\dot{m}_3 h_3 = \dot{W}_t + \dot{m}_4 h_4$$

$$2 \times 1528 = \dot{W}_t + 2 \times 880.8$$

$$\dot{W}_t = 1294.4\ kW$$

b) In order to calculate the compressor work, the EBE for compressor are written as follows:

$$\dot{m}_1 h_1 + \dot{W}_{comp} = \dot{m}_2 h_2$$

$$2 \times 305.5 + \dot{W}_c = 2 \times 692.1$$

$$\dot{W}_c = 773.2\ kW$$

c) In order to determine the flow rate, it is required to write the EBE for the combustion chamber:

$$\dot{m}_2 h_2 + \dot{m}_{fuel} LHV_{fuel} = \dot{m}_3 h_3$$

$$2 \times 692.1 + \dot{m}_{fuel} \times 40,320 = 2 \times 1528$$

$$\dot{m}_{fuel} \times 40,320 = 1671.8$$

Thus, the flow rate of the fuel is calculated to be $\dot{m}_{fuel} = 0.042$ kg/s.

d) The backflow rate is the ratio of the compressor work to turbine work. It can be calculated as:

$$r_{bw} = \frac{\dot{W}_{comp}}{\dot{W}_t} = \frac{773.2}{1671.8} = 0.462$$

e) The energy and exergy efficiencies are found from Eqs. 9.3 and 9.4, respectively.

$$\eta_{en} = \frac{\dot{W}_{net}}{\dot{m}_{fuel} LHW_{fuel}} = \frac{\dot{W}_t - \dot{W}_{comp}}{\dot{m}_{fuel} LHW_{fuel}} = \frac{1671.8 - 773.2}{0.042 \times 40,320} = \frac{898.4}{1693.4} = 0.5305 \; or \; 53.05\%$$

and

$$\eta_{en} = \frac{\dot{W}_{net}}{\dot{m}_{fuel} ex_{fuel}} = \frac{\dot{W}_t - \dot{W}_{comp}}{\dot{m}_{fuel} ex_{fuel}} = \frac{1671.8 - 773.2}{0.042 \times 36,520} = \frac{898.4}{1533.8} = 0.5857 \; or \; 58.57\%$$

Illustrative Example 9.7   Reconsider Illustrative Examples 9.6. In the biodigester, 10 kg of biomethane is produced from 600 kg of municipal waste per hour. For 24 hours of operation, calculate how much waste is required.

Solution:

In the previous example, the fuel flow rate is found to be 0.042 kg/s. For the 24 hours of operation, the total amount of fuel is found as:

$$m_{fuel} = 0.042 \times 24 \times 60 \times 60$$

Thus, the total fuel amount is found to be $m_{fuel} = 3628.8$ kg of biomethane is required.

As a 600 kg of municipal waste is required to produce 10 kg of biomethane, the amount of the total municipal waste is found to be:

$$m_{waste} = \frac{3628.8 \times 600}{10} = 217.728 \; kg \; or \; 217.7 \; tons$$

As another example, Figure 9.22 depicts a renewable fueled micro-gas turbine where a biogas generation system converts biomass to biogas. In the system, there are two main useful outputs which are electricity and heat. Also, the system integrated with fuel cell to produce hydrogen to be used as the promoter in the combustion process. Depending on the biomass type and operating conditions, the fuel cost of the biogas can be 50% lower than natural gas and propane [2]. Also depending on the temperature level of the heat recovered in the Brayton cycle, it can be used for steam production for electricity generation.

In this particular system shown in Figure 9.22, gasification is used for biogas production. It is also possible to biodigestion. A wastewater treatment plant can also utilize digester gas derived from anaerobic sludge digestion. Anaerobic sludge digestion is a common component of

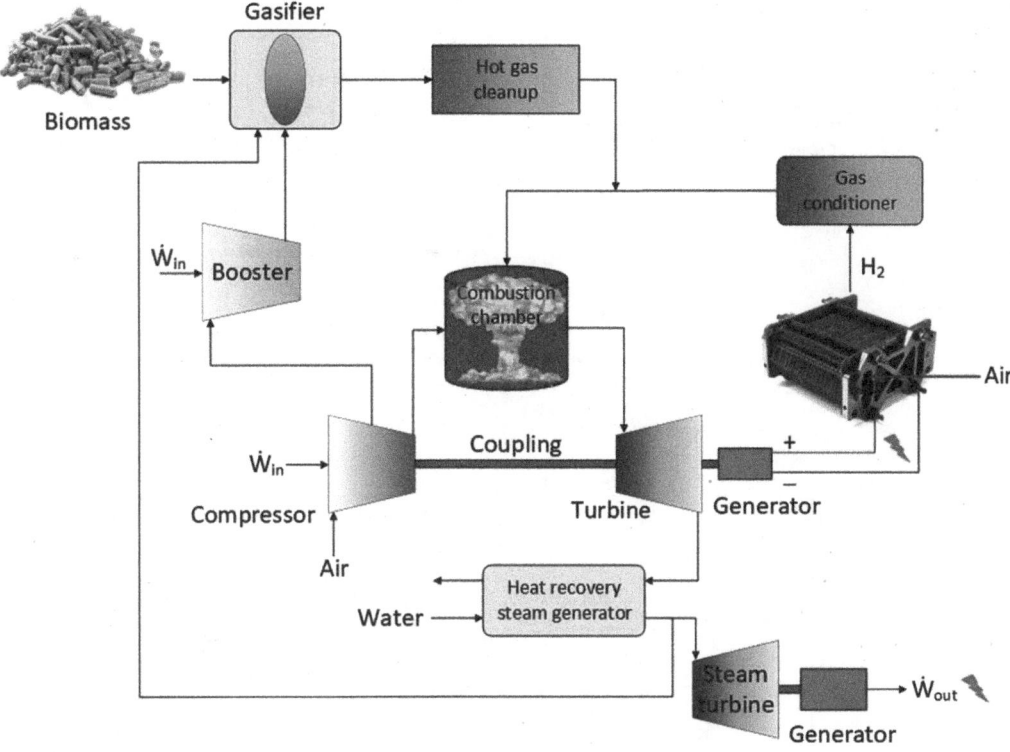

Figure 9.22 Schematic illustration of the gasified biomass powered micro-gas turbine system (Reproduced from [9]).

wastewater treatment plants in the world. By converting the methane-based biogas produced as a by-product of sludge digestion into electricity and thermal energy, microturbines are used instead of disposing of the biogas as waste. Technical and financial factors must be considered, as well as barriers to the adoption of microturbine technology, before a microturbine installation can be considered practical.

9.10 Case Studies

Biomass systems may be combined with any energy systems, especially waste-to-energy systems. In this section, various applications of some selected biomass-sources systems are introduced and discussed. In these systems, many useful outputs will be obtained by using wastes combining with other renewable energy sources.

Case Study 1
Development and analysis of a novel biomass-based integrated system for multigeneration with hydrogen production as summarized from [6]:   Anaerobic digestion of wastewater sludge from wastewater treatment plants (WWTP) for multigeneration is investigated in this study in terms of energy and exergy. Figure 9.23

(Continued)

Case Study 1 (Continued)

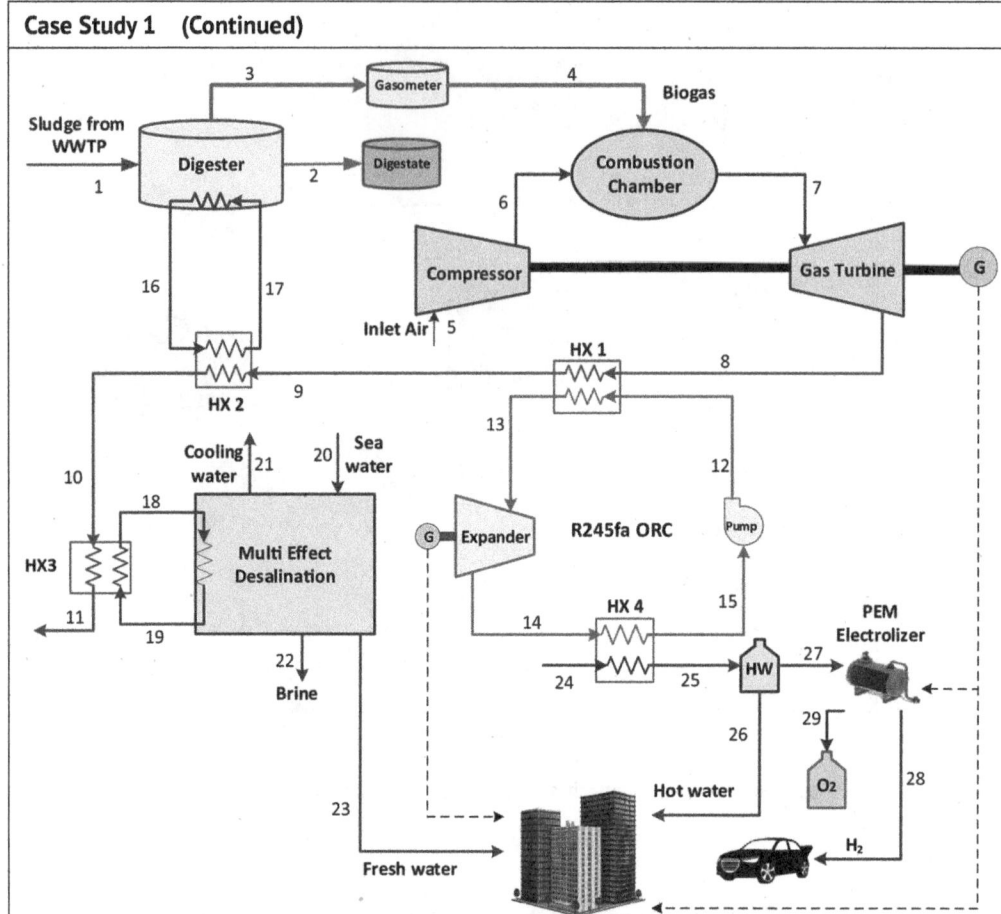

Figure 9.23 Schematic illustration of Case Study 1 (modified from [6]).

shows the process flow diagram for the proposed system. The proposed system aims to utilize anaerobic digestion of sludge to generate power, heat, desalinate, and produce hydrogen.

Biogas produced from digestion is used to operate the multigeneration system. Power, freshwater, heat, and hydrogen are the useful outputs of this system, while some heat can be recovered to increase efficiency. Power is generated using an open-air Brayton cycle and an organic Rankine cycle (ORC) with R-245fa as the working fluid. Through parallel/ cross multi-effect desalination systems, water is also desalinated using the waste heat of power generation units. Additionally, excess electricity can be converted into green hydrogen using proton exchange membrane electrolyzer (PEMs). The ORC's working fluid rejects heat, which is used for heating. Under the base case conditions, the production rates for power, freshwater, hydrogen, and hot water are 1102 kW, 0.94 kg/s, 0.347 kg/h, and 1.82 kg/s, respectively. Moreover, the developed system has 63.6% energy and 40% exergy efficiency, respectively.

Case Study 2

Development of biogas-driven multigenerational system as summarized from [7]:

An analysis of the energy, exergetic, and environmental impact of a biogas-powered multigeneration energy system (see Figure 9.24) is the objective of this study. Using chicken manure and maize silage as biogas, this multigeneration energy system generates electricity, heating, and cooling power. Additionally, the water content of the flue gas is separated for agricultural use in a water separator. The proposed system consists of various subsystems, such as a two-stage biomass digester, Brayton Cycle, single-effect absorption chiller, Organic Rankine Cycle (ORC), greenhouse heating, and water separation of flue gas.

As a biomass input for the biogas production subsystem, chicken manure and maize silage are considered, respectively, supplying 70,000 kg and 30,000 kg daily. Figure 9.24 demonstrates the proposed multigeneration system that consists of different subsystems: a two-stage

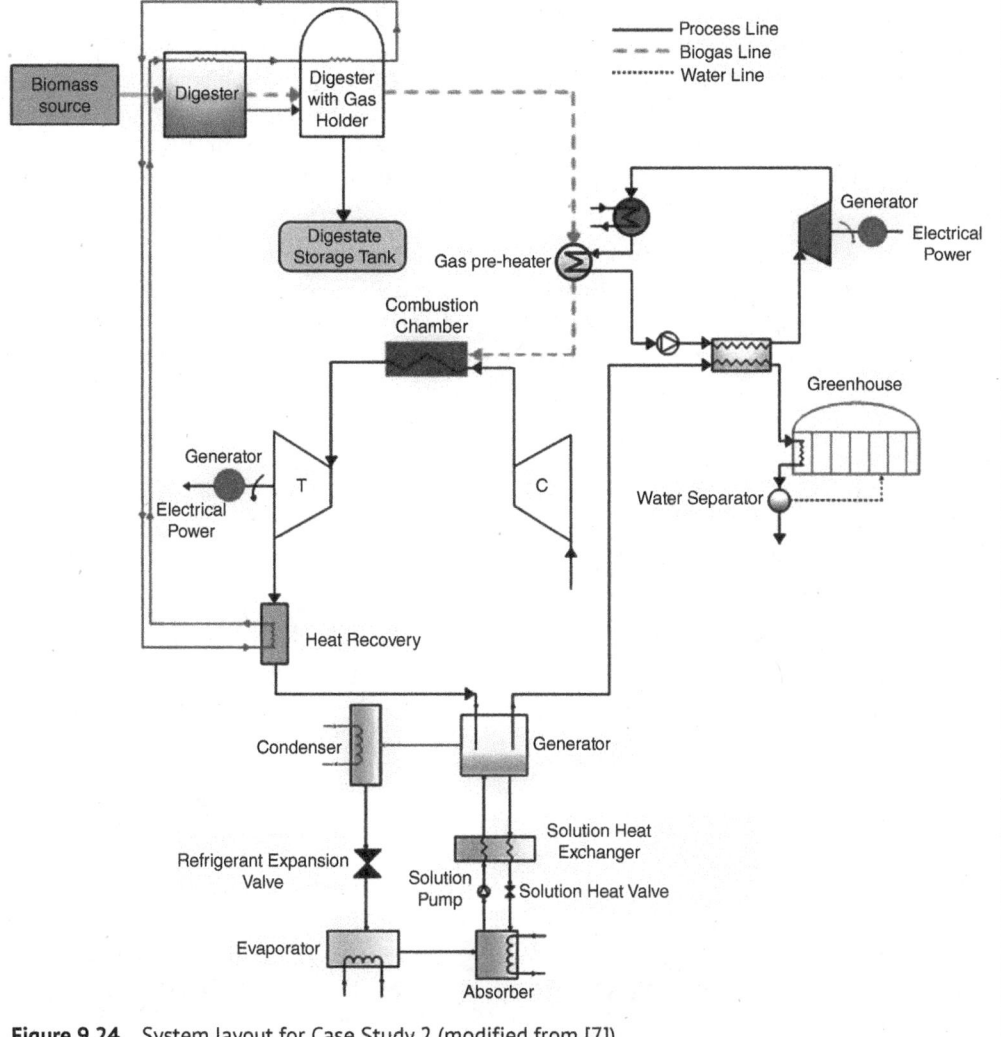

Figure 9.24 System layout for Case Study 2 (modified from [7]).

(Continued)

Case Study 2   (Continued)

biomass digester, open-type Brayton cycle, Organic Rankine Cycle (ORC), single-effect absorption chiller, heat recovery, and water separation unit. As part of this study, energy and energy-related analyses and environmental impact assessments are conducted for each component of the system. Multigeneration systems can generate electrical power for a minimum of 300 homes, heating power for 5 greenhouses, cooling power for greenhouses, and product water for agricultural use. With 1078 kW of electrical, 198 kW of heating, 87.54 kW of cooling power, and around 40 kg of water produced daily, the proposed system has an overall energy efficiency of 72.5%. A multigeneration system is found to have a maximum efficiency of 30.44%, with 65% of its exergy being destroyed in the combustion chamber.

Case Study 3

Development and assessment of a hybrid biomass and wind energy-based system for cleaner production of methanol with electricity, heat, and freshwater as summarized from [8]:

Different from the first two case studies, in the present case study, biomass is integrated with wind energy with energy storage options. Figure 9.25 shows the schematic illustration of the proposed system. There are seven subsystems in the system: CAES, MED, anaerobic digestion, biogas upgrading, Rankine cycle, alkaline electrolyzer, and methanol synthesis. The main objectives of the study are listed as follows:

- To address the mismatch between demand and supply, a wind-driven integrated energy system with a CAES option will be developed, as well as a seawater desalination system.
- The discharging phase of this system can be further enhanced by integrating a biomass energy-based anaerobic digestion system and biogas upgrading to increase biogas's calorific value. Additionally, it consists of a Brayton-Rankine cycle coupled with a CAES subsystem.

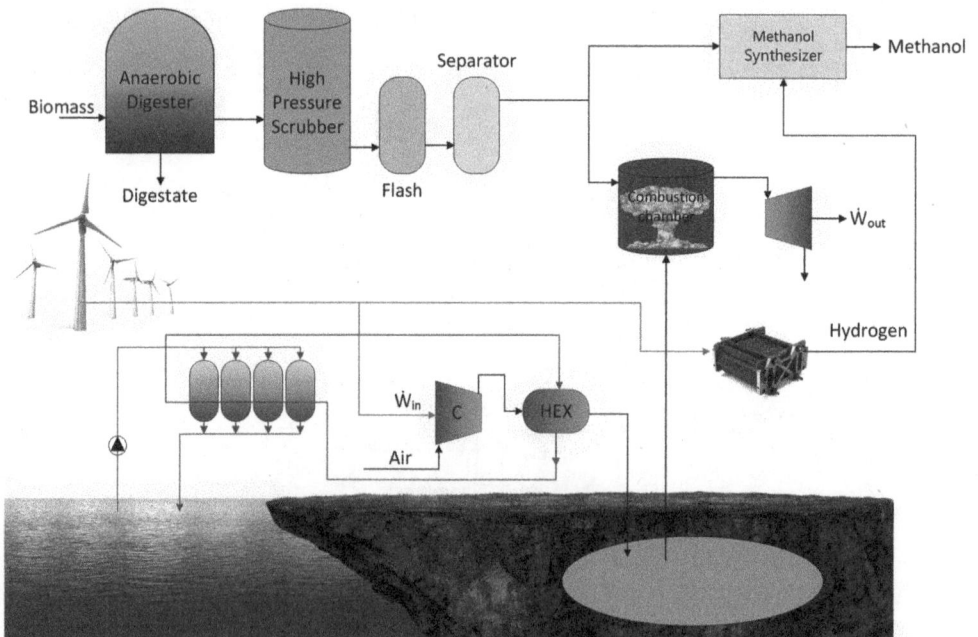

Figure 9.25   System layout for Case Study 3 (Reproduced from [8]).

Case Study 3 (Continued)

Using energy and exergy methods, this case study examines the development of a new biomass and wind energy-based system for the production of green methanol. In addition, a multi-effect desalination unit is proposed to be used in conjunction with compressed air energy storage (CAES) to store excess wind power during charging. Anaerobic digestion produces biogas during peak hours, which is upgraded into biomethane through water scrubbing. Electricity and heat are generated by burning compressed air and biomethane together. Green methanol is also synthesized via alkaline electrolysis. Freshwater and electricity outputs are 2684 kW and 7.9 kg/s, respectively, during charging. A total of 11,979 kW of electricity, 5186 kW of heating, and 0.03 kg/s of methanol were produced during the discharging period. It is determined that the overall energy and energy efficiency of the present system is 40.96% to 46.31%.

Case Study 4

Development and assessment of a new agrivoltaic-biogas energy system for sustainable communities as summarized from [9]:

Several subsystems are integrated synergistically in the proposed system shown in Figure 9.26. A district heating system, a biomethane power generation system, a lithium bromide absorption refrigeration chiller cycle, and an agrivoltaic system make up the integration. An agrivoltaic plant minimizes the land footprint of PV plants. The land can be used for both agricultural and PV plant purposes. For more solar energy to reach the crops, the height of PV modules is crucial. The transparency of bifacial PV modules makes them particularly

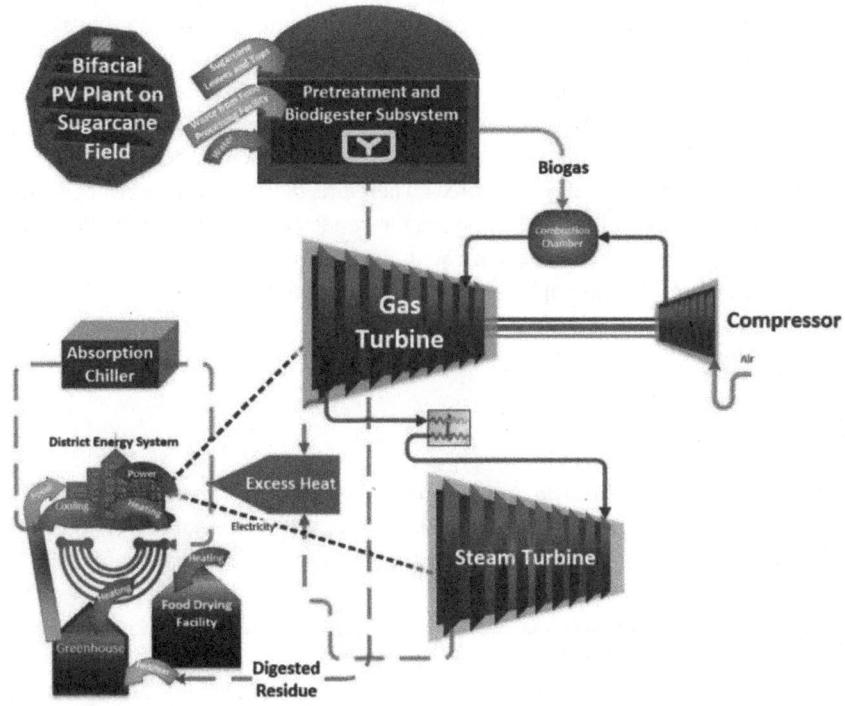

Figure 9.26 System layout of the biomass-based multigenerational system (modified from [9]).

(Continued)

Case Study 4   (Continued)

advantageous. More irradiations can be provided to crops with high transparency. However, in order to minimize shading on crops, the pitch distance between PV module rows and also the overall layout should be designed with unconventional gaps. As for the agrivoltaic system, it uses 8 m of pitch distance and 5 m of height.

This system is a unique energy system integrated with an agrivoltaic plant, a biogas production, and utilization facility, and a community-based energy system is analyzed in order to fully cover the needs of electricity, biogas, heat, and cooling for a community. Both sugarcane and cow manure are considered and used to generate biogas. This particular system is designed for a community where more than 34,000 people living in Nakhon Ratchasima, Thailand. The proposed system which has an 8 MWp of PV plant with 415 MWp of a biogas power plant as well as 105 MWp of high- and low-pressure steam turbines can meet 100% of 92 GWh non-thermal electricity, 63 GWh cooling, and 13 GWh heating loads in a year. The overall energy and exergy efficiencies of the overall system are found to be 33.64% and 22.96% under average ambient conditions and average load requirements.

The applications of biomass, waste-to-energy and their integration with other renewables, are not limited with these examples. It is possible to integrate biomass and waste-to-energy systems with any renewable energy sources to boost up the useful output. Any waste can be considered in the waste-to-energy systems. More examples will be discussed in Chapter 11.

9.11 Closing Remarks

This chapter is technically about biomass energy, biofuels and waste to energy systems and applications. These biomass and waste to energy systems present a significant contribution to waste management along with additional useful outputs, such as electricity, heat, treated water, hydrogen, and many more. Also, the chapter provides the basics of the biomass and waste to energy systems for readers. Biofuels are important for sustainable development as a large volume of biomasses are produced every day, and it is required to manage those wastes in a sustainable manner. Biomass sources are considered renewable and reliable energy resources. Therefore, biomass based power generating systems and micro gas turbines appear to be a key option for standby power – power quality and reliability, peak shaving, and cogeneration applications. With the developed technologies in the last few decades, lower costs and environmental impacts make them more attractive. Finally, integrated systems appear to be more efficient and more environmentally benign and hence more sustainable.

References

1 Biofuel energy production. *Our World in Data*. Available online https://ourworldindata.org/grapher/biofuel-production?country=MEX~DEU~BRA~USA~GBR~OWID_WRL (accessed 22 December 2022).

2 Farret, F.A. and Simoes, M.G. (2006). *Integration of Alternative Sources of Energy*. John Wiley & Sons, Ltd.

3 Rahmouni, C., Tazergut, M., and Le Corre, O. (2002). A method to determine biogas composition for combustion control. *Journal of Fuels and Lubricants* 111, Section 4: 700–709.

4 Waste-to-energy: how it works? Deltaway Waste and Biomass Power Plant Design and Operation. Available online https://deltawayenergy.com/2018/08/waste-to-energy-how-it-works (accessed 03 December 2023).

5 Dincer, I. (2020). *Thermodynamics: A Smart Approach*, 1e. John Wiley & Sons, Ltd.

6 Safari, F. and Dincer, I. (2019). Development and analysis of a novel biomass-based integrated system for multigeneration with hydrogen production. *International Journal of Hydrogen Energy* 44 (7): 3511–3526. doi: 10.1016/j.ijhydene.2018.12.101.

7 Sevinchan, E., Dincer, I., and Lang, H. (2019). Energy and exergy analyses of a biogas driven multigenerational system. *Energy* 166: 715–723. doi: 10.1016/j.energy.2018.10.085.

8 Oner, O. and Dincer, I. (2022). Development and assessment of a hybrid biomass and wind energy-based system for cleaner production of methanol with electricity, heat and freshwater. *Journal of Cleaner Production* 367: 132967. doi: 10.1016/j.jclepro.2022.132967.

9 Temiz, M., Sinbuathong, N., and Dincer, I. (2022). Development and assessment of a new agrivoltaic-biogas energy system for sustainable communities. *International Journal of Energy Research* 46 (13): 18663–18675. doi: 10.1002/er.8483.

10 Reed, T. and Desrosiers, R. (1979). The equivalence ratio: the key to understanding pyrolysis, combustion and gasification of fuels. SERI/TR33239.

Questions/Problems

Questions

1 Describe biomass in general terms and give examples.

2 Explain when the first biomass was as used by humans for long time.

3 Describe biofuel and give examples from daily life.

4 Discuss which countries lead in biomass sources and power generation from those sources.

5 Classify biomass sources and give an example for each category.

6 Describe what combustion, gasification, and pyrolysis are and give examples for each.

7 Define air-fuel ratio and equivalence ratio and give examples in specific to combustion, gasification, and pyrolysis.

8 What is biodigestion process and how it can be achieved?

9 Explain how to design an effective biodigester.

10 Explain why micro gas turbines are preferred and what capacities.

11 Why waste to energy field is important? Explain pros and cons.

12 Explain what types of biofuels are commonly used.

13 Explain why waste management is important.

14 Describe waste to energy systems and their role in managing wastes.

15 Explain if possible to use micro gas turbines beyond cogeneration.

Problems

1 One ton of municipal solid waste is burned directly in a boiler to generate saturated steam at 250°C, as illustrated in the figure below. Water enters the boiler at 40°C. The heating value of the MSW is 22 MJ/kg and the chemical exergy is 19 MJ/kg. The boiler efficiency is 80%. The steam produced in the boiler is run through a turbine-generator group to generate electricity. The steam exits from the turbine at 110 kPa of pressure. The isentropic efficiency of the turbine is 0.85. The generator efficiency is 90%.

   a Write the mass, energy, entropy and exergy balance equations for the system components.

   b Find the steam generation rate in the boiler.

   c Calculate the amount of electricity generated.

   d Determine the energy and exergy efficiencies of the system.

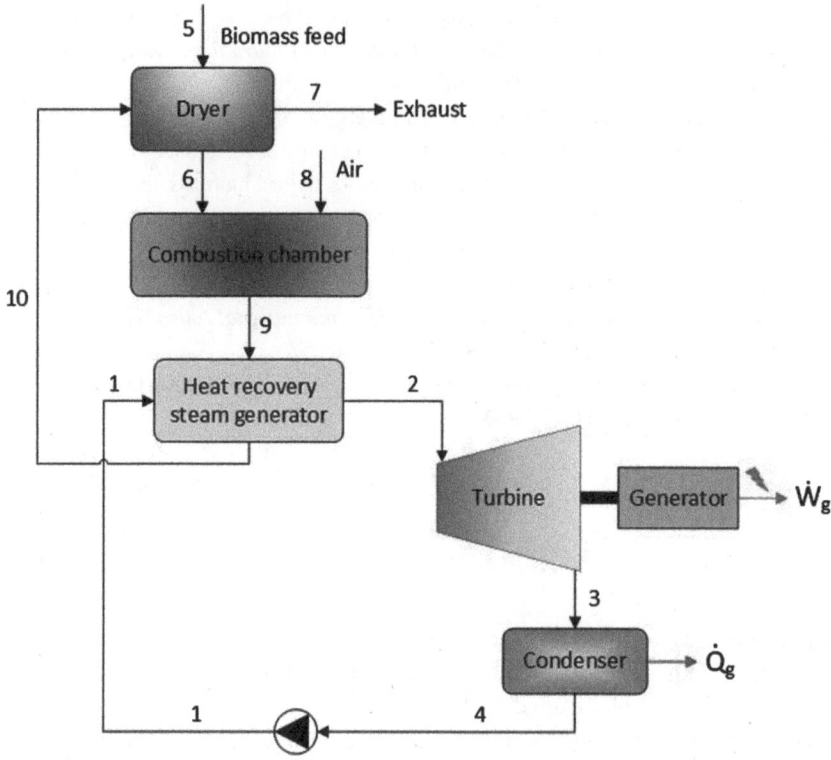

2 A total of 20 tons of yard is burned directly in a boiler to generate saturated steam at 225°C. Water enters the boiler at 40°C. The heating value of the MSW is 22 MJ/kg and the chemical exergy is 19 MJ/kg. The boiler efficiency is 80%. The steam produced in the boiler is run through a turbine-generator group to generate electricity. The steam exits from the turbine at 90 kPa of pressure. The isentropic efficiency of the turbine is 0.90. The generator efficiency is 85%.

  a Write the mass, energy, entropy and exergy balance equations for the system components.
  b Find the steam generation rate in the boiler.
  c Calculate the amount of electricity generated.
  d Determine the energy and exergy efficiencies of the system.

3 Reconsider the system given in Problem 1. Calculate CO_2 emissions to be emitted from the system for the same amount of electricity generation found in the previous example for different fuels. Use table given below for the fuel properties.

Details of the fuels used in Problem 3

Fuel	LHV (kWh/kg)	CO_2 Emissions (kg CO_2/kg fuel)	Fuel to Electricity Efficiency
MSW	5	1.1	0.10
Wood	4.28	1.65	0.05
Coal	7.9	2.69	0.3
Hydrogen	33.33	0	0.56
Natural gas	13.1	1.91	0.34

4 Reconsider Problem 2. Calculate CO_2 emissions per ton of fuel and compare the results. Use table data given in Problem 3 for the fuel properties.

5 Biomethane produced from sugar cane crops collected from a farm area is used as fuel in a micro gas turbine system, as shown below to meet the electricity demand of the farm. Air enters the compressor at the reference conditions which are $T_0 = T_1 = 20°C$ and $P_0 = P_1 = 100$ kPa with mass flow rate 3 kg/s. In the combustion chamber, the compressor compression ratio is 1:10. Air leaves from the compressor at 410°C. Air enters the turbine at 1000°C and exits at 485°C. The combustion temperature is $T_c = 1500°C$. The LHV of biomethane is 36,320 kJ/kg and its chemical exergy is 32,520 kJ/kg.
  a Calculate the turbine output.
  b Determine the compressor work.
  c Determine the flow rate of the fuel.
  d Find the back work ratio.
  e Calculate the energy and exergy efficiencies.

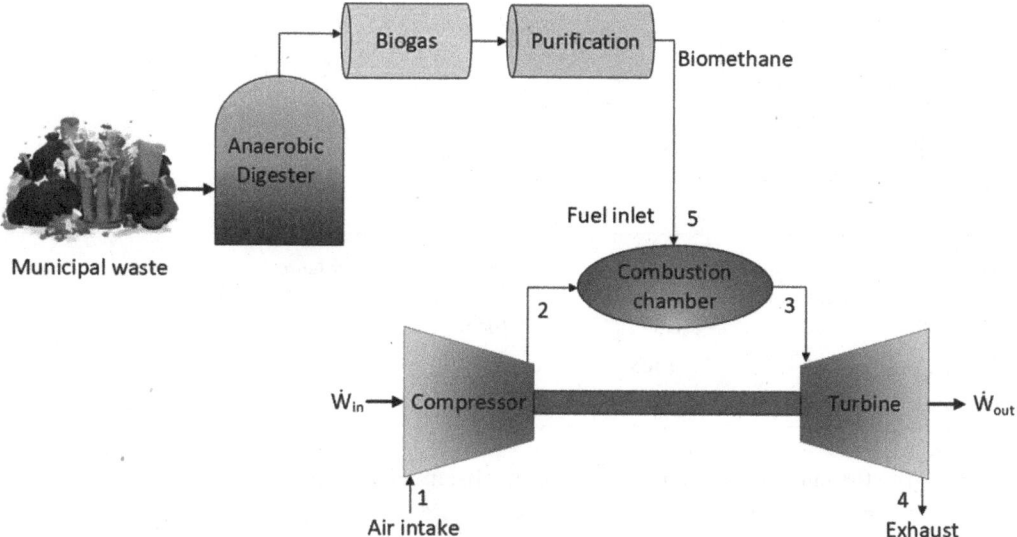

6 Reconsider Problem 5. The CO_2 emission of the biomethane used in the micro turbine is 0.85 kg CO_2/kg fuel. Considering the fuels given in Problem 3, recalculate CO_2 emissions. Compare the CO_2 emissions for the fuels and evaluate accordingly.

7 Reconsider Problem 5. In anaerobic digestion, 5 kg of biomethane is produced from 400 kg of sugar cane crops per hour. For 24 hours of operation, calculate how much waste is required.

8 Biomethane produced from corn crops collected from a farm area to produce biomethane. The produced biomethane is used in the micro turbine for power and heat generation. For the micro gas turbine, the following information is known: Air enters the compressor at the reference conditions which are $T_0 = T_1 = 20°C$ and $P_0 = P_1 = 100$ kPa with mass flow rate 5 kg/s. In the combustion chamber, the compressor compression ratio is 1:10. Air leaves from the compressor at 500°C. Air enters the turbine at 1125°C and exits at 510°C. The combustion temperature is $T_c = 1650°C$. The LHV of biomethane is 41,020 kJ/kg and its chemical exergy is 39,250 kJ/kg. In order to recover the heat from exhaust gases, a heat exchanger is used. In this heat exchanger aims to produce hot water. The water enters the heat exchanger at 10°C and leaves at 85°C.

a Calculate the turbine output.
b Determine the compressor work.
c Determine the flow rate of the fuel.
d Find the back work ratio.
e Calculate the flow rate of the water to be heated up.
f Calculate the energy and exergy efficiencies.

9 A biodigester used for producing biogas is shown in the figure below. Sludge from wastewater is entered into the digester. The thermodynamic properties of these substances are measured as listed in the table below.

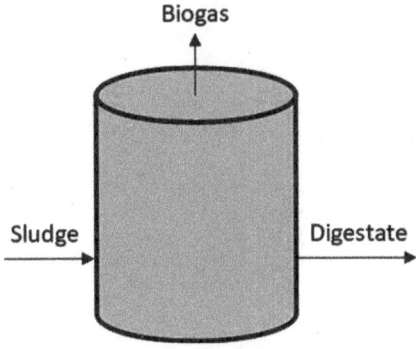

Substance	h (kJ/kg)	s (kJ/kg K)	ex (kJ/kg)	$\dot{m}$ (kg/s)	P (kPa)	T (°C)
Sludge	1308.3	0.37	104.8	12.06	100	25
Digestate	941.34	0.505	146.7	11.85	100	25
Biogas	18,570.7	7.282	357.6	?	100	25

a Write the mass, energy, entropy and exergy balance equations.
b Find to mass flow rate of the biogas to be produced.
c Determine the heat required for biodigestion.
d Calculate the energy and exergy efficiencies.

10 A biodigester used for producing biogas is shown in the figure below. Sludge from wastewater is entered into the digester. The thermodynamic properties of these substances are measured as listed in the table below. The flow rate of the sludge is 50% higher than the digestate.

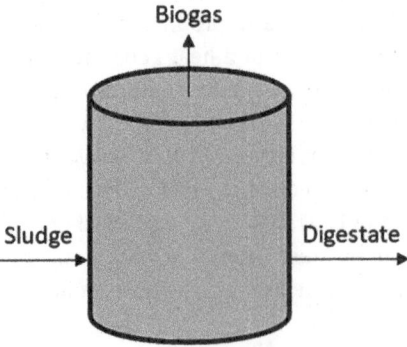

Substance	h (kJ/kg)	s (kJ/kg K)	ex (kJ/kg)	$\dot{m}$ (kg/s)	P (kPa)	T (°C)
Sludge	1308.3	0.37	104.8	?	100	25
Digestate	941.34	0.505	146.7	?	100	25
Biogas	18,570.7	7.282	357.6	15	100	25

a Write the mass, energy, entropy and exergy balance equations.
b Find mass flow rate of the sludge and digestate to be required for biodigester.
c Determine the heat required for biodigestion.
d Calculate the overall energy and exergy efficiencies.

11 A biomass gasifier is shown in the figure below. Corn crops biomass is entered into the gasifier with separate streams of oxygen and water. The thermodynamic properties of these substances are measured as listed in table below.

Substance	h (kJ/kg)	s (kJ/kg K)	ex (kJ/kg)	$\dot{m}$ (kg/s)	P (kPa)	T (°C)
Biomass	1510	3.15	1049.6	2.0	101.325	850
Oxygen	451.3	0.115	428.9	4.76	2431.8	483.8
Water	419.2	1.31	34.13	4.89	111.0	100
Syngas	4950	2.0	3071.6	?	506.6	2056

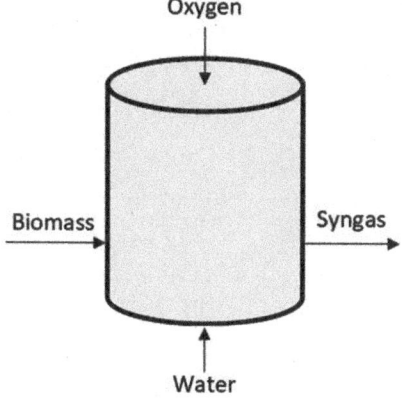

a Write the mass, energy, entropy, and exergy balance equations for the gasifier.
b Find the mass flow rate of output syngas.
c Determine the rate of heat input required.
d Find the exergy destruction rate if the source temperature is 3000°C.
e Find the energy and exergy efficiencies of the gasifier if the lower heating value of syngas is 2426 kJ/kg and the lower heating value of biomass is 14,690 kJ/kg.

12 Reconsider Case study 1 given in the chapter and make a similar study for your community and compare the results.

13 Reconsider Case study 2 given in the chapter and make a similar study for your community and compare the results.

14 Reconsider Case study 3 given in the chapter and make a similar study for your community and compare the results.

15 Reconsider Case study 4 given in the chapter and make a similar study for your community and compare the results.

10

Hydro and Ocean Energies

10.1 Introduction

In the previous chapters, nuclear energy and main renewable energy sources, including solar, wind, geothermal, biofuels and biomass, are presented in detail with illustrative examples. In this chapter, renewable energy sources related to water (hydro) energy, including hydro and ocean energy systems, are presented. The classification of these energy sources is shown in Figure 10.1. Hydro energy systems are generally classified according to their turbine types (Pelton, Francis, Michel-Banki, Kaplan, Deriaz, etc.) or power generation capacities of the systems. Ocean energies are classified as tides, tidal waves, currents, thermal gradient and salinity gradients.

Hydro energy is a form of renewable energy that harnesses the power of falling water to generate electricity. It is typically generated at hydroelectric power plants, which use the force of water falling through a dam to turn a turbine, which in turn generates electricity. Hydro energy is considered a clean and sustainable source of power, as it does not produce greenhouse gas emissions or other pollutants. It is also considered a reliable source of energy, as water flow can be controlled to meet changing electricity demands. Despite having all these advantages, large hydro plants (dams) cause some ecological concerns, due to local drastic chance in ecosystem and climate change, displacing people and communities, etc.

Ocean energy systems are technologies that generate electricity from the motion of ocean waves, tides, currents, and temperature gradients in the ocean. These systems fall into several categories: tidal, wave and current, and ocean thermal energy systems. Wave energy systems convert the kinetic energy of ocean waves into electricity. They can take many forms, including floating devices that move up and down with the waves, and submerged devices that are anchored to the seafloor. Ocean current energy systems harness the kinetic energy of ocean currents to generate electricity. They typically take the form of underwater turbines that are placed in areas with strong currents. Tidal energy systems harness the energy of the tide, which is the rise and fall of sea levels caused by the gravitational pull of the moon and sun. Tidal energy systems can take many forms, including barrages, which are dams that span a tidal estuary, and tidal turbines, which appear to be similar to wind turbines but are placed in water. Ocean thermal energy systems: These systems generate electricity by exploiting the temperature difference between the surface and deep waters of the ocean. Such systems use this temperature difference to power a turbine, which effectively generates electricity.

Hydro energy systems are well-established and mature renewable energy systems, despite the fact that large hydro is questioned in terms of damage to the ecosystem, environment and local communities. Today, they meet almost 20% of the global electricity generation. Ocean energy systems are considered renewable and sustainable sources of energy, but still in early stage of development and not yet widely adopted, due to the high cost of installation and maintenance, as well

Introduction to Energy Systems, First Edition. Ibrahim Dincer and Dogan Erdemir.
© 2023 John Wiley & Sons, Inc. Published 2023 by John Wiley & Sons, Inc.
Companion Website: www.wiley.com/go/Dincer/Introduction_to_Energy_Systems

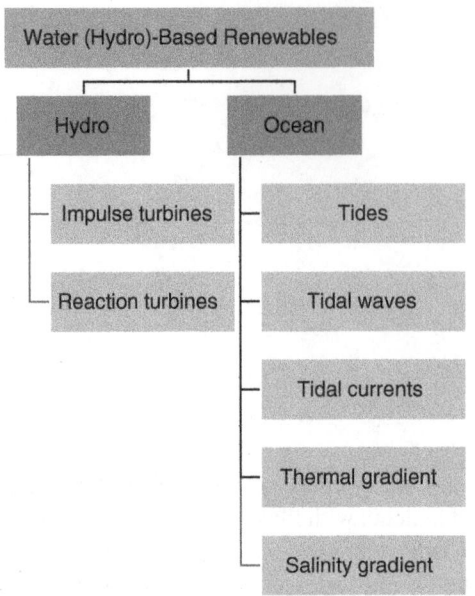

Figure 10.1 Classification of water (hydro)-based renewables.

as the technical challenges associated with building structures in the ocean.

Other renewable energy sources, which are hydro and ocean energies, are significant for a sustainable future. Hydro power systems have been used for many years for power generation. Today, they meet almost 20% of the global electricity generation. On the other hand, their capacities tend to decrease due to decreasing water resources all over the world. However, they will be critical for power generation to meet the electricity demands all over the world. Although there are many attempts to use the oceans as a clean energy source, they require high investment cost. Although their efficiencies are relatively lower than other renewable energy sources, their higher potential, reliability and predictability make them a significant candidate for clean energy sources.

In this chapter, the basics of hydro and ocean energy systems are introduced to the readers. It starts with the introductory information on hydro energy systems and their historical development and proceeds with the classification of hydropower and the type of turbines. There are numerous examples presented, and thermodynamic analysis methods are introduced and included performance assessment through energy and exergy efficiencies. Furthermore, ocean energy systems are discussed with their different applications. Moreover, ocean (marine) energy sources and systems are discussed, and the ocean energy systems are classified based on tidal, wave, ocean thermal and salinity gradient and covers their discussion with the examples and applications.

10.2 Hydro Energy

Hydro energy, or hydropower, or hydroelectric power, is one of the oldest, most mature, and largest sources of renewable energy, which is a result of moving (falling) water as a clean, renewable, and reliable source for electricity generation. The initial use of hydro energy has started with the utilization of the natural flow of water to grind grain and distribute irrigation water. It is a worldwide proven and potential method for electric power production and has gained significance. Small and micro hydropower systems now receive increasing attention due to environmental concerns.

Historical development of the hydro energy systems is illustrated in Figure 10.2. The first use of hydro energy was for grinding grain. Then, it was followed for irrigation purposes. The engineered hydropower use goes back to the era where there were studies performed by Al-Jazari, who was a polymath, mathematician, inventor and engineer, designed many mechanisms, robots, and many different systems in the late 1100s. The development of the first hydro turbine was completed by French engineer Bernard Forest de Bélidor in 1827. Its capacity was 4.5 kW. Next, James Francis invented the first modern water turbine for power generation in 1849, which is the most widely used water turbine in the world so far. In the years of 1870, Lester Allan Pelton invented the Pelton wheel. In 1878, the world's first hydroelectric project was achieved to power up a single lamp in the UK. The use of hydro energy for electricity generation was first achieved on September 30, 1882.

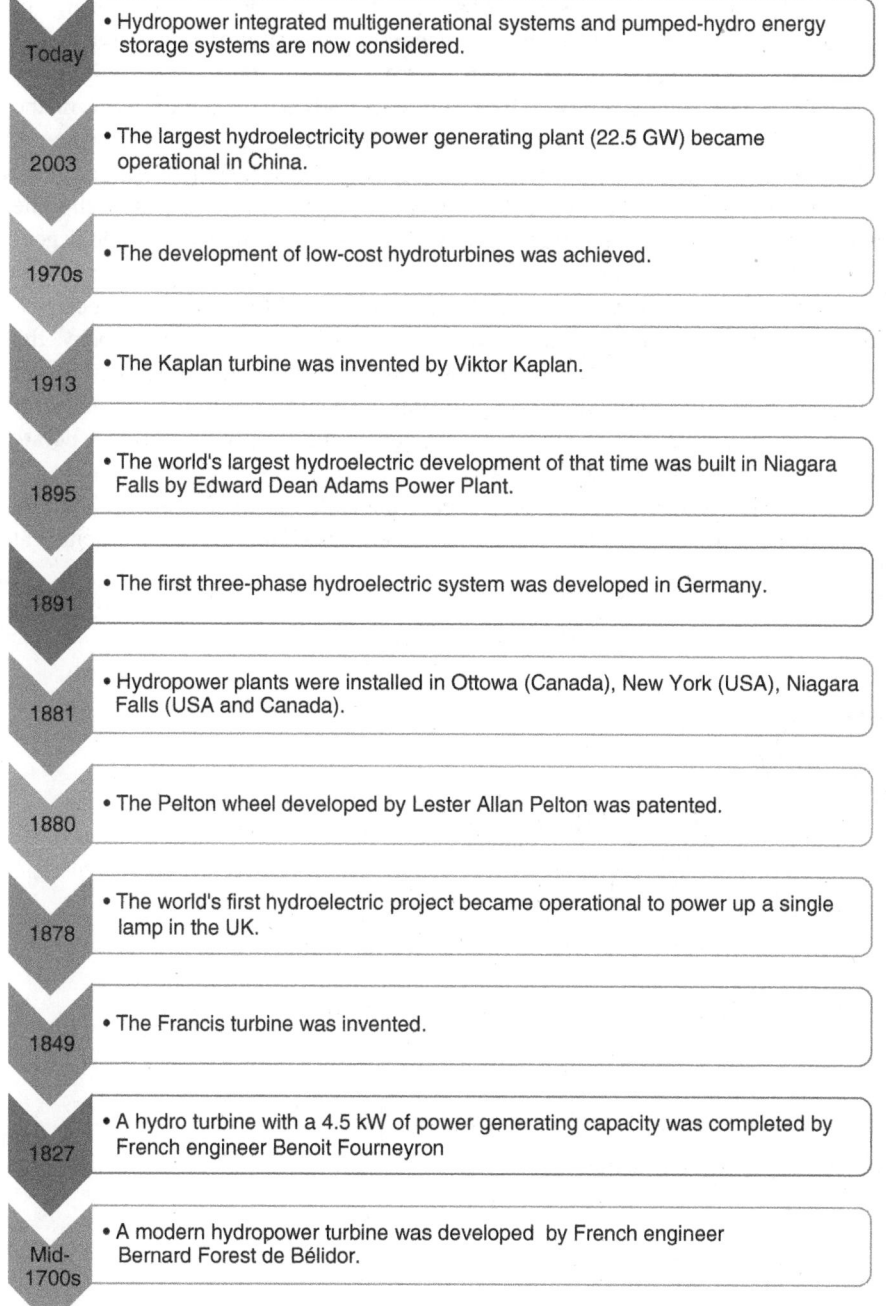

Figure 10.2 Historical development of hydropower.

The world's first hydroelectric power plant began operation on the Fox River in Appleton, Wisconsin. The first three-phase hydroelectric system was developed in Germany in 1891. The Kaplan turbine was invented by Viktor Kaplan in 1913. Today, hydropower plants are one of the significant clean power generation techniques. The world's largest capacity hydropower which has 22.5 GW of the capacity became operational in 2003 in China.

The working principle of hydro energy systems is based on the conversion of kinetic energy of the water to the shaft power in the hydro turbine to generate electricity in a generator that is attached to it. As illustrated in Figure 10.3, the main source of the hydropower plants includes dams, currents, tidal barrages, and wave power. The sources already have kinetic energy, or the potential energy that is ready to convert into kinetic energy. The kinetic energy of the water is converted into the shaft power in the hydro turbine. The turbine is connected to the generator for electricity production. The flow rate of the water is controlled automatically to maintain the safe operating conditions for turbine, generator, transmission used in the systems. The electricity generated can directly be connected to the grid.

Despite the capacity of hydropower systems vary, they have capability to achieve higher capacities. This feature is one of the most significant advantages of hydropower systems. Today, the world's highest capacity hydropower plant is 22,500 MW. There are many more which have lower capacities ranging from a few 1 kW to 1 MW for many purposes. Another advantage of hydropower generation is its fast response time. When a hydropower system is operated, it can start to generate electricity at its full capacity in a few minutes.

The capacity factor of the hydropower is relatively higher than other renewable energy sources. Figure 10.4 illustrates the capacity factors for the energy systems used in the USA. It is clear that

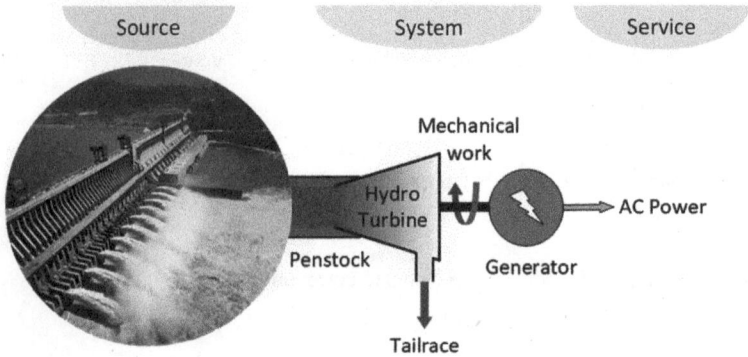

Figure 10.3 Schematic illustration of the working principle of hydropower.

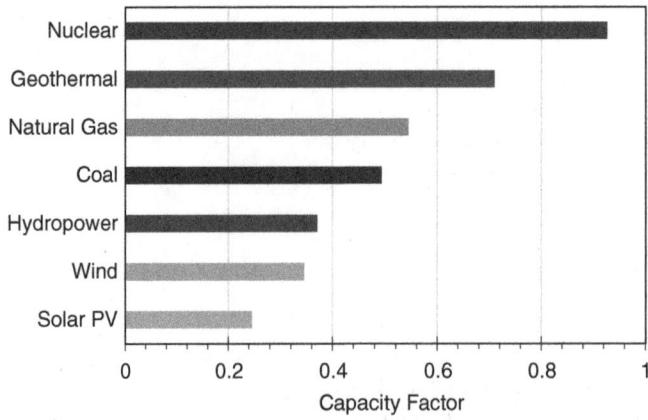

Figure 10.4 Capacity factor of electricity generation methods (data from [1]).

nuclear energy has the highest capacity factor for power generation due to its stability. It is followed by geothermal energy. Although hydro power presents a higher capacity factor, due to environmental issues in water resources their capacities tend to decrease.

Today, hydro power meets 15.5% of global power generation, as illustrated in Figure 10.5. In 1985, the share of the hydro power in the global energy generation was 20%; it has slightly dropped by 15.5%, due to the reducing capacity of the water resources and increasing the contribution of other renewable and other energy sources in the power generation.

Figure 10.6 shows a useful chart for the energy production methods used in Ontario, Canada on 19 January 2023 between 7:00 am and 8:00 am, respectively. These data have been retrieved

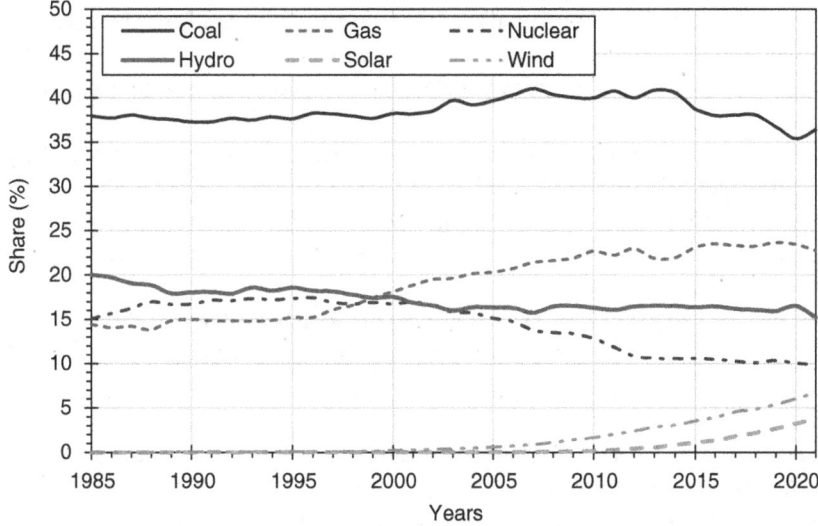

Figure 10.5 The share of energy sources in electricity generation (data from [2]).

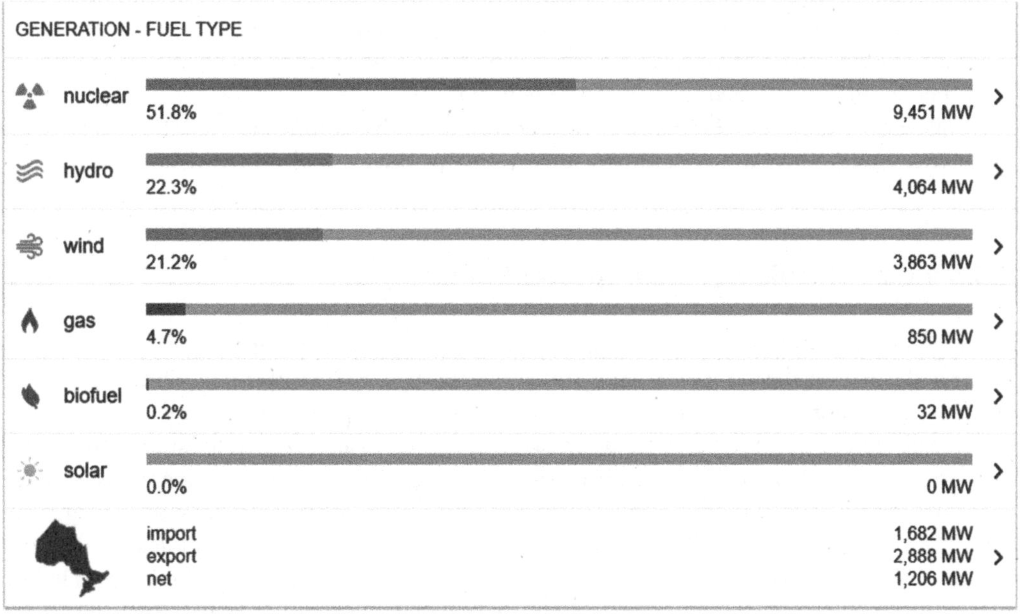

GENERATION - FUEL TYPE

nuclear	51.8%	9,451 MW
hydro	22.3%	4,064 MW
wind	21.2%	3,863 MW
gas	4.7%	850 MW
biofuel	0.2%	32 MW
solar	0.0%	0 MW
	import	1,682 MW
	export	2,888 MW
	net	1,206 MW

Figure 10.6 The shares of energy sources utilized in electricity generation in Ontario, Canada.

from the online platform, named the Grid Watch Ontario. The details of the power generation with its sources and power demand are shared. For the time data captured, the total power generation is 18,260 MW and the total demand is 17,054 MW. As seen from Figure 10.6, a total of 22.3% of the total electricity production is met by hydro energy systems. Also, only 4.7% of the total energy production is performed by fossil fuel-based sources. Emission value was calculated to be 365 tons and CO_2-eq intensity is indicated to be 20 g/kWh.

As a matter of fact, every energy choice has some advantages and disadvantages, just like the way every drug has some benefits and some unavoidable side effects. Hydro energy may be treated in a similar way. The following section provides some advantages and disadvantages. Here are some advantages of hydropower plants, including:

- Renewable and sustainable energy source: Hydroelectric power is generated from the energy of falling water, which is a renewable and sustainable energy source.
- Low operating costs: Once a hydroelectric power plant is built, the cost of generating electricity is relatively low.
- Long lifespan: Hydroelectric power plants have a lifespan of several decades, which means they can provide power for a long time.
- Low emissions: Hydroelectric power plants do not produce greenhouse gas emissions, which makes them an environmentally-friendly option. If life cycle emissions are considered, there will be very low emissions.
- Flexibility: Hydroelectric power plants can effectively be used for peak load management, meaning they can quickly adjust their output to meet changes in demand for electricity.

In addition, some disadvantages of hydropower plants are listed as follows:

- High initial costs: Building a hydroelectric power plant can be expensive, as it requires the construction of dams and other large infrastructure.
- Environmental impacts: Building a hydroelectric power plant can have negative impacts on the environment and hence change in local climate, in addition to the displacement of people and changes to natural habitats.
- Limited locations: Hydroelectric power can only be generated in locations where there is a sufficient supply of falling water, which limits the number of potential sites.
- Vulnerability to drought: Hydroelectric power plants rely on a steady supply of water, and during periods of drought, their output can be reduced.
- Reliance on water flow: Hydroelectric power plants require significant water flow to generate electricity, and during low water flow, they may not produce any electricity at all.

10.3 Classification of Hydropower Plants

Hydropower plants are facilities that generate electricity by harnessing the potential energy of water and converting into mechanical work output through hydro turbines. In order to increase the potential energy of the water, a barrier, dam, or set is set up in front of the water. This technique helps to increase the flow rate of the water. It also provides more reliability for operation with respect to the changing flow rate throughout the year. The water's kinetic energy is converted into the shaft power in a hydro turbine. That shaft power drives a generator for electricity generation. There are a few aspects to the classification of hydropower plants. As mentioned earlier, they can be classified based on the type of turbine used to generate electricity. The main types of hydropower turbines are impulse and reaction turbines. For example, impulse turbines use the kinetic energy of the water to move the blades. Pelton wheels are a common type of impulse turbines. Reaction turbines use the pressure of the water to move the blades. Francis and Kaplan turbines are examples of reaction turbines.

It can be classified into several different types, including run-of-river, impoundment, pumped-hydro storage, and micro-hydro power.

- Run-of-river: This type of hydropower uses the natural flow of a river to generate electricity, without the need for a large dam or reservoir.
- Impoundment: This type of hydropower uses a dam to create a reservoir of water, which is then released through turbines to generate electricity.
- Pumped storage: This type of hydropower uses excess electricity to pump water into a reservoir at a higher elevation, and then releases the water through turbines to generate electricity during periods of high demand.
- Micro hydropower: This type of hydropower uses small-scale hydroelectric systems, typically with a capacity of up to 100 kilowatts, to generate electricity for individual homes or small communities.

The capacity of the hydropower plants varies depending on the flow rate of the water, head height, and the community around the power plants. Therefore, it is also quite common to classify hydro power plants according to their capacities as follows:

- Large hydro (over 100 MW): usually feeding into a large electricity grid.
- Medium hydro (over 10 MW to 100 MW): usually feeding a grid.
- Small hydro (up to 10 MW): usually feeding a grid.
- Mini hydro (over 100 kW to 1 MW): either standalone schemes or more often feeding into the grid.
- Micro hydro (up to 100 kW): usually powering a small community or rural industry in remote areas away from the grid.

The world's highest capacity hydropower plant is China's Three Gorges Dam, which became operational in 2003. Its installed capacity is 22,500 MW, which is almost double the second highest one, Itaipu Dam. Hydropower plants which have an installed capacity higher than 100 MW are called large hydropower stations. Due to their higher capacity factors, they are a significant energy source to meet electricity demands. However, these systems require a large volume of water sources along with initial investment costs. So, it is directly dependent on the geographical and water resources conditions.

Note that 10 MW to 100 MW of installed capacities for hydropower plants are called medium hydropower plants. They are the most common hydropower station all over the world. They play a critical role in meeting the energy demand for countries. Less than 10 MW hydropower systems are named as mini hydropower stations. They are generally used for meeting electricity demands in the nearby communities where the power plant is found. Micro hydropower plants, which are less than 100 kW, are used for agricultural irrigation purposes.

Finally, hydropower plants are also classified into categories, depending on the head lengths, as follows:

- Low head power plants which generally range from 5 to 20 m.
- High head power plants which generally vary from 20 to 1000 m.

10.3.1 Hydro Turbines

A hydro turbine is a machine that converts the energy of flowing water into mechanical energy in terms of mechanical work, which is then be used to generate electricity. There are several types of

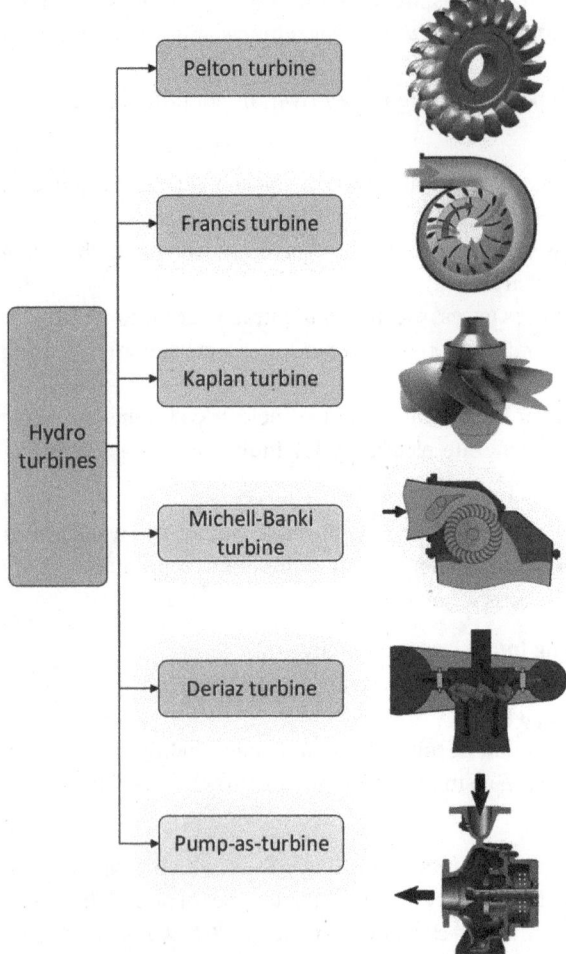

Figure 10.7 Common types of hydro turbines. *Source:* birhat / cleanpng.com

hydro turbines, each of which is designed to work best under specific conditions of water flow and head. There are two main types of hydro turbines: reaction and impulse. Hydropower turbines are selected according to the height of standing water, known as "head," and the water flows at the site over time. Turbine efficiency, cost and the depth at which the turbine must be placed also play a role in the decision. The most common types of hydro turbines are given in Figure 10.7. These hydro turbines are deployed for converting the fluid power into the mechanical work output. They essentially convert the kinetic energy of the water into shaft power. As indicated in Figure 10.7, there are six different turbines which are Pelton, Francis, Kaplan, Michell-Banki, Deriaz, and pump-as-turbine. The most common types are Pelton, Francis and Kaplan hydro turbines.

Pelton turbine is an impulse turbine that uses the kinetic energy of a high-velocity water jet to turn a wheel with a series of cups or buckets attached to its rim. It is typically used in high head, low flow applications. Francis turbine is a reaction turbine that uses the pressure and kinetic energy of water to turn a wheel with a series of radial vanes. It is typically used in medium head, medium flow applications. Kaplan turbine is a reaction turbine that has adjustable blades that can be optimized for a given water flow and head. It is typically used in low head, high flow applications. On the other hand, propeller turbine is similar to Kaplan turbine, but the blades are fixed and not adjustable. Deriaz turbine is a variation of the Pelton turbine which is an impulse turbine.

a) Pelton Turbines

A Pelton turbine is a type of water turbine that is commonly used in hydroelectric power plants. Figure 10.8 shows a general view of Pelton turbine. It is known as a reaction turbine, which means that it converts the kinetic energy of falling water into mechanical energy, which is then converted into electricity. Pelton turbine has a unique design that features one or more cups or buckets that are mounted on the periphery of a wheel. Water is directed onto the cups or buckets through a nozzle, which causes the wheel to rotate. The turbine's speed is controlled by adjusting the flow of water through the nozzle, which can be done using a governor or a speed control system.

Pelton turbines are typically used in high head hydroelectric power plants, which are those that have a large difference in elevation between the water source and the turbine. They are also commonly used in micro hydro power plants, which are small-scale hydroelectric power plants that

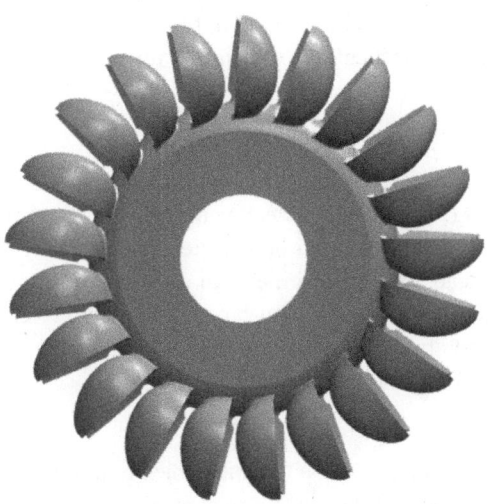

Figure 10.8 Some common views of Pelton turbines.

are used to generate electricity for homes, businesses, and other small-scale operations. Advantages of Pelton turbine are its high efficiency, compactness, easy to maintain, suitable for high head projects, and can handle wide range of water flow. However, they are relatively more expensive than other turbine types and may have high maintenance cost.

b) Francis Turbines

Francis turbine is a type of water turbine that is commonly used in hydroelectric power plants. The view of Francis turbine is shown in Figure 10.9. It is a reaction turbine, which means that it effectively utilizes the pressure of the incoming water to generate power, rather than the kinetic energy of the water. The turbine consists of a runner with curved vanes, which is enclosed in a spiral housing. As the water flows through the turbine, it pushes against the vanes, causing the runner to spin and generate power. Francis turbine is known for its high efficiency and ability to handle a wide range of water flow rates for practical applications.

c) Michell-Banki Turbines

A Michell-Banki turbine, also known as a Banki turbine or a crossflow turbine, is a type of water turbine that is characterized by its unique, crossflow design, as illustrated in Figure 10.10. It is an axial turbine, which means that the water flows parallel to the axis of rotation of the turbine blades.

Figure 10.9 Some common views of Francis turbines.

The turbine was invented by a Hungarian engineer, Sándor Michell-Banki in the 1920s. It is used for low head and low flow required applications and is particularly well-suited for use in small hydroelectric power plants.

d) Kaplan or Propeller Turbines

A Kaplan turbine, also known as a propeller turbine, is a type of water turbine that is characterized by its adjustable blades, as illustrated in Figure 10.11. The turbine was invented by an Austrian engineer, Viktor Kaplan in the 1913. The blades of the turbine can be adjusted to optimize the turbine's performance for a given water flow and head. Kaplan turbine is an axial flow turbine, which means that the water flows parallel to the axis of rotation of the turbine blades. It is a versatile turbine that can be used for a wide range of head and flow conditions, making it suitable for use in large hydroelectric power plants as well as in small and medium-sized plants.

e) Deriaz Turbines

Deriaz turbine is a type of water turbine; it is a variation of the Pelton turbine, which is an impulse turbine. It was developed by a Swiss engineer, Paul Deriaz, in the early 20th century. Figure 10.12 shows the view of a typical Deriaz turbine. The turbine uses the kinetic energy of a high-velocity

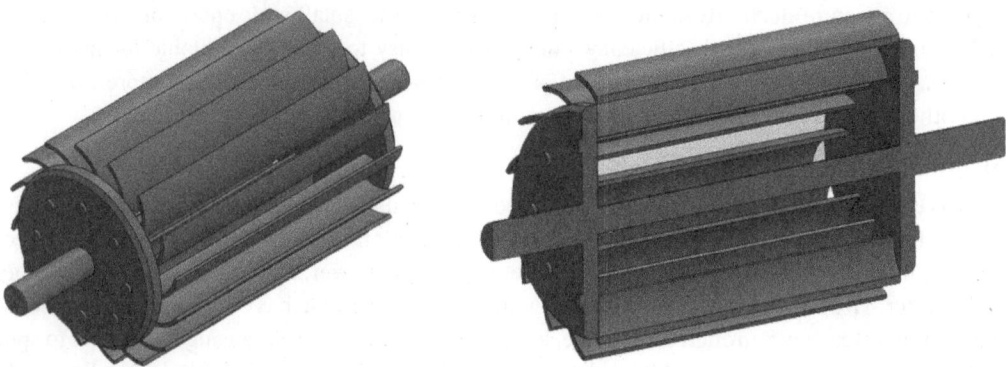

Figure 10.10 Some views of the Michell-Banki turbines.

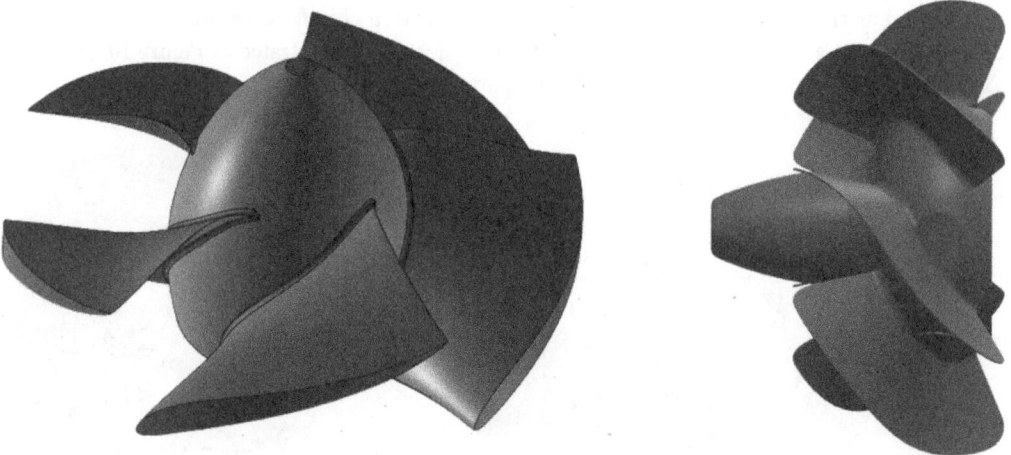

Figure 10.11 Some views of the Kaplan turbines.

water jet to turn a wheel with a series of cups or buckets attached to its rim. It is typically used in high head, low flow applications and is particularly well-suited for use in small hydroelectric power plants.

f) Using Reversed Water Pumps as Turbines

Water pumps can also be used as turbines, a process known as "pump-as-turbine" or "PAT" operation. This means that instead of using electricity to drive the pump and move water, the flow of water is used to turn the pump's impeller, which in turn generates electricity. This can be done by reversing the flow of water through the pump, so that it turns the impeller in the opposite direction. The view of the pump-as-turbine system is shown in Figure 10.13.

This method is particularly useful in micro-hydro power systems, where a small water flow can be used to generate electricity. It can also be used as a way to generate electricity during periods of high water flow, such as during a flood, when a hydroelectric dam would not be able to generate as much power. The efficiency of water pump as turbine is less compared to turbine design but it does have some advantages such as low cost, easy maintenance, and it can be used for both pumping and generating electricity.

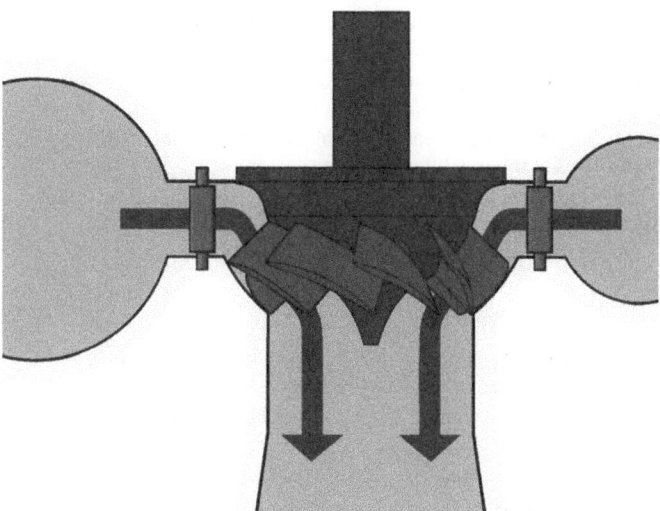

Figure 10.12 The view of Deriaz turbine.

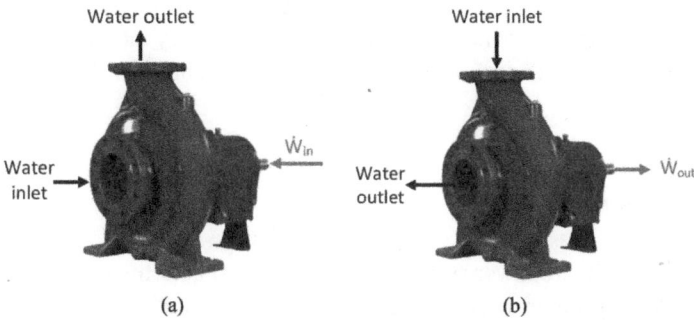

Figure 10.13 Some views of the pump-as-turbine system (a) pump mode and (b) turbine mode.

Table 10.1 Comparative technical details of the common hydro turbines.

Turbine	Details
Pelton turbine	• Pelton turbine is a high-head turbine, which means it is used in hydroelectric power plants where the water falls from a great height. • It is a type of impulse turbine, which converts the kinetic energy of the water into mechanical energy. • It is a simple design and easy to maintain. • It is suitable for applications with a head of more than 300 m.
Francis turbine	• Francis turbine is a medium- to low-head turbine, which means it is used in hydroelectric power plants where the water falls from a moderate to a low height. • It is a type of reaction turbine, which means it converts the potential and kinetic energy of the water into mechanical energy. • It has a more complex design than Pelton turbine, with adjustable blades that can change the turbine's output. • It is suitable for applications with a head between 10 and 200 m.
Kaplan turbine	• Kaplan turbine is a medium-head turbine, which means it is used in hydroelectric power plants where the water falls from a moderate height. • It is a type of reaction turbine, similar to Francis turbine, but it has a different design. • It is suitable for applications with a head between 20 and 70 m.
Michell-Banki turbine	• Michell-Banki turbine is a type of reaction turbine, similar to Kaplan, Francis and Deriaz turbines, but it has a different design. • It is a very efficient turbine, but it is also more complex to build and maintain than other types of turbines. • It is suitable for applications with a head between 5 and 20 m.
Deriaz turbine	• Deriaz turbine is a low-head turbine, which means it is used in hydroelectric power plants where the water falls from a low height. • It is a type of reaction turbine, similar to Francis, Kaplan, and Michell-Banki turbines, but it has a different design. • It is also more efficient than other types of turbines, but also more complex. • It is suitable for applications with a head between 5 and 20 m.

In general, the selection of a turbine depends on the site-specific conditions and the power plant design and requirements. Table 10.1 summarizes and compares the hydro turbines commonly used in practical applications. Practicing engineers in the field consider some key factors, such as the head and flow of water, the desired power output, and the cost and availability of the turbine when making the choice.

10.4 Analysis of Hydro Energy System

In order to understand the concept of hydropower systems, it is required to determine its energy capacity. The energy capacity of flowing water sources are defined through the following relation:

$$\dot{E}_{total} = \dot{m}_w(h + ke + pe) \tag{10.1}$$

where $\dot{m}_w$ is the flow rate of the water, h is the specific enthalpy, ke is the kinetic energy, and pe is the potential energy. When we consider the energy changes in a hydro power plant, as shown in Figure 10.15. The difference between the two reservoirs is then defined as the potential useful output from the system:

$$\Delta \dot{E}_{total} = \dot{m}_w \left[(h_A + ke_A + pe_A) - (h_B + ke_B + pe_B) \right] \tag{10.2}$$

When we consider a hydropower plant, there is no significant change in the kinetic energy content and enthalpy. The energy production potential of the hydropower plants comes from the potential energy of the water which directly depends on the height of the water reservoir and flow rate of the water.

$e_{total} = h + ke + pe$ and hence

$$e_{total} = \frac{P}{\rho_w} + gH + \frac{V^2}{2} \tag{10.3}$$

The total hydraulic head is found by modifying the above written equation as follows:

$$H = \frac{e_{total}}{g} = \frac{P}{\rho_w g} + z + \frac{V^2}{2g} \tag{10.4}$$

Here, $\dfrac{P}{\rho_w g}$ is called pressure term, z is the elevation term (referring to potential energy), and $\dfrac{V^2}{2g}$ is the kinetic energy term. It should be noted that unit of each term is in meter (m).

The total head or hydraulic head is then found as:

$$H = \frac{P}{\rho_w} + z + \frac{V^2}{2g} \tag{10.5}$$

Note that in Eq. 10.4, the potential energy part of the total energy is dominant parameter for hydro-power generation. Therefore, the power generation potential of a hydropower plant is calculated as follows:

$$\dot{W}_{out} = \rho \dot{V} g H$$

By rearranging this equation, the relation between the head and flow rate is written as follows:

$$H = \frac{\dot{W}_{out}}{\rho \dot{V} g} \tag{10.6}$$

Here, there is an inverse ratio between the flow rate and head for the constant work output. The effect of the volumetric flow rate and head on the work output is demonstrated in Figure 10.14a. As seen in Figure 10.14a, the work output increases with increasing head and flow rate of the water. The water flow rate in a hydropower system depends on the diameter of the penstock which is a connection pipe between the upper reservoir and turbine. Therefore, in order to show the relation between the flow rate and head height is often defined over the following equation:

$$\dot{V} = C_d A \sqrt{2gH} \tag{10.7}$$

where $\dot{V}$ is the flow rate of the water stream flows down from the dam through a hydro turbine, A is the cross-sectional area of the water flow in the connection pipe between the dam and hydro turbine which is called penstock. H is the head height, and g is the gravity. C_d is the discharge rate which is a value between 0.6 and 0.98, depending on the design of the penstock.

One has to remember that each type of turbine has its own advantages and disadvantages, and the choice of turbine depends on the specific site conditions and the power output required. There are various charts for selection of hydro turbines. Figure 10.14 illustrates a useful chart to select a hydro turbine depending on the water flow rates and head heights. As can be seen from Figure 10.14, Kaplan turbine looks like a promising turbine for lower flow rates and low water head conditions. Pelton turbines are suitable for a higher head height. The Francis turbine can serve a wide range of operating conditions comparing to the other types of turbines. As seen from Figure 10.14, there are two main drivers for hydro turbines, which are head height and flow rate of the water.

Illustrative Example 10.1 In a hydropower station, the head is designed to be 50 m. The discharge rate (C_d) is known to be 0.90. The diameter of the penstock is 1 m. Calculate the flow rate to be obtained in this hydropower station.

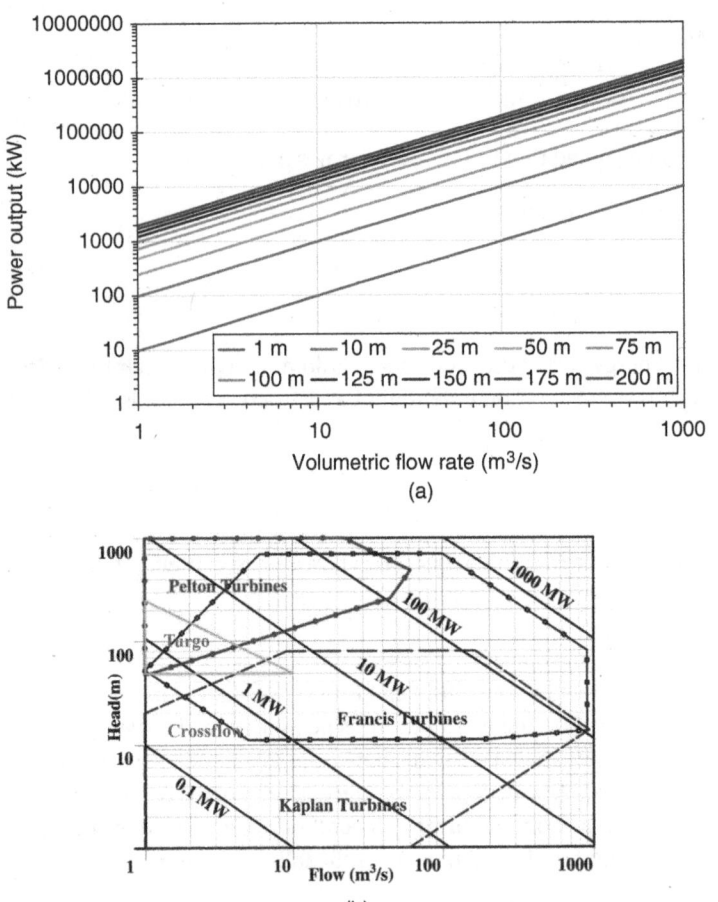

Figure 10.14 The useful charts for (a) variation of power output of a hydro turbine with volumetric flow rate at various heads and (b) variation of head of various hydro turbines versus volumetric flow rate (modified from [3]).

Solution:

Equation 10.7 is used for finding the flow rate depending on the head.

$$\dot{V} = C_d A \sqrt{2gH}$$

In order to use this equation, it is first required to determine the cross-sectional area of the penstock using

$$A = \pi \frac{D^2}{4}$$

$$A = \pi \times \frac{1^2}{4} = 0.785 \ m^2$$

Now, the flow rate can be calculated by substituting the area and other known parameters,

$$\dot{V} = 0.90 \times 0.785 \times \sqrt{2 \times 9.91 \times 50}$$

The volumetric flow rate is found to be $\dot{V} = \mathbf{490.1 \ m^3 / s}$.

As mentioned earlier, there are various types of hydro turbines to be able to be used in the hydro power generation power plants. Although they have different types, thermodynamic analysis of the hydro turbines and power plants are the same. In order to show this concept clearly, let's check Figure 10.15. It is seen in Figure 10.15 that the flow energy (due to potential energy of the water or water current) of the water is converted to the turbine work (shaft work).

In order to analyze the hydro turbine thermodynamically, see the system boundary for a hydro turbine provided in Figure 10.15. Like in any system, the thermodynamic analysis of the turbine is incorporated with inlets(s) and outlet(s) passing through the system boundary. For a hydro

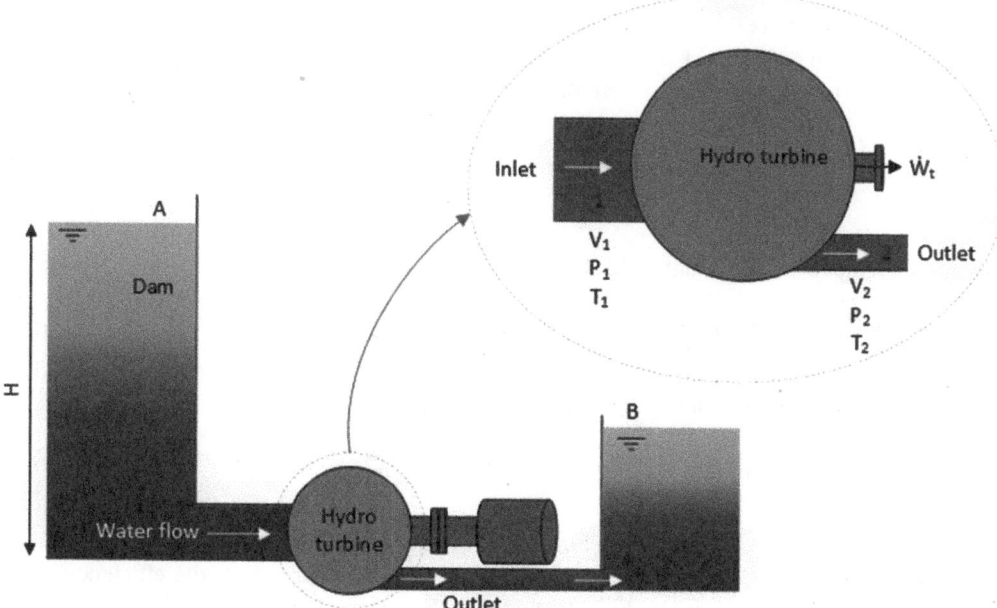

Figure 10.15 A thermodynamic illustration of a hydro turbine for analysis.

turbine, there is a water inlet with high pressure and high velocity. The outlet of the hydro turbine is a water outlet with low pressure and lower velocity, and the turbine works. So, the hydro turbine converts the kinetic energy of the water into the turbine work. The kinetic energy of the water can be created by the potential energy of the water coming from dams resulting in water currents that are referred to kinetic energy content.

The mass balance equation for the hydro turbine is written as follows:

$$\dot{m}_1 = \dot{m}_2 \tag{10.8}$$

Considering the continuity equation in the form of density, cross-sectional area and velocity, one can write the following:

$$\rho_1 A_1 V_1 = \rho_2 A_2 V_2 \tag{10.9}$$

where ρ is the density of water, A is the cross-sectional area of the turbine, and V is the velocity of water.

The energy balance equation of the hydro turbine can be written in general as follows:

$$\sum E_{in} = \sum E_{out} \tag{10.10}$$

Here, E_{in} and E_{out} represent the total energy input and output in the turbine. For a hydro turbine, the energy contents at the inlet and outlet of the turbine can be written by considering Bernoulli equation, which is energy balance for a steady-state fluid flow.

$$P_1 + \frac{1}{2}\rho V_1^2 + \rho g h_1 = P_2 + \frac{1}{2}\rho V_2^2 + \rho g h_2 + w_t \tag{10.11}$$

Here, P_1 and P_2 are the energy contents resulted out of pressure. $\frac{1}{2}\rho V_1^2$ and $\frac{1}{2}\rho V_2^2$ represent the kinetic energy content per unit volume. $\rho g h_1$ and $\rho g h_2$ denote the potential energy per unit volume. The first term of Eq. 10.11 shows the potential energy content of the hydro power plant. So, the maximum (ideal) power to be generated from a hydro turbine can be determined as the following:

$$w_{max} = P_1 + \frac{1}{2}\rho V_1^2 + \rho g h_1 \tag{10.12}$$

which is then modified for the dam which uses the potential energy of the water to generate power:

$$\dot{W}_{max} = \dot{m}_1 g H = \rho \dot{V}_1 g H \tag{10.13}$$

The electricity generation for a hydro power plant is calculated as follows:

$$\dot{W}_{electricity} = \dot{W}_{max} \eta_t \eta_g \eta_p \tag{10.14}$$

where η_t, η_g, and η_p are the efficiency for hydro turbine, generator, and piping. While the efficiency of the turbine and generator are related to the device performance, the efficiency of the piping is related to energy losses in the piping used in the power plant.

The energy and exergy efficiencies of the hydro power plant are found to be:

$$\eta_{en} = \frac{\dot{W}_{electricity}}{\dot{E}_{max}} \tag{10.15}$$

and

$$\eta_{ex} = \frac{\dot{W}_{electricity}}{\dot{Ex}_{max}} \tag{10.16}$$

In a hydro power plant, generally, the potential energy of the water is used as the energy input. As the potential energy equals to its exergy content. The energy and exergy efficiencies of the hydro power plant become identical as $\eta_{en} = \eta_{ex}$.

The basics of hydro power plants are significant for pumped hydro storage system which will be discussed later in Chapter 11.

Illustrative Example 10.2   A water reservoir is considered to generate electricity by the installation of a hydraulic turbine-generator at a location where the depth of the water is 100 m, as illustrated in Figure 10.16. The flow rate of the water is 4000 kg/s. Calculate the maximum turbine work rate generated in this dam.

Solution:
The maximum power that can be obtained in a dam is calculated using Eq. 10.13 as follows:

$$\dot{W}_{max} = \dot{m}_1 gH = \rho \dot{V}_1 gH$$

$$\dot{W}_{max} = 4000 \times 9.81 \times 100 = 3,924,000 \ W$$

The maximum turbine work rate is found to be $\dot{W}_{max} = 3.924$ MW.

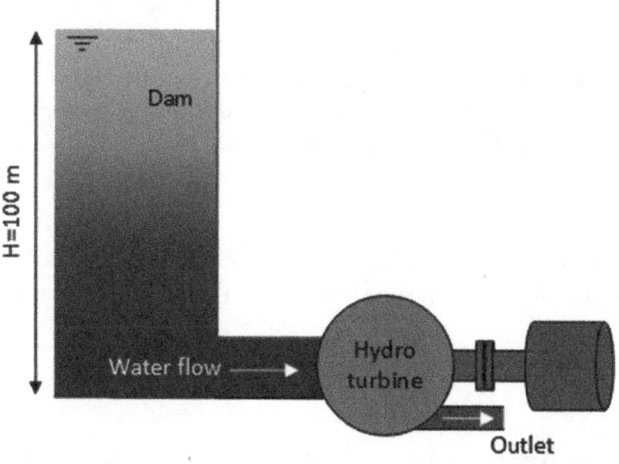

Figure 10.16   An illustration of the dam in Illustrative Example 10.2.

Illustrative Example 10.3 Consider the problem in Illustrative Example 10.1. In the hydro power plant, the turbine efficiency is 0.80, the generator efficiency is 0.90, and the piping efficiency is 0.95. Calculate the net electricity output from this hydro power plant and the energy and exergy efficiencies.

Solution:

The maximum turbine work rate was 3.924 MW as obtained earlier. The net electricity output is then calculated via Eq. 10.14.

$$\dot{W}_{electricity} = \dot{W}_{max} \eta_t \eta_g \eta_p$$

$$\dot{W}_{electricity} = 3.924 \times 0.80 \times 0.90 \times 0.95$$

Thus, the net electricity output is found to be 2.684 MW.

The energy and exergy efficiencies can be calculated as follows:

$$\eta_{en} = \frac{\dot{W}_{electricity}}{\dot{E}_{max}}$$

$$\eta_{en} = \frac{2.684}{3.924} = 0684 \text{ or } 68.4\%$$

As the energy and exergy contents are identical, the exergy efficiency is written as:

$$\eta_{ex} = \frac{\dot{W}_{electricity}}{\dot{E}x_{max}}$$

which results in the same result as

$$\eta_{ex} = \frac{2.684}{3.924} = 0.684 \text{ or } 68.4\%$$

Illustrative Example 10.4 By the international agreement, the Canadian power stations are able to draw 1599 m³/s of water from the Niagara River. In the Adam Beck Power Stations, 26 turbine units are available, and the average water flow rate through each of the turbines is 61.5 m³/s. The elevation difference between the free surfaces upstream and downstream of the dam is 51 m. If the turbine and generator efficiencies are 90% and 95%, determine the average power generation rate in the Adam Beck Power Stations. Calculate the overall efficiency of the power generation in the Station. The losses in the piping are neglected (g=9.81 m/s², ρ=997 kg/m³).

Solution:

In the problem, the volumetric flow rate is provided to be $\dot{V}_{total} = 1599$ m³/s. The volumetric flow rate for each turbine is $\dot{V}_t = 61.5$ m³/s. The elevation is given as 51 m. From Eq. 10.14, we proceed to make the necessary calculations as follows:

$$\dot{W}_{electricity} = \dot{W}_{max} \eta_t \eta_g \eta_p$$

where $\dot{W}_{max}$ can be included in this equation from Eq. 10.13. We therefore obtain

$$\dot{W}_{electricity} = \rho \dot{V}_1 g H \eta_t \, \eta_g \, \eta_p$$

$$\dot{W}_{electricity} = 997 \times 61.5 \times 9.81 \times 51 \times 0.90 \times 0.95 \times 1$$

$$\dot{W}_{electricity} = 26,228,627.9W = 26.2 \ MW$$

As there are 26 turbines in the facility, the total electricity production becomes

$$\dot{W}_{electricity,total} = 26.2 \times 26 = 681.2 \ MW$$

The energy efficiency of the power station is calculated from Eq. 10.15 as follows:

$$\eta_{en} = \frac{\dot{W}_{electricity}}{\dot{E}_{max}} = \frac{\dot{W}_{electricity}}{\rho \dot{V}_{total} \, gH} = \frac{681.2}{997 \times 1599 \times 9.81 \times 51} = \frac{681.2}{797.6} = 0.854$$

Illustrative Example 10.5   Consider the hydropower plant as shown in Figure 10.17. The head losses through the pipes are known to be $\dfrac{3V_1^2}{2g}$ and $\dfrac{2V_2^2}{2g}$ in pipes 1 and 2, respectively. The elevation at the given points is known as $H_a = 135$ m, $H_b = 45$ m, $H_c = H_d = 30$ m. The diameters are known as $D_1 = 0.3$ m and $D_2 = 0.6$ m. The volume flow rate at C is 0.7 m³/s. Determine the power output from the turbine. Calculate the overall efficiency ($g = 9.81$ m/s², $\rho = 1000$ kg/m³).

Solution:
The flow rate of the water is given as 0.7 m³/s. The mass flow rate of the water is calculated as:

$$\dot{m} = \rho \dot{V} = 1000 \times 0.7 = 700 \ kg/s$$

The turbine power is calculated from Eq. 10.11 as given below:

$$\dot{W}_{Turbine} = \dot{m}g\left(H_a - H_c - \frac{3V_1^2}{2g} - \frac{2V_2^2}{2g} - H_b \right)$$

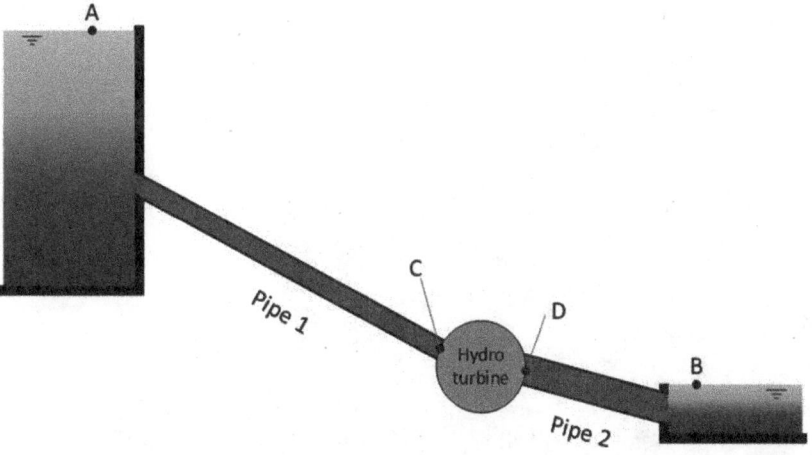

Figure 10.17   An illustration of the hydro power plant included in Illustrative Example 10.5.

In order to use this equation, it is required to calculate V_1 and V_2 accordingly from $\dot{V} = \rho A$.

$$V_1 = \frac{\dot{V}}{A_1} = \frac{0.7}{\left(\frac{\pi D_1^2}{4}\right)} = 9.9 \, m/s$$

and

$$V_2 = \frac{\dot{V}}{A_2} = \frac{0.7}{\left(\frac{\pi D_2^2}{4}\right)} = 2.5 \, m/s$$

The turbine work is found to be:

$$\dot{W}_{Turbine} = 700 \times 9.81 \times (135 - 14.9 - 0.64 - 45) = 511{,}317 \, W \text{ or } 511.3 \text{ kW}$$

The efficiency of the turbine can be found as

$$\eta = \frac{\dot{W}_{Turbine}}{\dot{m}g(H_a - H_b)} = \frac{511{,}317}{618{,}030} = 82.7\%$$

Illustrative Example 10.6   In a location, due to the local conditions and capacity of the water sources, there is a potential to build two hydropower plants on a river, as illustrated in Figure 10.18. In the first power station, the head is 60 m, and the flow rate is 5000 kg/s. For the second hydropower plant, the head is 40 m, and the flow rate is 2500 kg/s. The efficiencies for turbine and generator are 0.90 and 0.95, respectively. The piping losses are negligible. Calculate the total electricity power in these two power plants.

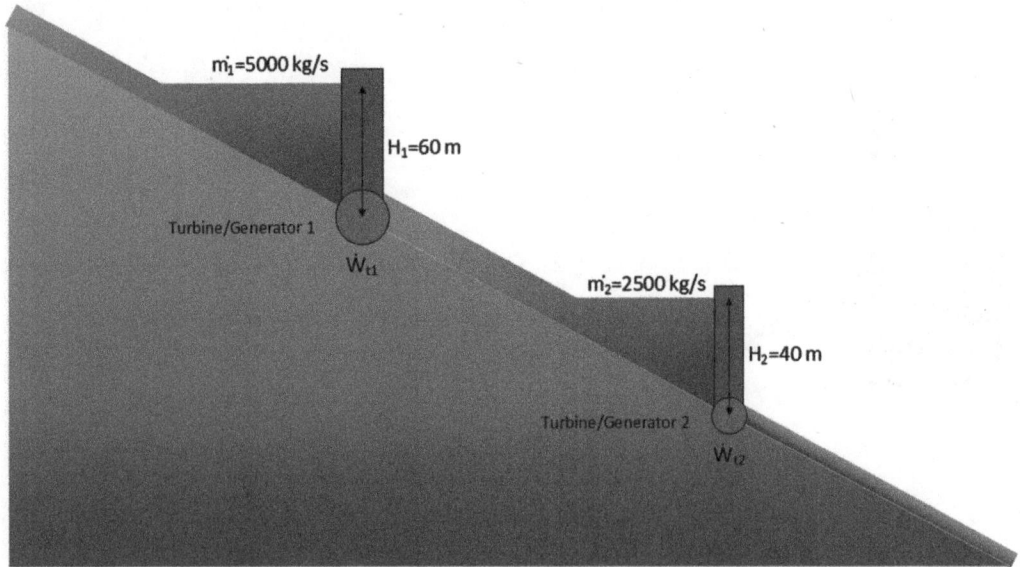

Figure 10.18   An illustration of two hydro power plants in Illustrative Example 10.6.

Solution:

The power generation in the first station is calculated as follows:

$$\dot{W}_{electricity,1} = \dot{m}_1 g \, H_1 \, \eta_t \, \eta_g \, \eta_p$$

$$\dot{W}_{electricity,1} = 5000 \times 9.81 \times 60 \times 0.90 \times 0.95 \times 1$$

$$\dot{W}_{electricity,1} = 2{,}516{,}265 W \; or \; 2{,}561.3 \, kW$$

The electricity generation in the second station is calculated as follows:

$$\dot{W}_{electricity,2} = \dot{m}_2 g \, H_2 \, \eta_t \, \eta_g \, \eta_p$$

$$\dot{W}_{electricity,2} = 2500 \times 9.81 \times 40 \times 0.90 \times 0.95 \times 1$$

$$\dot{W}_{electricity,2} = 838{,}755 W \; or \; 838.7 kW$$

Thus, the total electricity generation is found as

$$\dot{W}_{electricity,total} = \dot{W}_{electricity,1} + \dot{W}_{electricity,2}$$

$$\dot{W}_{electricity,total} = 2561.3 + 838.7 \, kW$$

The total electricity in these two power stations is found to be

$$\dot{W}_{electricity,total} = 3400 \, kW \; or \, 3.4 \, MW.$$

10.5 Ocean Energy

Ocean energy, which is also known as marine energy, refers to the energy that can be obtained from the ocean, such as tides, tidal waves, tidal currents, thermal gradient, and salinity gradient. These forms of energy are considered renewable and sustainable, as they do not produce greenhouse gas emissions or other pollutants. Wave energy is generated by the movement of ocean waves, while tidal energy is generated by the rise and fall of ocean tides. Ocean thermal energy is generated by the difference in temperature between the surface water and deep water, and ocean current energy is generated by the movement of ocean currents. The technology to harness these forms of energy is still in development and not yet widely used, but it has the potential to make a significant contribution to meeting our energy needs in the future.

In order to deploy ocean energy systems, there is strong need for the following aspects:

- Investment opportunities
- Incentives and tax credits
- Training and education for personnel

One may raise a question: Why ocean energy? Oceans have a significant potential to produce useful outputs in terms of energy generation. The ocean is the largest solar collector in the world and can provide huge amounts of energy (potential, kinetic, and thermal) which is absolutely clean. It is a stable and reliable source of energy. As mentioned earlier, salinity gradients, thermal gradients, tidal sea level change and powerful currents, or ocean waves can be used to generate electricity. The ocean energy can be classified as in Figure 10.1.

- Tides
- Tidal waves
- Tidal currents
- Thermal gradients
- Salinity gradients

10.5.1 Energy Production with Tides

Tidal energy is a form of hydropower that converts the energy of tides into electricity. Tides rise the water level in the ocean. The potential energy of the water risen is used for collecting the water in the basin called tidal basin. When the elevation of the ocean is back to normal level, the water is ejected from tidal basin to ocean via the turbine. Figure 10.19 demonstrates working principles of the power generation from tides. Tidal energy systems are divided into two main categories: tidal stream generators and tidal barrages. Tidal stream generators, also known as tidal turbines, are similar to wind turbines but are placed in the ocean to generate electricity from the movement of water. Such generators are placed on the ocean floor or floating on the surface.

Tidal barrages are large structures that are built across estuaries or bays to hold back water during high tide and release it during low tide. The resulting difference in water levels is used to generate electricity through turbines. Both tidal stream generators and tidal barrages have the potential to provide clean and reliable energy, but they also have some environmental impacts. Tidal barrages can affect the movement of fish and other marine animals, while tidal stream generators can disrupt the ocean's ecosystem and pose a risk to marine animals. Despite these challenges, many countries have begun to invest in tidal energy as a way to diversify their energy mix.

Tidal energy, which is driven by the moon and sun's gravitational pull, is predictable centuries in advance. In world's largest tidal power plant in the Rance estuary near St. Malo, France. The capacity of this power station is 240 MW, which is generated by 24 turbines. Also, there is forthcoming project which is candidate to be the world's largest tidal power plant, in Sihwa, South Korea. The capacity of the power station will be 260 MW and yearly power generation is expected to be 543 GWh.

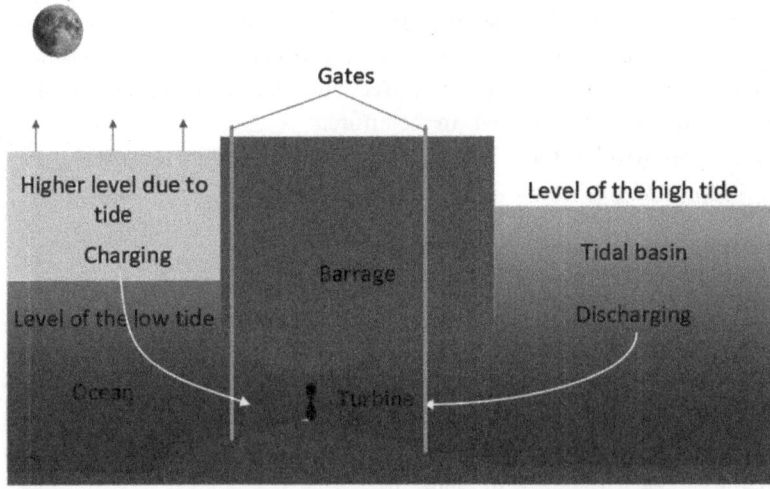

Figure 10.19 A schematic view of hydro power production from tides.

Illustrative Examples 10.7 In a tidal barrage power generation system, the ocean water is collected in the reservoir which is located 65 m higher than the sea level. A hydro turbine is used for power generation. The flow rate of the water is 5000 kg/s. If turbine, generator and piping efficiencies are 0.80, 0.95, and 0.90, calculate the electricity generation rate in this tidal barrage power generation system.

Solution:
Power generation potential in the tidal barrage comes from the potential energy of the water. Therefore, the electricity generation is calculated from Eq. 10.12.

$$\dot{W}_{electricity} = \dot{W}_{max}\eta_t\eta_g\eta_p$$

Here, the efficiencies are given in the problem, we need to calculate $\dot{W}_{max}$ by using Eq. 10.11.

$$\dot{W}_{max} = \dot{m}_1 gH$$
$$\dot{W}_{max} = 5000 \times 9.81 \times 65 = 3,188,250W \ or \ 3,188 kW$$

The electricity generation rate is found as follows:

$$\dot{W}_{electricity} = 3,188 \times 0.80 \times 0.95 \times 0.90$$

The electricity generation is found to be $\dot{W}_{electricity} = \textbf{2,180.6 kW}$.

10.5.2 Energy Production with Ocean Waves

Ocean waves are produced by winds blowing across the surface and undercurrents. Waves travel across the ocean, so their arrival time at the wave power facility may be more predictable than wind. Ocean wave energy systems are devices that convert the energy from ocean waves into usable electricity. Wave energy has the potential to be a significant source of clean, renewable energy, but the technology is still in the early stages of development. The classification of the wave energy systems can be classified as wave collection and wave current systems, as demonstrated in Figure 10.20. While wave collection works based on potential energy of the water, the wave current uses the kinetic energy of the water.

As seen in Figure 10.20, there are several different types of wave energy converters that are developed, including wave barrage, floating reservoir, oscillating water column, underwater turbine, and wave mat.

The working principles of the wave barrage are the same as the tidal barrage as illustrated in Figure 10.21. The high waves are collected in a higher reservoir. Then, the collected water is used for power generation with a hydro turbine. It is a quite useful technique to generate power when the location has higher wavelengths.

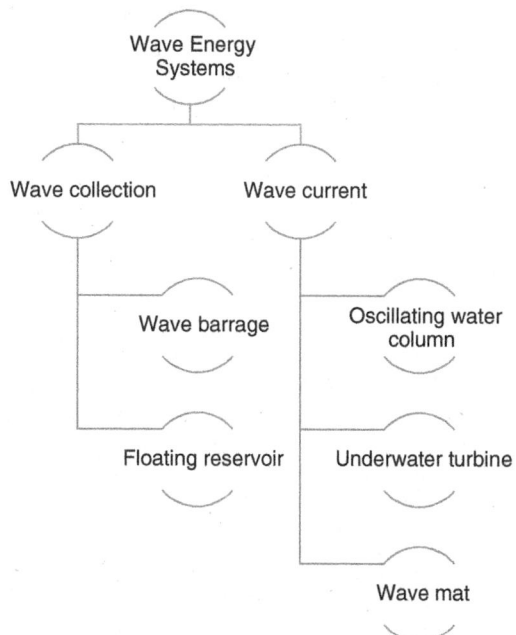

Figure 10.20 Classification of wave energies.

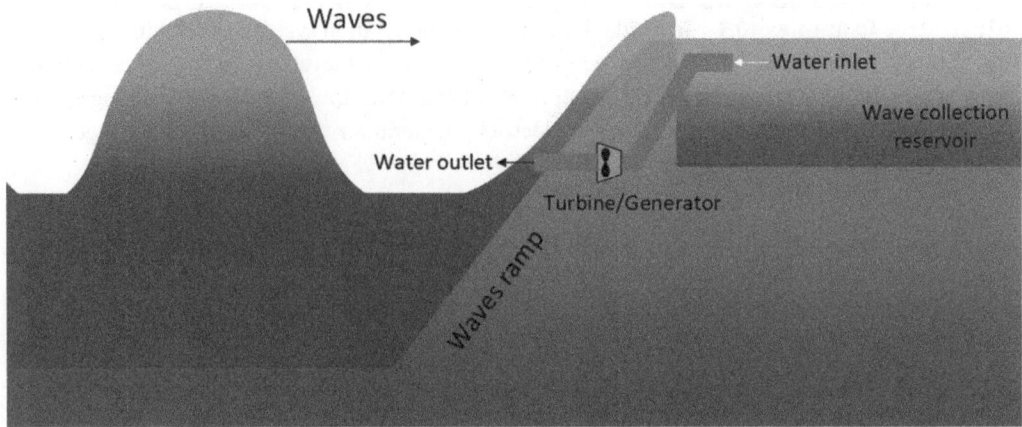

Figure 10.21 An illustration of power generation through wave collection in a reservoir.

The wavelength is generally higher in the middle of the open sea. In order to benefit from this source, floating reservoirs are used. This is also a type of wave collection system, as shown in Figure 10.22. The water is collected in the floating reservoir. Then, like in all hydro power generation systems, the collected water is for power generation in a hydro turbine.

Different from the wave collection systems which use the potential energy of the water, there are systems using the kinetic energy of the water includes: oscillating water column, underwater turbine, and wave mat. These systems convert the kinetic energy of the water into mechanical (shaft) work.

Oscillating water column systems have a chamber that is open to the ocean, and as waves pass through the chamber, the water level inside rises and falls, as illustrated in Figure 10.23. The rising water level compresses the air inside the control volume. The compressed air is expanded in the turbine to produce useful energy. This motion is used to drive a turbine, which generates electricity.

Another wave energy system is the system that uses the guided water through a hydro turbine, as shown in Figure 10.24. In this system, the kinetic energy of the wave is guided with a guide plate, the cross-sectional area of the water flow is reduced with guide plate. The water passes through a hydro turbine, generally Kaplan turbine, to convert the kinetic energy of the water to electricity.

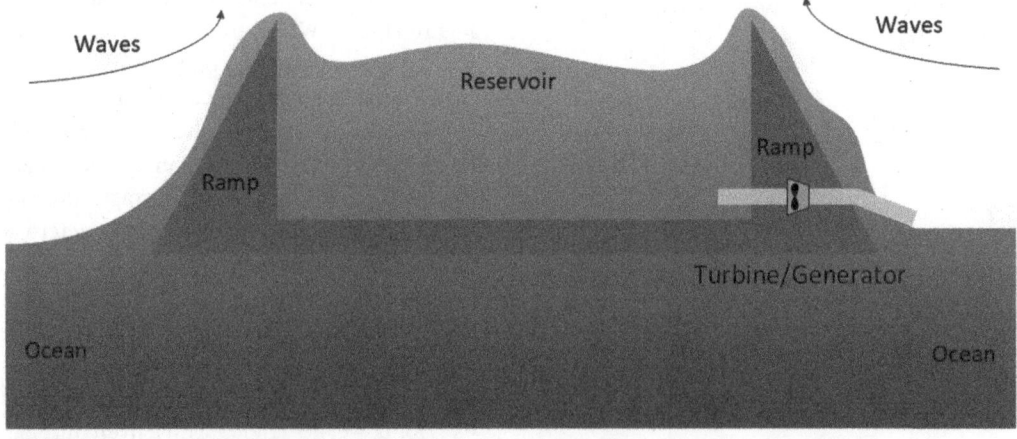

Figure 10.22 Another illustration of power generation with waves collected in a reservoir.

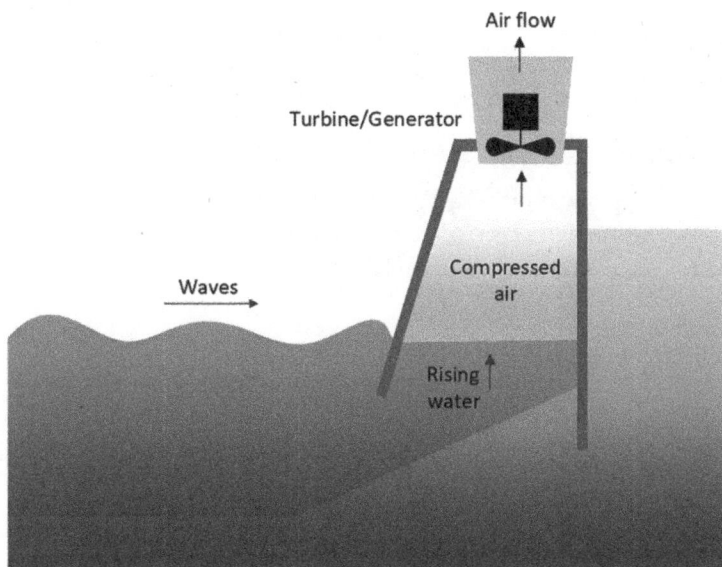

Figure 10.23 A schematic view of the oscillating water column system using rising water through the waves.

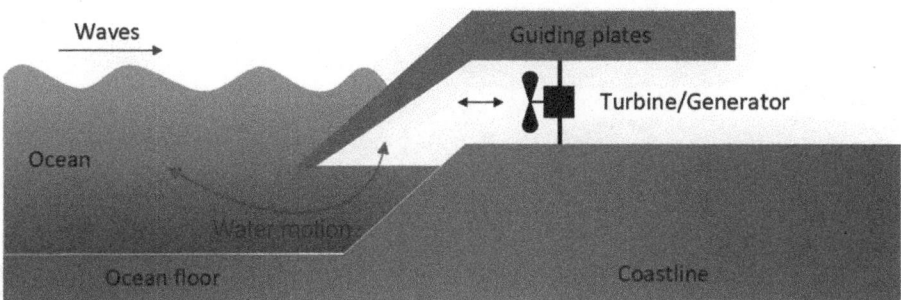

Figure 10.24 An illustration of the power generation with guided wave.

Other power generation systems which are using the kinetic energy of the water is the point absorber system (Figure 10.25a) and water map (Figure 10.25b). In the point absorber systems, the system consists of a buoy or other floating device that moves up and down with the waves, as shown in Figure 10.25a. The motion is used to drive a generator on the buoy, which produces electricity. Fluid pressure inside accumulators is created by the movement of the floaters. In this case, the rising waves create fluid pressure that turns a hydraulic motor. As a result, a generator is activated, sending electricity to the power grid.

The wave mats are wave energy converters, designed to capture surface waves. Figure 10.25b demonstrates the view of the wave map. As a result, the wave mat becomes one moving mass with the wave. The wave motions are converted to the mechanical work via mechanism. This mechanical work is for driving a generator for electricity production.

The Pelamis wave energy converter uses the kinetic energy of the waves to extract energy from the oscillations generated by wave motion on different sections of the tube. Figure 10.25c demonstrates the schematic view of the Pelamis wave energy converter. This is a semi-submerged

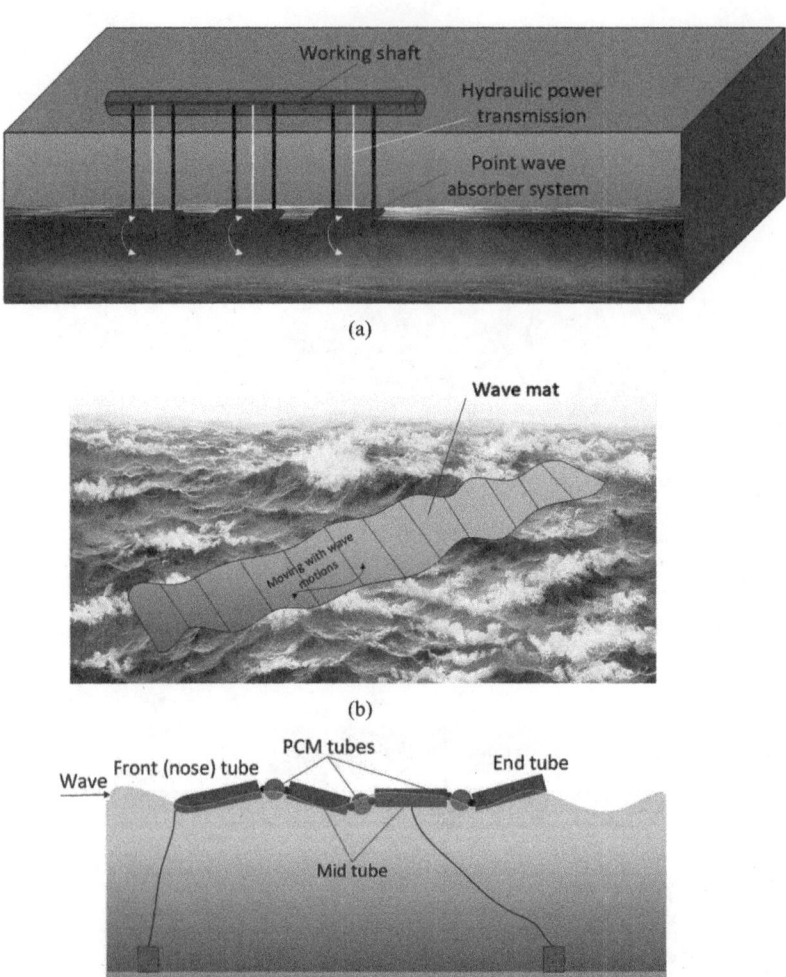

Figure 10.25 Illustrations for (a) a point absorber system, (b) the water mat, and (c) the Palamis type wave energy converters.

offshore device that looks like a snake, developed by the Pelamis Wave Power in Portugal. The main components of the device are shown in Figure 10.25c. In Figure 10.25c, four cylindrical steel tubes are hinged together by three power conversion modules which have PCMs (phase change materials). As the wave front passes, the Pelamis flexes vertically and horizontally in response to the moorings. A hydraulic ram resists this motion by pumping high-pressure oil into accumulators, which are drained through hydraulic motors to drive induction generators. As the PCMs contain the hydraulic power take-off, generators, and controls, the nose tube, tapered at one end, houses the switchgear and transformer to collect the power for export to shore.

10.5.3 Energy Production with Ocean Currents

It is exceptionally important to benefit from the powerful ocean currents. For this purpose, the propellers are placed under water, as illustrated in Figure 10.26. The propellers are deployed,

Figure 10.26 Schematic illustration of the energy production from ocean current.

similar to wind turbines, to convert the kinetic energy of the ocean currents into shaft power. Thus, the electricity is generated via generators that are attached to the propellers. The main advantage of ocean current power generation is the predictable current. Therefore, energy generation with ocean currents is quite similar to tidal power. The main difference is the source of the currents. While the tides are the source of the water current in the tidal power, natural water currents and wind are the main source of the water currents.

Illustrative Example 10.8 Water steadily flows over a tidal turbine at a velocity of 10 km/h, having a diameter of 20 m, as shown in Figure 10.27. The water exit velocity is assumed to be 8 km/h. The inlet water temperature and pressure are 30°C and 500 kPa and the outlet temperature is 29.992°C. The heat loss from the turbine is 3.5 kW. For a reference temperature of 25°C and pressure of 100 kPa.

a) Write the mass, energy, entropy and exergy balance equations for the turbine.
b) Calculate the power generated by the undersea turbine.
c) Calculate the energy and exergy efficiencies.

Take $h_1 = 126.2\dfrac{kJ}{kg}$, $s_1 = 0.4366\dfrac{kJ}{kgK}$, $h_0 = 104.9\dfrac{kJ}{kg}$, $s_0 = 0.3672\dfrac{kJ}{kg}K$.

The reference temperature is 25°C, the density of water is 1000 kg/m³, and specific heat of water is c_p=4.2 kJ/kg K.

Solution:

a) The thermodynamic balance equations for mass, energy, entropy and exergy are written as follows:

MBE: $\dot{m}_1 = \dot{m}_2$

EBE: $\dot{m}_1\left(h_1 + \dfrac{V_1^2}{2}\right) = \dot{m}_2\left(h_2 + \dfrac{V_2^2}{2}\right) + \dot{W}_{out} + \dot{Q}_{loss}$

EnBE: $\dot{m}_1 s_1 + \dot{S}_{gen} = \dot{m}_2 s_2 + \dfrac{\dot{Q}_{loss}}{T_{surr}}$

ExBE: $\dot{m}_1 ex_1 + \dot{m}_1 ex_{ke1} = \dot{m}_2 ex_2 + \dot{m}_2 ex_{ke2} + \dot{W}_{out} + \left(1 - \dfrac{T_0}{T_{ave}}\right)\dot{Q}_{loss}$

Figure 10.27 The view of the tidal turbine in Illustrative Example 10.8.

b) The turbine work can be calculated by using the EBE. Before using the EBE, it is required to calculate the flow rate:

$$\dot{m}_1 = \rho\left(\frac{\pi D^2}{4}\right)V = (1000)\left(\frac{\pi(20^2)}{4}\right)(2.78) = 873{,}476\frac{kg}{s}$$

Re-arranging the EBE:

$$\dot{W}_{out} = \dot{m}\left(h_1 - h_2 + \frac{V_1^2}{2} - \frac{V_2^2}{2}\right) - \dot{Q}_{loss}$$

$$W_{out} = m\left(c_p(T_1 - T_2) + \frac{V_1^2}{2} - \frac{V_2^2}{2}\right) - \dot{Q}_{loss}$$

$$\dot{W}_{out} = 873{,}476\times\left(4.2\times(30 - 29.992) + \frac{2.78^2}{2} - \frac{2.224^2}{2}\right)$$

Thus, $\dot{W}_{out}$ is found to be 3.057×10^7 W.

c) The energy and exergy efficiencies can be found to be

$$\eta_{en} = \frac{\dot{W}}{\dot{E}_{in}} = \frac{3.057\times10^7\ \text{W}}{\dot{m}\left(h_1 - h_0 + \frac{V_1^2}{2}\right)} = \frac{3.057\times10^7}{873{,}476\left(126{,}200 - 104{,}900 + \left(\frac{2.78^2}{2}\right)\right)} = 0.002\ \text{or}\ 0.2\%$$

and

$$\eta_{ex} = \frac{\dot{W}}{\dot{Ex}_{in}} = \frac{3.057\times10^7\ \text{W}}{\dot{m}\left(ex_1 + \frac{V_1^2}{2}\right)} = \frac{3.057\times10^7}{873{,}476\left(580 + \left(\frac{2.78^2}{2}\right)\right)} = 5.9\%$$

Illustrative Example 10.9   As given in the previous problem, the underwater current turbines work like wind turbines. Assuming the water velocity and wind velocity are same as 10 m/s. The turbine length is 25 m. The density of the air and water are 1.1 kg/m³ and 998 kg/m³, respectively. Calculate the electricity generation rate for both by assuming turbine-generator efficiency of 0.85.

Solution:

In order to calculate the electricity generation, we first need to determine the energy contents for both water and air. The total energy flow content is calculated with the following equation:

$$e_{fluid} = h + ke + pe = h + \frac{1}{2}\rho V^2 + \rho g H$$

Here, flow enthalpy and potential energy changes are not significant, while kinetic energy dominates. Therefore, the flow energy content becomes:

$$\dot{E}_{fluid} = \dot{m}_{fluid}\frac{1}{2}\rho V^2$$

In order to use this equation, we also need to find the mass flow rates for both air and water. The mass flow rates can be calculated as follows:

$$\dot{m}_{fluid} = \rho A V$$

where ρ is the density of fluid, A is the cross-sectional area of the turbine, and V is the velocity of the fluid stream which is 10 m/s for this problem.

$$\dot{m}_{air} = 1.1 \times \pi \times 25^2 \times 10 = 21{,}598.5 \ kg/s$$

$$\dot{m}_{water} = 998 \times \pi \times 25^2 \times 10 = 19{,}595{,}684.2 \ kg/s$$

We now calculate the energy contents of the flows as:

$$\dot{E}_{wind} = \dot{m}_{air}\frac{1}{2}\rho V^2 = 21{,}598.5 \times \frac{1}{2} \times 1.1 \times 10^2 = 1{,}187{,}917.5 W \ or \ 1.18 MW$$

and

$$\dot{E}_{water} = \dot{m}_{water}\frac{1}{2}\rho V^2 = 19{,}595{,}684.2 \times \frac{1}{2} \times 1.1 \times 10^2 = 1{,}077{,}762{,}631 W \ or \ 1{,}077.8 MW$$

As seen from these results, the energy content of the water current is 913.4 times higher than air flow. It clearly shows that the density of working fluid makes big difference.

Let's potentially calculate the electricity generation for these water current and wind turbines. The efficiency for the turbine-generator mechanism is provided to be 0.85 in the problem. Thus, the electricity generation rate is found as:

$$\dot{E}_{wind,elect} = \eta_{tg}\dot{E}_{wind} = 0.85 \times 1.18 = 1.003 MW$$

$$\dot{E}_{water,elect} = \eta_{tg}\dot{E}_{wind} = 0.85 \times 1077.8 = 916.13 MW$$

10.5.4 Energy Production with Ocean Thermal Energy

The Ocean Thermal Energy Conversion (OTEC) method utilizes the temperature difference between the warmer surface water and the colder deep ocean water to generate electricity. Figure 10.28 demonstrates the OTEC system depending on the ocean water cycles which are open-type and closed-type. The warm surface water is used to vaporize a working fluid, such as ammonia, which then drives a turbine to generate electricity. The vaporized working fluid is then cooled by the cold deep water and condensed back into a liquid, completing the thermodynamic cycle.

OTEC has the potential to provide a significant source of clean, renewable energy, particularly in tropical regions near the equator where the temperature difference between the surface and deep water is greatest. However, the technology is still in the early stages of development and has not yet been widely adopted for commercial power generation. Some challenges that need to be addressed may include the high capital costs of building OTEC plants, and the technical challenges of pumping large volumes of cold water from the deep ocean to the surface.

Illustrative Example 10.10 For an ocean thermal power plant, it is known that the surface temperature of the ocean is 32°C in the summer season, and it is 10°C in deep waters. A group of engineers claims that they will produce electricity with a heat engine that works with 10% efficiency between these two levels. Is it really possible to generate electricity with a 10% of efficiency for a heat engine under these conditions?

Solution:
For a heat engine, the Carnot efficiency defines the theoretical limit. Therefore, we can check this claim by going through the Carnot efficiency. The Carnot efficiency is now calculated from the following equation (see Chapter 3 for details):

$$\eta_{Carnot} = 1 - \frac{T_L}{T_H}$$

$$\eta_{Carnot} = 1 - \frac{10 + 273}{32 + 273} = 0.08 \; or \; 8\%$$

As seen from a heat engine works between 10°C and 32°C can have the highest 8% efficiency. Therefore, the engineer's claim is wrong.

Illustrative Example 10.11 For an ocean thermal power generation power plant, seen in Figure 10.29. The ocean water is used for power generation in an organic Rankine cycle using ammonia. The parameters for the working fluids are given in Table 10.2.

a) Write the mass, energy, entropy and exergy balance equations for each component of the system.
b) Find the power output of the turbine.
c) Calculate the energy and exergy efficiencies.

Solution:

a) The mass, energy, entropy, and exergy values can be written for each device in the system as follows:

For evaporator:
MBE: $\dot{m}_1 = \dot{m}_2$ and $\dot{m}_3 = \dot{m}_4$
EBE: $\dot{m}_1 h_1 + \dot{m}_3 h_3 = \dot{m}_2 h_2 + \dot{m}_4 h_4$
EnBE: $\dot{m}_1 s_1 + \dot{m}_3 s_3 + \dot{S}_{gen,eva} = \dot{m}_2 s_2 + \dot{m}_4 s_4$

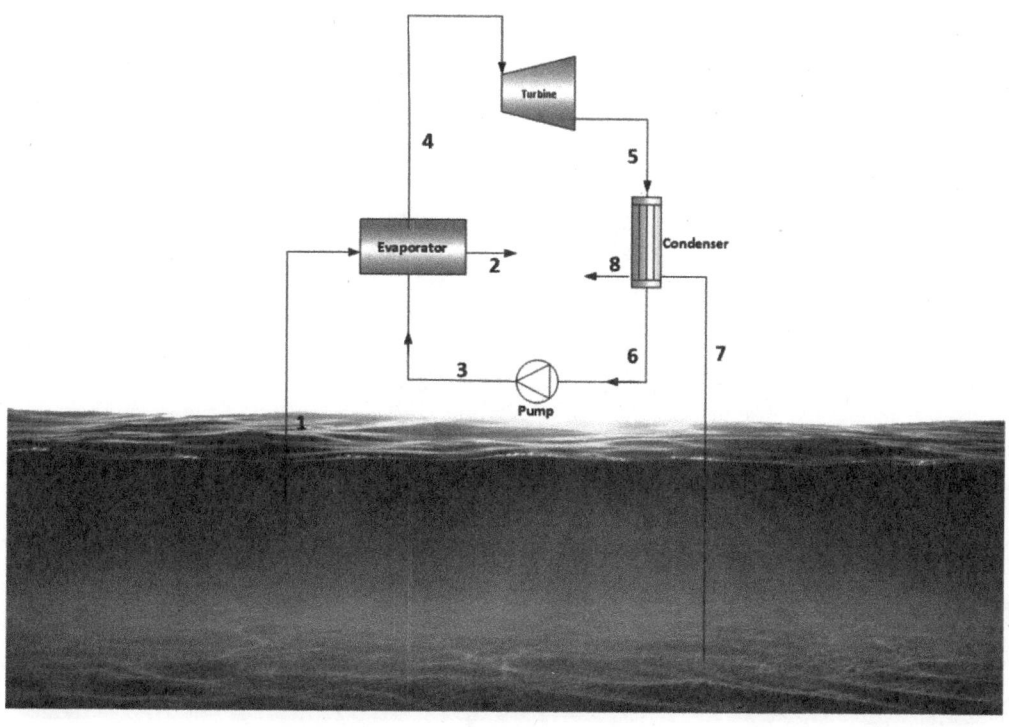

(a)

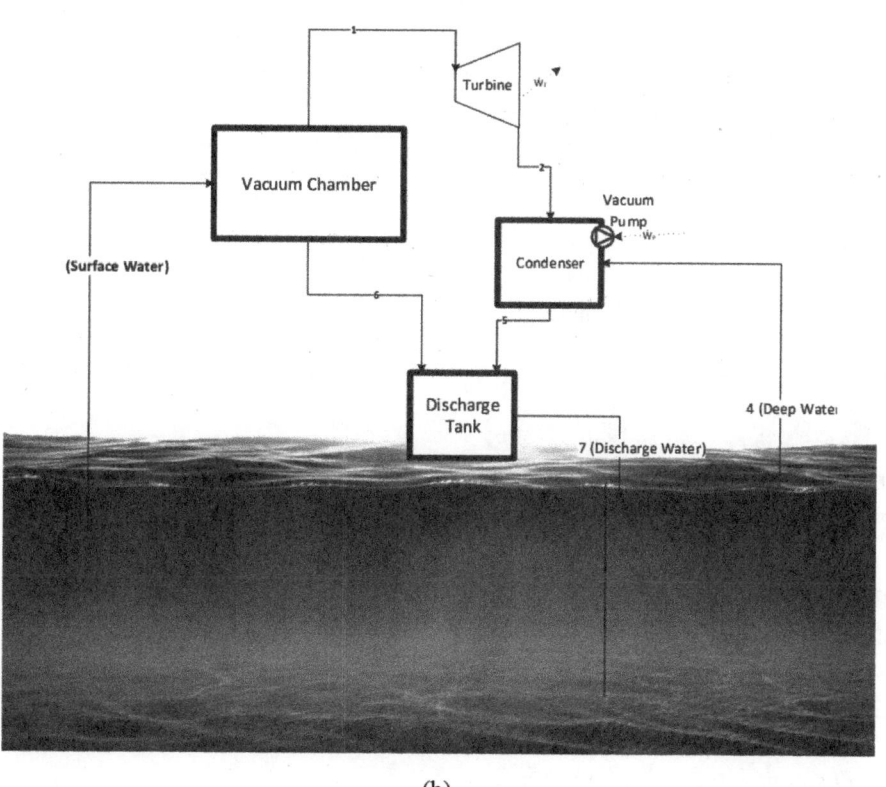

(b)

Figure 10.28 The view of the OTEC systems (a) closed-type OTEC and (b) open-type OTEC.

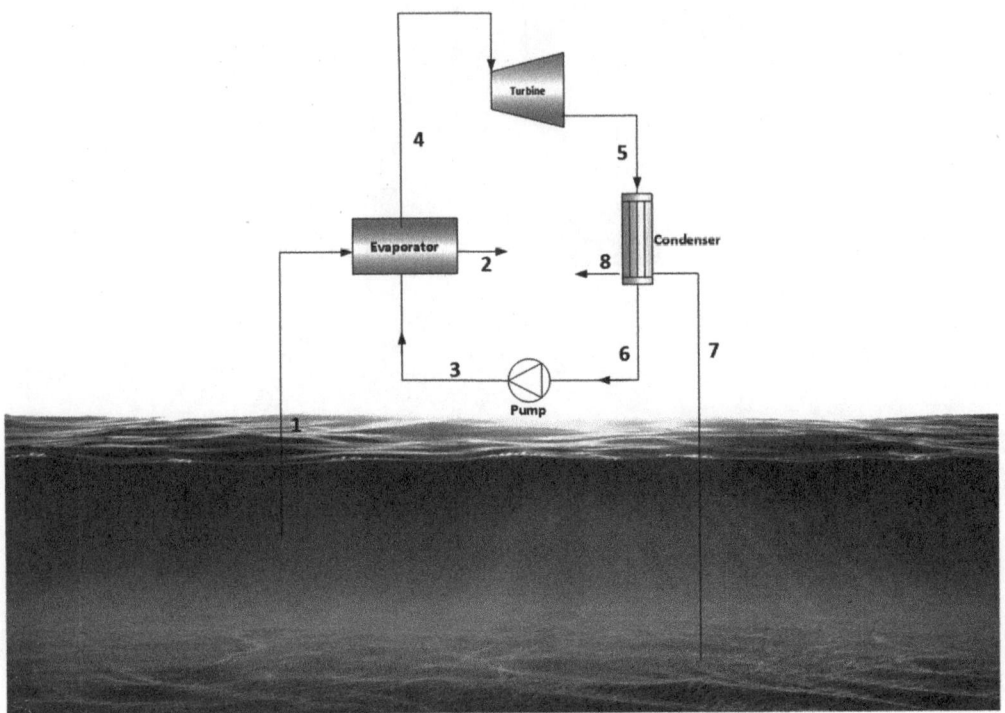

Figure 10.29 The view of the ocean thermal power plant in Illustrative Example 10.11.

Table 10.2 The properties of the working fluid in Illustrative Example 10.11.

State	Mass Flow Rate(kg/s)	Substance	T(°C)	h(kJ/kg)	ex(kJ/kg)
1	2	Water	35	146.7	417.3
2	2	Water	33	138.4	417.1
3	1	Ammonia	−25.2	1409	
4	1	Ammonia	−25.2	?	
5	1	Ammonia	−35	1418	

ExBE: $\dot{m}_1 ex_1 + \dot{m}_3 ex_3 = \dot{m}_2 ex_2 + \dot{m}_4 ex_4 + \dot{Ex}_{dest,eva}$

For turbine:

MBE: $\dot{m}_4 = \dot{m}_5$

EBE: $\dot{m}_4 h_4 = \dot{m}_5 h_5 + \dot{W}_T$

EnBE: $\dot{m}_4 s_4 + \dot{S}_{gen,T} = \dot{m}_5 s_5$

ExBE: $\dot{m}_4 ex_4 = \dot{m}_5 ex_5 + \dot{W}_T + \dot{Ex}_{dest,T}$

For condenser:

MBE: $\dot{m}_5 = \dot{m}_6$ and $\dot{m}_7 = \dot{m}_8$

EBE: $\dot{m}_5 h_5 + \dot{m}_7 h_7 = \dot{m}_6 h_6 + \dot{m}_8 h_8$

EnBE: $\dot{m}_5 s_5 + \dot{m}_7 s_7 + \dot{S}_{gen,con} = \dot{m}_6 s_6 + \dot{m}_8 s_8$

ExBE: $\dot{m}_5 ex_5 + \dot{m}_7 ex_7 = \dot{m}_6 ex_6 + \dot{m}_8 ex_8 + \dot{Ex}_{dest,con}$

For pump:

MBE: $\dot{m}_6 = \dot{m}_3$

EBE: $\dot{m}_6 h_6 + \dot{W}_P = \dot{m}_3 h_3$

EnBE: $\dot{m}_6 s_6 + \dot{S}_{gen,P} = \dot{m}_3 s_3$

ExBE: $\dot{m}_6 ex_6 + \dot{W}_P = \dot{m}_3 ex_3 + \dot{Ex}_{dest,P}$

b) In order to calculate the power output of the turbine, the EBE equation for the turbine is written:

$$\dot{m}_4 h_4 = \dot{m}_5 h_5 + \dot{W}_t$$

In order to use this equation, it is required to first determine h_4. h_4 can be obtained from the EBE for the evaporator.

$$\dot{m}_1 h_1 + \dot{m}_3 h_3 = \dot{m}_2 h_2 + \dot{m}_4 h_4$$

Substituting the values from Table 10.2, one may proceed with the calculation as:

$$2 \times 146.7 + 1 \times 1409 = 2 \times 138.4 + 1 \times h_4$$

resulting in $h_4 = 1426$ kJ/kg.

We now use the EBE for turbine to find the turbine work output as:

$$1 \times 1426 = 1 \times 1409 + \dot{W}_t$$

The turbine output is found to be $\dot{W}_t = 7.43$ kW.

c) The overall energy and exergy efficiencies are then calculated as follows:

$$\eta_{en} = \frac{\dot{W}_t}{\dot{E}_t} = \frac{\dot{W}_t}{\dot{m}_1 h_1} = \frac{7.43}{2 \times 146.7} = 0.025 \ or \ 2.5\%$$

and

$$\eta_{ex} = \frac{\dot{W}_t}{\dot{Ex}_t} = \frac{\dot{W}_t}{\dot{m}_1 h_1} = \frac{7.43}{2 \times 417.9} = 0.80 \ or \ 89\%$$

Illustrative Example 10.12   An open-cycle ocean thermal energy conversion system operates between 27°C surface water temperature and 4°C deep water temperature in order to generate power, as illustrated in Figure 10.30. A vacuum pump with surface water and deep water pumping requires 7 kW of power. The properties of the working fluids are listed in Table 10.3.

a) Find the work output of the turbine.
b) Find the energy efficiency of the power plant.

Solution:

a) When checking the parameters given in Problem 10.12 and Table 10.3, we first need to find h_4 to find the work output of the turbine:

The energy balance equation for the evaporator is then written as follows:

$$\dot{m}_3 h_3 = \dot{m}_6 h_6 + \dot{m}_1 h_1$$

$$100 \times 2519 = 98.88 \times 85.65 + 1.12 \times h_1$$

We use the table values to obtain $h_1 = 2551 \frac{kJ}{kg}$.

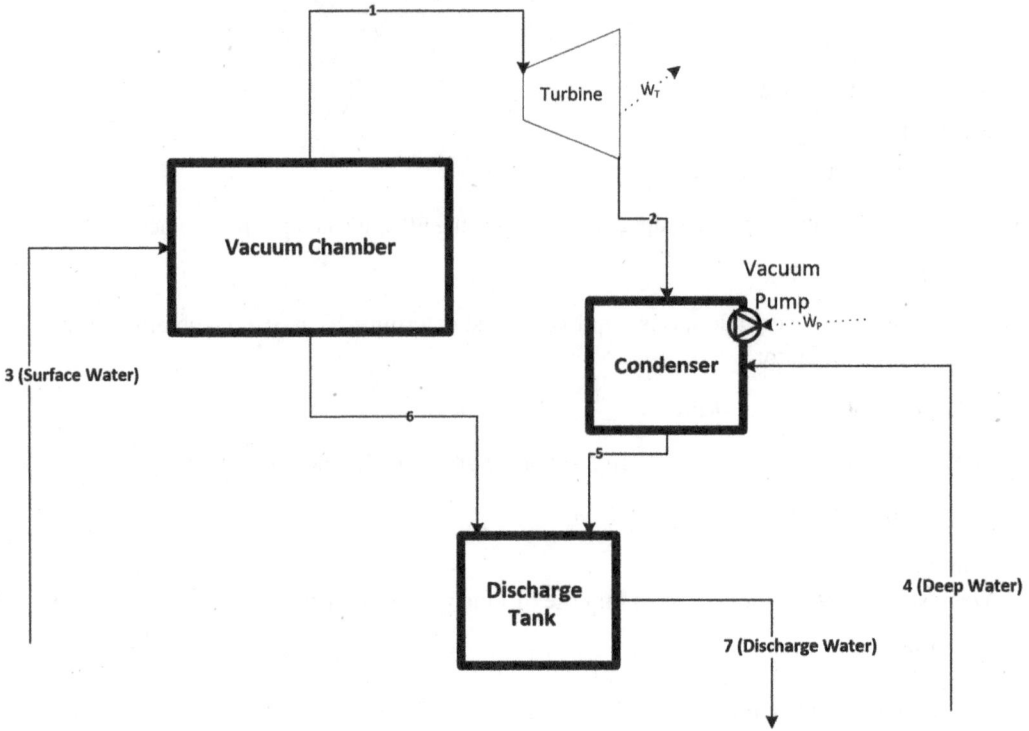

Figure 10.30 The system layout for the open-cycle ocean thermal energy conversion system in Illustrative Example 10.12.

Table 10.3 The properties of the working fluid in Illustrative Example 10.12.

State	Mass Flow Rate (kg/s)	Pressure (kPa)	Temperature (°C)	Specific Enthalpy (kJ/kg)
1	1.12	2.4	27	?
2	1.12	1	10	2519
3	100	100	27	113.3
6	98.88	2.4	27	85.65

Using the energy balance equation for the turbine as follows, we obtain

$$\dot{m}_1 h_1 = \dot{m}_2 h_2 + \dot{W}_T$$

$$1.12 \times 2551 = 1.12 \times 2519$$

The turbine work rate is found to be $\dot{W}_T = 35.84 \ kW$.
b) The energy efficiency of the power plant is obtained using

$$\eta_{en} = \frac{\dot{W}_{net}}{\dot{E}_{in}} = \frac{\dot{W}_T - \dot{W}_P}{\dot{E}_{in}} = \frac{\dot{W}_T - \dot{W}_P}{\dot{m}_1 h_1}$$

where $\dot{W}_{net}$ is the net work in the system. It is then calculated as follows:

$$\dot{W}_{net} = \dot{W}_T - \dot{W}_P$$

Thus, the energy efficiency is calculated as:

$$\eta_{en} = \frac{\dot{W}_{net}}{\dot{E}_{in}} = \frac{\dot{W}_T - \dot{W}_P}{\dot{m}_1 h_1} = \frac{35.84 - 7}{1.13 \times 2519} = 0.01 \ or \ 1\%$$

10.5.5 Energy Production with Salinity Gradient in Oceans

There is a unique process where electricity is generated by pressure difference between fresh and ocean water caused essentially by the difference in ionic concentrations. Figure 10.31 shows the schematic view of the power generation with salinity gradient of the oceans. Ocean water has a higher osmotic pressure than freshwater due to its high concentration of salt. Ocean salinity gradients can be used to generate electricity through a process known as osmotic power. This process involves the use of a membrane that separates saltwater and freshwater, with the natural osmotic pressure created by the difference in salt concentrations on either side of the membrane being used to drive a turbine and generate electricity. While the theoretical potential for this type of energy is significant, the technology is still in the early stages of development and has not yet been demonstrated on a commercial scale.

Most salinity gradient technologies are associated with impacts on the physical environment and changes in water quality. By mixing freshwater with seawater, nutrient-poor water is flushed out and nutrient- and oxygen-rich water is brought in, creating a brackish water habitat that contains some of the planet's most productive ecosystems. There is a wide variety of biological and physical life in these areas. The mixing process could be sped up, the balance between freshwater

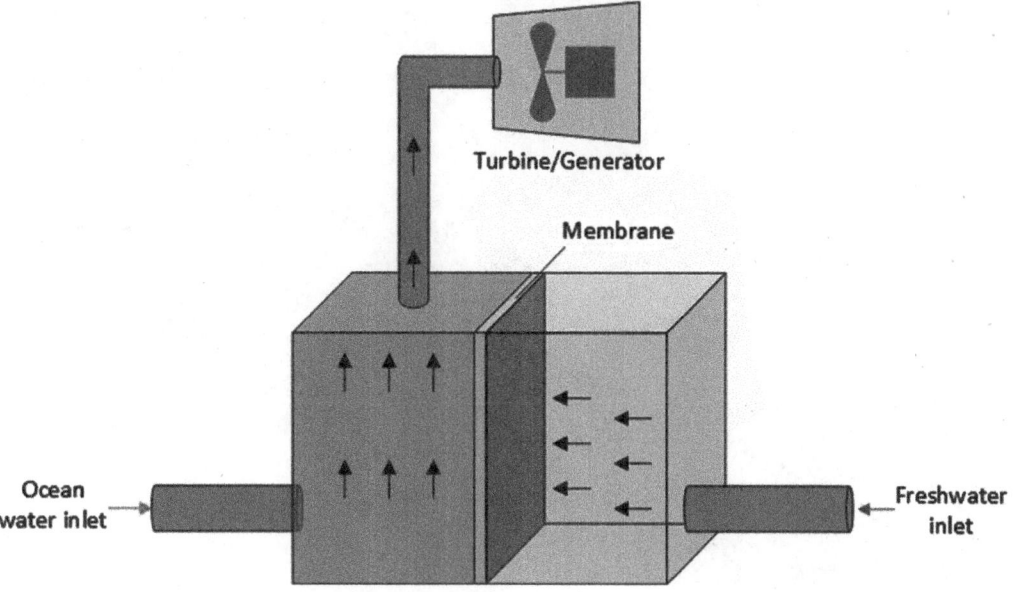

Figure 10.31 The schematic illustration of the power generation with salinity gradient of the oceans.

and saltwater could be altered, or organisms could be exposed at intake and discharge points to threats. The impacts of brackish water could be reduced by releasing it into the middle of the water column and covering the intake tubes with screens. It is essential to avoid water stress and scarce regions to prevent the main socioeconomic concern associated with salinity gradient technology, which involves diverting freshwater resources for power generation. As a result of limited deployments and information on these technologies, there is much uncertainty about their potential environmental impacts. Further research is required to understand these impacts better.

The world's first salinity gradient power generation plant has become operational in Tofte, Norway in 2015. The prototype plant has a designed capacity to generate 10 kW of electricity.

Illustrative Example 10.13   In a power generation plant, which uses the salinity gradient, shown in Figure 10.32, the power generated is used for the agricultural watering purposes to pump the water far location from the river, since there is no way to create a dam. The pumping power is required for irrigation is 10 kW. The salinity gradient is able to increase the water height up to 5 m in the saline side. If water pumps is used for 10 h, calculate the ejected freshwater volume from the river. The efficiencies of the turbine, generator and piping are 0.80, 0.90 and 0.95, respectively.

Solution:

In the power generation system, it is expected to generate 10 kW for 10 h. In other words, 100 kWh of energy should be generated via this power plant. In order to run the desired pumps in the system,

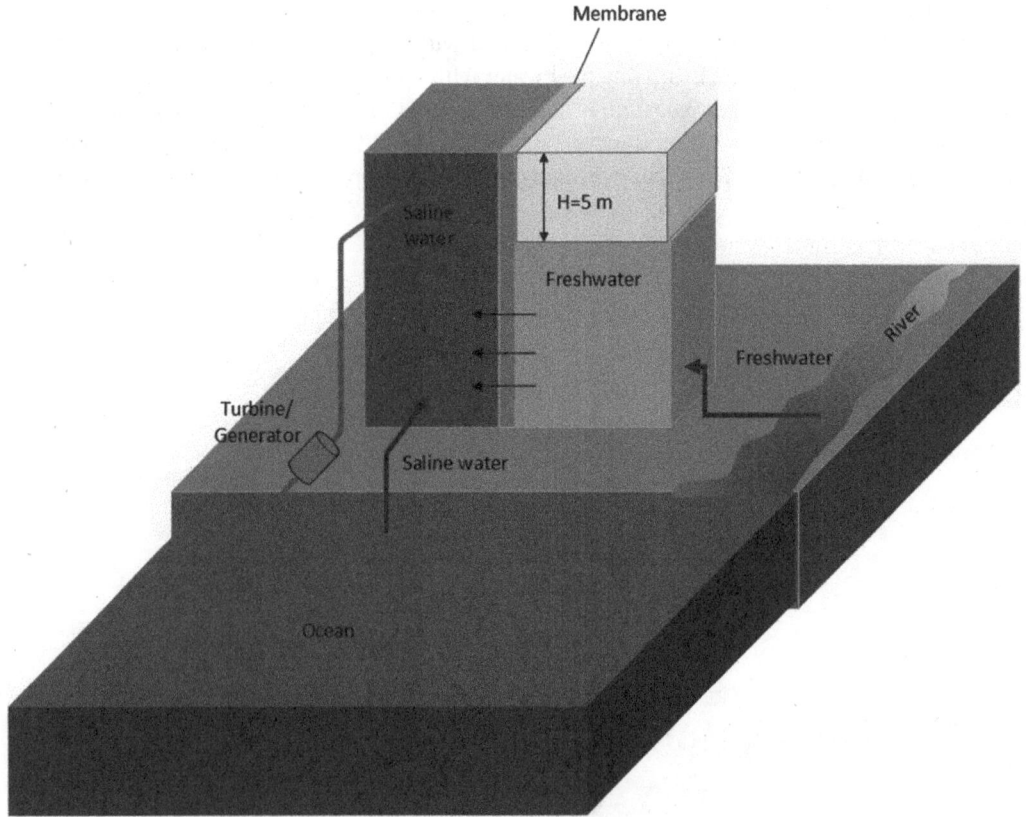

Figure 10.32   The schematic illustration of a salinity gradient-based power generation.

the turbine which has the capacity of 10 kW should be run by the water which is elevated by the salinity effect. The water height required by the salinity gradient is given to be 5 m.

In order to determine the water required for running the system, we can use Eq. 10.14 as:

$$\dot{W}_{electricity} = \dot{W}_{max}\eta_t\eta_g\eta_p$$

When we write $\dot{W}_{max}$ into this equation, we obtain

$$\dot{W}_{electricity} = \dot{m}_1 gH\eta_t\eta_g\eta_p$$

Here, $\dot{W}_{electricity}$ is 10 kW in this problem as it is the electricity needed by the water pumps. From this equation, we further calculate

$$10,000 = \dot{m}_1 \times 9.81 \times 5 \times 0.8 \times 0.9 \times 0.95$$

$$10,000 = \dot{m}_1 \times 33.55$$

The water flow rate is found to be $\dot{m}_1 = 298.06$ kg/s.
This system will run for 10 h. So, the total water mass required for the power generation is found to be:

$$m_{total} = \dot{m}_1 \times 10 \times 3600 = 10,730,160\,kg \ or \ 10,730.16\,tons$$

Note that a total of 10,730 tons of freshwater should be ejected from the river to run the system for 10 h.

10.6 Closing Remarks

This chapter dwells on water (hydro)-based renewable energy sources, including hydro and ocean energies. The first part of the chapter primarily focuses on hydro energy options, hydro turbines and hydro energy systems where thermodynamic analysis, based on balance equations for mass, energy entropy and exergy and performance assessment studies and examples presented for various applications. It also discusses pros and cons of hydropower plants and their utilization. The second part of the chapter is about ocean (marine) energy sources and systems, and the ocean energy systems are classified based on tidal, wave, ocean thermal, and salinity gradient and cover their discussion with the examples and applications.

References

1 What is generation capacity? Office of Nuclear Energy. Available online https://www.energy.gov/ne/articles/what-generation-capacity (accessed 28 December 2022).

2 Share of electricity production by source. Our World in Data. Available online https://ourworldindata.org/grapher/share-electricity-source-facet?country=~OWID_WRL (accessed 29 December 2022).

3 Chen, J., Yang, H.X., Liu, C.P. et al. (2013). A novel vertical axis water turbine for power generation from water pipelines. *Energy* 54: 184–193. doi: 10.1016/j.energy.2013.01.064.

Questions/Problems

Questions

1 Describe hydro sources and their potential utilization with examples.
2 Classify hydro sources and identify suitable applications.
3 Classify hydro turbines and discuss their pros and cons.
4 Discuss advantages and disadvantages of hydropower plants.
5 Explain how each type of hydro turbine works.
6 Develop a mathematical relationship between head and volumetric flow rate.
7 Explain how to use a pump as a hydro turbine and discuss potential advantages and disadvantages.
8 Describe ocean energy sources and their potential utilization with examples.
9 Classify ocean energy sources and identify suitable applications.
10 Classify ocean energy systems and discuss their pros and cons.
11 Discuss advantages and disadvantages of ocean thermal energy conversion plants.
12 Explain how salinity gradient works for power generation.
13 Identify hydro sources in your town and assess potentially for power generation.
14 Identify ocean energy sources in your country and asses the potential use for energy applications.
15 Discuss salinity gradient options for power generation and identify the theoretical limits.

Problems

1 The water in a dam is used to generate electricity by the installation of a hydraulic turbine-generator at a location where the depth of the water is 100 m, as illustrated in the figure below. The flow rate of the water is 4000 kg/s. Calculate the maximum power to be able to be generated in this dam.

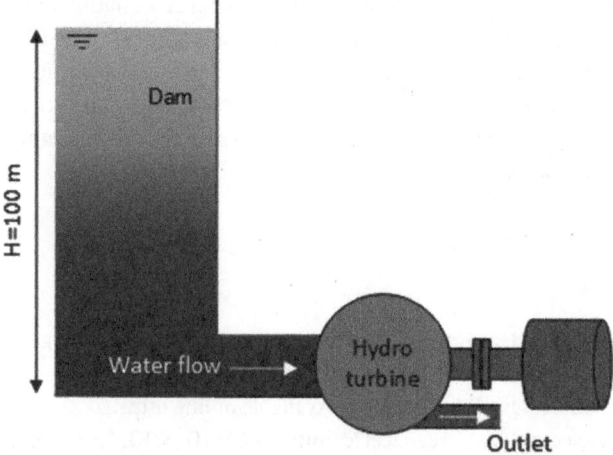

2 Consider the system in Problem 1. In the power plant, the turbine efficiency is 0.80, the generator efficiency is 0.90, and the piping efficiency is 0.95. Calculate the net electricity output from the power plant. Calculate the energy and exergy efficiencies for the power plant.

3 A hydro power station is expected to draw 2400 m³/s of water from the river. In the power station, 30 turbine units are available, and the average water flow rate through each of the turbines is 80 m³/s. The elevation difference between the free surfaces upstream and downstream of the dam is 60 m. If the turbine and generator efficiencies are 90% and 95%, determine the average power generation rate in the power station. Calculate the overall efficiency of the power generation. The losses in the piping are neglected (g=9.81 m/s², ρ=997 kg/m³).

4 Consider the hydropower plant shown in the figure below where the head losses through the pipes are known to be $\dfrac{3V_1^2}{2g}$ and $\dfrac{2V_2^2}{2g}$ in pipes 1 and 2 respectively. The elevations at the given points are known as H_a=145 m, H_b=35 m, H_c=H_d=40 m. The diameters are known as D_1=0.35 m and D_2=0.65 m. The volume flow rate at C is 1 m³/s. Determine the power output from the turbine. Calculate the overall energy efficiency (g=9.81 m/s², ρ=1000 kg/m³).

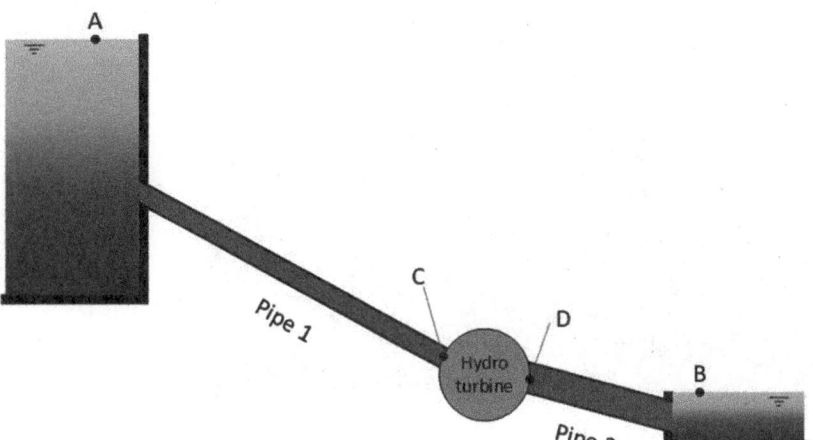

5 Restudy Problem 4 by considering that the outlet diameter is 1 m with the elevation H_a=200 m.

6 Repeat illustrative example 10.6 with some assumed values for a selected location.

7 Repeat illustrative example 10.7 with some assumed values for a selected location.

8 Repeat illustrative example 10.8 with some assumed values for a selected location.

9 Water steadily flows over a tidal turbine at a velocity of 15 km/h, having a diameter of 30 m. The water exit velocity is assumed to be 11 km/h. The inlet water temperature and pressure are 25°C and 600 kPa, and the outlet temperature is 24.992°C. The heat loss from the turbine is 4.5 kW. For a reference temperature of 25°C and pressure of 100 kPa.
   a Write the mass, energy, entropy and exergy balance equations for the turbine.
   b Calculate the power generated by the turbine.
   c Calculate the energy and exergy efficiencies.
   Take $h_1 = 126.2\dfrac{kJ}{kg}$, $s_1 = 0.4366\dfrac{kJ}{kgK}$, $h_0 = 104.9\dfrac{kJ}{kg}$, $s_0 = 0.3672\dfrac{kJ}{kg}K$.
   The reference temperature is 25°C, the density of water is 1000 kg/m³, and specific heat of water is c_p=4.2 kJ/kg K.

10 For an ocean thermal power plant, it is known that the temperature of the surface level is 29°C and 5°C in the summer time for the deep in the ocean. An engineer claims to generate electricity with a heat engine that works with 10% efficiency between these two temperature levels. Is this device possible by checking through the Carnot concept?

11 For an ocean thermal power plant, it is known that the temperature of the upper level is 10°C with 1°C at the lower level in winter time operation. An engineer claims to generate electricity with a heat engine that works with 5% efficiency between these two levels. Study this to see if this is conceptually possible?

12 In a tidal barrage power generation system, shown in the figure below, ocean water is collected in the reservoir where located 65 m higher than the sea level. A hydro turbine is used for power generation. The flow rate of the water is 5000 kg/s. If turbine, generator and piping efficiencies are 0.80, 0.95 and 0.90, respectively, calculate the electricity generation rate in this tidal barrage power generation system.

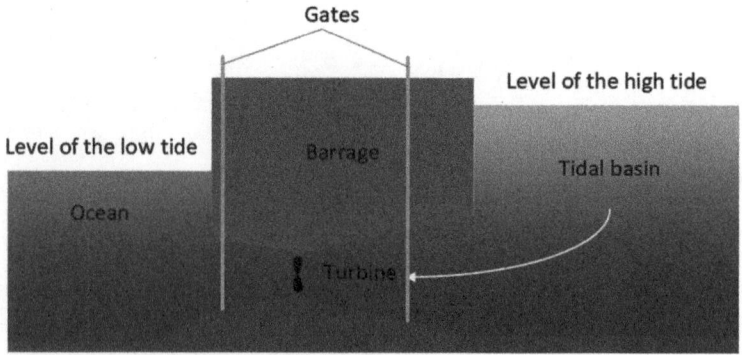

13 For an ocean thermal power generation power plant, as seen in the figure below, the ocean water is used for power generation in an organic Rankine cycle using ammonia. The parameters for the working fluids are given in the table below.

a Write the mass, energy, entropy and exergy balance equations for each component of the system.
b Find the work rate of the turbine.
c Calculate the energy and exergy efficiencies.

State	Mass Flow Rate(kg/s)	Substance	T(°C)	h(kJ/kg)	ex(kJ/kg)
1	2	Water	30	140.7	417.3
2	2	Water	28	130.4	417.1
3	1	Ammonia	−25.2	1409	
4	1	Ammonia	−25.2	?	
5	1	Ammonia	−35	1418	

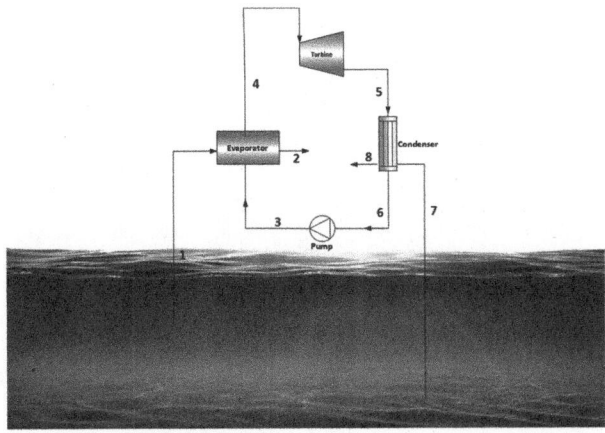

14 Repeat illustrative example 10.12 with some assumed values for a selected location.

15 In a power generation plant, which uses the salinity gradient, as shown in the figure below, the power generated is used for the agricultural watering purposes to pump the water to a remote location from the river, since there is no way to build a dam. The pumping power is required for irrigation is 500 kW. The salinity gradient is able to increase the water height up to 20 m in the saline side. If water pumps is used for 10 h, calculate the ejected freshwater volume from the river. The efficiencies of the turbine, generator and piping are 0.80, 0.90 and 0.95, respectively.

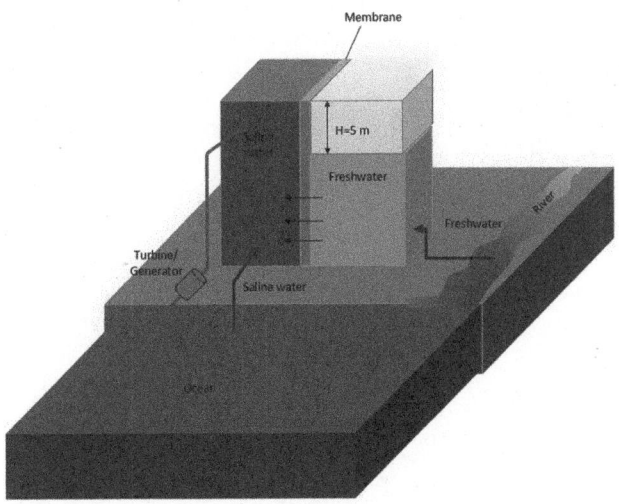

11

Energy Storage

11.1 Introduction

Mismatches between demand and supply profiles are common almost in every energy sector. The use of energy storage techniques is primarily aimed to help solve such a mismatch problem in a unique way. It is well known that this kind of mismatch occurs because of the availability of energy sources. For example, the availability of solar energy really depends on meteorological conditions, such as clouds, humidity, etc. There are also variations in its availability on a daily and yearly basis. In wind energy systems, there are fluctuations in their availability in terms of wind speeds and directions. Therefore, the capacity factors are used for energy systems to define their availability and power generation capacity, as mentioned in previous chapters. Furthermore, energy demands also vary with time in a daily, weekly or yearly basis. Such an unstable energy supply profiles also cause this dissonance. These mismatches may cause peak loads, additional power requirements, higher energy costs, using the energy devices in ineffective ways, etc.

Energy storage methods are considered a unique solution to solve mismatch between the energy supply and demand profiles, extend the availability of the energy sources (especially for renewables), provide lower cost operation, and run the energy systems in more effective conditions. Therefore, today, energy storage is a critical topic for a sustainable future to adjust energy supplies and demands and recover the losses and excess energy supplies. This can be more meaningful to explain the importance of energy storage systems over the 3S + 2S = 5S concept as illustrated in Figure 11.1. As seen in this figure, energy storage can be integrated into the energy systems between source(s) and system(s) or system(s) and service(s), or in both. The logic behind the energy storage techniques is the same for every case that it helps overcome the mismatch between demand and supply.

Energy demands also vary in a time frame (daily, weekly, seasonal, or yearly) since demand levels and capacities are affected by many factors. Figure 11.2 provides detailed information regarding the energy demand side issues. For example, the energy demands of the UK are in daily, weekly, and yearly basis in Figure 11.2. It is clear from Figure 11.2 that the energy demands fluctuate in all time frames considered. For example, in daily period (Figure 11.2a) which is the first days of January, March, May, July, September, and November, the energy demands are minimum during the night-time due to reducing activities in all sectors. It peaks between 16:00 and 19:00. When we consider the daily demands, the energy supply should be adjusted considering the daily peak demands.

Also, when the energy demand variation in a weekly demand is considered, it is clearly seen that the energy demands are lower on Fridays and at the weekends due to reducing activities in industrial and commercial sectors as seen from Figure 11.2b. Also, the energy demands in December

Introduction to Energy Systems, First Edition. Ibrahim Dincer and Dogan Erdemir.
© 2023 John Wiley & Sons, Inc. Published 2023 by John Wiley & Sons, Inc.
Companion Website: www.wiley.com/go/Dincer/Introduction_to_Energy_Systems

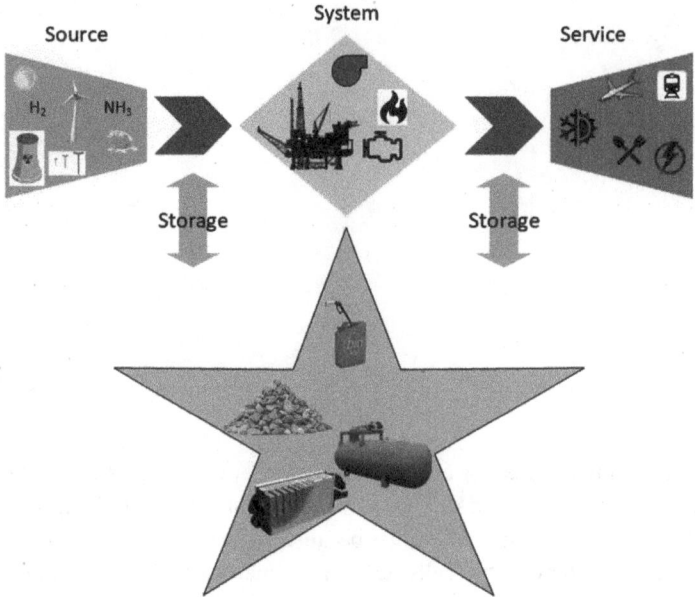

Figure 11.1 The 3S + 2S = 5S concept.

and June are compared to the energy demands in December are quite higher than June, due to increasing heating demand in the winter in the UK. However, for the countries that have wild summers, the energy demands are higher in summer than the winter. Also, since there are many sources of heating demand, cooling has more impact on the electricity grid load.

When yearly energy demand of the UK is considered in Figure 11.2c, there is a significant fluctuation in the energy demand between the winter and summer. It is evident in Figure 11.2 that, there is a serious fluctuation in daily, weekly, and yearly basis. When the average values are considered, there is a significant difference between the peak and average load. It occurs in a short time when compared with the average load. The main problem with the energy demand side is to meet the peak loads, which are quite a bit higher than the average load and needed for a short time.

As stated earlier, energy storage systems are unique solution to solve the problem due to the mismatch between energy supply and demand profiles in any scale, namely from a building to community, or from city to country. Energy storage can store energy when energy source is available to use later when energy source is not available. For example, in solar thermal systems, the solar energy can be stored in a medium in the heat form to use when solar energy is not available. This storage can be done for a short time (hourly, daily, etc.) or long period (seasonal or annual). Batteries are considered a significant option to store electricity in an electrochemical form. Hydrogen and ammonia are treated as unique energy storage options due to their high capacities and carbon-free nature.

Energy storage may further be defined "to store energy when energy source is available and excess capacity in order to use later when energy source is not available or adequate for meeting the demand." The availability of the energy source and demand profile are the decision factors for the energy storage systems as they are directly related to the operational periods of the energy storage, which are charging, storing, and discharging periods. Further details on energy storage techniques and their operational steps will be discussed in the following chapters.

The most widely deployed grid-scale storage technology today is pumped-storage hydropower. It is estimated that the total installed capacity will reach 160 GW by 2021. Around 8500 GWh of energy

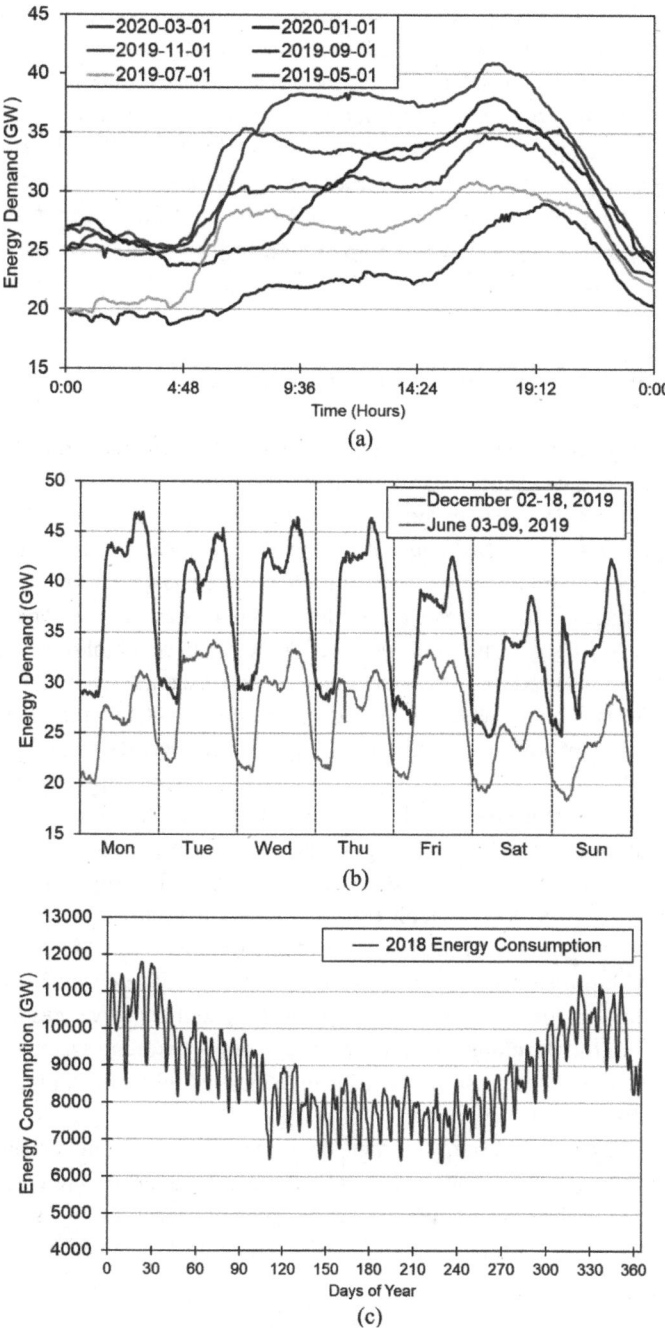

Figure 11.2 Energy demand in the UK: (a) daily, (b) weekly and (c) yearly basis (data from [1]).

was stored globally in 2020, accounting for over 90% of the world's total electricity storage capacity. As of 2021, the USA has the largest energy storage capacities. However, China has been investing heavily in energy storage and may surpass the USA in the near future. Most plants in operation today balance the supply and demand profiles on a daily basis [2]. In addition to grid-scale energy storage, there are millions of energy storage systems that are used globally for many purposes. For example, each solar domestic hot water system has a hot water storage tank(s) which is a cost-effective sensible energy storage method. Batteries are potentially considered for the electricity storage for buildings.

In this chapter, energy storage techniques are presented through various systems and processes with practical applications. Also, energy storage methods are classified and dicussed with historical developments. Furthermore, working principles of the energy storage systems are described through practical applications and examples. This is then followed by the analysis and performance evaluation of energy storage systems. Finally, each energy storage technique is specifically discussed from every perspective to provide technical and operational details.

11.2 Historical Development of Energy Storage Operations

The history of energy storage goes back to early days of humanity. People collected and stored the ice and snow for later cooling purposes, or wood was collected and stored in the summer to meet the heating needs in the winter when the ground was covered with snow and wood was wet to burn. Figure 11.3 illustrates some historical events and details of energy storage systems in a chronological manner. To store cooled or frozen goods, caves, underground pits or other naturally shaped formations have been used for centuries as thermal energy storage (TES) systems. The development of modern TES systems, however, began in the early 20th century. TES systems were developed primarily to cool buildings in the 1930s and 1940s. These early TES systems used ice storage as a means of thermal energy storage.

The first TES systems using sensible heat storage, which store heat in a material (such as water or a phase-change material) with a change in temperature, rather than a phase change, were developed in the 1960s. These systems were primarily used in industrial process heating and cooling applications. The first TES systems for use in power generation were developed in the 1970s and 1980s. These systems used molten salt as a storage medium and were primarily used in concentrated solar power plants.

In the 19th century, the invention of battery allowed for the storage of electrical energy for the first time. The lead-acid battery, invented in 1859, was widely used in automobiles and other applications. In the early 20th century, the development of the electric grid and the use of fossil fuels as a primary energy source led to the widespread use of large-scale energy storage systems, such as hydroelectric dams and pumped hydro storage. In the latter part of the 20th century, advances in materials science and electronics led to the development of new types of energy storage systems, such as lithium-ion batteries and fuel cells. These systems have become increasingly important as the use of renewable energy sources has grown.

Recent years have shown a renewed interest in energy storage, driven by advances in materials science and technology, as well as growing concerns about climate change and the need to reduce dependence on fossil fuels. This has led to the development of a wide range of energy storage systems, including advanced batteries, flywheels, compressed air energy storage, and thermal energy storage.

11.3 Energy Storage Methods

Despite the fact that the energy is able to be stored theoretically in any forms of energy that may be mechanical, thermal, chemical, electrochemical, magnetic, electromagnetic, or biological, it can be classified as illustrated in Figure 11.4, considering today's technologies and energy storage capacities. Today's energy storage capacities are taken into consideration, mechanical energy storage, especially

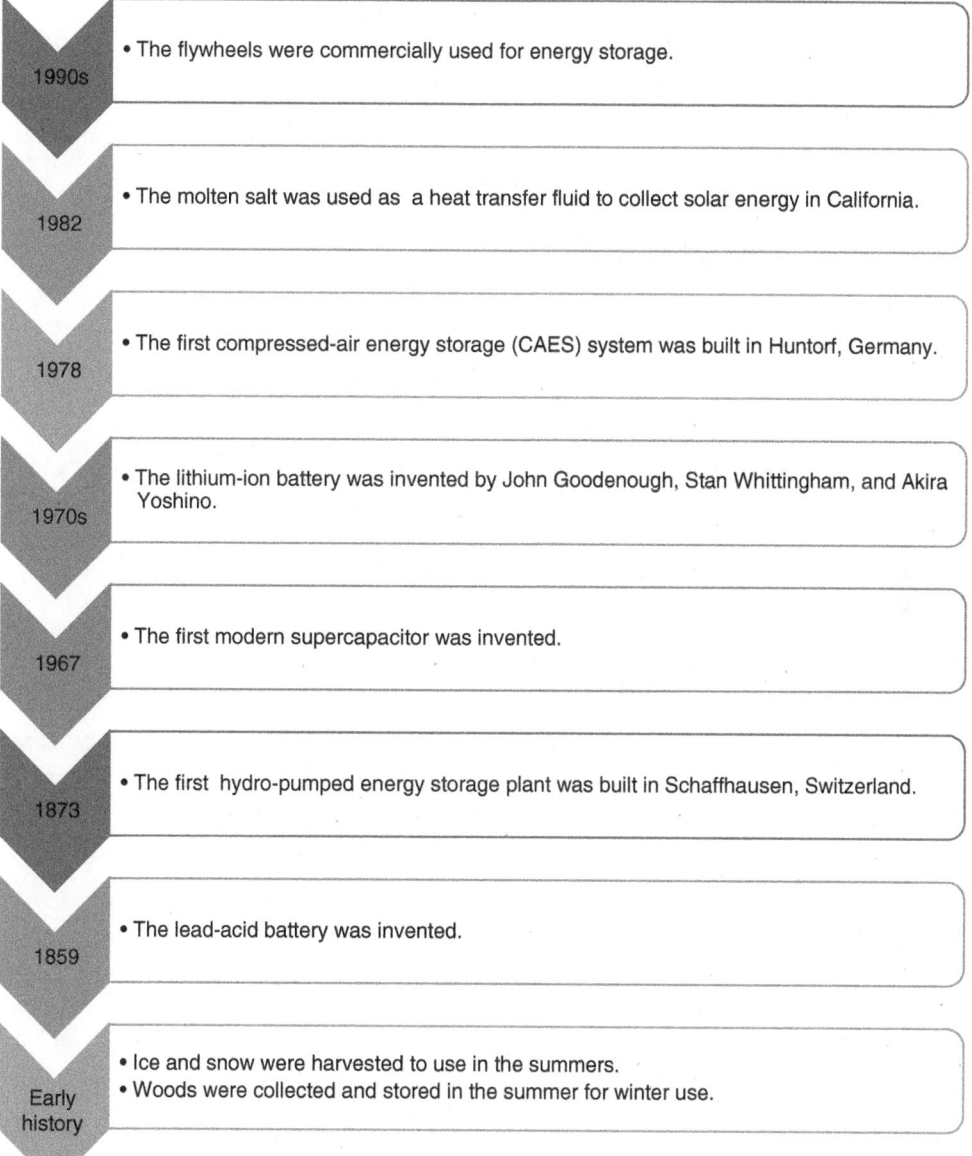

1990s
• The flywheels were commercially used for energy storage.

1982
• The molten salt was used as a heat transfer fluid to collect solar energy in California.

1978
• The first compressed-air energy storage (CAES) system was built in Huntorf, Germany.

1970s
• The lithium-ion battery was invented by John Goodenough, Stan Whittingham, and Akira Yoshino.

1967
• The first modern supercapacitor was invented.

1873
• The first hydro-pumped energy storage plant was built in Schaffhausen, Switzerland.

1859
• The lead-acid battery was invented.

Early history
• Ice and snow were harvested to use in the summers.
• Woods were collected and stored in the summer for winter use.

Figure 11.3 Some historical landmark type events energy storage methods and applications.

pumped-energy storage comes at the first place. However, chemical energy storage systems present promising performance due to their high energy contents, especially hydrogen and ammonia. On the other hand, the use of batteries is increasing to store electricity. Capacitors and supercapacitors receive increased attention with their high energy discharge potentials. Flywheels are considered a significant option to store electricity for a short time with a high energy discharge rate.

Each energy storage method has particular advantages and disadvantages depending on its specific area of use. Therefore, they have many different sub-applications as a part of energy systems. In Figure 11.4, energy formations and storage materials are classified according to energy storage methods.

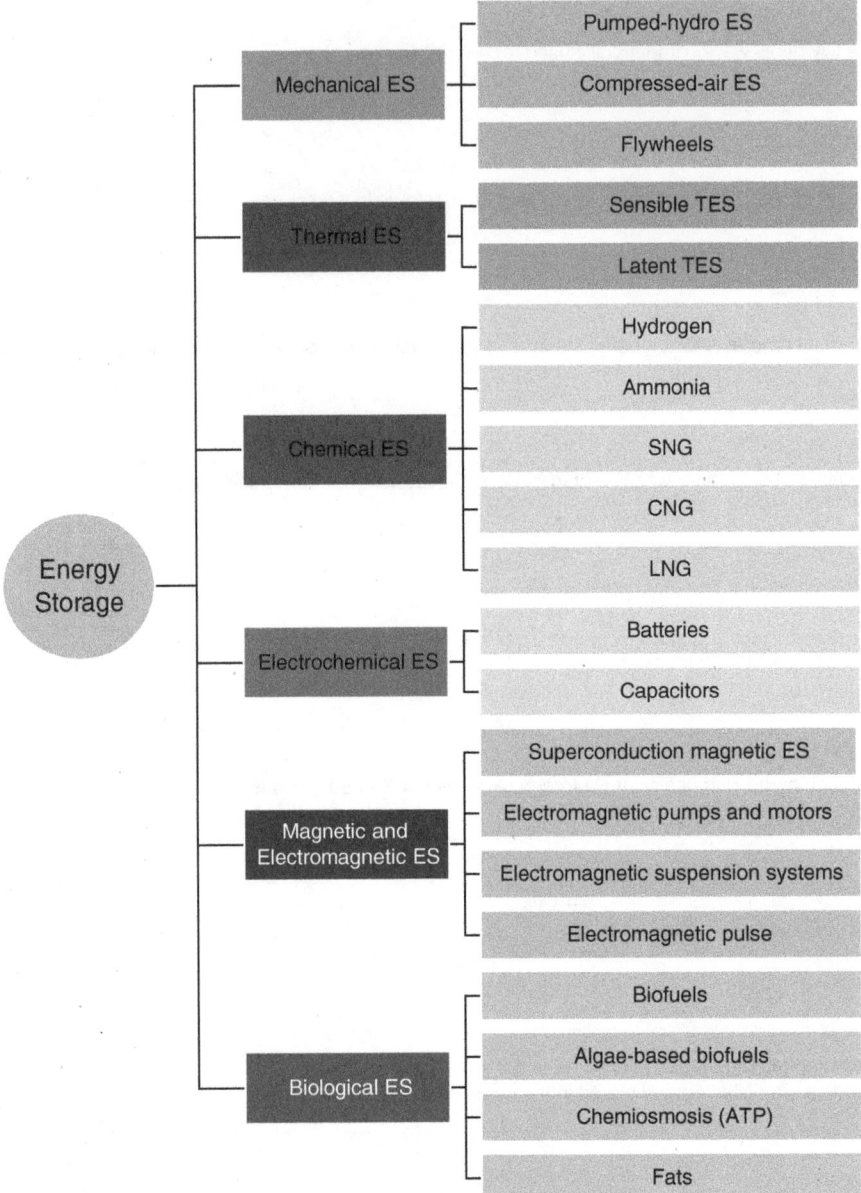

Figure 11.4 Classification of common energy storage methods.

Many energy storage methods are still under improvement to improve their performance or reduce their cost.

In this book, the common and promising energy storage techniques, which are pumped-hydro, compressed-air, flywheels, thermal energy storage methods, hydrogen, and batteries, are discussed with illustrative examples. The rests are briefly explained for readers under the section of other energy storage technologies.

11.4 Working Principles of Energy Storage Systems

The basic working principles of energy storage systems consist of three main processes that follow one another, which are charging, storing and discharging periods. Figure 11.5 generally exhibits the operating periods of energy storage systems. These periods are achieved in a cyclic manner. In the charging period, the energy source is charged into a storage medium either converting to another energy type or keeping in the same form with the source. For example, in order to store electricity in batteries, the electricity is converted into the electrochemical form. Also, electricity can be stored as kinetic energy in the flywheels by increasing the rotational velocity of the flywheel. As another example, it is a common way to store solar energy as the heat for both heating/cooling needs and power generation purposes by converting the solar energy into the heat in a storage medium.

The storing period aims to hold the energy stored for a while until the energy is needed. In order to increase performance of the energy storage, the losses should be minimized. In some cases, the storing period does not exist when there is no time gap between the availability of the energy source and the energy demand. In these cases, the stored energy is used right after the energy is charged.

In the discharging period, the stored energy is discharged to meet the energy demand, when energy source is not available or insufficient to meet the energy demand. The stored energy can be used for meeting the energy demand directly in the form of energy stored, or it can be converted into the other energy forms which are demanded.

The periods of these operating steps are adjusted according to the purpose of the energy storage technique to be applied. They vary from a few seconds to a year. Therefore, the energy storage systems are often classified according to the period of the energy storage, which are short-term energy storage and long-term energy storage. While the short-term is for the periods up to one week of period, the long-term mentions the seasonal energy storage.

Energy storage (ES) techniques are expected to provide an optimum way to benefit from energy sources by balancing energy supply and demand, minimizing the losses, extending the energy source active time, lowering capacities of the devices, etc. ES methods offer a great potential to enhance energy conversion processes efficiently and manage energy sources sustainably. Some advantages of ES techniques may be listed as follows:

- Reduces energy consumption costs
- Improves the efficiencies of the system and system components
- Extends the period of use of energy source
- Decreases the capacity and size of the system's equipment
- Reduces the initial investment and service costs
- Balances the energy supply and demand
- Provides resilient operation

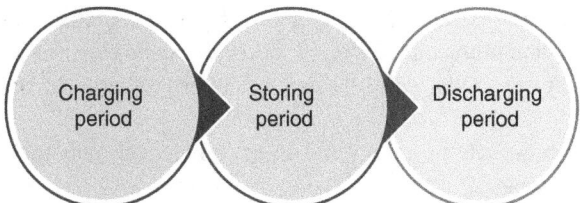

Figure 11.5 The operational periods of energy storage.

- Helps increase the generation capacity
- Increases the flexibility of working conditions and hours
- Reduces the fossil-based energy consumption
- Reduces the carbon emission and environmental impacts

The utilization strategy of an energy storage system may bring multiple benefits. Of course, the benefits may vary from one system to another. For example, in a building, it is expected to reduce its energy consumption and hence their costs or extend the duration of its benefit from an energy source. Ideally, these goals should be achieved with the lowest payback period or with the greatest cost savings. In summary, such economic benefits will help provide feasible and more appealing solutions.

11.5   Analysis of Energy Storage Systems

The analysis of any energy storage system is generally performed through the overall energy and exergy balance equations under the specific operating conditions. The overall energy balance equation may be written as follows:

$$\left(Energy\ charged\right) - \left[\left(Energy\ discharged\right) + \left(Energy\ loss\right)\right] = \left(Energy\ accumulation\right)$$

The exergy balance equation is written similarly as follows:

$$\left(Exergy\ charged\right) - \left[\left(Exergy\ discharged\right) + \left(Exergy\ loss\right)\right] + \left(Exergy\ destruction\right)$$
$$= \left(Energy\ accumulation\right)$$

A thermodynamic analysis of any energy storage system is performed by considering three periods of energy storage. As the duration of these periods is significant for the performance of energy storage, they should be considered in the analysis. The overall energy and exergy efficiencies for energy storage systems are generally defined to be:

$$\eta_{en} = \frac{\sum_{t=0}^{t_{disch}} \dot{E}_{disch}}{\sum_{t=0}^{t_{ch}} \dot{E}_{ch}} \tag{11.1}$$

and

$$\eta_{ex} = \frac{\sum_{t=0}^{t_{disch}} \dot{Ex}_{disch}}{\sum_{t=0}^{t_{ch}} \dot{Ex}_{ch}} \tag{11.2}$$

where $\dot{E}_{ch}$ and $\dot{E}_{disch}$ are the energy rates during the charging and discharging, respectively. "t_{ch}" and "t_{disch}" are the duration of the charging and discharging periods, respectively. $\dot{E}_{ch}$, the rate of energy to be charged, should be calculated depending on the energy source, it can be electricity, work, kinetic energy, potential energy, heat, etc. This may be the same for the energy to be discharged.

In order to improve the performance of each period, the energy and exergy efficiencies may specifically be written for each period of energy storage.

11.6 Mechanical Energy Storage Methods

Energy may be stored in the form of mechanical energy, such as kinetic and potential energies. While the kinetic energy-based storage is performed by changing the linear or rotational velocities of the objects, the potential energy-based energy storage is done by elevating of the materials or compressing the gasses. Since mechanical energy storage provides higher capacity storage than others through the well-known techniques, such as compression, pumping, or rotating, it is considered more advantageous. They are generally used for storing electricity. Mechanical energy storage techniques may generally be classified into three categories as follows:

- Pumped-hydro energy storage
- Compressed-air energy storage
- Flywheels

In addition to these techniques, there are various applications of mechanical energy storage such as Potter's wheel, springs, elevation of the objects. However, due to their energy storage capacities and practical applications, they will not be part of this book.

11.6.1 Pumped-Hydro Energy Storage

Pumped-hydro energy storage is a method of storing energy by pumping water from a lower elevation reservoir to an upper elevation reservoir. Figure 11.6 demonstrates the schematic illustration of the pumped-hydro energy storage systems. The stored energy can be converted into electrical energy when needed by releasing the water from the upper reservoir and allowing it to flow through a turbine, which drives a generator, as defined earlier in hydropower generation systems (see Chapter 10). This process is managed in reversed form and can be used for both energy storage and generation.

Pumped energy storage is a well-established technology and is currently one of the most widely used forms of energy storage in the world. It is particularly well-suited for use in conjunction with hydroelectric power plants but can also be used in conjunction with other forms of power generation. Pumped energy storage has several advantages over other forms of energy storage, such as batteries. It has a high energy density, meaning that it can store large amounts of energy in a relatively small space. It is also relatively inexpensive and has a long lifespan. However, it has several limitations as well such as the need for two reservoirs, one at a higher elevation than the other, and the environmental impact of building such reservoirs.

As indicated in Figure 11.6b, the lower reservoir can be a lake, sea, ocean, etc. or an underground source. Underground water reservoirs are a natural lower-level reservoir. When water is pumped to ground level artificial or natural reservoirs, energy can easily be stored thanks to distance between the underground and ground levels.

The working principles of pumped ES are the following:

- **Charging period**: During the off-peak hours, generally at night, when the energy demands are low and the unit cost of electricity is cheaper, pumps are operated to pump the water from a lower level to the upper level.
- **Storing period**: The elevated water is kept in the upper reservoir until there is a demand for use. It is important to benefit from this during the peak demand hours.
- **Discharging period**: During peak hours, when the energy demands are high and the unit cost of electricity is higher, the pumped water in the upper level flows to the lower level and this flow turns the turbine. A generator assembled to the turbine is used for generating electricity.

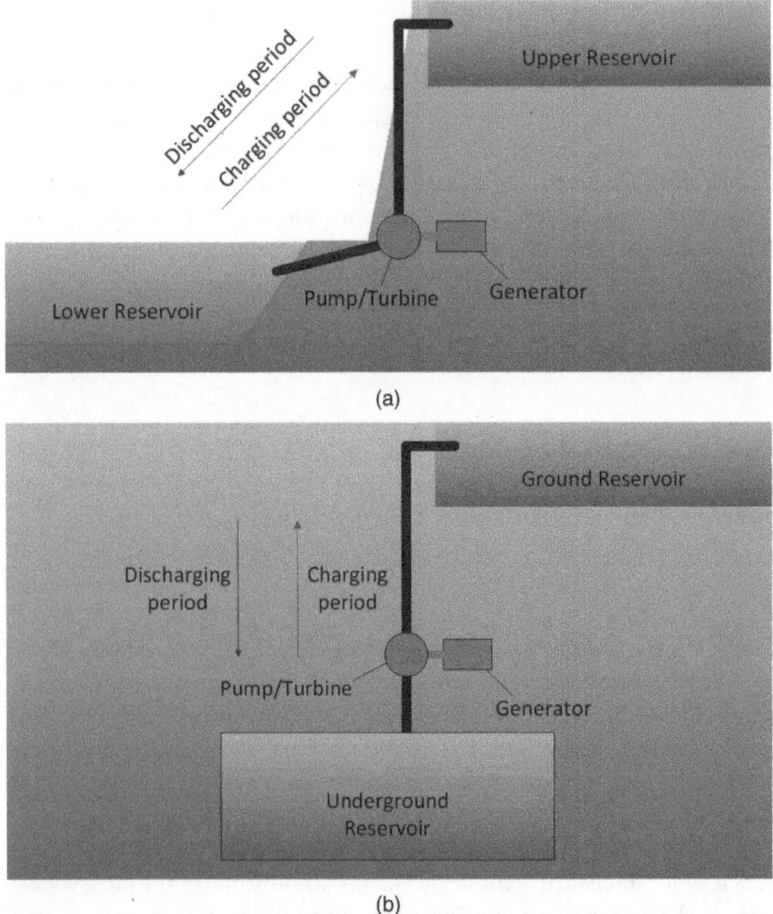

Figure 11.6 The schematic illustration for (a) a pumped-hydro energy storage system and (b) an underground pumped-hydro energy storage system.

As of 2021, there were approximately 50 GW of pumped-hydro energy storage capacity installed worldwide. Most of this capacity is located in the USA and China, with smaller amounts in other countries, such as the UK, Japan, Germany and South Korea. The capacity of pumped-hydro energy storage systems varies widely, with some systems having a capacity of just a few megawatts, while others have a capacity of several hundred megawatts.

The global pumped-hydro energy storage system market is expected to grow significantly in the coming years, as the need for energy storage increases with the growth of renewable energy sources such as wind and solar power. It is one of the most efficient and cost-effective ways to store energy on a large scale, and it is expected to play an important role in balancing the intermittent of renewable energy sources and maintaining grid stability.

For the pumped-hydro energy storage systems, there are two significant parameters one the power output to be able to be obtained from the hydro turbine during the discharging period and the pumping power to be consumed during the charging period. The maximum power to be generated from a pumped-hydro energy storage is calculated as follows:

$$\dot{W}_{max} = \rho \dot{V}_{down} gH \tag{11.3}$$

where ρ is density of the water, $\dot{V}$ is the downward flowrate of the water, g is the gravity, and H is the height difference between lower and upper reservoirs. Equation 11.1 defines the ideal power capacity of the pumped ES system. The electricity generation is calculated like in the hydropower generation systems as follows:

$$\dot{W}_{electric} = \eta_t \eta_g \eta_p \dot{W}_{max} \tag{11.4}$$

Here, η_t, η_g, and η_p are the efficiencies of the turbine, generator and piping.

Another significant parameter in the pumped-hydro storage systems is the pumping power that can be calculated as follows:

$$\dot{W}_{pump} = \frac{\rho g H \dot{V}_{up}}{\eta_{P1}} \tag{11.5}$$

where $\dot{V}_{up}$ is the upward flowrate of the water and η_{P1} is the efficiency of the pump.

The efficiency of the pumped-hydro energy storage is defined as follows:

$$\eta_{pumped\ ES} = \frac{\sum_{t=0}^{t_{disch}} \left[\dot{W}_{electric}\ t_{disch}\right]}{\sum_{t=0}^{t_{ch}} \left[\dot{W}_{pump}\ t_{ch}\right]} \tag{11.6}$$

Illustrative Example 11.1 A pumped-hydro storage system is used for storing the off-peak electricity. In the charging period, the water is pumped from the lower reservoir to upper reservoir, as depicted in Figure 11.7. The elevation is 50 m. It is desired to get 20,000 kWh of energy from this pumped-hydro energy storage facility. The charging period takes place between 22:00 and 06:00 (8 h). The discharge period also takes place between 15:00 and 19:00 (4 h). The efficiencies for the turbine, generator and pump are given as 0.85. The losses in the pipes are neglected. Take density of the water as 998 kg/m³.

a) In order to obtain 40,000 kWh of energy, calculate the amount of water required during the charging phase.
b) Calculate the volumetric flow rate of water during the charging phase.

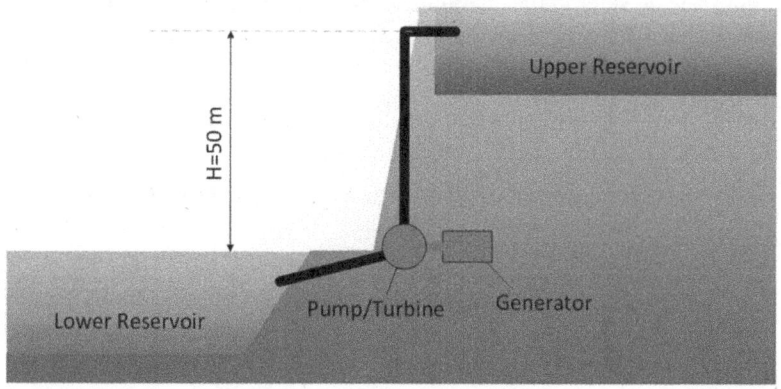

Figure 11.7 The schematic illustration of the pumped-hydro energy storage system in Illustrative Example 11.1.

c) Calculate the pumping power.

d) Calculate the overall energy efficiency of the system.

Solution:

a) As 40,000 kWh of energy will be generated during the discharging period which cover from 22:00 to 6:00, it is firstly required to calculate the turbine power as follows:

$$\dot{W}_{electric} = \frac{E_{total}}{t_{disch}} = \frac{20,000\,(kWh)}{4\,(h)} = 5000\ kW\ or\ 5\ MW$$

In order to calculate the flowrate of the water during the discharging period, the following equation (Eq. 11.4) is used:

$$\dot{W}_{electric} = \eta_t \eta_g \eta_p \rho \dot{V}_{down} gH$$

$$5,000,000 = 0.85 \times 0.85 \times 1 \times 998 \times \dot{V}_{down} \times 9.81 \times 50$$

$$5,000,000 = \dot{V}_{down} \times 353,677.48$$

$$\dot{V}_{down} = 14.14\ m^3/s$$

The water volume to be required to operate the turbine for 4 h can be calculated as follows:

$$V_{total} = \dot{V}_{down} t_{disch} = 14.14 \times 4 \times 3600 = 203,616\ m^3$$

b) In order to run the system, $203,616\ m^3$ of water should be pumped to a higher reservoir. In order to calculate flow rate in the charging period, we need to consider the duration of the charging period.

$$V_{total} = \dot{V}_{up} t_{ch}$$

$$203,616 = \dot{V}_{up} \times 8 \times 3600$$

Thus, the volumetric flow rate of water in the charging period is found to be $\dot{V}_{up} = 7.07\ m^3/s$.

c) The pumping power can be calculated by using Eq. 11.5, as follows:

$$\dot{W}_{pump} = \frac{\rho g H \dot{V}_{up}}{\eta_{P1}} = \frac{998 \times 9.81 \times 50 \times 7.07}{0.85} = 4,071,646.27\ W\ or\ 4,071.65\ kW$$

d) The overall energy efficiency of the pumped-hydro energy storage system is found by using Eq.11.1, as follows:

$$\eta_{en} = \frac{\sum_{t=0}^{t_{disch}} \dot{E}_{disch}}{\sum_{t=0}^{t_{ch}} \dot{E}_{ch}} = \frac{\dot{W}_{electric} t_{disch}}{\dot{W}_{pump} t_{ch}} = \frac{5,000 \times 4}{4,071.65 \times 8} = 0.6140\ or\ 61.40\%$$

Illustrative Example 11.2 In a geothermal-based underground pumped-hydro storage system, the geothermal fluid is pumped from the underground reservoir, which is located 150 m below the ground, as illustrated in Figure 11.8. The temperature of the geothermal fluid is 90°C. Durations of the charging and discharging periods are 6 h and 12 h. The flow rates in the charging and discharging periods are 15 and 30 m³/s. The efficiencies for the turbine, generator, pump and piping are taken 0.90. In addition to the electricity storage, the geothermal fluid is used for heating purposes in the upper reservoir. The temperature of the return water is 80°C. The geothermal fluid is used for heating purposes in the upper reservoir. Take the density of water as 998 kg/m³. The heat losses in the pipelines are negligible.

a) Calculate the volume of water required.
b) Find the energy to be charged and discharged.
c) Determine the amount of heat obtained.

Solution:

a) The volume of water which is pumped up and drawn down is calculated by using either charging or discharging phase. For the discharging phase, one may find the following:

$$V_{total} = \dot{V}_{down}t_{disch} = 30 \times 6 \times 3600 = 648{,}000 \ m^3$$

For the charging period, one may calculate the same below:

$$V_{total} = \dot{V}_{up}t_{ch} = 15 \times 12 \times 3600 = 648{,}000 \ m^3$$

b) The energy to be discharged is calculated from Eq. 11.4 as:

$$\dot{W}_{electric} = \eta_t \eta_g \eta_p \dot{W}_{max}$$

where $\dot{W}_{max} = \rho \dot{V}_{down}gH$. By substituting $\dot{W}_{max}$ to this equation,

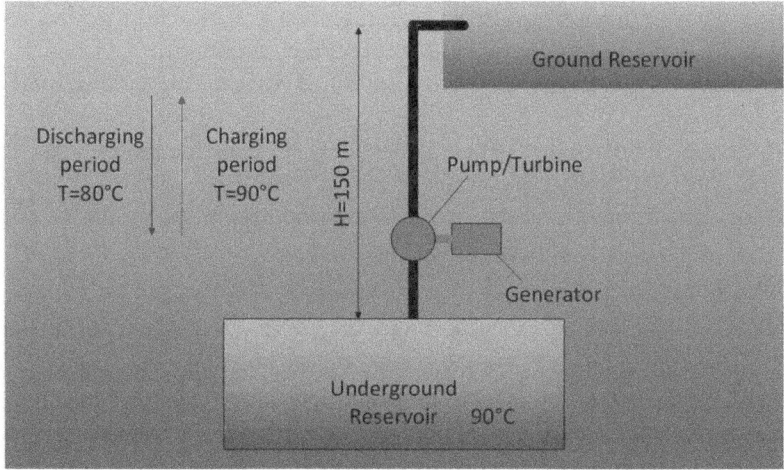

Figure 11.8 The schematic illustration of the underground pumped-hydro energy storage system in Illustrative Example 11.2.

$$\dot{W}_{electric} = \eta_t \eta_g \eta_p \rho \dot{V}_{down} gH$$

$$\dot{W}_{electric} = 0.90 \times 0.90 \times 1 \times 998 \times 30 \times 9.81 \times 150$$

The electricity generated is calculated as $\dot{W}_{electric}$=35,665,935.1 W or 35,665.9 kW.

The energy consumed in the charging period through the pump work is found by using Eq. 11.5 as:

$$\dot{W}_{pump} = \frac{\rho g H \dot{V}_{up}}{\eta_{P1}} = \frac{998 \times 9.81 \times 150 \times 15}{0.85}$$

The pump work is then found to be 25,915,711.8 W or 25,915.71 kW.

c) The heat amount of heat obtained is calculated as follows:

$$Q = m_{water} c_p \Delta T = \rho V_{water} c_p \Delta T = 998 \times 648,000 \times 4.18 \times (90 - 80)$$

Thus, the heat can be used as found to be Q=2.70 10^{10} kW.

11.6.2 Compressed-Air Energy Storage

A compressed-air energy storage (CAES) system is a method of storing energy generated at one time for use at a later time. Figure 11.9 shows two types of the compressed-air storage techniques, such as traditional CAES system and underwater CAES system. The system uses electricity to compress air and store it in underground caverns, depleted oil and gas fields, above-ground tanks, or underwater balloons. When electricity is needed, the compressed air is released and heated, which causes it to expand and drive a turbine to generate electricity. CAES systems can provide grid-scale energy storage and can help to balance the supply and demand of electricity on the grid. They have the potential to be an efficient and cost-effective way to store energy from renewable sources, such as wind and solar.

- **Charging period**: During the off-peak hours, generally at night, when the energy demands are low, and the unit cost of electricity is cheaper, high-capacity compressors are operated to compress the air into a reservoir. This process can also be operated when a renewable energy source is available if the aim of the storage is to store renewable energy for later usage.
- **Storing period**: The compressed air is then kept in the insulated reservoir until the energy is needed back during peak hours.
- **Discharging period**: At peak hours when the energy demands are high and the unit cost of electricity is higher or during periods when renewable energy is unavailable, the air is released into the atmosphere by passing through a turbine. A generator assembled to the turbine is used for generating electricity. Also, the pressurized air can also be used in gas turbine systems to reduce the compressor work.

Despite the fact that the working principles of a basic CAES seem easy, there are practically two significant issues comparing to the pumped-hydro energy storage, such as compressor efficiency and storage pressure. The compressor efficiency is the first point to consider. In contrast to pumps, compressors are quite irreversible. It is also crucial to find a convenient reservoir for storing compressed air, such as caverns, old oil or gas wells, porous rock formations, or pressure vessels. The storage pressure is the main performance criterion for the capacity and performance of the CAES. Higher storage pressure means higher stored energy. The reservoir should, however, be kept as airtight as possible during the storage period to minimize the losses. High amounts of energy can easily be stored by compressing the air in underground caverns.

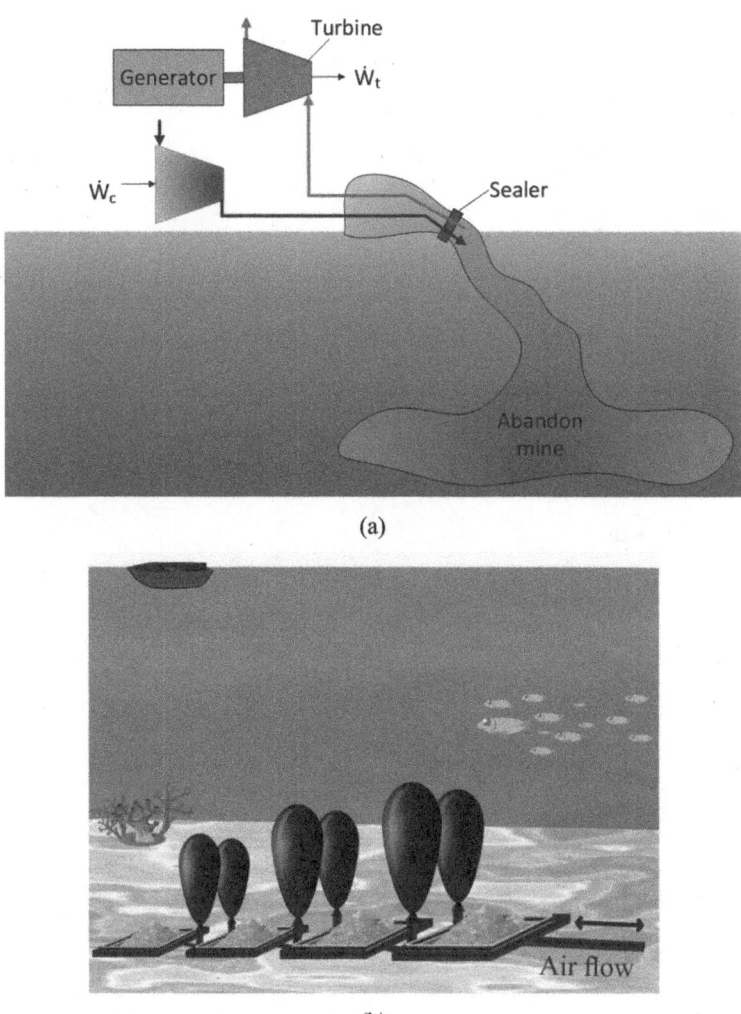

(a)

(b)

Figure 11.9 The schematic views for (a) the compressed-air storage and (b) underwater compressed-air storage system.

Illustrative Example 11.3 A compressed-air storage system is used for storing solar PV-based electricity. The excess PV output is known as 1 MW and available 4 h. In order to store electricity, the atmospheric air is compressed up to 100 bar in two stages which are 1:10 compression ratio. The isentropic efficiency of the compressors is 0.85. The compressed air is used in the turbine to generate power. The flowrate of air is 1.003 kg/s in the charging period. The flowrate of air is 0.5015 kg/s in the charging period. The efficiency of the turbine/generator is 0.85. Use the state point table given in Table 11.1.

a) Calculate the amount of energy for storage.
b) Find the power output in the system.
c) Determine the energy efficiency of the system.

Table 11.1 The state point table for Illustrative Example 11.3.

State	T (°C)	P (kPa)	h (kJ/kg)	s (kJ/kg K)
0	25	100	298.4	6.862
1	25	100	298.4	6.862
2	344.6	1000	626	6.945
3	935.6	10,000	1296	7.031
4	25	100	298.4	6.862

Solution:

a) The energy balance equation is written to calculate compressor work for each one as follows:

$$\dot{m}_1 h_1 + \dot{W}_{c1} = \dot{m}_2 h_2$$

$$1.003 \times 298.3 + \dot{W}_{c1} = 1.003 \times 626$$

$$\dot{W}_{c1} = 328.6 \; kW$$

For the second compressor, we obtain

$$\dot{m}_2 h_2 + \dot{W}_{c2} = \dot{m}_3 h_3$$

$$1.003 \times 626 + \dot{W}_{c2} = 1.003 \times 1296$$

$$\dot{W}_{c2} = 671.9 \; kW$$

Thus, the total energy consumed in the compressors is found as:

$$\dot{W}_{c,total} = \dot{W}_{c1} + \dot{W}_{c2}$$

$$\dot{W}_{c,total} = 328.6 + 671.9 = 1,000.5 \; kW$$

For 4 h of charging period, the total energy charged is determined to be:

$$\dot{E}_{ch,total} = 1,000.5 \times 4 = 4,004 \; kWh$$

b) In order to find the electricity generation, it is required to determine the turbine work output. From the energy balance equation of the turbine one may find

$$\dot{m}_3 h_3 = \dot{W}_t + \dot{m}_4 h_4$$

$$0.5015 \times 1296 = \dot{W}_t + 0.5015 \times 298.4$$

$$\dot{W}_t = 500.3 \; kW$$

c) The energy efficiency of the system is found as follows:

$$\eta_{en} = \frac{E_{disch}}{E_{ch}} = \frac{500.3 \times 4}{1005.5 \times 8} = 0.25 \; or \; 25\%$$

11.6.3 Flywheels

Flywheel energy storage is a method of storing energy in the form of kinetic energy in a spinning flywheel. Figure 11.10 demonstrates the view of the flywheel used as electricity storage technique. The flywheel is typically made of a lightweight and strong material, such as carbon fiber, and it is mounted on a low friction bearing. When electricity is available and the grid does not need it, the excess energy is used to spin the flywheel up to a high speed. When electricity is needed, the flywheel is used to drive an electrical generator, slowing the flywheel down and converting the kinetic energy back into electricity. Flywheel energy storage systems can provide short-term energy storage in the order of minutes to hours and can help to balance the supply and demand of electricity on the grid. They are typically used for frequency regulation and other grid stabilization services. They are more compact and fast response than batteries, but less energy dense.

- **Charging period:** The speed of the rotor is increased by the electric motor. The kinetic energy content in the flywheel is increased by increasing the angular velocity of the rotor. The charging period is completed when the angular velocity limit is reached.
- **Storing period:** Flywheel continues to rotate during the storing period, and the velocity of the flywheel and kinetic energy content reduce. In order to minimize the energy losses, the magnetic bearing and vacuum environment are used in the flywheel.
- **Discharging period:** When the energy is required, the generator is driven by the flywheel shaft, thus the stored kinetic energy is converted to electricity.

Note that the flywheel has a rotating mass, called as flywheel or rotor, generally axisymmetric, and it stores the energy in the form of kinetic energy by changing the angular velocity of the mass,

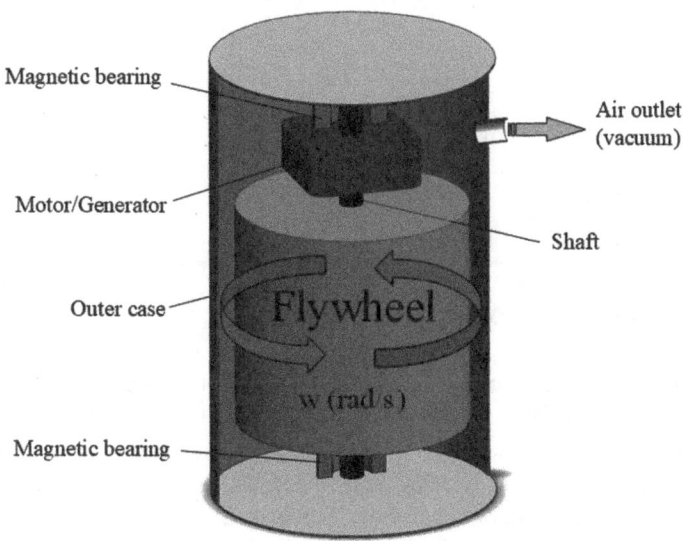

Figure 11.10 The schematic illustration of the flywheel as an energy storage technique.

as illustrated in Figure 11.10. The rotary type kinetic energy change in the flywheel is defined as follows:

$$\Delta E = \frac{1}{2} I \left(w_f^2 - w_i^2 \right) \tag{11.7}$$

where ΔE is the stored energy in the flywheel, I is the flywheel moment of inertia, and w_i and w_f are the initial and final angular velocities of the flywheel.

Illustrative Example 11.4   A set of flywheels are used for storing electricity in the charging station to reduce the peak loads. The charging period is 20 minutes while the discharging period is 5 minutes. In a flywheel, the rotational speed of the flywheel varies between 2000–40,000 rpm. The mass of the flywheel is 20 kg, and its moment of inertia is 16.9 kg m². If it is aimed to store 10,000 kWh of energy, how many flywheel units are required? The efficiencies for the motor and generator are taken as 0.95. Calculate the energy efficiency of the flywheel.

Solution:

In order to determine the energy needed for the increase the rotational speed of the flywheel, Eq. 11.7 is used:

$$E_{ch} = \frac{1}{2} I \left(w_f^2 - w_i^2 \right) = \frac{1}{2} \times 16.9 \times \left(4188.79^2 - 209.44^2 \right) = 147,892,715.9 \, J \ \text{or} \ 41.08 \, kWh$$

As motor works with the efficiency of 0.95, the electricity consumed by the motor is found to be:

$$E_{ch} = \frac{41.08}{0.85} = 48.33 \, kWh$$

The energy discharged from the flywheel is determined as follows:

$$E_{disch} = 41.08 \times 0.85 = 34.92 \, kWh$$

If 10,000 kWh of energy is stored, the number of flywheel units required is found as:

$$N = \frac{10,000}{34.92} \cong 287 \, units$$

The efficiency of the flywheel is calculated as follows:

$$\eta_{en} = \frac{E_{disch}}{E_{ch}} = \frac{34.92}{48.33} = 0.7225 \ \text{or} \ 72.25\%$$

11.7   Thermal Energy Storage Methods

Thermal energy storage (TES), also called heat storage, is recognized as the process of converting any incoming energy into heat and storing it in a storage medium. The TES method is one of the most advanced and mature methods of storing energy. Applications and capacities may be unlimited with TES techniques. Latent heat storage and sensible heat storage are the two main categories of TES systems. Sensible heat storage systems store heat in materials whose temperature changes with heat addition or removal. Storage tanks for hot water, rock beds and solar ponds are some examples. A latent heat storage system stores heat in a material that changes its phase when heat is added or removed. Molten salts, phase-change materials and ice storage are some typical examples. Table 11.2 shows common thermal energy storage mediums.

The working principles of thermal energy storage systems are shown in Figure 11.11. During the charging period, the incoming energy is converted to the heat and charged to the thermal energy medium via a heat transfer fluid. Then, the heat stored is kept in the storage unit in the storing unit.

Table 11.2 Common thermal energy storage mediums for sensible and latent TES systems (modified from [3]).

Sensible Short Term	Latent Short Term	Long Term
Rock beds	Inorganic materials	Rock beds
Earth beds	Organic materials	Earth beds
Water tanks	Fatty acids	Larger water tanks
	Aromatics	Aquifers
		Solar ponds

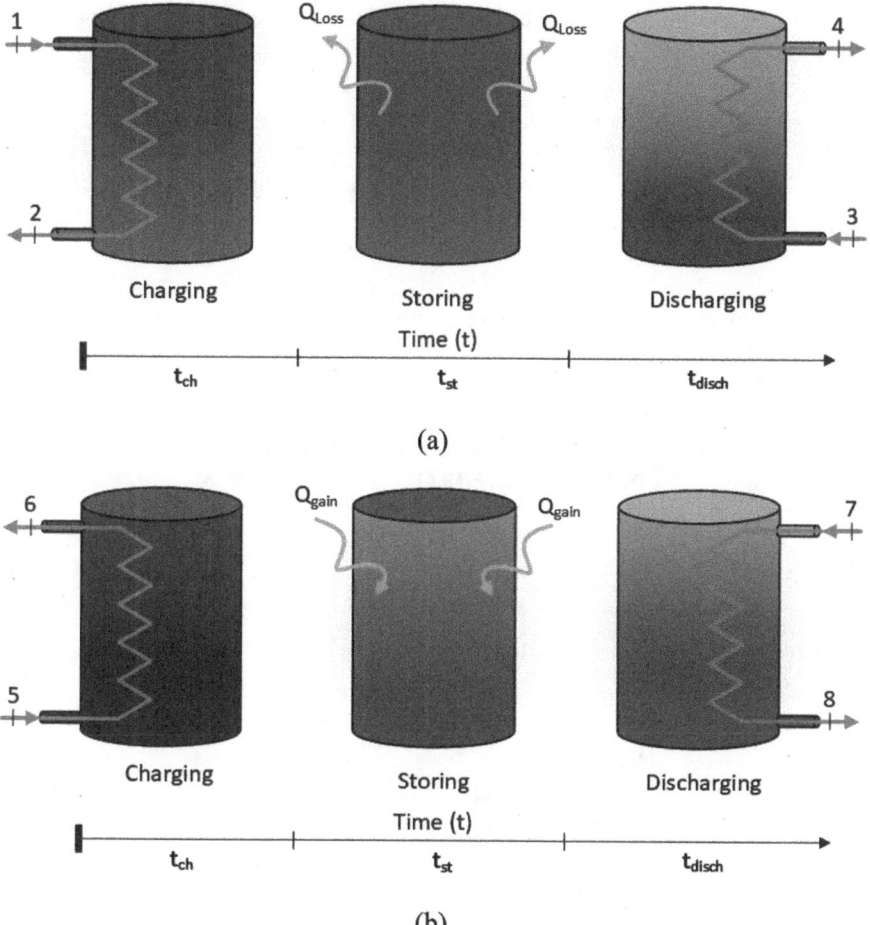

Figure 11.11 The working principles of the thermal energy storage systems: (a) heating purposes and (b) cooling purposes.

In the discharging period, the stored heat is recovered from the thermal energy storage unit via the heat transfer fluid. These periods run in a cyclic manner.

TES systems can be used to store energy from renewable sources, such as solar thermal and geothermal energy, and can be used to provide heating and cooling for buildings, industrial processes, and power generation. It is possible to shift energy consumption to times when renewable energy is available and energy prices are low using TES systems. Additionally, they can improve the efficiency of power plants by allowing them to operate at a constant temperature and reducing the need for peak demand power plants.

11.7.1 Sensible Thermal Energy Storage

Sensible thermal energy storage refers to the storage of heat in a material that changes temperature as energy is added or removed. This type of storage may be achieved through the use of materials such as water, molten salt, or concrete. These materials have a high heat capacity and are relatively inexpensive, making them a popular choice for thermal energy storage systems. These systems can be used in applications such as solar thermal power plants, where heat is collected during the day and stored for use at night, or in buildings to store heat generated by a heating system for later use.

The working principle of sensible thermal energy storage is based on the ability of a material to absorb and release heat without undergoing a phase change. This is completed by using materials with a high heat capacity, such as water, molten salt, or concrete. Table 11.3 lists some common materials used in the sensible thermal energy storage applications.

When heat is added to the storage material, its temperature increases. This heat can then be removed from the storage material and used for a variety of purposes, such as heating a building or generating electricity. The heat is removed from the storage material by circulating a fluid, such as water, through the storage material. As the fluid absorbs the heat, it becomes hot and can be used to generate steam to power a turbine and generate electricity.

Table 11.3 The sensible thermal energy storage features at 20°C for some common materials (modified from [3]).

Material	Density (kg/m^3)	Specific Heat (J/kg K)	Thermal Conductivity (W/kg K)	Volumetric Thermal Capacity (MJ/m^{3}K)
Aluminum	2710	896		2.43
Brick	1800	837	1.3	1.51
Clay	1458	879		1.28
Concrete	2000	880	1.5	1.76
Glass	2710	837		2.27
Gravelly earth	2050	1840		3.77
Iron	7900	452		3.57
Magnetite	5177	752	5.0	3.89
Sandstone	2200	712	1.0	1.57
Steel	7840	465		3.68
Water	988	4182		4.17
Wood	700	837		1.51

When the heat is removed from the storage material, its temperature decreases. This process can be reversed by adding heat to the storage material. In this way, the storage material can be used as a heat sink, absorbing excess heat during periods of high demand, and releasing it during periods of low demand.

A typical sensible heat storage system consists of a storage material, a container, and input/output devices. The storage tanks should be well insulated to prevent heat losses. The common sensible heat storage system used in practical applications can be listed as

- Thermally stratified heat storage tanks
- Rock and water/rock beds heat storage
- Borehole storage
- Aquifer heat storage
- Solar ponds
- Concrete heat storage

The efficiency of sensible heat storage is related to the heat capacity of the storage material, the temperature difference between the storage material and the heat source or sink and the storage volume. The specific heat and internal energy of the materials are recognized as the key parameters for sensible heat storage systems.

$$\sum_{t=0}^{t_{ch}}\left(\dot{m}_1\left(h_1-h_2\right)\right)t = m_{sm}\left(u_{final}-u_{initial}\right) \tag{11.8}$$

$$\sum_{t=0}^{t_{ch}}\left(\dot{m}_3\left(h_3-h_4\right)\right)t = m_{sm}\left(u_{initial}-u_{final}\right) \tag{11.9}$$

The exergy balance may be written as follows:

$$\sum_{t=0}^{t_{ch}}\left(\dot{m}_1\left(ex_1-ex_2\right)\right)t - Ex_{dest} = m_{sm}\left(ex_{final}-ex_{initial}\right) \tag{11.10}$$

$$\sum_{t=0}^{t_{ch}}\left(\dot{m}_3\left(ex_3-ex_4\right)\right)t - Ex_{dest} = m_{sm}\left(ex_{initial}-ex_{final}\right) \tag{11.11}$$

The energy efficiency for a sensible thermal energy storage system can be written as follows:

$$\eta = \frac{Net\ discharged\ energy\left(net\ energy\ output\right)}{Charged\ energy\left(energy\ input\right)}$$

Therefore, the overall energy efficiency is written as follows:

$$\eta_{overall} = \frac{Q_{disch}}{Q_{ch}} = \frac{\sum_{t=0}^{t_{ch}}\left(\dot{m}_1\left(h_1-h_2\right)\right)t}{\sum_{t=0}^{t_{disch}}\left(\dot{m}_3\left(h_3-h_4\right)\right)t} \tag{11.12}$$

The energy efficiencies for the operational periods are written as follows:

$$\eta_{ch} = \frac{m_{sm}\left(u_{ch@final}-u_{ch@initial}\right)}{\sum_{t=0}^{t_{ch}}\left(\dot{m}_1\left(h_1-h_2\right)\right)t} \tag{11.13}$$

$$\eta_{st} = \frac{m_{sm}\left(u_{st@final} - u_{st@initial}\right) - Q_{loss}}{m_{sm}\left(u_{st@final} - u_{st@initial}\right)} \qquad (11.14)$$

$$\eta_{disch} = \frac{\sum_{t=0}^{t_{ch}}\left(\dot{m}_3\left(h_3 - h_4\right)\right)t}{m_{sm}\left(u_{ch@initial} - u_{ch@final}\right)} \qquad (11.15)$$

The exergy efficiency for a sensible thermal energy storage system is defined as:

$$\eta = \frac{Net\ discharged\ exergy\left(net\ exergy\ output\right)}{Charged\ exergy\left(exergy\ input\right)}$$

Thus, the overall exergy efficiency can be written as follows:

$$\eta_{overall} = \frac{Ex_{disch}}{Ex_{ch}} = \frac{\sum_{t=0}^{t_{ch}}\left(\dot{m}_1\left(ex_1 - ex_2\right)\right)t}{\sum_{t=0}^{t_{disch}}\left(\dot{m}_3\left(ex_3 - ex_4\right)\right)t} \qquad (11.16)$$

The exergy efficiencies for the operating periods, which are charging, storing, and discharging periods may be written as:

$$\eta_{ch} = \frac{m_{sm}\left(ex_{ch@final} - ex_{ch@initial}\right)}{\sum_{t=0}^{t_{ch}}\left(\dot{m}_1\left(ex_1 - ex_2\right)\right)t} \qquad (11.17)$$

$$\eta_{st} = \frac{m_{sm}\left(ex_{st@final} - ex_{st@initial}\right) - Q_{loss}}{m_{sm}\left(ex_{st@final} - ex_{st@initial}\right)} \qquad (11.18)$$

and

$$\eta_{disch} = \frac{\sum_{t=0}^{t_{ch}}\left(\dot{m}_3\left(ex_3 - ex_4\right)\right)t}{m_{sm}\left(ex_{ch@initial} - ex_{ch@final}\right)} \qquad (11.19)$$

Illustrative Example 11.5 A rock bed thermal energy storage system is used for storing solar energy. The air is heated through the solar air collectors and forced flow through the rocks which are kept in a well-insulated environment. The average solar radiation in the location is 750 W/m². The air enters the collector at 30°C and exits at 90°C. The total area of the solar collectors is 500 m². The temperatures at the inlet and outlet of the storage units are 85°C and 35°C, respectively. The total mass of rock bed is 50,000 kg. Take the density of the air as 1.1 kg/m³. The specific heat of the rock is 1.48 kJ/kg°C. Consider the steady-state flow and constant solar radiation conditions for calculations.

a) Find the air flowrate circulated in the system.
b) Calculate the final temperature of the rock bed after 4 h of charging period.
c) Calculate the amount of the stored energy.

Solution:
a) In order to calculate the air flow rate circulated through the solar collector, there is a need to write the energy balance equation as follows:

$$\dot{I}_{solar} A_{coll} = \dot{m}_{air} c_p \left(T_{in} - T_{out} \right)$$

$$0.750 \times 500 = \dot{m}_{air} \times 1.009 \times \left(90 - 30 \right)$$

Thus, the mass flow rate is found to be $\dot{m}_{air}$=6.19 kg/s.

b) In order to find the final temperature of the rock bed, the energy balance equation is now written for the storage unit.

$$\dot{m}_{air} c_p \left(T_{in} - T_{out} \right) t_{ch} = m_{rock} c_p \left(T_f - T_i \right)$$

$$6.19 \times 1.009 \times \left(85 - 35 \right) \times 4 \times 3600 = 50{,}000 \times 1.48 \times \left(T_f - 25 \right)$$

$$T_f = 60.77°C$$

The stored energy is then calculated as follows:

$$Q_{stored} = 50{,}000 \times 1.48 \times \left(60.77 - 25 \right)$$

$$Q_{stored} = 2{,}646{,}980 \ kJ$$

11.7.2 Latent Thermal Energy Storage

Latent thermal energy storage refers to the storage of heat energy by means of a phase change of a material, such as the melting or freezing of a substance. The material used for this type of thermal energy storage is usually a solid or liquid with a high latent heat of fusion or solidification. When heat is added to the material, it melts, and when heat is removed, it solidifies.

The main advantage of latent thermal energy storage is that a large amount of heat can be stored in a relatively small volume of material due to large enthalpy differences. This is because the heat energy is stored in the form of the potential energy of the material, rather than as an increase in temperature. This allows for more efficient storage and greater energy density. Some examples of materials used for latent thermal energy storage include ice, paraffin wax, and eutectic alloys. These materials have a high latent heat of fusion, which means that they can absorb or release a large amount of heat energy without a significant change in temperature. Latent thermal energy storage systems can be used in applications such as air conditioning and refrigeration, where heat is removed from a space during the day and stored for use at night, or in buildings to store energy generated by a solar thermal system for later use. Figure 11.12 illustrates the classification of latent heat storage materials.

The working principle of latent thermal energy storage is based on the ability of a material to absorb or release a large amount of heat energy without a significant change in temperature. This is achieved by using materials with a high latent heat of fusion or solidification, such as ice, paraffin wax, and eutectic alloys. When heat is added to a material with a high latent heat of fusion, it melts. The heat energy is stored in the material in the form of the potential energy of the material, rather than as an increase in temperature. This process is known as charging the storage system. When heat is removed from the material, it solidifies. During this process, the stored heat energy is released and can be used for a variety of purposes, such as heating a building or generating electricity. This process is known as discharging the storage system.

A common example of latent thermal energy storage is the use of ice storage systems in air conditioning, where ice is made at night when electricity demand is low, and the stored ice is used to cool the building during the day. The ice is made by circulating a refrigerant through a heat

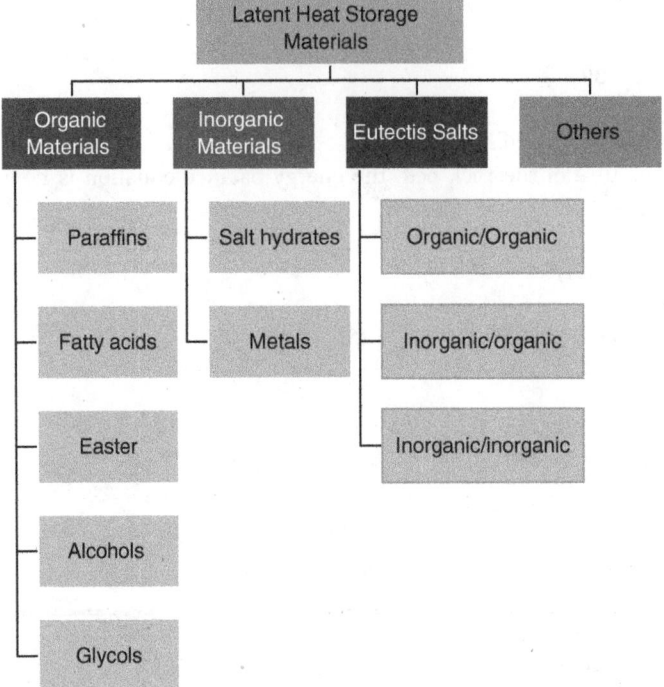

Figure 11.12 A general classification of phase change materials.

exchanger filled with water. The refrigerant absorbs heat from the water, causing it to freeze into ice. The ice is then stored in an insulated tank. During the day, when the building needs to be cooled, the ice is melted by circulating warm water through the heat exchanger. The heat absorbed by the ice is transferred to the water, which is then used to cool the building.

The efficiency of the latent heat storage is related to the heat energy stored per unit volume of the storage material, the temperature difference between the storage material and the heat source or sink and the storage volume.

The energy change in the latent thermal energy storage system is defined as follows:

$$\Delta E_{sm} = m_{sm}\left(u_{inital} - u_{pcm,f}\right) + \left(h_{sf}\right) + c_s\left(T_{pcm,s} - T_{final}\right) \tag{11.20}$$

$$\Delta E_{sm} = m_{sm}\left[\left(c_f\left(T_{inital} - T_{pcm}\right)\right) + \left(h_{sf}\right) + \left(c_s\left(T_{pcm} - T_{final}\right)\right)\right] \tag{11.21}$$

where s and f subscripts denote solid and fluid phases of the PCM. h_{sf} is the enthalpy of phase change.

When the PCM consists of a mixture of solid and liquid phases, the energy change in the PCM is defined as:

$$\Delta E_{sm} = m_{sm}\left[\left(1 - F\right)\left(h_{final,s} - h_{initial}\right) + F\left(h_{final,f} - h_{initial}\right)\right] \tag{11.22}$$

Illustrative Example 11.6 In a latent thermal energy storage system, paraffin has 120°C of the melting point and 250 kJ/kg of the latent heat capacity. 1000 kg of paraffin is used for storing solar thermal energy via parabolic through collectors. The initial temperature of the paraffin is 25°C and final temperature is 150°C. The specific heat of paraffin for solid and liquid phases is 2.07 kJ/kg K and 2.21 kJ/kg K, respectively. Calculate the amount of thermal energy stored in the paraffin.

Solution:

The thermal energy stored in the paraffin calculated by using Eq. 11.21.

$$E_{sm} = m_{sm}\left[c_{solid}\left(T_{pcm} - T_i\right) + h_{sf} + c_{fluid}\left(T_f - T_{pcm}\right)\right]$$

where $T_i = 25°C$, $T_{pcm} = 120°C$, $T_f = 150°C$, by substituting these values in this equation:

$$E_{sm} = 1000 \times \left[2.07 \times \left(120 - 25\right) + 250 + 2.21 \times \left(150 - 120\right)\right]$$

$$E_{sm} = 512,950 \ kJ$$

Under these conditions, the amount of the stored thermal energy is found to be 512,950 kJ.

11.7.3 Cold Thermal Energy Storage Systems

The purpose of cold heat storage is to store cold energy (cooling capacities) for later use when needed. Cold heat storage is primarily intended to reduce cooling costs and energy consumption depending on the cooling process. Currently, cold heat storage techniques are used not only to reduce cooling costs, but also to manage grid load and store renewable energy. When heat is stored for cooling purposes, it is done at a temperature lower than that of the surrounding environment. Off-peak electricity and renewable energy sources can be stored using heat storage. As wind turbines are increasingly used and their contribution to nighttime electricity generation increases, a nighttime surplus of electricity is created, which contributes to the attraction of shifting power consumption to the night. As a result, it becomes more attractive to shift power demands to night hours, thus increasing interest in energy storage applications, such as batteries and cold heat storage.

The capacity and working duration of air conditioning (AC) systems are ever-increasing because of increasing environmental temperature due to global warming, spending a more extended time indoors, and using many devices producing heat indoor. Therefore, AC systems have become important operating and initial investment costs for buildings. When the daily, seasonal, and yearly peak cooling load distributions of buildings are considered, the peak cooling loads are seen in a limited time period for each period. High-capacity AC system are required to meet the on-peak cooling demands which brings up high initial cost of HVAC systems. Additionally, these high-capacity devices increase the capacities and costs of auxiliary devices such as grid connection transformers, power generator, the diameter of cables, etc. Consequently, a substantial amount of initial cost is required to meet the peak cooling loads. Cold heat storage techniques are an essential method to reduce peak cooling loads and the operating cost of AC units. These also have a significant impact on the peak electricity loads of the grid since they are operated at the same time. In many countries, the peak electricity loads are seen in the hottest days of the summer due to the use of AC units since electricity is the main energy source of these units. Meeting the peak electricity loads is an essential issue for the energy supplier and generation companies. In order to meet the peak electricity load, extra power plants are built or electricity imported directly. Therefore, meeting peak electricity loads require a high cost. Cold heat storage systems play a critical role in shifting the peak electricity loads dependent by cooling systems. Thus, both the operating cost of AC units and the peak electricity loads are reduced by cold heat storage systems.

Cold heat storage methods can also be used for meeting emergency cooling needs. In buildings where cooling is very critical such as supercomputer centers, data processing centers, and stem cell centers, there are a few back-up chillers, and generators used in power cut-off. Also, today, we have a critical issue with AC systems. During the COVID-19 pandemic, we have seen that almost all intensive care units and hospitals are working with overcapacity conditions. These overcapacity working conditions have caused poorly meeting the clean air and air conditioning demands. Additionally, the capacity increases in these systems are required due to converting regular patient services into an intensive care unit. Besides, fresh air needs have become a critical issue for the buildings which have high people circulation such as malls, hypermarkets, etc. especially in current extraordinary days. These needs require the capacity increasing in the AC systems. Cold heat storage systems are a unique solution to meet the increase in cooling capacities. While the cold heat storage systems are used for reducing cooling costs during normal days, they can be used for meeting emergency cooling capacities in tough days like a pandemic, long-duration power cut-off.

The cooling load of the buildings is stored for later use by taking advantage of the latent heat required or released during the water/ice phase change in the ice storage systems. Figure 11.13 shows the energy storage capacity of the cold water and water/ice phase changing. As seen in Figure 11.13, using ice as an energy storage medium provides a significant energy storage density. Practically, the main purpose of the ice storage systems is to reduce the energy consumption costs of ACs by benefiting electricity tariffs. The working principles of ice storage systems are similar to other energy storage systems; energy charging, storing, and discharging periods. In the charging period, ice is produced during the off-peak tariff hours which the electricity unit price is the cheapest during the day. When the off-peak period is completed, the produced ice is kept until it is used in the insulated storage tank in the storing period.

During the discharging period, the cooling demand of the building is carried out by using the produced ice during the mid-peak and on-peak hours. Thus, the air conditioning of the buildings is done at a lower cost. Besides lowering the operating costs of AC units, the main advantages of energy storage systems, as shown in Figure 11.14, can be listed as follows:

- Reduces the operating costs of cooling systems: By reducing the use of air conditioning systems during the hours when the electricity unit price is high, it reduces the cooling costs of the buildings. The economic performance of an ice storage system depends on the price change and distribution of the electricity tariff during the day.
- Shifts the peak cooling loads to off-peak hours: Peak cooling loads correspond to several hours of the day. Thanks to ice storage systems, it will be possible to shift peak cooling loads to off-peak hours. Thus, it also helps to reduce the peak loads by providing a more balanced consumption of energy.

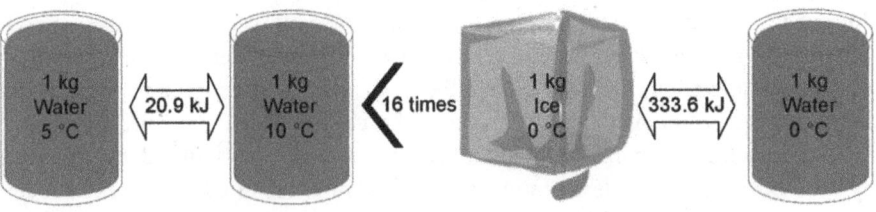

Figure 11.13 The latent heat change of 1 kg water/ice.

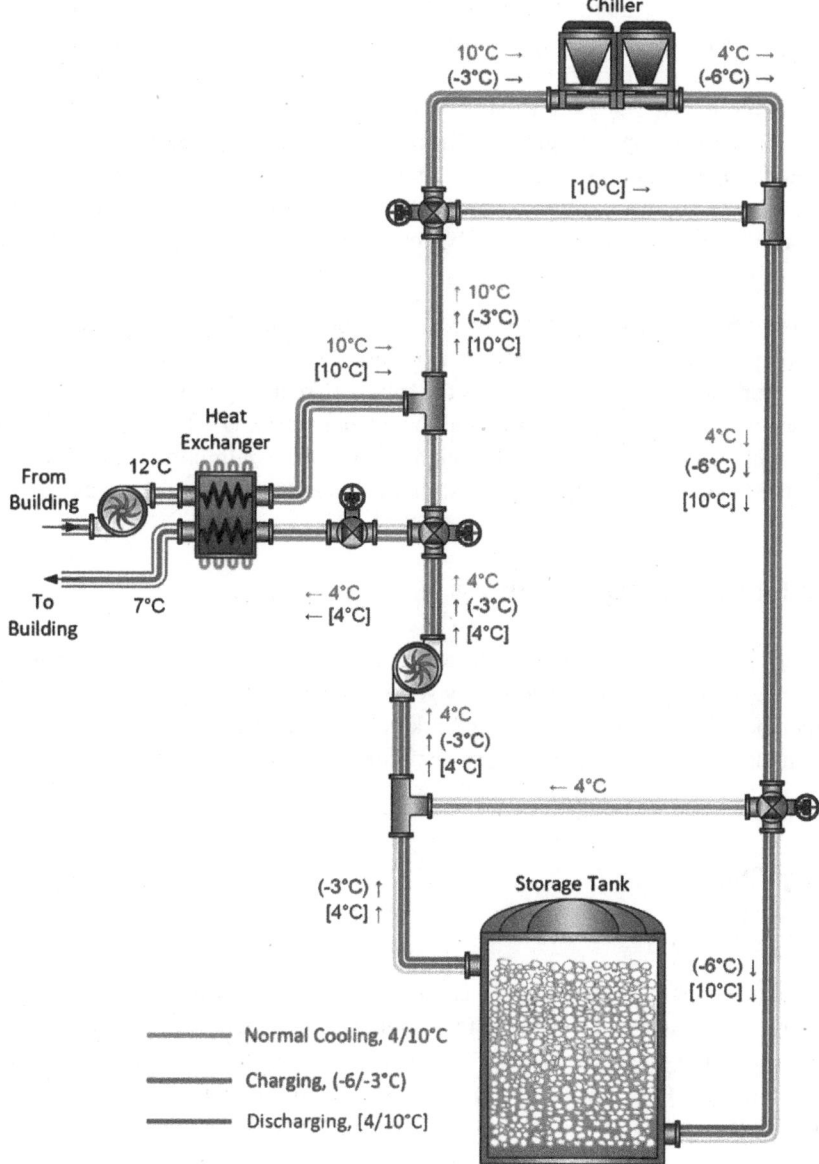

Figure 11.14 The schematic view of the ice storage integrated AC system (modified from [4]).

- Reduces chiller capacities: AC systems are designed according to the peak cooling loads of the buildings. Thanks to the ice storage systems, the capacity required to meet the peak load can be reduced by shifting these peak loads to off-peak hours.
- Provides chillers to work with higher COP: Air conditioning systems are used extensively in the summer months. Especially due to the increasing outdoor temperature, solar radiation, and relative humidity in the afternoon, the working environment to which the chillers are exposed significantly reduces the COP of the chillers. Thanks to the energy storage systems, by using the chillers at night, the COP values of chillers can increase significantly.

- Reduces the capacities and sizes of auxiliary equipment like a transformer, cable, generator etc. required for building power. Lowering the peak cooling load provides a lower power demand from the grid. Thus, power can be provided to the building with a lower cable cross-section. It will also be possible to reduce the generator capacity.
- Meets emergency demands: Cooling is a significant issue in buildings like medical laboratories, data processing centers, super-computer facilities, etc. In such places, generator groups are used to continue cooling in cases of power failure. There are backups of these generator groups too. Ice storage systems can be used to reduce the number of generator groups or provide more effective cooling backup.

In addition to the benefits listed above, it helps to consume energy more efficiently because it provides more efficient operation of chillers and helps to reduce the energy peak load. Due to the higher energy efficiency provided, ice storage systems contribute to reducing fossil fuel consumption and carbon emissions, thereby they help to eliminate global warming effects. In addition, reducing the peak electricity consumption will also decrease the amount of imported energy, thus contributing to the reduction of the country's current account deficit due to energy imports.

11.8 Chemical Energy Storage Methods

Chemical energy storage generally refers to the storage of energy in the chemical bonds of a substance. This type of energy storage is achieved by converting any energy source into chemical energy, which can then be stored and later converted back into energy desired as a useful output. There are several methods of chemical energy storage, including, but not limited to:

- Hydrogen
- Ammonia
- Synthetic natural gas (SNG)
- Liquid natural gas (LNG)
- Compressed natural gas (CNG)

The main objective for all these types of chemical energy storage is to store any energy sources by obtaining the materials listed above. Chemical energy storage is particularly useful for storing large amounts of energy for long periods of time. The efficiency of chemical energy storage is related to the energy stored per unit mass of the storage material and the energy losses during charge and discharge cycles.

11.8.1 Hydrogen

The clean, nontoxic, and high energy content of hydrogen makes it a significant energy source. Hydrogen contains 142 kJ/kg of chemical energy, which is higher than that of other hydrocarbon-based fuels. Hydrogen can be produced through steam reforming and gasification of fossil fuels, thermochemical water splitting and high-temperature electrolysis from nuclear energy, and electrolysis and gasification of renewable energy sources. Therefore, hydrogen is an effective energy storage medium to store any energy source from renewables to nuclear, from electricity to biogas. After combustion or utilization, it emits only water vapor. Details of hydrogen as an energy storage medium will be discussed in Chapter 12 in detail with some practical applications and illustrative examples.

As a quick review of hydrogen as an energy storage medium, let's consider water electricity powered by off-peak electricity or renewable energy sources. The off-peak electricity can be stored by producing hydrogen with water electrolysis. During the peak period, hydrogen can be used for power generation in the fuel cells. Thus, it can help to compensate for the electricity supply and demand profiles.

Illustrative Example 11.7   Hydrogen is used for off-peak electricity storage where electricity is first utilized to produce hydrogen, which is stored and converted into electricity when needed. For this purpose, the PEM-based electrolyzer is employed for hydrogen production and later fuel cells are used for power generation from hydrogen. It is known that in the PEM electrolyzer, a total of 55 kWh of energy is required for 1 kg of hydrogen production. On the other hand, it is possible to generate 15 kWh of energy with 1 kg of hydrogen in the PEM fuel cell. In the energy storage systems, it is aimed to store 10 MWh of energy in the off-peak hours from 22:00 to 6:00. The stored hydrogen is for power generation during the peak hours from 14:00 to 18:00.

a) Calculate how much hydrogen is needed for this storage.
b) Calculate the energy and exergy efficiency for this energy storage system.

Solution:

a) With hydrogen, it is aimed to store 10 MWh of energy. Therefore, the total power output of the energy storage unit is 10 MWh in 4 h. In order to find the amount of hydrogen to be needed to achieve this energy storage, the following approach can be used. As 15 kWh of energy can be produced with 1 kg of hydrogen. In order to get 10 MWh of energy,

$$10,000 \, (kWh) = m_{H_2} \, (kg) \times 15 \left(\frac{kWh}{kg} \right)$$

Thus, the hydrogen to be needed for this energy storage can found as $m_{H_2} = 666.67 \, kg$.

b) In order to calculate the efficiency of the energy storage systems, it is first required to calculate how much energy is consumed to obtain that much hydrogen. It is calculated as follows:

$$E_{ch} \, (kWh) = m_{H_2} \, (kg) \times 55 \left(\frac{kWh}{kg} \right)$$

Thus, the total energy needed for hydrogen production is found as E_{ch}=36,666.85 kWh. The energy efficiency of the system is therefore found as follows:

$$\eta_{en} = \frac{E_{disch}}{E_{ch}} = \frac{10,000}{36,666.85} = 0.2727 \text{ or } 27.27\%$$

As only energy input and output in the system is electricity, the energy and exergy efficiencies become identical.

11.8.2   Ammonia

Ammonia (NH_3), a carbon-free fuel, has approximately two times the energy density per unit volume of liquid hydrogen. Due to its high energy density and ease of storage, shipping, and distribution, it is a remarkable energy material. Hydrogen makes up 17.8% of its composition. From renewable hydrogen and nitrogen separated from the air, it can be produced easily and converted to hydrogen and nitrogen. As an alternative fuel, ammonia can also be used. There are, however, some barriers for its use, including low flammability, high NOx emissions, and low radiation intensity. Currently, ammonia can be burned in turbines, co-fired with pulverized coal, and in furnaces. It is also used for obtaining H_2 to use in fuel cells by cracking ammonia into nitrogen and hydrogen.

Today, production of ammonia requires carbon-intensive due to hydrogen production techniques which is steam-reforming used in the ammonia production. Greenization of ammonia production is a significant task for a sustainable future. Because ammonia is one the most produced chemical in the world.

It is high energy and hydrogen contents make it a promising energy storage medium and fuel. Off-peak electricity and renewable energy sources can be used for running the systems for hydrogen production and air separation. Then, ammonia can be used for power generation or directly used for chemicals.

11.8.3 Synthetic Natural Gas (SNG)

Synthetic natural gas, also known as substitute natural gas, is one of the most promising technologies in that field. SNG is an alternative to natural gas with similar properties. Therefore, it plays a critical role in energy systems. In natural gas production, the partial conversion of solid feedstock is achieved through gasification, followed by gas conditioning, SNG synthesis, and upgrading, or similar processes. Gasification is a non-combustion heating process that converts solid carbon fuels into hydrogen, CO_2, and CO using solid carbon fuels. Electrolysis creates hydrogen, which can then be converted into e-gas or syngas by Power-to-Gas or Power-to-X. Renewable energy surpluses are processed in this way. As long as the raw material is plant cellulose, the process results in thermochemical SNG production and bio-SNG is the result. The result of natural anaerobic digestion of organic materials is biochemical SNG or biogas if the feedstock used is natural anaerobic digestion of organic materials. Fuels like SNG are critical, as they are flexible, storable, transportable, and can be produced with renewable resources. It is therefore proposed as a potential choice for storing renewables.

11.8.4 Liquid Natural Gas (LNG)

A significant benefit of natural gas is that it emits fewer carbon dioxide emissions than other fossil fuels. Natural gas is usually transported in liquid form. Figure 11.15 illustrates how natural gas is liquefied and regasified to produce energy. The following are three periods of LNG which is considered a potential energy storage option:

- Charging Period: Natural gas is compressed by the cryogenic compression process during the electricity off-peak periods or when the renewable energy source is active. Thus, natural gas is liquefied.
- Storing Period: LNG is kept in an insulated pressurized tank depending on its utilization requirement.
- Discharging Period: The stored LNG is used to generate electricity in gas turbines when the energy is needed during peak periods.

Figure 11.15 The schematic view of the LNG production, storage, and regasification processes.

11.9 Electrochemical Energy Storage Systems

Electrochemical power sources convert chemical energy to electrical energy and heat. Electrochemical storage is an ES method used to store electricity under a chemical form. This storage technique benefits from the fact that both electrical and chemical energy share the same carrier, the electron. This common point allows limiting the losses due to the conversion from one form to another. In the electrochemical ES, at least two reaction partners undergo a chemical process. The result of this reaction is obtained as electric current and voltage. When practical applications are considered, electrochemical ES have two common applications which are rechargeable batteries and flow batteries. Batteries generally consist of one or more electrochemical cells in series. Flow batteries consist of the energy storage material that is dissolved in the electrolyte as a liquid. Capacitors are also a significant energy storage technique for short-period and higher-rate electricity storage.

11.9.1 Batteries

Today, batteries usually come to mind first when considering the storage of electricity, as they are widely used for electricity storage in daily life. Energy storage in batteries involves the conversion of electrical energy into chemical energy and vice versa. This is achieved by the movement of ions through an electrolyte between two electrodes, known as the anode and cathode. The anode is where oxidation occurs, while the cathode is where reduction takes place. The most common type of battery is the lithium-ion battery, which is widely used in consumer electronics, electric vehicles, and renewable energy systems. Other types of batteries include lead-acid, nickel-cadmium, and sodium-sulfur batteries. Their specific energy on mass basis range between 172.8 and 828 kJ/kg. Among batteries, in recent years, lithium-ion batteries are commonly preferred due to having higher energy density, bad memory effect, lower mass density, lower self-discharge rates. Figure 11.16 shows the views of various types of li-Ion batteries, such as coin (button) cell, cylindrical cell, prismatic cell, and pouch cell which are practically deployed for various applications.

Heat generation in the batteries is known as the most significant issue due to increasing temperature reducing the capacity of the lithium batteries. Therefore, the temperature of batteries should be kept under control during the charging and discharging conditions. The capacity fade of battery cells accelerates at higher battery operation temperatures. High operating temperatures can cause overheating, which leads to thermal runaway. High temperature on batteries may cause an explosion or fire. Two main ways of solving the thermal issues of the batteries are to reduce the heat generation rate or to increase the heat dissipation rate. However, the high energy demand and fast charging conditions are considered, these processes are not possible to achieve. Therefore, batteries should be cooled in an effective way. There are many studies in the open literature that develop battery thermal management systems (BTMSs).

A flow battery is a type of electrochemical energy storage that consists of two chemical components dissolved in liquid separated by a membrane. Batteries charge and discharge by transferring ions through their membranes. A major advantage of flow batteries is their ability to pack in large quantities. With the increase in renewable energy storage needs, interest in flow batteries has increased considerably. High-capacity flow batteries can store large amounts of electricity because they have giant electrolyte tanks. Vanadium, however, is one of the most expensive materials used in flow batteries. Flow batteries have been shown to be useful in some recent studies. An integrated energy storage system using a flow battery for storing renewable energy is shown in Figure 11.17.

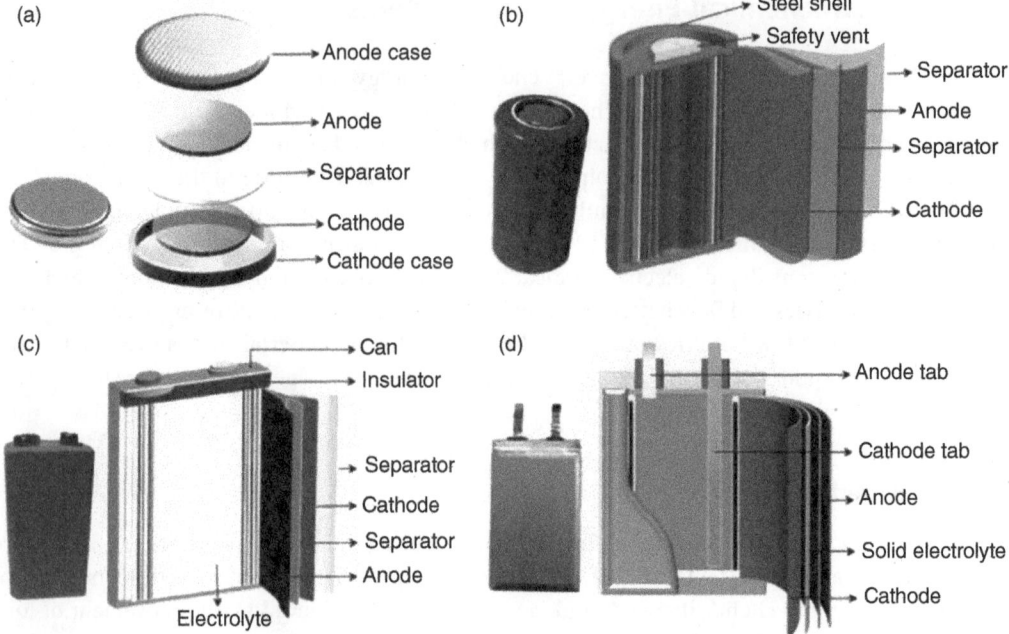

Figure 11.16 The view of various types of Li-Ion batteries (a) coin (button) cell, (b) cylindrical cell, (c) prismatic cell, and (d) pouch cell.

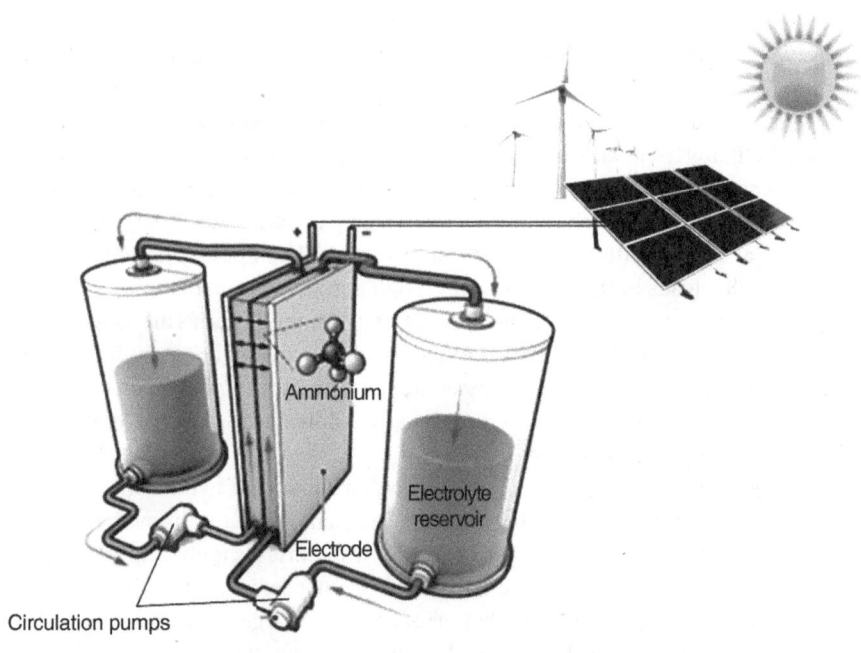

Figure 11.17 Schematic illustration of the flow batteries.

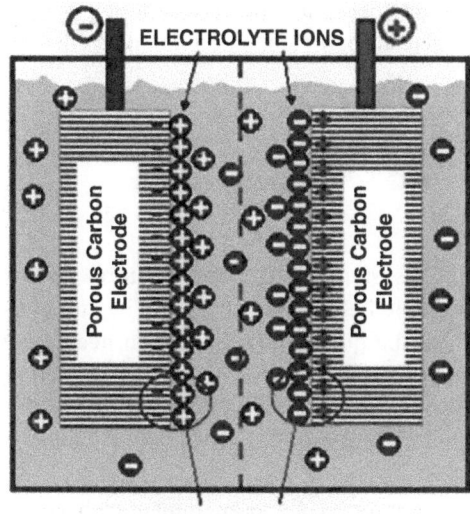

Double Layer Capacitors
(Adsorbed layers of ions and solvated ions)

Figure 11.18 A schematic view of the capacitor.

11.9.2 Capacitors

Capacitors are considered another type of energy storage medium that store energy in an electric field. They consist of two conductive plates separated by an insulating material known as a dielectric, as shown in Figure 11.18. When a voltage is applied to the plates, an electric charge builds up on each plate, creating an electric field between them. Capacitors may store and release energy quickly, making them useful for applications such as power conditioning, voltage stabilization, and energy storage in regenerative braking systems in electric vehicles. They are also used in power electronic applications, to smooth the output of power supplies.

Capacitors are not as energy-dense as batteries, which means they typically can't store as much energy in a given volume or weight. However, they are able to handle large amounts of power, meaning they can release their stored energy very quickly when needed. Therefore, they are used for short-time high-demand rate applications.

11.10 Other Energy Storage Techniques

In the previous sections, common energy storage techniques are widely used in the practical applications are presented. The remaining energy storage techniques listed in Figure 11.4 are presented under this caption.

11.10.1 Magnetic and Electromagnetic Energy Storage

Magnetic energy storage is a method of storing energy in a magnetic field. This can be achieved through the use of devices such as superconducting magnetic energy storage systems and magnetic air-core flywheels. Electromagnetic energy storage refers to the storage of energy in an electromagnetic field. This may be achieved through the use of devices such as electromagnetic flywheels and electromagnetic linear generators. The following techniques are considered some common types of magnetic and electromagnetic methods which are assumed as energy storage techniques.

- Superconducting magnetic energy storage
- Electromagnetic pumps and motors
- Electromagnetic suspension systems
- Electromagnetic pulse systems

a) Superconducting Magnetic Energy Storage

A superconducting magnetic energy storage (SMES) system is a device that stores energy in a superconducting coil by creating a strong magnetic field within the coil. The energy is stored in the form of magnetic field energy, and can be quickly released when needed by passing a current through the coil. An SMES system typically consists of a superconducting coil, a cryogenic cooling system to maintain the superconductivity of the coil, and power electronic converters to control the flow of energy into and out of the system, as shown in Figure 11.19.

The SMES systems have several advantages over other forms of energy storage, such as batteries. They have a very high energy density, so they can store a large amount of energy in a small volume. They also have very fast response times, so they can quickly release stored energy when needed. Additionally, the SMES systems do not degrade over time like batteries do, which means they have a long lifespan. However, the SMES systems also have some disadvantages. They require cryogenic cooling, which can be expensive and complex. They also have relatively high cost, which can make them less economically viable for some applications.

The SMES systems are mainly used in the power industry to provide a quick response to changes in power demand and to balance the grid. They can also be used in industrial and commercial settings to smooth out power fluctuations and provide backup power during outages. Superconducting magnetic energy storage systems store energy in a superconducting coil by creating a strong magnetic field within the coil.

b) Electromagnetic Pumps and Motors

Electromagnetic pumps and motors may also be used as energy storage devices. An electromagnetic pump uses an electromagnetic field to move a fluid, while an electromagnetic motor uses an electromagnetic field to generate mechanical motion. An electromagnetic pump can be used to store energy by pumping a fluid into a high-pressure storage tank when excess energy is available, and then releasing the stored energy by allowing the fluid to flow through a turbine to generate electricity when needed.

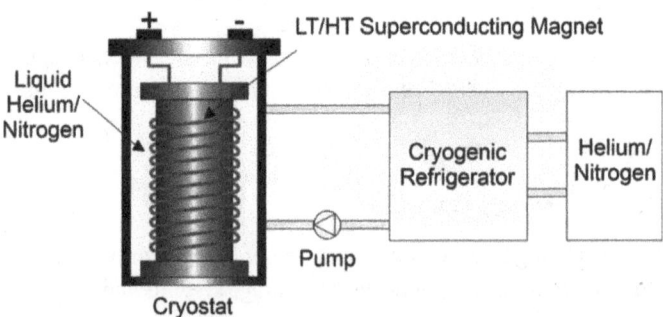

Figure 11.19   The schematic view of superconducting magnetic energy storage system.

An electromagnetic motor can also be used to store energy by using excess energy to drive the motor and generate mechanical motion, which can be stored in the form of kinetic energy. The stored energy can then be released by using the motor as a generator to convert the kinetic energy back into electricity when needed. Both of these technologies can be used to quickly release stored energy when needed, making them useful for applications such as load leveling and frequency regulation in power grids. Electromagnetic pumps and motors are also cheaper and more durable than batteries. However, they have lower energy density and thus require larger space for energy storage.

Both technologies are still in the research and development phase and have not yet been widely adopted as commercial energy storage technology.

c) Electromagnetic Suspension System

An electromagnetic suspension (EMS) system is a device that uses an electromagnetic field to levitate and control the position of a magnetically suspended object, such as a train or a wind turbine blade. An EMS system can also be used as an energy storage device by using the motion of the suspended object to store and release energy.

One way to use an EMS system for energy storage is to use excess energy to drive the electromagnets and cause the suspended object to move against a load, such as a spring or a hydraulic cylinder. The energy is stored as potential energy in the load. When energy is needed, the electromagnets can be turned off and the suspended object will move back to its original position, releasing the stored energy in the form of kinetic energy.

Another way to use an EMS system for energy storage is to use excess energy to drive the electromagnets and cause the suspended object to move in a circular motion, such as a flywheel. The energy is stored as kinetic energy in the flywheel. When energy is needed, the electromagnets can be used as generators to convert the kinetic energy back into electricity.

The EMS systems have the potential to be highly efficient energy storage devices, as they can convert energy very quickly and with minimal loss. However, they are also relatively complex and have not yet been widely adopted as commercial energy storage technology.

d) Electromagnetic Pulse System

An electromagnetic pulse (EMP) system is a device that generates a high-power electromagnetic pulse, which can be used for a variety of applications such as testing electronic devices for EMP susceptibility or for military purposes as a weapon. However, the EMP systems are not commonly used as energy storage techniques. The energy for the EMPs is usually generated by a high-energy capacitor bank, which stores a large amount of electrical energy and then rapidly discharges it to create the EMP. But the energy from the EMPs is usually released in a very short time, less than a microsecond, and the energy density is extremely high, making it not suitable for energy storage applications.

Additionally, the EMP systems can cause severe damage to electronic devices and infrastructure and are therefore heavily regulated and controlled. Therefore, it is not a safe or viable method for energy storage. There are other more suitable technologies that can store energy, such as batteries, flywheels, compressed air energy storage, and pumped hydro storage. These technologies have been tested and proven to be safe, reliable, and efficient for energy storage purposes.

11.10.2 Biological Energy Storage

Biological energy storage refers to the storage of energy in biofuels and living organisms, typically in the form of carbohydrates, such as glycogen and glucose, in animals and plants respectively. Carbohydrates are molecules made up of sugar units that can be easily broken down to release energy in the form of glucose when needed. In animals, glycogen is stored in the liver and muscle tissue, and is quickly broken down into glucose to provide energy for the body's needs. This process is regulated by hormones such as insulin and glucagon, which control the conversion of glycogen to glucose. In plants, the most common form of energy storage is in the form of starch, a polysaccharide that is stored in chloroplasts and amyloplasts, which can be broken down into glucose when light intensity is low or during the night.

Biological energy storage is a natural and efficient method of storing energy, as it is a part of the organisms' normal metabolic processes. However, it is not a method of energy storage that can be used on a large scale for human use. Additionally, the energy stored in biological systems is not directly usable for most human energy needs and thus it needs to be transformed into other forms of energy such as electricity, heat, or fuels.

a) Biofuels

Biofuels are a type of renewable energy that is derived from biomass, which is any organic material that comes from plants or animals. The most common types of biofuels are ethanol and biodiesel, which are produced from crops such as corn, sugarcane, and soybeans. These biofuels can be used in place of fossil fuels, such as gasoline and diesel, in transportation. Ethanol, also known as ethyl alcohol, is a biofuel that can be blended with gasoline in varying concentrations, up to a maximum of 10% (E10) in most countries. It is typically made from corn, sugarcane, and other crops that are rich in starch or sugar.

Biodiesel, on the other hand, is made from vegetable oils and animal fats, such as soybean oil and used cooking oil. It can be used in place of diesel fuel in vehicles and machinery. Biodiesel can also be blended with diesel fuel in varying concentrations, usually up to a maximum of 20% (B20) in most countries. Biofuels have several advantages over fossil fuels, such as being renewable and producing fewer greenhouse gas emissions. However, they also have some disadvantages, such as the potential competition with food production and the need for large areas of land to grow the crops used to produce biofuels.

In addition, there are other forms of biofuels such as biomethane, hydrogen, and synthetic biofuels that are produced from biomass through different processes and have different properties. Some details are presented in Chapter 9.

b) Algae-based Biofuels

Algae-based biofuels are a type of biofuel that is derived from microalgae, which are tiny aquatic plants that can grow in a wide range of environments. Algae are considered as one of the most promising sources of biofuel because they can be grown in non-arable land and using non-potable water, and they have a high growth rate and a high oil content. There are different ways to produce algae-based biofuels, but one of the most common methods is to grow the algae in large ponds or photobioreactors, which are closed systems that control the light, temperature, and other conditions to optimize the growth of the algae. Once the algae have grown, the oil can be extracted and processed to produce biofuel.

Algae-based biofuels have several advantages over traditional biofuels, such as higher yield per area, lower water consumption, and lower greenhouse gas emissions. Additionally, they do not compete with food production as they are not grown on land used for agricultural purposes. However, algae-based biofuels are still in the early stages of development and have not yet been widely adopted as a commercial energy source. There are still some challenges that need to be overcome, such as developing more efficient and cost-effective methods of growing and processing the algae and scaling up production to meet the demand for biofuels.

Research is ongoing to improve the efficiency and scalability of algal biofuel production, and it's expected to play an important role as a sustainable biofuel source in the future.

c) Chemiosmosis

Chemiosmosis is a biological process that occurs in the mitochondria of eukaryotic cells, and also in bacteria, where energy is stored in the form of a proton gradient across a membrane. This process is used by cells to generate adenosine triphosphate (ATP), the primary energy currency of cells. The basic principle of chemiosmosis is that protons are pumped across a membrane, creating a proton gradient or electrochemical potential. This proton gradient is used to drive the production of ATP through the action of an enzyme called ATP synthase.

Chemiosmosis can also be used as a way to store energy in the form of a proton gradient. For example, in photosynthesis, light energy is used to pump protons across a membrane creating a proton gradient that is stored as potential energy. This stored energy can be used later on by cells to generate ATP. Chemiosmosis energy storage has not been developed yet as a commercial energy storage technology, but it is an area of research. It has some advantages over other energy storage systems, such as high energy density and high efficiency. However, it also has some challenges such as the need to maintain a proton gradient across a membrane and the need to develop efficient ways of harnessing the stored energy. Overall, chemiosmosis is a biological process that can be used for energy storage, but it is not a technology that has been developed for use on a large scale.

d) Fats

Biological fats, also known as lipids, are a type of biomolecule that can be used for energy storage in living organisms. Lipids are composed of a glycerol molecule and one, two, or three fatty acid molecules, and are found in all living organisms. They are an important source of energy storage because they have a high energy density, are easily transportable, and can be stored for long periods of time. In animals, lipids are stored in adipose tissue, also known as fat, and can be broken down into fatty acids and glycerol when energy is needed. These fatty acids can be used as fuel for cells, providing energy for the body's needs. In plants, lipids are stored in the form of oils and can be found in seeds, nuts, and fruits. These oils can be extracted and used as a biofuel.

Biological fats can be used as a biofuel by extracting the oils from plants or animals and processing them to produce biodiesel, which can be used in place of diesel fuel in vehicles and machinery. Additionally, lipids can be converted to other forms of biofuels such as jet fuel, and can also be used to produce chemical products such as lubricants and plastics.

Biological fats are a renewable and sustainable source of energy, but the extraction and processing of the oils require a significant amount of energy, which can make them less efficient as an energy storage method. Additionally, the use of animal fats as biofuels raises ethical concerns. However, the use of plant-based oils as biofuels can be a more sustainable and eco-friendly alternative.

11.11 Closing Remarks

Energy storage appears to be a significant dimension of energy spectrum and plays a critical role to offset the mismatch between demand of supply or energy for applications in various sectors. It also helps achieving more resilient systems, reduced environmental impact and increased sustainability. This chapter, in this regard, classifies and discusses energy storage methods, namely mechanical, thermal, chemical, electrochemical, and others which may cover magnetic, electromagnetic, and biological methods. There are illustrative examples presented to provide some examples about the systems and their performance assessments through energy and exergy efficiencies. It is also noted that energy storage systems are expected to play a major role in overcoming the fluctuating challenges which some renewables, such as solar and wind, have.

References

1 Grid Watch. GB electricity national grid demand and output. Available online https://gridwatch.co.uk (accessed 04 January 2020).
2 Grid scale energy storage. International Energy Agency (IEA), Paris. Available online https://www.iea.org/reports/grid-scale-storage (accessed 21 January 2023).
3 Dincer, I. and Rosen, M.A. (2021). *Thermal Energy Storage: Systems and Applications*, 3rd ed. John Wiley and Sons.
4 Erdemir, D., Altuntop, N., and Cengel, Y.A. (2021). Experimental investigation on the effect of ice storage system on electricity consumption cost for a hypermarket. *Energy and Buildings* 251: 111368. doi: 10.1016/j.enbuild.2021.111368.

Questions/Problems

Questions

1 What is energy storage? Briefly explain.
2 What is the role of energy storage introduced by the 5S concept? Explain with examples.
3 Describe why energy storage is important for energy supply and demand management and provide examples from various economic sectors.
4 Classify energy storage techniques and describe each with an example.
5 What are the operational periods of any energy storage system?
6 Classify mechanical energy storage methods. Explain their working principles.
7 Classify thermal energy storage methods. Explain their working principles.
8 Classify chemical energy storage methods. Explain their working principles.
9 Classify electrochemical energy storage methods. Explain their working principles.
10 Classify magnetic and electromagnetic energy storage methods. Explain their working principles.
11 Discuss the role of biofuels in energy storage with an example.
12 Explain how the performance of any energy storage is evaluated.
13 Identify some commercially viable phase change materials.
14 Discuss what types of phase change materials are suitable for high temperature applications.
15 Explain how to improve the storage efficiency and what criteria are important.

Problems

1 A pumped-hydro storage system is designed to use for storing the off-peak electricity. In the charging period, the water is pumped from a lower reservoir to an upper reservoir, as depicted in the figure below. The elevation is 50 m. It is desired to get 30,000 kWh of energy from this pumped-hydro energy storage facility. The charging period is between 22:00 and 06:00 (8 h). The discharge period is between 15:00 and 19:00 (4 h). The efficiencies for the turbine, generator and pump are 0.85. The pipeline losses are negligible. Take density of the water as 998 kg/m^3.
   a In order to obtain 70,000 kWh of energy, calculate the amount of water to pump up.
   b Find the volumetric flow rate of water for charging period.
   c Calculate the pumping power.
   d Calculate the overall energy efficiency of the system.

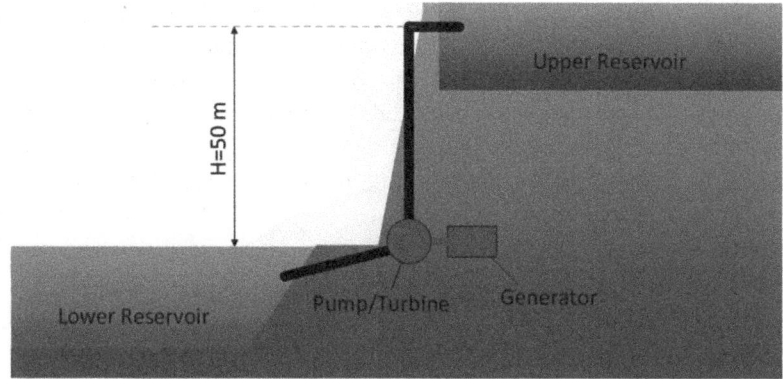

2 Restudy problem 1 for H = 200 m. It is desired to get 50,000 kWh of energy from this pumped-hydro energy storage facility. The charging period is 10 h. The discharging period is 4 h. The efficiencies for the turbine, generator, and pump are 0.85. All losses are negligible. Take density of the water as 998 kg/m³.

   a In order to obtain 100,000 kWh of energy, calculate the amount of water to pump up.
   b Find the volumetric flow rate during the charging period.
   c Calculate the pumping power.
   d Calculate the overall efficiency of the system.

3 Repeat Problem 2 for a charging period of 8 hours and a discharging period of 6 hours.

4 In a geothermal-based underground pumped-hydro storage system, the geothermal fluid is pumped from the underground reservoir, which is located 100 m below the ground. The temperature of the geothermal fluid is 80°C. The durations of the charging and discharging periods are 6 h and 12 h. The flow rates in the charging and discharging periods are 15 and 30 m³/s. The efficiencies for the turbine, generator, pump, and piping are to be taken as 0.90. In addition to the electricity storage, geothermal fluid is used for heating purposes in the upper reservoir. The temperature of the temperature of the return water is 80°C. The geothermal fluid is used for heating purposes in the upper reservoir. Take density of the water as 998 kg/m³. The heat losses are negligible.

   a Calculate the water volume for pumping.
   b Determine the charging and discharging energy capacities.
   c Determine the potential heat capacity obtained.

5 Repeat Problem 4 for a charging period of 8 hours and a discharging period of 6 hours.

6 In a geothermal-based underground pumped-hydro storage system, the geothermal fluid is pumped from the underground reservoir, which is located 50 m below the ground. The temperature of the geothermal fluid is 60°C. The durations of the charging and discharging periods are 4 h and 8 h. The flow rates in the charging and discharging periods are 15 and 30 m³/s. The efficiencies for the turbine, generator, pump, and piping are given as 0.85. In addition to the electricity storage, geothermal fluid is used for heating purposes in the upper reservoir. The temperature of the return water is 40°C. Consider negligible heat losses. Geothermal fluid is used for heating purposes in the upper reservoir. Take density of the water as 998 kg/m³.

   a Calculate the volume of water for pumping.
   b Determine the charging and discharging energies.
   c Determine the potential heat capacity obtained.

7 Repeat Problem 6 for a charging period of 8 hours and a discharging period of 6 hours.

8 A compressed-air storage system is used for storing wind electricity. The excess electricity output is known as 3 MW and available 3 h per day. In order to store the excess electricity, the atmospheric air is compressed up to 100 bars in two stages which are 1:10 compression ratio. The isentropic efficiency of the compressors is 0.85. The compressed air is used in the turbine to generate power. The flowrate of the air is 2.006 kg/s in the charging period. The flowrate of the air is 1.003 kg/s in the charging period. The efficiency of the turbine/generator is taken as 0.95. Use the state point table given below.

   a Calculate the energy stored.

   b Find the power output in the system.

   c Determine the energy efficiency of the system.

State	T (°C)	P (kPa)	h (kJ/kg)	s (kJ/kg K)
0	25	100	298.4	6.862
1	25	100	298.4	6.862
2	344.6	1000	626	6.945
3	935.6	10,000	1296	7.031
4	25	100	298.4	6.862

9 Repeat Problem 8 for an air flow rate of 3 kg/s during charging and an air flow rate of 2 kg/s during discharging.

10 A flywheel is used for storing electricity in the charging station to reduce the peak loads. In the flywheel, the rotational speed of the flywheel varies from 1000 rpm to 100,000 rpm. The mass of the flywheel is 50 kg, and its moment of inertia is 35 kg m². Calculate the energy stored.

11 Repeat Problem 10 for a flywheel mass of 100 kg and an inertia of 50 kgm².

12 A set of flywheels are used for storing electricity in the charging station to reduce the peak loads. The charging period is 30 minutes and discharging period is 8 minutes. In a flywheel, the rotational speed of the flywheel is 1000–50,000 rpm. The mass of the flywheel is 50 kg, and its moment of inertia is 35 kg m². If it is aimed to store 20,000 kWh of energy, how many flywheel unit are required? The efficiencies for the motor and generator are 0.95. Calculate the energy efficiency of the flywheel.

13 In a latent thermal energy storage system, paraffin has 90°C of the melting point and 245 kJ/kg of the latent heat capacity. 1000 kg of paraffin is used for storing solar thermal energy via parabolic through collectors. The initial temperature of the paraffin is 30°C and final temperature is 135°C. The specific heat of paraffin for solid and liquid phases is 2.01 kJ/kg K and 2.45 kJ/kg K, respectively. Calculate the amount of thermal energy stored in the paraffin.

14 Consider an ice thermal energy storage system. The water is at 20°C initially and cooed down up to –10°C. Calculate the amount of stored cold capacity per kg of the storage material.

15 Hydrogen is used for off-peak electricity storage. For this purpose, the PEM-based electrolyzer and fuel cells are used for hydrogen production and power generation from hydrogen. It is known that in the PEM electrolyzer, 45 kWh of energy is required for 1 kg of hydrogen production. On the other hand, it is possible to generate 12 kWh of energy with 1 kg of hydrogen in the PEM fuel cell. In the energy storage systems, it is aimed to store 20 MWh of energy in the off-peak hours from 22:00 to 6:00. The stored hydrogen is for power generation during the peak hours from 14:00 to 18:00.

   a Calculate how much hydrogen is needed for this storage.

   b Calculate the energy and exergy efficiencies for this energy storage system.

12

Hydrogen Energy Systems

12.1 Introduction

Global economic development and population growth have caused ever-increasing energy demand. Power generation has been a vital need since the industrial revolution where it was primarily in railways through the use of steam engines. Global energy production has been mainly met by fossil-based sources, nuclear energy, and hydropower generation, which are coal, natural gas, oil, etc. These traditional energy production methods using fossil fuels are expected to cover the majority of this energy demand, while renewable energy sources are expected to contribute significantly to global energy demand. The dominance of fossil resources in energy generation causes serious issues such as environmental problems, energy supply problems, and even distribution of resources. As indicated in Figure 12.1a, especially after the Industrial Revolution, the use of fossil fuels, and their negative impacts have increased substantially. Today, fossil fuels are responsible for the majority of major disasters like environmental and economic. In order to reduce the effects of these issues in the near future and eliminate them in the medium and long term, renewable energy sources have come into the picture to contribute to the global energy supply, which are solar, wind, geothermal, biofuels, innovative nuclear technologies, hydro and ocean energies, as demonstrated in Figure 12.1b.

Hydrogen is one of the key solutions to reaching a carbon-free future with the sustainability solution. Although there was some resistance in considering hydrogen energy as a prime solution, it has been pretty recently accepted as a fuel, energy storage medium, energy carrier, and feedstock. Its high energy content, carbon-free nature, and carbon-free production potential with renewables make it a popular resource.

The simplest member of the chemical family, hydrogen (H), is a colorless, odorless, tasteless, and flammable gas. An atom of hydrogen consists of a proton with one unit of positive electrical charge and an electron with one unit of negative electrical charge. The hydrogen gas formed by hydrogen molecules is a loose aggregation, each containing a pair of atoms, called a diatomic molecule (H_2). It is known that hydrogen burns with oxygen to form water (H_2O), which is the earliest chemical property of hydrogen. Although hydrogen is the most abundant element in the universe, it makes up only less than 1% of Earth's weight. It is available in the water form in oceans, ice packs, rivers, lakes, and the atmosphere. It is also the main part of the hydrocarbons which are all animal and vegetable residues and in petroleum.

Introduction to Energy Systems, First Edition. Ibrahim Dincer and Dogan Erdemir.
© 2023 John Wiley & Sons, Inc. Published 2023 by John Wiley & Sons, Inc.
Companion Website: www.wiley.com/go/Dincer/Introduction_to_Energy_Systems

(a)   (b)

Figure 12.1 (a) The past of the world issues and (b) a clean future with hydrogen.

As a matter of fact, hydrogen is an essential need for many sectors. Hydrogen can potentially be used as a fuel, an energy carrier, an energy storage medium, and a feedstock which are very critical in developing a low-carbon or carbon-free economy locally and globally.

Its higher energy density, lower heating value, and combustibility make hydrogen an important carbon-free fuel. Therefore, there are many attempts to use hydrogen as a fuel in various applications such as internal combustion engines, gas turbines, burners, furnaces, and industrial combustion purposes. Thus, hydrogen can be an environmentally-friendly fuel for transportation, energy generation, aviation, space, industrial, and residential sectors.

One of the main problems in the energy sector is the distribution and storage of the energy sources due to the mismatch between energy supply and demand profiles. Due to its high energy density and production from many energy sources, hydrogen is a critical energy carrier and storage medium. It is an important energy storage option for storing renewables. The stored hydrogen can be used for fuel, power generation, and feedstock purposes.

As hydrogen is a feedstock of various industrial applications, it is one of the highly demanded commodities. For example, along with nitrogen, it is one of the two raw material inputs of ammonia which is mostly produced chemicals in the world. Therefore, even to reach the clean commodities which consisted of hydrogen, the production of hydrogen is critical.

As mentioned before, hydrogen has been a significant feedstock for synthesizing many chemicals and fuels in various sectors where hydrogen was produced out of fossil fuels. In the last decade, the interest in hydrogen has increased due to the increasing contribution of renewables, the need for energy storage and rising demand for commodities containing hydrogen. Current development needs, issues, pandemics and other challenges indicate the need for change in almost everything. Therefore, the year of 2020 is called the turning point for better future as indicated in Figure 12.2, as introduced by Dincer [1]. The COVID-19 coronavirus pandemic, in this regard, showed that the people should change their habits for almost everything. Hydrogen is therefore recognized as a critical option for solving the common problems.

This chapter provides the basics of hydrogen and hydrogen energy systems and discuss their advantages and disadvantages as well as their deployment for various applications. Also, the historical development of hydrogen energy systems is discussed. Furthermore, hydrogen production methods are introduced and evaluated for various utilization opportunities. This is then followed by hydrogen storage techniques. Moreover, hydrogen utilization system and methods are discussed with some illustrative examples.

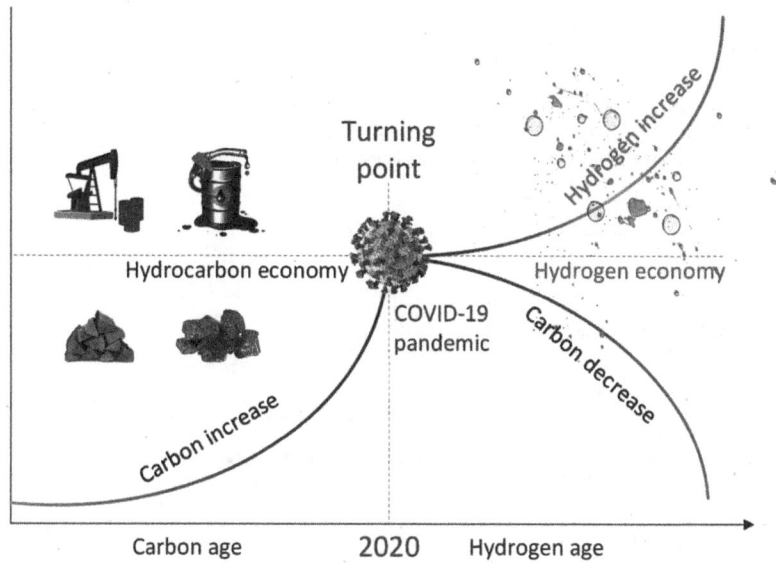

Figure 12.2 COVID-19 pandemic as a historical turning point for the hydrogen age.

12.2 Historical Development of Hydrogen Energy Systems

Hydrogen was identified as an element by Henry Cavendish in 1776. French chemist, Antoine Lavoisier, gave hydrogen its name, which was derived from the Greek words "hydro" and "genes," meaning "water" and "born of" or "maker of water." Figure 12.3 presents some historical milestones about hydrogen and hydrogen energy systems and applications. The water electrolysis concept was demonstrated by Jan Rudolph Deiman and Adrian Pates van Troostwijk in 1789. In 1828, the fuel cell concept used hydrogen and oxygen to generate voltage by releasing water.

The use of hydrogen as a fuel was introduced by a French author, Jules Verne, in 1874. In his book entitled *The Mysterious Island*, hydrogen was used as a fuel in the submarine named Nautilus. In the second quarter of the 1900s, researchers and engineers focused on the use of hydrogen and hydrogen/fuel blends in internal combustion engines. One of the critical uses of hydrogen is space technologies. In 1958, NASA started to use liquid hydrogen in rocket propulsion systems.

In 1970, the terminology, "hydrogen economy" was declared in the USA. Now, all around the world, micro and macro economies, countries, local governments have included hydrogen and hydrogen energy systems. Today, the countries have started declaring their hydrogen road maps, strategies and action plans with short-, medium-, and long-term objectives.

The first international conference on hydrogen technologies was organized under Prof. Nejat Veziroglu's leadership in 1974. With this conference, the International Association for Hydrogen Energy was founded. The participants became members of this foundation. After this organization, hydrogen turned into a critical solution for a clean and sustainable future. In 1990, the first solar-based hydrogen production facility became operational. Another milestone in hydrogen economy is the PEM-based fuel cell for automotive sector by the Ballard in Canada.

As mentioned earlier, the year of 2020 was the turning point for the world by Dincer [1]. During the COVID-19 pandemic, it became essential for people to have clean air, clean water, clean food,

2023	• Hydrogen 1.0 has been initiated by Dincer [2].
2020	• It was a declaration of closing carbon age, but opening hydrogen age in the month of May by Dincer [1]. • The European Community announced their green hydrogen deal in the month of July.
2000	• The first hydrogen refueling station became operational in Dearborn, USA.
1990	• The world's first solar-based hydrogen production plant was built for operation.
1974	• The world's first hydrogen energy conference was organized by Prof. Nejat Veziroglu and the term "hydrogen economy" was declared. • The International Association for Hydrogen Energy was established by Prof. Nejat Veziroglu and his colleagues.
1966	• The first hydrogen fuel cell car was developed.
1958	• The liquid hydrogen was first used in the NASA's space programs.
1920	• The uses of hydrogen and hydrogen/fuel blends in internal combustion engines were tested.
1874	• Jules Verne wrote the potential use of hydrogen as a fuel in his book entitled "*The Mysterious Island*".
1839	• The first fuel cell was developed by Sir William Robert Grove.
1838	• The fuel cell effect was first used through hydrogen and oxygen to generate voltage while releasing water.
1789	• The water electrolysis was first demonstrated by Jan Rudolph Deiman and Adriaan Pates van Troostwijk.
1788	• Antoine Lavoisier named it as hydrogen, meaning maker of water.
1783	• Hydrogen was used in a hot-air baloon in France.
1776	• Hydrogen was identified as a distinct element by Henry Cavendish.

Figure 12.3 Historical landmarks of hydrogen and related developments.

and clean energy where clean hydrogen was the key target. In addition, the people focused on the balance of the world starting from the energy sector which is responsible for almost all issues in the world. That's why it is called the Hydrogen era. More recently, Dincer [2] has initiated the Hydrogen 1.0 concept which covers better resource use, better environment, better energy security, better economy, better design, and better efficiency, as illustrated in Figure 12.4. Hydrogen can provide a significant contribution to each item. That's why it is often called a key solution for a better future.

Figure 12.4 Concept of hydrogen 1.0 (modified from [2]).

12.3 Hydrogen Production

It is true that hydrogen is the most common element on earth, but it is not really available on its own. The separation of pure hydrogen must therefore be carried out in an environmentally-friendly manner, in order to dissociate it from hydrogen-containing sources. There is no doubt that this separation or cracking process requires energy; however, hydrogen production can be performed with any primary energy source since hydrogen is an energy carrier, not an energy source itself. Hydrogen is produced using different sources of energy, such as fossil fuels, nuclear, natural gas, nuclear, solar, biomass, wind, hydro, or geothermal. There are a variety of energy resources that can be used to produce hydrogen, including fossil fuels such as coal and natural gas, nuclear energy, and renewable energy sources such as solar, wind, geothermal, hydro, ocean energy and biomass. Additionally, hydrogen is a promising energy carrier due to its diversity and range of alternative energy sources.

There are many hydrogen production methods available. While many of them are commercially available in the market and utilized, many more are under development. Figure 12.5 demonstrates the available hydrogen production techniques according to the source. However, renewable electricity can also be utilized to produce green hydrogen that has a zero carbon footprint, even though steam methane reforming is the CO_2-intensive process used globally for hydrogen production today. Electrolysis is another conventional method of splitting water into oxygen and hydrogen using electricity. Electricity can be harvested from renewable sources to produce green hydrogen through electrolysis.

Hydrogen production costs are currently considered a significant concern. Steam reforming-based hydrogen production costs roughly three times higher than the natural gas cost per unit of produced energy. Similarly, electrolysis for hydrogen production using 5 cents/kWh of electricity will cost somewhat less than natural gas-based hydrogen production. Recent reports state the USA will offer wind energy at the lowest recorded price of 2.5 cents/kWh, so the electricity will cost somewhat less than four times the cost of natural gas-based hydrogen production.

In addition to the hydrogen production cost, the environmental impact of hydrogen production methods is essential for the systems which use hydrogen. These systems will be as clean as hydrogen is clean. Therefore, CO_2 emissions of hydrogen production methods are quite important. Figure 12.6 shows CO_2 emission values of hydrogen production according to the source. Although they provide higher production rates, the fossil-based hydrogen production methods cause the

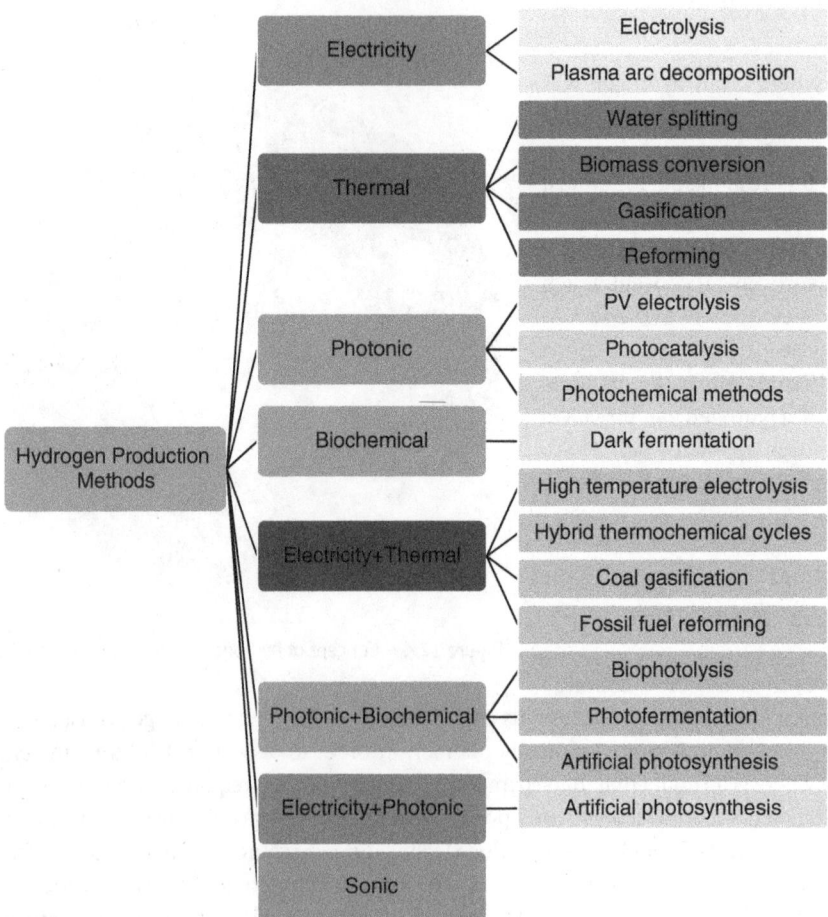

Figure 12.5 Hydrogen production methods by sources.

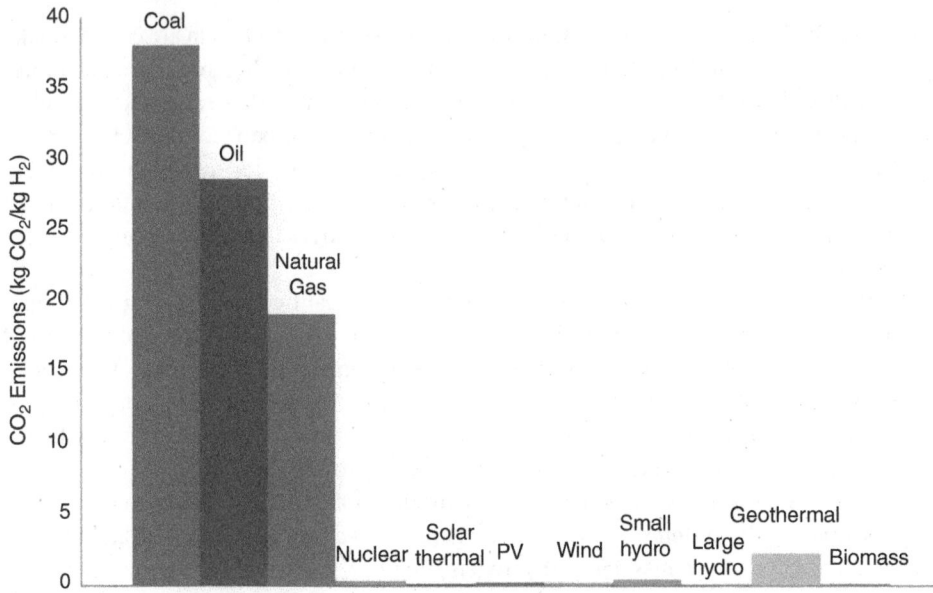

Figure 12.6 CO₂ emissions of hydrogen production according to the source.

Table 12.1 Color codes of hydrogen production methods.

Source and Method of Hydrogen Production	Color Code
Renewable	**Green**
Nuclear	**Yellow**
Fossil fuel based (with carbon capturing)	**Blue**
Fossil gas based (with gasification/pyrolysis)	**Light Green**
Fossil gas based (with no capturing)	**Gray**
Lignite based	**Brown**
Coal based	**Black**

highest CO_2 emissions. While coal-based production emits about 37.5 kg CO_2 per kg H_2, oil-based production causes almost 29 kg CO_2 per kg of hydrogen production. Natural gas emits approximately 19 kg CO_2 per kg H_2 production.

As expected, the CO_2 emissions are quite lower with using the renewables in hydrogen production. Geothermal energy-based hydrogen production emits higher CO_2 emissions comparing to other renewable and nuclear energy sources. The emissions of the solar thermal- and biomass-based hydrogen production methods are almost none.

In order to illustrate the cleanliness levels of hydrogen production, the color codes are commonly used, as illustrated in Table 12.1. Coal-based hydrogen production is indicated with black as it emits the highest CO_2 emissions. The yellow color indicates nuclear-based hydrogen production. In order to reduce the environmental impact of hydrogen production with fossil-based sources, carbon capturing is recently integrated into the system. As these systems reduce the carbon footprint of the methods, they are illustrated with the blue color. The green color indicates hydrogen production with renewable sources. Therefore, hydrogen produced with renewable energy sources is called "green hydrogen."

12.3.1 Electrolysis

Water electrolysis is a process that involves splitting water (H_2O) into its two basic components, hydrogen (H_2) and oxygen (O_2), by applying an electrical current. Figure 12.7 shows the illustration of basic water electrolysis. The electrical current passes through a water-based solution or through a solid-state electrolyte and causes the water molecules to break down into hydrogen and oxygen ions. The hydrogen ions then migrate to the cathode, where they combine with electrons to form hydrogen gas, and the oxygen ions migrate to the anode, where they combine with electrons to form oxygen gas. The two gases are then collected separately and can be used for various purposes, such as fuel in hydrogen fuel cells, industrial processes, or as a source of oxygen for breathing. The efficiency and cost-effectiveness of water electrolysis depend on

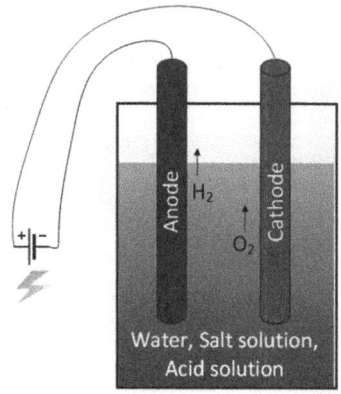

Figure 12.7 Schematic illustration of the water electrolysis.

several factors, including the type of electrolyzer used, the source of the electrical current, and the operating conditions.

The chemical reaction of water electrolysis can be represented as follows:

$$2H_2O + \text{electricity} \rightarrow 2H_2 + O_2$$

In this reaction, water (H_2O) is split into hydrogen (H_2) and oxygen (O_2) through the application of an electric current. This reaction occurs at the electrodes of the electrolysis cell, with hydrogen being produced at the cathode and oxygen being produced at the anode. The hydrogen and oxygen produced in this reaction are typically separated by a membrane in the electrolysis cell. Some details of water electrolysis and their types are further discussed in Section 12.4.

12.3.2 Plasma Arc Decomposition

Plasma arc decomposition is a method of producing hydrogen through the decomposition of natural gas or other hydrocarbons, as illustrated in Figure 12.8. In this process, an electric arc is used to heat the hydrocarbon feedstock to a high temperature, causing it to decompose into hydrogen and carbon. The hydrogen produced in this reaction can then be separated from the carbon and other by-products through a series of filters and condensers. Plasma arc decomposition is a highly efficient method of producing hydrogen, but it requires a significant amount of energy input, which can make it cost-prohibitive in some cases.

Furthermore , it is important to note that the plasma state of matter is an ionized state containing electrons in an excited state and atomic species. The presence of electrically charged particles in plasma makes it suitable as a medium for high-voltage electric current release. As a result of thermal plasma activity, natural gas (mostly methane) dissociates into hydrogen and carbon black (soot). The solid phase of carbon black is collected at the bottom, while the gas phase of hydrogen is collected at the top. Methane decomposes into hydrogen and carbon as follows:

$$CH_4 \rightarrow C_s + 2H_2 \rightarrow \Delta H = 74.6 \frac{MJ}{kmol}$$

12.3.3 Thermolysis

Thermolysis is a process of hydrogen production that involves heating water to high temperatures to dissociate its molecules into hydrogen and oxygen directly. The heat energy causes the water molecules to break apart into hydrogen and oxygen through a chemical reaction, which can be captured and stored for use as a fuel. The process is typically performed using an electrical heating element and can be used to generate hydrogen on-demand for various applications. Water thermolysis, also known as single step thermal dissociation of water, reaction can be written as

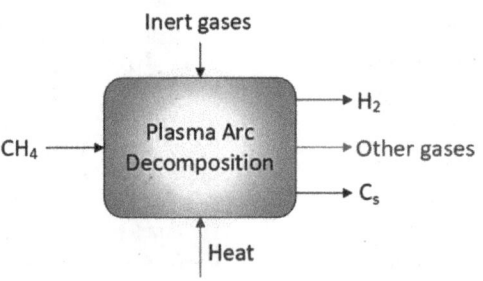

Figure 12.8 Schematic illustration of hydrogen production with plasma arc decomposition.

$$H_2O \xrightarrow{heat} H_2 + \frac{1}{2}O_2$$

Figure 12.9 demonstrates the schematic view of water thermolysis process. In order to accomplish a reasonable degree of dissociation, the reaction requires a heat source which could provide temperatures above 2500 K. For instance, at 3000 K and 1 bar, the degree of dissociation may remain about 64%. One of the challenges of this production method is the separation of H_2 and O_2. The existing semipermeable membranes can be used at temperatures up to 2500 K. Therefore, the mixture needs to be cooled down before being sent to the separation process.

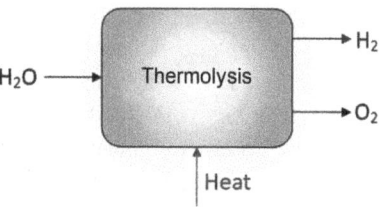

Figure 12.9 Schematic illustration of thermolysis process.

12.3.4 Thermochemical Water Splitting

Thermochemical water splitting is a process for producing hydrogen through a series of chemical reactions that occur in the presence of heat. The process typically involves the use of a metal oxide or hydroxide, which reacts with water to generate hydrogen and oxygen. The

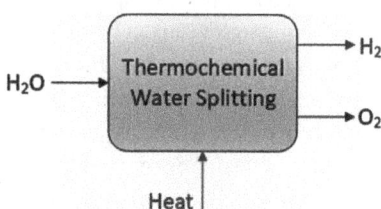

Figure 12.10 Schematic illustration of thermochemical water splitting process.

heat energy required for the reaction can come from a variety of sources, including solar, nuclear, or conventional combustion. The hydrogen produced through thermochemical water splitting can be used as a fuel in various applications, including power generation, transportation, and industrial processes.

Figure 12.10 depicts the view of the thermochemical water splitting method. A major advantage of thermochemical water splitting cycles is that no catalytic reactions are required. Other advantages of thermochemical water splitting cycles can be listed as: (i) no need for O_2–H_2 separation membranes, (ii) reasonable temperature requirement range of 600–1200 K, and (iii) zero or low electrical energy requirement. There are some challenges related to materials and recovery of chemicals.

12.3.5 Thermochemical Biomass Conversion

Thermochemical biomass conversion is a process that uses heat and chemical reactions to convert biomass, such as wood chips or agricultural waste, into hydrogen. The process typically involves the thermolysis of biomass, which breaks down the organic matter into its constituent elements, including hydrogen. The heat for the reaction can come from a variety of sources, including solar, nuclear, or conventional combustion. The hydrogen produced through thermochemical biomass conversion can be used as a fuel in various applications, including power generation, transportation, and industrial processes. The process has the potential to provide a renewable source of hydrogen, as the biomass used as a feedstock can be sustainably grown and replenished.

Thermochemical biomass conversion for hydrogen production can be performed using two main processes: pyrolysis and gasification. In pyrolysis, the biomass is heated in the absence of oxygen, causing it to break down into a mixture of gases, liquids, and solids, including hydrogen. In gasification, the biomass is heated in the presence of a controlled amount of oxygen, which causes it to break down into a mixture of gases, including hydrogen and carbon monoxide. Hydrogen and carbon monoxide can be separated and purified, and hydrogen may be used as a fuel or as an energy carrier or a feedstock.

The advantages of using thermochemical biomass conversion for hydrogen production include the ability to use a renewable, sustainable feedstock and the potential to reduce greenhouse gas emissions by replacing fossil fuels with hydrogen as a fuel source. However, the process is currently limited by high production costs and low hydrogen yields. Despite these limitations, thermochemical biomass conversion remains an attractive option for hydrogen production, particularly in regions where biomass resources are abundant. Research and development in the field continues to focus on improving the efficiency and economics of the process, with the goal of making it a viable source of renewable hydrogen for various applications.

Figure 12.11 shows the thermochemical biomass-based hydrogen production with gasification. Thermochemical gasification is a process for producing hydrogen from a range of feedstocks, including coal, natural gas, and biomass, through a series of chemical reactions that occur in the presence of heat. The process involves heating the feedstock in the presence of a controlled amount of oxygen, which causes it to break down into a mixture of gases, including hydrogen and carbon monoxide. In thermochemical gasification, the feedstock is first converted into a synthetic gas (syngas) through a series of chemical reactions that occur in the presence of heat and a controlled amount of oxygen. The syngas is then cleaned and purified, and the hydrogen can be separated and captured for use as a fuel. Thermochemical gasification is a well-established technology with a long history of use in various industrial applications, including power generation and the production of chemicals. However, the process is also associated with several environmental and economic challenges, including the need for high-temperature processing, the production of greenhouse gases, and the high costs associated with purifying and capturing hydrogen. Despite these challenges, thermochemical gasification remains an important technology for producing hydrogen, particularly in regions where abundant supplies of fossil fuels or biomass are available. Efforts to improve efficiency and reduce the costs of the process continue to be an active area of research and development.

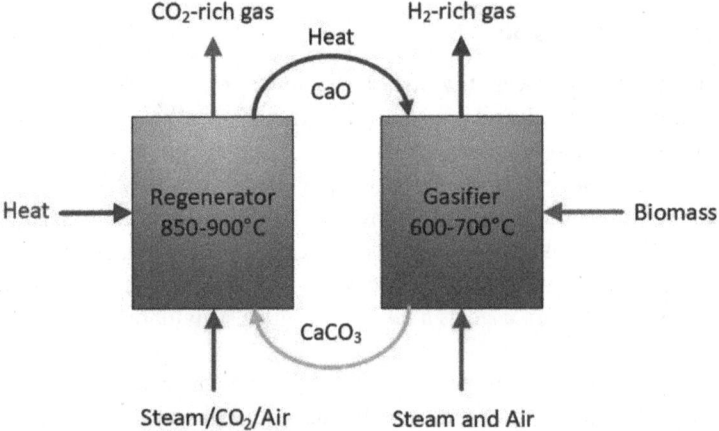

Figure 12.11 Illustration of thermochemical biomass conversion-based hydrogen production.

12.3.6 Thermochemical Reforming

Thermochemical reforming is a process for producing hydrogen through a series of chemical reactions that occur in the presence of heat. Figure 12.12 demonstrates the schematic view of the thermochemical reforming in hydrogen production. The process typically involves the reaction of a hydrocarbon feedstock, such as natural gas

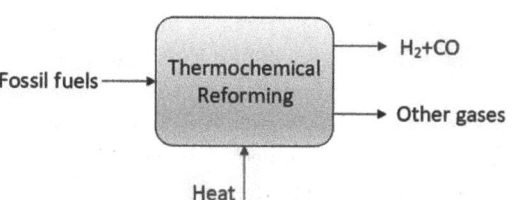

Figure 12.12 Schematic view of the hydrogen production with thermochemical reforming.

or methanol, with steam at high temperatures to produce a mixture of gases, including hydrogen and carbon monoxide. In thermochemical reforming, the feedstock is first converted into a synthetic gas (syngas) through a series of chemical reactions that occur in the presence of heat and steam. The syngas is then cleaned and purified, and the hydrogen can be separated and captured for use as a fuel.

Thermochemical reforming is a mature technology as they are used in many sectors. The process is typically performed using high-temperature steam reformers, which are capable of producing high volumes of hydrogen with high levels of purity. One of the main advantages of thermochemical reforming is its ability to produce hydrogen from a range of feedstocks, including natural gas, methanol, and biomass. However, the process is also associated with several environmental and economic challenges, including the need for high-temperature processing, the production of greenhouse gases, and the high costs associated with purifying and capturing hydrogen. Despite these challenges, thermochemical reforming remains an important technology for producing hydrogen, particularly in regions where abundant supplies of fossil fuels or biomass are available. There are efforts to made improve process efficiency and reduce the costs.

12.3.7 Photocatalysis

Photocatalytic hydrogen production is defined as a process for producing hydrogen by splitting water molecules into hydrogen and oxygen using light energy and a photocatalyst. The process involves exposing water to light in the presence of a photocatalyst, which triggers a chemical reaction that splits the water molecules into hydrogen and oxygen, as shown in Figure 12.13. The main advantage of photocatalytic hydrogen production is that it is a clean and renewable source of hydrogen, as it relies on light energy rather than fossil fuels. The process also has the potential to be highly efficient, as it can harness the energy from a wide range of light sources, including sunlight and artificial light. However, photocatalytic hydrogen production also has several disadvantages, including the need for specialized equipment and the limited efficiency of the process, particularly when

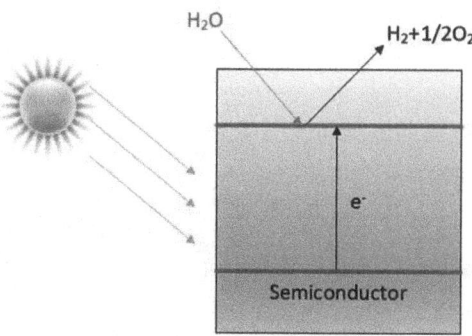

Figure 12.13 A schematic representation of photocatalytic hydrogen production.

compared to other methods of producing hydrogen. Additionally, the process is limited by the availability of light, which means that it may not be able to produce hydrogen on a continuous basis, particularly in regions with low levels of sun exposure. Consequently, photocatalytic hydrogen production is a promising technology for producing hydrogen in a clean and renewable manner, but it remains subject to several challenges that must be addressed in order to increase its viability as a source of energy.

12.3.8 Photoelectrochemical Method

Photoelectrochemical (PEC) hydrogen production is a process for producing hydrogen by splitting water molecules into hydrogen and oxygen using light energy and an electrically conductive material. The process involves exposing water to light in the presence of a photoelectrode, which triggers a chemical reaction that splits the water molecules into hydrogen and oxygen, as illustrated in Figure 12.14. The main advantage of PEC hydrogen production is that it is a clean and renewable source of hydrogen, as it relies on light energy rather than fossil fuels. The process also has the potential to be highly efficient, as it can harness the energy from a wide range of light sources, including sunlight and artificial light. However, PEC hydrogen production also has several disadvantages, including the need for specialized equipment and the limited efficiency of the process, particularly when compared to other methods of producing hydrogen. Additionally, the process is limited by the availability of light, which means that it may not be able to produce hydrogen on a continuous basis, particularly in regions with low levels of sun exposure. PEC hydrogen production is a promising technology for producing hydrogen in a clean and renewable manner, but it remains subject to several challenges that must be addressed in order to increase its viability as a source of energy.

12.3.9 Dark Fermentation

Dark fermentation is a process for producing hydrogen by utilizing microorganisms, such as bacteria and fungi to break down any organic matter in the absence of light. The process involves introducing a mixture of organic matter, such as food waste or agricultural waste, into a bioreactor, where microorganisms

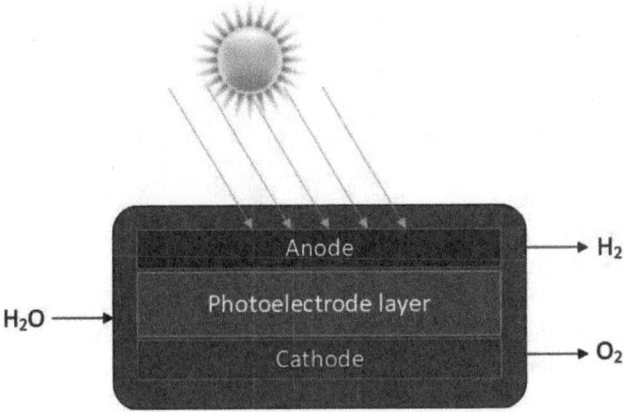

Figure 12.14 A schematic illustration of photoelectrochemical hydrogen production.

break down the organic matter and produce hydrogen gas as a by-product. Figure 12.15 demonstrates the view of the hydrogen production with dark fermentation.

The main advantage of dark fermentation is that it can utilize a wide range of organic waste materials that would otherwise be discarded, making it a sustainable and low-cost method of producing hydrogen. Additionally, dark fermentation can be performed in a continuous manner, allowing for the production of hydrogen on a large scale. However, dark fermentation also has several disadvantages, including the limited efficiency of the process and the need for specialized equipment and conditions to maintain the microorganisms in the bioreactor. Additionally, the process can produce other by-products, such as methane, which may limit the overall efficiency of the process. Dark fermentation is a promising technology for producing hydrogen in a sustainable and low-cost manner, but it remains subject to several challenges that must be addressed in order to increase its viability as a source of energy.

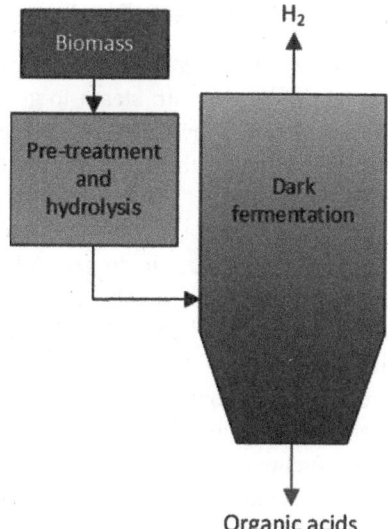

Figure 12.15 A schematic view of hydrogen production with dark fermentation.

12.3.10 Coal Gasification

Coal gasification is a process for producing hydrogen by reacting coal with steam and oxygen to generate a mixture of gases, including hydrogen. Figure 12.16 demonstrates the hydrogen production with coal gasification. Coal gasification involves the partial oxidation of coal to produce a mixture of gases, including hydrogen. The process typically starts with pulverized coal that is mixed with steam and oxygen, which are then fed into a gasifier. The gasifier is a high-temperature reactor that heats the coal mixture to temperatures of 700°C to 1200°C, causing a chemical reaction to occur that converts the coal into a mixture of gases, including hydrogen, carbon monoxide, carbon dioxide, and others. After leaving the gasifier, the gas mixture is then cooled and cleaned to remove impurities and particles, resulting in a synthesis gas, or syngas, which is primarily composed of hydrogen and carbon monoxide. The syngas can then be used as a fuel in various applications, including power generation, heating, or further processed to produce hydrogen.

The main advantage of coal gasification is that it can utilize coal, a readily available and abundant energy source, to produce hydrogen. Additionally, the process can potentially capture and utilize the carbon dioxide generated by the reaction, making it a potentially sustainable source of hydrogen. However, coal gasification also has several disadvantages, including the need for high-temperature materials and specialized equipment, which can increase the cost and complexity of the process. Additionally, the process generates large amounts of carbon dioxide, which is the most critical greenhouse gas and must be captured and stored in order to avoid contributing to climate change.

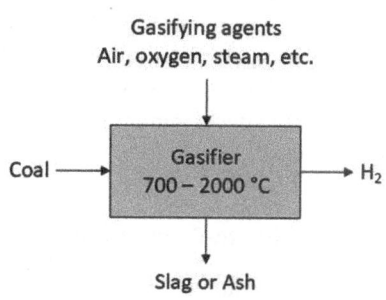

Figure 12.16 A schematic illustration of hydrogen production with coal gasification.

12.3.11 Fossil Fuel Reforming

Fossil fuel reforming is a process for producing hydrogen by reacting fossil fuels, such as natural gas or petroleum, with steam to generate a mixture of gases, including hydrogen. The process involves heating the fossil fuel to high temperatures in the presence of steam, which triggers a chemical reaction that converts the fossil fuel into a mixture of gases, including hydrogen and carbon monoxide.

The main advantage of fossil fuel reforming is that it can utilize readily available and abundant fossil fuels to produce hydrogen. Additionally, the process can potentially capture and utilize the carbon dioxide generated by the reaction, making it a potentially sustainable source of hydrogen. However, fossil fuel reforming also has several disadvantages, including the need for high-temperature materials and specialized equipment, which can increase the cost and complexity of the process. Additionally, the process generates large amounts of carbon dioxide, which is a potent greenhouse gas and must be captured and stored in order to avoid contributing to climate change. Fossil fuel reforming is a potentially viable source of hydrogen, but it remains subject to several challenges that must be addressed in order to increase its sustainability and viability as a source of energy. The process must be improved to minimize its impact on the environment and make it a more sustainable source of hydrogen.

12.3.12 Bio-Photolysis

Bio-photolysis is recognized as a process for producing hydrogen using photosynthetic organisms, such as algae or bacteria. In bio-photolysis, photosynthetic organisms are exposed to light, which triggers a series of reactions that produce hydrogen as a by-product. During photosynthesis, photosynthetic organisms convert light energy into chemical energy, which is stored in the form of organic molecules, such as sugars. In bio-photolysis, the organic molecules produced during photosynthesis are broken down through a process known as dark fermentation, which releases hydrogen as a by-product.

One advantage of bio-photolysis is that it can utilize renewable energy sources, such as sunlight, to produce hydrogen, making it a potentially sustainable source of energy. Additionally, bio-photolysis can be implemented in a variety of settings, including terrestrial and aquatic environments, and can utilize a variety of photosynthetic organisms, including algae, bacteria, and plants. However, there are also several challenges associated with bio-photolysis, including the need for specialized equipment and the need to optimize the growth conditions of the photosynthetic organisms. Additionally, the process remains in the early stages of development and further research and development is needed to increase its efficiency and scalability. Bio-photolysis is a promising technology for producing hydrogen using renewable energy sources, but further research and development is needed to optimize the process and increase its viability as a source of energy.

12.3.13 Photo-Fermentation

Photo-fermentation is a process for producing hydrogen using photosynthetic microorganisms, such as algae or bacteria. In photo-fermentation, photosynthetic microorganisms are exposed to light, which triggers a series of reactions that produce hydrogen as a by-product. During photosynthesis, photosynthetic microorganisms convert light energy into chemical energy, which is stored in the form of organic molecules, such as sugars. In photo-fermentation, the organic molecules produced during photosynthesis are broken down through a process known as fermentation, which releases hydrogen as a by-product.

One advantage of photo-fermentation is that it can utilize renewable energy sources, such as sunlight, to produce hydrogen, making it a potentially sustainable source of energy. Additionally, photo-fermentation can be implemented in a variety of settings, including terrestrial and aquatic environments, and can utilize a variety of photosynthetic microorganisms, including algae, bacteria, and plants. However, there are also several challenges associated with photo-fermentation, including the need for specialized equipment and the need to optimize the growth conditions of photosynthetic microorganisms. Additionally, the process remains in the early stages of development and further research and development is needed to increase its efficiency and scalability. Photo-fermentation is a promising technology for producing hydrogen using renewable energy sources, but further research and development is needed to optimize the process and increase its viability as a source of energy.

12.3.14 Artificial Photosynthesis

Artificial photosynthesis is a process for producing hydrogen using artificial systems that mimic the process of photosynthesis that occurs in plants and other photosynthetic organisms. In artificial photosynthesis, light energy is used to trigger chemical reactions that produce hydrogen. Artificial photosynthesis systems can be designed using a variety of materials, including semiconductors and catalysts. These systems can utilize different light-absorbing materials, such as dyes or pigments, to convert light energy into chemical energy. One advantage of artificial photosynthesis is that it has the potential to produce hydrogen more efficiently and cost-effectively than other methods of hydrogen production. Additionally, artificial photosynthesis can utilize renewable energy sources, such as sunlight, to produce hydrogen, making it a potentially sustainable source of energy. However, there are also several challenges associated with artificial photosynthesis, including the need for specialized equipment and the need to optimize the efficiency of the light-absorbing materials and catalysts used in the system. Furthermore, the process remains in the early stages of development, and further research and development is needed to increase its efficiency and scalability.

12.3.15 Sonic Hydrogen Production

The energy content of the sonic power can be used in the water splitting for hydrogen production. Hydrogen production using sonic power, also known as ultrasonic hydrogen generation or sono-hydrogen production, involves using high-frequency sound waves to generate hydrogen. In this process, sound waves are used to agitate the water molecules, which then dissociate into hydrogen and oxygen. Figure 12.17 illustrates the ultrasonic hydrogen production. The process of ultrasonic hydrogen generation can be carried out in a variety of configurations, including batch and continuous systems. In some cases, catalysts may be added to the water to increase the efficiency of the hydrogen generation process.

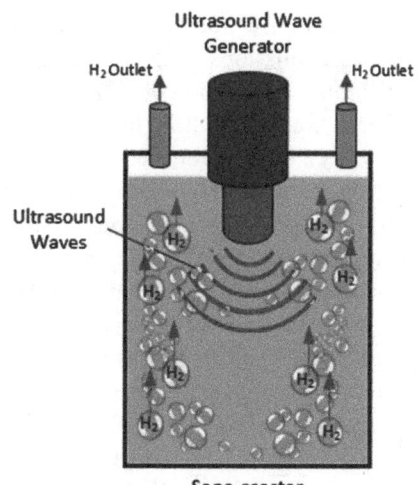

Figure 12.17 A simplified view of the ultrasonic hydrogen production.

The main advantage of ultrasonic hydrogen generation is that it has the potential to be more efficient and cost-effective than other methods of hydrogen production, such as steam reforming or electrolysis. Additionally, ultrasonic hydrogen generation can utilize renewable energy sources, such as wind or solar power, to generate the sound waves needed for the process. However, there are also several challenges associated with ultrasonic hydrogen generation, including the need for specialized equipment and the need to optimize the efficiency of the process. Additionally, the technology remains in the early stages of development and further research and development is needed to increase its viability as a source of energy.

Case Study 1   Comparison of Hydrogen Production Methods

Here, various potential hydrogen production methods using renewable and nonrenewable sources are compared in terms of the environmental impact, cost, energy efficiency, and exergy efficiency. This case study considers a number of potential primary energy sources, including electrical, thermal, biochemical, photonic, electro-thermal, photo-electric, and photo-biochemical. Nineteen hydrogen production methods have been compared based on data published by Dincer and Acar [3].

As a result of their adverse effects on the environment and human health, CO_2 emissions are considered to be the primary source of greenhouse gases. CO_2 emissions can be mitigated by capturing and storing the CO_2, managing the CO_2 as waste or as a commodity, etc. One of the most heavily researched topics in the literature is switching to a carbon neutral economy. Hydrogen can reduce CO_2 emissions significantly when produced from renewable energy sources. Life cycle assessments (LCAs) are necessary to fully understand and assess CO_2 emissions. One may see Chapter 14 for more details on life cycle assessment.

In this case study, hydrogen production methods are assessed based on this operational guide provided by Dincer and Acar [3]. Table 12.2 lists the hydrogen production methods studied and shows the overall comparison of these methods over the normalized values. In terms of the ranking, there is a scale from 0 to 10, where 0 represents poor performance, and 10 represents the ideal case (zero costs and zero emissions). A higher ranking is given to the companies that have lower costs and emissions. A value of "0" is assigned to the categories with the highest costs and emissions. As an example, the coal gasification method emits the highest number of greenhouse gases; therefore, it is assigned a GWP ranking of zero. The ranking range is again between 0 and 10, with 0 indicating poor performance and 10 indicating the ideal case (100% efficiency). The higher the ranking, the greater the efficiency. Table 12.2 presents the normalized emissions, cost, and efficiency rankings. In the hypothetical ideal case, there would be no costs or emissions, which would also mean that there would be no SCC. In this ideal scenario, energy and energy efficiency are both 100%. Fossil fuel reforming and biomass gasification provide the closest performance to the ideal case in terms of energy efficiency and energy output. Nevertheless, biomass gasification has a considerable AP (low AP ranking) when compared to other selected methods.

From Figure 12.18, it is possible to observe that fossil fuel reforming, plasma arc decomposition, and the combustion and gasification of coal and biomass are the most energy and energy efficient methods of hydrogen production, which have an advantage over other methods. As compared to the other methods, photonic energy-based hydrogen production methods show poorer performance compared to the other methods that have been selected.

Table 12.2 Overall comparisons of selected hydrogen production methods over normalized values, ranging between 0 and 10 where 0 stands for the worst and 10 represents the best.

Method		Energy Efficiency	Exergy Efficiency	Cost	SCC	GWP	AP
M1	Electrolysis	5.30	2.50	7.34	3.33	3.33	8.86
M2	Plasma arc decomposition	7.00	3.20	9.18	0.83	0.83	5.14
M3	Thermolysis	5.00	4.00	6.12	7.50	7.50	7.43
M4	Thermochemical water splitting	4.20	3.00	8.06	9.17	9.17	9.43
M5	Biomass conversion	5.60	4.50	8.10	6.67	6.67	2.00
M6	Biomass gasification	6.50	6.00	8.25	5.83	5.83	0.00
M7	Biomass reforming	3.90	2.80	7.93	6.25	6.25	0.86
M8	PV electrolysis	1.24	0.70	4.50	7.50	7.50	7.71
M9	Photocatalysis	0.20	0.10	5.19	9.58	9.58	9.71
M10	Photoelectrochemical process	0.70	0.15	0.00	9.58	9.58	9.71
M11	Dark fermentation	1.30	1.10	7.52	9.58	9.58	9.71
M12	High temperature electrolysis	2.90	2.60	5.54	7.92	7.92	8.57
M13	Hybrid thermochemical cycles	5.30	4.80	7.41	9.43	9.43	9.02
M14	Coal gasification	6.30	4.60	9.11	0.00	0.00	1.31
M15	Fossil fuel reforming	8.30	4.60	9.28	2.50	2.50	5.71
M16	Biophotolysis	1.40	1.30	7.27	7.50	7.50	9.71
M17	Photo-fermentation	1.50	1.40	7.61	9.58	9.58	9.71
M18	Artificial photosynthesis	0.90	0.80	7.54	9.58	9.58	9.71
M19	Photoelectrolysis	0.78	0.34	7.09	8.33	8.33	9.71
Ideal	No emissions. lowest cost, and highest efficiency	10.00	10.00	10.00	10.00	10.00	10.00

Source: Modified from [3].

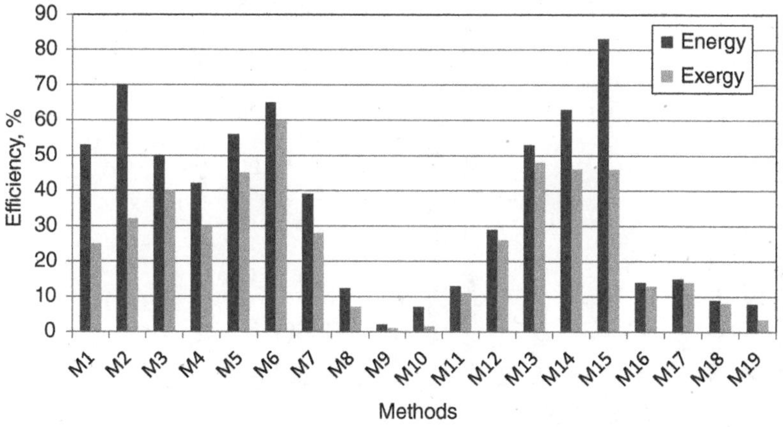

Figure 12.18 Energy and exergy efficiencies for the selected hydrogen production methods (per kg of hydrogen) (data from [3]).

Figure 12.19 shows the average cost of hydrogen production (per kilogram of hydrogen). Based on this figure, steam methane reforming, coal and biomass gasification, and plasma arc decomposition are the most economically feasible methods of producing hydrogen. It appears that the prices of thermochemical cycles, biomass conversion, and hybrid thermochemical cycles are competitive with those of fossil fuels and biomass. It is important to note that the average cost of production in this study is based on literature. In terms of cost per kilogram of hydrogen produced, photoelectrochemical systems are the most expensive. Nevertheless, this method is still in the early stages of research and development, and its ability to be applied locally is one of its major advantages. Due to the advancement in the technology of PEC systems, it is expected that production costs related to PEC operation will decrease in the future.

The environmental impacts of hydrogen production methods are evaluated using global warming potential (GWP) and acidification potential (AP). GWP (kg CO_2 eq.) measures CO_2 emissions. AP (g SO_2 eq.) measures the change in acidity resulting from SO_2 discharge into soil and water. Figure 12.20 shows the GWP and AP results of selected hydrogen production methods.

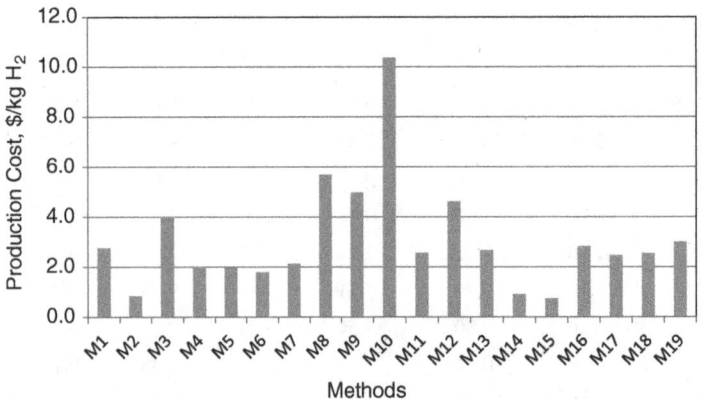

Figure 12.19 Production costs of the selected hydrogen production methods (per kg of hydrogen) (data from [3]).

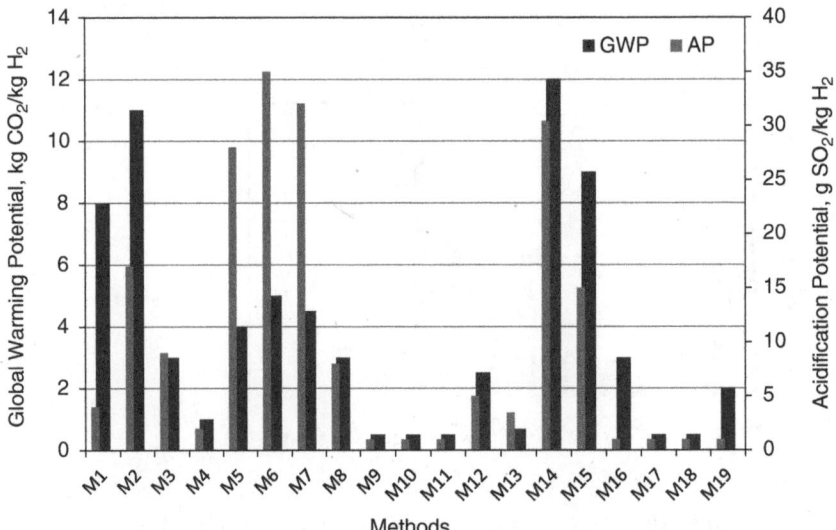

Figure 12.20 GWP and AP values of the selected hydrogen production methods (data from [3]).

Most environmentally harmful methods are coal gasification, fossil fuel reforming, and plasma arc decomposition. As compared to the other selected methods, biomass gasification has the highest AP even though it has the lowest GWP. As shown in Figure 12.20, the photonic energy and hybrid thermochemical cycle-based hydrogen production have the lowest CO_2 emissions and acidification potentials of the selected methods.

12.4 Electrolysis

The most essential industrial process for almost pure hydrogen production is water electrolysis. During electrolysis, water is decomposed into oxygen and hydrogen gas by using electricity. This process is also known as electrochemical water splitting. Electrolysis involves the movement of electrons supported by an external circuit. Figure 12.7 illustrates the view of the water electrolysis. The minimum potential difference required for the electrolysis of water is 1.23 V, as well as external heat. Under practical conditions, higher voltage values are required. In spite of the fact that water electrolysis has simple operation principles, it is rarely used in industrial applications, as fossil fuels can be used to produce it more cheaply. The importance of water electrolysis is expected to grow with the growth of renewable electricity.

Alkaline, polymer membrane, and solid oxide electrolyzers are the key electrochemical hydrogen production technologies. Figure 12.21 demonstrates the view of the common water electrolysis methods. Catalysts are generally used in order to increase current density and rate of electrolysis reactions. Platinum is one of the most commonly used heterogeneous catalysts – applied to the surface of the electrodes. Homogeneous catalysts can also be used during electrolysis. Due to their high turnover

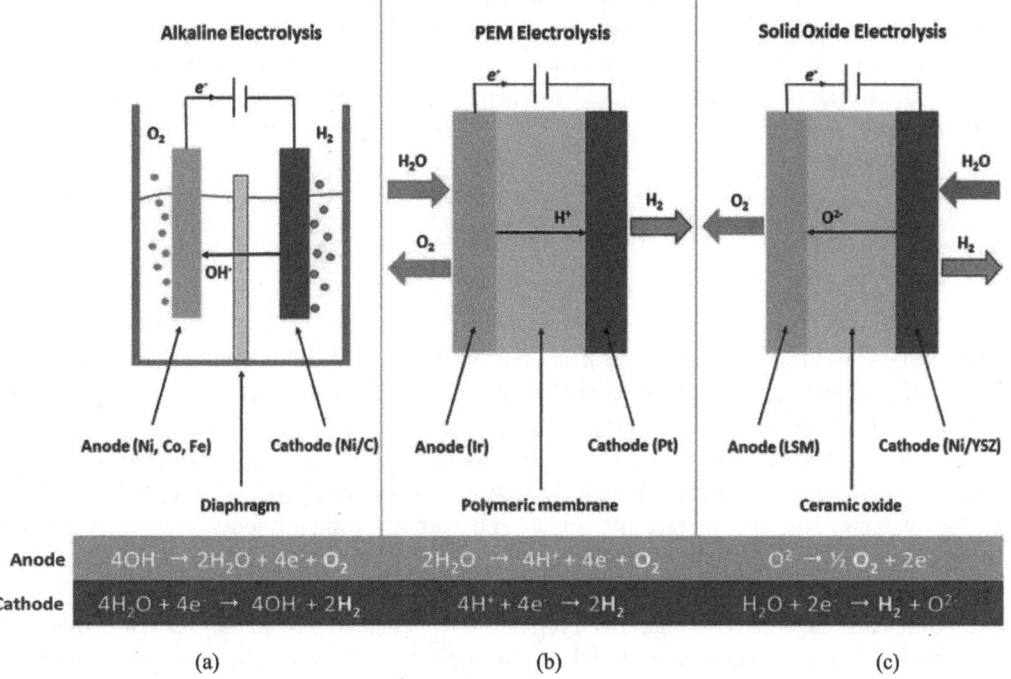

Figure 12.21 Schematic illustration of the common water electrolysis methods: (a) alkaline electrolysis, (b) PEM electrolysis and (c) solid oxide electrolysis.

rates, homogeneous catalysts are less expensive than the heterogeneous ones. In the literature, there are some homogeneous catalysts with turnover rates of 2.4 mol of hydrogen per mole of catalyst and second.

Since electrolyzers (especially PEM electrolyzers) are highly sensitive to the purity of water, desalination and demineralization must be applied before electrolysis process. For instance, if brine (or sea water) is supplied to an electrolyzer, it is more likely to produce chlorine rather than oxygen. There are several methods available in the literature to stop side reactions (like chlorine evolving reaction) during electrolysis; one of them is utilization of ion-selective membranes to desalinate water.

An alkaline electrolyzer produces hydrogen by transporting hydroxide ions (OH-) through the electrolyte from the cathode to the anode, as illustrated in Figure 12.21a. It has been commercially available for many years to use a liquid alkaline solution of sodium or potassium hydroxide as the electrolyte. Several new approaches using solid alkaline exchange membranes (AEM) as the electrolyte are showing promise in the lab. The reaction that takes place at the anode and cathode in alkaline water electrolysis is as follows:

$$\text{Anode: } 2H_2O\,(l) \rightarrow O_2\,(g) + 4\,H+(aq) + 4e^-$$

$$\text{Cathode: } 2H^+\,(aq) + 2\,e^- \rightarrow H_2\,(g)$$

At the anode, water molecules are oxidized, releasing oxygen gas, hydrogen ions, and electrons. The hydrogen ions then migrate to the cathode, where they react with electrons to form hydrogen gas. The oxygen gas is collected in one chamber and the hydrogen gas in another, allowing for the separation of the two gases.

Solid specialty plastic materials are used as electrolytes in polymer electrolyte membrane electrolyzers (PEMs). Figure 12.21b illustrates the schematic illustration of the PEM electrolyzer. The anode reacts with water to produce oxygen and positively charged hydrogen ions (protons). A PEM is used to selectively transfer hydrogen ions to the cathode by passing electrons through an external circuit. An external circuit combines hydrogen ions with electrons to form hydrogen gas at the cathode. The reaction that takes place at the anode and cathode in PEM (Proton Exchange Membrane) electrolysis are as follows:

$$\text{Anode: } 2H_2O\,(l) + 2\,e^- \rightarrow O_2\,(g) + 4\,H^+\,(aq)$$

$$\text{Cathode: } 4H^+\,(aq) + 4\,e^- \rightarrow 2\,H_2\,(g)$$

At the anode, water molecules are oxidized, releasing oxygen gas and hydrogen ions, which are positively charged. The positively charged hydrogen ions then migrate through the proton exchange membrane to the cathode, where they react with electrons to form hydrogen gas. The oxygen gas is collected in one chamber and the hydrogen gas in another, allowing for the separation of the two gases. Unlike in alkaline electrolysis, PEM electrolysis uses a special type of ion-exchange membrane to separate the anode and cathode compartments, ensuring that only hydrogen ions pass through the membrane, thus avoiding the formation of hydroxide ions that can corrode the electrodes and other components.

In solid oxide electrolyzers (Figure 12.21c), negatively charged oxygen ions (O_2-) are selectively conducted by a solid ceramic material as the electrolyte to generate hydrogen at elevated temperatures. Hydrogen gas and negatively charged oxygen ions are formed at the cathode when steam is combined with electrons from the external circuit. The oxygen ions pass through solid ceramic membranes and react with oxygen gas at the anode to produce electrons. The reaction that takes place at the anode and cathode in solid oxide electrolyzers are as follows:

Anode: O_2^- (doped) $\rightarrow O_2$ (g) $+ 2\,e^-$

Cathode: $2\,H_2O$ (l) $+ 2\,e^- \rightarrow H_2$ (g) $+ O_2^-$ (doped)

At the anode, oxygen ions in the solid oxide electrolyte are reduced to oxygen gas and electrons. The electrons are then transported through an external circuit to the cathode, where they react with water molecules, forming hydrogen gas and oxygen ions. The oxygen ions then return to the anode, completing the electrical circuit and enabling the electrolysis process to continue. Solid oxide electrolyzers use a solid-state electrolyte made of a ceramic material, such as yttria-stabilized zirconia, which provides high ionic conductivity and stability at high temperatures. This allows for high efficiency and long-term stability of the electrolysis process.

The solid oxide membranes in solid oxide electrolyzers must be heated to 700–800°C for proper operation (versus 70–90°C for PEM electrolyzers and less than 100°C for commercial alkaline electrolyzers). Using proton-conducting ceramic electrolytes, advanced solid oxide electrolyzers can lower the operating temperature to 500–600°C. Hydrogen can be produced from water using solid oxide electrolyzers by using heat available at these elevated temperatures.

High-temperature electrolysis (HTE) is a process for producing hydrogen by splitting water molecules into hydrogen and oxygen using heat and an electrical current. The process involves heating water to high temperatures (typically above 700°C) and passing an electrical current through the water, which triggers a chemical reaction that splits the water molecules into hydrogen and oxygen. The main advantage of HTE is that it can achieve higher hydrogen production efficiencies than other methods of producing hydrogen, as the high temperatures enable a more efficient ionization of the water molecules. Additionally, HTE can utilize the waste heat generated by other industrial processes, making it a potentially sustainable source of hydrogen. However, HTE also has several disadvantages, including the need for high-temperature materials and specialized equipment, which can increase the cost and complexity of the process. Additionally, the process generates large amounts of heat, which may pose a safety risk and require significant cooling systems to maintain safe operating conditions. HTE is a promising technology for producing hydrogen with high efficiency, but it remains subject to several challenges that must be addressed in order to increase its viability as a source of energy.

Like in every energy system, it is possible to apply a complete thermodynamic analysis based on four main balance equations: mass, energy, entropy, and exergy balance equations. Before applying to the balance equations, it is required to write the chemical reaction and balance it for water splitting. It can be written as follows:

$$H_2O \rightarrow H_2 + \frac{1}{2}O_2$$

Based on this chemical equation, the mass balance equation for an electrolyzer may be written as follows:

$$\dot{m}_{H_2O} = \dot{m}_{H_2} + \dot{m}_{O_2} \tag{12.1}$$

The energy balance equation is defined as:

$$\dot{m}_{H_2O}h_{H_2O} + \dot{P}_{in} = \dot{m}_{H_2}h_{H_2} + \dot{m}_{O_2}h_{O_2} \tag{12.2}$$

The entropy balance equation for an electrolysis process is expressed as:

$$\dot{m}_{H_2O}s_{H_2O} + \dot{S}_{gen} = \dot{m}_{H_2}s_{H_2} + \dot{m}_{O_2}s_{O_2} \tag{12.3}$$

Lastly, the exergy balance equation is written as follows:

$$\dot{m}_{H_2O}ex_{H_2O} + \dot{P}_{in} = \dot{m}_{H_2}ex_{H_2} + \dot{m}_{O_2}ex_{O_2} + \dot{Ex}_{dest} \tag{12.4}$$

Thus, the energy efficiency of electrolysis can be defined as the ratio of the energy content of hydrogen to total electricity input. This is formulated as follows:

$$\eta_{en} = \frac{\dot{N}_{totalH_2}LHV_{H_2}}{\dot{P}_{in}} \tag{12.5}$$

The exergy efficiency of electrolysis can be defined as the ratio of the chemical exergy content of the hydrogen to total energy input as formulated below:

$$\eta_{ex} = \frac{\dot{N}_{totalH_2}ex_{H_2}}{\dot{P}_{in}} \tag{12.6}$$

Illustrative Example 12.1 A 3-cell electrolyzer stack shown in Figure 12.22 is provided. If the current through the cells is measured to be 20 A. The ohmic, activation, and concentration polarization voltages are given as ($V_{activation} + V_{ohmic} + V_{concentration} = 0.77$ V). Refer to Table 12.3 for thermochemical properties.

a) Write the chemical reaction and balance it first, and list the mass, energy, entropy and exergy balance equations for the electrolyzer.
b) Find the total power input to the electrolyzer stack.
c) Find the hydrogen production rate.
d) Find the energy and exergy efficiencies of the electrolyzer.

Solution:

a) The balance chemical equation can be written as follows:

$$H_2O = H_2 + \frac{1}{2}O_2$$

For the electrolyzer, one may write the balance equations accordingly for energy, entropy, and exergy as follows:

$$\dot{N}_{H_2O}\bar{h}_{H_2O} + \Delta H = \dot{N}_{H_2}\bar{h}_{H_2} + \dot{N}_{O_2}\bar{h}_{O_2}.$$

$$\dot{N}_{H_2O}\bar{s}_{H_2O} + \Delta S = \dot{N}_{H_2}\bar{s}_{H_2} + \dot{N}_{O_2}\bar{s}_{O_2}$$

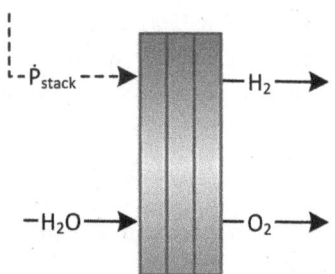

Figure 12.22 A view of the electrolyzer given in Illustrative Example 12.1.

Table 12.3 Thermochemical properties of hydrogen, oxygen, and water.

Substance	Formula	Molar Mass, M (kg/kmol)	Enthalpy of Formation, $\bar{h}°f$, (kJ/kmol)	Gibbs Function of Formation, $\bar{g}°f$, (kJ/kmol)	Absolute Entropy, $\bar{s}°$, (kJ/kmol K)	HHV (kJ/mol)	ex_{ch} (kJ/mol)
Hydrogen	H_2 (g)	2.02	0	0	130.57	285,828	236,090
Oxygen	O_2 (g)	32.00	0	0	205.03	0	0
Water	H_2O (g)	18.02	−241.82	−228.59	0	0	0
Water	H_2O (l)	18.02	−285.83	−237.18	0	0	0

and

$$\dot{N}_{H_2O}ex_{H_2O} = \dot{N}_{H_2}ex_{H_2} + \dot{N}_{O_2}ex_{O_2} + \dot{Ex}_d$$

The enthalpy change for the hydrogen reaction can be expressed as:

$$\Delta H = \Delta G + T\Delta S$$

The Gibbs free energy needs to be negative to perform the reaction, which is represented with ΔG. ΔH represents enthalpy change and ΔS represents entropy change and T represents temperature in Kelvin.

b) In order to find the minimum voltage input (V_{min}) to split water, the following equation is used:

$$\Delta G = -2F \, V_{min}$$

where F stands for the Faraday constant which is the charge on 1 mol of electrons (F=96.48 kJ/mol).

Therefore, the minimum voltage input is found as follows:

$$V_{min} = \frac{-\Delta G}{2F} = \frac{237.18}{192.96} = 1.23 \ V$$

The actual electrolyzer voltage for each cell can be calculated after adding ohmic, activation, and concentration polarizations into the reversible open circuit potential, as follows:

$$V_{actual} = V_{min} + V_{activation} + V_{ohmic} + V_{concentration}$$

$$V_{actual} = 1.23 \ V + 0.77 \ V = 2 \ V$$

In order to find the actual power input for each cell, the following equation is writen:

$$\dot{P}_{actual,cell} = V_{actual} \times I$$

$$\dot{P}_{actual,cell} = 2 \ V \times 20 \ A = 40 \ W$$

In order to find the total power input for the stack, the following calculation is made:

$$\dot{P}_{stack} = n_{cell} \times \dot{P}_{actual,cell}$$

$$\dot{P}_{stack} = 3 \times 40 = 120 \ W$$

c) The hydrogen production rate is calculated as follows:

$$\dot{N}_{H_2} = \frac{I}{2F} = \frac{20}{2(96500)} = 0.000103 \ mol/s$$

As there are three cells in the stack, the total hydrogen production rate is found to be:

$$\dot{N}_{totalH_2} = 0.000103 \times 3 = 0.000309 \ \frac{mol}{s}$$

d) The energy and exergy efficiencies for this hydrogen production can be calculated as follows:

$$\eta_{en} = \frac{\dot{N}_{totalH_2} \left(HHV_{H_2} \right)}{\dot{P}_{in}} = \frac{0.000309(285828)}{120} = 0.736$$

and

$$\eta_{ex} = \frac{\dot{N}_{totalH_2} \left(ex_{ch_{H_2}} \right)}{\dot{P}_{in}} = \frac{0.000309(236,090)}{120} = 0.608$$

Illustrative Example 12.2   It is further aimed to produce 1 g of hydrogen gas with the electrolyzer stack given in the previous problem. Calculate the number of cells in the stack to achieve the desired hydrogen production. Calculate the energy and exergy efficiencies of the electrolyzer (heating value of $H_2 = 286$ kJ/mol, chemical exergy of $H_2 = 236.09$ kJ/mol).

Solution:
For this electrolyzer cell, the hydrogen production rate per cell is

$$\dot{m}_{H_2,per\,cell} = 0.000207 \frac{g}{s}$$

As 1 g of H_2 production is aimed, the total number of cells in the stack can be found to be:

$$\dot{m}_{totalH_2} = N \times \dot{m}_{H_2,per\,cell}$$

$$1 = N \times 0.000207$$

Thus, the number of cells is calculated to be 4831. With this electrolyzer, 3600 kg of hydrogen can be produced per hour.

12.5   Hydrogen Storage Methods

Storage of hydrogen is a critical topic for the hydrogen economy, especially when it is used as an energy carrier and energy storage medium. Generally, unless it is not used as an industrial gas, it is produced whenever it is needed. Today's common hydrogen production techniques, which are mainly fossil-based production methods, are utilized to produce hydrogen before synthesizing it into various other fuels and chemicals. Only some portion of it is kept for industrial usages. However, the storage of hydrogen will be more critical with the increase utilization and demand of hydrogen, and deployment of the green hydrogen techniques. Figure 12.23 shows the classification of hydrogen storage methods, which are mainly divided into two main sections: physical-based and material-based hydrogen storage. While the hydrogen is stored in the form of H_2 by changing its form as compressed-gas, cryogenic gas, or liquid in the physical-based hydrogen storage techniques, it can be stored various chemical forms which are in adsorbents, liquid organic materials, interstitial hydride, complex hydride, and some chemicals.

The methods are given in Figure 12.23 have some advantages and disadvantaged according to each other. However, when energy storage techniques are considered, the energy storage density, which indicates the ratio of the total amount of the stored materials to its volume, is the critical performance parameter. Figure 12.24 compares the energy storage methods in the terms of hydrogen storage density per liter. When hydrogen is stored in atmospheric conditions, one liter of storage volume will contain only 0.3 g of hydrogen. In order to increase the hydrogen storage density, to store hydrogen in compressed gas form is a common technique. If the storage pressure of the hydrogen is increased to 150, 350, and 700 bar, the storage density will be 10 g/L, 28 g/L, and 40 g/L, respectively. The hydrogen amount of hydrogen in unit storage volume increases with increasing storage pressure. However, it should be noted that higher-pressure means higher energy consumption for compression and cooling. Hydrogen should be cooled down after compression due to maintain safety conditions. In hydrogen is stored in its liquid form at 20 K, it is possible to store approximately 70 g of hydrogen per liter. The main advantage of the physical-based hydrogen storage methods is to require well-known techniques which are compression and cooling. However, in order to enhance the economic feasibility of hydrogen storage, there are numerous attempts on them.

On the other hand, materials-based hydrogen storage provides more energy storage density comparing to the physical-based ones. The main problem with materials-based storage is the response rate of the hydrogen storage method when it is needed. Due to chemical reactions kinetics, obtaining pure hydrogen rate during the discharging period is quite lower than physical-based methods. They are very useful when there is a storage volume restriction, and a lower discharging rate is possible.

12.5.1   Physical-based Hydrogen Storage

Physical-based hydrogen storage covers the storage techniques of hydrogen in the various form of pure hydrogen (H_2). Compressed hydrogen gas is stored at high pressures, typically in the range of 35–700 bar. It is stored in high-pressure tanks made of composite materials or metal

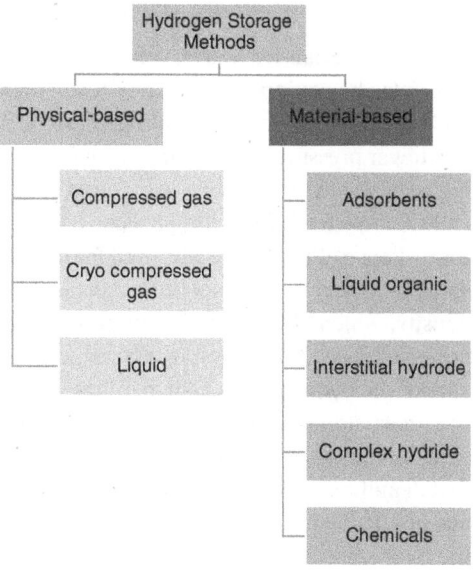

Figure 12.23   Classification of hydrogen storage methods.

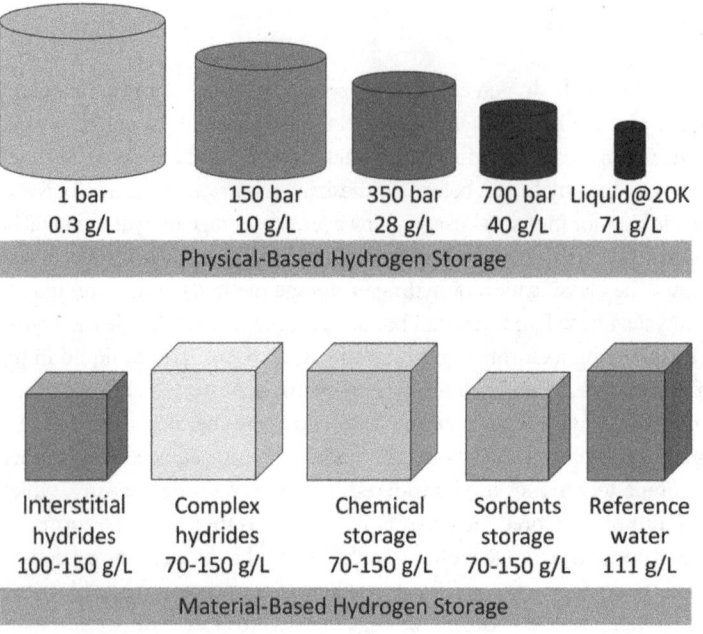

Figure 12.24 A comparison of hydrogen storage methods.

hydrides. The storage tanks need to be designed to withstand high pressure and prevent hydrogen leaks. The tanks can be refilled at hydrogen fueling stations and are used in hydrogen fuel cell vehicles. The following materials are widely used in the compressed-gas hydrogen storage:

- Carbon fiber reinforced plastic (CFRP)
- Steel
- Aluminum
- Composite material with a polymer liner

These materials need to be lightweight, durable, and able to withstand high pressures. The tanks must also be equipped with safety features such as pressure relief valves to prevent overpressure incidents. The compressed hydrogen can be dispensed by releasing it through a pressure regulator to a lower pressure for use, such as in fuel cell vehicles.

Cryogenic hydrogen storage involves cooling hydrogen gas to extremely low temperatures, typically below –100°C, and compressing it to high pressures in order to store it in a compact and dense form. This method is typically used for large-scale hydrogen storage, such as in hydrogen refueling stations and hydrogen pipelines. Advantages of cryogenic hydrogen storage include high storage density, which allows for efficient transportation and storage of hydrogen, and the ability to store large volumes of hydrogen in a compact space. Additionally, the low temperature and high pressure of the hydrogen gas help to minimize the risk of leaks or hydrogen spills. However, cryogenic hydrogen storage also has some disadvantages. It requires specialized equipment and infrastructure, including cryogenic storage tanks and specialized piping systems, which can be expensive to install and maintain. Additionally, the low temperatures and high pressures involved in this storage method can pose safety and health hazards for personnel involved in handling and storage of the hydrogen.

In summary, cryogenic compressed-gas hydrogen storage is a method of storing hydrogen at low temperatures and high pressures, offering high storage density and the ability to store large volumes of hydrogen in a compact space. However, it requires specialized equipment and infrastructure, and can pose safety and health hazards, making it more expensive than other hydrogen storage methods.

Liquid hydrogen storage is a method of storing hydrogen in its liquid state at extremely low temperatures, typically below −253 °C. The hydrogen is kept in a cryogenic storage tank, which is typically made of materials with low thermal conductivity and high insulation properties, such as stainless steel or vacuum-insulated panels. One of the main advantages of liquid hydrogen storage is its high storage density, which allows for efficient transportation and storage of hydrogen in a compact space. This makes it a popular method of storage for applications where large volumes of hydrogen are needed, such as in the aerospace and cryogenic refrigeration industries. Additionally, liquid hydrogen has a low boiling point, which allows it to be stored at relatively low pressures, reducing the risk of leaks or hydrogen spills. However, there are also some disadvantages to liquid hydrogen storage. It requires specialized equipment and infrastructure, including cryogenic storage tanks and specialized piping systems, which can be expensive to install and maintain. Additionally, the low temperatures involved in this storage method can pose safety and health hazards for personnel involved in handling and storage of the hydrogen. The hydrogen must also be kept at extremely low temperatures to remain in its liquid state, which requires specialized cooling equipment and large amounts of energy.

The liquid hydrogen storage method offers high storage density and the ability to store large volumes of hydrogen in a compact space, as hydrogen is stored in its liquid state at low temperatures. As a result, it is more expensive than other methods of hydrogen storage because it requires specialized equipment and infrastructure. It also poses potential safety and health hazards.

12.5.2 Material-based Hydrogen Storage

Material-based hydrogen storage refers to the storage of hydrogen in solid materials, such as metal hydrides, chemical hydrides, and carbon-based (organic) materials. This method of hydrogen storage offers several advantages over other storage methods, including high hydrogen storage capacity, stability, and low-pressure storage requirements. In metal hydrides, hydrogen is stored in the form of a chemical compound formed between hydrogen and a metal. These materials offer high hydrogen storage capacities and are relatively safe, as they are stable and do not pose a risk of leaks or spills. In chemical hydrides, hydrogen is stored in the form of a chemical compound with a metal or nonmetal element. These materials are also relatively safe and offer high hydrogen storage capacities, although they typically require high pressures and/or high temperatures for hydrogen release. In organic materials, hydrogen is stored as hydrogen gas or as a chemical compound in the pores or defects of carbon materials, such as activated carbon or carbon nanotubes. These materials offer high hydrogen storage capacities and are relatively safe, but they typically require high pressures and/or high temperatures for hydrogen release.

Consequently, material-based hydrogen storage refers to the storage of hydrogen in solid materials, such as metal hydrides, chemical hydrides, and carbon-based materials. These materials offer high hydrogen storage capacities and stability, and can store hydrogen at low pressures, but may require high pressures and/or high temperatures for hydrogen release.

a) Adsorbents

Hydrogen storage in adsorbents refers to the storage of hydrogen in porous materials, such as activated carbon, zeolites, and metal-organic frameworks. These materials have high surface areas and porous structures, which allow them to adsorb (or store) large amounts of hydrogen gas in their pores.

Adsorbent-based hydrogen storage offers several advantages over other storage methods, including high hydrogen storage capacities, stability, and low-pressure storage requirements. Additionally, adsorbent materials can be regenerated and reused multiple times, making them a more sustainable option for hydrogen storage. One of the main challenges in adsorbent-based hydrogen storage is the requirement for high temperatures and/or pressures to release the hydrogen from the adsorbent material. However, researchers are working to develop new adsorbent materials with improved hydrogen release characteristics, as well as new methods for hydrogen release that do not require high temperatures or pressures.

Hydrogen storage in adsorbents refers to the storage of hydrogen in porous materials, such as activated carbon, zeolites, and metal-organic frameworks. These materials offer high hydrogen storage capacities and stability, and can store hydrogen at low pressures, but may require high temperatures and/or pressures for hydrogen release.

b) Liquid Organics

Among liquid organic compounds, hydrogen can be stored in formic acid, ethanol, and hydrocarbons. Hydrogen can be absorbed and released by these compounds, allowing them to serve as hydrogen storage materials. A major advantage of hydrogen storage in liquid organics is that it can be stored and transported at ambient temperatures and pressures, making it a more practical and safer solution. Moreover, liquid organic compounds are easy to handle, transport, and widely available. In addition to the advantages of hydrogen storage in liquid organics, there are also some disadvantages. Due to the relatively low hydrogen storage capacity of these materials, more volumes of liquid organic compounds are required to store the same amount of hydrogen as other storage methods. Additionally, the process of releasing hydrogen from the liquid organic compounds often requires high temperatures or pressures, making it less energy-efficient than other storage methods.

Essentially, hydrogen storage in liquid organics refers to storing hydrogen in formic acid, ethanol, or hydrocarbons. However, these materials have relatively low hydrogen storage capacities and may require high temperatures or pressures to release hydrogen. While they are easy to handle and transport at ambient temperatures and pressures, they also have relatively low hydrogen storage capacities.

c) Interstitial Hydride

Solid materials can store hydrogen in the interstitial spaces between molecules or atoms, called interstitial hydrides. Metal hydrides, such as AB5, AB2, and AB2-type hydrides, can store hydrogen gas by absorption into their interstitial sites. Interstitial hydrides are excellent hydrogen storage materials for portable devices and vehicles because they can store large amounts of hydrogen in compact, dense form. Furthermore, interstitial hydrides can be easily controlled for hydrogen release by changing their temperature or pressure, allowing them to be a flexible hydrogen storage solution. There are, however, some challenges associated with hydrogen storage in interstitial hydrides. The hydrogen must be released at high temperatures and/or pressures in order to be stored, which is energy-intensive and limits the efficiency of the process. The durability and stability of the interstitial hydrides can also be an issue, since repeated hydrogen uptakes and releases can degrade the material. Interstitial hydrides are solid materials, such as metal hydrides, that store hydrogen in their interstices. Despite providing a high hydrogen storage capacity in a compact, dense form, this method of hydrogen storage may have stability and durability issues due to repeated hydrogen uptake and release cycles that may require high temperatures and/or pressures for hydrogen release.

d) Complex Hydrides

Hydrogen storage in complex hydrides refers to the storage of hydrogen in complex hydride compounds, which are materials that contain hydrogen and metal atoms. Complex hydrides can store hydrogen through chemical bonds between the hydrogen and metal atoms, allowing them to act as hydrogen storage materials. The main advantage of hydrogen storage in complex hydrides is their high hydrogen storage capacities, which can be several times greater than those of other hydrogen storage methods. Additionally, the hydrogen release characteristics of complex hydrides can be easily controlled by changing the temperature or pressure, making them a flexible option for hydrogen storage. However, there are also some challenges associated with hydrogen storage in complex hydrides. One of the main challenges is the requirement for high temperatures and/or pressures to release the hydrogen from the complex hydride compounds, which can be energy-intensive and limit the efficiency of the storage process. Additionally, the stability and durability of the complex hydrides can be an issue, as repeated hydrogen uptake and release cycles can lead to degradation of the material.

In summary, hydrogen storage in complex hydrides refers to the storage of hydrogen in complex hydride compounds, which contain hydrogen and metal atoms. This method of hydrogen storage offers the advantage of high hydrogen storage capacities and the ability to control hydrogen release by changing temperature or pressure but may require high temperatures and/or pressures for hydrogen release and may have stability and durability issues associated with repeated hydrogen uptake and release cycles.

e) Chemicals

Chemical compounds store hydrogen within their chemical bonds. In this method of hydrogen storage, hydrogen is absorbed into the chemical structure of the storage material, allowing it to be stored in a stable form. Hydrogen storage in chemical compounds offers several advantages, including high hydrogen storage capacities, compact and dense storage, and easy transport and handling. Furthermore, hydrogen storage in chemical compounds is an established technology, with many established storage materials and technologies. The storage of hydrogen in chemical compounds, however, also presents some challenges. The release of hydrogen from some chemical storage materials requires high temperatures and/or pressures, limiting its efficiency. In addition, certain chemical storage materials can be degraded by repeated hydrogen uptakes and releases.

Hydrogen storage in a chemical compound refers to storing hydrogen within the chemical bonds of specific compounds through chemical reactions that absorb hydrogen into the structure of the material. There are many advantages to this method of hydrogen storage, including its high hydrogen storage capacity, compact and dense storage capacity, and ease of transport and handling. As hydrogen is released at high temperatures and/or pressures, stability and durability concerns may occur as a result of repeated hydrogen uptake and release cycles.

12.6 Sectoral Hydrogen Utilization

As mentioned earlier, hydrogen is a remarkable choice of energy which can be used as fuel, energy carrier and storage medium, and feedstock. Hydrogen is used in various sectors due to its unique properties as a clean and versatile energy carrier. Some of the sectors where hydrogen may primarily be used are:

1) Energy sector: Hydrogen can be used as a fuel in power generation, particularly in fuel cells where it can generate electricity with high efficiency and low emissions. Also, it is possible to use for electricity generation in gas turbines and power generators.

2) Transportation sector: Hydrogen can be used as a fuel for vehicles, either in fuel cell vehicles (FCVs) or in internal combustion engine vehicles. FCVs have zero tailpipe emissions, while hydrogen-powered internal combustion engines emit only water vapor.

3) Industrial sector: Hydrogen is used in various industrial processes, such as oil refining, ammonia production, and metal processing, as a reducing agent or as a source of energy.

4) Space sector: Hydrogen is used as a fuel for rocket engines and for power generation in satellites and other spacecrafts.

5) Residential sector: Hydrogen can be used in household appliances for residential heating and cooking, either in fuel cells or through conventional units as a carbon-free fuel.

These are just a few examples of the many sectors where hydrogen is used. Hydrogen's versatility and cleanliness make it a promising energy carrier for a wide range of applications, and its use is expected to grow in the coming years as new technologies and applications are developed.

12.6.1 Fuel Cells

It is crystal clear to everyone that hydrogen is a key commodity for sustainable future, as a fuel, an energy carrier, an energy storage medium and a feedstock. Therefore, clean hydrogen production techniques are essential for a better future. However, the sustainability of hydrogen is not limited to hydrogen production, but depends essentially upon the entire ecosystem. Therefore, better hydrogen storage methods are needed in terms of efficiency, storage density, and cost. Also, the utilization of hydrogen is another topic for hydrogen economy. The cost effective, higher performance, and long lifetime fuel cells are required for power generation from hydrogen.

Hydrogen fuel cells are devices that generate electricity through a chemical reaction between hydrogen and oxygen. Figure 12.25 shows the schematic view of a fuel cell. They work by passing hydrogen gas over an anode, which splits the hydrogen into protons and electrons. The protons pass through a proton exchange membrane (PEM), while the electrons are conducted through an external circuit, generating an electric current. On the other side of the PEM, the protons and electrons combine with oxygen from the air to produce water and heat.

Fuel cells have several advantages over traditional power sources, such as internal combustion engines or batteries. They are highly efficient, with electrical conversion efficiencies of up to 60% or more, and they produce only water as a by-product, making them a clean source of power. Additionally, fuel cells can be used as a stationary power source, or as a power source for vehicles, and they can be easily refueled with hydrogen. However, there are also some challenges associated with hydrogen fuel cells. One of the main challenges is the high cost of fuel cells and the associated infrastructure, as well as the limited availability of hydrogen fuel. Additionally, fuel cells are still relatively new technology, and there is a need for further research and development to improve their performance, reliability, and durability. Some common types of the fuel cells are listed as follows:

- Polymer electrolyte membrane fuel cell (PEMFC)
- Direct methanol fuel cell (DFMC)
- Alkaline fuel cell (AFC)
- Solid oxide fuel cell (SOFC)
- Phosphoric acid fuel cell (PAFC)
- Molten carbonate fuel cell (MCFC)

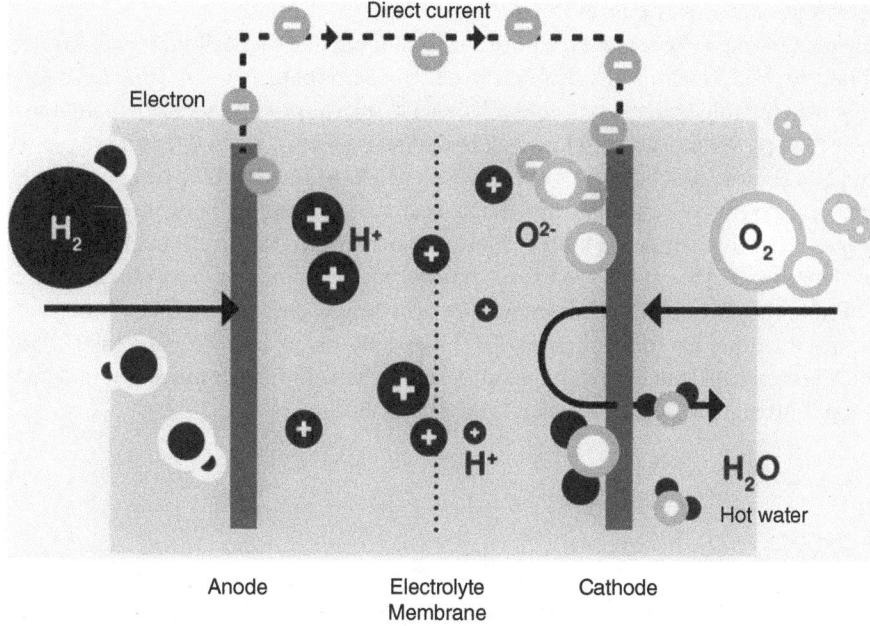

Figure 12.25 The schematic illustration of a fuel cell.

Table 12.4 demonstrates the comparison of fuel cells given above in the terms of operating temperature, power output range, and cell efficiency. Depending on the applications, operating conditions, and fuel sources, there are advantages and disadvantages according to each other. Solid oxide fuel cells and molten carbonate fuel cells can reach up to 100 MW of power. PEMFC can provide more flexible operation due to its power output range, working temperature, and cell efficiency.

Table 12.4 Comparison of the fuel cells.

Type of Fuel Cell	Anode Gas	Cathode Gas	Electrolyte	Working Temperature (°C)	Power Output Range	Cell Efficiency (%)
PEMFC	Hydrogen	Air, oxygen	Polymer membrane	20–80	Up to 500 (kW)	50%–70%
DFMC	Methanol	Air, oxygen	Polymer membrane	20–130	Up to 100 (kW)	20%–30%
AFC	Hydrogen	Oxygen	Potassium hydroxide solution	20–90	Up to 100 (kW)	60%–70%
SOFC	Natural gas, coal, biogas	Air, oxygen	Yttrium-stabilized zirconium oxide	800–1000	Up to 100 MW	60%–65%
PAFC	Hydrogen, natural gas, or biogas	Air, oxygen	Phosphoric acid	160–220	Up to 10 MW	55%
MCFC	Natural gas, coal, biogas	Air, oxygen	Alkali carbonate melting	620–660	Up to 100 MW	65%

Source: modified from [4].

a) Polymer Electrolyte Membrane Fuel Cells

Polymer electrolyte membrane (PEM) fuel cells are a type of hydrogen fuel cell that use a proton exchange membrane to produce electricity through an electrochemical reaction. Among the many advantages of polymer electrolyte membrane fuel cells, also known as proton exchange membrane fuel cells, are their high power density and low weight and volume as compared to other types of fuel cells. Figure 12.26 shows the schematic representation of a PEM fuel cell. The reaction occurs between hydrogen and oxygen, producing water as a by-product and emitting only heat and water vapor. Solid polymer electrolytes are used in PEM fuel cells, and porous carbon electrodes containing platinum or platinum alloy catalysts are used. The only requirements for their operation are hydrogen, oxygen from the atmosphere, and water. Pure hydrogen is generally supplied from storage tanks or reformers as fuel for these engines. PEM fuel cells have a high power density, fast start-up, and quick response to load changes, making them suitable for use in transportation and portable power applications. The overall reaction is then described as:

Anode: $H_2 \rightarrow 2H^+ + 2e^-$

Cathode: $2H^+ + 2e^- + O_2 \rightarrow H_2O$

Overall reaction: $H_2 + O_2 \rightarrow H_2O$

The PEM fuel cells work by electrochemically reacting hydrogen and oxygen to produce electricity. The fuel cell consists of two electrodes (the anode and cathode) separated by a proton exchange membrane. The hydrogen fuel is fed to the anode, where it undergoes a process called oxidation and releases electrons. These electrons flow through an external circuit, providing electrical power,

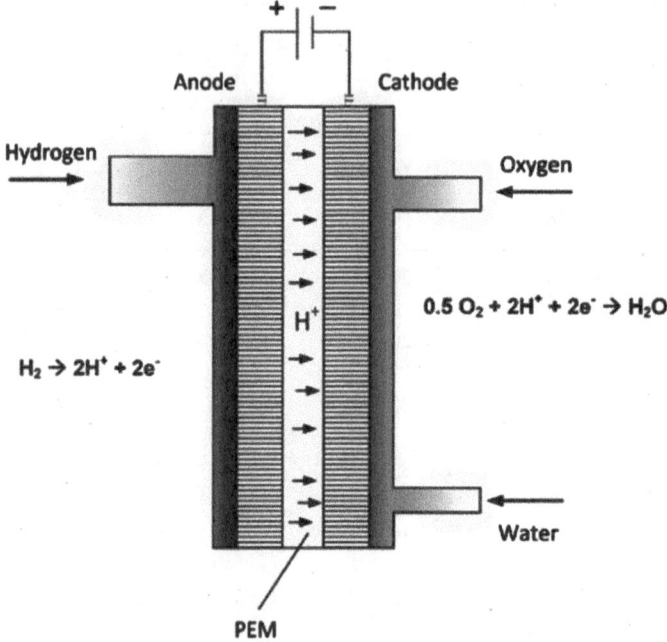

Figure 12.26 The illustration of a polymer electrolyte membrane (PEM) fuel cell.

while the positively charged hydrogen ions (protons) are conducted through the PEM to the cathode. At the cathode, the hydrogen ions combine with oxygen and electrons to form water, producing heat and water vapor as by-products. This process generates a direct current that can be used as an electricity source. The fuel cell will continue to produce electricity as long as hydrogen and oxygen are supplied to the anode and cathode, respectively.

In PEM fuel cells, temperatures are relatively low, around 80°C. The operation of these systems at low temperatures enables them to start quickly (less warm-up time) and results in less wear on the components of the system, which enhances their durability. Despite these advantages, it requires the use of a noble metal catalyst (typically platinum) in order to separate the electrons and protons of hydrogen, increasing the cost of the system. Additionally, platinum catalysts are extremely sensitive to carbon monoxide poisoning, requiring an additional reactor to reduce carbon monoxide in the fuel gas if the hydrogen is derived from hydrocarbons. As a result of this reactor, there is an additional cost involved.

Fuel cells based on PEM technology are primarily used in transportation and some stationary applications. Vehicle applications, such as cars, buses, and heavy-duty trucks, are particularly well suited to PEM fuel cells.

b) Direct methanol fuel cells

Direct methanol fuel cells (DFMCs) are devices that convert chemical energy from methanol into electrical energy through a direct electrochemical reaction. They typically consist of an anode and a cathode separated by a proton exchange membrane, with methanol and oxygen as reactants, as illustrated in Figure 12.27. The DMFCs are attractive for portable power applications due to their high energy density, low operating temperature, and simple system design. However, limitations such as low power density, methanol crossover, and low fuel utilization efficiency have hindered their widespread commercialization.

In a DMFC, methanol and oxygen are supplied to the anode and cathode, respectively. At the anode, methanol is oxidized to produce carbon dioxide and hydrogen ions (protons), which are transported through the proton exchange membrane to the cathode. At the cathode, the protons react with oxygen to form water, while electrons are supplied to an external circuit to generate an electrical current. The overall reaction can be described as:

Anode: $CH_3OH + H_2O \rightarrow CO_2 + 6H^+ + 6e^-$

Cathode: $6H^+ + 6e^- + 3O_2 \rightarrow 6H_2O$

Overall reaction: $CH_3OH + 3O_2 \rightarrow CO_2 + 2H_2O$

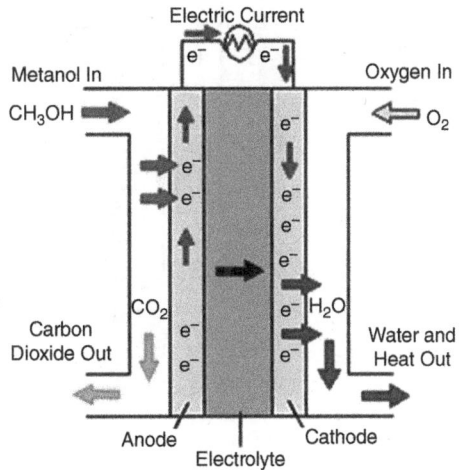

Figure 12.27 The view of a direct methanol fuel cell.

The electrical energy generated by the DMFC can be used to power various devices or stored in a battery. The efficiency of a DMFC is dependent on several factors, including the type of proton exchange membrane used, the operating temperature, and the methanol concentration.

c) Alkaline fuel cells

Alkaline fuel cells are a type of fuel cell that uses an alkaline electrolyte solution, typically potassium hydroxide (KOH), to facilitate the transfer of ions between the anode and cathode. They can run on hydrogen, methanol, or other fuels and are known for their high power density and long operational lifetime. Figure 12.28 demonstrates an alkaline fuel cell. In an AFC, hydrogen is supplied to the anode, where it is oxidized to produce protons and electrons. The protons diffuse through the KOH solution to the cathode, while the electrons are supplied to an external circuit to generate an electrical current. At the cathode, the protons react with oxygen to form water, completing the overall reaction. The overall reaction can be described as:

Anode: $H_2 \rightarrow 2H^+ + 2e^-$

Cathode: $2H^+ + 2e^- + O_2 \rightarrow H_2O$

Overall reaction: $H_2 + O_2 \rightarrow 2H_2O$

The AFCs have been used in various applications, including space missions, backup power systems, and stationary power generation. They are known for their high efficiency, durability, and low cost, making them an attractive option for various power generation applications.

d) Solid oxide fuel cells

Solid oxide fuel cells (SOFCs) are a type of fuel cell that uses a solid ceramic electrolyte, typically yttria-stabilized zirconia (YSZ), to facilitate the transfer of ions between the anode and cathode. Figure 12.29 displays the view of the solid oxide fuel cells. They can run on hydrogen, natural gas, or other fuels and are known for their high operating temperature, high efficiency, and long operational lifetime. In an SOFC, fuel (typically hydrogen or natural gas) is supplied to the anode, where it is oxidized to produce protons and electrons. The protons diffuse through the YSZ electrolyte to the

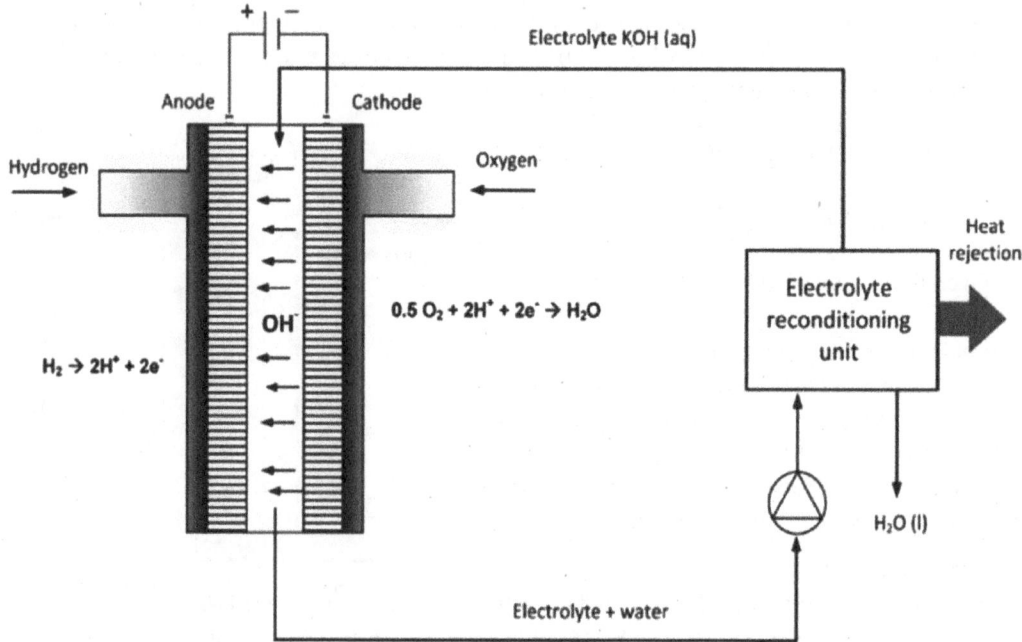

Figure 12.28 The schematic view of an alkaline fuel cell.

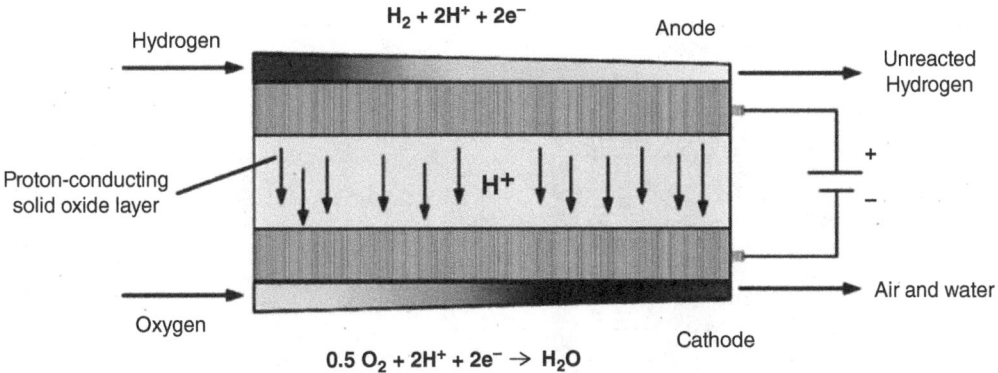

Figure 12.29 The schematic view of a solid oxide fuel cell.

cathode, while the electrons are supplied to an external circuit to generate an electrical current. At the cathode, the protons react with oxygen to form water, completing the overall reaction. The overall reaction can be described as:

Anode: Fuel (H_2 or CH_4) $\rightarrow$ Proton + Electron

Cathode: Proton + Electron + O_2 $\rightarrow$ H_2O

Overall reaction: Fuel + O_2 $\rightarrow$ H_2O

Note that SOFCs are used in various applications, including stationary power generation, cogeneration systems, and hybrid power systems. They are known for their high efficiency, high power density, and long operational lifetime, making them an attractive option for various power generation applications. However, their high operating temperature and complex fabrication process can make them more expensive to manufacture than other types of fuel cells.

e) Phosphoric acid fuel cells

Phosphoric acid fuel cells (PAFCs) are a type of fuel cell that uses a phosphoric acid electrolyte to facilitate the transfer of ions between the anode and cathode. Figure 12.30 demonstrates the schematic illustration of phosphoric acid fuel cell. They can run on hydrogen, natural gas, or other fuels and are known for their relatively high efficiency and stability compared to other types of fuel cells. In a PAFC, fuel (typically hydrogen or natural gas) is supplied to the anode, where it is oxidized to produce protons and electrons. The protons diffuse through the phosphoric acid electrolyte to the cathode, while the electrons are supplied to an external circuit to generate an electrical current. At the cathode, the protons react with oxygen to form water, completing the overall reaction. The overall reaction can be described as:

Anode: Fuel (H_2 or CH_4) $\rightarrow$ Proton + Electron

Cathode: Proton + Electron + O_2 $\rightarrow$ H_2O

Overall reaction: Fuel + O_2 $\rightarrow$ H_2O

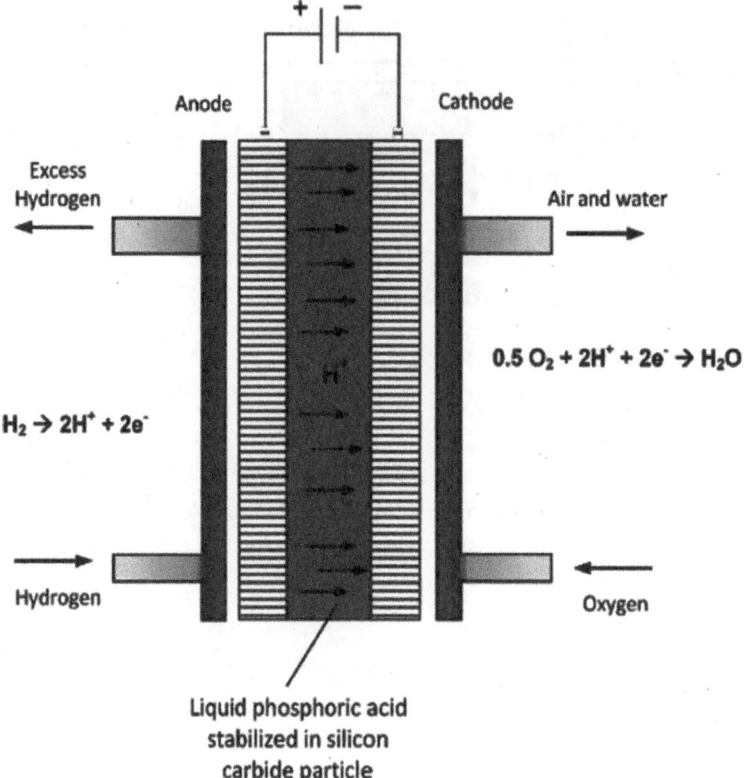

Figure 12.30 The illustration of a phosphoric acid fuel cell.

Note that PAFCs have been used in various applications, including stationary power generation, cogeneration systems, and backup power systems. They are known for their high efficiency and stability, making them an attractive option for various power generation applications. However, they are also relatively expensive to manufacture compared to other types of fuel cells, and their relatively low operating temperature limits their ability to generate high power densities.

f) Molten carbonate fuel cells

Molten carbonate fuel cells (MCFCs) are a type of fuel cell that uses a molten carbonate electrolyte to facilitate the transfer of ions between the anode and cathode. They can run on hydrogen, natural gas, or other fuels and are known for their high operating temperature and high efficiency, as shown in Figure 12.31. In an MCFC, fuel (typically hydrogen or natural gas) is supplied to the anode, where it is oxidized to produce protons and electrons. The protons diffuse through the molten carbonate electrolyte to the cathode, while the electrons are supplied to an external circuit to generate an electrical current. At the cathode, the protons react with oxygen to form water, completing the overall reaction. The overall reaction is described as:

Anode: Fuel (H_2 or CH_4) $\rightarrow$ Proton + Electron

Cathode: Proton + Electron + O_2 $\rightarrow$ H_2O

Overall reaction: Fuel + O_2 $\rightarrow$ H_2O

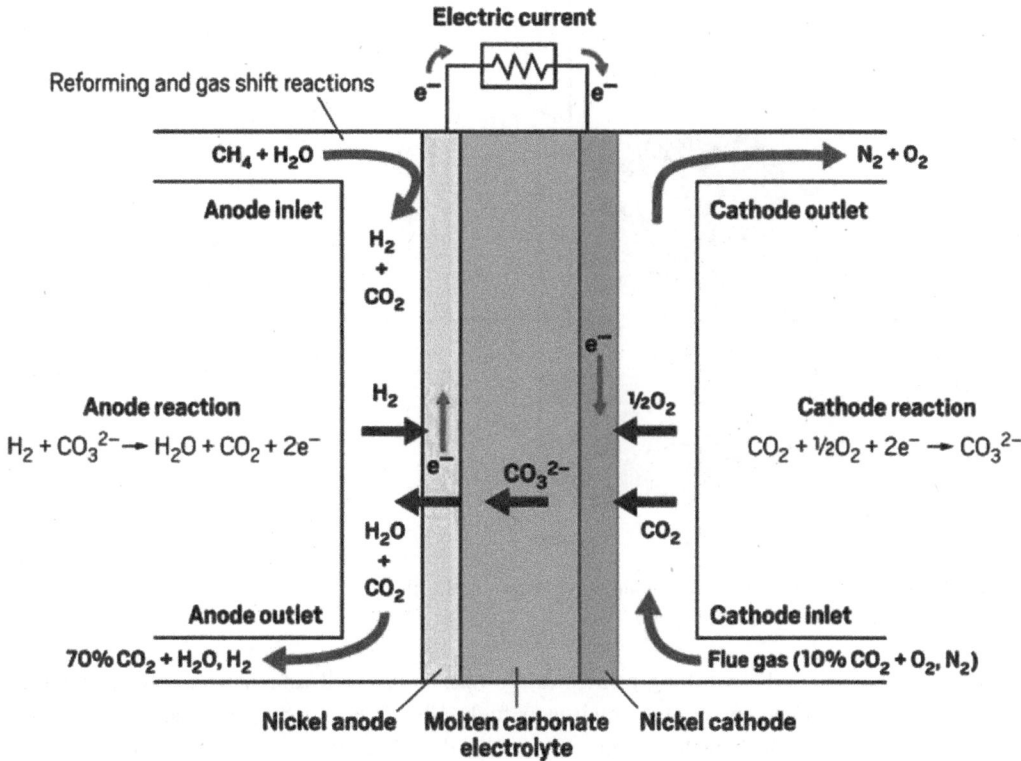

Figure 12.31 The schematic view of a molten carbonate fuel cell.

Note that the MCFCs are used in various applications, including stationary power generation, cogeneration systems, and hybrid power systems. They are known for their high efficiency and high operating temperature, making them an attractive option for various power generation applications. However, their high operating temperature and complex fabrication process can make them more expensive to manufacture than other types of fuel cells. Additionally, the use of a high-temperature molten electrolyte can also pose technical challenges in terms of material compatibility and durability.

g) Ammonia Fuel Cell

The working principle of an ammonia fuel cell is similar to the hydrogen fuel cell involving electrode reactions in addition to membrane electrolytes. Additionally, hydrogen can be mixed with ammonia and fed to the fuel cell that improves the experimental performance and efficiency of the ammonia fuel cell. Nevertheless, alkaline electrolyte-based ammonia fuel cell involves some differences in comparison with hydrogen-fueled PEM fuel cell. Figure 12.32 displays a general schematic of the alkaline electrolyte-based direct ammonia fuel cell (DAFC) displaying the anodic and cathodic reactants and products.

The input fuel feed in a DAFC contains direct ammonia (NH_3) inlet. In alkaline-based DAFC, this fuel input feed reacts at the anode electrochemically in the existence of catalyst through negatively charged hydroxyl ions. The anodic reaction can be expressed as

$$NH_3 + \frac{3}{4}O_2 \rightarrow \frac{1}{2}N_2 + 1.5\,H_2O$$

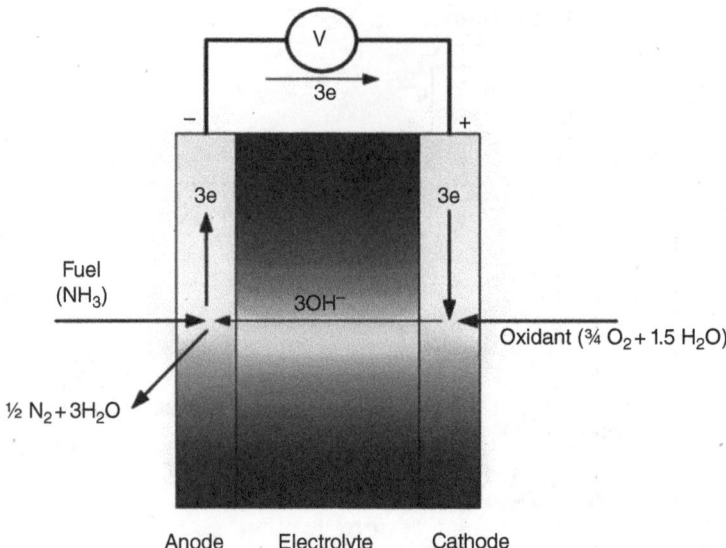

Figure 12.32 The schematic view of a direct ammonia fuel cell.

Even though the DAFC working principles are reasonably well established, and their performance is comparatively lower than hydrogen fuel cells. This is principally accredited to the electro-oxidation of unsatisfactory ammonia molecules. The catalyst activity practical in the hydrogen oxidation case for fuel cells is not established for ammonia oxidation case yet. This remains a key factor that prohibits the high-performance DAFC development.

Illustrative Example 12.3   A 5-cell fuel cell stack shown in Figure 12.33 is provided. If the current through the cells is measured to be 25 A. The ohmic, activation and concentration polarization voltages are given as ($V_{activation} + V_{ohmic} + V_{concentration} = 0.43$ V). Refer to Table 12.3 for thermochemical properties.

a) Write the chemical reaction and balance it accodingly, and write the mass, energy, entropy, and exergy balance equations for the fuel cell.
b) Find the current passing through the fuel cell.
c) Find the amount of hydrogen supply required.
d) Find the energy and exergy efficiencies of the fuel cell (heating value of $H_2 = 286$ kJ/mol, chemical exergy of $H_2 = 236.09$ kJ/mol).

Solution:

a) The overall chemical equation is written as follows:

$$H_2 + \frac{1}{2}O_2 = H_2O$$

For the fuel cell, one may write enthalpy, entropy, and exergy balance equations accordingly as follows:

$$\dot{N}_{H_2}\bar{h}_{H_2} + \dot{N}_{O_2}\bar{h}_{O_2} = \dot{N}_{H_2O}\bar{h}_{H_2O} + \Delta H$$

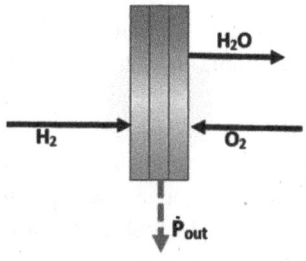

Figure 12.33 Schematic view of the fuel cells in the illustrative example.

and

$$\dot{N}_{H_2}\bar{s}_{H_2} + \dot{N}_{O_2}\bar{s}_{O_2} = \dot{N}_{H_2O}\bar{s}_{H_2O} + \Delta S$$

and

$$\dot{N}_{H_2}ex_{H_2} + \dot{N}_{O_2}ex_{O_2} = \dot{N}_{H_2O}ex_{H_2O} + \dot{Ex}_d$$

The enthalpy change for the hydrogen reaction is expressed as:

$$\Delta G = \Delta H - T\Delta S$$

The Gibbs free energy is represented with ΔG. ΔH represents enthalpy change and ΔS represents entropy change and T represents temperature in Kelvin.

b) In order to find the maximum voltage output (V_{max}), the following equation is used:

$$\Delta G = -2F\,V_{max}$$

where F stands for the Faraday constant which is the charge on 1 mol of electrons (F=96.48 kJ/mol).

Therefore, the maximum voltage output is found as follows:

$$V_{max} = \frac{-\Delta G}{2F} = \frac{237.18}{192.96} = 1.23\ V$$

The actual fuel cell voltage for each cell can be calculated after adding ohmic, activation, and concentration polarizations into the reversible open circuit potential, as follows:

$$V_{actual} = V_{max} - (V_{activation} + V_{ohmic} + V_{concentration})$$

$$V_{actual} = 1.23\ V - 0.43\ V = 0.8\ V$$

In order to find actual power output for each cell, the following equation is then used:

$$\dot{P}_{actual,cell} = V_{actual} \times I$$

$$\dot{P}_{actual,cell} = 0.8\ V \times 25\ A = 20\ W$$

In order to find the total power output for the stack, the following calculation is made:

$$\dot{P}_{stack} = n_{cell} \times \dot{P}_{actual,cell}$$

$$\dot{P}_{stack} = 5 \times 20 = 100\ W$$

c) The amount of hydrogen supply required can be calculated as follows:

$$\dot{N}_{H_2} = \frac{I}{2F} = \frac{25}{2(96,500)} = 0.000129\ mol/s$$

As there are five cells in the stack, the total hydrogen consumption rate can be found to be:

$$\dot{N}_{totalH_2} = 0.000129 \times 5 = 0.000645 \frac{mol}{s}$$

d) The energy and exergy efficiencies can be calculated as follows:

$$\eta_{en} = \frac{\dot{P}_{out}}{\dot{N}_{totalH_2}\left(HHV_{H_2}\right)} = \frac{100}{(0.000645 \ mol/s)\left(285,828\frac{kJ}{mol}\right)} = 0.542 \ or \ 54.2\%$$

and

$$\eta_{ex} = \frac{\dot{P}_{out}}{\dot{N}_{totalH_2}\left(ex_{H_2}\right)} = \frac{100}{(0.000645 \ mol/s)\left(236,090\frac{kJ}{mol}\right)} = 0.657 \ or \ 65.7\%$$

12.7 Closing Remarks

Humanity is experiencing a remarkable transition to from hydrocarbon economy, depending heavily upon fossil fuels, to hydrogen economy where renewable energy sources are widely deployed to produce it in a green/clean manner. Hydrogen appears to be a remarkable energy solution to serve primarily as a fuel, an energy carrier, and a feedstock and will shape the future by providing environmentally-benign and sustainable solutions to the global community. This chapter discusses hydrogen energy options by covering its entire spectrum from production to storage and from storage to utilization, particularly where fuel cells are employed. There are examples presented about electrolyzers and fuel cells in addition to a case study where numerous hydrogen production methods are comparatively evaluated.

References

1 Dincer, I. (2020). Covid-19 coronavirus: closing carbon age, but opening hydrogen age. *International Journal of Energy Research* 44: 6093–6097. doi: 10.1002/er.5569.
2 Dincer, I. (2023). Hydrogen 1.0: a new age. *International Journal of Hydrogen Energy* 48 (43): 16143–16147. doi: 10.1016/j.ijhydene.2023.01.124.
3 Dincer, I. and Acar, C. (2015). Review and evaluation of hydrogen production methods for better sustainability. *International Journal of Hydrogen Energy* 40 (34): 11094–11111. doi: 10.1016/j.ijhydene.2014.12.035.
4 Hydrogen fuel cell: function & types. TUVNORD. Available online https://www.tuev-nord.de/en/company/energy/hydrogen/hydrogen-fuel-cell (accessed 30 January 2023).

Questions/Problems

Questions

1 Discuss why hydrogen is important for a carbon-free future with examples.
2 What is the Hydrogen 1.0 concept? Write an essay about it.
3 Select some hydrogen production methods and compare CO_2 emissions.
4 List hydrogen production methods and discuss each briefly.

5 Explain how a basic electrolyzer works. Compare it with some other types of hydrogen production devices.

6 Explain why hydrogen storage is needed and what methods are practically available and feasible.

7 Classify the physical-based hydrogen storage techniques. Explain each in brief.

8 What are the materials-based hydrogen storage methods? Briefly explain.

9 How does a fuel cell work? Briefly explain.

10 What are the common types of fuel cells? Discuss their pros and cons.

11 Discuss how hydrogen will change every economic sector.

12 Explain the role of hydrogen in ammonia making.

13 Identify some potential fuel cells for household usages.

14 Discuss the role of hydrogen as a feedstock.

15 Discuss the key criteria in evaluating both electrolysers and fuel cells.

Problems

1 A 10-cell electrolyzer stack shown below is provided with a power input of 200 W. If the current through the cells is measured to be 30 A.
   a Write the chemical reaction and balance it accordingly, and write the mass, energy, entropy, and exergy balance equations for the electrolyzer.
   b Find the voltage across each cell.
   c Find the hydrogen production rate.

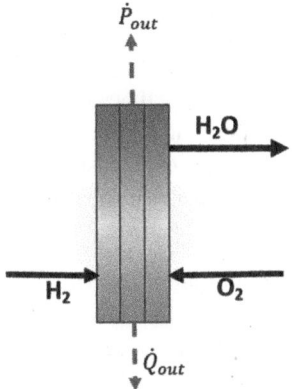

d Find the energy and exergy efficiencies of the electrolyzer (heating value of H_2=286 kJ/mol, chemical exergy of H_2=236.09 kJ/mol).

2 Recalculate Problem 1 by considering power input as 150 W and 250 W. Compare the results.

3 A 100-cell electrolyzer stack shown below is provided with a power input of 2 kW. If the current through the cells is measured to be 300 A.
   a Write the chemical reaction and balance it accordingly, and write the mass, energy, entropy, and exergy balance equations for the electrolyzer.
   b Find the voltage across each cell.
   c Find the hydrogen production rate.
   d Find the energy and exergy efficiencies of the electrolyzer (heating value of H_2=286 kJ/mol, chemical exergy of H_2=236.09 kJ/mol).

4 It is aimed to produce 1 kg of hydrogen gas per hour with the electrolyzer stack. The stack has the following details a power input per cell 35 W and current through the cell is 7 A. Calculate the number of cells in the stack to achieve the desired hydrogen production. Calculate the energy and exergy efficiencies of the electrolyzer (heating value of H_2=286 kJ/mol, chemical exergy of H_2=236.09 kJ/mol).

5 Consider a 3-cell electrolyzer stack as shown in the figure below. If the current through the cells is measured to be 20 A. Ohmic, activation, and concentration polarizations voltages are given as ($V_{activation} + V_{ohmic} + V_{concentration} = 0.77$ V). Refer to Table 12.3 for thermochemical properties.

    a Write the chemical reaction and balance it accordingly, and write the mass, energy, entropy, and exergy balance equations for the electrolyzer.

    b Find the total power input to the electrolyzer stack.

    c Find the hydrogen production rate.

    d Find the energy and exergy efficiencies of the electrolyzer.

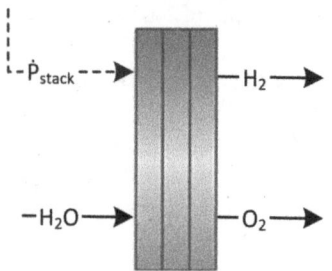

6 With a set of electrolyzer stuck, it is aimed to produce 100 kg of hydrogen per day with the electrolyzer stack. The stack has the following details a power input per cell 25 W and current through the cell is 5 A. Calculate the number of cells in the stack to achieve the desired hydrogen production. Calculate the energy and exergy efficiencies of the electrolyzer (heating value of H_2=286 kJ/mol, chemical exergy of H_2=236.09 kJ/mol).

7 A 10-cell fuel cell stack shown in the figure below provides a power output of 200 W. The voltage across each cell is measured to be 0.8 V.

    a Write the chemical reaction and balance it accordingly, and write the mass, energy, entropy, and exergy balance equations for the fuel cell.

    b Find the current passing through the fuel cell.

    c Find the amount of hydrogen supply required.

    e Find the energy and exergy efficiencies of the fuel cell (heating value of H_2=286 kJ/mol, chemical exergy of H_2 = 236.09 kJ/mol).

8 With a fuel cell stuck, it is aimed to generate 1 kW of electricity. One cell provides a power output of 10 W. The voltage across each cell is measured to be 0.8 V.

    a Write the chemical reaction and balance it accordingly, and write the mass, energy, entropy, and exergy balance equations for the fuel cell.

    b Calculate the number of cells to generate 1 kW of power.

    c Find the current passing through the fuel cell.

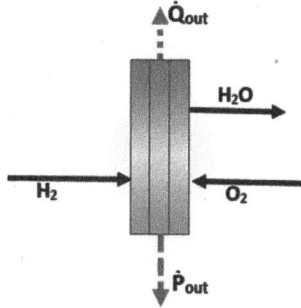

d Find the amount of hydrogen supply required.

e Find the energy and exergy efficiencies of the fuel cell (heating value of H_2=286 kJ/mol, chemical exergy of $H_2 = 236.09$ kJ/mol).

9 With a fuel cell stack, it is aimed to generate 30 kWh of energy. One cell provides a power output of 10 W. The voltage across each cell is measured to be 0.8 V. Considering 10 hours of operation,

a Write the chemical reaction and balance it accordingly, and write the mass, energy, entropy, and exergy balance equations for the fuel cell.

b Calculate the number of cells to generate 30 kWh of energy.

c Find the current passing through the fuel cell.

d Find the amount of hydrogen supply required.

e Find the energy and exergy efficiencies of the fuel cell (heating value of H_2=286 kJ/mol, chemical exergy of $H_2 = 236.09$ kJ/mol).

10 A 5-cell fuel cell stack is shown in the figure below. If the current through the cells is measured to be 25 A. Ohmic, activation, and concentration polarization voltages are given as ($V_{activation} + V_{ohmic} + V_{concentration} = 0.43$ V). Refer to Table 12.3 for thermochemical properties.

a Write the chemical reaction and balance it accordingly, and write the mass, energy, entropy, and exergy balance equations for the fuel cell.

b Find the current passing through the fuel cell.

c Find the amount of hydrogen supply required.

d Find the energy and exergy efficiencies of the fuel cell (heating value of H_2=286 kJ/mol, chemical exergy of $H_2 = 236.09$ kJ/mol)

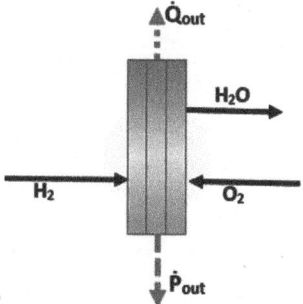

13

Integrated Energy Systems

13.1 Introduction

Everything around us is getting integrated, such as integrated plan, integrated action, integrated services, integrated users, and many more. When we look at the food sector in specific, they make products where two or more ingredients are put together to produce integrated products. Since everything is getting more integrated in almost every sector, we need to resemble this in energy sector and come up with integrated energy systems for multigeneration, where multiple useful outputs, such as power, heat, cooling, freshwater, hot water, hot air, hydrogen, ammonia, other alternative fuels, etc. are generated simultaneously. Therefore, designing and building integrated energy systems have to be a primary goal.

Integration of energy systems refers to combining several components, units, and subsystems to form a larger, integrated energy system. The ultimate goal is really to achieve better sustainability by combining energy systems in order to have multiple useful outputs and multiple benefits: better efficiency, better resource utilization, better cost effectiveness, better environment, better energy security, better management, better design and analysis (see Figure 13.1). The key question is that possible to develop integrated energy systems to be able to address some of these or all of these benefits. Especially from the energy aspect, to address and to cover the above-mentioned goal and multiple benefits becomes very critical in order to meet essential needs of future generations such as heat, cooling, freshwater, fresh air, food, etc. Numerous industries and sectors already have integrated systems, such as in food industry, food and beverage companies have products as three in one, four in one, ten in one, etc., where they combine multiple ingredients to form a single product. In order to have a similar concept in energy applications, how to achieve having multiple useful outputs and multiple benefits from the same input is the key.

Figure 13.2 illustrates the global electricity production in 2007 when the conventional methods are vast majority to make single generation. It is clear that the majority of primary energy is wasted with conventional single generation systems. There are numerous possibilities of integrated energy systems which use the same input to generate multiple useful outputs, such as a high temperature thermal energy input can be used to produce power, heat, cooling, hot water, hydrogen, freshwater, dry air, refrigeration, or air conditioning, etc. Many more can be added by using energy subsystems such as heat recovery units, heat pumps, desalination units, organic Rankine cycle units, hydrogen production units, ammonia generation units, absorption refrigeration cycle units, in an

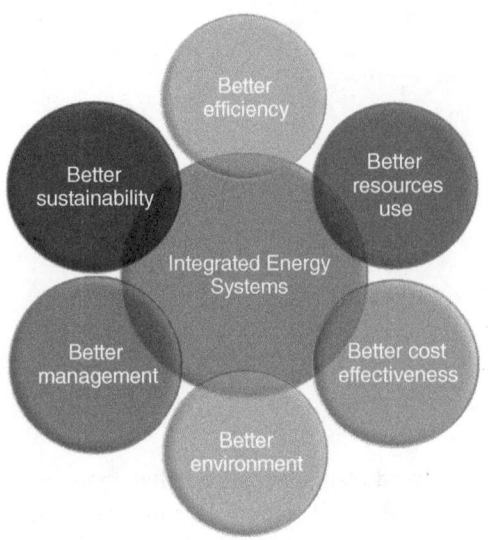

Figure 13.1 Multiple benefits with the ultimate goal to achieve better sustainability by integrating energy systems.

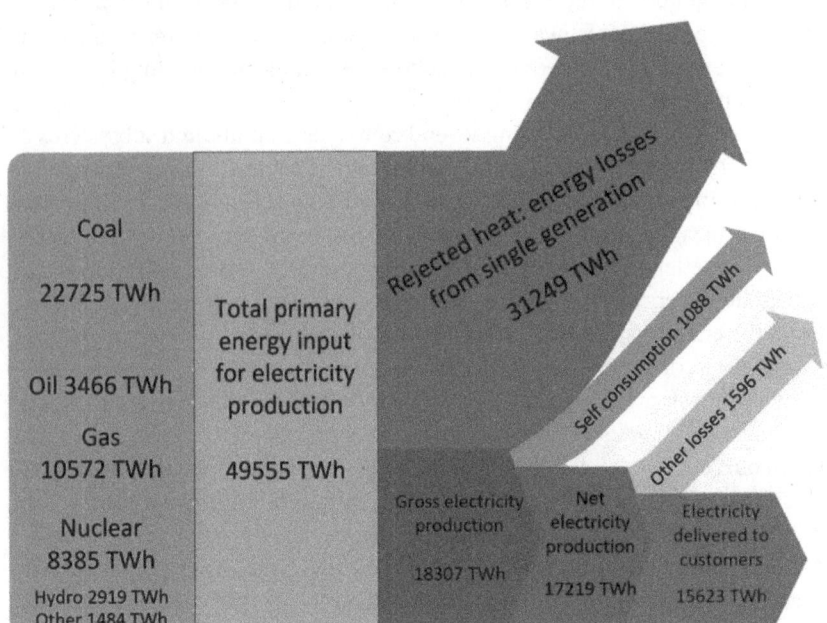

Figure 13.2 Global electricity production processes and energy flows (data from [1]).

integrated manner. When its performance is compared with the conventional approach which focuses on single generation, efficiency will be very different. Conventional power plants use high temperature thermal energy to have only one useful output, electricity, which will be used as numerator and will be divided by the input in order to assess the efficiency. On the other hand, a cogeneration, trigeneration, or multigeneration plant can generate multiple useful outputs which will be added into numerator as output one, output two, output three, etc., and then divided by the same input. From single generation to cogeneration, trigeneration, or multigeneration, the efficiency will be increasing very significantly.

The concept of integrated energy systems may depend on in specific how exploit the wasted energy which is commonly rejected into the environment as low temperature heat, by using integrated units, components, and subsystems. Exploiting the wasted energy will reduce the production of waste and will convert it into variety of useful commodities which will improve the sustainability. Especially conventional systems heavily deploy on fossil fuels which is less sustainable. Integrating subsystems and reducing the losses improve the sustainability. Furthermore, having much more sustainable systems depends on incorporating with renewable energy sources, reducing waste production, reducing emissions and making it with multiple useful outputs. The recipe for energy technologies to achieve sustainability, to switch from conventional systems with single generation to renewables and do the system integration to produce multiple useful outputs.

It is further to note that every energy source has different nature and brings some challenges and opportunities. So, it is really necessary to better design, analyze, and evaluate every energy system. When it is based on exergy, different natures of various types of commodities can be more insightfully and rationally assessed. This chapter formulates and presents integrated energy systems and how to assess their performance, efficiency, and sustainability. The differences between single and multigenerational systems are comparatively assessed and provided within examples and case studies.

13.2 System Integration

Various sectors and industries deploy numerous types of systems to generate many different products which can benefit from system integration where several components, units, subsystems are integrated to form a larger system for better effectiveness and efficiency, reduced environmental impact and increased resilience. In energy systems, many system integration practices are potentially available, such as:

- Combined heat and power systems recover heat from gas turbine cycles for cogeneration of power and heat.
- Combined cycle power plants recover heat from gas turbine cycles and exploit the recovered heat with Rankine cycles in order to enhance the utilization of fuel and improve efficiency.
- Integrating multi-effect desalination unit along with district heating system into Rankine cycles in order to exploit the excess heat for trigeneration of power, heat, and freshwater from single input and to improve efficiency.

There are numerous combinations and examples to integrate components, units, and subsystems in order to form an integrated system but there are a few key points which need to be highlighted in order to design and develop such systems. Heat, thermal energy, is one of the most common forms of energy which is used as primary energy input to generate various useful outputs. Heat grade refers to the temperature level of heat, which is the key in order to generate specific outputs. There are specific heat grade levels which enables certain processes to be executed. Higher grade of heat mostly enables more variety of processes such as power generation with Rankine cycles, hydrogen generation with thermochemical cycles. It is a common approach to exploit excess heat from power generation processes along with other forms of excess energy in order to produce wide variety of useful outputs. When there are more outputs, efficiency tends to be higher.

Note that the efficiency is generally defined represents the system performance by means of the ratio of useful outputs to the consumption of primary energy input which can be expressed as follows:

$$\eta_{single\ generation} = \frac{Output_1}{Input}$$

$$\eta_{cogeneration} = \frac{Output_1 + Output_2}{Input}$$

$$\eta_{trigeneration} = \frac{Output_1 + Output_2 + Output_3}{Input}$$

$$\eta_{quadgeneration} = \frac{Output_1 + Output_2 + Output_3 + Output_4}{Input}$$

and

$$\eta_{n-generation} = \frac{Output_1 + Output_2 + Output_3 + Output_4 + \ldots + Output_n}{Input}$$

where each equation involves a number of outputs with the same input, according to the system type.

13.3   Multigeneration

The main objective of designing integrated systems is to minimize the waste and increase the number of outputs. It is possible that to generate the same useful output with excess energy such as combined Rankine and Brayton cycles or combined Rankine and organic Rankine cycles which exploits different grades of heat to generate power. The combination of various components to generate same useful output is still considered as single generation. According to the customer or location, the integrated system can be designed to meet various requirements and demands. Residential heating, cooling, desalination, and domestic hot water are some of the useful outputs which can be more suitable for residential applications or community systems. Sustainable communities need clean electricity, clean air, clean water; however, those are not always available or not clean. Integrated energy systems especially with low carbon sources can meet those requirements by generating multiple useful outputs in a sustainable and integrated fashion. For the industrial applications, other requirements such as chemical substances, feedstocks, process heat, steam can be needed. Integrated energy systems for industrial applications can meet specific requirements or can support auxiliary systems to meet those needs.

An energy system with single generation is shown in Figure 13.3. It has only single output and the rest of the energy will be wasted in various forms of energy such as waste heat. Therefore, a single generation system can be mentioned as less sustainable with higher inefficiencies. In order to calculate its efficiency, the numerator has only output 1 and the denominator has the input.

A cogeneration system is shown in Figure 13.4. In order to reduce the production of waste, cogeneration system uses the excess from the first process and utilizes it to generate a useful output. Cogeneration process multiplies the outputs into two and tends to be more efficient compared

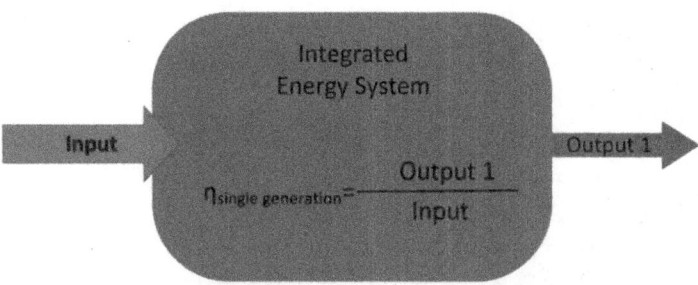

Figure 13.3 Illustration of an energy system for single generation.

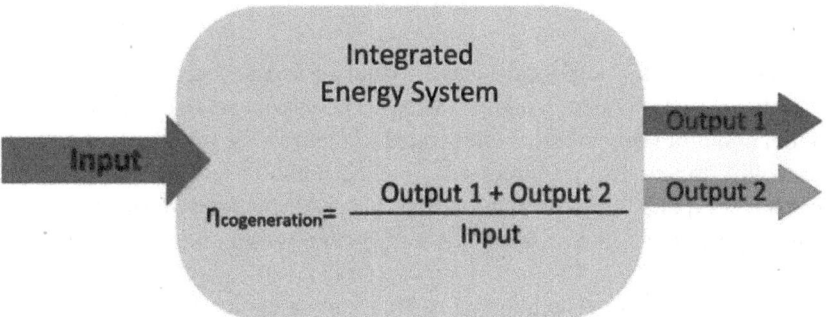

Figure 13.4 Illustration of an integrated energy system for cogeneration.

to the single generation process. Cogeneration systems can be mentioned as more sustainable with higher efficiencies in reference to the single generation systems. However, inefficiencies still can be mitigated by exploiting the further waste with trigeneration, quadrogeneration, or more generation systems.

For further, it is possible to reduce wastes by increasing the utilization and adding subsystems to generate variety of useful outputs. A trigeneration system (see Figure 13.5) can further exploits the excess and multiplies useful outputs to three. In order to analyze the performance, the numerator increases to three for the same input, which improves the efficiency and sustainability compared to single and cogeneration systems.

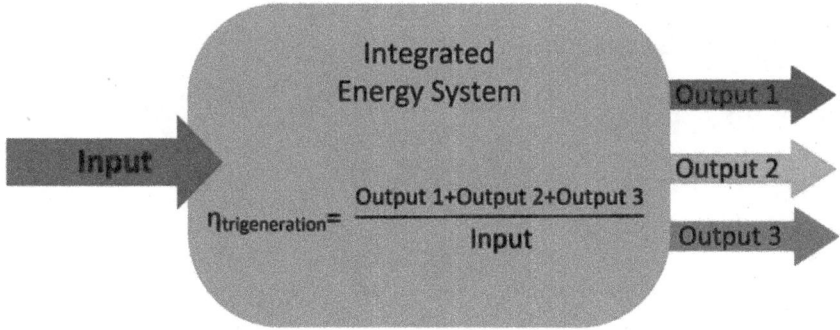

Figure 13.5 Illustration of an integrated energy system for trigeneration.

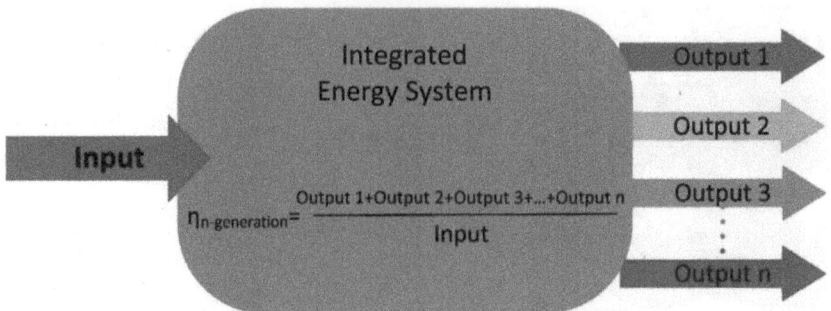

Figure 13.6 Illustration of sustainability improved by multigeneration.

   The possibilities to increase useful outputs are just a matter of imagination of the designer. Numerous useful outputs can be generated by adding the correct components to integrate subsystems. Quadrogeneration and more generation are illustrated in Figure 13.6. In order to reach the ultimate goal, sustainability can be further improved until the limits of thermodynamic laws. Integrated energy systems for quadro- and more generation possesses higher performances compared to tri-, co-, and single generation systems by including more useful outputs in its performance numerator.

   In order to meet needs of the present generation on earth, future generations' needs should be considered. Multigeneration systems are potential solutions to improve sustainability by improving the ratio of useful outputs to the primary energy input. As illustrated in Figure 13.7, more useful outputs tend to increase sustainability. The process of multiplying outputs need to be done by converting the energy source into similar forms of energy such as process heat or power, thereafter, different kinds of outputs can be generated with the integrated systems. The assessment of performance can be done by calculating the energy contents of the outputs and inputs. However, exergy content can provide more insightful results since it quantifies different kinds of outputs by considering their qualities.

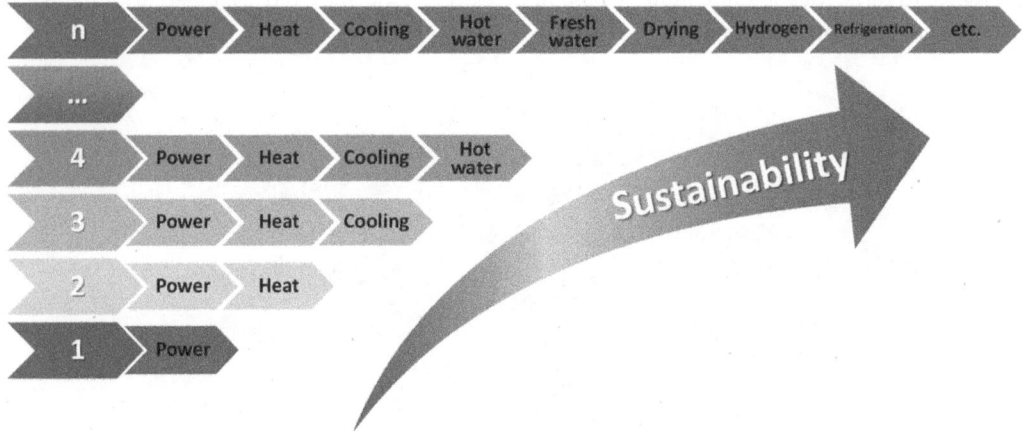

Figure 13.7 Illustration of sustainability improved by multigeneration.

Example 13.1 An integrated system is designed to make trigeneration with single input where biomass is used as primary energy input in order to generate power, heat, and freshwater (see Figure 13.8). Biomass-driven system evaporates n-octane working fluid of organic Rankine cycle at the evaporator where 90% of the lower heating value of biomass transferred to n-octane. The working fluid in the system at 11 kg/s mass flow rate enters organic Rankine cycle turbine at 275.85°C and 2000 kPa in order to generate electricity with the generator which has 90% efficiency and leaves the turbine at 159.7°C and 35.7 kPa. In heat exchanger 1, n-octane rejects heat and cools down to 91.85°C and condenses until the vapor quality X = 0.4. Thereafter, it becomes saturated liquid at 91.85°C before entering the pump.

A residential heating system works with liquid water which enters heat exchanger 1 at 70°C and leaves at 85°C in atmospheric pressure.

A multi-effect desalination system is fed by 72°C 33 kPa water which enters as saturated vapor to the unit and leaves as saturated liquid from the unit in order to desalinate sea water and produce freshwater with 2.68 kg/s mass flow rate at 56°C in atmospheric pressure.
(take $LHV_{biomass} = 18,000$ kJ/kg, $ex_{biomass} = 12,500$ kJ/kg, $v_1 = 0.001935$ m³/kg)

a) Design and discuss the single generation, cogeneration and trigeneration configurations of the integrated system separately.
b) Write the mass, energy, entropy and exergy balance equations for the organic Rankine cycle.
c) By using the given parameters, prepare a thermophysical properties table.
d) Calculate the power generation, freshwater production rate, and heat rate.
e) Calculate the single generation (power), cogeneration (power+heat), and trigeneration (power+heat+freshwater) efficiencies.

Solution:

a) Single generation configuration is shown in Figure 13.9. The main purpose of the single generation system is to produce power by using biomass energy. Biomass is combusted with air

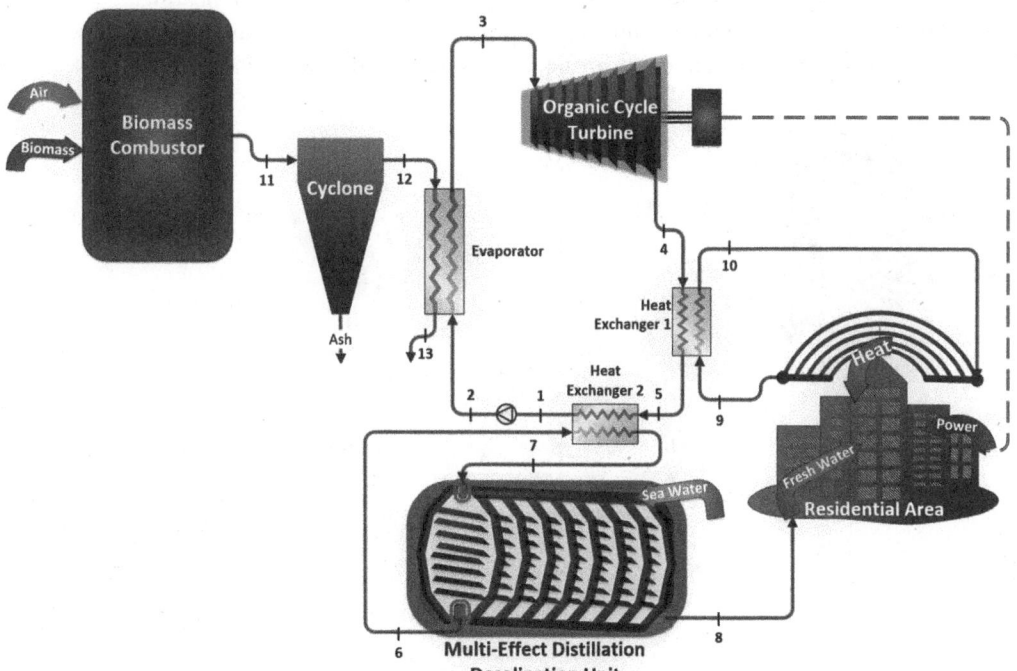

Figure 13.8 Illustration of an integrated energy system layout for Example 13.1.

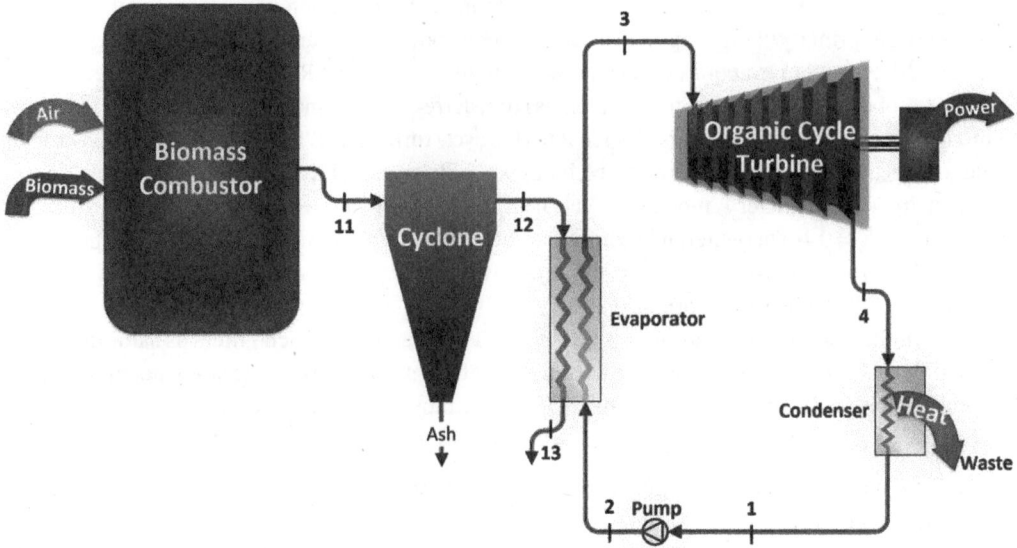

Figure 13.9 Layout of a single generation system configuration for Example 13.1.

and then transferred to the evaporator to run an organic Rankine cycle. The working fluid is selected as n-octane which can exploit lower temperatures compared to the steam Rankine cycle. Between state points 4 and 1, the process rejects heat to the atmosphere. This single generation process is similar to one of the common power generation processes. A conventional power generation system consists of Rankine cycle where they generate heat from different types of fuels and then converts the generated heat into power. Around 70% of the generated heat gets lost from the condenser. It is released to the atmosphere in most of the cases. However, this heat can be exploited for various purposes, such as residential heating.

The cogeneration system, as shown in Figure 13.10, exploits the waste heat and uses it for residential heating. Conventionally, residential heating and power generation processes have

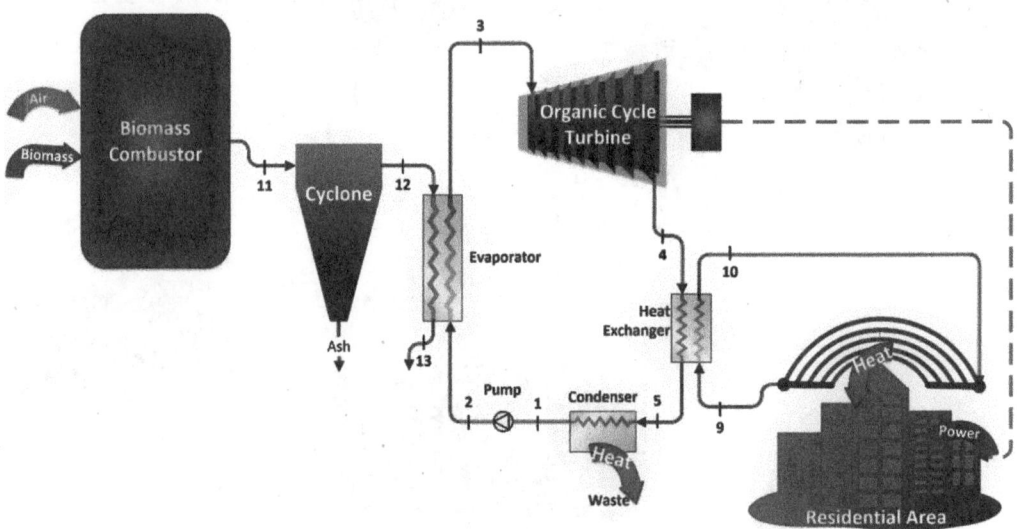

Figure 13.10 Layout of a cogeneration system configuration for Example 13.1.

separate systems. It can be estimated that around 70% of the primary energy input is wasted for conventional power generation processes with Rankine cycle, and 25% of the primary energy input is wasted for residential heating processes. As a potential alternative, cogeneration systems, such as the configuration above, can generate power and exploit the waste heat for residential heating purposes. This can further improve the utilization of primary energy input and as well as sustainability.

But there are still rooms for improvement, since condenser in cogeneration configuration rejects heat to the atmosphere, which generates further waste. In order to exploit the waste, trigeneration system can be designed which exploits the rejected heat between state points 1 and 5 and generates freshwater for the community (see Figure 13.11).

b) The working fluid n-octane flows into pump as a saturated liquid in order to get pressurized. For pump component, the mass, energy, entropy and exergy balance equations are expressed as follows:

$$\dot{m}_1 = \dot{m}_2$$

$$\dot{m}_1 h_1 + \dot{W}_P = \dot{m}_2 h_2$$

$$\dot{m}_1 s_1 + \dot{S}_{gen,P} = \dot{m}_2 s_2$$

$$\dot{m}_1 ex_1 + \dot{W}_P = \dot{m}_2 ex_2 + \dot{Ex}_{dest,P1}$$

The pressurized n-octane enters evaporator to get superheated before entering organic cycle turbine. For the evaporator component, the mass, energy, entropy and exergy balance equations can be expressed as follows:

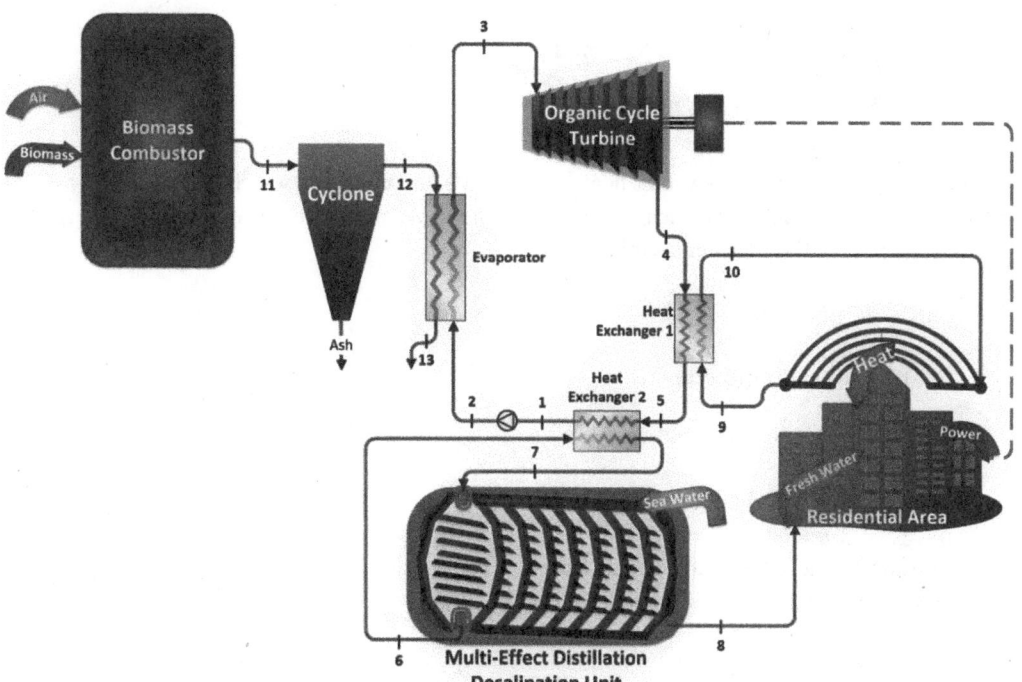

Figure 13.11   Layout of a trigeneration system configuration for Example 13.1.

$$\dot{m}_2 = \dot{m}_3 \; , \dot{m}_{12} = \dot{m}_{13}$$

$$\dot{m}_2 h_2 + \dot{m}_{12} h_{12} = \dot{m}_3 h_3 + \dot{m}_{13} h_{13}$$

$$\dot{m}_2 s_2 + \dot{m}_{12} s_{12} + \dot{S}_{gen,eva} = \dot{m}_3 s_3 + \dot{m}_{13} s_{13}$$

$$\dot{m}_2 ex_2 + \dot{m}_{12} ex_{12} = \dot{m}_3 ex_3 + \dot{m}_{13} ex_{13} + \dot{Ex}_{dest,eva}$$

In the organic Rankine cycle turbine, superheated n-octane expands in order to generate power. For the organic cycle turbine, the mass, energy, entropy and exergy balance equations are written as follows:

$$\dot{m}_3 = \dot{m}_4$$

$$\dot{m}_3 h_3 = \dot{m}_4 h_4 + \dot{W}_{T1}$$

$$\dot{m}_3 s_3 + \dot{S}_{gen,T1} = \dot{m}_4 s_4$$

$$\dot{m}_3 ex_3 = \dot{m}_4 ex_4 + \dot{W}_{T1} + \dot{Ex}_{dest,T1}$$

Heat exchanger 1 recovers the excess heat and injects it into the water which will heat the residential area. For heat exchanger 1, the mass, mass, energy, entropy, and exergy balance equations can be expressed as follows:

$$\dot{m}_4 = \dot{m}_5 \; , \dot{m}_9 = \dot{m}_{10}$$

$$\dot{m}_4 h_4 + \dot{m}_9 h_9 = \dot{m}_5 h_5 + \dot{m}_{10} h_{10}$$

$$\dot{m}_4 s_4 + \dot{m}_9 s_9 + \dot{S}_{gen,HX1} = \dot{m}_5 s_5 + \dot{m}_{10} s_{10}$$

$$\dot{m}_4 ex_4 + \dot{m}_9 ex_9 = \dot{m}_5 ex_5 + \dot{m}_{10} ex_{10} + \dot{Ex}_{dest,HX1}$$

Heat exchanger 2 recovers the excess heat and injects it into the water which will be used for multi-effect desalination process. For heat exchanger 2, the mass, mass, energy, entropy and exergy balance equations are then written as follows:

$$\dot{m}_5 = \dot{m}_1 \; , \dot{m}_6 = \dot{m}_7$$

$$\dot{m}_5 h_5 + \dot{m}_6 h_6 = \dot{m}_1 h_1 + \dot{m}_7 h_7$$

$$\dot{m}_5 s_5 + \dot{m}_6 s_6 + \dot{S}_{gen,HX2} = \dot{m}_1 s_1 + \dot{m}_7 s_7$$

$$\dot{m}_5 ex_5 + \dot{m}_6 ex_6 = \dot{m}_1 ex_1 + \dot{m}_7 ex_7 + \dot{Ex}_{dest,HX2}$$

c) According to the given parameters, enthalpy values can be found by using the engineering equation solver (EES) by finding the thermophysical properties of n-octane:

$$\left. T_1 = 91.85°C \atop P_1 = 35.7kPa \right\} \begin{array}{l} h_1 = 157.5 \dfrac{kJ}{kg} \\[2mm] s_2 = 0.476 \dfrac{kJ}{kg\ K} \end{array}$$

$$w_{pump} = \nu_1(P_2 - P_1) = 0.001935 \dfrac{m^3}{kg}(2000\,kPa - 35.7\,kPa) = 3.8\dfrac{kJ}{kg}$$

$$h_1 + w_{pump} = h_2$$

$$h_2 = 157.5\dfrac{kJ}{kg} + 3.8\dfrac{kJ}{kg} = 161.3\dfrac{kJ}{kg}$$

$$\left. h_2 = 275.85°C \atop P_2 = 2000kPa \right\} \begin{array}{l} h_2 = 275.85°C \\[2mm] s_2 = 0.4781\dfrac{kJ}{kg\ K} \end{array}$$

$$P_3 = P_2 = 2000kPa$$

$$\left. T_3 = 275.85°C \atop P_3 = 2000kPa \right\} \begin{array}{l} h_3 = 708.4\dfrac{kJ}{kg} \\[2mm] s_3 = 1.674\dfrac{kJ}{kg\ K} \end{array}$$

$$\left. T_4 = 159.7°C \atop P_4 = 35.7kPa \right\} \begin{array}{l} h_4 = 625\dfrac{kJ}{kg} \\[2mm] s_4 = 1.723\dfrac{kJ}{kg\ K} \end{array}$$

$$\left. x_5 = 0.4 \atop T_5 = 91.85°C \right\} \begin{array}{l} h_5 = 287.3\dfrac{kJ}{kg} \\[2mm] s_5 = 0.5433\dfrac{kJ}{kg\ K} \end{array}$$

$$\left. T_6 = 72°C \atop x_6 = 0 \right\} \begin{array}{l} h_6 = 301\dfrac{kJ}{kg} \\[2mm] s_6 = 0.971\dfrac{kJ}{kg\ K} \end{array}$$

$$\left. T_7 = 72°C \atop x_7 = 1 \right\} \begin{array}{l} h_7 = 2630\dfrac{kJ}{kg} \\[2mm] s_7 = 7.739\dfrac{kJ}{kg\ K} \end{array}$$

$$\left. T_8 = 56°C \atop P_8 = 101.3kPa \right\} \begin{array}{l} h_8 = 235\dfrac{kJ}{kg} \\[2mm] s_8 = 0.7807\dfrac{kJ}{kg\ K} \end{array}$$

$$T_9 = 70°C \atop P_9 = 101.3 \, kPa \Bigg\} \quad h_9 = 293.1\frac{kJ}{kg} \atop s_9 = 0.9551\frac{kJ}{kg \, K}$$

$$T_{10} = 85°C \atop P_{10} = 101.3 \, kPa \Bigg\} \quad h_{10} = 356\frac{kJ}{kg} \atop s_{10} = 1.135\frac{kJ}{kg \, K}$$

In order to find the specific exergy values, the following equation is used:

$$ex_n = h_n - h_0 - T_0(s_n - s_0)$$

If the reference state point is considered as 25°C and 101.3 kPa, the enthalpy and entropy values are obtained for n-octane as:

$$T_0 = 25°C \atop P_0 = 101.3 \, kPa \Bigg\} \quad h_0 = 5.24 \times 10^{-5} \frac{kJ}{kg} \atop s_0 = 7.02 \times 10^{-5} \frac{kJ}{kg \, K}$$

Therefore, a thermophysical properties table, as shown in Table 13.1, is prepared as follows:

d) To find the power generation, we use the following balance equations:

$$\dot{m}_3 h_3 = \dot{m}_4 h_4 + \dot{W}_{T1}$$

$$\dot{m}_3 = \dot{m}_4$$

Table 13.1 Thermophysical properties of substances at each state point for Example 13.1.

State Point	Substance	T (°C)	P (kPa)	h (kJ/kg)	ex (kJ/kg)
1	n-octane	91.85	35.7	157.5	15.6
2	n-octane	92.75	2000	161.3	18.8
3	n-octane	275.85	2000	708.4	209.2
4	n-octane	159.7	35.7	625	111.2
5	n-octane	91.85	35.7	287.3	20.8
6	water	72	33	301	23.0
7	water	72	33	2630	332.0
8	water	56	101.3	235	6.4
9	water	70	101.3	293.1	12.9
10	water	85	101.3	356	22.2

The work output for the organic Rankine cycle turbine is found as follows:

$$\dot{W}_{T1} = \dot{m}_3(h_3 - h_4)$$

$$\dot{W}_{T1} = 917.4\frac{kJ}{s} = 917.4\,kW$$

Furthermore, the generator efficiency is used to find electricity output from the generator as:

$$\dot{W}_{el} = \eta_{gen}\dot{W}_{T1}$$

$$\dot{W}_{el} = 0.9 \times 917.4\,kW = 825.66\,kW$$

The net power generation is found by subtracting pump work from the generated electricity as follows:

$$\dot{W}_{net} = \dot{W}_{el,T_1} - \dot{W}_p$$

where $\dot{W}_p$ can be find by multiplying w_{pump} with mass flow rate as follows:

$$\dot{W}_p = \dot{m}_1 w_{pump} 3 \dot{W}_p = \dot{m}_1(h_2 - h_4)$$

$$\dot{W}_p = 3.8\frac{kJ}{kg} \times 11 = 41.8\,kW$$

Therefore, it results in

$$\dot{W}_{net} = 825.66 - 41.8 = 783.86\,kW$$

In order to find the mass flow rate of water which carries heat to residential area, the following equations are used:

$$\dot{m}_4 = \dot{m}_5,\ \dot{m}_9 = \dot{m}_{10}$$

$$\dot{m}_4 h_4 + \dot{m}_9 h_9 = \dot{m}_5 h_5 + \dot{m}_{10} h_{10}$$

$$\dot{m}_4(h_4 - h_5) = \dot{m}_9(h_{10} - h_9)$$

$$\dot{m}_9 = 59.1\,kg/s$$

To find the residential heating capacity, the following calculation is made:

$$\dot{Q}_{RH} = \dot{m}_9\left(h_{10} - h_9\right) = 3714.7\frac{kJ}{s} = 3714.7\,kW$$

In order to find the third useful output, the enthalpy and exergy values of the freshwater are considered and calculated as follows:

$$\dot{m}_{fw}h_{fw} = 628.75\frac{kJ}{kg}$$

$$\dot{m}_{fw}ex_{fw} = 17.12\frac{kJ}{kg}$$

For the biomass input, the problem statement, as given above, indicates 90% of LHV from biomass is transferred into n-octane at the evaporator. The following expression is used to find the mass flow rate of biomass as follows:

$$\eta_{gen}\dot{m}_{biomass}LHV_{biomass} = \dot{m}_2(h_3 - h_2)$$

$$\dot{m}_{biomass} = \frac{11\frac{kg}{s}\left(708.4\frac{kJ}{kg} - 161.3\frac{kJ}{kg}\right)}{0.9 \times 18000\frac{kJ}{kg}} = 0.37\,kg\,/\,s$$

The total energy and exergy inputs are then obtained as follows:

$$\dot{m}_{biomass}LHV_{biomass} = 6686.8\frac{kJ}{s} = 6686.8kW$$

$$\dot{m}_{biomass}ex_{biomass} = 4643.6\frac{kJ}{s} = 4643.6kW$$

e) The energy efficiency of single generation is calculated by dividing net electricity generation into total energy input as follows:

$$\eta_{en,single\ generation} = \frac{\dot{W}_{net}}{\dot{m}_{biomass}LHV_{biomass}}$$

$$\eta_{en,single\ generation} = \frac{783.9kW}{6686.8kW} = \mathbf{11.72\%}$$

The exergy efficiency of single generation is calculated by dividing net electricity generation into total exergy input as follows:

$$\eta_{ex,single\ generation} = \frac{\dot{W}_{net}}{\dot{m}_{biomass}ex_{biomass}}$$

$$\eta_{ex,single\ generation} = \frac{783.9kW}{4643.6kW} = \mathbf{16.88\%}$$

The single generation system intends to generate power from biomass; however, majority of the energy input is wasted which could be recovered by various processes (see Figure 13.12). Exergy efficiency of the single generation system found higher than the energy efficiency of the single generation system since the exergy value of biomass is lower than its energy value, while exergy value of power equals to its energy value.

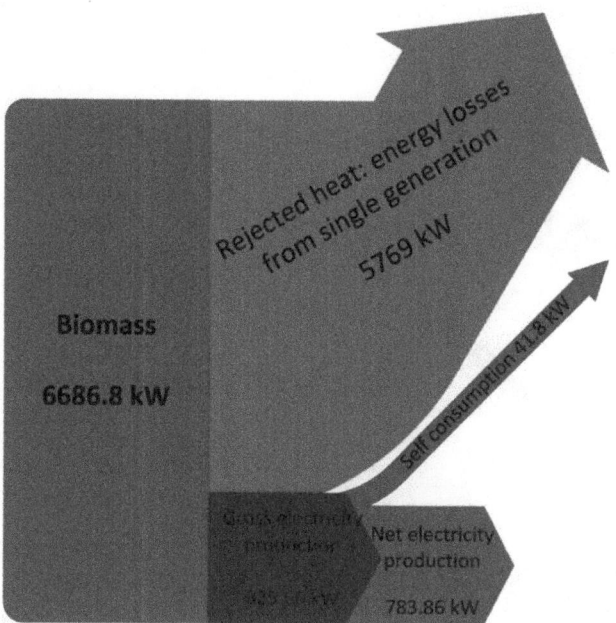

Figure 13.12 Illustration of energy flows for single generation configuration of Example 13.1.

The energy efficiency of cogeneration is calculated by adding residential heating to net electricity generation and dividing the input into total energy input as follows:

$$\eta_{en,cogeneration} = \frac{\dot{W}_{net} + \dot{Q}_{RH}}{\dot{m}_{biomass}LHV_{biomass}}$$

Therefore, it results in

$$\eta_{en,cogeneration} = \frac{4498.6kW}{6686.8kW} = \textbf{67.28\%}$$

The exergy efficiency of cogeneration is calculated by adding residential heating to net electricity generation and dividing the input into total exergy input as follows:

$$\eta_{ex,cogeneration} = \frac{\dot{W}_{net} + \dot{Ex}_{QRH}}{\dot{m}_{biomass}ex_{biomass}}$$

Therefore, it results in

$$\eta_{ex,cogeneration} = \frac{1457.1kW}{4643.6kW} = \textbf{31.38\%}$$

The cogeneration configuration enables the integrated system to recover more than half of the waste heat and use it for residential heating purposes (see Figure 13.13). While the energy efficiency is increased to 67.28%, exergy efficiency is found 31.38% since the exergy value of residential heating is significantly lower than its energy value due to its heat grade and ambient temperature.

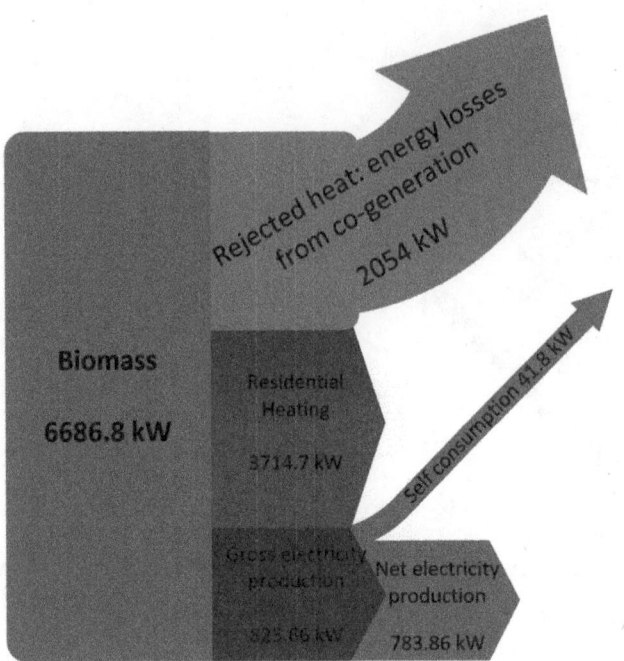

Figure 13.13 Illustration of energy flows for cogeneration system configuration of Example 13.1.

The energy efficiency of trigeneration can be calculated by adding residential heating to net electricity generation and dividing the input into total energy input as follows:

$$\eta_{en,trigeneration} = \frac{\dot{W}_{net} + \dot{Q}_{DH} + \dot{m}_{FW}h_{FW}}{\dot{m}_{biomass}LHV_{biomass}}$$

Thus, it becomes

$$\eta_{en,trigeneration} = \frac{5127.3kW}{6686.8kW} = \mathbf{76.68\%}$$

The exergy efficiency of trigeneration can be calculated by adding residential heating to net electricity generation and dividing the input into total exergy input as follows:

$$\eta_{ex,single\ generation} = \frac{\dot{W}_{net} + \dot{Ex}_{QDH} + \dot{m}_{FW}ex_{FW}}{\dot{m}_{biomass}ex_{biomass}}$$

Thus, it results in

$$\eta_{ex,trigeneration} = \frac{1474.3kW}{4643.6kW} = \mathbf{31.75\%}$$

Trigeneration system enables more heat recovery by adding a multi-effect desalination system (see Figure 13.14). Even the increase in energy and exergy efficiencies are not significant, freshwater is a crucial commodity especially for communities which has water scarcity and drought.

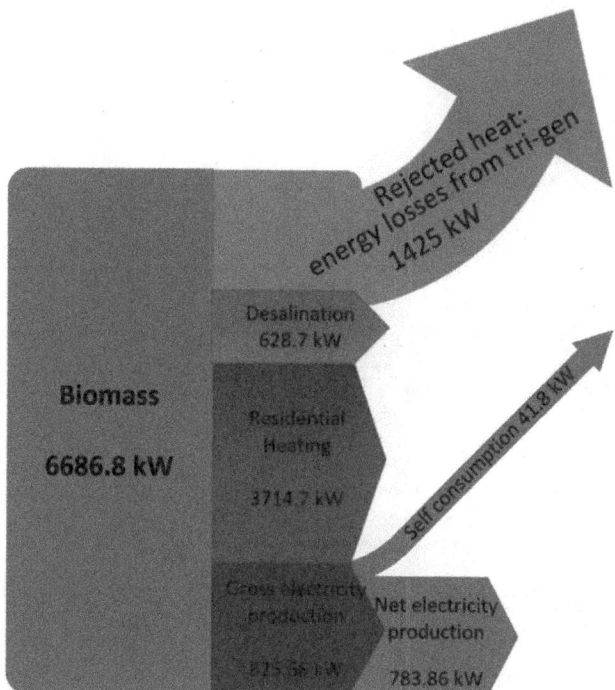

Figure 13.14 Illustration of energy flows for trigeneration configuration of Example 13.1.

Example 13.2 A solar-driven integrated system (see Figure 13.15) designed to generate power and provide residential heating as well as domestic hot water. The pressurized water with 16 kg/s at 180.6°C and 7700 kPa is heated with concentrated solar plant which is able to transfer 65% solar irradiation heat to generate high temperature steam at 383.1°C which is expanded until 180°C as saturated vapor in a steam turbine to generate power. Heat exchanger 1 transfers heat from water to fluoromethane (R41) until water becomes saturated liquid.

At the second stage, an organic Rankine cycle produces power with fluoromethane (R41) working fluid. The pressurized R41 at 37°C and 8000 kPa is superheated until 152°C and 8000 kPa and entered to the organic cycle turbine to generate power and expanded until 102.5°C and 4292 kPa. In heat exchanger 2, R41 is cooled down until 70°C to heat water from 45°C to 65°C at atmospheric pressure for residential heating purposes. Thereafter, R41 is cooled down until 30° and it became saturated liquid while transferring the remaining heat into the domestic hot water unit. (Take $T_{sun} = 5780$ K, $v_4 = 0.001004$ m^3/kg)

a) Provide a solution methodology, and design the single generation, cogeneration and trigeneration configurations of the integrated system, separately.
b) Write the mass, energy, entropy and exergy balance equations for each component.
c) By using the given parameters, prepare a thermophysical properties table.
d) Calculate the power generation, residential heating and domestic hot water production capacities.
e) Calculate both energy and exergy efficiencies for each case of single generation (power), cogeneration (power+heat), and trigeneration (power+heat+domestic hot water) efficiencies.

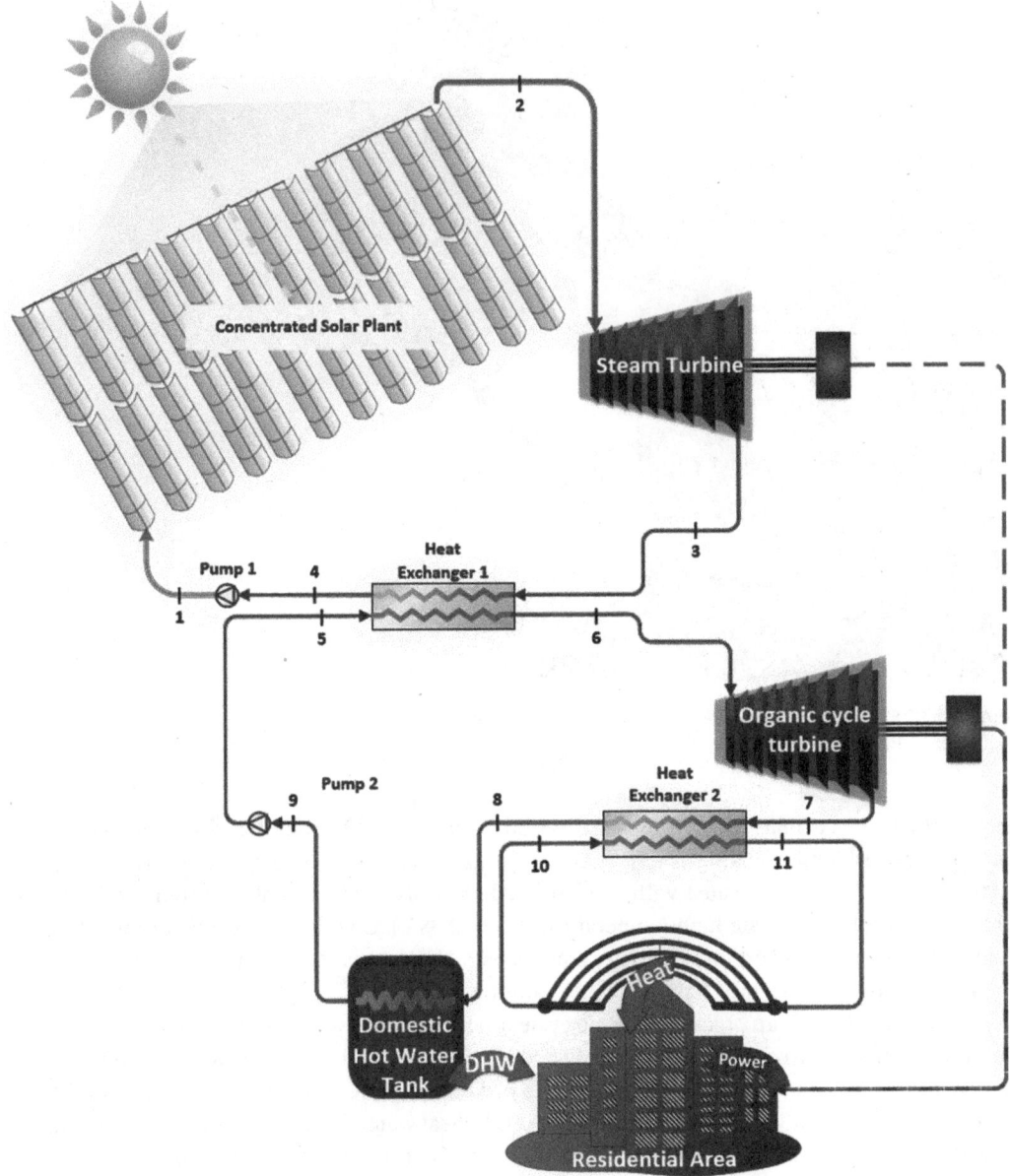

Figure 13.15 Solar-driven integrated energy system layout for trigeneration.

Solution:

a) Before starting the problem solution, there is a need to follow a solution methodology as a key
    guidance which is very important. A solution methodology is given in Figure 13.16 to help for
    better understanding.

    The configuration in Figure 13.17 shows the system layout for single generation. The concen-
    trated solar plant directly heats steam to the desired point without using any heat transfer fluid. It is
    able to transfer 65% of the heat from solar irradiation into the water. Only one-stage power genera-
    tion system leads losses which is mainly wasted from the condenser and released to the atmosphere.

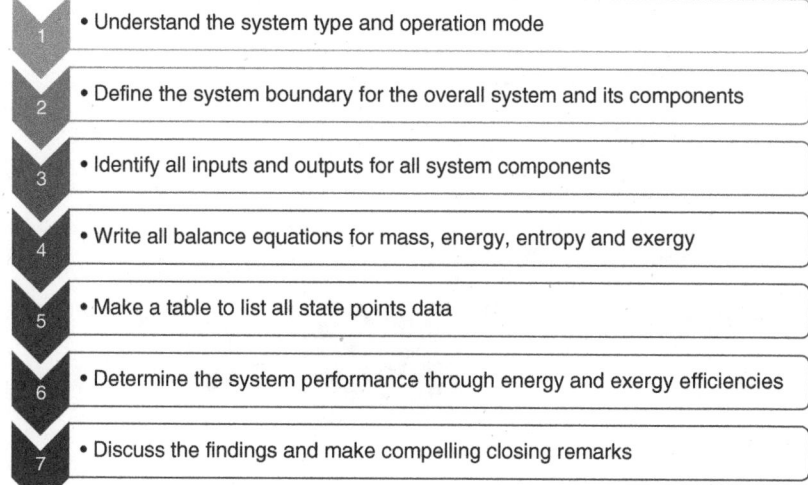

1 • Understand the system type and operation mode

2 • Define the system boundary for the overall system and its components

3 • Identify all inputs and outputs for all system components

4 • Write all balance equations for mass, energy, entropy and exergy

5 • Make a table to list all state points data

6 • Determine the system performance through energy and exergy efficiencies

7 • Discuss the findings and make compelling closing remarks

Figure 13.16 The solution procedure which can be used in solving such problems.

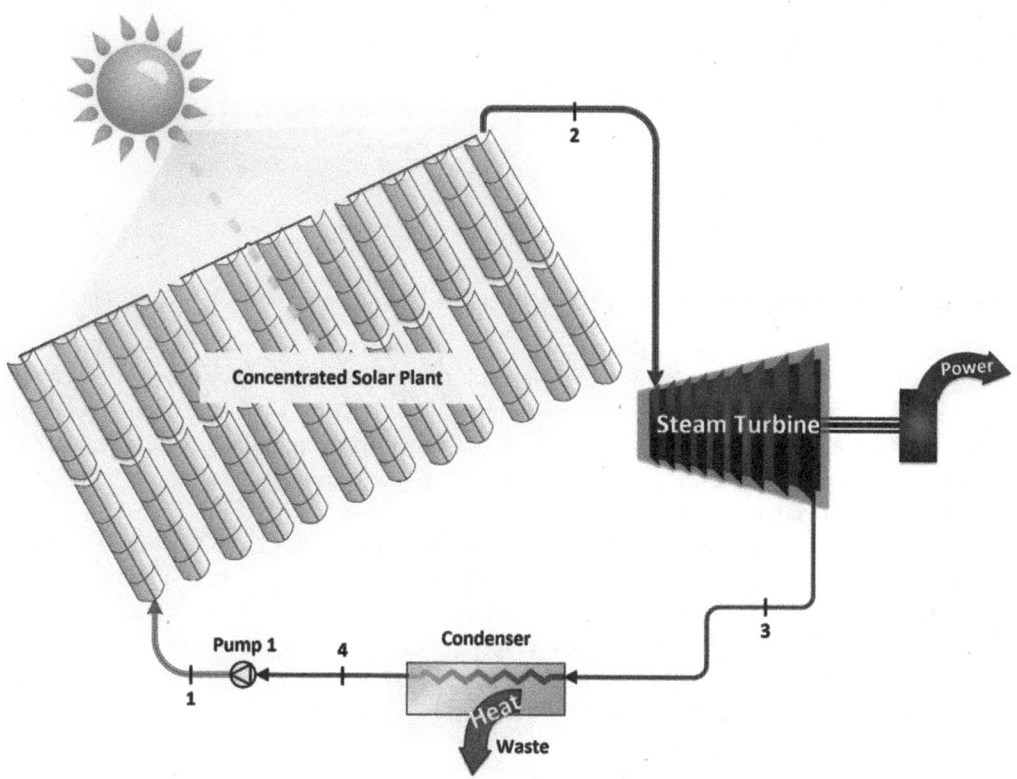

Figure 13.17 Layout of the one-stage single generation system configuration for Example 13.2.

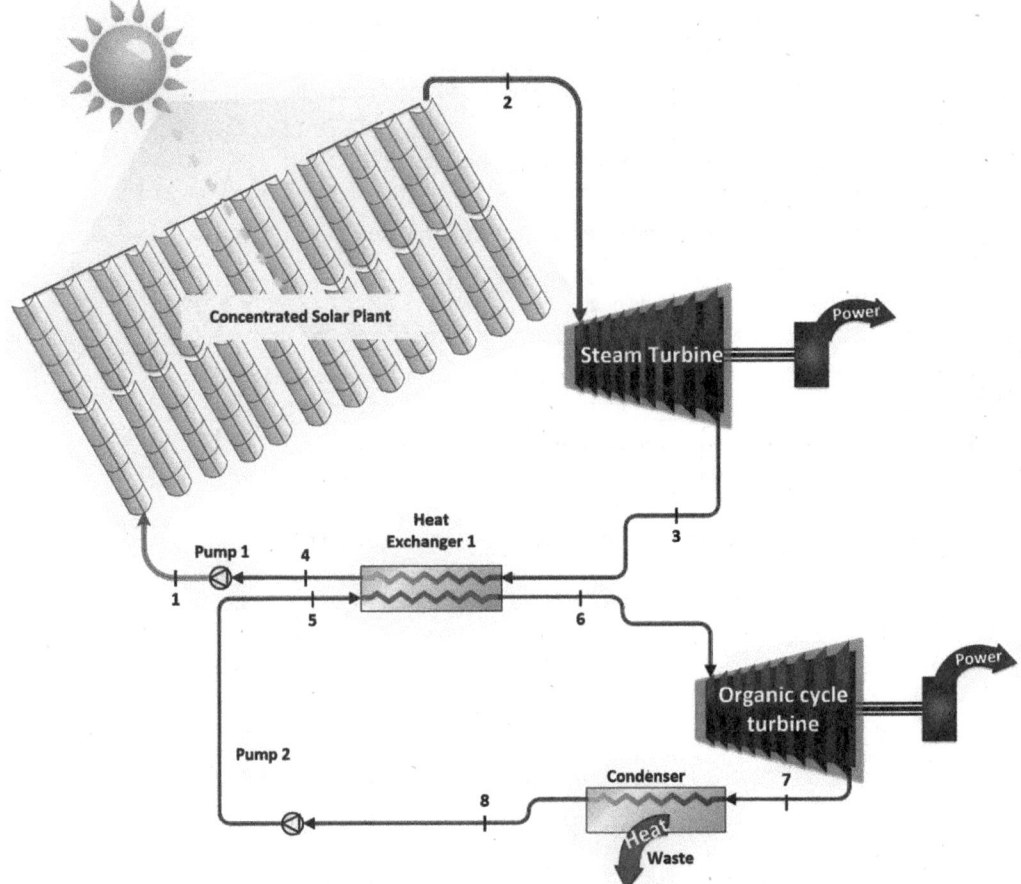

Figure 13.18 Layout of the combined cycle single generation system configuration for Example 13.2.

The second configuration is also a single-generation system; however, the second stage power generation system improves the system performance (see Figure 13.18). The organic Rankine cycle in the second stage is able to run with lower temperatures compared to the steam Rankine cycle. This enables the integrated system to exploit lower grade heat for power generation purpose, as well as to improve performance, and sustainability.

There is further waste which can be exploited between state points 7 and 8, where a residential heating system is integrated in the third configuration. A cogeneration system is designed to exploit heat from solar source, after the first and second stage power generation cycles (see Figure 13.19). In the first example, higher temperature heat was used to heat residential area; however, the current configuration would struggle to generate higher temperatures due to the waste heat's grade at state point 7 and 8. Therefore 65°C water is used for residential heating purposes in the cogeneration system.

A trigeneration system can be configured by adding the domestic hot water system to the previous configuration (see Figure 13.20). Low temperature heat can be used for domestic hot water applications, which can reduce the further production of waste. Between state points 8 and 9, a domestic water heater and tank is added and the remaining heat is used in there.

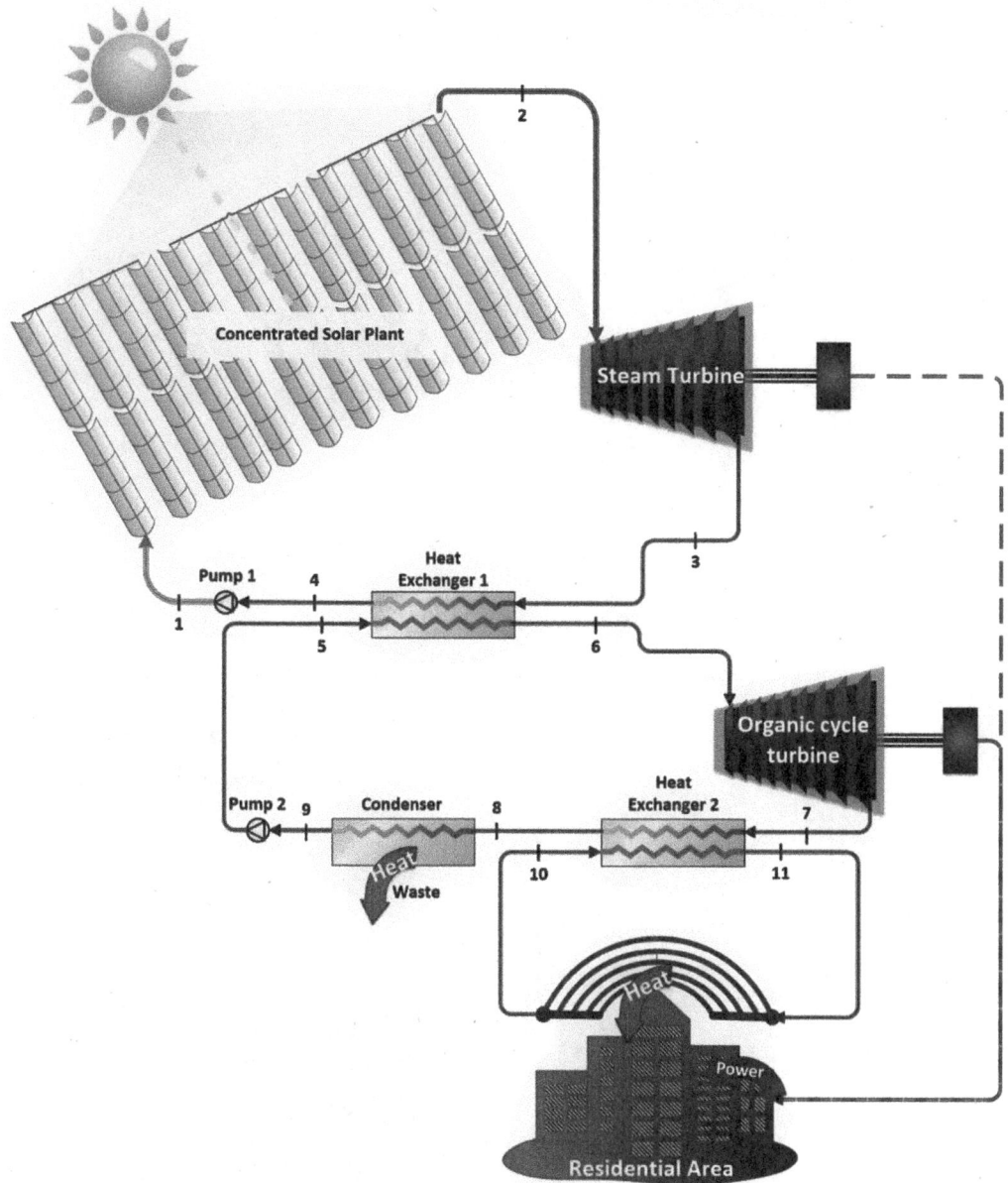

Figure 13.19 Layout of the cogeneration system configuration for Example 13.2.

b) The working fluid flows into pump as a saturated liquid in order to get pressurized. For pump 1 component, the mass, energy, entropy and exergy balance equations are expressed as follows:

$$\dot{m}_4 = \dot{m}_1$$

$$\dot{m}_4 h_4 + \dot{W}_{P1} = \dot{m}_1 h_1$$

$$\dot{m}_4 s_4 + \dot{S}_{gen,P1} = \dot{m}_1 s_1$$

$$\dot{m}_4 ex_4 + \dot{W}_{P1} = \dot{m}_1 ex_1 + \dot{Ex}_{dest,P1}$$

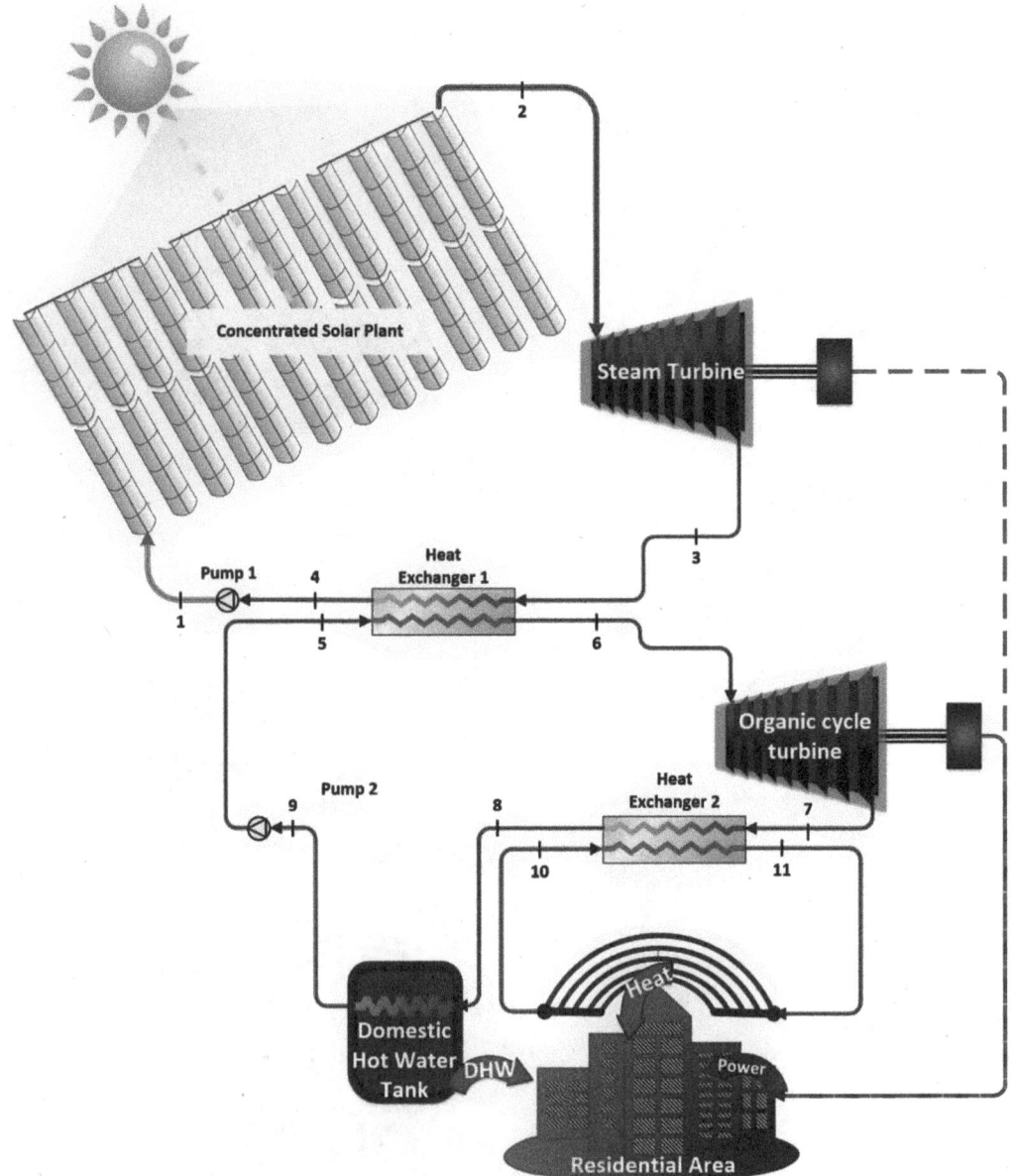

Figure 13.20 Layout of the trigeneration system configuration for Example 13.2.

The pressurized water enters concentrated solar plant to get superheated before entering steam turbine. For the concentrated solar plant component, the mass, energy, entropy and exergy balance equations are written as follows:

$$\dot{m}_1 = \dot{m}_2$$

$$\dot{m}_1 h_1 + \dot{Q}_{CSP} = \dot{m}_2 h_2$$

$$\dot{m}_1 s_1 + \frac{\dot{Q}_{CSP}}{T_{b,CSP}} + \dot{S}_{gen,CSP} = \dot{m}_2 s_2$$

$$\dot{m}_1 ex_1 + + \dot{Ex}_{Q_{CSP}} = \dot{m}_2 ex_2 + \dot{Ex}_{dest,CSP}$$

In the steam turbine, the superheated water expands in order to generate power. For the steam turbine, the mass, energy, entropy and exergy balance equations are written as follows:

$$\dot{m}_2 = \dot{m}_3$$

$$\dot{m}_2 h_2 = \dot{m}_3 h_3 + \dot{W}_{T1}$$

$$\dot{m}_2 s_2 + \dot{S}_{gen,T1} = \dot{m}_3 s_3$$

$$\dot{m}_2 ex_2 = \dot{m}_3 ex_3 + \dot{W}_{T1} + \dot{Ex}_{dest,T1}$$

Heat exchanger 1 recovers the excess heat and injects it into R41 which will be used in organic Rankine cycle. For heat exchanger 1, the mass, energy, entropy and exergy balance equations are expressed as follows:

$$\dot{m}_3 = \dot{m}_4 \;, \dot{m}_5 = \dot{m}_6$$

$$\dot{m}_3 h_3 + \dot{m}_5 h_5 = \dot{m}_4 h_4 + \dot{m}_6 h_6$$

$$\dot{m}_3 s_3 + \dot{m}_5 s_5 + \dot{S}_{gen,HX1} = \dot{m}_4 s_4 + \dot{m}_6 s_6$$

$$\dot{m}_3 ex_3 + \dot{m}_5 ex_5 = \dot{m}_4 ex_4 + \dot{m}_6 ex_6 + \dot{Ex}_{dest,HX1}$$

In the organic cycle turbine, the superheated R41 expands in order to generate power. For the organic cycle turbine, the mass, energy, entropy and exergy balance equations are written as follows:

$$\dot{m}_6 = \dot{m}_7$$

$$\dot{m}_6 h_6 = \dot{m}_7 h_7 + \dot{W}_{T2}$$

$$\dot{m}_6 s_6 + \dot{S}_{gen,T2} = \dot{m}_7 s_7$$

$$\dot{m}_6 ex_6 = \dot{m}_7 ex_7 + \dot{W}_{T2} + \dot{Ex}_{dest,T2}$$

Heat exchanger 2 recovers the excess heat and injects it into the water which will be used for residential heating. For heat exchanger 2, the mass, mass, energy, entropy and exergy balance equations are written as follows:

$$\dot{m}_{10} = \dot{m}_{11} \;, \dot{m}_7 = \dot{m}_8$$

$$\dot{m}_{10} h_{10} + \dot{m}_7 h_7 = \dot{m}_{11} h_{11} + \dot{m}_8 h_8$$

$$\dot{m}_{10}s_{10} + \dot{m}_7s_7 + \dot{S}_{gen,HX2} = \dot{m}_{11}s_{11} + \dot{m}_8s_8$$

$$\dot{m}_{10}ex_{10} + \dot{m}_7ex_7 = \dot{m}_{11}ex_{11} + \dot{m}_8ex_8 + \dot{Ex}_{dest,HX2}$$

c) According to the given parameters, the enthalpy and entropy values are obtained by using the EES:

$$w_{P1} = v_4(P_1 - P_4) = 0.001004\frac{m^3}{kg}(7700kPa - 1000kPa) = 6.73\frac{kJ}{kg}$$

$$h_1 = h_4 + w_{P1}$$

$$h_1 = 762.5 + 6.7 = 769.2\frac{kJ}{kg}$$

$$\left.\begin{array}{l} h_1 = 769.2\dfrac{kJ}{kg} \\ P_1 = 7700kPa \end{array}\right| \begin{array}{l} T_1 = 180.6°C \\ s_2 = 2.13\dfrac{kJ}{kg\ K} \end{array}$$

$$P_2 = P_1 = 7700kPa$$

$$\left.\begin{array}{l} T_2 = 383.1°C \\ P_2 = 7700kPa \end{array}\right| \begin{array}{l} h_2 = 3098\dfrac{kJ}{kg} \\ s_2 = 6.56\dfrac{kJ}{kg\ K} \end{array}$$

At state point 3, the water is indicated as saturated vapor; therefore, the quality is taken as $X_3 = 1$,

$$\left.\begin{array}{l} T_3 = 180°C \\ x_4 = 1 \end{array}\right| \begin{array}{l} h_3 = 2777\dfrac{kJ}{kg} \\ s_3 = 6.56\dfrac{kJ}{kg\ K} \end{array}$$

At state point 4, the water is indicated as saturated liquid; therefore, the quality is taken as $X_4 = 0$,

$$\left.\begin{array}{l} T_4 = 180°C \\ x_4 = 0 \end{array}\right| \begin{array}{l} h_4 = 762.5\dfrac{kJ}{kg} \\ s_4 = 2.13\dfrac{kJ}{kg\ K} \end{array}$$

At the second stage, the R41 working fluid is used for organic Rankine cycle,

$$\left.\begin{array}{l} T_5 = 37°C \\ P_5 = 8000kPa \end{array}\right| \begin{array}{l} h_5 = 297.8\dfrac{kJ}{kg} \\ s_5 = 1.302\dfrac{kJ}{kg\ K} \end{array}$$

$$\left.\begin{array}{l} T_6 = 152°C \\ P_6 = 8000\ kPa \end{array}\right| \begin{array}{l} h_6 = 705.7\dfrac{kJ}{kg} \\ s_6 = 2.465\dfrac{kJ}{kg\ K} \end{array}$$

$$
\left.\begin{array}{l} T_7 = 102.5°C \\ P_7 = 4292 kPa \end{array}\right\} \begin{array}{l} h_7 = 663.7 \dfrac{kJ}{kg} \\ s_7 = 2.484 \dfrac{kJ}{kg\ K} \end{array}
$$

$$
\left.\begin{array}{l} T_8 = 70°C \\ P_8 = 4292 kPa \end{array}\right\} \begin{array}{l} h_8 = 606.9 \dfrac{kJ}{kg} \\ s_8 = 2.326 \dfrac{kJ}{kg\ K} \end{array}
$$

$$
\left.\begin{array}{l} T_9 = 30°C \\ P_9 = 4292 kPa \end{array}\right\} \begin{array}{l} h_9 = 289.7 \dfrac{kJ}{kg} \\ s_9 = 1.298 \dfrac{kJ}{kg\ K} \end{array}
$$

$$
\left.\begin{array}{l} T_{10} = 45°C \\ P_{10} = 101.3 kPa \end{array}\right\} \begin{array}{l} h_{10} = 188.5 \dfrac{kJ}{kg} \\ s_{10} = 0.6386 \dfrac{kJ}{kg\ K} \end{array}
$$

$$
\left.\begin{array}{l} T_{11} = 65°C \\ P_{11} = 101.3 kPa \end{array}\right\} \begin{array}{l} h_{11} = 272.2 \dfrac{kJ}{kg} \\ s_{11} = 0.8936 \dfrac{kJ}{kg\ K} \end{array}
$$

In order to find specific exergy values, the following equation can be used:

$$
ex_n = h_n - h_0 - T_0(s_n - s_0)
$$

If the reference point is considered as 25°C and 101.3 kPa, the enthalpy and entropy values of water are obtained as:

$$
\left.\begin{array}{l} T_0 = 25°C \\ P_0 = 101.3 kPa \end{array}\right\} \begin{array}{l} h_0 = 104.9 \dfrac{kJ}{kg} \\ s_0 = 0.3672 \dfrac{kJ}{kg\ K} \end{array}
$$

For R41, the reference enthalpy and entropy values are taken as:

$$
\left.\begin{array}{l} T_0 = 25°C \\ P_0 = 101.3 kPa \end{array}\right\} \begin{array}{l} h_0 = 626.2 \dfrac{kJ}{kg} \\ s_0 = 3.231 \dfrac{kJ}{kg\ K} \end{array}
$$

In order to find the mass flow rate of R41, the following calculation is made:

$$
\dot{m}_3(h_3 - h_4) = \dot{m}_5(h_6 - h_5)
$$

$$
\dot{m}_5 = 79 kg/s
$$

In order to find the mass flow rate of residential heating water, the balance equations for heat exchanger 2 are used as follows:

$$\dot{m}_7(h_7 - h_8) = \dot{m}_{10}(h_{11} - h_{10})$$

$$\dot{m}_{10} = 53.6\, kg\,/\,s$$

Therefore, a thermophysical properties table, as shown in Table 13.2, is prepared using the given, calculated, and taken data accordingly.

d) To find the power generation, it is necessary to use the following balance equations:

$$\dot{m}_2 h_2 = \dot{m}_1 h_1 + \dot{W}_{T1}$$

$$\dot{m}_2 = \dot{m}_1$$

The work output in rate form for the steam turbine can be obtained as follows:

$$\dot{W}_{T1} = \dot{m}_2(h_2 - h_1)$$

$$\dot{W}_{T1} = 5136\frac{kJ}{s} = 5136\,kW$$

To find the work output from the organic Rankine cycle, the calculation is made as follows:

$$\dot{m}_6 = \dot{m}_7$$

$$\dot{m}_6 h_6 = \dot{m}_7 h_7 + \dot{W}_{T2}$$

$$\dot{W}_{T2} = \dot{m}_6(h_6 - h_7)$$

$$\dot{W}_{T2} = 3318.8\frac{kJ}{s} = 3318.8\,kW$$

Table 13.2 Thermophysical properties of substances at each state point.

State Point	Substance	$\dot{m}$(kg/s)	T (°C)	P (kPa)	h (kJ/kg)	ex (kJ/kg)
1	Water	16	180.6	7700	769.2	138.7
2	Water	16	383.1	7700	3098	1218.3
3	Water	16	180	1000	2777	825.7
4	Water	16	180	1000	762.5	132
5	Fluoromethane	79	37	8000	297.8	246.7
6	Fluoromethane	79	152	8000	705.7	307.9
7	Fluoromethane	79	102.5	4292	663.7	260.2
8	Fluoromethane	79	70	4292	606.9	250.5
9	Fluoromethane	79	30	4292	289.8	239.9
10	Water	53.6	45	101.3	188.5	2.7
11	Water	53.6	65	101.3	272.2	10.4

The net work rate, which is used for power generation, is found by subtracting pump work from the generated work as follows:

$$\dot{W}_{net,1} = \dot{W}_{T_1} - \dot{W}_{p_1}$$

$$\dot{W}_{net,2} = \dot{W}_{T_2} - \dot{W}_{p_2}$$

$$\dot{W}_{net,total} = \dot{W}_{net,1} + \dot{W}_{net,2}$$

where $\dot{W}_{p1}$ can be obtained by multiplying w_{P_1} with mass flow rate as follows:

$$\dot{W}_{p_1} = \dot{m}_1 w_{p_1}$$

$$\dot{W}_{p_1} = \dot{m}_1 (h_4 - h_1)$$

$$\dot{W}_{p_1} = 107.2 \frac{kJ}{s} = 107.2 \, kW$$

To find $\dot{W}_{p_2}$, w_{P_2} should be multiplied with mass flow rate as follows:

$$\dot{W}_{p_2} = \dot{m}_6 w_{p_2}$$

$$\dot{W}_{p_2} = \dot{m}_6 (h_6 - h_7)$$

$$\dot{W}_{p_2} = 632 \frac{kJ}{s} = 632 \, kW$$

The net work rate is then found as follows:

$$\dot{W}_{net,1} = 5028.8 \, kW$$

$$\dot{W}_{net,2} = 2686.7 \, kW$$

$$\dot{W}_{net,total} = 7715.5 \, kW$$

In order to find the mass flow rate of water which carries heat to residential area, the following calculation is made:

$$\dot{m}_4 (h_4 - h_5) = \dot{m}_9 (h_{10} - h_9)$$

$$\dot{m}_9 = 59.1 \, kg/s$$

To find the residential heating capacity, the following calculation is made:

$$\dot{Q}_{RH} = \dot{m}_{10} \left(h_{11} - h_{10} \right) = 4488.3 \frac{kJ}{s} = 4488.3 \, kW$$

$$\dot{Ex}_{Q_{RH}} = \dot{Q}_{RH} \left(1 - \frac{T_0}{T_{s,RH}} \right) = 588.5 \frac{kJ}{s} = 588.5 \, kW.$$

To find the domestic hot water capacity, the following calculation is made:

$$\dot{Q}_{DHW} = \dot{m}_8(h_8 - h_9) = 25057\frac{kJ}{s} = 25057\,kW$$

$$\dot{Ex}_{Q_{DHW}} = \dot{Q}_{DHW}\left(1 - \frac{T_0}{T_{s,DHW}}\right) = 413.3\frac{kJ}{s} = 413.3\,kW$$

The solar energy input is therefore calculated as follows:

$$\dot{Q}_{CSP} \times 0.65 = \dot{m}_1(h_2 - h_1)$$

$$\dot{Q}_{CSP} = 57324\,kW$$

$$\dot{Ex}_{QCSP} = \dot{Q}_{CSP}\left(1 - \frac{T_0}{T_{sun}}\right) = 54367\,kW$$

In this regard, based on the calculations performed above for the organic Rankine cycle and combined cycle single generation configurations, energy flows are given in Figure 13.21, where total energy input is converted into electricity, the rest of the energy is wasted.

On the other hand, both cogeneration and trigeneration configurations exploit the excess heat between state points 7 and 8 by using water as working fluid at state points 10 and 11 and transfers the heat for residential heating purpose. For further, a domestic hot water tank is used in order to be heated by the remaining heat at state point 8. Figure 13.22 shows cogeneration and trigeneration configurations' energy flows, where waste energy is reduced significantly.

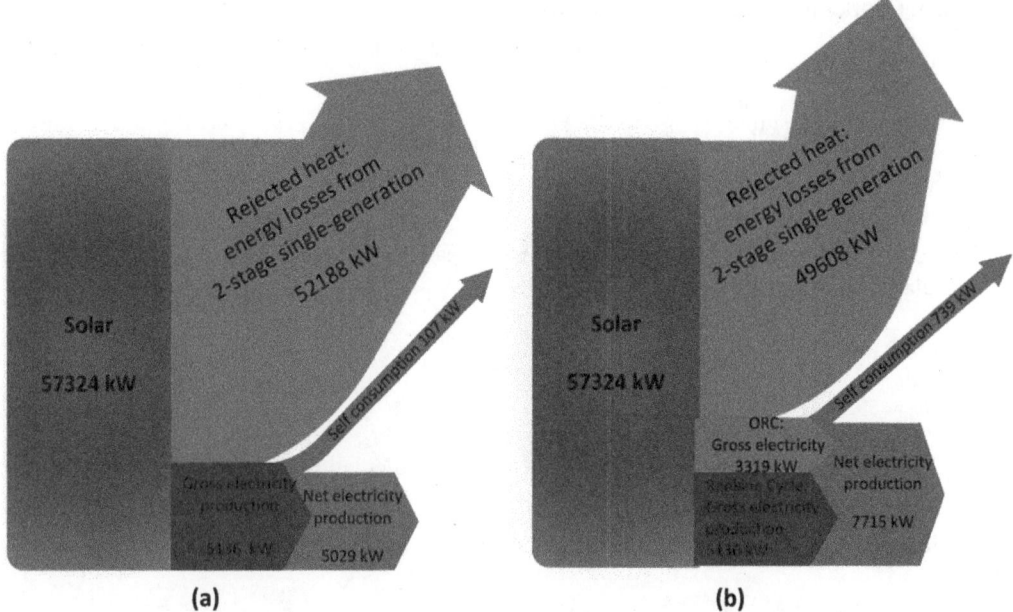

Figure 13.21 Illustration of energy flows for Rankine cycle (a) and combined cycle (b) single generation configurations of Example 13.2.

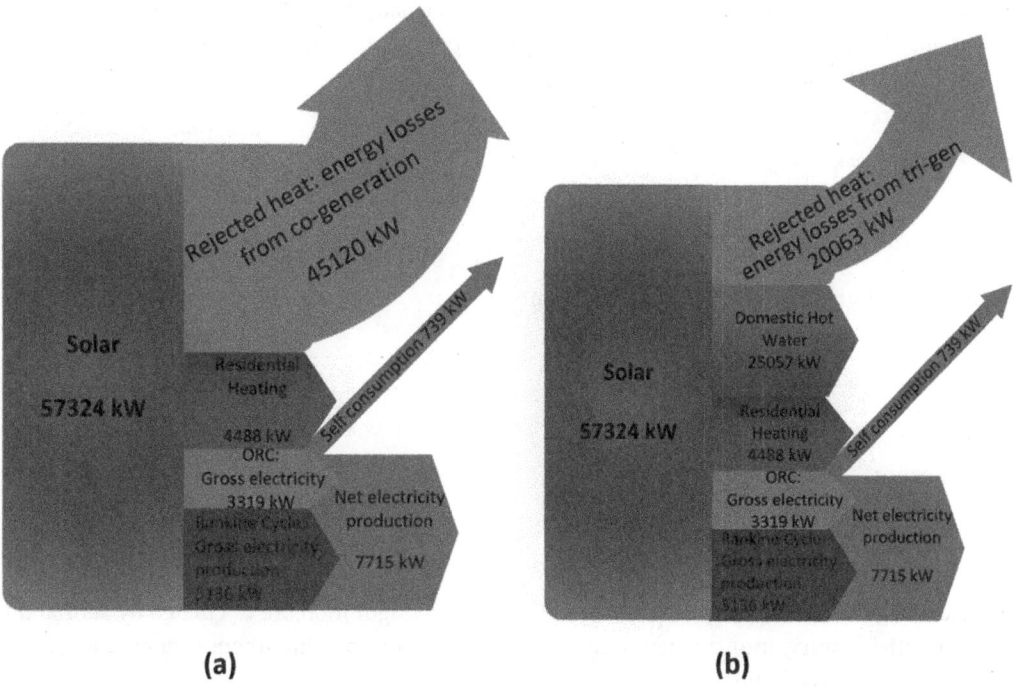

Figure 13.22 Illustration of energy flows for cogeneration (a) and trigeneration (b) configurations of Example 13.2.

e) The energy efficiency of Rankine cycle single generation configuration can be calculated by dividing net electricity generation of the steam Rankine cycle to the total energy input as follows:

$$\eta_{en, single\ generation} = \frac{\dot{W}_{net,1}}{\dot{Q}_{CSP}}$$

which results in

$$\eta_{en, single\ generation} = \textbf{8.77\%}$$

The exergy efficiency of Rankine cycle single generation configuration can be calculated by dividing net electricity generation of the steam Rankine cycle into total exergy input as follows:

$$\eta_{ex, single\ generation} = \frac{\dot{W}_{net,1}}{\dot{Ex}_{QCSP}}$$

which results in

$$\eta_{ex, single\ generation} = \textbf{9.25\%}$$

The energy and exergy efficiencies of combined cycle single generation configuration can be calculated by dividing net electricity generation of both steam Rankine cycle and organic Rankine cycle into total energy input as follows:

$$\eta_{en,\,single\,generation,2} = \frac{\dot{W}_{net,total}}{\dot{Q}_{CSP}}$$

$$\eta_{en,\,single\,generation,2} = \mathbf{13.46\%}$$

In addition, the exergy efficiency then results in

$$\eta_{ex,\,single\,generation,2} = \frac{\dot{W}_{net,total}}{\dot{Ex}_{Q_{CSP}}}$$

and hence

$$\eta_{ex,\,single\,generation,2} = \mathbf{14.19\%}$$

The energy and exergy efficiencies of cogeneration configuration are calculated by adding the residential heating into the net electricity generation and dividing them to total energy and exergy inputs as follows:

$$\eta_{en,\,cogeneration} = \frac{\dot{W}_{net,total} + \dot{Q}_{RH}}{\dot{Q}_{CSP}}$$

resulting in

$$\eta_{en,\,cogeneration} = \mathbf{21.29\%}$$

and

$$\eta_{ex,\,cogeneration} = \frac{\dot{W}_{net,total} + \dot{Ex}_{Q_{RH}}}{\dot{Ex}_{Q_{CSP}}}$$

resulting in

$$\eta_{ex,\,cogeneration} = \mathbf{15.27\%}$$

The energy and exergy efficiencies of trigeneration configuration are calculated by adding the residential heating and domestic hot water into the net electricity generation and dividing them to total energy and exergy inputs as follows:

$$\eta_{en,\,trigeneration} = \frac{\dot{W}_{net,total} + \dot{Q}_{RH} + \dot{Q}_{DHW}}{\dot{Q}_{CSP}}$$

resulting in

$$\eta_{en,\,trigeneration} = \mathbf{65\%}$$

$$\eta_{ex,trigeneration} = \frac{\dot{W}_{net,total} + \dot{Ex}_{Q_{RH}} + \dot{Ex}_{Q_{DHW}}}{\dot{Ex}_{Q_{CSP}}}$$

resulting in

$$\eta_{ex,trigeneration} = \textbf{16.03}\%$$

By integrating three different subsystems, namely, the organic Rankine cycle, residential heating unit, domestic hot water unit, the utilization of solar energy is increased from 8.77% to 65%, and the utilization of exergy value of solar input is increased from 9.25% to 16.03%.

Case Study 1

A case study is adapted from [2] proposed by Temiz and Dincer. Solar and nuclear sources are employed with small modular reactor and bifacial PV plant in order to generate hydrogen, power, and freshwater in a sustainable and integrated fashion (see Figure 13.23).

The proposed system can be divided into five parts; two of these parts are related with source utilization and three of these parts can be related with outputs. A bifacial PV plant and a high-temperature gas cooled reactor are used in order to utilize solar and nuclear sources. Bifacial PV plants generate electricity for the system and nuclear source generate process heat for solid oxide electrolysis, Rankine cycle, and multi-effect distillation desalination processes.

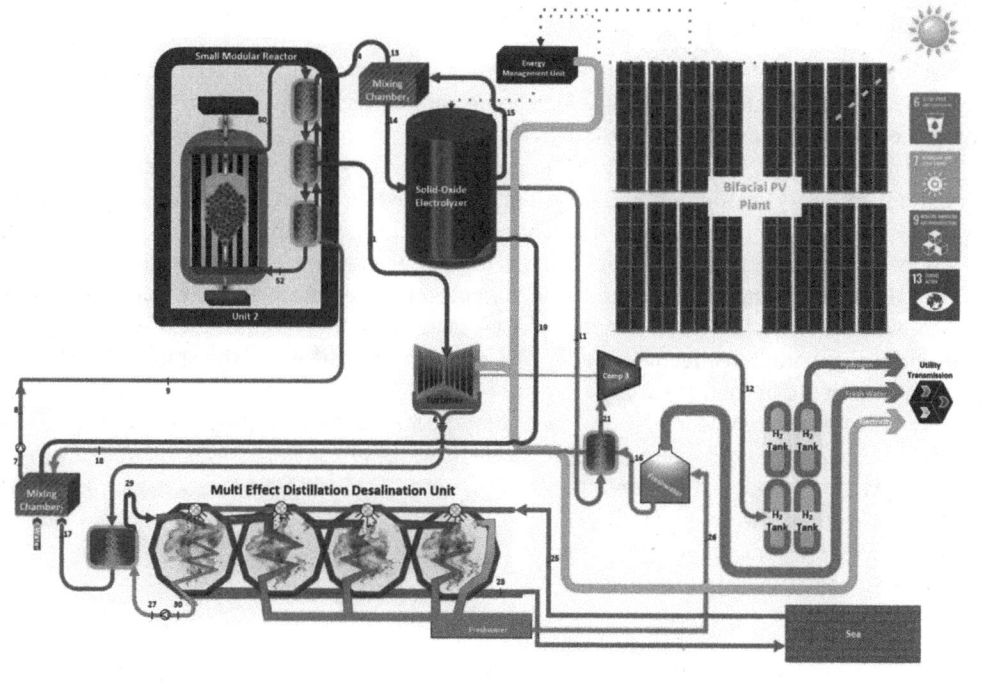

Figure 13.23 Illustration of system layout for Case Study 13.1.

(Continued)

Case Study 1   (Continued)

Nuclear source is employed with two small modular reactor units each at 10 MW thermal energy generation capacity, which can generate 20 MW thermal energy in total. Each reactor can employ 27,000 spherical fuel elements with 60 mm of diameter. High temperature gas cooled pebble bed reactor uses helium gas as coolant. Helium gas reaches 950°C at 30 bar and rejects its heat to the water via heat exchangers. Two water circuits leave from the heat exchangers at different temperatures, in order to feed solid-oxide electrolyzer and steam turbine. The excess heat then employed by the multi-effect desalination unit.

Bifacial PV plant is integrated mainly for its lower levelized cost of electricity compared to the small modular reactor. When we look at the solid-oxide electrolysis process, it employs both heat and electricity in order to generate hydrogen. But to produce electricity, nuclear system should use the high-grade heat in Rankine cycle, which will reduce the high grade heat flow from nuclear source into solid-oxide electrolyzer. To maximize the hydrogen production capacity, electricity from bifacial PV plant can be used in solid-oxide electrolyzer, which enables to exploit more heat for solid-oxide electrolysis process rather than Rankine cycle.

The proposed system is analyzed in a time-dependent manner for six different locations:

- Polaris, Nunavut, Canada: 75°N 96.8°W
- Kangirsuk, Quebec, Canada: 60°N 70°W
- Richardson, New Brunswick, Canada: 45°N 64.8°W
- Ponte Vedra Beach, Florida, the United States of America: 30°N 81.4°W
- Aztlan, Chiapas, Mexico: 15°N 92.7°W
- Coaque, Ecuador: 0.1°N 80.1°W

Each 15th latitude from Northern Hemisphere is selected to run and analyze the proposed system in order to increase generality. Also, some parametric studies are carried out to see the effects of various parameters on system performance.

In system analysis, it is important to write the mass, energy, entropy, and exergy balance equations for every component of the system. These have to be correctly written by considering all inputs and outputs with the boundaries placed for the components. The balance equations for system units are then listed in Table 13.3 for energy, entropy, and exergy accordingly.

It is also equally important to prepare a statepoints table where all state points are listed along with the mass flow rate, temperature, pressure, specific enthalpy, specific entropy, and specific exergy values, respectively. Table 13.4 lists such properties of the state points for the system considered in the case study.

In order to assess single, co-, and trigeneration performances, the proposed system will be simplified by using single energy input as nuclear process heat to generate hydrogen, power, and freshwater in steady-state conditions. For the single, co-, and trigeneration configurations, energy flows are shown in Figure 13.24.

The energy efficiency of single generation system can be calculated by dividing hydrogen energy into heat input as follows:

$$\eta_{en,single\ generation} = \frac{\dot{m}_{H_2} LHV_{H_2}}{\dot{Q}_{SMR}}$$

$$\eta_{en,single\ generation} = \frac{3624kW}{19936kW} = \textbf{18.18\%}$$

The energy efficiency of cogeneration system can be calculated by adding power generation with hydrogen energy, and dividing them into heat input as follows:

Table 13.3 Energy, entropy, and exergy balance equations of the proposed system for Case Study 13.1.

Component	Energy Balance Equations	Entropy Balance Equations	Exergy Balance Equations
Each pump	$\dot{m}_1 h_1 + \dot{W}_p = \dot{m}_o h_o$	$\dot{m}_1 s_1 + \dot{S}_{gen,p} = \dot{m}_o s_o$	$\dot{m}_1 ex_1 + \dot{W}_p = \dot{m}_o ex_o + \dot{Ex}_{d,p}$
Steam turbine, N	$\dot{m}_1 h_1 = \dot{W}_{ST,N} + \dot{m}_6 h_6$	$\dot{m}_1 s_1 + \dot{S}_{gen,ST,N} = \dot{m}_6 s_6$	$\dot{m}_1 ex_1 = \dot{W}_{ST,N} + \dot{m}_6 ex_6 + \dot{E}_{d,ST,N}$
Heat exchanger 2	$\dot{m}_{16} h_{16} + \dot{m}_{11} h_{11} = \dot{m}_{21} h_{21} + \dot{m}_{18} h_{18}$	$\dot{m}_{16} s_{16} + \dot{m}_{11} s_{11} + \dot{S}_{gen,HEx2} = \dot{m}_{21} s_{21} + \dot{m}_{18} s_{18}$	$\dot{m}_{16} ex_{16} + \dot{m}_{11} ex_{11} = \dot{m}_{21} ex_{21} + \dot{m}_{18} ex_{18} + \dot{E}_{d,HEx2}$
Mixing Chamber 1	$\dot{m}_{19} h_{19} + \dot{m}_{17} h_{17} + \dot{m}_{18} h_{18} = \dot{m}_7 h_7$	$\dot{m}_{19} s_{19} + \dot{m}_{18} s_{18} + \dot{m}_{17} s_{17} + \dot{S}_{gen,MC2} = \dot{m}_7 s_7$	$\dot{m}_{19} ex_{19} + \dot{m}_{18} ex_{18} + \dot{m}_{17} ex_{17} = \dot{m}_7 ex_7 + \dot{E}_{d,MC2}$
Small modular reactor	$\dot{m}_{52} h_{52} + \dot{Q}_{N2} = +\dot{m}_{50} h_{50}$	$\dot{m}_{52} s_{52} + \dot{Q}_{N2}/T_{bN2} + \dot{S}_{gen,N2} = \dot{m}_{50} s_{50}$	$\dot{m}_{52} ex_{52} + \dot{Ex}_{Q_{N2}} = \dot{m}_{50} ex_{50} + \dot{X}_{d,N2}$
Nuclear steam generator 1	$\dot{m}_{50} h_{50} + \dot{m}_9 h_9 = \dot{m}_{52} h_{52} + \dot{m}_4 h_4$	$\dot{m}_{50} s_{50} + \dot{m}_9 s_9 + \dot{S}_{gen,NSG1} = \dot{m}_{52} s_{52} + \dot{m}_4 s_4$	$\dot{m}_{50} ex_{50} + \dot{m}_9 ex_9 = \dot{m}_{52} ex_{52} + \dot{m}_4 ex_4 + \dot{E}_{d,NSG1}$
Mixing Chamber 2	$\dot{m}_{13} h_{13} + \dot{m}_{15} h_{15} = \dot{m}_{14} h_{14}$	$\dot{m}_{13} s_{13} + \dot{m}_{15} s_{15} + \dot{S}_{gen,MC1} = \dot{m}_{14} s_{14}$	$\dot{m}_{13} ex_{13} + \dot{m}_{15} ex_{15} = \dot{m}_{14} ex_{14} + \dot{E}_{d,MC1}$
Each compressor	$\dot{m}_1 h_1 + \dot{W}_c = \dot{m}_o h_o$	$\dot{m}_1 s_1 + \dot{S}_{gen,c} = \dot{m}_o s_o$	$\dot{m}_1 ex_1 + \dot{W}_c = \dot{m}_o ex_o + \dot{Ex}_{d,c}$
MED desalination	$\dot{m}_{25} h_{25} + \dot{m}_{29} h_{29} = \dot{m}_{30} h_{30} + \dot{m}_{23} h_{23} + \dot{m}_{26} h_{26}$	$\dot{m}_{25} s_{25} + \dot{m}_{29} s_{29} + \dot{S}_{gen,MED} = \dot{m}_{30} s_{30} + \dot{m}_{23} s_{23} + \dot{m}_{26} s_{26}$	$\dot{m}_{25} ex_{25} + \dot{m}_{29} ex_{29} = \dot{m}_{30} ex_{30} + \dot{m}_{23} ex_{23} + \dot{m}_{26} ex_{26} + \dot{Ex}_{d,MED}$

Case Study 1 (Continued)

Table 13.4 Thermohysical properties of state points for steady-state conditions for 45°N case at Richardson, New Brunswick, Canada, location without solar enhancement.

State Point	Substance	Mass Flow Rate (kg/s)	Temperature (°C)	Pressure (bar)	h (kJ/kg)	s (kJ/kg K)	ex (kJ/kg)
0	Water	–	297	1.0	98.65	0.346	–
0	Hydrogen	–	297	1.0	3910	53.310	–
0	Helium	–	297	1.0	1546	27.960	–
1	Water	2.8	550.0	18.0	3541.0	7.031	1677.5
2	Water	2.8	550.0	18.0	3541.0	7.031	1677.5
3	Water	5.5	550.0	18.0	3541.0	7.031	1677.5
4	Water	0.2	700.0	18.0	3919.0	8.000	1799.7
5	Water	0.2	700.0	18.0	3919.0	8.000	1799.7
6	Water	5.5	76.0	0.4	2686.0	7.389	728.0
7	Water	5.5	95.0	1.0	398.1	1.250	60.8
8	Water	5.5	95.0	18.0	399.4	1.249	62.4
9	Water	0.3	95.0	18.0	399.4	1.249	62.4
10	Water	5.5	95.0	18.0	399.4	1.249	62.4
11	Hydrogen	0.036	700.0	17.6	13,807.0	58.830	8439.7
12	Hydrogen	0.036	75.0	360.0	4879.0	31.240	6795.5
13	Water	0.3	700.0	18.0	3919.0	8.000	1799.7
14	Water	0.4	700.0	18.0	3919.0	8.000	1799.7
15	Water	0.0	700.0	17.6	3920.0	8.011	1797.8
16	Water	0.3	26.0	1.0	109.0	0.381	1.2
17	Water	5.5	76.0	0.4	319.0	1.030	39.8
18	Water	0.3	85.0	1.0	356.0	1.135	49.1
19	Water	0.034	700.0	17.6	3920.0	8.011	1797.8
21	Hydrogen	0.036	75.0	17.6	4659.0	43.820	3254.4
23	Brine	10.0	34.0	1.0	142.5	0.492	5.5
25	Brine	11.1	26.0	1.0	109.0	0.381	1.2
26	Freshwater	1.1	56.1	1.0	235.2	0.783	6.4
27	Water	5.3	75.9	0.8	319.0	1.030	39.8
29	Water	5.3	75.9	0.4	2636.0	7.669	604.1
30	Water	5.3	75.9	0.4	318.1	1.028	39.4
50	Helium	3.2	900.0	30.0	4552.0	0.077	10,367.1
51	Helium	3.2	900.0	30.0	4552.0	0.077	10,367.1
52	Helium	3.2	300.0	30.0	1437.0	−3.642	8233.9

Case Study 1   (Continued)

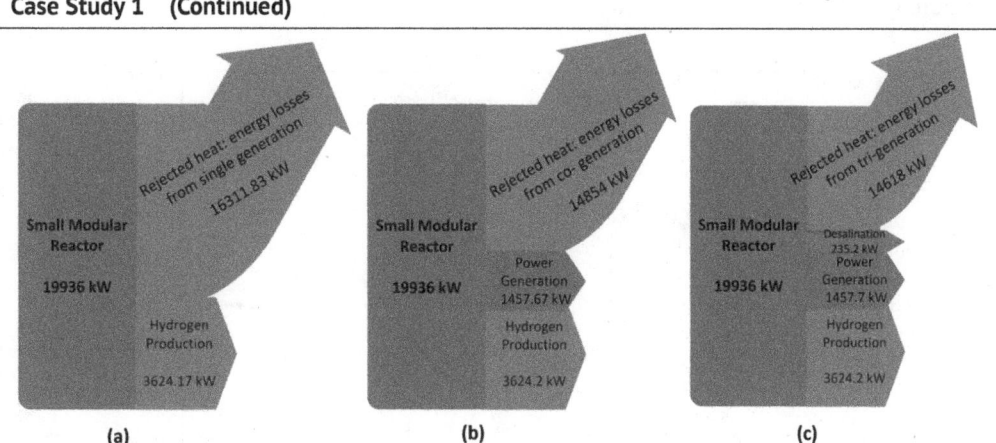

Figure 13.24   Illustration of energy flows for single generation (a), cogeneration (b), and trigeneration (b) configurations of Case Study 13.1.

$$\eta_{en,cogeneration} = \frac{\dot{W}_{net} + \dot{m}_{H_2} LHV_{H_2}}{\dot{Q}_{SMR}}$$

$$\eta_{en,cogeneration} = \frac{5082 kW}{19,936 kW} = \textbf{25.49\%}$$

The energy efficiency of trigeneration system can be calculated by adding freshwater output with power generation and hydrogen energy, and dividing them into heat input as follows:

$$\eta_{en,trigeneration} = \frac{\dot{W}_{net} + \dot{m}_{FW} h_{FW} + \dot{m}_{H_2} LHV_{H_2}}{\dot{Q}_{SMR}}$$

$$\eta_{en,trigeneration} = \frac{5317 kW}{19,936 kW} = \textbf{26.67\%}$$

While considering the time-dependent analysis with dynamic parameters, a proposed system along with a bifacial PV plant can have 22.7% energy and 19.8% exergy efficiencies. The overall system efficiencies are lower with above-mentioned steady-state analyses, because of the fact that bifacial PV plant generates work with lower efficiencies and to use this work for hydrogen generation further decreases the overall efficiencies. However, bifacial PV plants can generate lower levelized cost of electricity, which will improve the cost-effectiveness. Figure 13.25 shows overall energy and exergy efficiencies for Richardson, NB, Canada case as monthly averages during a typical meteorological year.

(Continued)

Case Study 1   (Continued)

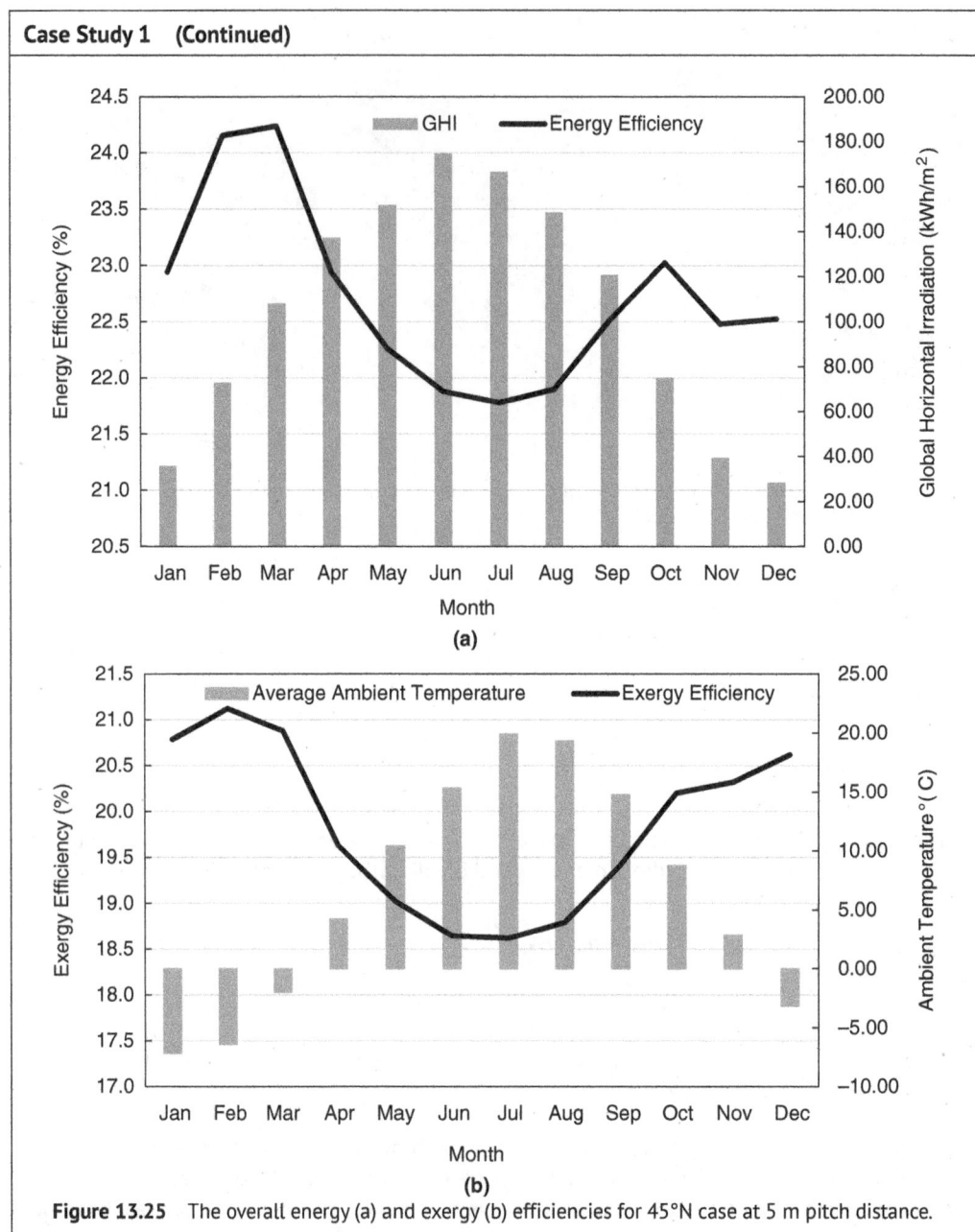

Figure 13.25   The overall energy (a) and exergy (b) efficiencies for 45°N case at 5 m pitch distance.

Case Study 2

A cogeneration system is studied by Dincer et al. [3]. Their proposed system generates DC electric current by using a solid-oxide fuel cell stack, shaft rotation power by using a gas turbine, and low temperature heat by recovering the heat from steam condenser.

Their proposed system can be analyzed in three main subsystems, from bottom to top, the first one is the compression section, the second one is the heat recovery section, and the third one is the power generation section. A Brayton cycle kind system is integrated with the solid-oxide fuel cell, which can be used as active preheater while consuming a part of the primary fuel.

Case Study 2   (Continued)

The reference environmental conditions for T_0, P_0 are taken as standard. The inlet and outlet temperatures for the solid-oxide fuel cell are considered as 1073 K and 1273 K, respectively. The outlet temperature of the gas turbine is taken as 1123 K. Further assumptions are made as isentropic efficiencies of turbine and compressors are taken as 93% and 85%, respectively. Their proposed system and simplified operational modules are shown in Figure 13.26.

The chemical reaction in the solid-oxide fuel cell stacks can be expressed as follows:

$$CH_4 + 2O_2 \rightarrow CO_2 + 2H_2O(g) - 802 \text{ kJ/mo}$$

$$CH_4 + H_2O \leftrightarrow CO + 3H_2 + 206 \text{ kJ/mol}$$

$$CO + H_2O \leftrightarrow CO_2 + H_2 + 41 \text{ kJ/mol}$$

In order to analyze the energy and exergy efficiencies of the overall system, following equations can be used:

$$\eta_{en} = \frac{\dot{W}_{FC} + \dot{W}_T + \dot{Q}_C}{\dot{m}_{CH_4} LHV_{CH_4}}$$

$$\eta_{ex} = \frac{\dot{W}_{FC} + \dot{W}_T + \dot{Q}_C (1 - T_0 / T_C)}{\dot{m}_{CH_4} \dot{Ex}_{CH_4}^{ch}}$$

When we look at the exergy destructions and overall outputs (see Figure 13.27), the single generation for one-stage gas turbine energy and exergy efficiencies can be calculated by dividing the work output from gas turbine into the energy and exergy value of primary fuel input as follows:

$$\eta_{en} = \frac{\dot{W}_T}{\dot{m}_{CH_4} LHV_{CH_4}}$$

$$\eta_{en} = \frac{345}{802} = \mathbf{43.02\%}$$

$$\eta_{ex} = \frac{\dot{W}_T}{\dot{m}_{CH_4} \dot{Ex}_{CH_4}^{ch}}$$

$$\eta_{ex} = \frac{345}{818} = \mathbf{42.2\%}$$

When the second stage work generation is considered the single generation for combined cycle gas turbine and solid-oxide fuel cell energy and exergy efficiencies can be calculated by dividing total work output into the energy and exergy value of primary fuel input as follows:

$$\eta_{en} = \frac{\dot{W}_{FC} + \dot{W}_T}{\dot{m}_{CH_4} LHV_{CH_4}}$$

$$\eta_{en} = \frac{275 + 345}{802} = \mathbf{77.3\%}$$

$$\eta_{ex} = \frac{\dot{W}_{FC} + \dot{W}_T}{\dot{m}_{CH_4} \dot{Ex}_{CH_4}^{ch}}$$

(Continued)

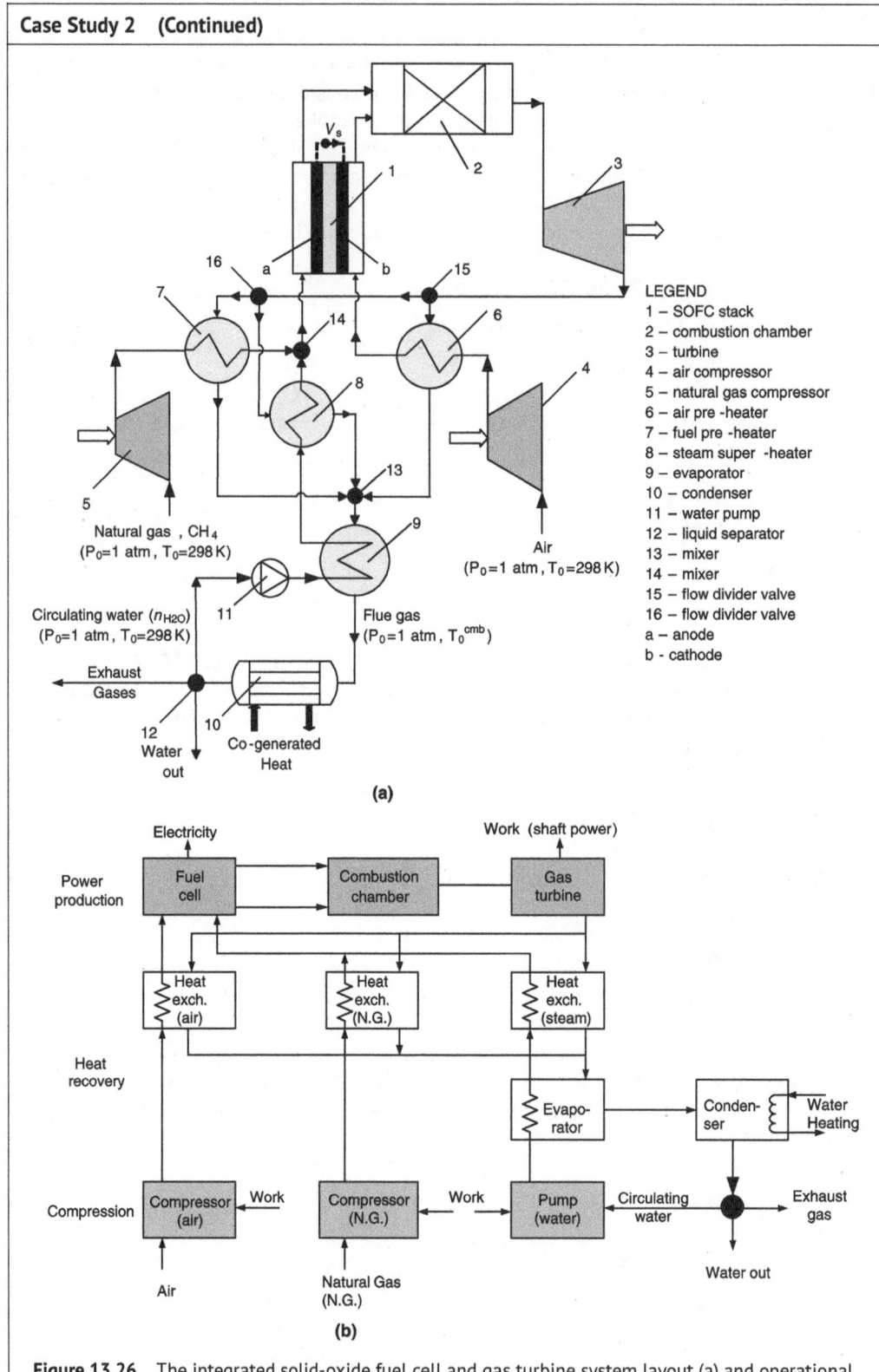

Figure 13.26 The integrated solid-oxide fuel cell and gas turbine system layout (a) and operational modules (b) for cogeneration (modified from [3]).

Case Study 2   (Continued)

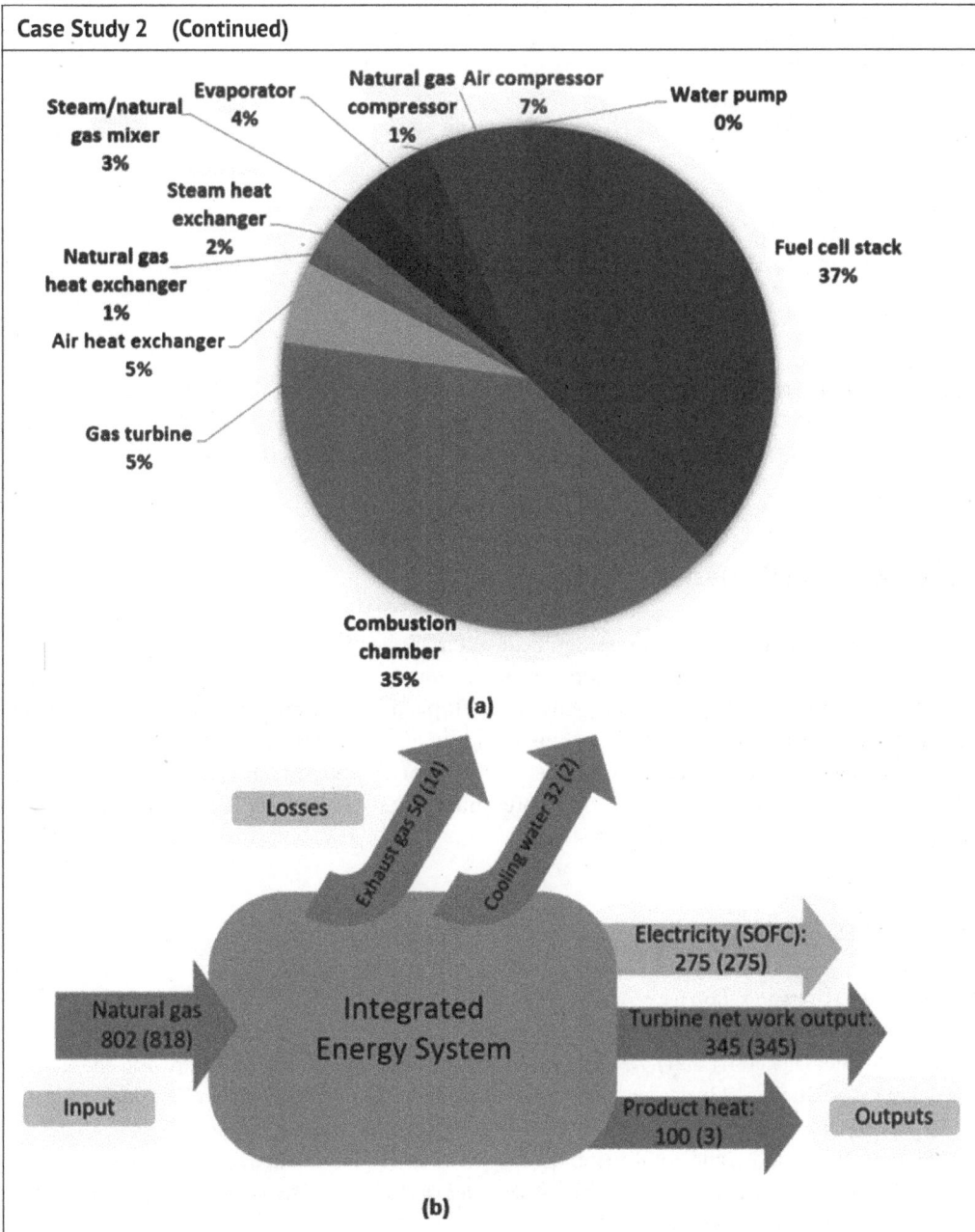

Figure 13.27 The destruction of exergy in various components (a) and the overall outputs (b) of the SOFC/GT system (data from [3]).

$$\eta_{ex} = \frac{275+345}{818} = 75.8\%$$

For the cogeneration system, in addition to solid-oxide fuel cell and gas turbine outputs, heat output from the condenser will be also considered and divided into the energy and exergy values of primary fuel input as follows:

Case Study 2 (Continued)

$$\eta_{en} = \frac{\dot{W}_{FC} + \dot{W}_T + \dot{Q}_C}{\dot{m}_{CH_4} LHV_{CH_4}}$$

$$\eta_{ex} = \frac{720}{802} = 90\%$$

$$\eta_{en} = \frac{\dot{W}_{FC} + \dot{W}_T + \dot{Q}_C}{\dot{m}_{CH_4} LHV_{CH_4}}$$

$$\eta_{ex} = \frac{623}{818} = 76\%$$

These examples and case studies demonstrate that benefit of integrated energy systems for multigeneration can achieve better efficiencies as well as better sustainability. For further details and information about integrated energy systems, source [4] has extensive information and illustrative examples.

13.4 Closing Remarks

Integrated energy systems are of great significance for applications, where we really want to bring more efficient, more cost-effective, more resilient, more environmentally-benign and more sustainable solutions, with a prime aim to generate multiple useful outputs as needed for the targeted applications. This chapter provides a unique perspective on integrated energy systems for multi-generational applications and discuss the benefits obtained through such systems. There are multiple illustrative examples and case studies presented to clearly indicate the advantages of using multigenerational integrated energy systems.

References

1 OECD (2007). Organisation for economic co-operation and development/international energy agency. World Energy Outlook 2007 – Analysis. ISSN: 20725302 (online) doi: 10.1787/20725302.
2 Temiz, M. and Dincer, I. (2021). Design and analysis of a new renewable-nuclear hybrid energy system for production of hydrogen, fresh water and power. *e-Prime-Advances in Electrical Engineering, Electronics and Energy* 1: 100021. doi: 10.1016/j.prime.2021.100021.
3 Dincer, I., Rosen, M.A., and Zamfirescu, C. (2009). Exergetic performance analysis of a gas turbine cycle integrated with solid oxide fuel cells. ASME. *Journal of Energy Resources Technology* 131 (3): 032001. doi: 10.1115/1.3185348.
4 Dincer, I. and Bicer, Y. (2019). *Integrated Energy Systems for Multigeneration*. Elsevier. doi: 10.1016/C2015-0-06233-5.

Questions/Problems

Questions

1 Explain the goals to achieve better sustainable development in regards to energy systems.
2 Discuss the potential improvements in currently dominating power generation methods.

3 What is the main objective of system integration?

4 Classify the integrated energy systems and explain the effectiveness in terms of energy and exergy aspects.

5 What is the key role of renewable sources in integrated energy systems?

6 Explain the importance of exergy efficiency while assessing the integrated energy systems.

7 Write mass, energy, entropy and exergy balance equations for the system in Case Study 13.2 with a system boundary and all inputs and outputs indicated.

8 Explain why domestic hot water or residential heating outputs lower exergy values have compared to their energy values, in contrast to work output.

9 Describe cogeneration and trigeneration systems and discuss their advantages and disadvantages.

10 Discuss how organic Rankine cycle can improve the system performance in terms of efficiencies, and discuss their advantages and disadvantages.

11 What is combined heat and power? Give examples.

12 What is combined cooling, heat, and power? Give examples.

13 Discuss the role of energy and exergy efficiencies about multigeneration systems.

14 Identify your own needs in the the community and come up with an integrated energy system.

15 Write an essay about how integrated energy systems help achieve better sustainability.

Problems

1 Consider an integrated energy system as given in the figure below, using coal, natural gas and hydrogen fuels in order to provide the necessary heat to boiler for the working fluid. Follow the steps below and compare the results for three different fuels.

   a Design and discuss the single generation, cogeneration and trigeneration configurations of the integrated system, separately.

   b Write the mass, energy, entropy and exergy balance equations for each component of the integrated system.

   c By using the given parameters and your design parameters, prepare a thermophysical properties table.

   d Calculate the power generation, residential heating capacity, and domestic hot water production capacity.

   e Calculate the single generation (power), cogeneration (power + heat), and trigeneration (power + heat + domestic hot water) efficiencies.

   f Make a comparative evaluation of these fuel options about their potential environmental impacts.

2 Consider the system given in Example 13.1 and replace the biomass input with municipal solid wastes. Take the necessary lower heating and exergy values from the literature, and solve again by following the steps below:

   a Design and discuss the single generation, cogeneration and trigeneration configurations of the integrated system, separately.

   b Write the mass, energy, entropy and exergy balance equations for each component.

   c By using the given parameters and your design parameters, prepare a thermophysical properties table.

   d Calculate the power generation, residential heating load, and freshwater production capacity.

   e Calculate the single generation (power), cogeneration (power + heat), and trigeneration (power + heat + freshwater) efficiencies.

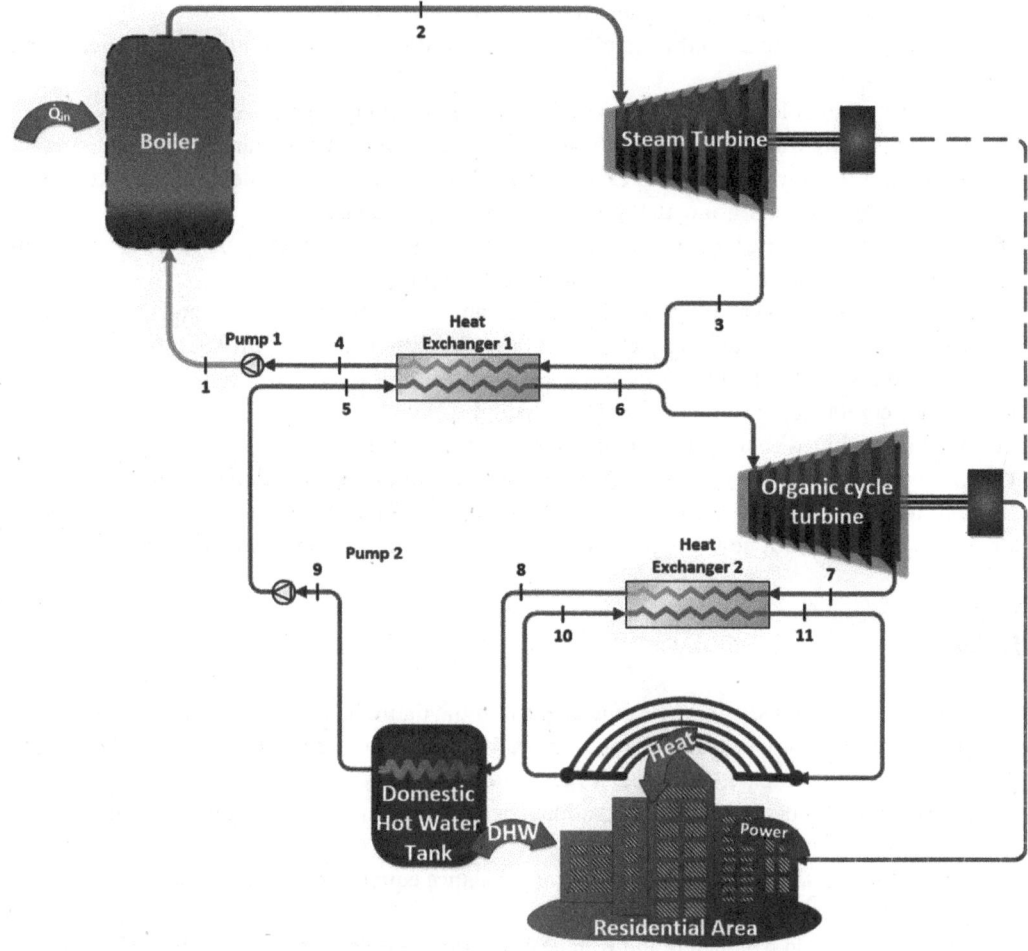

3 Consider the system given in Example 13.2 and the location you live. Redesign the system again based on the capacities required in your location.

   a Design and discuss the single generation, cogeneration and trigeneration configurations of the integrated system, separately.

   b Write the mass, energy, entropy and exergy balance equations for each component.

   c By using the given parameters and your design parameters, prepare a thermophysical properties table.

   d Calculate the power generation, residential heating load, and domestic hot water production capacity.

   e Calculate the single generation (power), cogeneration (power + heat), and trigeneration (power + heat + domestic hot water) efficiencies.

4 Consider the system presented in (Al-Sulaiman, F. A., Dincer, I., & Hamdullahpur, F. (2012). Energy and exergy analyses of a biomass trigeneration system using an organic Rankine cycle. Energy, 45(1), 975–985. https://doi.org/10.1016/j.energy.2012.06.060)

   a Design and discuss the single generation, cogeneration and trigeneration configurations of the integrated system given in the journal paper, separately.

   b Write the mass, energy, entropy and exergy balance equations for each component.

   c By using the given parameters and your design parameters, prepare a thermophysical properties table.

 d Calculate the power generation, residential heating and cooling production capacities.
 e Calculate the single generation (power), cogeneration (power + heat), and trigeneration
   (power + heat + cooling) efficiencies.

5 Consider the given single generation system above. 10 tons of municipal solid waste (MSW) is
  burned directly in a boiler to generate saturated steam at 200°C. The water enters the boiler at
  20°C. The heating value of the MSW is 18 MJ/kg and the chemical exergy is 16 MJ/kg. The
  boiler efficiency is 75%. The steam produced in the boiler is run through a turbine-generator
  group to generate electricity. The steam exits from the turbine at 100 kPa of pressure. The isentropic efficiency of the turbine is 0.85. The generator efficiency is 95%.
  Modify the system and make it a cogeneration system, in order to recover the heat from condenser for residential heating purpose in order to meet requirements in your location.
  a Write the mass, energy, entropy and exergy balance equations for each component.
  b By using the given parameters and your design parameters, prepare a thermophysical prop-
    erties table.
  c Calculate the power generation, residential heating and freshwater production capacities.
  d Calculate the single generation (power), cogeneration (power + heat), and trigeneration
    (power + heat + freshwater) efficiencies.

6 Biomethane produced from municipal waste collected from residential units is used as fuel in
  a micro turbine in a small community. Air enters the compressor at the reference conditions
  which are T0=T1=20°C and P0=P1=100 kPa with mass flow rate 2 kg/s. In the combustion
  chamber, the compressor compression ratio is 1:12. Air leaves from the compressor at 510°C.
  Air enters the turbine at 1100°C and exits at 595°C. The combustion temperature is Tc=1650°C.
  The LHV of biomethane is 40,320 kJ/kg and its chemical exergy is 36,520 kJ/kg.
  Modify the above system and make it a trigeneration system, in order to recover the heat from
  exhaust gases for cooling with absorption cooling system and freshwater with multi-effect desalination system.
  a Design and discuss the single generation, cogeneration and trigeneration configurations of
    the integrated system, separately.
  b Write the mass, energy, entropy and exergy balance equations for each component.

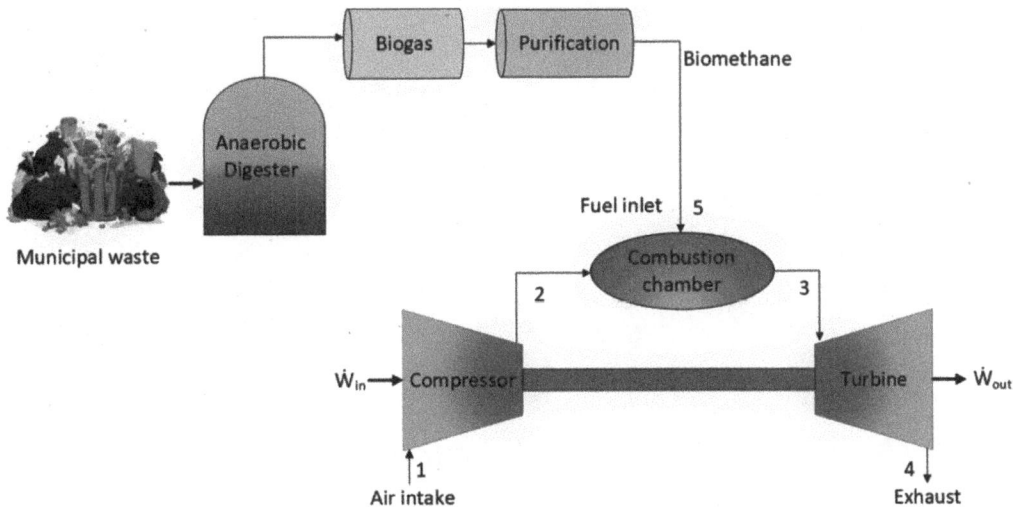

    c By using the given parameters and your design parameters, prepare a thermophysical properties table.

    d Calculate the power generation, residential heating and freshwater production capacities.

    e Calculate the single generation (power), cogeneration (power + cooling), and trigeneration (power + cooling + freshwater) efficiencies.

7 Repeat Example 13.2 with a parametric study by changing the ambient temperature between 10°C and 30°C. Discuss the exergy value of domestic hot water at 35°C ambient temperature.

8 Consider a geothermal source with 270°C hot water. Design a trigeneration system to generate power, residential heat, and freshwater.

    a Design and discuss the single generation, cogeneration and trigeneration configurations of the integrated system, separately.

    b Write the mass, energy, entropy and exergy balance equations for each component.

    c By using the given parameters and your design parameters, prepare a thermophysical properties table.

    d Calculate the power generation, residential heating and freshwater production capacities. Calculate the single generation (power), cogeneration (power + heat), and trigeneration (power + heat + freshwater) efficiencies.

9 Consider the system given in Case Study 13.1 and replace the nuclear reactor part with a biomass-based energy system where heat is produced for the power generating system. Identify the capacities based the needs potentially in the location to be selected and cover the needs for the community by providing three useful outputs, such as electricity, heat and freshwater.

10 Consider the system given in Case Study 13.2 and apply it to a targeted application in a specific location by replacing natural gas supply with hydrogen supply. Calculate the requested capacities and compare the results as well discuss the environmental impact dimensions.

14

Life Cycle Assessment of Energy Systems

Life cycle assessment is considered a holistic approach to study the entire spectrum of anything, including material development, processing and utilization stages for any service, product, or specific device or unit depending on the targeted application. During the past several decades, there has been an attempt to implement life cycle assessment tools due to the increased environmental impacts, energy needs, necessities for daily utilization and, of course, increase in the population, which has affected this in a way expedited the global warming causing local and global environmental and ecosystem-related issue. This has brought us to a level where we need solutions at every level of application, and this has then become a critical issue for anything everything we produce and for anything everything we use. Therefore, this life cycle assessment tool is essential because its coverage makes it a circular approach in dealing with all services, systems, products, and applications around us as we interact daily. In every sector, it is now essential to implement the life cycle assessment tools for any service or product or system or application considered. When we bring this life cycle assessment subject to energy systems, it even becomes more critical since energy systems are comprehensive and may use any kinds of materials, fuels and then many other commodities to be able to provide useful outputs. Therefore, this is an important subject to consider and treat accordingly, as systems are analyzed, designed, built, and implemented for the desired applications in various sectors and our communities.

Life cycle assessment (LCA) is recognized in simple terms as a tool for compiling and examining the inputs and outputs of materials and energy and the associated environmental impacts directly attributable to the functioning of a product or service system throughout its life cycle. LCA aims minimizing the production cost and environmental impact and maximizing the product quality and efficiency. LCA should be considered in any engineering decisions since it leads to better environment and sustainability. In this chapter, methodology of LCA, steps, and procedures that need to be executed for a high-standard LCA analysis are presented in depth. Various case studies and scenarios are presented to illustrate the importance of LCA studies and show how effectively they can be used.

14.1   Introduction

Life cycle assessment (LCA) is further recognized as a science-based methodology that allows quantifying and evaluating the lifetime environmental impacts of a product or process and used as a critical tool for environmental impact assessment studies [1]. An LCA analysis can help to determine areas of

Introduction to Energy Systems, First Edition. Ibrahim Dincer and Dogan Erdemir.
© 2023 John Wiley & Sons, Inc. Published 2023 by John Wiley & Sons, Inc.
Companion Website: www.wiley.com/go/Dincer/Introduction_to_Energy_Systems

production with high environmental burden, aiding the product improvement procedure and decision making with comparative analysis of alternative pathways [2]. The results of an LCA are indicated per functional unit, e.g., 1 MJ of energy, 1 kg of hydrogen, or transportation of a passenger by 1 km. Even if alternatives exist presently in open literature, following three common approaches are considered widely by many and deployed for various systems and applications as potential tools of LCA (see [3] for further details):

- Cradle to gate approach: This is recognized as an LCA methodology that considers raw material extraction, production, manufacturing, packaging, and transportation procedures. This variant of LCA does not take into account distribution, end use, and disposal phases.
- Cradle to grave approach: This is recognized as the most comprehensive LCA approach that considers all processes and stages within the life cycle of a product or process. In addition to material extraction and manufacturing, this approach also considers the emissions associated with distribution, end-use, and disposal phases. Figure 14.1 presents a cradle to grave type LCA framework along with process and stages that are considered within such type of analysis.
- Cradle to cradle approach: This is technically a cradle to grave type assessment which considers the end-of-life stage as recycling rather than disposal.
- Well to wheels approach: This is recognized as a comprehensive methodology to evaluate the LCA of fuels from extraction to their use.
- Well to tank approach: This is technically known as a cradle to gate type LCA to evaluate the environmental impacts of fuels that covers extraction of resources, production, and distribution phases but leaves out the end-use.
- Tank to wheels approach: This is recognized as another methodology, dealing with environmental impacts associated with consumption of fuels on board, but leaves out extraction, production, manufacturing, and distribution phases.

Distinctions between LCA types and the presented results in accordance with the scope of the considered LCA approach matter. For instance, electric vehicles are often referred as not

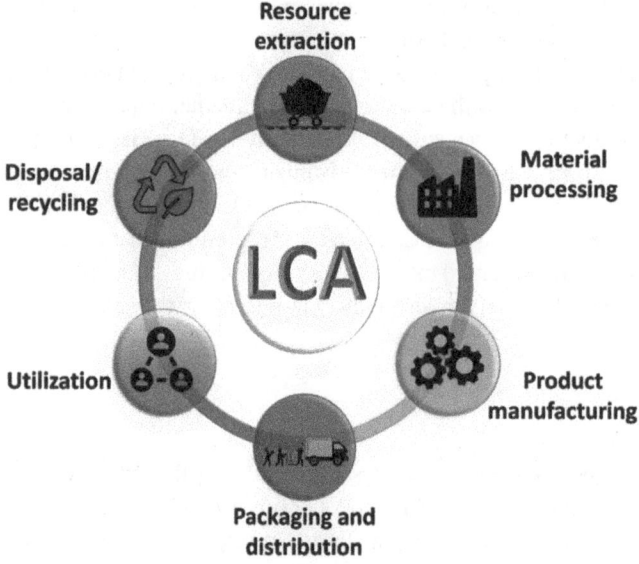

Figure 14.1 A cradle to grave approach of life cycle assessment.

producing any greenhouse gas emissions since they do not produce any tailpipe, in another word, tank-to-wheel emissions. Nevertheless, well-to-wheel LCA analysis of these vehicles shows that they have considerable impacts on environment from the production and manufacturing phases. Therefore, fair and actual environmental impacts of fuel types and vehicle technologies can be evaluated by conducting a well-to-wheel type LCA. Table 14.1 presents some strengths and limitations of LCA methods. These are not expected to be seen as concrete reasons for use or not using. These pros and cons are provided just to give an idea about these important LCA methods where there may be some advantages and disadvantages just like any other methods for any other type of analysis or assessment.

The LCA framework, distinguished by the ISO 14040 standard, comprises four separate phases, namely *(1)* goal and scope definition of the targeted study, *(2)* inventory analysis of all inputs and outputs (primarily materials and energy) for the process/system studied, *(3)* impact assessment of every step in the process/system, and *(4)* interpretation of the results which is also considered improvement analysis [4]. Figure 14.2 presents LCA framework standardized according to ISO 14040. LCA method allows identify the main product or process stages making the highest contribution to the environmental impact, so that environmental footprints can be mitigated with determination of problem and development of solution throughout improvement analyses.

14.1.1 Goal and Scope Definition

A careful definition of the goal and scope of an LCA study is very critical to ensure that the performed LCA consistent and product or procedure left outside the scope would not influence results considerably. An LCA practitioner should be able to understand the potential impacts of considered simplifications on the results. Therefore, the best way to avoid any misleading outcomes is to define the product and its life cycle precisely as well as describing the system boundaries carefully at the first stage of an LCA study, which also helps to determine the type of conducted LCA study. In consistent with the goal and scope of conducted LCA, a proper functional unit should be defined in this stage.

Table 14.1 Strengths and limitations of LCA methods.

Strengths	Limitations
• It provides a comprehensive analysis of product, service, process, application, or any activity.	• It may be costly and expensive in labor time.
• It is overall a transparent method for accounting for any environmental damages or impacts directly or indirectly.	• It may take lots of time for analysis, assessment, and improvement.
• It potentially helps determine potential environmental trade-offs.	• There are many different types of LCA packages available.
• It gives specific details and portrays the complete pictures about every step considered.	• There are numerous LCA approaches to select the right one for a specific purpose.
• It delivers right information and knowledge to help decision makers about any initiative.	• Not sufficient data may be available for every process, system, service, application, or activity.
• It fosters better communication among decision and policymakers.	• Combining impacts into a single score may not be fully scientific, but requiring proper judgments.
• It gives details about opportunities and challenges related to any product.	• There is a need to well document all specific assumptions.
	• There is strong need to make justifications and assumptions in a correct manner for decision makers.
	• The LCA doer is expected to have enough experience to interpret the results.

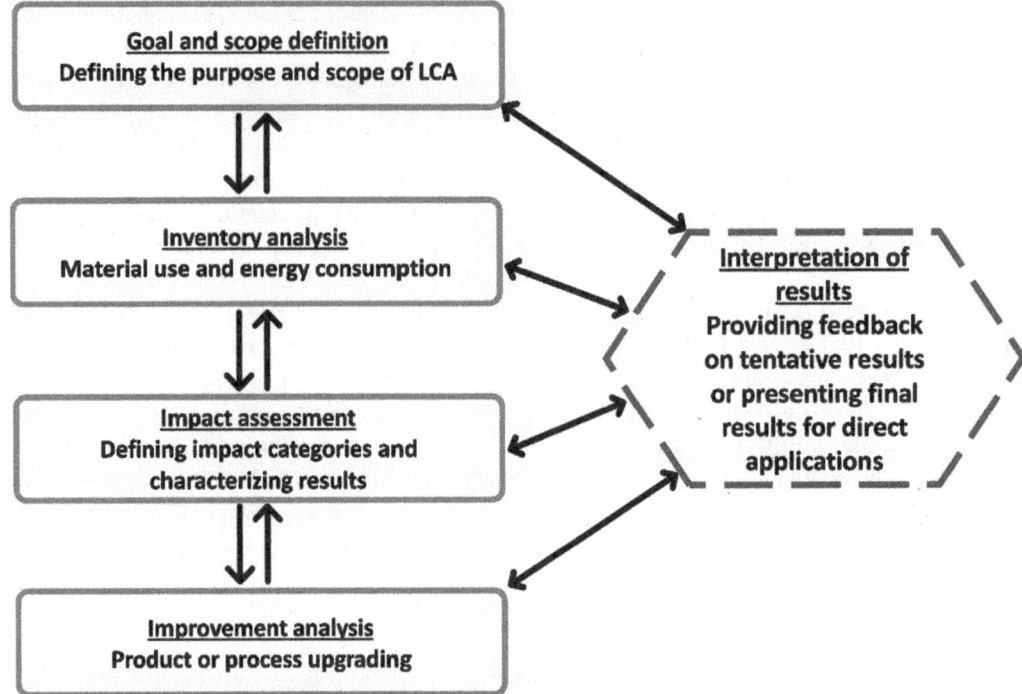

Figure 14.2 A modified LCA framework to consider (Modified from [5]).

14.1.2 Life Cycle Inventory Analysis

In this stage, data regarding material and energy inputs and outputs required for product or service are collected, calculated, related and quantified as well as assessed accordingly. Of course, the proper definitions and quantifications of life cycle inventory (LCI) are expected to improve the precision and reliability of the obtained LCA results.

14.1.3 Life Cycle Impact Assessment

In this phase of LCA, the environmental impacts of a product or process are quantified by multiplying the results of inventory analysis with environmental impact factors [6]. There are many different types of life cycle impact assessment (LCIA) methods that can be adapted in an LCA study. Among these, the CML baseline is one of the early developed LCIA methods. Depending on the environmental problems of the regions or countries, institute developed different LCIA methods. Each method provides different impact categories for different fields of environmental problems. Table 14.2 presents some of commonly utilized LCIA methods.

The CML baseline is known by many as one of the most used impact assessment methods for LCA studies that contain various impact categories. The method is developed by the University of Leiden in the Netherlands in 2001 [8]. There are various impact categories included in the method CML (baseline), among which abiotic depletion of sources, acidification potential, eutrophication potential, global warming potential, and ozone depletion potential are investigated in numerous LCA studies available in the open literature [9]. Table 14.3 presents CML baseline impact categories along with their units shown in equivalence factor. The compounds contributing to a specific environmental impact category are multiplied with a characterization factor to express the additive of the compound; thus, it is defined as an equivalence factor.

Table 14.2 Overview of life cycle impact assessment methods (adapted from [7]).

LCIA Methods	Developer	Origin
CML 2001	Institute of Environmental Sciences, Leiden University	Netherlands
Eco-Indicator 99	Pre-Consultants	Netherlands
EDIP-2003	IPU, Technical University of Denmark	Denmark
ReCiPe	RIVM, CML, Pre-Consultants, Radboud Universities Nijmegen and CE Delft	Netherlands
LIME	National Institute of Advanced Industrial Science and Technology	Japan
EPS 2000	Swedish Environmental Research Institute	Sweden
Ecological Scarcity Method	Federal Office for the Environment	Switzerland
TRACI	US Environmental Protection Agency	USA

Table 14.3 The impact categories and their units [10].

Impact Category	Acronym	Unit	Unit Definition
Abiotic depletion potential	ADP	kg Sb-eq	Kilogram of antimony equivalent
Acidification potential	AP	kg SO_2-eq	Kilogram of sulfur equivalent
Eutrophication potential	EP	kg PO_4-eq	Kilogram of phosphate equivalent
Global warming potential	GWP	kg CO_2-eq	Kilogram of carbon dioxide equivalent
Ozone depletion potential	ODP	kg CFC 11-eq	Kilogram of trichlorofluoromethane equivalent

14.1.4 Improvement Analysis

The step of improvement analysis of LCA seeks for opportunities to improve efficiencies, reduce energy consumptions, replace, or reduce material inputs, and all others to mitigate environmental impacts of product or process within life cycle.

14.1.4.1 Interpretation of Results

The step of interpretation of LCA results can be intermediate based on tentative LCA results, where redefinition of goal and scope or revaluation of inventory analysis may be requested. On the other hand, interpretation of results in the decision-making phase, the LCA results must be consistent and reliable for direct application.

14.2 Case Studies

In this section, LCA analyses of different case studies are presented. Case Study 1 evaluates the environmental impacts of 1 kWh of electricity for different production scenarios. Case Study 2 comparatively evaluates various PEM water electrolysis-based hydrogen production scenarios. In

Case Study 3, environmental impacts of different road transport options are comparatively investigated.

14.2.1 Case Study 1 – LCA Analysis of Different Electricity Production Scenarios

In this case study, a comprehensive analysis and assessment is presented where the environmental impacts of electricity production for a small city are comparatively evaluated for different scenarios.

Step-1 Goal and scope definition: In this case study, environmental impacts resulted from electricity production for a small city are comparatively evaluated for different electricity production scenarios, including combined power cycle plant, PV solar power plant, and wind power plant. The functional unit of this LCA analysis is 1 kWh of electricity production. Figure 14.3 presents the selected electricity production method in this comparative LCA analysis, namely combined power cycle (a), wind power plant (b), and PV solar power plant (c).

Step-2 Inventory analysis: The required inventory for the LCA analysis is obtained from SimaPro software databases. Power plant construction, material extraction, and production activities are included in the LCA scenarios. Table 14.4 presents information regarding the considered plants and details of databases that are utilized while developing the LCA cases.

Step-3 Life cycle impact analysis: The selected three different electricity production methods are comparatively evaluated in terms of abiotic depletion potential, acidification potential, eutrophication potential, global warming potential, and ozone depletion potential. Note that the LCA does not include the consumption phase of the produced electricity. In this regard, the conducted LCA analysis can be classified as cradle to gate type.

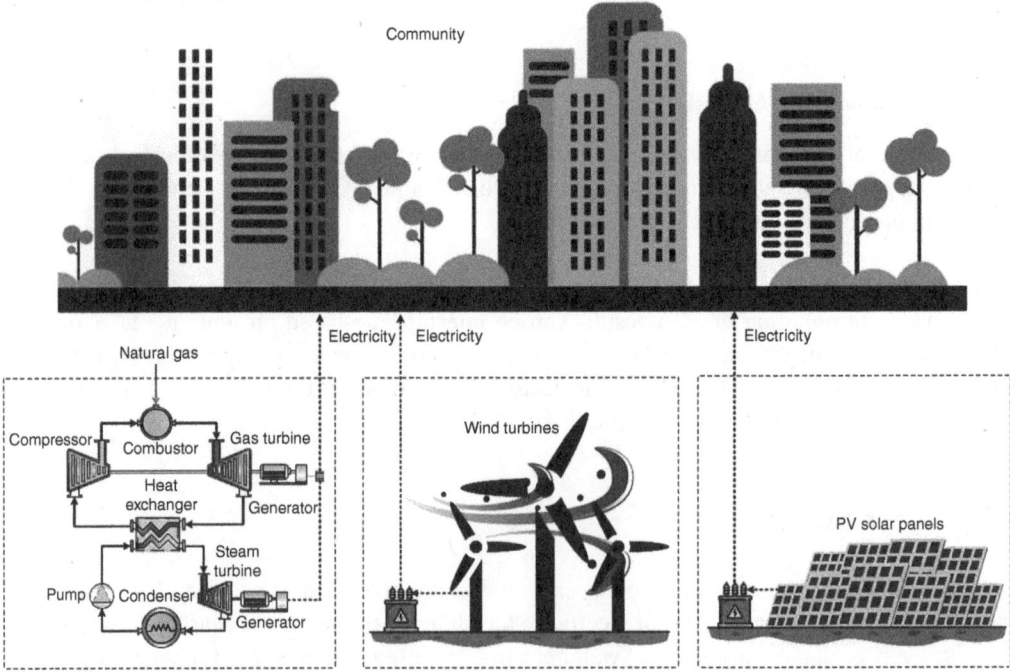

Figure 14.3 Electricity production scenarios for a small city.

Table 14.4 Details of the inventory selected from the LCA software database.

Plant Type	Characteristics	Location	Library
Combined gas-steam power plant	Electricity, high voltage {CA-ON} I electricity production, natural gas, combined cycle power plant I APOS, U	Ontario, Canada	Ecoinvent-3
Wind power plant	Electricity production, wind >3 MW, onshore, APOS, U	Ontario, Canada	Ecoinvent-3
PV solar power plant	570 kWp open ground installation, multi-Si, APOS, U	Ontario, Canada	Ecoinvent-3

Figure 14.4 presents the abiotic depletion potentials (ADP) of the selected electricity production scenarios. This abiotic depletion potential (ADP) is considered as the measure of the use of nonrenewable natural resources for human being's livelihood and activities [11]. Water and land uses are separated from ADP and evaluated as individual environmental impact categories [12]. In this regard, the impact category of ADP in an LCA study considers the degradation of abiotic finite resources such as fossil fuels, minerals, and metals [13]. According to the LCA results, electricity production from combined cycle power plant performs the lowest ADP in this LCA (1.06EE-07 kg Sb eq/kWh). ADP of electricity from PV solar plant is evaluated as 4.17E-06 kg Sb eq/kWh. The wind-based electricity production has the highest ADP in this comparative analysis corresponding to a value of 5.88E-06 kg Sb eq/kWh. Fossil fuel consumption during copper extraction and manufacturing procedure appears to be the main ADP contributor of wind-based electricity production.

Figure 14.5 presents acidification potentials of the selected electricity production scenarios. This acidification potential (AP) refers to releasing chemical compounds to air that may cause acid precursors and eventually acid rains [14]. The most important acidifying human-made chemical compounds are noted as SO_x, NO_x, HCl, NH_3, and HF [15]. The release of these compounds may form new hydrogen ions (H^+) and greatly reduce the quantity of substances in the ecosystem capable of neutralizing hydrogen ions. With a value of 0.000407 kg SO_2 eq/kWh, PV solar-based electricity production is evaluated as the most AP-dense electricity production, which

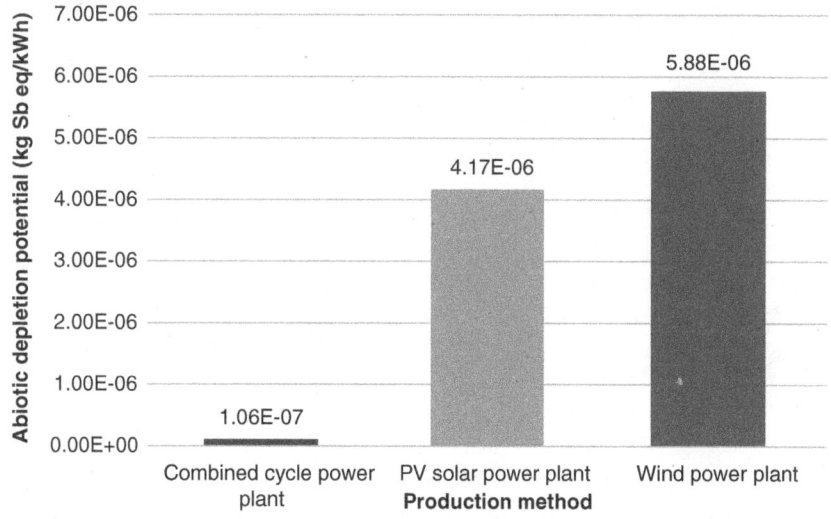

Figure 14.4 Comparative ADP evaluation of different electricity production scenarios.

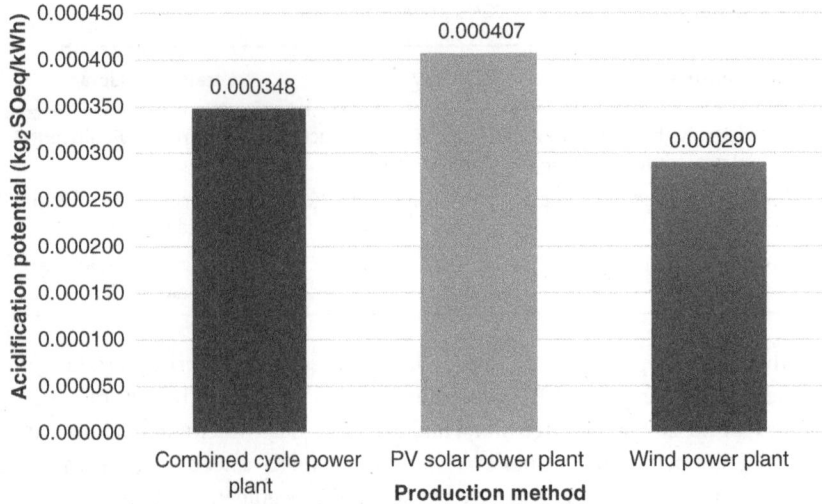

Figure 14.5 AP of different electricity production methods.

is followed by combined power cycle and wind-based electricity productions with values of 0.000348 and 0.000290 kg SO_2 eq/kWh, respectively. Multi-Si wafer production for manufacturing of PV modules is the highest AP contributor of this electricity production method.

Figure 14.6 presents the eutrophication potentials of the selected electricity production methods. This eutrophication potential (EP) is the environmental impact responsible for the contamination of soil and groundwater due to overabundance of nitrogen and phosphorus coming from polluting emissions, e.g., wastewater or fertilizers [16]. These excessive polluting compounds originate aggressive development of algae and plants, decreasing the rate of solar energy and oxygen, and resulting in the contamination of plants and groundwater consequently. In this environmental impact category as well, the EP of electricity production from PV solar power plant is evaluated as the highest corresponding to

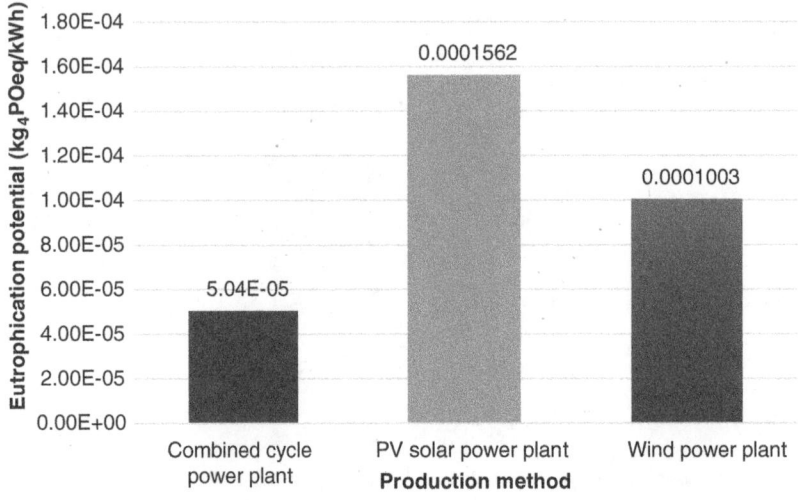

Figure 14.6 Comparative EP analysis of the selected electricity production methods.

a value of 0.0001562 kg PO_4 eq/kWh. After careful analysis of the subprocesses taking place in this electricity production method, chemical substances and fossil fuels consumed during multi-Si wafer production process appears as the main EP contributor. EP of electricity from wind power plant is evaluated as 0.0001003 kg PO_4 eq/kWh. On the other hand, the EP of electricity production from combined cycle power plant is evaluated as the lowest in this analysis (5.04E-05 kg PO_4 eq/kWh). High efficiency, production capacity, and high operational capacity factors of this plant can be shown as main reasons for this result. The capacity factor is the ratio of actual electrical energy output of a power plant to plant's theoretical energy production capacity. Even though they utilize from renewable resources, solar and wind power plants operate with much lower capacity factors compared to power cycles, which may result in higher results in some environmental impact categories.

Figure 14.7 presents the global warming potential results evaluated from the selected electricity production method. Global warming potential (GWP) was originally developed for the comparison of the ability of different greenhouse gases in the atmosphere relative to another gas. In this regard, CO_2 was chosen as the reference gas due to its long duration in the atmosphere that increases the concentration over a certain period [17]. A 100-year period is commonly used for GWP evaluation. According to Intergovernmental Panel on Climate Change, GWP is defined as a quantified measure of the globally averaged radiative forcing impacts of a unit mass of well-mixed greenhouse gas over a chosen time horizon relative to that of a unit mass of CO_2 defined as a reference gas [18]. In terms of this impact category, electricity produced by wind power plant has the lowest GWP with a value of 0.028 kg CO_2 eq/kWh, which is followed by electricity from PV solar (0.073 kg CO_2 eq/kWh). The highest GWP is evaluated from electricity produced by combined cycle power plant (0.425 kg CO_2 eq/kWh). This is mainly because of natural gas consumption during the electricity production process. Materials and energy are two primary inputs, where type of energy consumed during the production is the primary parameter determining the GWP results. From extraction to consumption, fossil fuels are CO_2-dense energy sources. Consumption of these energy resources during a production phase results in high GWP values.

Figure 14.8 presents the ozone depletion potentials of the evaluated electricity production methods. The ozone depletion potential (ODP) is another impact category considered in the CML baseline method. The concept of ODP is commonly used as a measure of the impact of a compound in

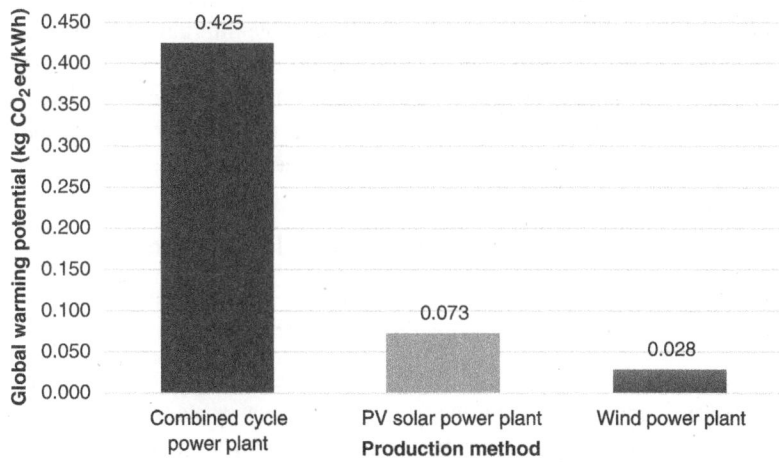

Figure 14.7 GWP of different electricity production scenarios.

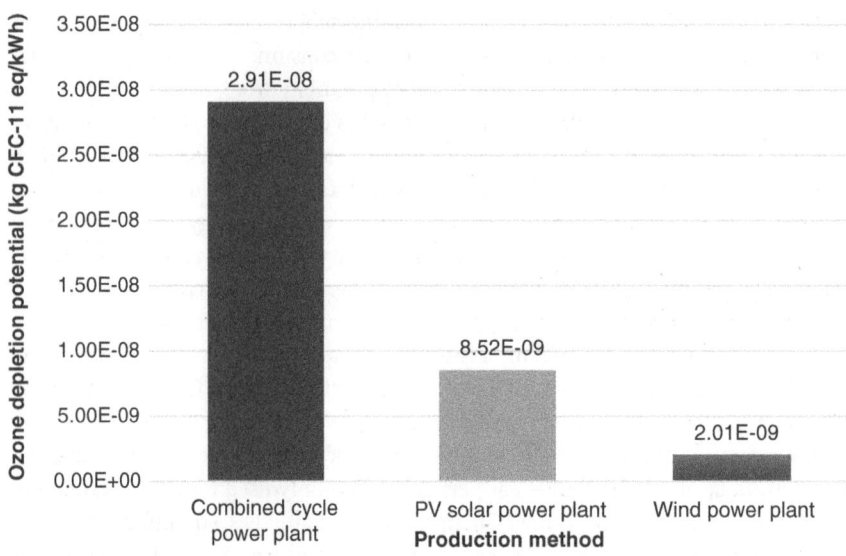

Figure 14.8 ODP of various electricity production scenarios.

depleting the ozone layer relative to CFC-11 as a standard compound [19]. The stratospheric ozone layer protects the planet earth against UV-B radiations that may harm living species, biochemical cycles, and materials. ODP is evaluated at a global-scale geographical scope for an infinite time horizon. Electricity from the combined cycle power plant has the highest environmental impact in this impact category due to fossil fuel consumption for electricity production. In the analysis, ODP of the combined cycle-based electricity generation is evaluated as 2.91E-08 kg CFC-11 eq/kWh. Electricity production from wind has the lowest ODP with a value of 2.01E-09 kg CFC-11 eq/kWh that is followed by PV solar-based electricity (8.52E-09 kg CFC-11 eq/kWh).

Step-4 Improvement analysis and interpretation of results: Efficiency is treated as one of the most influential parameters on the environmental impacts of an energy system. In this regard, combined cycle power plants operate with high efficiencies and capacity factors. Nevertheless, electricity production by this technology has high impacts on the environment as long as consuming fossil fuels in the production process. Cleaner and locally producible alternative fuels, e.g., biogas, biomethane, hydrogen-blended fuels, can help to obtain much better environmental performance from this technology. Even though it brings intermittency and low efficiency-related concerns, electricity production from renewable energy resources offers much lower environmental impacts than conventional fossil-fuel-based electricity production methods.

14.2.2 Case Study 2 – LCA Analysis of Different Hydrogen Production Scenarios

In this case study, a comprehensive analysis and assessment is presented where the environmental impacts of hydrogen production are comparatively evaluated for different production scenarios.

Step-1 Goal and scope: Hydrogen is produced from a 1 MW proton exchange membrane type electrolyzer. The required electrical energy is considered from different power plant types including photovoltaic (PV) solar, wind, nuclear, and hydro plants, and the environmental impacts of hydrogen production fed-by corresponding plants are comparatively evaluated. The functional unit of the LCA study is determined to be 1 kg of hydrogen production. Since the scope of the conducted LCA does not consider the end-use, in another word, the distribution and consumption phases of hydrogen, the performed LCA can be defined as cradle to gate type which may be accomplished accordingly depending on the specific tasks.

Step-2 Inventory analysis: Table 14.5 presents characteristics and operational parameters of the PEM electrolyzer considered in the life cycle inventory analysis. Figure 14.9 presents the system boundaries of the selected hydrogen production scenarios.

First thing is to determine the required electrical energy to produce 1 kg H_2. As depicted in Table 14.3, it is considered that the PEM electrolyzer operates with an efficiency of 60%. With this information, the electricity consumption of PEM electrolyzer per kg of H_2 production can easily be evaluated by the following equation:

$$\eta_{PEM} = \frac{m_{H_2} \times LHV_{H_2}}{E_{El}}$$

where η_{PEM} represents the efficiency of the electrolyzer (60%), LHV_{H_2} is the lower heating value of hydrogen (120 MJ/kg or 33.3 kWh/kg), m_{H_2} is the mass of produced hydrogen (1 kg), and E_{El} is the required electrical energy (MJ or kWh) for 1 kg hydrogen production. According to the given data, the electricity consumption of the PEM per kg H_2 is evaluated as 55.6 kWh/kg H_2. With a plant capacity of 90% and lifetime of 20 years, the lifetime hydrogen production from the PEM electrolyzer can be calculated by using the following equation:

$$m_{LF} = \frac{0.9 \times CP_{PEM} \times 8760}{E_{El}} \times 20$$

Where CP_{PEM} represents the PEM electrolyzer capacity (1000 kW or 1 MW). 8760 is the number of hours in year. From the equation, annual hydrogen production from the PEM electrolyzer is evaluated as around 142 tons (141,798 kg exactly), which corresponds to 2836 tons of hydrogen production over a lifetime operation of 20 years of PEM. The chemical reaction of electrolysis can be given as follows:

$$H_2O \rightarrow H_2 + \frac{1}{2}O_2$$

Table 14.5 Operational parameters of the 1 MW PEM electrolyzer.

Parameter	Value	
Functional unit of LCA	1 kg hydrogen production	
Source of electricity	Nuclear- Wind- PV solar- Hydro- Biomass	
Water consumption	9 kg deionized water per kg H_2 production	
Capacity	1 MW	
Plant capacity factor (%)		90%
Lifetime operation	20 years	
PEM efficiency		60%
Lower heating value of hydrogen	120 MJ/kg – 33.3 kWh/kg	
Annual hydrogen production	142 tons	
Lifetime hydrogen production	2836 tons	
Electricity consumption	55.6 kWh/kg H_2	

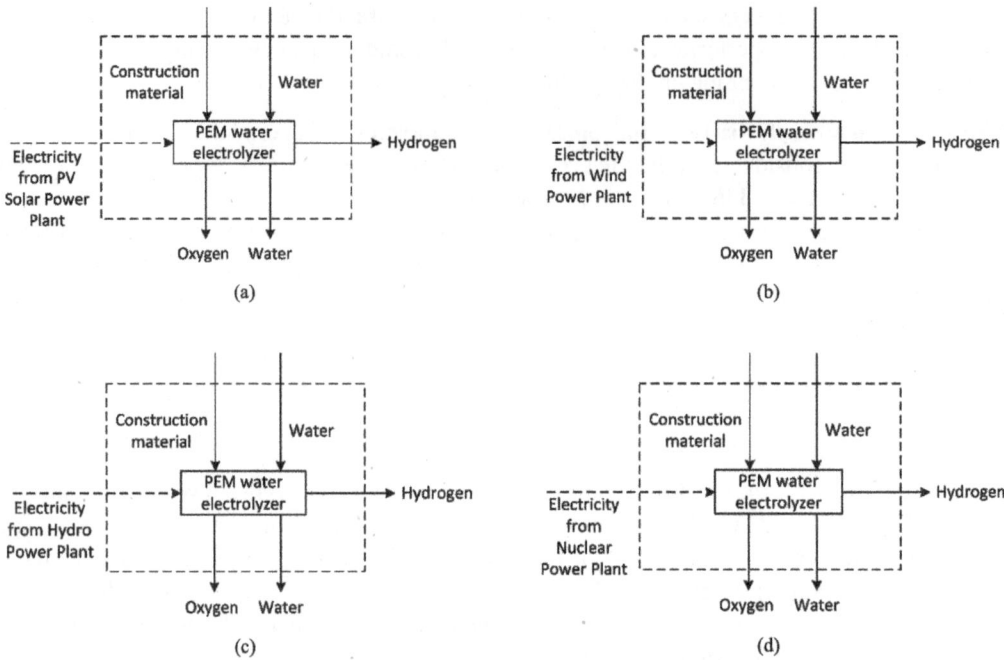

Figure 14.9 System boundaries of PV solar (a), wind (b), hydro (c), and nuclear (d) based PEM water electrolysis.

By considering the molecular weight of the substances taking place in the chemical reaction (18 g of H_2O, 1 g of H_2, 16 g of O_2), the required amount of water for 1 kg of hydrogen production can be calculated as 9 kg according to the preservation of mass principle. Note that a complete dissociation of water is assumed in the analysis. Preferably, the consumed water is to be distilled to protect the decency of the PEM electrolyzer, which is selected accordingly in the LCA inventory. The next step in the LCI of the conducted LCA analysis is to determine the materials required to build the PEM electrolyzer. Table 14.6 presents the inventory and quantity of the materials needed to build a 1 MW PEM electrolyzer along with detailed information of data utilized to build LCA case in SimaPro software.

In this case study, 1 kg of hydrogen production from the PEM electrolyzer is comparatively assessed for different scenarios, where the required electricity for the water electrolysis is provided from different power plants including PV solar, onshore wind, hydro, and nuclear power plants. Table 14.7 presents the details and locations of the selected power plants while building the LCA cases in SimaPro software.

Step-3 Life cycle impact analysis: CML baseline is considered as impact assessment method. Environmental impacts of hydrogen production scenarios are assessed in terms of abiotic depletion potential (ADP), acidification potential (AP), eutrophication potential (EP), global warming potential (GWP), human toxicity potential (HTP), and ozone depletion potential (ODP) impact categories. Figure 14.10 presents the ADP values of various PEM-based hydrogen production scenarios. According to the LCA results presented, the hydro-based PEM water electrolysis has the lowest ADP among the selected hydrogen production pathways corresponding to the value of 1.95E-06 kg Sb eq per kilogram of produced hydrogen. In the analysis, the

Table 14.6 Materials for a 1 MW PEM water electrolyzer stack (data from [20]).

Material	Quantity (kg)	Brief Info	Library
Titanium	528	Inputs from nature	Ecoinvent-3
Aluminum	27	Inputs from nature	Ecoinvent-3
Copper	4.5	Inputs from nature	Ecoinvent-3
Stainless steel	100	Steel hot dip galvanized, including recycling	Agri-footprint- mass allocation
Nafion	16	Dummy process – Tetrafluoroethylene	USLCI
Activated carbon	9	Activated carbon (APOS, S)	Ecoinvent-3
Iridium	0.75	Inputs from nature	Ecoinvent-3
Platinum	0.075	Inputs from nature	Ecoinvent-3

Table 14.7 The power plants' characteristics.

Plant Type	Characteristics	Location	Library
PV solar	570 kWp open ground installation, multi-Si, APOS, S	Ontario, Canada	Ecoinvent-3
Wind	Electricity production, wind >3 MW, onshore, APOS, S	Ontario, Canada	Ecoinvent-3
Hydro	Electricity production, hydro, run-of-river, APOS, S	Ontario, Canada	Ecoinvent-3
Nuclear	Pressure water reactor, heavy water moderated, APOS, S	Ontario, Canada	Ecoinvent-3

second lowest ADP is obtained from the nuclear-based production (3.00E-05 kg Sb eq/kg H_2) that is followed by PV solar (0.000232 kg Sb eq/kg H_2) and wind-based (0.000327 kg Sb eq/kg H_2) PEM water electrolysis.

As it can be seen, wind-based PEM water electrolysis has the highest ADP among the selected production methods. Potential reasons for this result can be evaluated by using network analysis. Network analysis shows the contribution of each product and processes taking place in the life cycle to the evaluated impact category. Figure 14.11 presents the network analysis results of wind-based PEM water electrolysis where the contribution of each sub process to ADP of the corresponding hydrogen production method is weighted. Here, the contributors can be evaluated from the materials used for the construction of the PEM electrolyzer, and energy and material used per kg of produced hydrogen. Environmental impacts coming from material production used in the system construction are distributed over the lifetime production which is calculated as around 2836 tons as indicated in Table 14.2. Therefore, environmental impacts from the construction material per kg of hydrogen production is evaluated as very low compared to the energy (electricity) and material (water) consumption required for 1 kg of hydrogen production. In parallel to this, the electricity generated from wind power plant appears as the main ADP contributor to this impact category.

Figure 14.12 presents the comparative acidification potential results of the selected hydrogen production pathways. According to the results, hydro-based PEM water electrolysis has the lowest

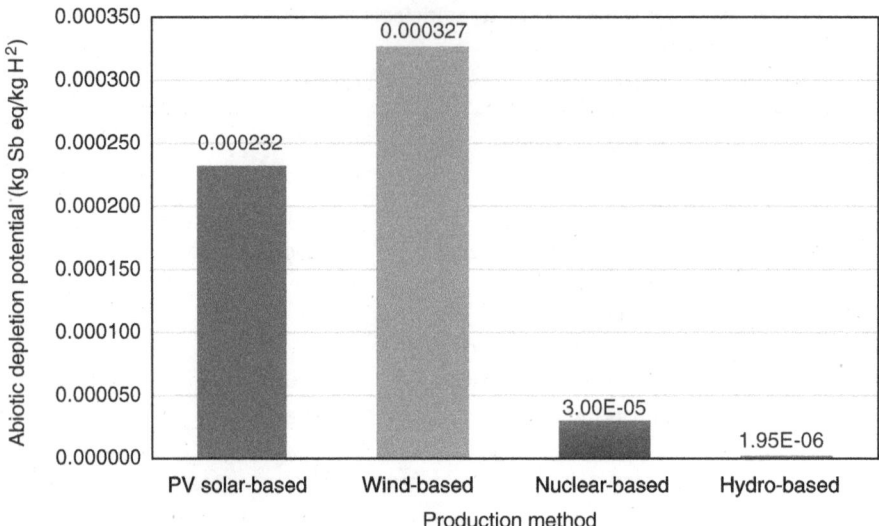

Figure 14.10 ADP of various PEM-based hydrogen production pathways.

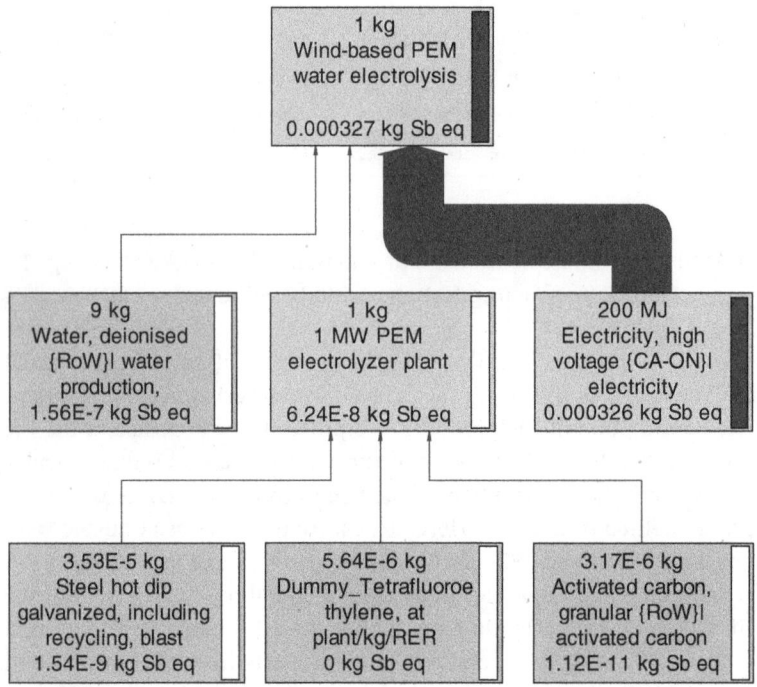

Figure 14.11 Main ADP contributing processes in wind-based PEM water electrolysis.

AP among the evaluated methods corresponding to the value of 0.00095 kg SO_2 eq per kg of produced hydrogen, whereas PV solar-based method performs the highest AP with a value of 0.02261 kg SO_2/kg H_2. Figure 14.13 presents the network analysis results regarding the main AP contributors of PV solar-PEM water electrolysis.

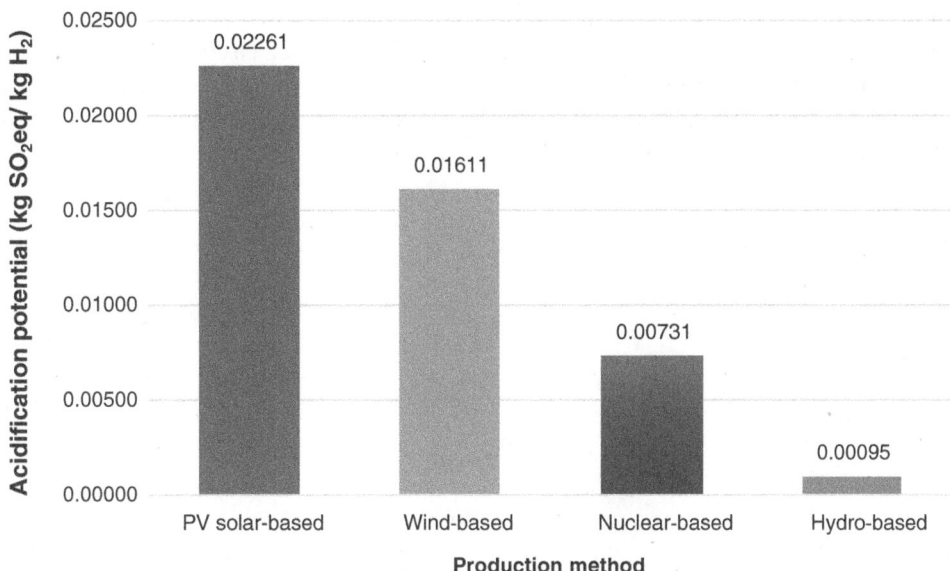

Figure 14.12 AP of various PEM-based hydrogen production pathways.

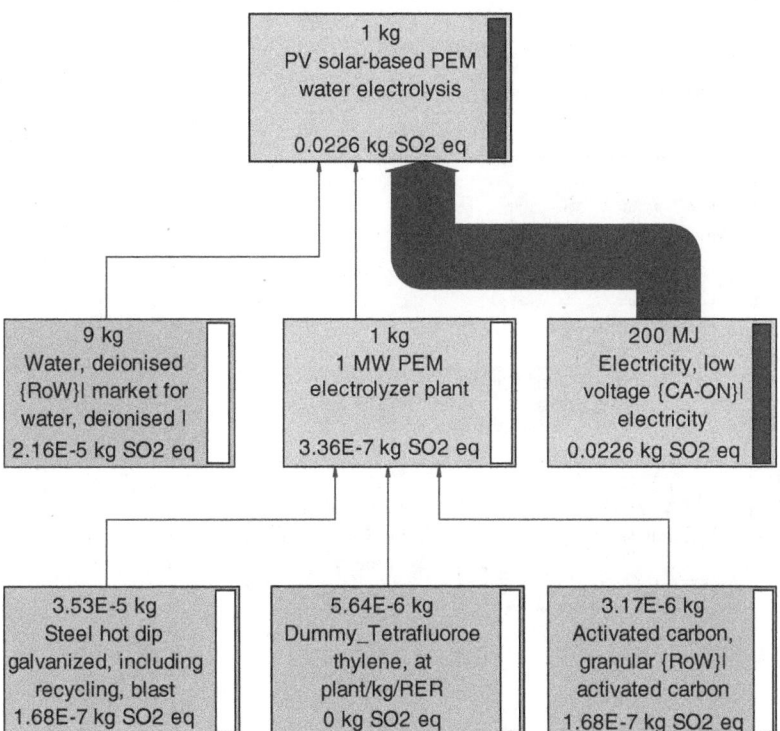

Figure 14.13 Main AP contributor of PV solar-based PEM water electrolysis.

Figure 14.14 presents the eutrophication potentials of selected PEM water electrolysis production pathways. In the analysis, the EP of nuclear-based PEM water electrolysis is evaluated 0.003218 kg PO_4 eq per kilogram of produced hydrogen. The lowest EP is evaluated from the hydro-based production with a value of 0.000317 kg PO_4 eq/kg H_2, whereas the PV solar-based production has the highest EP value in this analysis (0.008683 kg PO_4 eq/kg H_2). Figure 14.15

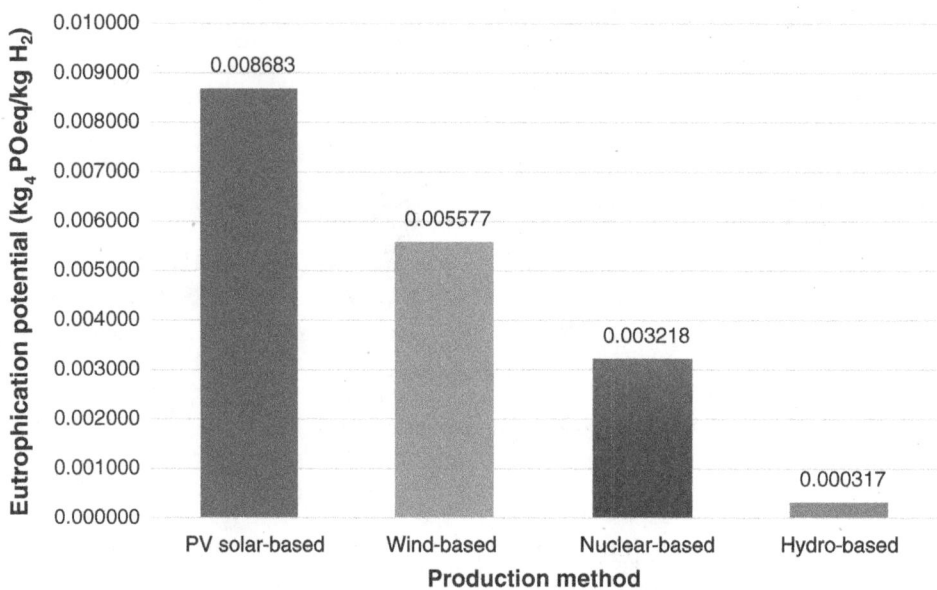

Figure 14.14 EP of various PEM-based hydrogen production pathways.

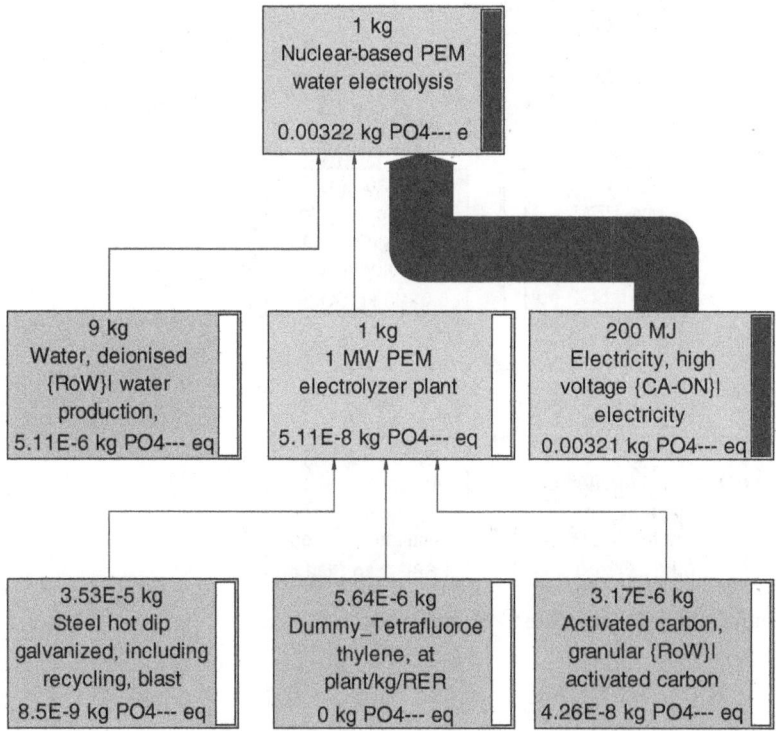

Figure 14.15 Main EP contributors of nuclear-based PEM water electrolysis.

presents the main EP contributors of nuclear-based PEM water electrolysis pathway for hydrogen production.

Figure 14.16 presents the comparative global warming potential evaluation of various PEM water electrolysis pathways. The results of the LCA study conducted on the selected hydrogen production methods indicate that hydro-based PEM water electrolysis has the lowest GWP value (0.245 kg CO_2 eq/kg H_2), which is followed by the nuclear-based, wind-based, and PV solar-based as from lowest to highest, respectively. Multi-Si wafer production for PV solar panels appears as the high GWP contributing production process making GWP of this hydrogen production method poorest in this case study. It should be noted that PV solar based is accepted as one of the clean, so-called green, hydrogen production pathway. This is totally true, because the GWP of PV solar-based PEM water electrolysis is almost three times less than that of conventional steam methane reforming having a GWP value of around 10.28 kg CO_2/kg H_2 [21]. Figure 14.17 presents the main GWP contributors of hydro-based PEM water electrolysis.

Figure 14.18 presents the ozone depletion potentials of the investigated PEM water electrolysis-based hydrogen production pathways. Among all evaluated methods, ODP of PV solar-based PEM water electrolysis is evaluated as the highest with a value of 4.74E-07 kg CFC-11 eq per kilogram of produced hydrogen. The wind-based production has the second highest ODP corresponding to the value of 1.13E-07 kg CFC-11 eq/kg H_2. In this impact category, the lowest ODP is evaluated from hydro-based PEM water electrolysis (1.64E-08 kg CFC-11 eq/kg H_2), which is followed by the nuclear-based hydrogen production (7.32E-08 kg CFC-11 eq/kg H_2) in the analysis. Figure 14.19 presents the ODP contributors of hydro power-based PEM water electrolysis process.

Step-4 Improvement analysis and interpretation of results: The main environmental impact contributors of the production are the active energy and material consumption during the production phase. Depending on the efficiency and capacity factor of the system where the required energy provided from, environmental impacts due to consumed energy decrease or increase. In this regard, electrical energy from hydro and nuclear power plants has low environmental impacts due to their high energy production capacities, which significantly lowers the environmental impacts per kWh of produced electricity. In parallel to this, feeding hydrogen production via PEM

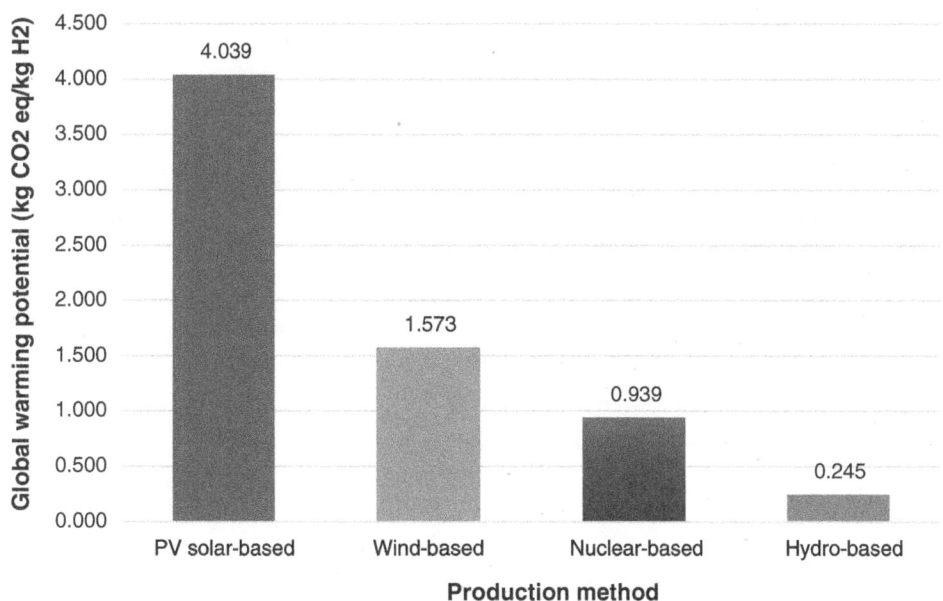

Figure 14.16 GWP of various PEM-based hydrogen production pathways.

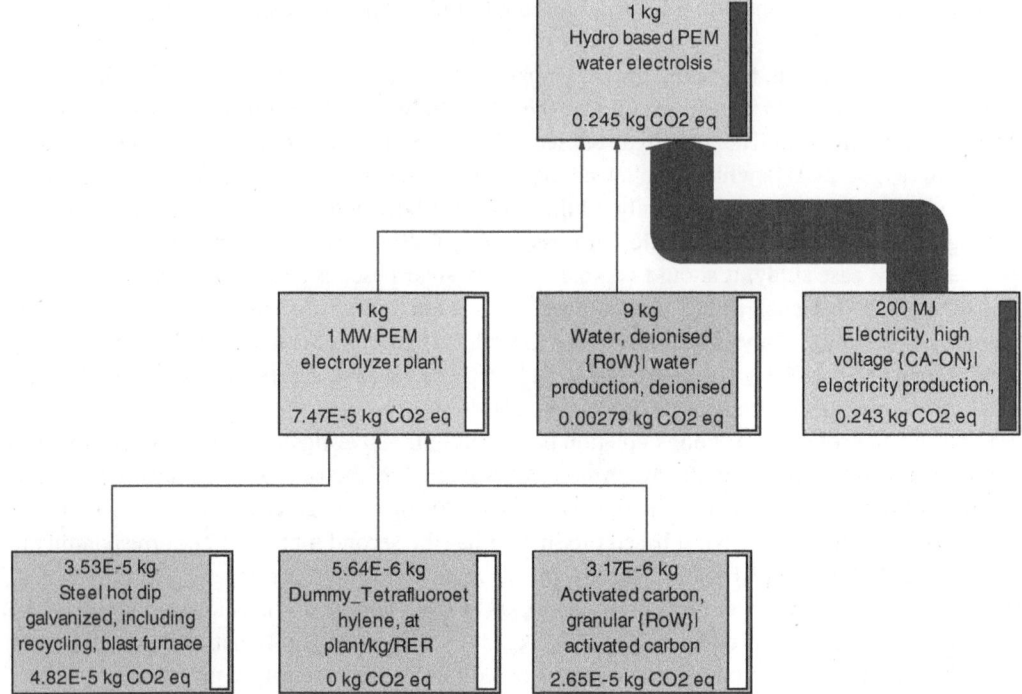

Figure 14.17 GWP contributing subprocesses in hydro-based PEM water electrolysis.

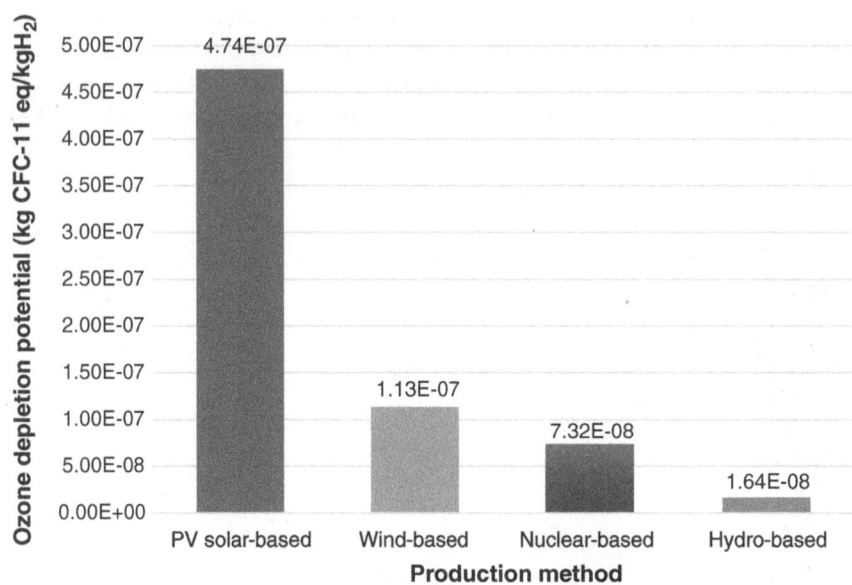

Figure 14.18 ODP of various PEM-based hydrogen production pathways.

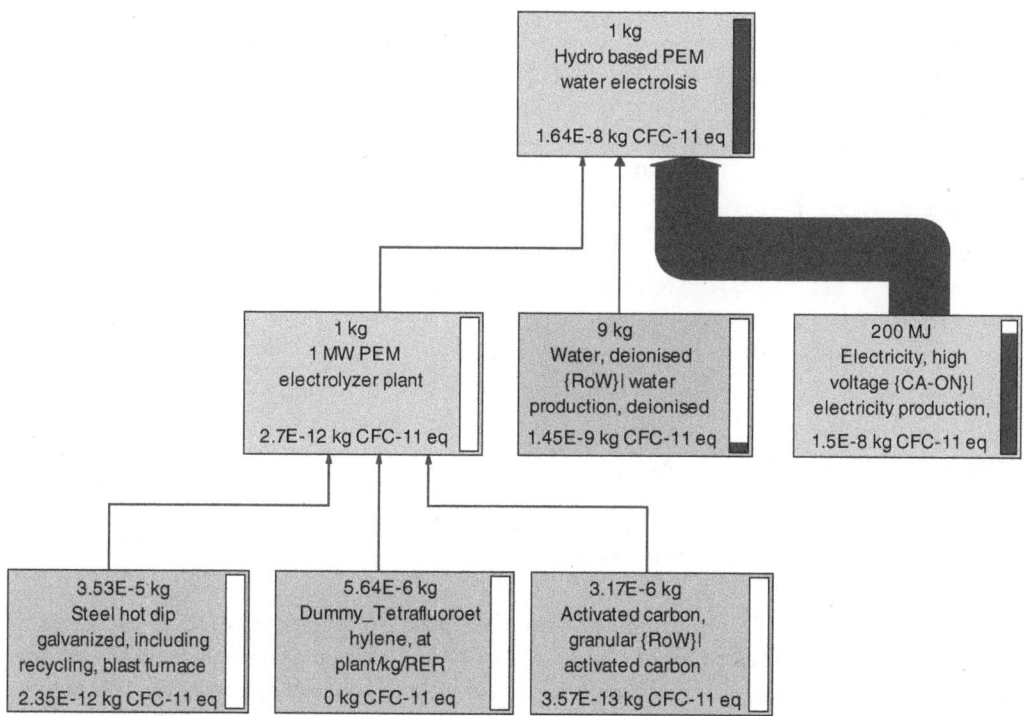

Figure 14.19 Main ODP contributors to the hydro power-based PEM water electrolysis.

water electrolysis with electrical energy from hydro power plants and nuclear power plants appears as the most environmentally production pathways among the evaluated ones in this comparative analysis.

14.2.3 Case Study 3 – GWP Analysis of Different Road Transportation Scenarios

In this case study, a comprehensive analysis and assessment is presented where the environmental impacts of public transit bus configurations are comparatively evaluated with respect to power-train technology and consumed fuel types.

Step-1 Goal and scope definition: The objective of this environmental impact analysis is to evaluate the CO_2 emissions of public transit buses with respect to power train and consumed fuel type. The functional unit of the analysis is determined as 1 km of city driving. It should be noted that emissions associated with construction of vehicle body is not considered in the analysis.

Step-2 Inventory analysis: In the analysis, the city driving speed range is considered from 0 to 11.11 m/s (40 km/h). For a fair fuel consumption-based environmental impact analysis, power demand and fuel consumption ratios are determined by simulating a 40-foot bus configuration. Further details regarding the simulation of bus configuration can be found in the source [22]. Table 14.8 presents some specifications of the considered bus configuration in the analysis.

Since it depends on numerous factors, e.g. driver, weight, and terrain, it is difficult to estimate the acceleration time. In the analysis, 10 s is considered as an acceleration time for reaching the top

Table 14.8 The parameters and numerical values considered in the analysis of the system.

Parameters	Value
Curb weight of the bus	13,000 kg
Fully loaded weight of the bus	16,000 kg
Bus length-height-width	12 m^{-2}.6 m^{-2}.5 m
Frontal area	6.5 m^2
Driving range	500 km
City driving speed range	0–40 km/h (0–11.11 m/s)
Drag coefficient	0.7
Rolling resistance coefficient	0.006
Gravitational acceleration	9.81 m/s^2
Density of air*	1.184 kg/m^3

* At a temperature of 25°C and 101.3 kPa pressure.

city driving speed of 11.11 m/s from rest position. Thus, the acceleration rate is evaluated as 1.111 m/s^2, and assumed as constant. There are three main forces that the vehicle needs to overcome to travel at desired speeds; acceleration force, drag force, and rolling resistance. It should be noted that asphalt road and flat terrain conditions are considered in the analysis. Table 14.9 presents energy consumption ratios of the vehicle and other parameters considered in the comparative environmental impact analysis.

After calculating energy required for travel of the 40-foot bus over 1000 m at an average city driving speed of 11.11 m/s (40 km/h), the fuel consumption ratio per km with respect to powertrain types can be calculated by applying proper equations.

Hydrogen consumption of hydrogen fuel cell electric bus

$$E_{km_a} = m_{H_2} \times LHV_{H_2} \times \eta_{fc} \times \eta_{em}$$

where η_{fc} represents the electrical efficiency of PEM fuel cell (53%), and η_{em} is the efficiency of electric motor (75%).

Table 14.9 The values considered in the comparative analysis.

Parameters	Value
Energy consumption with acceleration (E_{km_a})	3.325 MJ/km
Energy consumption at constant speed (E_{km_c}).	1.279 MJ/km
Transmission efficiency	87%
Diesel engine thermal efficiency	35%
Gas engine thermal efficiency	30%
PEM fuel cell electrical efficiency	53%
Electric motor efficiency	75%

Electricity consumption of battery electric bus

$$E_{km_a} = E_e \times \eta_{em}$$

Biomethane or diesel consumption of the bus can be evaluated by using following equation.

$$E_{km_a} = m_{fuel} \times LHV_{fuel} \times \eta_{ge} \times \eta_{gear}$$

where η_{ge} is the thermal efficiency of the gas engine, and η_{gear} is the transmission efficiency. The CO_2 emission ratio can be calculated by

$$EM_{CO_2} = m_{fuel} \times LHV_{fuel} \times EF_{fuel}$$

where EF is the emission factor (kg CO_2 eq/MJ) of fuel. Table 14.10 presents the results considering powertrains with respect to fuel type.

After consumed in the vehicle, the obtained results represent the environmental impacts from well-to-wheel point of view. Emission factor for electricity represents the emission intensity of grid electricity in Ontario [23]. Emission factor of hydrogen is from hydro-based proton exchange membrane water electrolysis.

Step-3 Life cycle impact analysis: Figure 14.20 presents the well-to-wheel CO_2 emissions of the public transit bus for different fuel consumption scenarios. According to this, a public transit bus powered by hydrogen fuel has the lowest emission rate with a value of 0.020 kg CO_2 eq per km of city driving. As it can be anticipated, the consumption of diesel fuel has the highest emission rate with a value of 0.943 kg CO_2/km. Besides being a fossil fuel by itself that emits significant amount of CO_2 emission when combusted, extraction, refining, long-distance transportation, and distribution steps of this fuel are energy-dense and mostly fossil fuel consuming processes. Due to this, consumption of diesel fuel ends up with high emission rates from well-to-wheel point of view. Battery electric bus has second lowest emission rate with a value of 0.038 kg CO_2 eq/km. Note that the emissions associated with battery pack manufacturing, which brings significant concerns regarding a sustainable supply chain as well as social and environmental costs, is not considered in the analysis. Another important clean fuel option is biomethane, which has an emission rate of 0.191 kg CO_2/km. From the sense that organic and inorganic wastes are produced as long as the life exist, biomethane, which is obtained by cleaning biogas, is considered as a sustainable fuel. By using proper systems with respect to the waste type, biogas can be produced locally. Another advantage is that using waste to produce biogas prevents the release of methane from landfills. So that, cleaner environment and sustainable fuel cycles can be obtained by using domestic resources.

Table 14.10 The values considered in the comparative environmental impact analysis.

Fuel Type	Fuel Consumption (kg or MJ/km)	Lower Heating Value (MJ/kg)	Emission Factor (kg CO_2 eq/MJ)
Diesel	0.2563	42.3	0.087
Biomethane	0.2548	50.0	0.015
Hydrogen	0.0697	120.0	0.0024
Electricity	4.432	–	0.0086

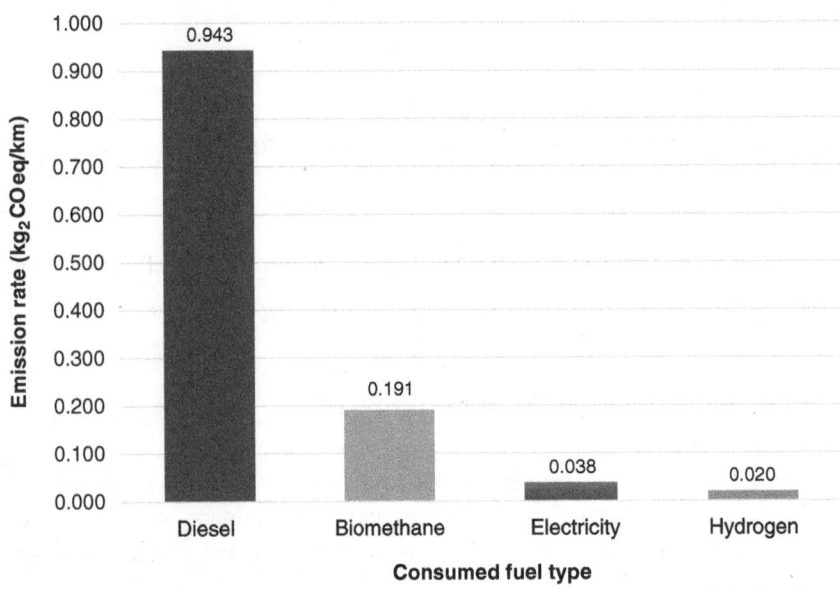

Figure 14.20 Well-to-wheel emission rate of public transit bus with respect to consumed fuel type.

Step-4 Improvement analysis and interpretation of results: In closing, hydrogen fuel cell electric vehicles for public transit are the lowest CO_2 emitting platforms, which is followed by battery electric and biomethane. The results indicate that emissions from using fossil fuels in road transport are almost 40 times higher than hydrogen, which highlights the importance of hydrogen for achieving low-emission road transportation. Nevertheless, production method of hydrogen and electricity are critical to obtain low emissions from these platforms. Therefore, clean hydrogen and electricity produced from renewable resources are necessary to promote these vehicles for road transport.

14.2.4 Case Study 4 – Well-to-Tank GHG Emissions of Various Fuel Types

In this case study, a comparative LCA study is performed on coal and natural gas-based combined heat and power cogeneration plants. The publicly available GREET, developed by Argonne National Laboratory, LCA tool is utilized in the present study. This tool employs attributional LCA methodology and considers the primary life cycle stages upstream and downstream of a well to pump LCA study. The Greet LCA tool can be accessed through the following link https://www.lifecycleassociates.com/lca-tools/greet-model or by the source [24].

The GREET LCA tool has accommodations for a variety of fuels, vehicles, and feedstocks. More than 100 fuel paths are included in GREET, comprising those for gasoline, natural gas, biofuels, hydrogen, and electricity generated from a variety of energy feedstocks. The main sheets in the GREET workbook are a summary of the key input parameters for all fuel pathways, a fuel production time series sheet with input parameters that vary by target year, fuel pathway sheets that calculate results, and a results sheet that summarizes the well-to-wheels (WTW) results in g/mi for the majority of fuel pathways. Several input sheets are used to execute inputs into the model. The fuel economy and emission factors for vehicles are organized on the "Vehicle" sheets as available elsewhere [24].

Step-1 Goal and scope definition: The objective of this case study is to perform an LCA study on both natural gas and coal-based power plants. Specifically, focusing on CHP type plants that provide both power and heat useful outputs. The geographical focus of the current study comprises the US. The functional unit of the LCA is determined as 1 MJ of energy production. Further the specific details of this comparative LCA analysis can be found in the source [25]. The schematic of coal-based steam cycle is shown in Figure 14.21a, and Figure 14.21a presents the natural-based combined cycle power plant.

Step-2 Inventory analysis: The GREET software is used for performing the inventory analysis as it specifies the emissions of different types of power generation methodologies. The system inputs that are inventoried are the resource usage of several fossil fuel and nonfossil fuel-based energy inputs. The water consumption for each power plant type is also considered. However, since the GREET package only provides the life cycle emissions, the GaBi LCA software is deployed to convert these emission values into the corresponding environmental impact categories to better reflect the results.

The initial stage of development of well infrastructure is comprehensively analyzed by GREET through considering several parameters as well as resource inputs and outputs. Firstly, the drilling phase is considered. This includes the utilization of materials, fuel, and drill rigs. Further, the water usage for the construction of the well that is used for several purposes including fracturing, case cementing, and for drilling fluid are also considered. Moreover, the shale wells considered include the 4-6H Carol Baker wells. The drill rig is assumed to run by an engine of 2000 hp. Also, fuel consumption is taken to be 0.06 gallon/hp h and a capacity factor of 45% is applied on the drill rig . Once the drilling phase is completed, the well completion is performed. This includes the preparation and insertion of production tubing followed by piercing or fracturing the well hydraulically.

Next, the installation stage of appropriate pipelines that transport the produced natural gas to the compression stations is considered. After the natural gas is extracted from the well, the next life cycle stage comprises the processing phase. In this phase, the obtained raw gas is treated to eliminate unwanted products such as other fluids. After processing the extracted gas, it is passed to the transmission and distribution stage, where it is compressed to high pressures and transported through pipelines. Furthermore, the processed natural gas is passed to the power generation or CHP life cycle stage. For the inventory data, the fuel combusted for power and thermal energy generation is considered for each type of plant along with the energy efficiencies values. Once the electricity is generated, the transmission and distribution life cycle stage are considered. In this phase, the grid losses that arise as a result of heat generation are assessed. Figures 14.22a and 14.22b present the LCA boundaries of the coal-based and the natural gas-based combined power plants.

Step-3 Life cycle impact analysis: After determining the environmental emissions from the GREET software, the associated impact assessment categories and their magnitudes per unit functional unit are determined as

$$Impact = \sum_{b} e_b.Cfr_{a,b}$$

where e_b is denoted by emission and Cf represents the classification factor. The Cf value of each pollutant varies according to the impact category considered. In the present study, the impact assessment is conducted according to CML 2001. From the pollutants inventoried, the impact

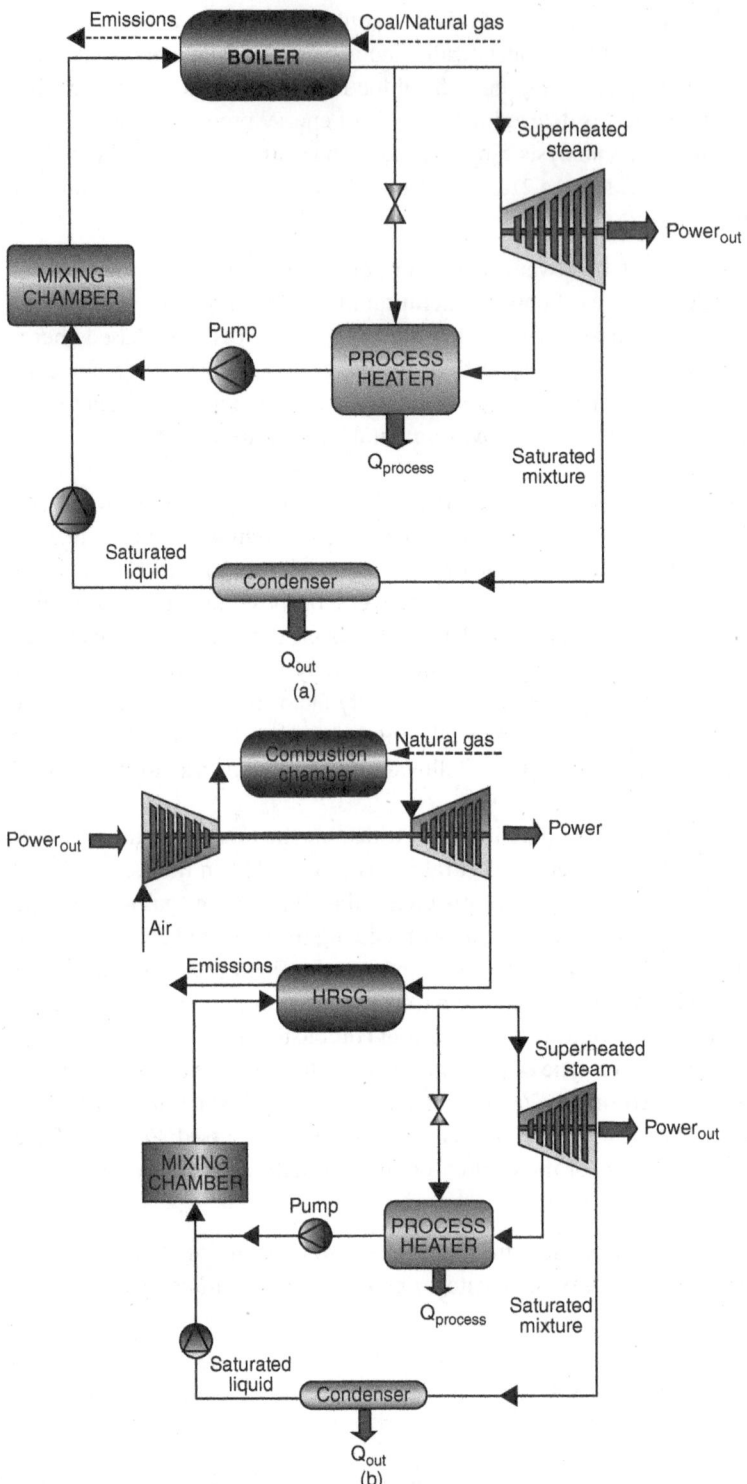

Figure 14.21 Schematic diagrams of (a) a coal driven steam Rankine cycle and (b) natural gas-based the combined cycle power plant.

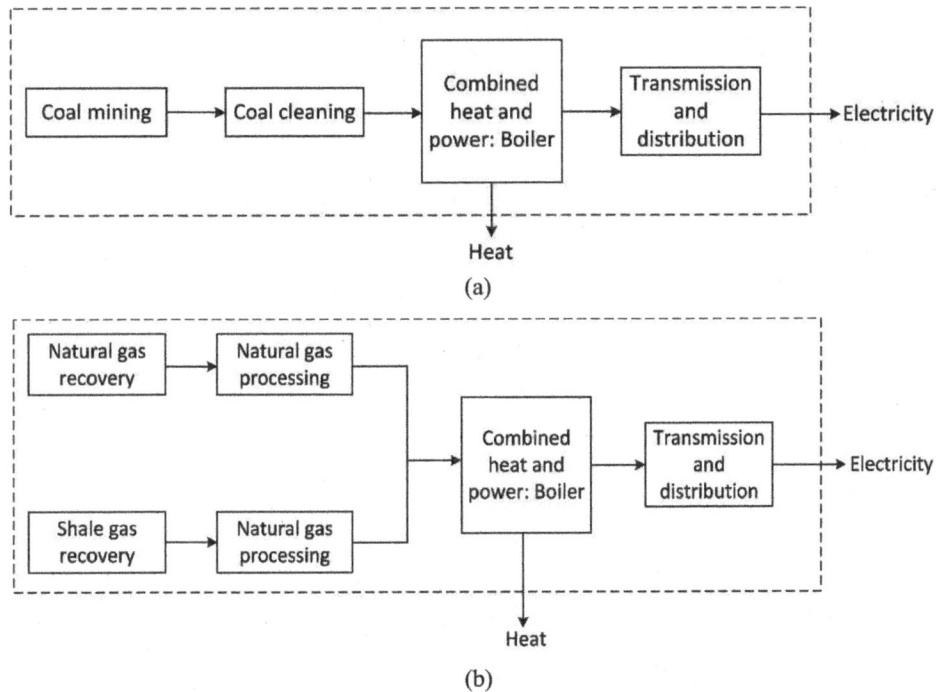

Figure 14.22 Schematics of system boundaries considered for LCA analyses of coal-based (a) and natural gas-based (b) power plants.

categories evaluated according to the CML 2001 methodology are global warming potential (GWP100), acidification potential (AP), freshwater aquatic ecotoxicity potential (FAETP), and eutrophication potential (EP). Figure 14.23 presents GWP and AP results.

The coal-based CHP plant entails the highest life cycle fossil fuel-based energy input of 2289 kJ. Further, the coal-fired boiler power plant is found to have the lowest comparative life cycle efficiency of 43.6%. The global warming potential (GWP100) is found to be comparatively higher for the coal-based CHP (Coal: CHP: B) life cycle. This life cycle is found to have a GWP100 of 0.23 kg CO_{2eq}/MJ. This is considerably higher than the other natural-based life cycles. The natural gas-based CC plant (NG:CC) is found to have a GWP100 of 0.13 kg CO_{2eq}/MJ. Furthermore, the AP of the coal-based plant is also estimated to be the highest among the considered plant types (0.0007 kg SO_{2eq}/MJ), followed by the natural gas-based boiler CHP plant (NG: CHP: B) that is estimated to have an AP of 0.00017 kg SO_{2eq}/MJ. Hence, in terms of GWP100 and AP, the coal-based cycle is found to be most environmentally detrimental. Furthermore, the natural gas-based combined cycle with CHP (NG: CHP:CC) is found to entail the lowest AP and GWP100 of 0.102 kg CO_{2eq}/MJ and 5.5 E-5 kg SO_{2eq}/MJ.

The natural gas-based combined cycle plant with CHP is found to have comparatively higher life cycle energy efficiency of 59.6%. This is attributed to the dual usage of waste heat. Firstly, the waste thermal energy entailed in the gas turbine exhaust is utilized to operate on a steam Rankine cycle. Secondly, the available thermal energy is also used to provide heat as a useful system output. Figure 14.24 presents FAETP and EP results. The NG: CHP:B-based life cycle is found to have the highest FAETP of 7.6E-7 kg DCB eq/MJ and also the highest EP of 2.78 E-5 kg PO_4 eq/MJ. Also, this is followed by the coal-based life cycle that entails a FAETP of 6.19 E-7 kg DCB eq/MJ and an EP value of 1.52E-5 kg PO_{4eq}/MJ. Moreover, the NG: CHP:CC-based life cycle is found to have the lowest FAETP and EP values of 4.49E-7 kg DCB eq/MJ and 9.27E-6 kg PO_{4eq}/MJ. respectively.

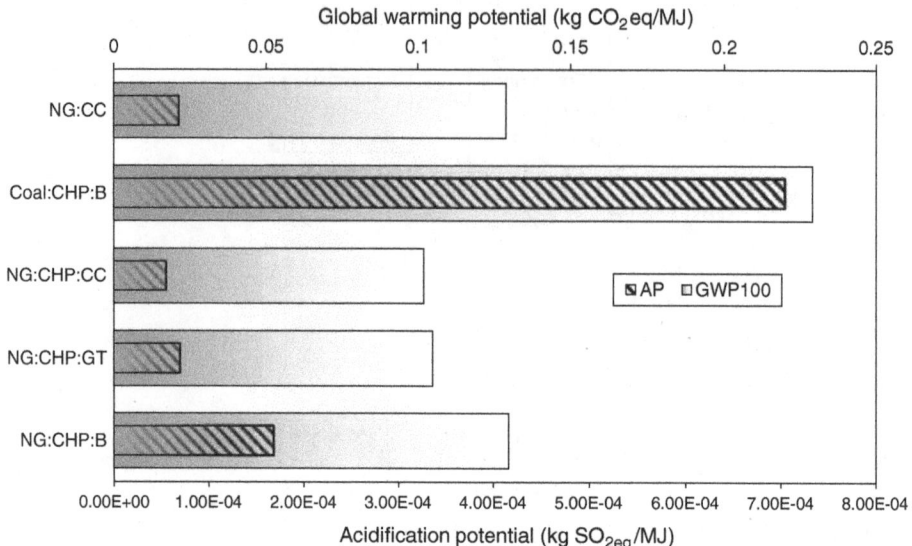

Figure 14.23 Global warming potential and acidification potential results.

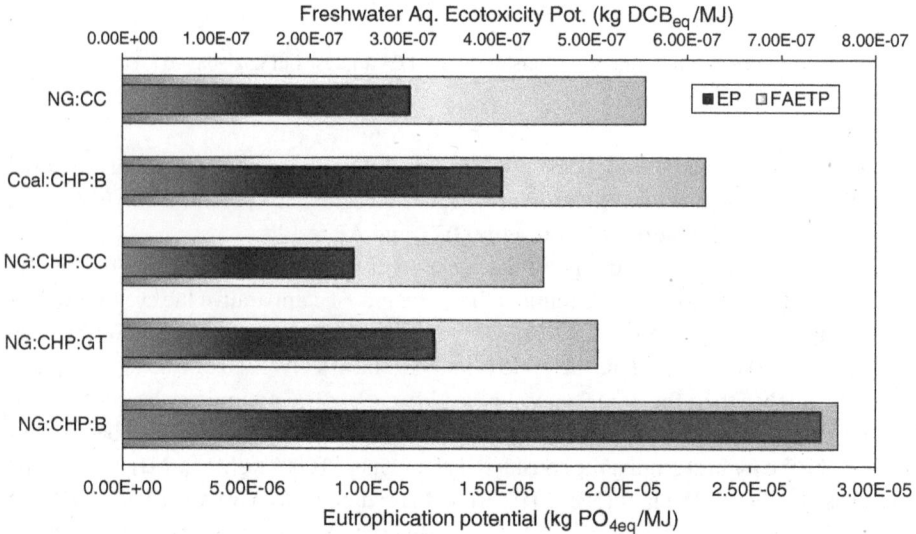

Figure 14.24 Freshwater ecotoxicity potential and eutrophication potential results of the considered plants according to CML 2001.

***Step-4 Improvement analysis and interpretation of results*:** The high AP associated with the coal-based life cycle can be attributed to the high life cycle SOx and NOx emissions. Furthermore, the high GWP100 is attributed to the high emission factors of the GHG emissions including CO_2 and CH_4. Hence, it is recommended firstly to employ carbon capture technologies for coal-based power plants. Further, the sulfur content in the coal plays an important role in the AP; thus, it is recommended to utilize coal that entails a low sulfur content to ensure minimum SOx emissions occur during the life cycle.

14.3   Closing Remarks

Using life cycle Assessment (LCA) tools is a necessity to better understand the systems, services, products, applications, and activities and their relationships with the environment through their lifetime and the consequences. Such tasks are important to achieve for reaching carbon neutral or net zero carbon energy infrastructure goals. It is also important to note that there are specific standards (ISO 14040 and ISO 14044) about LCA methods and their utilization which make the tools and study results are recognized and trusted for implementation. Since LCA takes into account the entire life cycle and provides a holistic overview on systems, it helps to reduce environmental impacts from life cycle point of view rather than being reduced at one stage of the life cycle while growing on the other. This chapter covers the LCA methods, the specific steps considered in any LCA analysis and assessment study, solution methodologies and environmental impact categories and discusses their roles and outcomes as related to various energy systems and applications. There are multiple case studies presented to better highlight the importance of LCA studies and their role in achieving better environment and sustainable development.

References

1 Hyrkäs, T.B. (2018). *7 Steps guide to building life cycle assessment.* 16.
2 Algren, M., Fisher, W., and Landis, A.E. (2021 January). Machine learning in life cycle assessment. In: *Data Science Applied to Sustainability Analysis*, 167–190. doi: 10.1016/B978-0-12-817976-5.00009-7.
3 Muthu, S.S. (2020 January). Ways of measuring the environmental impact of textile processing: an overview. In: *Assessing the Environmental Impact of Textiles and the Clothing Supply Chain*, 33–56. doi: 10.1016/B978-0-12-819783-7.00002-8.
4 *ISO 14040:2006(en), environmental management — life cycle assessment — principles and framework.* https://www.iso.org/obp/ui#iso:std:iso:14040:ed-2:v1:en (accessed 09 July 2021).
5 Karaca, A.E., Dincer, I., and Gu, J. (2020 October). A comparative life cycle assessment on nuclear-based clean ammonia synthesis methods. *Journal of Energy Resources Technology, Transactions of the ASME* 142 (10). doi: 10.1115/1.4047310/1083981.
6 Kikuchi, Y. (2016 January). Life cycle assessment. In: *Plant Factory: An Indoor Vertical Farming System for Efficient Quality Food Production*, 321–329. doi: 10.1016/B978-0-12-801775-3.00024-X.
7 Hollberg, A. Parametric Life Cycle Assessment: introducing a time-efficient method for environmental building design optimization DBU: intuitive communication and visualization of whole-building LCA and hazardous building materials for decision support in the digital planning process (BIM) View project Parametric Life Cycle Assessment View project. doi: 10.25643/bauhaus-universitaet.3800.
8 Handbook on life cycle assessment: operational guide to the ISO standards – Google Books. https://books.google.ca/books?hl=en&lr=&id=Q1VYuV5vc8UC&oi=fnd&pg=PR8&dq=Handbook+on+Life+Cycle+Assessment&ots=mX80pqBMOP&sig=XxGNveJg0m7F9f5KxYeiSjurrkY#v=onepage&q=Handbook on Life Cycle Assessment&f=false (accessed 09 July 2021).
9 Valente, A., Iribarren, D., and Dufour, J. (2017). Life cycle assessment of hydrogen energy systems: a review of methodological choices. *International Journal of Life Cycle Assessment* 22 (3): 346–363. doi: 10.1007/s11367-016-1156-z.
10 Karaca, A.E., Dincer, I., and Gu, J. (2020 August). Life cycle assessment study on nuclear based sustainable hydrogen production options. *International Journal of Hydrogen Energy* 45 (41): 22148–22159. doi: 10.1016/J.IJHYDENE.2020.06.030.

11 Swart, P., Alvarenga, R.A.F., and Dewulf, J. (2015). Abiotic resource use. 247–269. doi: 10.1007/978-94-017-9744-3_13.

12 Hauschild, M.Z., Rosenbaum, R.K., and Olsen, S.I. (2017 September). Life cycle assessment: theory and practice. In: *Life Cycle Assessment: Theory and Practice*, 1–1216. doi: 10.1007/978-3-319-56475-3.

13 *Life Cycle Assessment (LCA): a guide to best practice | Wiley*. https://www.wiley.com/en-ca/Life+Cycle+Assessment+%28LCA%29%3A+A+Guide+to+Best+Practice-p-9783527329861 (accessed 09 July 2021).

14 Saur, K. (1997). *Life cycle impact assessment*. 2 (2). doi: 10.1007/BF02978760.

15 Azapagic, A., Emsley, A., and Hamerton, L. (2003). Definition of environmental impacts. In: *Polymers, the Environment and Sustainable Development*, vol. 9, 197–200. doi: 10.1002/0470865172.app2.

16 Farinha, C., de Brito, J., and do Veiga, M. (2021 January). Life cycle assessment. In: *Eco-Efficient Rendering Mortars* 205–234. doi: 10.1016/B978-0-12-818494-3.00008-8.

17 Understanding global warming potentials | US EPA. https://www.epa.gov/ghgemissions/understanding-global-warming-potentials (accessed 17 July 2021).

18 Shine, K.P. (2000). Radiative forcing of climate change. *Space Science Reviews* 94 (1–2): 363–373. doi: 10.1023/A:1026752230256.

19 Brune, W.H. (1991). Stratospheric chemistry. *Reviews of Geophysics* (Suppl): 12–24. doi: 10.1002/rog.1991.29.s1.12.

20 Bareiß, K., de la Rua, C., Möckl, M., and Hamacher, T. (2019 March).Life cycle assessment of hydrogen from proton exchange membrane water electrolysis in future energy systems. *Apply Energy* 237: 862–872. doi: 10.1016/J.APENERGY.2019.01.001.

21 Cetinkaya, E., Dincer, I., and Naterer, G.F. (2012 February). Life cycle assessment of various hydrogen production methods. *International Journal of Hydrogen Energy* 37 (3): 2071–2080. doi: 10.1016/J.IJHYDENE.2011.10.064.

22 Karaca, A.E., Dincer, I., and Nitefor, M. (2022 June). Analysis of a newly developed hybrid pneumatic powertrain configuration for transit bus applications. *Energy* 248: 123557. doi: 10.1016/J.ENERGY.2022.123557.

23 Gridwatch | Web App. https://live.gridwatch.ca/home-page.html (accessed 20 January 2023).

24 GREET Model – Life Cycle Associates, LLC. https://www.lifecycleassociates.com/lca-tools/greet-model (accessed 20 January 2023).

25 Siddiqui, O. and Dincer, I. (2020 November). A comparative life-cycle assessment of two cogeneration plants. *Energy Technology* 8 (11): 1900425. doi: 10.1002/ENTE.201900425.

Questions/Problems

Questions

1 Describe the need for life cycle assessment study and present a procedure on how to apply it.

2 Discuss the specific LCA approaches and give an example for each.

3 Illustrate the cradle to grave type approach and discuss possible domains in it.

4 Discuss possible pros and cons of LCA methods.

5 Illustrate the LCA framework and describe each step with an example.

6 List some potential environmental impact categories and discuss each.

7 List some potential improvement studies to combat global warming.

8 What are the key outcomes of the first case study presented in this chapter?

9 What are the key outcomes of the second case study presented in this chapter?

10 What are the key outcomes of the third case study presented in this chapter?

11 Identify all openly available LCA softwares in the literature and compare them.

12 Use GREET LCA software and apply to an energy system.

13 Identify all environmental impact categories available in the literature and compare them for possible use.

14 Compare the listed LCA methods by providing their utilization spectrum.

15 Write an essay about exergetic LCA for energy systems and explain what additional benefits it brings in.

Problems

1 Repeat the study presented in Case Study 1 by using the GREET LCA package and discuss the results comparatively for different environmental impact categories.

2 Repeat the study presented in Case Study 2 by using the GREET LCA package and discuss the results comparatively for different environmental impact categories.

3 Repeat the study presented in Case Study 3 by using the GREET LCA package and discuss the results comparatively for different environmental impact categories.

4 Repeat the study presented in Case Study 4 this time by considering solar and geothermal driven systems and discuss the results comparatively for different environmental impact categories.

5 Consider a house where one is allowed to use only natural gas, fuel oil, and coal. Conduct an LCA study to investigate the environmental impact through CO_2 emissions.

6 Consider a thermal power running on a steam Rankine cycle where three types of fuels, such as natural gas, diesel, and coal are potentially utilized. Conduct an LCA study to investigate the environmental impact through CO_2 emissions.

7 Consider three types of personal vehicles, namely with gasoline, electric, and hydrogen with fuel cell. Conduct an LCA study to investigate the environmental impact through CO_2 emissions for a total of driving range of 10,000 km.

8 Consider the hydrogen production methods available in GREET LCA Package and compare the environmental impact for each kilogram of hydrogen product.

9 Consider three different power generators running on diesel, gasoline, and hydrogen and apply an LCA method to study the environmental impact categories comparatively for each kWh of electricity generated.

10 Conduct a life cycle assessment study on all possible electricity generation options available in your community and evaluate the CO_2 emissions comparatively.

Index

Introduction to Energy Systems, First Edition. Ibrahim Dincer and Dogan Erdemir.
© 2023 John Wiley & Sons, Inc. Published 2023 by John Wiley & Sons, Inc.
Companion Website: www.wiley.com/go/Dincer/Introduction_to_Energy_Systems